ANNUAL REVIEW OF CELL BIOLOGY

EDITORIAL COMMITTEE (1985)

ANNUAL REVIEW OF CELL BIOLOGY

VOLUME 1, 1985

GEORGE E. PALADE, *Editor*
Yale University School of Medicine

BRUCE M. ALBERTS, *Associate Editor*
University of California, San Francisco

JAMES A. SPUDICH, *Associate Editor*
Stanford University

ANNUAL REVIEWS INC. 4139 EL CAMINO WAY PALO ALTO, CALIFORNIA 94306 USA

Ⓡ ANNUAL REVIEWS INC.
Palo Alto, California, USA

International Standard Serial Number : 0743-4634
International Standard Book Number : 0-8243-3101-X

TYPESETTING BY AUP TYPESETTERS (GLASGOW) LTD., SCOTLAND
PRINTED AND BOUND IN THE UNITED STATES OF AMERICA

PREFACE

The first volume of this new series of Annual Reviews came into being as a result of the convergence of two major currents of interest connected with the relatively young but rapidly advancing field of Cell Biology. One of these currents had its source in Annual Reviews Inc. (ARI), whose mission is to review the literature of significant areas of contemporary research. Individual proposals for an annual review of Cell Biology had been received from a number of scientists as early as 1969, and these led to the decision in 1983 by the Board of Directors of Annual Reviews Inc. that the time had come for a new review in this field. The other current came from the American Society for Cell Biology whose 1982 president, Marilyn Farquhar, and publication committee, led by Richard McIntosh, felt that the degree of development of the field required an annual review in addition to regular journals. Fortunately the two currents converged and, after preliminary discussions, an organizing meeting was held in October 1983 at the Annual Reviews headquarters in Palo Alto, California. Those in attendance included Winslow Briggs (Stanford University), William Brinkley (Baylor University), Marilyn Farquhar (Yale University), Peter Hepler (University of Massachusetts), Richard Hynes (MIT), Marc Kirschner (UCSF), David Luck (Rockefeller University), George Palade (Yale University), David Prescott (University of Colorado), Jean Paul Revel (California Institute of Technology), James Spudich (Stanford University), and several ARI staff members.

This meeting, which was chaired by Winslow Briggs, representing the Board of Directors of Annual Reviews, worked out the list of authors and chapters for the present volume. On that occasion, I was asked to become Editor of the Annual Review of Cell Biology. I accepted and was fortunate in obtaining the assistance of a competent and helpful Editorial Committee as well as the support of the editorial staff of Annual Reviews. From the beginning, I decided that close contact with the American Society for Cell Biology should be maintained, so as to benefit from the Society's suggestions as to appropriate topics and authors for future volumes.

The currents mentioned above were effective in giving birth to this new Annual Reviews series, but the real forces behind the move were the spectacular growth of Cell Biology over the last two or three decades, and the broad and close contacts it has established with the equally young fields of Molecular Biology and Molecular Genetics, as well as with such old and traditionally established disciplines as Biochemistry, Genetics, and Physiology.

The broadest and most intimate of these contacts are leading, in fact, to a gradual merger of the fields of Cellular and Molecular Biology, Molecular Genetics included. We are witnessing a powerful syncretic movement that has already generated a common conceptual and methodological ground for many basic biological sciences, which until recently were separated into distinct disciplines on the basis of approaches used or domains covered. We can consider this syncretic process as the logical expression of our recently acquired knowledge that the unity of organization of living systems extends far beyond the small and large molecules of Biochemistry to reach the level of the macromolecular assemblies, cell organs, and cells of Cellular and Molecular Biology. The fact that so many pieces of complex bio-chemical equipment that perform highly integrated metabolic reactions have been extensively conserved through evolution makes even more impressive than originally thought the unity of the foundation on which all living systems are built.

About three decades ago, the contact established and maintained with Biochemistry was particularly important for the subsequent development of Cell Biology. Over the last decade the nascent merger with Molecular Biology has given every sign of being even more important and fruitful than the first close contact. The new development is rooted in more than common technology; it stems from the realization that the genome can function only within the structural and chemical framework of the cell, a framework produced and modulated by the genes of distant ancestors, preserved by temporal continuity through cell generations, and inherited together with a daughter genome by each daughter cell at each cell division. The framework cannot be maintained without the genome, and the genome cannot function or even survive without the cellular framework. In the future, cellular and molecular biology, molecular genetics included, promise to be as inextricably dependent on one another as the genome is on the cellular framework that acts as the equipment for its expression, as its protective prison, and as its access to potential immortality. The *Annual Review of Cell Biology* will make a sustained effort to cover the merging fields of Cellular and Molecular Biology to the widest extent practically possible.

The importance assumed by Cell Biology in all fundamental biological sciences is clearly attested by the fact that topics developed (or under development) in Cell Biology have been reviewed in recent years in the Annual Reviews of Biochemistry, Genetics, Physiology, Plant Physiology, and Biophysics and Bioengineering. The publication of a separate *Annual*

Review of Cell Biology is justified on two accounts. First, the research community in Cellular and Molecular Biology is large enough and diverse enough in its specific interests to need an annual review that covers the whole field. Second, there is enough interest in Cellular and Molecular Biology in adjacent areas to warrant convenient access to a review of the entire field as it develops year by year. These grounds do not restrict in any way the publication of reviews on Cell Biology topics in other Annual Reviews series. Since the object is efficient communication among research scientists in the general area of basic biological sciences, all outlets should be useful. The *Annual Review of Cell Biology*, however, may prove to be the most useful means of communication within the field and immediately outside it. This is the goal of the new series.

GEORGE E. PALADE
EDITOR

Annual Review of Cell Biology
Volume 1, 1985

CONTENTS

SOME RELATED ARTICLES IN OTHER *ANNUAL REVIEWS*

From the *Annual Review of Biochemistry*, Volume 54 (1985):

> *Genetic Analysis of Protein Export in* Escherichia Coli *K12*, S. A. Benson, M. N. Hall, and T. J. Silhavy
>
> *Cell Adhesion and the Molecular Processes of Morphogenesis*, G. M. Edelman
>
> *Receptors and Phosphoinositide-Generated Second Messengers*, L. E. Hokin
>
> *The Membrane Skeleton of Human Erythrocytes and its Implications for More Complex Cells*, V. Bennett
>
> *Molecular Biology and Genetics of Tubulin*, D. W. Cleveland and K. F. Sullivan
>
> *Evolving Ribosome Structure*, J. A. Lake
>
> *Assembly of Asparagine-Linked Oligosaccharides*, R. Kornfeld and S. Kornfeld
>
> *DNA Topoisomerases*, J. C. Wang
>
> *Protein-Tyrosine Kinases*, T. Hunter and J. A. Cooper
>
> *Eukaryotic Protein Synthesis*, K. Moldave

From the *Annual Review of Biophysics and Biophysical Chemistry*, Volume 14 (1985):

> *Photoelectron Imaging in Cell Biology*, O. H. Griffith and G. F. Rempfer
>
> *Actin and Tubulin Polymerization*, C. Frieden
>
> *New Observations on Cell Architecture and Dynamics by Video-Enhanced Contrast Optical Microscopy*, R. D. Allen
>
> *Structure of Muscle Filaments Studied by Electron Microscopy*, L. A. Amos
>
> *Organization of Glycosphingolipids in Bilayers and Plasma Membranes of Mammalian Cells*, T. E. Thompson and T. W. Tillack

From the *Annual Review of Genetics*, Volume 19 (1985):

> *Structure and Evolution of the Insulin Gene*, D. F. Steiner et al
>
> *The Structure and Function of Yeast Centromeres*, L. Clarke and J. Carbon
>
> *The Genetic Analysis of Mammalian Cell-Cycle Mutants*, M. Marcus, A. Fainsod, and G. Diamond
>
> *Genetics of* Drosophila *Embryogenesis*, A. P. Mahowald and P. A. Hardy
>
> *Processed Pseudogenes*, E. Vanin
>
> *Steroid Receptor Regulated Transcription of Specific Genes and Gene Networks*, K. R. Yamamoto

From the *Annual Review of Immunology*, Volume 3 (1985):

*T-Lymphocyte Recognition of Antigen in Association with Gene Products of the
Major Histocompatibility Complex*, R. H. Schwartz
Human T-Cell Leukemia Virus (HTLV), S. Broder and R. C. Gallo
Genetics and Expression of Murine Ia Antigens, L. Mengle-Gaw and H. O.
McDevitt
The Coming of Age of the Immunoglobulin J Chain, M. E. Koshland
The Atomic Mobility Component of Protein Antigenicity, J. A. Tainer, E. D.
Getzoff, Y. Paterson, A. J. Olson, R. A. Lerner

From the *Annual Review of Microbiology*, Volume 39 (1985):

Oncogenes, L. Ratner, S. F. Josephs, and F. Wong-Staal
Viral Taxonomy for the Nonvirologist, R. E. F. Matthews
Antigenic Variation in African Trypanosomes, J. C. Boothroyd
Protein Secretion in Escherichia coli, D. Oliver
*Plant and Fungal Protein and Glycoprotein Toxins Inhibiting Eukaryote Protein
Synthesis*, A. Jiménez and D. Vázquez

From the *Annual Review of Physiology*, Volume 47 (1985):

Growth and Development of Gastrointestinal Cells, M. Lipkin
Chemistry of Gap Junctions, J.-P. Revel, B. J. Nicholson, and S. B. Yancey
Physiology and Pharmacology of Gap Junctions, D. C. Spray
Antibody Probes in the Study of Gap Junctional Communications, E. L.
Hertzberg
The Nature and Regulation of the Insulin Receptor, M. P. Czech
Intracellular Mediators of Insulin Action, K. Cheng and J. Larner
The Glucose Transporter of Mammalian Cells, T. J. Wheeler and P. C. Hinkle
Kinetic Properties of the Plasma Membrane Na^+-H^+ Exchanger, P. S. Aronson
Mechanism of Calcium Transport, G. Inesi

From the *Annual Review of Plant Physiology*, Volume 36 (1985):

Plant Lectins, M. E. Etzler
Plant Chromatin Structure, S. Spiker
*Topographic Aspects of Biosynthesis, Extracellular Secretion, and Intracellular
Storage of Proteins in Plant Cells*, T. Akazawa and I. Hara-Nishimura
H^+ Translocating ATPases, H. Sze
Cell-Cell Interactions in Chlamydomonas, W. J. Snell

ANNUAL REVIEWS INC. is a nonprofit scientific publisher established to promote the advancement of the sciences. Beginning in 1932 with the *Annual Review of Biochemistry*, the Company has pursued as its principal function the publication of high quality, reasonably priced *Annual Review* volumes. The volumes are organized by Editors and Editorial Committees who invite qualified authors to contribute critical articles reviewing significant developments within each major discipline. The Editor-in-Chief invites those interested in serving as future Editorial Committee members to communicate directly with him. Annual Reviews Inc. is administered by a Board of Directors, whose members serve without compensation.

ANNUAL REVIEWS OF		SPECIAL PUBLICATIONS
Anthropology	Medicine	Annual Reviews Reprints:
Astronomy and Astrophysics	Microbiology	Cell Membranes, 1975–1977
Biochemistry	Neuroscience	Cell Membranes, 1978–1980
Biophysics and Biophysical Chemistry	Nuclear and Particle Science	Immunology, 1977–1979
Cell Biology	Nutrition	
Earth and Planetary Sciences	Pharmacology and Toxicology	Excitement and Fascination
Ecology and Systematics	Physical Chemistry	of Science, Vols. 1 and 2
Energy	Physiology	
Entomology	Phytopathology	History of Entomology
Fluid Mechanics	Plant Physiology	
Genetics	Psychology	Intelligence and Affectivity,
Immunology	Public Health	by Jean Piaget
Materials Science	Sociology	Telescopes for the 1980s

A detachable order form/envelope is bound into the back of this volume.

Ann. Rev. Cell Biol. 1985. 1 : 1–39

RECEPTOR-MEDIATED ENDOCYTOSIS: Concepts Emerging from the LDL Receptor System

Joseph L. Goldstein, Michael S. Brown, Richard G. W. Anderson, David W. Russell, and Wolfgang J. Schneider

Departments of Molecular Genetics, Cell Biology, and Internal Medicine, University of Texas Health Science Center at Dallas, Dallas, Texas 75235

CONTENTS

INTRODUCTION

The concept of *receptor-mediated endocytosis* was formulated in 1974 to explain the observation that regulation of cellular cholesterol metabolism

1

0743–4634/85/1115–0001$02.00

depended on the sequential cell surface binding, internalization, and intracellular degradation of plasma low density lipoprotein (LDL) (Goldstein & Brown 1974, Goldstein et al 1976). This uptake mechanism was postulated on the basis of biochemical studies; it was soon verified morphologically when the receptors for LDL were observed to be clustered in coated pits that pinched off from the surface to form coated vesicles that carried the LDL into the cell (Anderson et al 1976, 1977a).

Coated pits and coated vesicles had been recognized by electron microscopy in the mid-1960s (Roth & Porter 1964, Fawcett 1965, Friend & Farquhar 1967). Their role in receptor-mediated endocytosis was appreciated a decade later as a result of the convergence of twin events: 1) the demonstration that coated vesicles were the sites at which LDL receptors were concentrated, and 2) the demonstration by Pearse (1975) that a single protein, clathrin, formed the cytoplasmic coat, an observation that provided a biochemical definition of coated vesicles. The biological implications of receptor-mediated endocytosis were vividly underscored by the finding that genetic defects in the LDL receptor preclude cellular uptake of LDL, producing hypercholesterolemia and heart attacks (Brown & Goldstein 1984).

During the last decade, receptor-mediated endocytosis was recognized as a mechanism by which animal cells internalize many macromolecules in addition to LDL (Goldstein et al 1979a, Pastan & Willingham 1981, Bretscher & Pearse 1984). The process is initiated when receptors on the cell surface bind macromolecules and slide laterally into clathrin-coated pits. Within minutes the coated pits invaginate into the cell and pinch off to form coated endocytic vesicles. After shedding their clathrin coats the vesicles fuse with one another to form endosomes whose contents are acidified by ATP-driven proton pumps (Tycko & Maxfield 1982, Helenius et al 1983, Pastan & Willingham 1983). Within the endosome the ligand and receptor part company. Often, but not always, the ligand is carried to lysosomes for degradation, while the receptor cycles back to the cell surface to bind new ligand (Brown et al 1983).

More than 25 specific receptors have been observed to participate in receptor-mediated endocytosis. These include receptors for transport proteins that deliver nutrients to cells, such as the cholesterol-carrying lipoprotein LDL, the iron transport protein transferrin, and the vitamin B_{12} transport protein transcobalamin II. Receptor-mediated endocytosis also applies to many nontransport plasma proteins, including asialoglyco-proteins, α-2-macroglobulin, and immune complexes. Moreover, the process mediates the cellular uptake of lysosomal enzymes, which occurs when these enzymes bind to receptors that recognize mannose-6-phosphate residues uniquely attached to this class of proteins. Certain viruses and toxins use receptor-mediated endocytosis to enter cells, apparently by binding opportunistically to receptors that normally function in the uptake of other substances.

Protein growth factors, such as epidermal growth factor (EGF) and platelet-derived growth factor (PDGF), as well as classic polypeptide hormones, such as insulin and luteinizing hormone, also enter cells by receptor-mediated endocytosis. The same receptors that mediate endocytosis of these proteins mediate their physiologic actions. However, frequently cellular entry of the ligand does not seem to be required for the action of the growth factor or the hormone. Rather, the entry mechanism functions in the rapid control of receptor number and in the removal of the growth factor or hormone from the circulation (Carpenter & Cohen 1979, Terris et al 1979).

Progress in this field has been rapid. Within a single year—1984—complementary DNAs (cDNAs) for five different coated pit receptors were isolated, and their nucleotide and corresponding amino acid sequences were determined (Mostov et al 1984, Ullrich et al 1984, Russell et al 1984, Yamamoto et al 1984, McClelland et al 1984, Schneider et al 1984, Holland et al 1984). The cDNA cloning and structure of a sixth coated pit receptor, the insulin receptor, was reported early in 1985 (Ullrich et al 1985, Ebina et al 1985). In this review we summarize the information that is emerging from study of the amino acid sequences of the receptor proteins, with emphasis on the LDL receptor.

PATHWAYS OF RECEPTOR-MEDIATED ENDOCYTOSIS

Entry Into Coated Pits

The various pathways of receptor-mediated endocytosis share one common feature: in each case the receptors move to coated pits and coated vesicles. However, there are differences in the mechanisms that trigger movement to coated pits as well as differences in the routes the ligands and receptors follow after entering the cell. We can divide the process of receptor-mediated endocytosis into subcategories according to these differences, as described below.

The first distinction is whether the receptors spontaneously move to coated pits and enter cells continuously (even in the absence of ligand), or whether the receptors wait on the surface until a ligand is bound, whereupon they are captured by coated pits. The receptors in the first category include those for LDL (Anderson et al 1982, Basu et al 1981), transferrin (Hopkins & Trowbridge 1983, Hopkins 1985), α-2-macroglobulin (Hopkins 1982, Via et al 1982), asialoglycoproteins (Wall et al 1980, Berg et al 1983), and insulin (Krupp & Lane 1982). Conversely, the receptor for EGF is diffusely distributed on the cell surface, and is not trapped in coated pits unless it is occupied with ligand (Schlessinger 1980, Dunn & Hubbard 1984).

The propulsive force for movement of receptors to coated pits may be simple diffusion, or it may involve a more directed type of propulsion (Bretscher 1984). The rate of diffusion of receptors on cell surfaces is sufficiently fast in itself to explain movement into coated pits (Goldstein et al 1981, Barak & Webb 1982). However, considerable evidence suggests that membrane lipids are continuously flowing toward coated pits (Bretscher 1984). This lipid flow may carry membrane proteins along passively (Bretscher & Pearse 1984, Hopkins 1985), but why are only certain cell surface proteins trapped in coated pits? One possibility is that receptors are marked for such entry by the attachment of prosthetic groups. Many receptors (such as those for transferrin, asialoglycoproteins, EGF, PDGF, and insulin) have phosphate groups attached to serine, threonine, or tyrosine residues in their cytoplasmic domains (see Table 1 in Brown et al 1983).

Recent attention has focused on phosphorylation or dephosphorylation as a potential mechanism for signaling entry, perhaps through induction of receptor binding to clathrin, the protein that covers the cytoplasmic surface of coated structures. Phosphorylation of the receptors for EGF (Hunter 1984), transferrin (Klausner et al 1984), and insulin (Jacobs et al 1983) can be enhanced by treatment of cells with phorbol esters, which activate protein kinase C. Phorbol esters cause transferrin receptors in K562 cells to become trapped within the cell, which suggests that phosphorylation either increases the rate of their cellular entry or slows their return to the cell surface, or both (Klausner et al 1984).

A few receptors undergo acylation of cysteine residues with fatty acids, but this modification does not apply to all receptors that participate in endocytosis. Moreover, in the one case that has been studied in detail, that of the transferrin receptor, the turnover of the fatty acid moiety is much slower than the internalization rate (Omary & Trowbridge 1981), which implies that acylation-deacylation does not occur during each recycling event.

Intracellular Routes

A second variation in the systems of receptor-mediated endocytosis is the fate of the ligand and receptor. It appears that all endocytotic receptors enter cells in the same coated pits, and are delivered to the same acidified endosomes (Pastan & Willingham 1981, Via et al 1982, Carpentier et al 1982). Thereafter, the pathways diverge. The receptor-ligand complex may follow one of four routes, which are discussed below and illustrated schematically in Figure 1.

ROUTE 1: RECEPTOR RECYCLES, LIGAND DEGRADED   This pathway is the classic one described for the endocytosis of LDL, asialoglycoproteins, and α-2-macroglobulin, as well as for insulin and luteinizing hormone. In following this route, ligands dissociate from their receptors within the endosome, apparently as a result of the drop in pH (Brown et al 1983, Helenius et al 1983). The ligand is carried further to lysosomes, where it is degraded. The receptor leaves the endosome, apparently via incorporation into the membrane of a vesicle that buds from the endosome surface. These recycling vesicles may originate as tubular extensions of the endosome, which gather receptors and then pinch off from the main body of the endosome (Geuze et al 1983, 1984). After their return to the surface, LDL receptors are said to remain clustered so that they can be incorporated rapidly into newly formed coated pits (Robenek & Hesz 1983). Conversely,

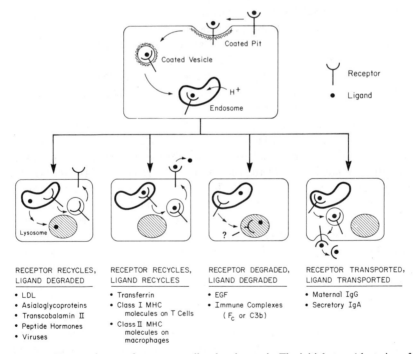

Figure 1   Four pathways of receptor-mediated endocytosis. The initial steps (clustering of receptors in coated pits, internalization of coated vesicles, and fusion of vesicles to form endosomes) are common to the four pathways. After entry into acidic endosomes, a receptor-ligand complex can follow any of the four pathways shown in the figure.

recycling transferrin receptors seem to go through a transient phase of monomolecular dispersion on the cell surface before clustering and internalizing again (Hopkins 1985).

Route 1 seems ideally adapted for use by receptors that transport ligands into cells at a high rate: It allows reuse of receptors once every 10–20 min. Thus, one receptor can mediate the uptake of hundreds of ligands during its usual lifespan of 10–30 hr. Recycling requires that the receptors have a stable structure that will permit them to pass repeatedly through the acidic environment of the endosome without denaturation. In the acidic endosome the receptors must undergo sufficient conformational change to release their ligands (DiPaola & Maxfield 1984), but they must not become irreversibly denatured. The LDL receptor, for example, can make up to 150 trips through the endosome without losing its function (Goldstein et al 1979a, Brown et al 1982). Maintenance of stability may require unusual protein structures, some of which are detailed below.

ROUTE 2: RECEPTOR RECYCLES, LIGAND RECYCLES   This pathway was originally described for the transferrin receptor (Octave et al 1983). When the transferrin/receptor complex reaches the endosome the two proteins do not dissociate. In vitro binding studies show that the transferrin receptor, in contrast to the receptors for LDL, asialoglycoproteins, and EGF, fails to dissociate from its ligand at pH 5 (Klausner et al 1983, Dautry-Varsat et al 1983). However, iron does dissociate from transferrin at acidic pH. Thus, in the endosome the iron is stripped from transferrin while the apo-transferrin remains attached to the receptor. The apo-transferrin/receptor complex then returns to the cell surface. The recycling transferrin receptor seems to leave the endosome by a network of membrane tubules and vesicles that eventually leads it back to the cell surface (Geuze et al 1984). Once on the surface and again at neutral pH, the apo-transferrin dissociates from the receptor. (Iron-containing transferrin binds to the receptor at neutral pH, but apo-transferrin dissociates from the receptor at this pH). The transferrin receptor is now free to bind another molecule of iron-containing transferrin and to repeat the cycle. Like the LDL receptor, the transferrin receptor is degraded very slowly with a half life of > 30 hr (Omary & Trowbridge 1981), even though it enters the cell every 15–20 min (Bleil & Bretscher 1982, Ciechanover et al 1983).

Recent studies suggest a new role for internalization and recycling of ligands by this route. Such recycling may provide the mechanism by which cells of the immune system process antigen and "present" it to effector cells (for reviews see Unanue 1984, Pernis 1985, Pernis & Tse 1985). Macrophages and certain B lymphocytes internalize foreign antigens by receptor-mediated endocytosis. The receptors responsible for this uptake are poorly characterized. Their function is to deliver the antigen to an acidic intracellular compartment where the antigen undergoes partial proteolysis. The proteolytic fragments are transported back to the cell surface where they are presented to the well-characterized antigen receptors on neighboring T lymphocytes. Presentation requires that the surface of the antigen-presenting cell express Class II major histocompatibility (MHC) proteins of the same genotype as the T lymphocyte. The Class II MHC proteins are dimers of two nonidentical transmembrane glycoproteins, which are continuously internalized and recycled without degradation. One reason for such recycling may be that after proteolysis the antigen and the Class II molecules must pass through the same acidic compartment so they can form a complex that presents the fragmented antigen to the T-cell antigen receptor (Brodsky 1984).

Once a responding T cell is stimulated by exposure to antigen, it begins to internalize and recycle its own MHC molecules, which are of the Class I type (Pernis 1985). These Class I MHC molecules are a complex of a transmembrane glycoprotein and another protein (β_2-microglobulin) that is adherent to the outer surface of the cell membrane. The Class I MHC molecules are not internalized by the T cells unless the cells are activated (i.e. stimulated by antigen). After activation the internalized Class I molecules are delivered to acidic endosomes and cycled back to the cell surface every 15 min (Pernis 1985). Each Class I molecule makes many trips in and out of the activated T cell during its half-life of 14 hr (Tse & Pernis 1984, Tse et al 1985). Monoclonal antibodies directed against the Class I molecules do not affect this recycling. In fact, the bound antibody enters the cell and cycles back to the cell surface with the Class I molecule still attached, in a manner analogous to the co-recycling of apo-transferrin and the transferrin receptor (Tse et al 1985).

The specificity of regulation of internalization of Class I MHC molecules is striking. Internalization and recycling occur only on activated T lymphocytes and not on resting T lymphocytes, B lymphocytes (resting or activated), or on any other known cell type. It is not possible to induce rapid internalization of these molecules in B cells, even when the Class I molecules have been cross-linked by exposure to a monoclonal anti-Class I antibody followed by a second layer of polyclonal antibodies (Pernis & Tse 1985). Moreover, nonlymphoid cells, such as fibroblasts and mouse L cells, do not rapidly internalize their Class I MHC molecules even though they rapidly internalize other membrane molecules, such as the receptors for LDL and transferrin (Pernis & Tse 1985). Internalization of Class I molecules by mouse L cells or fibroblasts can be observed when the molecules have been cross-linked by a double layer of antibodies, but the rate is considerably slower than receptor-mediated endocytosis—it takes hours rather than minutes. Moreover, this internalization leads to lysosomal degradation of the Class I molecules rather than to recycling (Pernis & Tse 1985, Huet et al 1980).

Do the internalized antigens and MHC molecules follow the classic coated pit to coated vesicle to endosome to lysosome pathway described for LDL? There are no electron microscopic data that address this question; however, there is circumstantial evidence that coated pits and vesicles are involved. For instance, recycling of MHC molecules is strictly regulated, has rapid kinetics, and is inhibited by drugs that raise the pH of endosomes (Tse & Pernis 1984, Tse et al 1985, Unanue 1984), all of which are features of endocytosis via coated pits and vesicles (Brown et al 1983).

ROUTE 3: RECEPTOR DEGRADED, LIGAND DEGRADED   This pathway has been described in greatest detail for EGF. After the EGF/receptor complex reaches the endosome both components are degraded, probably as a result of subsequent cotransport to the lysosome (Carpenter & Cohen 1979). The mechanism for this cotransport is unclear. Since EGF dissociates from its receptor at acidic pH (DiPaola & Maxfield 1984), it would presumably dissociate in the endosome. Somehow this dissociation does not allow the EGF receptor to return to the surface, but rather it is carried further into lysosomes. If the EGF receptor is delivered to lysosomes by vesicular fusion then the cytoplasmic domain of the receptor would remain outside of the lysosome, facing the cytoplasm. This domain of the EGF receptor consists of 542 amino acids and contains tyrosine kinase activity (Hunter 1984, Ullrich et al 1984). It is tempting to speculate that this tyrosine kinase domain of the receptor might be liberated from its hydrophobic anchor through proteolytic cleavage and then migrate elsewhere in the cell, where it could phosphorylate proteins that trigger cell division. The selective release of such a cytoplasmic fragment has not yet been demonstrated.

In certain cells, one population of EGF receptors may escape degradation and recycle. In cultured fibroblasts (Carpenter & Cohen 1979) and in the perfused rat liver (Dunn & Hubbard 1984) the addition of EGF causes a decrease of up to 80% in EGF receptors, apparently due to ligand-induced internalization and degradation. However, the remaining 20% of EGF receptors continue to bind, internalize, and degrade EGF with kinetics that suggest recycling.

ROUTE 4: RECEPTOR TRANSPORTED, LIGAND TRANSPORTED   This pathway has been described most clearly for the receptor that carries polymeric immunoglobulin A (IgA) and immunoglobulin M (IgM) across epithelial surfaces, such as across liver cells for excretion into the bile, and across mammary epithelia for excretion into milk (Solari & Kraehenbuhl 1984, Mostov et al 1984). In the liver the newly synthesized receptor appears on the sinusoidal surface of the hepatocyte, where it binds dimeric IgA. The receptor/immunoglobulin complex is incorporated into vesicles and carried into the cell (Renston et al 1980). Coated vesicles have not been implicated formally, but such involvement seems likely. At some point after internalization, the receptor is clipped proteolytically so that part of the receptor with the immunoglobulin still bound to it is released from the membrane. This released receptor fragment is the so-called secretory component. The IgA-containing vesicle eventually migrates to the bile canalicular face of the hepatocyte, where it discharges the IgA/secretory component adduct into the bile.

In neonatal animals, transepithelial transport of maternal IgG from the lumen of the intestine to the interstitial space is mediated by a receptor that binds to the Fc domain of the IgG. This transport probably does not involve cleavage of the receptor, since a secretory component has not been identified (Rodewald & Abrahamson 1982).

The delineation of four routes for disposal of receptors and ligands (Figure 1) implies that cells have multiple mechanisms for sorting receptors after they enter the cell. These mechanisms must be regulated so as to allow different cells to process the same receptor by different routes or a single cell to process the same receptor by different routes at different times. In some cases sorting involves passage of the receptors through vesicles that are located near the Golgi complex. However, the receptors do not seem to transit through classic Golgi stacks, which are the sites of sorting in the exocytotic pathway. Rather, they pass through nearby vesicles that may or may not contain Golgi-associated enzymes (Dunn & Hubbard 1984, Hanover et al 1984).

To assure accuracy of the multiple sorting and targeting events, each receptor must contain multiple functional domains. It must contain a binding domain that is specific for a given set of ligands, and regions that allow it to interact with other macromolecules so it can be transported to various sites within the cell. Often a receptor will proceed successively from one compartment to another, at each stage being sorted from other membrane molecules that are stationary or are moving to different sites. Therefore, each receptor must contain multiple sorting signals that act sequentially. These signals will be revealed only when the complete structures of the receptors are known.

THE LDL RECEPTOR: STRUCTURE-FUNCTION RELATIONSHIPS

We recently carried out detailed studies of the structure and biosynthesis of the LDL receptor. This receptor performs a simple function: it carries cholesterol-bearing lipoproteins into cells. To accomplish this task, the receptor must move from its site of synthesis in the membranes of the endoplasmic reticulum (ER) through the Golgi complex to the cell surface, where it is targeted to coated pits. It must then recycle from the endosome to the cell surface. Naturally occurring mutations in the gene for the LDL receptor disrupt several of these transport steps and produce a clinical condition of receptor deficiency known as familial hypercholesterolemia (FH).

Protein Purification and cDNA Cloning

The LDL receptor was purified from the bovine adrenal cortex, an organ that contains a relative abundance of LDL receptors ($\sim 10^5$ molecules per cell), which it uses to supply cholesterol for conversion to steroid hormones (Schneider et al 1982). Biochemical tools were developed that permitted cloning of cDNAs for the receptor. Thus, polyclonal antibodies raised against the purified bovine protein were used to enrich for the rare LDL-receptor mRNA by polysome immune purification. A cDNA library was constructed from the purified mRNA of bovine adrenal cortex, and was screened with two families of oligonucleotides derived from the amino acid sequence of a cyanogen bromide fragment of the bovine protein. These methods led to the isolation of a partial cDNA for the bovine receptor (Russell et al 1983).

The bovine cDNA was used as a probe to isolate a fragment of the human LDL-receptor gene. In turn, an exon probe from this genomic fragment was employed to isolate a cDNA clone representing the complete 5.3-kilobase (kb) human LDL-receptor mRNA. Transfection studies indicated that this cDNA could direct the expression of functional human LDL receptors in simian COS cells (Yamamoto et al 1984).

The nucleotide sequence of this cDNA was used to derive the complete amino acid sequence of the human LDL receptor (Yamamoto et al 1984). This sequence, together with biochemical experiments (Russell et al 1984), revealed that the mature receptor is divided into five distinguishable domains. A model of the domain structure is shown in Figure 2.

A Single Polypeptide Chain With Five Domains

FIRST DOMAIN: LIGAND BINDING   The extreme NH_2-terminus of the LDL receptor consists of a hydrophobic sequence of 21 amino acids that is cleaved from the receptor and is not present in the mature protein. This segment presumably functions as a classic signal sequence to direct the receptor-synthesizing ribosomes to the ER membrane. Because it does not appear in the mature receptor, the signal sequence is omitted from the numbering system that is described below.

The mature receptor (without the signal sequence) consists of 839 amino acids. The first domain of the mature LDL receptor consists of the NH_2-terminal 292 amino acids, which are extremely rich in cysteines (42 out of 292 amino acids). Studies with anti-peptide antibodies revealed that this domain is located on the external surface of the plasma membrane (Schneider et al 1983b). The cysteines are spaced at intervals of 4–7 amino acids (Figure 3). An initial computer analysis suggested that the first domain was made up of eight repeat sequences (Yamamoto et al 1984). More recent analysis of the sequence, considered together with exon/intron mapping data (see below and Figure 2), suggests that the number of repeats is only seven, as shown in Figure 3.

Each of the seven repeats consists of ~ 40 amino acids and contains 6 cysteine residues, which are essentially in register for all of the repeats. The receptor cannot be labeled with [^{3}H]iodoacetamide without prior reduction, suggesting that all of these cysteines are involved in disulfide bonds. This region of the receptor must therefore exist in a tightly cross-linked, convoluted state.

A striking feature of the COOH-terminus of each repeat sequence is a cluster of negatively-charged amino acids (Figure 3). These sequences are complementary to positively-charged sequences in the best-characterized ligand for the LDL receptor, apolipoprotein E (apo E), a 33-kilodalton protein component of the plasma lipid transport system. Apo E contains a cluster of positively charged residues that are believed to face one side of a single α-helix (Innerarity et al 1984). Studies with mutant and proteolyzed forms of apo E, and with monoclonal antibodies against different regions of apo E showed that the positively charged region contains the site by which

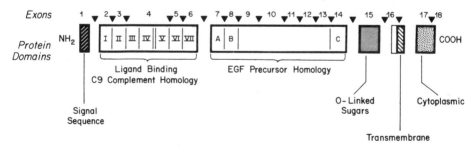

Figure 2 Exon organization and protein domains in the human LDL receptor. The domains of the protein are delimited by thick black lines and are labeled in the lower portion of the figure. The 7 cysteine-rich, 40–amino acid repeats in the LDL binding domain (see also Figure 3) are assigned roman numerals I–VII. Repeats IV and V are separated by 8 amino acids. The 3 cysteine-rich repeats in the EGF precursor homology domain are lettered A–C. The positions at which introns interrupt the coding region are indicated by arrow heads. Exon numbers are shown between the arrow heads. (Reprinted from Südhof et al 1985a with permission.)

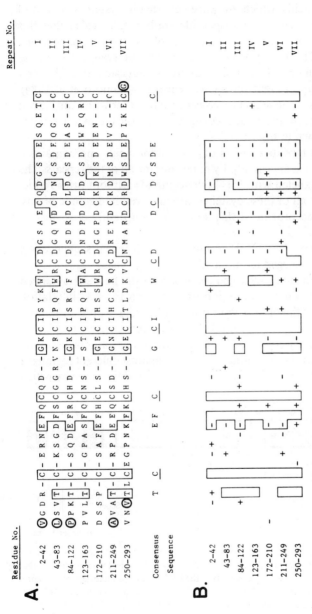

Figure 3   The first domain of the human LDL receptor is composed of seven cysteine-rich repeats. The amino acids constituting each repeat unit are numbered in the left column according to the sequence data of Yamamoto et al (1984). *Panel A*: Optimal alignment was made by the computer programs ALIGN and RELATE with slight modifications based on the sequence data of Yamamoto et al (1984, Südhof et al 1985a). Amino acids that are present at a given position in more than 50% of the repeats are boxed and shown as a consensus on the bottom line. Cysteine residues (C) are underlined. The positions at which intervening sequences interrupt the coding sequence of the gene are denoted by the encircled amino acids. The single letter amino acid code translates to the three letter code as follows: A = Ala; C = Cys; D = Asp; E = Glu; F = Phe; G = Gly; H = His; I = Ile; K = Lys; L = Leu; M = Met; N = Asn; P = Pro; Q = Gln; R = Arg; S = Ser; T = Thr; V = Val; W = Trp; Y = Tyr. *Panel B*: The net charge of each of the amino acids in Panel A is shown. All of the charged amino acids that are conserved bear a negative charge; none are positively charged. (Reprinted from Südhof et al 1985a with permission.)

this protein binds to the LDL receptor (Innerarity et al 1984). It is therefore tempting to speculate that the negatively charged clusters of amino acids within the cysteine-rich repeat sequence of the LDL receptor constitute multiple binding sites, each of which binds a single apo E molecule by attaching to its positively-charged α-helix (Table 1).

This speculation is supported by several observations. First, Innerarity & Mahley (1978) and Pitas et al (1980) showed through kinetic analysis that multiple apo E molecules (4–8) bind to a single LDL receptor. Second, receptor binding of apo E and of apo B (the other ligand for the LDL receptor) is blocked by modification of lysine or arginine residues of the two ligands, a reaction that is achieved with acetylation or cyclohexanedione treatment, respectively (Basu et al 1976, Mahley & Innerarity 1983). Third, although both apo E and apo B have a net negative charge, they nevertheless bind tightly to polyanions, such as heparin, which suggests that both have exposed clusters of basic residues. Fourth, the binding of ^{125}I-LDL to the LDL receptor is inhibited by negatively and positively charged molecules such as heparin, suramin, protamine, and platelet factor 4 (Goldstein et al 1976, Brown et al 1978, Schneider et al 1982). Fifth, the binding activity of the LDL receptor is destroyed by reduction of the disulfide bonds. Sixth, proteolytic treatment of the purified bovine LDL receptor with thrombin yields a 60-kilodalton fragment (isolated on SDS polyacrylamide gels) that is recognized by an anti-peptide antibody directed against the NH_2-terminus of the receptor, and that also specifically binds LDL (Schneider et al, manuscript in preparation). Thus, the LDL binding site is located within the NH_2-terminal 60 kilodaltons of protein, which includes the cysteine-rich acidic region.

The disulfide bonds confer great stability upon the binding site of the

Table 1 Complementarity between amino acid sequences in the LDL receptor and in one of its ligands

LDL Receptor	
	(-Cys-*Asp*-X-X-X-*Asp*-Cys-X-*Asp*-Gly-Ser-*Asp*-*Glu*-)₇
Apo E	
	-*His*-Leu-*Arg*-Lys-Leu-*Arg*-Lys-*Arg*-Leu-Leu-*Arg*-
	140 150

The LDL receptor sequence corresponds to the consensus from the most conserved part of the seven repeat units shown in Figure 3. The apo E sequence (amino acids 140–150) has been identified as being responsible for binding to the LDL receptor (McLean et al 1984, Innerarity et al 1984). Negatively charged amino acids in the LDL receptor sequence and positively charged amino acids in the apo E sequence are italicized. No amino acid sequence data are available for apo B, the other ligand for the LDL receptor.

receptor. The receptor can be boiled in SDS or guanidine and still retain binding activity as long as the disulfide bonds are not reduced (Daniel et al 1983). The disulfide bonds in this region may preserve the stability of the receptor when it delivers LDL to endosomes. In this acidic environment, the negatively-charged residues of the receptor become protonated and lose their charges, allowing the LDL to be released. (LDL is known to dissociate from the receptor in vitro when the pH falls below 6.5) (Basu et al 1978). Despite the titration of its carboxyl groups, the receptor is not irreversibly denatured by this acid exposure, apparently because of the structural stability afforded by the multiple disulfide bonds.

Each of the seven 40–amino acid repeats in the LDL receptor is strongly homologous to a single 40-residue sequence that occurs within the cysteine-rich region of human complement component C9, a plasma protein of 537 amino acids (Stanley et al 1985, DiScipio et al 1984). Of the 19 conserved amino acids in the LDL receptor repeats, 14 are found in the C9 sequence, including the highly conserved negatively charged cluster:

LDL Receptor Consensus (Figure 3):

```
x x T C x x x x E F x C x x G x C I x x x x W x C D x x x D C x D G S D E x x C
E D D C - G N D F Q C S T G R C I K M R L R C N G D N D C G D F S D E D D C
```

Complement factor C9 (residues 77–113) (above).

This finding raises the possibility that C9 might have measurable binding affinity for lipoproteins containing apo B or E, the two ligands for the LDL receptor.

SECOND DOMAIN: HOMOLOGY WITH THE EGF PRECURSOR   Epidermal growth factor (EGF) is a peptide of 53 amino acids that is synthesized as a large precursor of 1217 amino acids (Scott et al 1983, Gray et al 1983). Analysis of the amino acid sequence, as revealed from the sequence of the cloned cDNA, suggests that the EGF precursor is synthesized as a membrane-bound molecule (Doolittle et al 1984). During synthesis, the first 1038 amino acids of the precursor penetrate into the lumen of the ER, whereupon a stretch of 22 hydrophobic amino acids is encountered. According to the current view of protein synthesis in the ER, such a hydrophobic stretch should become anchored in the membrane and stop further transfer (Sabatini et al 1982). Upon completion of translation a short tail of 158 amino acids would face the cytoplasm, constituting the cytoplasmic domain of the precursor. The 53–amino acid EGF sequence lies just outside the membrane-spanning region in the external domain of 1038 amino acids. This external sequence also contains multiple repeats of the EGF sequence that have diverged during evolution, as well as spacer sequences that are not related to EGF. EGF is presumably liberated from this putative membrane-bound precursor by proteolysis, and the peptide is then released, and eventually gains access to receptors on epithelial cells, which it stimulates to divide.

The second domain of the LDL receptor, consisting of ∼400 amino acids (Figure 2), is homologous to a portion of the extracellular domain of the EGF precursor (Russell et al 1984, Yamamoto et al 1984, Südhof et al 1985b). Within this region approximately 35% of the amino acids are identical, with a few short gaps (Russell et al 1984, Yamamoto et al 1984). This overall domain is flanked by several short repetitive sequences of ∼40 amino acids each that are designated A, B, and C in Figure 2. Each of these repeats contains six cysteine residues spaced at similar intervals. The A, B, and C sequences are homologous to four repeat sequences in the EGF precursor (Scott et al 1983, Doolittle et al 1984, Südhof et al 1985a). Repeats A, B, and C in the LDL receptor are also homologous to certain proteins of the blood clotting system, including Factor IX, Factor X, and protein C (Doolittle et al 1984, Südhof et al 1985a).

The existence of these homologies implies that all of these proteins contain regions derived from a common ancestral protein. Does this observation bear any further significance? Does the homology imply that the functions of these regions are conserved among the proteins? It is difficult to imagine functions that would be conserved among a blood clotting factor, the EGF precursor, and the LDL receptor. All of these proteins are made in the ER and reach the cell surface, but it is unlikely that the homology relates to this shared characteristic since other proteins that follow similar routes do not have homologous regions.

Another possibility is that the EGF precursor and the LDL receptor evolved from the duplication of a single ancestral gene that played roles in growth control as well as nutrient delivery (Russell et al 1984). Comparison of the structures of the genes for the human LDL receptor and the human EGF precursor reveals that the region of homology is encoded by eight contiguous exons in each respective gene (Südhof et at 1985b). Of the nine introns that separate these exons, five are located in identical positions in the two protein sequences. This finding strongly suggests that the homologous region arose by a duplication of an ancestral gene. Each copy of the duplicated gene would have further evolved by recruitment of exons from other genes, which provided specialized functions, i.e. the provision of a nutrient (cholesterol) or the signaling of cell growth via the secretion of a peptide (EGF). It is even possible that at some phase of its life cycle the EGF precursor exists in the intact form on the cell surface, where it functions as a receptor, thus increasing the analogy with the LDL receptor. In this regard, Rall et al (1985) have recently shown that the EGF precursor is synthesized in the distal tubules of the mouse kidney where it accumulates in the intact form and is not detectably processed to EGF.

THIRD DOMAIN : O-LINKED SUGARS   Immediately external to the membrane-spanning domain of the human LDL receptor is a sequence of 58 amino acids that contains 18 serine or threonine residues (Yamamoto et al 1984). This domain is encoded within a single exon (see below). Proteolysis studies reveal that this region contains carbohydrate chains attached by O-linkage (Cummings et al 1983, Russell et al 1984). Each O-linked sugar chain consists of a core N-acetylgalactosamine, plus a single galactose and one or two sialic acids. In this respect, the LDL receptor resembles glycophorin, a red-cell membrane protein that contains short O-linked sugar chains attached to clusters of serines and threonines (Marchesi et al 1976). Another cell surface receptor, that for interleukin-2 (IL-2) on T lymphocytes, contains O-linked sugars and a cluster of serine and threonine residues immediately external to the membrane-spanning region (Leonard et al 1984, Nikaido et al 1984).

What is the function of the clustered O-linked sugars? Their similar location in two plasma membrane receptors suggests that these sugars may function as struts to keep the receptors extended from the membrane surface so they can bind their ligands. Why only certain receptors require such struts is not known.

FOURTH DOMAIN : MEMBRANE-SPANNING REGION   This domain consists of a stretch of 22 hydrophobic amino acids. Proteolysis experiments (see below) confirmed that this domain spans the membrane (Russell et al 1984). Comparison of the amino acid sequences of the bovine and human LDL receptors reveals that the membrane-spanning region is relatively poorly conserved (Figure 4). Of the 22 amino acids in this region, 7 differ between human and cow, but all of the substitutions are also hydrophobic. The human receptor contains a single cysteine in the membrane-spanning region. In the bovine receptor this cysteine is replaced by an alanine (Figure 4). Since the bovine and the human receptors function similarly, it seems likely that this intramembraneous cysteine exists in a reduced state in the human LDL receptor.

FIFTH DOMAIN : CYTOPLASMIC TAIL   The human and bovine LDL receptors both contain a COOH-terminal segment of 50 amino acids that projects into the cytoplasm. This sequence is strongly conserved; only four of the 50 amino acids differ between the two species, and each of these substitutions is conservative with respect to the charge of the amino acid (Figure 4). Localization of this domain to the cytoplasmic side of the membrane was determined through use of an anti-peptide antibody directed against the

```
Human   1        H₂N - A V G D R C E R N E F Q C Q D G . . . . . . . .   542 Amino Acids . . . . . .
Bovine           H₂N - A V E D N C G R N E F E C Q D G . . . . . . . ?              ?   ?   ?

Human   559      S K L H S I S S I D V N G G N R K T I L E D E K R L A H P F S L A V F E D K V F W T
Bovine  (1)      S K L H S I S S I D V N G G N R K T V L E D K K K L A H P F S L A I F E D K V F W T

Human   601      D I I N E A I F S A N R L T G S D V N L L A E N L L S P E D M V L F H N L T Q P R G
Bovine  (43)     D V I N E A I F S A N R L T G S D I S L M A E N L L S P E D I V L F H N L T Q P R G

Human   643      V N W C E R T T L S N G G C Q Y L C L P A P Q I N P H S P K F T C A C P D G M L L A
Bovine  (85)     V N W C E R T A L R N G G C Q Y L C L P A P Q I N P R S P K F T C A C P D G M L L A

Human   685      R D M R S C L T E A E A A V A T Q E T S T V R L K V S S T A V R T Q H T T T R P V F
Bovine  (127)    K D M R S C L T E S E S A V T T R G P S T V - - S S T A V G P K R T - - - - -

Human   727      D T S R L P G A T P G L T T V E I V T M S H Q A L G D V A G R G N E K K P S S V R A
Bovine  (159)    - - - - - A S P E L T T A E S V T M S Q Q G Q G D V A S Q A D T E R P G S V G A

Human   769      L S I V L P I V L L V F L C L G V F L L W K N W R L K N I N S I N F D N P V Y Q K T
Bovine  (194)    L Y I V L P I A L L L L A F G T F L L W K N W R L K S I N S I N F D N P V Y Q K T

Human   811      T E D E V H I C H N Q D G Y S Y P S R Q M V S L E D D V A - COOH   839
Bovine  (236)    T E D E V H I C R S Q D G Y T Y P S R Q M V S L E D D V A - COOH   (264)
```

Figure 4 Comparison of the amino acid sequences of the bovine and human LDL receptors. The single letter amino acid code is used (see legend Figure 3). Regions of identity are boxed. The amino acid sequence of the human receptor was deduced from the nucleotide sequence of a full-length cDNA (Yamamoto et al 1984). The NH₂-terminal sequence (16 amino acids) of the bovine receptor was obtained by chemical sequencing of purified receptor protein (Schneider et al 1983b). The COOH-terminal sequence of the bovine receptor (264 amino acids) was deduced from the nucleotide sequence of a partial cDNA (Russell et al 1984). The membrane-spanning region of the two receptors is indicated by the heavy underline.

COOH-terminal sequence (Russell et al 1984). When inside-out membrane vesicles containing receptor were digested with pronase, the antibody-reactive material was removed, and the molecular weight of the receptor was reduced by approximately 5000.

Immediately internal to the membrane the cytoplasmic tail contains a cluster of positively charged amino acids (3 of the first 6 residues are lysines or arginines). This is a frequent feature of plasma membrane proteins (Sabatini et al 1982). Near the COOH-terminal end of the receptor lies a cluster of negatively charged residues (glutamic-aspartic-aspartic) (Figure 4). The cytoplasmic segment also contains several serine and threonine residues and three tyrosines, which may be sites for phosphorylation. This domain also contains a single cysteine, which may be a site for disulfide bond formation or for fatty acylation. None of these modifications have been detected as yet.

The cytoplasmic domain of the LDL receptor plays an important role in clustering in coated pits, either through interaction with clathrin itself or with some protein associated with clathrin on the cytoplasmic side of the membrane (see below). For this reason, it is important to compare the sequences of the cytoplasmic domain of the LDL receptor with those of other receptors known to enter coated pits.

Comparison With Five Other Coated Pit Receptors

In 1984 and early 1985 complete cDNA sequences for 6 coated pit receptors were reported. When the predicted amino acid sequences are compared, no obvious conserved feature is apparent (Figure 5). In particular, the cytoplasmic domains show tremendous differences, varying in size from 38 amino acids (asialoglycoprotein receptor) to 542 amino acids with tyrosine kinase activity (EGF receptor). In addition, the orientations of the receptors are different. Four receptors, those for LDL, EGF, insulin, and polymeric IgA/IgM, are oriented with their NH_2-termini outside the cell and their COOH-termini in the cytoplasm. Two receptors, for transferrin and asialoglycoprotein, exhibit an inverted orientation with their NH_2-termini in the cytoplasm and their COOH-termini outside the cell. The cytoplasmic domains of the 4 receptors that lack tyrosine kinase domains contain clusters of negatively charged amino acids (glutamic and aspartic), generally in regions predicted to have an α-helical conformation. These acidic residues may play some role in interaction with coated pits.

The transferrin receptor exists as a homodimer, linked by a disulfide bond between two cysteine residues that are immediately external to the plasma membrane (McClelland et al 1984, Schneider et al 1984). The insulin receptor is composed of two α-subunits (ligand binding domain) and two β-subunits (tyrosine kinase domain) linked by disulfide bonds. The α and β

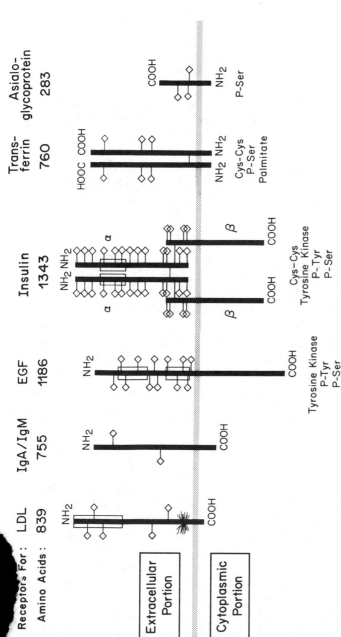

Figure 5 Comparison of the structures of the LDL receptor and 5 other receptors that enter coated pits. Each protein spans the plasma membrane once and has the indicated orientation. The number of amino acids in the cytoplasmic portion of each receptor is as follows: human LDL receptor, 50 residues; rabbit IgA/IgM receptor, 105; human EGF receptor, 542; human insulin receptor, 402; human transferrin receptor, 61; rat asialoglycoprotein receptor, 38. The potential sites of attachment of N-linked oligosaccharide chains are indicated by diamond symbols. The sites of attachment of O-linked oligosaccharide chains in the LDL receptor are shown by a cluster of horizontal lines. Regions rich in cysteine residues in the receptor for LDL, EGF, and insulin are boxed. The insulin receptor is shown as a disulfide-linked heterodimer (Cys-Cys). The transferrin receptor is shown as a disulfide-linked homodimer (Cys-Cys). The cytoplasmic domain of the receptors for EGF and insulin express tyrosine kinase activity. P-Tyr (phosphotyrosine), P-Ser (phosphoserine), and palmitate denote posttranslational covalent modifications that occur on the cytoplasmic domain of the indicated receptor. (For original sequence data see Yamamoto et al 1984; Mostov et al 1984; Ullrich et al 1984, 1985; Ebina et al 1985; Schneider et al 1984; McClelland et al 1984; Drickamer et al 1984; Chiacchia & Drickamer 1984.)

chains are derived from a single precursor molecule that undergoes proteolytic cleavage to assume the configuration shown in Figure 5 (Ullrich et al 1985, Ebina et al 1985). There is no published evidence that the other 4 coated pit receptors form disulfide-linked dimers. Four coated pit receptors (those for LDL, IgA/IgM, transferrin, and asialoglycoproteins) have a single cysteine residue in their cytoplasmic regions; the EGF and insulin receptors have several such residues. This observation suggests that a novel cytoplasmic interchain disulfide bond may play a role in clustering in coated pits.

As mentioned above, the amino acid sequences of the cytoplasmic domains of the bovine and human LDL receptors are highly conserved. In contrast, no significant conservation is observed between the cytoplasmic domains of the asialoglycoprotein receptor in the rat and the analogous receptor in the chicken, even though the extracellular domains of these two receptors are conserved (Drickamer et al 1984).

THE LDL RECEPTOR AT A GENETIC LEVEL

mRNA Structure

In the human tissues studied so far (cultured diploid fibroblasts, SV40-transformed fibroblasts, A-431 epidermal carcinoma cells, fetal and adult adrenal glands, and fetal and adult liver) the LDL-receptor mRNA appears on Northern blots as a single species of approximately 5.3 kb (Yamamoto et al 1984, unpublished observations). (Minor heterogeneity in size cannot be excluded by this technique.) About half of the mRNA consists of an unusually long 3' untranslated region of 2.5 kb. It terminates with a poly $(A)^+$ tract that is about 15 nucleotides downstream from a likely polyadenylation signal (AAUUAAA).

An unusual feature of the 3' untranslated region is the presence of $2\frac{1}{2}$ RNA copies of a middle repetitive sequence present in mammalian genomes. This sequence, designated *Alu*, occurs on average once in every 5000 base pairs (bp) in the human genome, for a total of $\sim 300,000$ copies (Schmid & Jelinek 1982). Each *Alu* repeat is about 300 bp long, and consists of a tandem repeat of two monomeric units—a left monomer of 130 bp and a right monomer 160 bp long, owing to a 30 bp insertion. The human LDL-receptor mRNA has two complete *Alu* sequences and an extra right monomeric unit, all clustered together within a region of about 750 nucleotides (Yamamoto et al 1984).

The bovine mRNA does not contain these *Alu* sequences or any other repetitive sequences (Hobbs et al, manuscript in preparation). The sequences on either side of the *Alu* repeats are conserved in the human and the cow, suggesting that the *Alu* sequences were inserted after the human and bovine evolutionary lines diverged. Restriction maps of genomic DNA suggest that the *Alu* repeats are present at the same location in the 3' untranslated regions of the LDL-receptor genes of the gorilla and chimpanzee, but not the baboon (unpublished observations). If this finding is confirmed by direct cloning studies, it would suggest that the *Alu* sequences have inserted into this location late in the evolution of the primates. Whether this insertion has any functional consequence for the processing, translation, or stability of the receptor mRNA is unknown.

Gene Structure

Southern blotting of genomic DNA demonstrated that the haploid human genome contains a single copy of the LDL-receptor gene (Lehrman et al 1985). This gene resides on chromosome 19, as determined by somatic cell genetic techniques (Francke et al 1984). The gene spans more than 45 kb. Sequences representing almost the entire gene have been isolated from bacteriophage lambda and cosmid libraries (Südhof et al 1985a,b). The position of each intron within the gene has been mapped, and the sequence of each exon-intron junction has been determined.

These studies reveal that the receptor gene is made up of 18 exons. The sites of the introns in relation to the protein sequence are indicated in Figure 2 (Südhof et al 1985a,b). Most of the introns separate regions of the protein that correspond to domains that were identified through the protein chemistry studies described above. The first intron is located just at the end of the DNA encoding the cleaved signal sequence of the protein. Within the binding domain of the receptor (which contains the seven cysteine-rich repeats), introns occur precisely between repeats I and II; II and III; V and VI; and VI and VII (Figures 2, 3). Repeats III, IV, and V are included in one exon. The binding domain is terminated by an intron at amino acid 292, the last residue in the seventh repeat.

The next domain, the region of homology with the EGF precursor, is encoded in 8 contiguous exons. Within this 400–amino acid region of homology are located 3 copies of a repeated sequence (repeats A, B, and C in Figure 2), each of which is encoded by a single exon (Südhof et al 1985a,b). (The striking similarity in the exon-intron organization of this region of the LDL receptor gene and the EGF precursor gene is discussed in the preceding section).

The O-linked sugar domain is also demarcated neatly by two introns (Figure 2). However, not all domains of the protein are encoded by single exons. Thus, the membrane-spanning region is interrupted by an intron. Another intron interrupts the coding region for the cytoplasmic tail 11 amino acids from the COOH-terminus.

The placement of the introns is consistent with the notion that the human

LDL receptor gene was constructed by the stepwise assembly of exons that encode useful protein sequences. Thirteen of the 18 exons comprising the LDL receptor gene encode protein sequences that are homologous to sequences in other proteins: 5 of these exons encode a sequence similar to one in the C9 component of complement; 3 exons encode a sequence similar to a repeat sequence in the EGF precursor and in 3 proteins of the blood clotting system; and 5 other exons encode nonrepeated sequences that are shared only with the EGF precursor. The LDL receptor thus appears to be a mosaic protein built up of exons shared with different proteins (Südhof et al 1985a,b).

The 5' untranslated region of the receptor gene is less than 100 base pairs, and it is not interrupted by an intron (Südhof et al 1985a). Two TATA-like boxes occur 20–30 base pairs to the 5' side of the two major sites of transcription initiation located between nucleotides -79 to -93. Analysis of the upstream promoter region will be of interest because transcription of the gene into mRNA appears to be regulated by a feedback mechanism. When cholesterol accumulates in cells, the level of cytoplasmic mRNA for the receptor declines dramatically (Russell et al 1983, Yamamoto et al 1984), and this leads to a decrease in the rate of synthesis of the receptor protein (Goldstein & Brown 1977). It is likely, but not yet proven, that the decrease in cytoplasmic mRNA is due to a cholesterol-mediated suppression of transcription of the receptor gene.

BIOSYNTHESIS OF THE HUMAN LDL RECEPTOR

The presence of a cleaved NH_2-terminal hydrophobic signal sequence suggests that the LDL receptor is synthesized on membrane-bound ribosomes. At the earliest time point that can be studied (15 min after the addition of $[^{35}S]$methionine to cultured human fibroblasts) the receptor appears in immunoprecipitates as a protein with an apparent molecular weight of 120,000, as estimated on SDS polyacrylamide gels (Tolleshaug et al 1982). This precursor contains asparagine-linked (N-linked), high-mannose oligosaccharide chains, which are sensitive to endoglycosidase-H (endo-H) (Tolleshaug et al 1983). According to the best estimates available, there are two N-linked sugar chains on the purified bovine LDL receptor (Cummings et al 1983). Although the protein sequence for the human receptor shows five potential N-linked glycosylation sites (Figure 5), it is possible that the three N-linked sites in the cysteine-rich, disulfide-linked region of the receptor are not glycosylated (Yamamoto et al 1984).

The earliest detectable receptor precursor also contains N-acetylgalactosamine (GalNAc) residues attached to serines and threonines by O linkage. This finding emerged from experiments in which the [^{3}H]glucosamine-labeled 120-kilodalton precursor of human A-431 epidermal carcinoma cells was isolated by SDS gel electrophoresis and digested with pronase. Multiple GalNAc residues were found on a single pronase-resistant fragment (Cummings et al 1983).

The presence of O-linked GalNAc residues at a time when the N-linked sugar chains are still in the high-mannose (endo-H sensitive) configuration implies that the GalNAc transferase that initiates synthesis of O-linked sugar chains is proximal to the *cis*-Golgi stacks. This follows from the observation that once a protein reaches the *cis*-Golgi the mannose residues are trimmed from the N-linked sugars and the chains become endo-H resistant (Hubbard & Ivatt 1981). Whether the O-linked GalNAc residues are added in the ER, or whether they are added in some transitional zone between the ER and the *cis*-Golgi is not known.

Between 30 and 60 min after synthesis, the LDL receptor precursor undergoes a sudden shift in apparent molecular weight from 120,000 to 160,000 (Tolleshaug et al 1982, 1983; Schneider et al 1983a). The timing of this shift coincides with the maturation of the N-linked and O-linked chains. The shift is not the result of the alteration in N-linked sugars, because a change of nearly equal magnitude occurs in cells that are treated with tunicamycin, which blocks the addition of N-linked chains. Conversely, the increase in apparent molecular weight is minimized in a mutant strain of Chinese hamster ovary (CHO) cells that is unable to add galactose to the core GalNAc residues of the O-linked chains (Cummings et al 1983). These findings suggest that the 40,000 change in apparent molecular weight is attributable to the elongation of the O-linked chains. This elongation consists of the addition of a single galactose and one or two sialic acids to the GalNAc core sugar of each O-linked chain.

The apparent molecular weight of the mature receptor is reduced by only about 10,000 when the sialic acids are completely removed with neuraminidase (Schneider et al 1982, Cummings et al 1983). Thus, most of the change from 120,000 to 160,000 daltons is contributed not by the sialic acids, but by the simple addition of galactose residues to the GalNAc core sugars. This change selectively retards the mobility of the receptor on SDS gels, so that the mature receptor migrates more slowly than would be appropriate for its true molecular weight. The calculated molecular weight of the protein component of the receptor is 93,102. When the mature carbohydrates are included, the molecular weight will be ~115,000, not 160,000 as observed on SDS gels. Aberrant migration of other membrane glycoproteins that contain clustered O-linked sugars has been previously documented (Marchesi et al 1976).

The increase in apparent molecular weight of the LDL receptor is partially blocked when cells are incubated with monensin, an ionophore that blocks vesicular transport in the Golgi complex (unpublished observations). Under these conditions there is no longer a discrete jump from apparent molecular weight of 120,000 to 160,000. Rather, the receptor appears as a smear between these two extremes.

The LDL receptor in human fibroblasts can be labeled with ^{35}S-sulfate, which attaches to N-linked sugars; incorporation is blocked by tunicamycin (Cummings et al 1983). The receptor synthesized in tunicamycin-treated fibroblasts appears to undergo normal internalization and recycling, but subtle changes in receptor half-life have not been ruled out (unpublished observations). Since these tunicamycin-treated receptors do not contain sulfate, it is unlikely that sulfate performs a crucial function in the LDL receptor. Incorporation of ^{35}S-sulfate into the receptor in human A-431 carcinoma cells could not be demonstrated (Cummings et al 1983).

NATURALLY OCCURRING MUTATIONS IN THE LDL RECEPTOR

The power of the LDL receptor as a system for the study of receptor-mediated endocytosis derives from the existence of many naturally occurring mutations in the LDL receptor gene that disrupt receptor function in revealing ways. The mutations occur in individuals with familial hypercholesterolemia (FH) (Goldstein & Brown 1983). Those who inherit one mutant LDL-receptor gene produce half the usual number of normal receptors. In tissue culture their cells degrade LDL at about half the normal rate. In the body the receptor deficiency causes LDL to build up in plasma to levels about twofold greater than normal. Eventually, the high plasma LDL levels lead to atherosclerosis and heart attacks as early as 40 years of age (Brown & Goldstein 1984).

Individuals with two mutant LDL-receptor genes are termed FH homozygotes. Their cells produce few or no functional LDL receptors. As a result, plasma LDL accumulates to levels eight to ten times greater than normal, and they develop atherosclerosis and heart attacks in childhood.

At least ten different mutant alleles at the LDL-receptor locus have been described (Goldstein & Brown 1983, Tolleshaug 1982, 1983, Schneider et al 1983a, Lehrman et al 1985). Many of the phenotypic FH homozygotes actually represent compound heterozygotes who inherit different mutant alleles of the receptor gene from each parent. Study of cultured skin fibroblasts from 104 FH homozygotes revealed that the mutations could be divided into four broad classes based upon their effects on receptor structure and function. These mutations are summarized in Table 2 and discussed below.

Class 1 Mutations: No Detectable Precursor

These alleles, designated R-0 for "receptor-zero," are the most frequent of the mutant alleles. It is difficult to determine their frequency directly (since they cannot be identified unequivocally in the heterozygous state), but they probably account for about one-third to one-half of all mutant alleles at the LDL-receptor locus. Class 1 alleles fail to express receptor proteins as measured by functional assays (binding of ^{125}I-LDL) or immunological assays (immunoblotting or precipitation by a variety of monoclonal and polyclonal antibodies directed against the LDL receptor). It is likely that this class includes nonsense mutations, which introduce termination codons early in the protein coding region. It may also include: point mutations in the promoter that block transcription of mRNA; point mutations in intron-exon junctions that alter the splicing of mRNA; and large deletions.

Class 2 Mutations: Precursor Not Processed

These alleles encode receptor precursors that are synthesized in normal or reduced amounts, but that do not undergo any apparent increase in molecular weight after synthesis. These receptors remain in the endo-H sensitive form, and they do not receive sialic acid, as indicated by their lack of susceptibility to neuraminidase. Thus, we believe these receptors are not transported to the Golgi complex. Receptors specified by these alleles never reach the cell surface, and hence they are protected when the surface of intact cells is treated with pronase (Tolleshaug et al 1983).

Most of the alleles in this class encode receptors with apparent molecular weights on SDS gels of 120,000, which is similar to the apparent molecular weight of the normal precursor. These alleles are designated R-120 (Tolleshaug et al 1982, 1983). Detailed structural analysis of the oligosaccharide chains of one mutant receptor encoded by the R-120 allele showed it to contain N-linked high mannose chains and O-linked core GalNAc residues indistinguishable from the normal 120-kilodalton receptor precursor (Cummings et al 1983). We have also observed precursor proteins with abnormal apparent molecular weights of 100,000 and 135,000 that fall into this class (designated R-100 and R-135 alleles, respectively). These molecular weight abnormalities may result from alterations in the length of the protein chain, rather than from alterations in carbohydrate, since the molecular weight remains abnormal after endo-H treatment (Tolleshaug et al 1983).

Variants of the Class 2 mutation were observed in a consanguineous black American family, in several Afrikaners, and in a strain of rabbits that has a syndrome similar to FH, i.e. Watanabe heritable hyperlipidemic

Table 2 Mutations at the LDL receptor locus that produce familial hypercholesterolemia (FH)

Class of mutation	Allele designation[1]	Apparent receptor mass on SDS gels (kDa)		Receptor location				LDL binding to intact cells	Frequency in FH patients
		Precursor	Mature	Intra-cellular	Plasma membrane		Extra-cellular		
					Coated pits	Noncoated regions			
Class 1: No detectable precursor	*R-0*	None	None					None	Common
Class 2: Precursor not processed	*R-100*	100	100	+				None	Rare; found in Lebanese
	R-120	120	120	+				None	Common
	R-135	135	135	+				None	Rare
Class 2 variant: Precursor processed slowly, mature receptor binds LDL poorly	*R-120^{slow}⇌160 b$^-$*	120	160	+	$(+)^2$			Reduced	Rare; found in Afrikaners and in WHHL rabbits

	Precursor	Mature				Frequency
Class 3: Precursor processed normally, mature receptor binds LDL poorly						
R-140 b⁻	100	140	+		Reduced	Rare
R-160 b⁻	120	160	+		Reduced	Common
R-210 b⁻	170	210	+		Reduced	Rare
Class 4: Precursor processed normally, mature receptor binds LDL normally, but does not enter coated pits						
R-150 i⁻, sec	110	150	(+)	(+)	Normal binding; defective internalization	Rare
R-160 i⁻	120	160		+	Normal binding; defective internalization	Rare
R-155 i⁻	115	155		+		Rare

[1] Allele designations are based on the apparent molecular weight (in kilodaltons) on SDS polyacrylamide gel electrophoresis of the mature form of the receptor, i.e. the predominant form observed after a 2 hr pulse followed by a 2 hr chase. b^- denotes defective LDL binding; i^- denotes defective internalization; *sec* denotes secretion from the cell.

[2] Symbols in parentheses refer to minor populations.

(WHHL) rabbits (Schneider et al 1983a, unpublished observations). In these variants the receptor is produced as a 120-kilodalton precursor that is processed to the mature form at a slow but finite rate. Eventually about 10% of the receptors appear on the cell surface as 160-kilodalton mature proteins. Even after they reach the surface, these receptors have a reduced ability to bind LDL. Normal receptors bind equimolar amounts of IgG-C7 (a monoclonal antibody against the external domain of the LDL receptor) and LDL protein (Beisiegel et al 1981). In the Class 2 variants the ratio of LDL binding to monoclonal antibody binding is reduced, suggesting that these receptors have an abnormality in the LDL binding site as well as a slower rate of transport to the surface (Schneider et al 1983a).

The molecular basis of the defect in the Class 2 mutations is not known. These receptors are all recognized by monoclonal and polyclonal antibodies against the receptor, so their structures are not drastically different from the normal receptor. It seems likely that the failure of transport arises from some subtle alteration in structure. Elucidation of this change should lead to new insights into the signals that govern transport of proteins from the ER to the *cis*-Golgi.

Scheckman and co-workers have described a mutation in yeast invertase, a secreted enzyme, that is analogous to the Class 2 mutations (Schauer et al 1985). The defect in invertase results from the alteration of a single amino acid at the site at which the hydrophobic NH_2-terminal signal sequence is cleaved from the protein. In the absence of cleavage, invertase is transported to the Golgi at 2% of the normal rate. A similar transport defect has been created in yeast by in vitro mutagenesis of the gene for acid phosphatase (Haguenauer-Tsapis & Hinnen 1984). It seems likely that some of the Class 2 mutations in FH may result from the failure to cleave the signal sequence from the protein.

Class 3 Mutations: Precursor Processed, Abnormal Binding of LDL by Receptor

Receptors specified by Class 3 mutant alleles reach the surface at a normal rate and are recognized on the surface by monoclonal anti-receptor antibody (IgG-C7). However, these receptors bind less than 15% of the normal amount of ^{125}I-LDL (Goldstein & Brown 1983, Beisiegel et al 1981, Tolleshaug et al 1983).

Most commonly, the receptors produced by the Class 3 alleles have a normal molecular weight on SDS gel electrophoresis. This allele is designated *R-160 b⁻*. Receptors with molecular weights of 140,000 (*R-140 b⁻*) (Tolleshaug et al 1983) and 210,000 (*R-210 b⁻*) (Tolleshaug et al 1982) have also been described. Both of these proteins originate as precursors with apparent molecular weights that are 40,000 less than their mature species, i.e. 100,000 and 170,000, respectively. The correct increase in apparent molecular weight suggests that the carbohydrate processing reactions occur normally. Structural analysis of the carbohydrates of the receptor specified by the *R-210 b⁻* allele showed no abnormality (Cummings et al 1983). We believe, therefore, that the abnormal molecular weight is due to alterations in the amino acid sequence, and not to changes in carbohydrate content.

An explanation for the abnormally sized receptors is suggested by the structure of the binding domain for LDL. As discussed above, this binding domain is made up of seven repeats of a 40 amino acid sequence. Because of this internal homology, the DNA encoding such a repeat structure would be susceptible to deletion or duplication following "slipped mispairing" and recombination during meiosis. Such duplications or deletions would change the size of the receptor and might reduce LDL binding without affecting the binding of monoclonal antibody IgG-C7 (which does not recognize the LDL binding site). This hypothesis is presently being tested with the available cDNA and genomic probes.

Class 4 Mutations: Precursor Processed, Receptor Binds LDL But Does Not Cluster In Coated Pits

These are the so-called internalization-defective mutations. The original example was patient JD (Brown & Goldstein 1976). Biochemical studies showed that patient JD is a compound heterozygote (Goldstein et al 1977). From his mother he inherited a gene that produces a nonfunctional receptor (*R-0* allele). From his father he inherited a gene that produces a receptor of normal size that reaches the surface and binds LDL normally, but is not able to carry the bound LDL into the cell (*R-160 i⁻* allele). Electron microscopic studies revealed that the internalization-defective receptors in JD and his father are present on the surface in small clusters, but they are not sequestered in coated pits, even though coated pits are present in these cells (Anderson et al 1977b) and the coated pits function normally in the receptor-mediated endocytosis of other ligands such as EGF (Goldstein et al 1978).

Subsequently, four other FH patients with internalization defects have been identified. One is a young man from Minnesota (designated FH 274 or BH) who is also a compound heterozygote with an internalization-defective allele inherited from his mother and a nonfunctional allele from his father (Goldstein et al 1982, Lehrman et al 1985). Another is a patient from Japan, born of consanguineous parents, who appears to be homozygous for an internalization-defective allele (Miyake et al 1981). The third and fourth patients are Arab siblings from a consanguineous marriage, who appear to be homozygous for an internalization-defective allele (Lehrman et al 1985a).

We have recently elucidated the molecular defect in the internalization-defective allele in patient FH 274 (Lehrman et al 1985b). Protein chemistry studies demonstrated that his mutant receptor has two abnormal properties. First, it is about 10,000 daltons smaller in apparent molecular weight than the normal receptor. Second, about 80% of the receptors are secreted into the culture medium, and only about 20% remain associated with the cell. (In normal fibroblasts no detectable amounts of receptor are secreted.) The allele giving rise to this abnormal receptor was inherited from the mother. Since the mother's cells had the internalization-defective phenotype, the shortened receptor must be responsible for the internalization-defective state.

Through restriction endonuclease mapping of genomic DNA and subsequent cloning of the relevant genomic fragments into bacteriophage lambda, Lehrman et al (1985b) demonstrated that the mutant internalization-defective allele in FH 274 cells (designated R-150 i^-, sec) had undergone a large deletion. The deletion resulted from a recombination between one of the *Alu* sequences in the 3'-untranslated region of the mRNA and an *Alu* sequence that is located 5 kb upstream in the intervening sequence that separates the exon encoding the O-linked sugar region from the exon encoding the membrane-spanning domain. This deletion eliminated the DNA sequences that encode the membrane-spanning region and the cytoplasmic domain of the receptor.

According to the above results the protein produced by the deleted gene is expected to have a normal sequence from the NH_2-terminus through the O-linked sugar region. Thereafter, the protein should terminate because the deletion joint should produce a random sequence of nucleotides in the mRNA. By chance, a protein termination codon is expected to be reached within 20 codons. These predictions derived from the genomic cloning were confirmed by studies of the FH 274 protein. The truncated receptor precursor (110 kilodaltons) showed the normal 40,000 increase in molecular weight after synthesis, indicating that the O-linked sugars were present (Lehrman et al 1985b). Moreover, the FH 274 protein reacted with anti-peptide antibodies directed against the NH_2-terminal domain of the receptor, but failed to react with anti-peptide antibodies directed against the COOH-terminal cytoplasmic tail, which confirms that the cytoplasmic domain was eliminated (Lehrman et al 1985b).

The secretion of the truncated receptor from the FH 274 cells is an expected result based on earlier studies with truncated mutants for viral envelope proteins, such as that of the influenza hemagglutinin (Gething & Sambrook 1982) and the G protein of vesicular stomatitis virus (Florkiewicz et al 1983). When such truncated proteins are synthesized on membrane-bound ribosomes, the lack of a hydrophobic membrane-spanning region allows the entire protein to translocate across the ER membrane and to appear in the lumen of the ER, from which it is eventually incorporated into secretory vesicles.

Why are some of the mutant receptors produced by the FH 274 cells found attached to the cell surface? These molecules appear to be firmly attached since they cannot be washed off easily by high salt, EDTA, highly charged polymers, or reducing agents (Lehrman et al 1985b, unpublished observations). The membrane-adherent receptors are found in noncoated regions of the cell surface where they bind LDL (Anderson et al, unpublished observations) and give rise to the internalization-defective phenotype by which the FH 274 cells were originally identified (Goldstein et al 1982). It is possible that these truncated receptors bind with high affinity to some other protein that keeps them anchored to the membrane. Alternatively, perhaps some of the mutant receptors have acquired a hydrophobic sequence by chance, as a result of alternative splicing of the mRNA within the intervening sequence that contains the deletion joint. Such splicing might lead to a random sequence of nucleotides that happens to encode a stretch of hydrophobic amino acids prior to the termination codon. Thus, some of the receptors might have sticky hydrophobic COOH-terminal tails that make them adhere to the membrane.

We have recently identified defects in two other internalization-defective LDL receptor mutations (Lehrman et al 1985a and manuscript in preparation). Both of these mutations involve single base substitutions in the exon encoding the majority of the cytoplasmic domain of the receptor (exon 17 in Figure 2), and both were identified through cloning and sequencing of genomic DNA fragments that contain this exon. One of these point mutations, found in the Arab family mentioned above, results from a guanosine to adenosine transition, which changes a tryptophan codon (UGG) to a termination codon (UGA). The receptor produced by this gene terminates at a position corresponding to the tryptophan that is just at the beginning of the cytoplasmic domain of the receptor (Figure 4). The resulting protein has a truncated cytoplasmic domain of only 2 (rather than 50) amino acids. This protein moves to the cell surface, but the lack of a cytoplasmic domain renders it incapable of clustering in coated pits.

A different point mutation occurs in JD, the first internalization-defective FH patient to be described (Brown & Goldstein 1976, Goldstein et al 1977). In this case, an adenosine to guanosine transition converts a codon for tyrosine (UAU) into a codon for cysteine (UGU). This mutation occurs in the cytoplasmic domain 33 residues from the COOH-terminus (Lehrman et al, manuscript in preparation). The mutant receptor has 2 cysteine residues in its cytoplasmic domain: the normal one at position 818 and a new one at position 807. We do not yet know whether the failure to cluster in coated pits and the abnormal internalization are attributable to the acquisition of an extra cysteine or to the loss of a crucial tyrosine.

The finding of defects in the cytoplasmic domain in three internalization-defective mutations (one deletion, one nonsense mutation, and one missense mutation) supports our earlier proposal that this domain is crucial in directing the LDL receptor to coated pits (Anderson et al 1977, Goldstein et al 1979).

EXPERIMENTALLY INDUCED MUTATIONS IN THE LDL RECEPTOR

To increase the repertoire of available mutations, Krieger et al (1981, 1983) developed two methods for creating LDL-receptor mutations in tissue culture cells. In the first procedure the cholesteryl esters of LDL are extracted and replaced with hydrophobic molecules that convert the LDL into either toxic or fluorescent particles (Krieger et al 1981). Mutagen-treated Chinese hamster ovary (CHO) cells are incubated with re-constituted LDL-containing toxic 25-hydroxycholesteryl oleate. Wild-type cells take up this lipoprotein via the LDL receptor, liberate the 25-hydroxycholesterol in lysosomes, and die. The few surviving clones are incubated with LDL reconstituted with fluorescent cholesteryl ester, and the colonies that fail to accumulate fluorescence are picked. This two-step isolation procedure yielded LDL-receptor-deficient cells at a frequency of 1×10^{-5} (Krieger et al 1981).

The second selection procedure takes advantage of two fungal metabolites: compactin, a potent inhibitor of cholesterol biosynthesis, and amphotericin B, a polyene antibiotic that forms toxic complexes with sterols in membranes (Krieger et al 1983). Mutagen-treated CHO cells are preincubated in a medium containing compactin, LDL, and small amounts of mevalonate, a combination that makes CHO cells dependent on the LDL receptor for obtaining cholesterol (Goldstein et al 1979b). After the preincubation, mutant cells that cannot utilize the cholesterol of LDL become cholesterol-deficient. Subsequent incubation with amphotericin B kills the wild-type cells through formation of complexes with membrane cholesterol. The receptor-deficient clones, which are depleted in cholesterol, do not bind amphotericin B and therefore they survive. With this procedure mutant cells are isolated at a frequency of approximately 2.6×10^{-5} (Krieger et al 1983).

All of the mutants obtained to date, by either of the two procedures, express an LDL-receptor-deficient or negative phenotype, i.e. there is a proportional reduction in the binding, internalization, and degradation of LDL. Complementation studies suggest that the mutants fall into four groups, designated ldlA, ldlB, ldlC, and ldlD (Kingsley & Krieger 1984). The ldlA locus appears to be the structural gene for the LDL receptor on the basis of two observations: 1) fusion of these cells with normal human fibroblasts, but not FH homozygote fibroblasts, leads to complementation; and 2) mutants in the ldlA group show abnormal LDL receptors by immunoprecipitation and SDS gel electrophoresis (Kozarsky & Krieger, manuscript in preparation). The ldlA locus appears to be diploid in CHO cells, as defined by the isolation of a heterozygous revertant of a homozygous mutant (Kingsley & Krieger 1984).

The biochemical basis of the LDL-receptor deficiency in the ldlB, ldlC, and ldlD mutants is unclear at the present time; it may involve defects in the posttranslational processing of the LDL receptor. Most of these mutants exhibit altered sensitivity to plant lectins, which suggests they harbor a pleiotropic abnormality in glycoprotein processing (Krieger et al, unpublished observations). The ldlD mutants are unique in that the receptor-negative phenotype can be corrected by co-cultivation with other mammalian cells (Krieger 1983). LDL-receptor activity can also be induced in the ldlD cells by addition of a factor found in human or bovine serum (Krieger 1983).

Sege et al (1984) recently used the technique of calcium phosphate–mediated gene transfer to introduce human genomic DNA into the ldlA cells. After transfection the ldlA cells were incubated in the presence of compactin, LDL, and mevalonate, a combination that selects for functional LDL receptors. One of the clones that survived this selection was shown to express the human LDL receptor on its cell surface, as determined by immunoprecipitation with monoclonal antibody IgG-C7, which reacts with the receptor of human but not of hamster origin (Beisiegel et al 1981). The transfected human receptor gene was functional in the binding, uptake, and degradation of LDL, and its expression was suppressed by cholesterol (Sege et al 1984).

The ability to introduce a functional human LDL-receptor gene into mutant CHO cells illustrates the power of the LDL-receptor system for use in studies of experimentally induced mutations. Not only are efficient selection systems available for the production of LDL-receptor-deficient mutants, but efficient systems are also now available for selecting revertants of these mutants, and for introducing transfected genes into these cells. Finally, the ability to express functional LDL receptors from cloned cDNAs following transfection of cultured cells (Yamamoto et al 1984), together with the techniques of in vitro mutagenesis, should provide much exciting information regarding the LDL receptor in particular, and the process of receptor-mediated endocytosis in general.

ACKNOWLEDGMENT

The original research described in this article was supported by grants from the National Institutes of Health (HL 20948 and HL 31346). D.W.R. is the recipient of an NIH Research Career Development Award (HL 01287). W.J.S. is an Established Investigator of the American Heart Association.

Literature Cited

Anderson, R. G. W., Brown, M. S., Beisiegel, U., Goldstein, J. L. 1982. Surface distribution and recycling of the LDL receptor as visualized by anti-receptor antibodies. *J. Cell. Biol.* 93:523–31

Anderson, R. G. W., Brown, M. S., Goldstein, J. L. 1977a. Role of the coated endocytic vesicle in the uptake of receptor-bound low density lipoprotein in human fibroblasts. *Cell* 10:351–64

Anderson, R. G. W., Goldstein, J. L., Brown, M. S. 1977b. A mutation that impairs the ability of lipoprotein receptors to localise in coated pits on the cell surface of human fibroblasts. *Nature* 270:695–99

Anderson, R. G. W., Goldstein, J. L., Brown, M. S. 1976. Localization of low density lipoprotein receptors on plasma membrane of normal human fibroblasts and their absence in cells from a familial hypercholesterolemia homozygote. *Proc. Natl. Acad. Sci. USA* 73:2434–38

Barak, L. S., Webb, W. W. 1982. Diffusion of low density lipoprotein-receptor complex on human fibroblasts. *J. Cell Biol.* 95:846–52

Basu, S. K., Goldstein, J. L., Anderson, R. G. W., Brown, M. S. 1976. Degradation of cationized low density lipoprotein and regulation of cholesterol metabolism in homozygous familial hypercholesterolemia fibroblasts. *Proc. Natl. Acad. Sci. USA* 73:3178–82

Basu, S. K., Goldstein, J. L., Anderson, R. G. W., Brown, M. S. 1981. Monensin interrupts the recycling of low density lipoprotein receptors in human fibroblasts.

Cell 24:493–502

Basu, S. K., Goldstein, J. L., Brown, M. S. 1978. Characterization of the low density lipoprotein receptor in membranes prepared from human fibroblasts. *J. Biol. Chem.* 253:3852–56

Beisiegel, U., Schneider, W. J., Brown, M. S., Goldstein, J. L. 1982. Immunoblot analysis of low density lipoprotein receptors in fibroblasts from subjects with familial hypercholesterolemia. *J. Biol. Chem.* 257:13150–56

Beisiegel, U., Schneider, W. J., Goldstein, J. L., Anderson, R. G. W., Brown, M. S. 1981. Monoclonal antibodies to the low density lipoprotein receptor as probes for study of receptor-mediated endocytosis and the genetics of familial hypercholesterolemia. *J. Biol. Chem.* 256:11923–31

Berg, T., Blomhoff, R., Naess, L., Tolleshaug, H., Drevon, C. A. 1983. Monensin inhibits receptor-mediated endocytosis of asialoglycoproteins in rat hepatocytes. *Exp. Cell Res.* 148:319–30

Bleil, J. D., Bretscher, M. S. 1982. Transferrin receptor and its recycling in HeLa cells. *EMBO J.* 1:351–55

Bretscher, M. S. 1984. Endocytosis: Relation to capping and cell locomotion. *Science* 244:681–86

Bretscher, M. S., Pearse, B. M. F. 1984. Coated pits in action. *Cell* 38:3–4

Brodsky, F. M. 1984. The intracellular traffic of immunologically active molecules. *Immunol. Today* 5:350–57

Brown, M. S., Anderson, R. G. W., Basu, S. K., Goldstein, J. L. 1982. Recycling of cell surface receptors: Observations from the LDL receptor system. *Cold Spring Harbor Symp. Quant. Biol.* 46: 713–21

Brown, M. S., Anderson, R. G. W., Goldstein, J. L. 1983. Recycling receptors: The round-trip itinerary of migrant membrane proteins. *Cell* 32: 663–67

Brown, M. S., Deuel, T. F., Basu, S. K., Goldstein, J. L. 1978. Inhibition of the binding of low-density lipoprotein to its cell surface receptor in human fibroblasts by positively charged proteins. *J. Supramol. Struct.* 8: 223–34

Brown, M. S., Goldstein, J. L. 1976. Analysis of a mutant strain of human fibroblasts with a defect in the internalization of receptor-bound low density lipoprotein. *Cell* 9: 663–74

Brown, M. S., Goldstein, J. L. 1984. How LDL receptors influence cholesterol and atherosclerosis. *Sci. Am.* 251: 58–66

Carpenter, G., Cohen, S. 1979. Epidermal growth factor. *Ann. Rev. Biochem.* 48: 193–216

Carpentier, J.-L., Gorden, P., Anderson, R. G. W., Goldstein, J. L., Brown, M. S., et al. 1982. Co-localization of ^{125}I-epidermal growth factor and ferritin-low density lipoprotein in coated pits: A quantitative electron microscopic study in normal and mutant fibroblasts. *J. Cell Biol.* 95: 73–77

Chiacchia, K. B., Drickamer, K. 1984. Direct evidence for the transmembrane orientation of the hepatic glycoprotein receptors. *J. Biol. Chem.* 259: 15440–46

Ciechanover, A., Schwartz, A. L., Dautry-Varsat, A., Lodish, H. F. 1983. Kinetics of internalization and recycling of transferrin and the transferrin receptor in a human hepatoma cell line. Effect of lysosomotropic agents. *J. Biol. Chem.* 258: 9681–89

Cummings, R. D., Kornfeld, S., Schneider, W. J., Hobgood, K. K., Tolleshaug, H., et al. 1983. Biosynthesis of the N- and O-linked oligosaccharides of the low density lipoprotein receptor. *J. Biol. Chem.* 258: 15261–73

Daniel, T. O., Schneider, W. J., Goldstein, J. L., Brown, M. S. 1983. Visualization of lipoprotein receptors by ligand blotting. *J. Biol. Chem.* 258: 4606–11

Dautry-Varsat, A., Ciechanover, A., Lodish, H. F. 1983. pH and the recycling of transferrin during receptor-mediated endocytosis. *Proc. Natl. Acad. Sci. USA* 80: 2258–62

DiPaola, M., Maxfield, F. R. 1984. Conformational changes in the receptors for epidermal growth factor and asialoglycoproteins induced by the mildly acidic pH

found in endocytic vesicles. *J. Biol. Chem.* 259: 9163–71

DiScipio, R. G., Gehring, M. R., Podack, E. R., Kan, C. C., Hugli, T. E., et al. 1984. Nucleotide sequence of cDNA and derived amino acid sequence of human complement component C9. *Proc. Natl. Acad. Sci. USA* 81: 7298–7302

Doolittle, R. F., Feng, D.-F., Johnson, M. S. 1984. Computer-based characterization of epidermal growth factor precursor. *Nature* 307: 558–66

Drickamer, K., Mamon, J. F., Binns, G., Leung, J. O. 1984. Primary structure of the rat liver asialoglycoprotein receptor: Structural evidence for multiple polypeptide species. *J. Biol. Chem.* 259: 770–78

Dunn, W. A., Hubbard, A. L. 1984. Receptor-mediated endocytosis of epidermal growth factor by hepatocytes in the perfused rat liver: Ligand and receptor dynamics. *J. Cell Biol.* 98: 2148–59

Ebina, Y., Ellis, L., Jarnagin, K., Edery, M., Graf, L., et al. 1985. The human insulin receptor cDNA: The structural basis for hormone-activated transmembrane signalling. *Cell* 40: 747–58

Fawcett, D. W. 1965. Surface specializations of absorbing cells. *J. Histochem. Cytochem.* 13: 75–91

Florkiewicz, R. Z., Smith, A., Bergmann, J. E., Rose, J. K. 1983. Isolation of stable mouse cell lines that express cell surface and secreted forms of the vesicular stomatitis virus glycoprotein. *J. Cell. Biol.* 97: 1381–88

Francke, U., Brown, M. S., Goldstein, J. L. 1984. Assignment of the human gene for the low density lipoprotein receptor to chromosome 19: Synteny of a receptor, a ligand, and a genetic disease. *Proc. Natl. Acad. Sci. USA* 81: 2826–30

Friend, D. S., Farquhar, M. G. 1967. Functions of coated vesicles during protein absorption in the rat vas deferens. *J. Cell Biol.* 35: 357–76

Gething, M.-J., Sambrook, J. 1982. Construction of influenza haemagglutinin genes that code for intracellular and secreted forms of the protein. *Nature* 300: 598–603

Geuze, H. J., Slot, J. W., Strous, G. J. A. M., Lodish, H. F., Schwartz, A. L. 1983. Intracellular site of asialoglycoprotein receptor-ligand uncoupling: Double-label immunoelectronmicroscopy during receptor mediated endocytosis. *Cell* 32: 277–87

Geuze, H. J., Slot, J. W., Strous, G. J. A. M., Peppard, J., von Figura, K., et al. 1984. Intracellular receptor sorting during endocytosis: Comparative immunoelectron microscopy of multiple receptors in rat liver. *Cell* 37:195–204

Goldstein, B., Wofsy, C., Bell, G. 1981. Interactions of low density lipoprotein receptors with coated pits on human fibroblasts: Estimate of the forward rate constant and comparison with the diffusion limit. *Proc. Natl. Acad. Sci. USA* 78:5695–98

Goldstein, J. L., Anderson, R. G. W., Brown, M. S. 1979a. Coated pits, coated vesicles, and receptor-mediated endocytosis. *Nature* 279:679–85

Goldstein, J. L., Basu, S. K., Brunschede, G. Y., Brown, M. S. 1976. Release of low density lipoprotein from its cell surface receptor by sulfated glycosaminoglycans. *Cell* 7:85–95

Goldstein, J. L., Brown, M. S. 1974. Binding and degradation of low density lipoproteins by cultured human fibroblasts: Comparison of cells from a normal subject and from a patient with homozygous familial hypercholesterolemia. *J. Biol. Chem.* 249:5153–62

Goldstein, J. L., Brown, M. S. 1977. The low-density lipoprotein pathway and its relation to atherosclerosis. *Ann. Rev. Biochem.* 46:897–930

Goldstein, J. L., Brown, M. S. 1979. The LDL receptor locus and the genetics of familial hypercholesterolemia. *Ann. Rev. Genet.* 13:259–89

Goldstein, J. L., Brown, M. S. 1983. Familial hypercholesterolemia. In *The Metabolic Basis of Inherited Disease*, ed. J. B. Stanbury, J. B. Wyngaarden, D. S. Fredrickson, J. L. Goldstein, M. S. Brown, Chapter 33, pp. 672–712. New York: McGraw-Hill. 2032 pp. 5th ed.

Goldstein, J. L., Brown, M. S., Stone, N. J. 1977. Genetics of the LDL receptor: Evidence that the mutations affecting binding and internalization are allelic. *Cell* 12:629–41

Goldstein, J. L., Buja, L. M., Anderson, R. G. W., Brown, M. S. 1978. Receptor-mediated uptake of macromolecules and their delivery to lysosomes in human fibroblasts. In *Protein Turnover and Lysosome Function*, ed. H. L. Segal, D. F. Doyle, pp. 455–77. New York: Academic. 790 pp.

Goldstein, J. L., Helgeson, J. A. S., Brown, M. S. 1979b. Inhibition of cholesterol synthesis with compactin renders growth of cultured cells dependent on the low density lipoprotein receptor. *J. Biol. Chem.* 254:5403–9

Goldstein, J. L., Kottke, B. A., Brown, M. S. 1982. Biochemical genetics of LDL receptor mutations in familial hypercholestero-

lemia. In *Human Genetics, Part B: Medical Aspects*, ed. B. Bonne-Tamir, T. Cohen, R. M. Goodman, pp. 161–76. New York: Alan R. Liss. 619 pp.

Gray, A., Dull, T. J., Ullrich, A. 1983. Nucleotide sequence of epidermal growth factor cDNA predicts a 128,000-molecular weight protein precursor. *Nature* 303:722–25

Haguenauer-Tsapis, R., Hinnen, A. 1984. A deletion that includes the signal peptidase cleavage site impairs processing, glycosylation, and secretion of cell surface yeast acid phosphatase. *Mol. Cell. Biol.* 4:2668–75

Hanover, J. A., Willingham, M. C., Pastan, I. 1984. Kinetics of transit of transferrin and epidermal growth factor through clathrin-coated membranes. *Cell* 39:283–93

Helenius, A., Mellman, I., Wall, D., Hubbard, A. 1983. Endosomes. *Trends Biochem. Sci.* 8:245–50

Holland, E. C., Leung, J. O., Drickamer, K. 1984. Rat liver asialoglycoprotein receptor lacks a cleavable NH_2-terminal signal sequence. *Proc. Natl. Acad. Sci. USA* 81:7338–42

Hopkins, C. R. 1982. Membrane recycling. *Ciba Foundation Symposium* 92:239–42

Hopkins, C. R. 1985. The appearance and internalization of transferrin receptors at the margins of spreading human tumor cells. *Cell* 40:199–208

Hopkins, C. R., Trowbridge, I. S. 1983. Internalization and processing of transferrin and the transferrin receptor in human carcinoma A431 cells. *J. Cell Biol.* 97:508–21

Hubbard, S. C., Ivatt, R. J. 1981. Synthesis and processing of asparagine-linked oligosaccharides. *Ann. Rev. Biochem.* 50:555–83

Huet, C., Ash, J. F., Singer, S. J. 1980. The antibody-induced clustering and endocytosis of HLA antigens on cultured human fibroblasts. *Cell* 21:429–38

Hunter, T. 1984. The epidermal growth factor receptor gene and its product. *Nature* 311:414–16

Innerarity, T. L., Mahley, R. W. 1978. Enhanced binding by cultured human fibroblasts of apo-E-containing lipoproteins as compared with low density lipoproteins. *Biochemistry* 17:1440–47

Innerarity, T. L., Weisgraber, K. H., Arnold, K. S., Rall, S. C. Jr., Mahley, R. W. 1984. Normalization of receptor binding of apolipoprotein E2: Evidence for modulation of the binding site conformation. *J. Biol. Chem.* 259:7261–67

Jacobs, S., Sahyoun, N. E., Saltiel, A. R., Cuatrecasas, P. 1983. Phorbol esters stimulate the phosphorylation of receptors for insulin and somatomedin C. *Proc. Natl. Acad. Sci. USA* 80:6211–13

Kingsley, D. M., Krieger, M. 1984. Receptor-mediated endocytosis of low density lipoprotein: Somatic cell mutants define multiple genes required for expression of surface-receptor activity. *Proc. Natl. Acad. Sci. USA* 81:5454–58

Klausner, R. D., Ashwell, G., Van Renswoude, J., Harford, J. B., Bridges, K. R. 1983. Binding of apotransferrin to K562 cells: Explanation of the transferrin cycle. *Proc. Natl. Acad. Sci. USA* 80:2263–66

Klausner, R. D., Harford, J., van Renswoude, J. 1984. Rapid internalization of the transferrin receptor in K562 cells is triggered by ligand binding or treatment with a phorbol ester. *Proc. Natl. Acad. Sci. USA* 81:3005–9

Krieger, M. 1983. Complementation of mutations in the LDL pathway of receptor-mediated endocytosis by cocultivation of LDL receptor-defective hamster cell mutants. *Cell* 33:413–22

Krieger, M., Brown, M. S., Goldstein, J. L. 1981. Isolation of Chinese hamster cell mutants defective in the receptor-mediated endocytosis of low density lipoprotein. *J. Mol. Biol.* 150:167–84

Krieger, M., Martin, J., Segal, M., Kingsley, D. 1983. Amphotericin B selection of mutant Chinese hamster cells with defects in the receptor-mediated endocytosis of low density lipoprotein and cholesterol biosynthesis. *Proc. Natl. Acad. Sci. USA* 80:5607–11

Krupp, M. N., Lane, M. D. 1982. Evidence for different pathways for the degradation of insulin and insulin receptor in the chick liver cell. *J. Biol. Chem.* 257:1372–77

Lehrman, M. A., Goldstein, J. L., Brown, M. S., Russell, D. W., Schneider, W. J. 1985a. Internalization-defective LDL receptors produced by genes with nonsense and frameshift mutations that truncate the cytoplasmic domain. *Cell.* In press

Lehrman, M. A., Schneider, W. J., Südhof, T. C., Brown, M. S., Goldstein, J. L., et al. 1985b. Mutation in LDL receptor: *Alu-Alu* recombination deletes exons encoding transmembrane and cytoplasmic domains. *Science* 227:140–46

Leonard, W. J., Depper, J. M., Crabtree, G. R., Rudikoff, S., Pumphrey, J., et al. 1984. Molecular cloning and expression of cDNAs for the human interleukin-2 receptor. *Nature* 311:626–31

McClelland, A., Kuhn, L. C., Ruddle, F. H. 1984. The human transferrin receptor gene: Genomic organization, and the complete primary structure of the receptor deduced from a cDNA sequence. *Cell* 39:267–74

McLean, J. W., Elshourbagy, N. A., Chang, D. J., Mahley, R. W., Taylor, J. M. 1984. Human apolipoprotein E mRNA: cDNA cloning and nucleotide sequencing of a new variant. *J. Biol. Chem.* 259:6498–6504

Mahley, R. W., Innerarity, T. L. 1983. Lipoprotein receptors and cholesterol homeostatis. *Biochim. Biophys. Acta* 737:197–222

Marchesi, V. T., Furthmayr, H., Tomita, M. 1976. The red cell membrane. *Ann. Rev. Biochem.* 45:667–98

Miyake, Y., Tajima, S., Yamamura, T., Yamamoto, A. 1981. Homozygous familial hypercholesterolemia mutant with a defect in internalization of low density lipoprotein. *Proc. Natl. Acad. Sci. USA* 78:5151–55

Mostov, K. E., Friedlander, M., Blobel, G. 1984. The receptor for transepithelial transport of IgA and IgM contains multiple immunoglobulin-like domains. *Nature* 308:37–43

Nikaido, T., Shimizu, A., Ishida, N., Sabe, H., Teshigawara, K., et al. 1984. Molecular cloning of cDNA encoding human interleukin-2 receptor. *Nature* 311:631–35

Octave, J.-N., Schneider, Y.-J., Trouet, A., Crichton, R. R. 1983. Iron uptake and utilization by mammalian cells. I: Cellular uptake of transferrin and iron. *Trends Biochem. Sci.* 8:217–20

Omary, M. B., Trowbridge, I. S. 1981. Biosynthesis of the human transferrin receptor in cultured cells. *J. Biol. Chem.* 256:12888–92

Pastan, I. H., Willingham, M. C. 1981. Journey to the center of the cell: Role of the receptosome. *Science* 214:504–9

Pastan, I., Willingham, M. C. 1983. Receptor-mediated endocytosis: Coated pits, receptosomes and the Golgi. *Trends Biochem. Sci.* 8:250–54

Pearse, B. M. F. 1975. Coated vesicles from pig brain: Purification and biochemical characterization. *J. Mol. Biol.* 97:93–98

Pernis, B. 1985. Internalization of lymphocyte membrane components. *Immunol. Today.* 6:45–49

Pernis, B., Tse, D. B. 1985. Dynamics of MHC molecules in lymphoid cells: Facts and speculations. In *The Cell Biology of the Major Histocompatibility Complex*, ed. B. Pernis, H. J. Vogel. New York: Academic. P & S Biomedical Sciences Symposium, pp. 152–64

Pitas, R. E., Innerarity, T. L., Mahley, R. W. 1980. Cell surface receptor binding of phospholipid: protein complexes containing different ratios of receptor-active and -inactive E apoprotein. *J. Biol. Chem.* 255: 5454-60

Rall, L. B., Scott, J., Bell, G. I., Crawford, R. J., Penschow, J. D., et al. 1985. Mouse prepro-epidermal growth factor synthesis by the kidney and mouse tissues. *Nature* 313:228-31

Renston, R. H., Jones, A. L., Christiansen, W. D., Hradek, G. T., Underdown, B. J. 1980. Evidence for a vesicular transport mechanism in hepatocytes for biliary secretion of immunoglobulin A. *Science* 208:1276-78

Robenek, H., Hesz, A. 1983. Dynamics of low-density lipoprotein receptors in the plasma membrane of cultured human skin fibroblasts as visualized by colloidal gold in conjunction with surface replicas. *Eur. J. Cell Biol.* 31:275-82

Rodewald, R., Abrahamson, D. R. 1982. Receptor-mediated transport of IgG across the intestinal epithelium of the neonatal rat. *Ciba Foundation Symp.* 92:209-32

Roth, T. F., Porter, K. R. 1964. Yolk protein uptake in the oocyte of the mosquito *Aedes aegypti. L. J. Cell. Biol.* 20:313-32

Russell, D. W., Schneider, W. J., Yamamoto, T., Luskey, K. L., Brown, M. S., et al. 1984. Domain map of the LDL receptor: Sequence homology with the epidermal growth factor precursor. *Cell* 37:577-85

Russell, D. W., Yamamoto, T., Schneider, W. J., Slaughter, C. J., Brown, M. S., et al. 1983. cDNA cloning of the bovine low density lipoprotein receptor: Feedback regulation of a receptor mRNA. *Proc. Natl. Acad. Sci. USA* 80:7501-5

Sabatini, D. D., Kriebich, G., Morimoto, T., Adesnik, M. 1982. Mechanisms for the incorporation of protein in membranes and organelles. *J. Cell Biol.* 92:1-21

Schauer, I., Emr, S., Gross, C., Schekman, R. 1985. Invertase signal and mature sequence substitutions that delay transport of active enzyme from the endoplasmic reticulum. *J. Cell Biol.* 100:1664-75

Schlessinger, J. 1980. The mechanism and role of hormone-induced clustering of membrane receptors. *Trends Biochem. Sci.* 5:210-14

Schmid, C. W., Jelinek, W. R. 1982. The *Alu* family of dispersed repetitive sequences. *Science* 216:1065-70

Schneider, C., Owen, M. J., Banville, D., Williams, J. G. 1984. Primary structure of human transferrin receptor deduced from the mRNA sequence. *Nature* 311:675-78

Schneider, W. J., Beisiegel, U., Goldstein, J. L., Brown, M. S. 1982. Purification of the low density lipoprotein receptor, an acidic glycoprotein of 164,000 molecular weight.

J. Biol. Chem. 257:2664-73

Schneider, W. J., Brown, M. S., Goldstein, J. L. 1983a. Kinetic defects in the processing of the LDL receptor in fibroblasts from WHHL rabbits and a family with familial hypercholesterolemia. *Mol. Biol. Med.* 1:353-67

Schneider, W. J., Slaughter, C. J., Goldstein, J. L., Anderson, R. G. W., Capra, D. J., et al. 1983b. Use of anti-peptide antibodies to demonstrate external orientation of NH_2-terminus of the LDL receptor in the plasma membrane of fibroblasts. *J. Cell Biol.* 97:1635-40

Scott, J., Urdea, M., Quiroga, M., Sanchez-Pescador, R., Fong, N., et al. 1983. Structure of a mouse submaxillary messenger RNA encoding epidermal growth factor and seven related proteins. *Science* 221:236-40

Sege, R. D., Kozarsky, K., Nelson, D. L., Krieger, M. 1984. Expression and regulation of human low-density lipoprotein receptors in Chinese hamster ovary cells. *Nature* 307:742-45

Solari, R., Kraehenbuhl, J.-P. 1984. Biosynthesis of the IgA antibody receptor: A model for the transepithelial sorting of a membrane glycoprotein. *Cell* 36:61-71

Stanley, K. K., Kocher, H.-P., Luzio, J. P., Jackson, P., Tschopp, J. 1985. The sequence and topology of human complement component C9. *EMBO J.* 4:375-82

Südhof, T. C., Russell, D. W., Brown, M. S., Russell, D. W. 1985a. The LDL receptor gene: A mosaic of exons shared with different proteins. *Science* 228:815-22

Südhof, T. C., Russell, D. W., Goldstein, J. L., Brown, M. S., Sanchez-Pescador, R., et al. 1985b. Cassette of eight exons shared by genes for LDL receptor and EGF precursor. *Science* 228:815-22

Terris, S., Hofmann, C., Steiner, D. F. 1979. Mode of uptake and degradation of ^{125}I-labelled insulin by isolated hepatocytes and H4 hepatoma cells. *Can. J. Biochem.* 57:459-68

Tolleshaug, H., Goldstein, J. L., Schneider, W. J., Brown, M. S. 1982. Posttranslational processing of the LDL receptor and its genetic disruption in familial hypercholesterolemia. *Cell* 30:715-24

Tolleshaug, H., Hobgood, K. K., Brown, M. S., Goldstein, J. L. 1983. The LDL receptor locus in familial hypercholesterolemia: Multiple mutations disrupting the transport and processing of a membrane receptor. *Cell* 32:941-51

Tse, D. B., Cantor, C. R., McDowell, J., Pernis, B. 1985. Recycling Class I MHC antigens: Dynamics of internalization, acidification and ligand-degradation in murine T lymphoblasts. Submitted for publication

Tse, D. B., Pernis, B. 1984. Spontaneous internalization of Class I major histocompatibility complex molecules in T lymphoid cells. *J. Exp. Med.* 159:193–207

Tycko, B., Maxfield, F. R. 1982. Rapid acidification of endocytic vesicles containing α_2-macroglobulin. *Cell* 28:643–51

Ullrich, A., Bell, J. R., Chen, E. Y., Herrera, R., Petruzzelli, L. M., et al. 1985. Human insulin receptor and its relationship to the tyrosine kinase family of oncogenes. *Nature* 313:756–61

Ullrich, A., Coussens, L., Hayflick, J. S., Dull, T. J., Gray, A., et al. 1984. Human epidermal growth factor receptor cDNA sequence and aberrant expression of the amplified gene in A431 epidermoid carcinoma cells. *Nature* 309:418–25

Unanue, E. R. 1984. Antigen-presenting function of the macrophage. *Ann. Rev. Immunol.* 2:395–428

Via, D. P., Willingham, M. C., Pastan, I., Gotto, A. M. Jr., Smith, L. C. 1982. Co-clustering and internalization of low-density lipoproteins and α_2-macroglobulin in human skin fibroblasts. *Exp. Cell Res.* 141:15–22

Wall, D. A., Wilson, G., Hubbard, A. L. 1980. The galactose-specific recognition system of mammalian liver: The route of ligand internalization in rat hepatocytes. *Cell* 21:79–93

Yamamoto, T., Davis, C. G., Brown, M. S., Schneider, W. J., Casey, M. L., et al. 1984. The human LDL receptor: A cysteine-rich protein with multiple *Alu* sequences in its mRNA. *Cell* 39:27–38

Ann. Rev. Cell Biol. 1985. 1 : 41–65

INTERMEDIATE FILAMENTS:
Conformity and Diversity of
Expression and Structure[1]

Peter M. Steinert

Dermatology Branch, National Cancer Institute, National Institutes of Health, Bethesda, Maryland, 20205

David A. D. Parry

Department of Physics and Biophysics, Massey University, Palmerston North, New Zealand

CONTENTS

INTRODUCTION: INTERMEDIATE FILAMENTS ARE UBIQUITOUS

Since the 1930s, the properties and structures of the filamentous component of wool and related keratins have been the subject of various studies (Astbury & Street, 1931). However, only within the last few years has it become apparent that these proteins are members of a larger class of fibrous proteins, related by their common morphologies, that form a component of the cytoskeleton of virtually all vertebrate cells. These are now known as intermediate filaments (IF), and consist of a heterogeneous population of protein subunits. Based on the patterns of tissue distribution, and on immunological and biochemical properties, the current trend is to use a classification system for IF that defines five major subclasses of subunits that can form IF in vitro and in vivo (Lazarides 1980, 1982; Zackroff et al 1981; Steinert 1981; Steinert et al 1984a, 1985a; Yang et al 1985). These subclasses include one, a large family of keratin subunits of about 40–70 kilodaltons (kDa) two or more members of which are coexpressed in most epithelial cells. A second subclass is found in myogenic cells and consists of a single protein of about 52 kDa called desmin (or skeletin). A third subclass consists of the 50 kDa glial fibrillary acidic protein, which is found in astroglial tissues. Many cells of mesenchymal origin, or cell lines established in culture, contain a protein, vimentin, of about 53 kDa, which defines the fourth subclass. The fifth subclass is found in neuronal tissues and consists of varying proportions of three "triplet" subunits of about 60–70 kDa (the NF-L subunit), 105–110 kDa (NF-M), and 135–150 kDa (NF-H). In addition, some cells can also express vimentin in addition to their more specific IF subclass.

There is growing evidence that invertebrate cells also contain IF or IF-like proteins, e.g. as neurofilaments in the squid brain, or in the giant axons of marine worms (Zackroff et al 1981); in *Drosophila* cells (Walter & Biessmann 1984); and in *Dictyostelium discoideum* (Koury & Eckert 1984). Thus, IF may be ubiquitous constituents of the cytoskeletons of all eukaryotic cells. Further, there is a growing list of proteins associated with IF (IFAP), or that share certain properties with the subclasses enumerated above (Lazarides 1982; Wiche et al 1982; Zackroff et al 1984; Yang et al 1985). Accordingly, as more is learned about IF, it may eventually be necessary to adopt a more systematic nomenclature system for all of these proteins.

Implicit in the ubiquity of IF and the chemical heterogeneity of their subunits is the realization that IF must be a functionally diverse component of the cytoskeletons of cells. In recent years there has been an extraordinary expansion in research in this field in an attempt to understand both the apparent different roles of IF in the biology of different specialized cells, and the molecular bases of this functional diversity. Unfortunately, few data are yet available that permit precise definition of IF functions, although several roles such as nuclear centration, cell shape definition, movement of organelles, and mechanical coordination of the cytoskeleton have been ascribed to them (Lazarides 1980, 1982; Jones et al 1983, 1985; Fey et al 1984; Steinert et al 1984a). On the other hand, a clearer picture of the molecular structure of IF subunits has emerged which can explain the morphological similarity yet chemical and functional diversity of IF in a simple way. We now know that all IF are morphologically similar because their subunits contain a central α-helical rod domain of conserved secondary structure that forms the basic framework or core of all IF. IF subunits differ from one another in the sizes and amino acid sequences of their N-terminal and C-terminal domains. These portions of the subunits protrude from the IF core, and are thought to be intimately involved in determining the functions of IF. In this review, we explore the data that led to these realizations, and the emerging evidence that the subunit terminal domain sequences are closely coordinated with their expression and with the functions of IF in cells.

SUBUNIT STRUCTURE

General Features

Partial or complete amino acid sequence information is now available for each of the five subclasses of IF (see Weber & Geisler 1984 and Parry & Fraser 1985 for lists of published sequences). Using these data in conjunction with secondary structure prediction methods (Chou & Fasman 1978; Garnier et al 1978) it has been shown that each IF subunit is composed of a central α-helix-rich rod domain about 311–314 residues long, and N- and C-terminal domains of variable size and chemical character, which are generally not α-helical (Figure 1) (Geisler & Weber 1982; Steinert et al 1983, 1984a,b, 1985a,b,c; Weber & Geisler 1984; Parry & Fraser 1985). The variations in molecular weight of IF subunits are due almost entirely to the different sizes of the end domains. Although diversity is a striking feature of the end domains, the central rod domain exhibits considerable structural homology between IF subunits from different sources, and is largely responsible for the axial coherence of all IF.

Nomenclature

Detailed analyses of the sequences of the central rod domains of various IF subunits have shown they may be further divided into four types (Types I–IV). This nomenclature arose initially because of the similarity observed

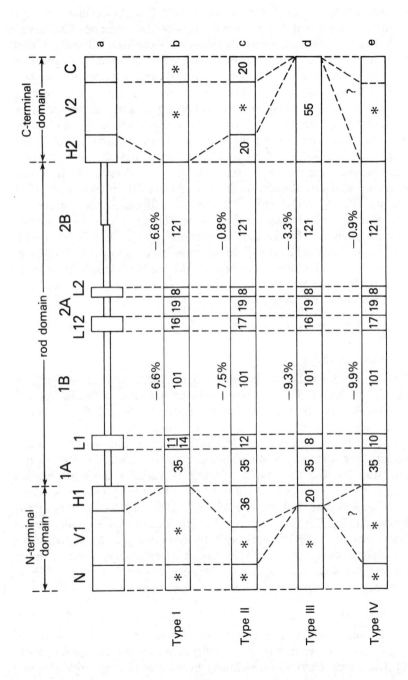

between the sequences of the Type I and Type II α-helical fragments from the 8c-1 and 7c subunits of wool keratin (Crewther et al 1978; Gough et al 1978) and the sequences found in an acidic (Hanukoglu & Fuchs 1982) and a basic (Hanukoglu & Fuchs 1983) epidermal keratin IF subunit. Accordingly, acidic keratins are now known as Type I IF subunits and neutral or basic keratins are Type II IF subunits. Since representatives of each Type are required for keratin IF assembly in vitro and in vivo (see Table 1), Type I and Type II keratin IF subunits are obligate heteropolymers. In contrast, three other IF subunits, vimentin, desmin, and glial fibrillary acidic protein, contain structurally similar Type III rod domain sequences that are homologous to each other but distinctly different from the keratins (Crewther et al 1983; Steinert et al 1984a; Parry & Fraser 1985). The molecules in these cases are homopolymers that require but a single subunit for IF formation in vivo and in vitro (Steinert et al 1982). More limited sequence data on mammalian neurofilament subunits (Geisler et al 1984) reveal that their rod domain sequences constitute a different Type IV group (Parry et al 1985; Parry & Fraser 1985). Although the NF-L subunits can form homopolymer IF, existing data are inconclusive as to whether the NF-M and NF-H subunits are also capable of homopolymerization (Steinert et al 1982). Interestingly, any of the Type III

Figure 1   IF subunit structure. (*a*) Universal model. All IF subunits possess a central α-helical rod domain flanked by end domains. The rod domain in all cases consists of four segments (horizontal rods) of invariant size. These are composed of a 7-residue (heptad) repeating peptide, which can participate in the formation of a coiled-coil with another subunit. Segment 1A is 35 residues long, 1B is 101, 2A is 19, and 2B is 121. These segments are separated by short regions or linkers (*vertical rectangles*) that do not possess the heptad repeat, and therefore cannot form a coiled-coil. Linker L1 varies from 8–14 residues in different subunits as shown; L12 is 16 or 17; and L2 is always 8. At the middle of segment 2B, the polarity of the progression of heptads is abruptly reversed, a feature that imposes a "stutter" in the regularity of the coiled-coil at this point. The large segments 1B and 2B have a net acidic charge, which is expressed here as a percentage of their total length. Despite the conserved structural features, the exact sequences of the heptads in the segments and of the linkers vary, which permits classification of all IF subunits into four distinct sequence Types. Each sequence Type of rod domain is coupled with a specific set of end domains, which themselves may be further subdivided into subdomains based on their basic charge (N or C), sequence homology (H), or sequence variability (V). In Type I (*b*) and Type II (*c*) keratin IF subunits, the N and C subdomains vary in size (*asterisk*), but are usually 15–30 residues long; the highly variable V1 and V2 subdomains may vary from 0 to 130 residues; and H1 and H2 subdomains of Type II keratins are 36 and 20 residues. In Type III IF subunits (*d*), H1 and H2 are 20 and 55 residues, and their N subdomains vary from 60 to 75 residues. (*e*) Full specification of the subdomainal organization of the end domains of Type IV neurofilaments must await more complete sequence information, but they appear to have N, V1, V2, and possibly C subdomains of variable size and sequence.

IF subunits are capable of copolymerization; for example, desmin and vimentin copolymerize in vitro (Steinert et al 1982), and in BHK-21 cells or developing myotubes they form coincident arrays of IF networks (Lazarides 1982). Apart from Type I and Type II keratins, members of different types do not copolymerize in vivo, rather, when present in the same cell [for example, Types I and II keratins and Type III vimentin in HeLa cells (Aynardi et al 1984)] they form separate IF networks.

Most of the available amino acid sequences of IF subunits have been deduced from the nucleotide sequences of cDNA clones. Members within a designated Type show 50–90% homology of their nucleotide sequences in the regions encoding the rod domain. IF subunits of different types show only limited ($<30\%$) homology. In the case of keratins, some workers have reported that cDNA clones encoding an IF subunit of one type can be used to identify clones for other subunits of the same type (Fuchs et al 1981; Fuchs & Marchuk 1983; Kim et al 1983, 1984), but other workers have not confirmed this (Roop et al 1983; Steinert et al 1985b).

Table 1　Expression rules of some keratin IF subunits[a]

Cell type	Type I (acidic)	Type II (neutral-basic)	V subdomain amino acid sequences[b]
Simple	40	52 or 54(?)	Sequences unknown,
	45	52	but predicted to be
	46	54	short or absent
Stratified squamous	48	56	V1, 50–90 residues
	50	58	long, closely related, and about 40% glycine
Hyperproliferating stratified squamous	48	56	V2, 48 residues long, high homology, low glycine
Terminally-differentiated:			
cornea	55	64	Sequences unknown
epidermis	56.5	65–67	V1 and V2 sequences, 104–130 residues long, 60% glycine, variable tandem peptide repeats
wool (sheep)[c]	48	54	25–55 residues long, cysteine-rich
lens	vimentin only		

The subunit molecular weights (kDa) are arranged horizontally in Type I–Type II coexpression pairs. Data are from: [a] Moll et al (1982) and Sun et al (1984) for human keratins, except sheep wool; [b] Steinert et al (1984b, 1985b); [c] Gillespie (1983).

Structural Features of the α-Helical Rod Domain

X-ray diffraction patterns from both native and reconstituted IF contain meridional reflections with spacings of 0.15 and 0.51 nm (Fraser et al 1976; Steinert et al 1976; Renner et al 1981). These reflections are typical not only of an α-helical structure oriented approximately parallel to the axis of the IF, but also of a grouping of α-helices in a coiled-coil rope structure of the form first described by Crick (1953). In these structures the amino acid sequences have a quasi-repeating heptad structure of the form $(a\text{-}b\text{-}c\text{-}d\text{-}e\text{-}f\text{-}g)_n$, where a and d are frequently apolar residues. This repeating pattern of residues gives rise to an inclined stripe of apolar residues running around each right-hand α-helix. Since individual α-helices are unstable in an aqueous environment, two (or more) such α-helices must aggregate to form a stable structure. This structure is a coiled-coil rope with the apolar residues buried along its axis.

Crewther et al (1978) provided the first chemical evidence that Type I and Type II wool keratin IF subunits do indeed have such a heptad substructure, which is now known to be characteristic of the rod domain of all IF amino acid sequences (Figure 1). Four distinct regions or segments that contain this heptad repeat, known as 1A, 1B, 2A, and 2B, have been recognized in the rod domain of all IF subunits (Geisler & Weber 1982; Crewther et al 1983; Hanukoglu & Fuchs 1983; Steinert et al 1983; Parry & Fraser 1985). These are interspersed by short linker sequences that cannot form a coiled coil: Linker L1 connects 1A (5 heptads long) to 1B (14–15 heptads); and linker L2 joins 2A (2–3 heptads) to 2B (17 heptads), thus forming segments 1 and 2, respectively, each about 21–22 nm long. A linker L12 connects segments 1 and 2. Finally, another feature that has been precisely conserved among all IF subunits occurs near the center of segment 2B: the regularity of the progression of heptads is reversed to form a distinct discontinuity or "stutter" (Figure 1).

The overall length of the rod domain of IF subunits is about 45 nm (Steinert et al 1983; Crewther et al 1983), a value close to that of the 47 nm axial repeat adduced by Fraser & MacRae (1973) from X-ray diffraction measurements.

In addition to the conservation of the structure of the central α-helical rod domain, several other features of these amino acid sequences are noteworthy. Despite the classification into distinct sequence types as described above, in all IF the sequences of the second and third heptads of segment 1A, and of the last 4–5 heptads of segment 2B are highly conserved (Steinert et al 1984b; Weber & Geisler 1984; Parry & Fraser 1985). The net charge of the four segments in each IF subunit type is distinctly acidic (Figure 1), except in Type II keratins and Type IV neurofilament subunits, in which the 2B segment is nearly neutral. In the keratins this may account for the overall basic charge of Type II keratin subunits (Steinert et al 1984b). These subtle differences in sequence and charge distribution may explain the specificity of antisera that cross-react with all IF subunits (Pruss et al 1981), or with only Type I or II keratin subunits (Sun et al 1983, 1984). Furthermore, as discussed below, there are statistically significant periodicities in the distributions of acidic and basic amino acid residues in the coiled-coil sequences of segments 1B and segment 2 that are also conserved among IF subunits. Also, the linker L12 has a sequence of the form (hydrophilic-apolar)$_4$, which has the potential to form a β-sheet structure.

Structural Features of the End Domains

A study of the sequences of Types I and II epidermal keratin IF subunits has revealed that subdivision of their N- and C-terminal domains is justified on the basis of (a) sequence homology (H subdomains), (b) variability (V subdomains), and (c) charge distribution (N or C subdomains). These sequences are distributed with bilateral symmetry with respect to the central α-helical rod domain (Steinert et al 1985b) (Figure 1). Type II, but not Type I, keratins contain short homologous sequences, H1 and H2, immediately adjacent to either side of the rod domain. The H1 sequences are highly conserved among all Type II keratins, but are different from the H2 sequences, which have likewise been highly conserved. The absence of these H1 and H2 sequences in Type I keratins accounts for most of the difference in mass between coexpressed Type I and Type II keratins (see Table 1).

These sequences are flanked by V1 and V2 subdomains that vary in both length and sequence, often contain tandem peptide repeats, and are conspicuously rich in glycines and serines. The repeat sequences conform to XY$_n$, where X is one of several apolar residues, not uncommonly tyrosine or phenylalanine, Y is glycine or serine, and $n = 1$–9. Coexpressed keratin subunits have generally similar V1 and V2 subdomains, although the exact sequence repeat(s) usually varies. Among keratin subunits of a given type, their mass variability is almost entirely due to variations in the size of their V1 and V2 subdomains. In large keratins they are up to 130 residues long, whereas small keratins possess only diminutive V subdomains, or lack them altogether (Steinert et al 1985b).

Finally, at the termini of the keratin subunits are basic subdomains, N and C, which are highly variable in sequence. This sequence variability has been exploited in the production of keratin IF subunit specific antisera (Roop et al 1984).

Analyses of the end domain sequences of Type III IF subunits also reveal a subdomainal organization (Figure 1). Type III subunits have H1 and H2 subdomains (though these are not homologous to each other or to those of Type II subunits) about 20 and 55 residues long, respectively. The H2 subdomain thus comprises the entire C-terminal end domain. The remainder of the amino-terminal domain consists of N, a region whose length varies somewhat among Type III subunits, is hypervariable in sequence, and is strongly basic. These sequences are thought to adopt random coil conformations (Weber & Geisler 1984).

Incomplete data on neurofilament subunit sequences suggest a subdomainal organization within their end domains as well. They appear to contain VI sequences in their N-terminal end domains, and unusual very glutamate-rich sequences of variable length in the C-terminal domains (Weber & Geisler 1984).

Based on these considerations it is apparent that each IF subunit type has both a characteristic α-helical rod domain and a set of end subdomains. We recommend, therefore, that as future sequence information on IF subunits, IF-like proteins, and IFAP becomes available, a nomenclature system should be retained that is consistent with the emerging pattern described above.

MODELS FOR IF STRUCTURE

The Two-Chain Coiled-Coil Molecule

Although early data suggested that the basic chemical unit in IF contained three subunits (Crewther & Harrap 1967; Crewther & Dowling 1971; Skerrow et al 1973; Steinert 1978; Steinert et al 1980), more recent evidence strongly favors four subunits, consisting of a pair of two-stranded coiled-coil molecules (Geisler & Weber 1982; Gruen & Woods 1983; Woods & Inglis 1984; Parry et al 1985). From the different lengths of the heptad repeats in segments 1A, 1B, 2A, and 2B of the rod domain, as well as from calculations of the number of possible ionic interactions between subunits determined as a function of relative subunit stagger and polarity, it has been predicted that the subunits in the two-chain coiled-coil molecule are parallel (rather than antiparallel), and that they are in exact axial register (Parry et al 1977; Steinert et al 1984b, 1985c; Parry & Fraser 1985; Parry et al 1985). By use of sequence analyses on two- and four-chain particle species isolated from enzymic digests of IF, it has now been confirmed that the subunits are indeed parallel in both wool keratin (Woods & Inglis 1984) and epidermal keratin IF (Parry et al 1985). These studies have also shown that each coiled-coil molecule is a heterodimer that consists of a Type I and a Type II subunit. The structural and functional significance of the requirement for the heterodimer for keratin IF is not yet understood.

Type III subunits each contain a single cysteine residue in the same location, which can be oxidized by use of 1,10-phenanthroline cupric ion complexes, to form disulfide-bonded cross-linked homodimers, or a heterodimer with another Type III subunit (Quinlan & Franke 1982, 1983). These data, together with those obtained from identical oxidation experiments of Type IV neurofilament subunits in which the cysteine residue(s) do not occur in the same position(s) as the Type III subunits (Carden & Eagles 1983), establish that the two-chain molecules in all IF consist of parallel subunits in exact axial register (Parry et al 1985).

The Fundamental Four-Chain Coiled-Coil Unit of All IF

The way in which a pair of two-stranded coiled-coil molecules aggregate to form the fundamental four-chain chemical unit is less certain. An initial important observation was reported by Parry et al (1977), who showed that segment 1B in Type I and II keratin subunits has a significant periodicity of about 9.5 residues in the linear distributions of both the acidic and the basic amino acid residues. The phases of the acidic and basic 9.5 residue period differ by about 180°; that is, clumps of acidic and basic residues form alternate stripes along the coiled-coil. This work has now been extended (McLachlan & Stewart 1982; Crewther et al 1983; Dowling et al 1983; Parry & Fraser 1985) to show that a statistically significantly different periodicity of 9.84 residues exists for the same sorts of residues in segment 2 (2A + L2 + 2B) of all IF subunits.

The most logical explanation for these periodicities is that intersegmental ionic interactions are critically important in the aggregation of neighboring two-chain coiled-coil molecules to form the four-chain units, and in the aggregation of these units to form the complete IF. Based on this assumption, and the slight difference in the periods of the charged residues, Parry & Fraser (1985) have demonstrated that the highest number of favorable ionic interactions occur when either two 1B segments or two 2 segments on neighboring coiled-coils are aligned. However, adjacent 1B and 2 segments would also give rise to a large number of favorable ionic interactions. Thus 1B–1B, 2–2, and 1B–2 segment interactions are predicted to be the most important determinants of IF structure. Two possibilities then exist for molecular aggregation: the molecules are either in axial register or are approximately half-staggered. Since ionic interactions can be made as easily between parallel segments as between antiparallel segments, the calculations of Fraser et al (1985) could not readily distinguish between the two possibilities, although the number of interactions between antiparallel segments was always greater than those between parallel subunits. However, Crewther et al (1983) have presented several lines of indirect evidence suggesting that the pairs of two-chain coiled-coil molecules in the four-chain unit (and hence in the IF) are antiparallel. It is important to note that no direct experimental evidence is yet available to establish this point unequivocally.

Possibilities for Packing of Four-Chain Units within IF

Detailed meridional and near-meridional X-ray diffraction data from the only known highly oriented form of IF, hard α-keratin, have revealed that the true axial rise or period is 47 nm, that there are seven quasi-equivalent units in the period, and that these units lie on a basic helix with a mean pitch of 22 nm (Fraser et al 1976; Fraser & MacRae 1983) (Figure 2a). The surface lattice, refined from a quantitative study of the distribution of meridional intensities, shows a shear plane along one of the so-called seven-start helices. (See Figure 2 for an explanation of the terms three-start, four-start, and seven-start helices.)

There is now considerable electron microscope evidence in support of this X-ray diffraction data. A 47 nm period has been visualized for keratin and vimentin IF by several workers (Steven et al 1982; Aebi et al 1983), and a 22 nm period has been observed for keratin, vimentin, desmin, and neurofilament IF. Shadowed preparations of keratin IF have revealed that the basic helix may be left-handed (Milam & Erickson 1982), but conflicting evidence, together with inconsistent observations of left-handed or right-handed subfilaments in various IF (Krishnan et al 1979; Henderson et al 1982; Aebi et al 1983), suggests that the definitive handedness of the helix with the 22 nm pitch length has yet to be established. In addition, it is noteworthy that this 22 nm length closely corresponds to the calculated length of segment 1 and segment 2 in the rod domain of all IF subunits.

Scanning transmission electron microscopy of unstained preparations of native and reassembled IF has shown that all IF are polymorphic, consisting of two or three mass forms (Steven et al 1982, 1983a,b, 1985). In most IF samples, the major mass class consists of about 32 subunits per 47 nm. All samples contain "light-weight" IF, which consist of about 22 subunits per 47 nm. These are thought to be incompletely assembled in that they contain the minimum amount of mass needed for IF elongation, since most such measurements occur either on short filaments (< 1 μm), or at the ends of long IF, to which more protein can be added to form the major or mature mass class. Some reassembled IF contain a more massive class of IF of about 44 subunits per 47 nm; these are considered "aberrant" assembly products. Interestingly, the three classes have masses in the ratio of 2:3:4. However, within each mass class all IF, irrespective of the size of their constituent subunits, contain the same numbers of subunits per unit length of IF. These data argue strongly that all IF are built according to a common plan.

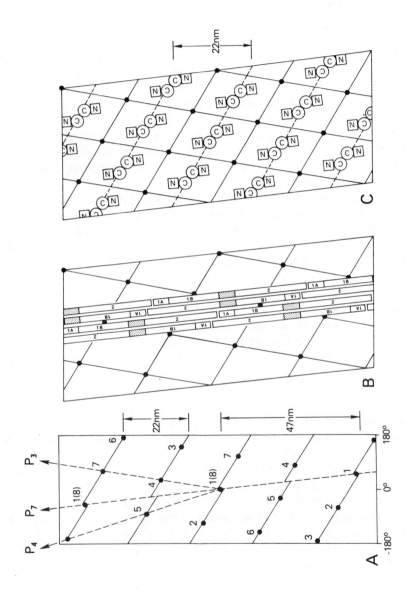

In conjunction with the X-ray diffraction data of Fraser & MacRae (1976) and Fraser et al (1973, 1983, 1985), the mass data corresponding to the major mass class of IF suggest that each of the seven quasi-equivalent units in the 47 nm repeat contains about 4.5 subunits, specifically, one four-chain chemical unit composed of a pair of two-stranded coiled-coils.

Fraser et al (1985) recently attempted to synthesize a model for IF structure that could accommodate all of these data, using calculations of the possible ionic interactions between segments 1B and 2, determined for all possible combinations, relative polarities, and relative axial staggers. Their analyses revealed that there are indeed a number of possible structures for IF that obey the mass criteria, retain a stable four-chain unit, allow a 47 nm unit height in the IF, and explain the symmetry of the basic helix adduced from X-ray diffraction data (Figures 2 and 3). However, additional biochemical, biophysical, and electron microscope or X-ray diffraction analyses are necessary to provide further constraints in these models, and hence resolve the structure unequivocally.

One of a family of closely related structures, based on all of these concepts, is illustrated in Figures 2 and 3. This particular model is attractive

---

Figure 2   (*a*) Surface lattice of keratin IF based on the X-ray diffraction data of Fraser & MacRae (1983). Such a lattice is generated by conceptually wrapping a sheet of paper around an IF and marking on it every point that appears similar. The "sheet of paper" is then cut down a line parallel to the axis of the IF along the edges marked −180° and 180°, and laid out flat. As a group, the seven quasi-equivalent chemical units (labeled 1–7) have a true axial repeat of 47 nm. Each unit, however, lies on a basic helix of mean pitch length 22 nm. From electron microscope observations (Milam & Erickson 1982) this helix is believed to be left-handed. The X-ray diffraction data are unable to distinguish between the two possible hands, so we have constructed this drawing in line with the original conclusion of Milam & Erickson (1982). The three-start (P_3), four-start (P_4), and seven-start (P_7) helices may be generated by connecting the first lattice point (1) to the one lattice point (4) above it, to the lattice point (5) above it, and to the lattice point (8 = 1), above it, respectively. The surface lattice does not form a continuous helix linking the seven quasi-equivalent lattice points to one another, thus the IF surface lattice suffers a sort of dislocation along one of the seven-start helices (Fraser & MacRae 1983). (*b*) The same lattice as in (*a*) except that the line of dislocation now lies along the left- and right-hand edges of the surface lattice. A topological map illustrating one of a number of possible arrangements of the coiled-coil segments deduced for IF by Fraser et al (1985) has, for the sake of clarity, been laid out on only a selection of the lattice points. Although the involvement of segment 1A in ionic interactions is unlikely (Parry & Fraser 1985; Fraser et al 1985), it is included in the drawing for the sake of clarity. The L12 regions, which may form a β-sheet conformation, are shaded. (*c*) In the model illustrated in (*b*) the N-terminal (*squares*) and C-terminal (*circles*) domains of the IF subunits are in close proximity to each other, and lie along the basic helix of mean pitch length 22 nm. Since these regions of the molecules lie on the periphery of the IF, they may be the delineating feature of the pitch length when IF are shadowed and observed in the electron microscope (Fraser et al 1985).

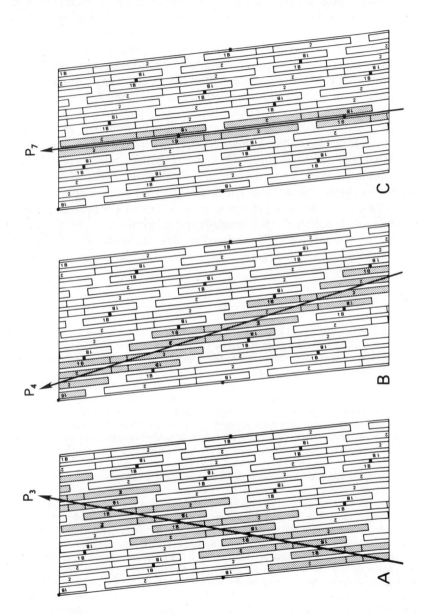

because it naturally incorporates an antiparallel arrangements of neighboring two-chain coiled-coil molecules, and the three most favorable modes of ionic interactions. Further, the antiparallel in-register alignment of the L12 segments has a length compatible with a four-stranded β-sheet, a possibility consistent with the amino acid sequence data. Interestingly, this model also offers an explanation for one of the nettlesome areas of controversy in the field of IF structure. Using a variety of assembly and disassembly procedures, subfilamentous "protofibrils" (Krishnan et al 1979; Aebi et al 1983), "protofilaments" (Quinlan et al 1984; Steinert et al 1984a; Sauk et al 1984), or "prekeratin" (Skerrow 1974) of different sizes, conformations, and handedness have been reported for the substructure of IF. The model in Figure 3 shows clearly how widely different types of subfilamentous particles could all be generated from the same IF structure by disassembly or partial assembly along the three-start (Figure 3a), four-start (Figure 3b), or seven-start helix (Figure 3c). All of these variations consist of different combinations of four-chain molecules. Similarly, the "light-weight" and heavy "aberrant" mass forms of IF could be generated by deletion or insertion of a strand in the three-stranded subfilament (Fraser et al 1985).

Experimental approaches to solving the packing arrangement of the coiled-coil molecules within the IF have already been described in the literature. Quinlan et al (1984) reported that in 4 M urea, keratins form a four-chain molecule in which the four subunits are exactly aligned (~ 48 nm long). The four-chain α-helix-enriched peptides generated by proteolytic digestion of keratin and other IF are about 20 nm long (Steinert 1978; Steinert et al 1980; Gruen & Woods 1983; Woods & Inglis 1984; Parry et al 1985), and consist of pairs of two-stranded coiled coils that are equally likely to have arisen from half-staggered or in-register, parallel or antiparallel molecules. Thus, determination of the exact structures of these various particles might offer solutions to IF structure.

These considerations all suggest that there is a "core" in the IF in which

---

Figure 3   The structure of IF, based on one of the models of Fraser et al (1985) as in Figure 2b, may be broken up into subfilaments of varying number, size, and handedness, depending on the sets of interactions considered to be broken. (a) If interactions between antiparallel 2 segments are broken, then dislocations will occur along the three-start directions, generating three right-handed subfilaments with a mean pitch length of 238.6 nm. (b) If interactions between antiparallel 1B segments are broken, then dislocations will occur along the four-start directions, generating four left-handed subfilaments with a mean pitch length of 121.6 nm. (c) If interactions between antiparallel 1B and 2 segments are broken, then dislocations will occur along the seven-start directions, generating seven left-handed subfilaments with a mean pitch length of 344.7 nm. [Redrawn with permission from Fraser et al (1985).]

the coiled-coil segments are predominantly located. In wool α-keratin the low-angle X-ray equatorial diffraction pattern reveals maxima at 4.5 and 2.8 nm (Fraser et al 1976); more limited data on epidermal keratin, desmin, and vimentin IF also yield equatorial spacings at 2.7 nm (Renner et al 1981). These spacings may be interpreted as a cylinder transform, which yields a calculated core diameter of 7.5–8.0 nm (Fraser & MacRae 1973; Fraser et al 1976), within which the coiled-coil segments are close-packed while they maintain the sets of interactions discussed above.

Peripheral Locations of Functional End Domains

From the evidence of electron microscope studies, IF are believed to have diameters in the range of 7–15 nm. Clearly, these values differ both with the technique used to find the "edge" of the IF, and with the definition of "diameter" that is used. Nevertheless, the range of sizes reported is largely explicable according to the variability of the sizes of the N- and C-terminal domains. These are believed to be predominantly located on the surface of the IF (Steinert et al 1980, 1983; Crewther et al 1983; Fraser & MacRae 1983; Geisler et al 1983; Weber & Geisler 1984), simply because there is insufficient space for them within the core of the IF. In addition, these end domains are probably responsible for delineating the 22 nm pitch length seen in the electron microscope (Milam & Erickson 1982) (see Figure 2c). Accordingly, the end domains would contribute a lower density "fringe" region about the IF periphery (Steven et al 1983a,b, 1985). Depending on the size of the end domains and their conformations, it is conceivable that the full IF diameter could be 3–6 nm greater than that of the α-helical core. Thus, in the helix lattice illustrated in Figure 2c, these end domain sequences would project outwards from the plane of the page. In the IF, therefore, the coiled-coil portions of the molecules would be close-packed with the end domains projecting from the cylindrical surface.

Direct evidence in support of the peripheral location of N- and C-terminal domains was obtained by Steinert et al (1983, 1985b) who showed that more than 60% of the glycine-rich portions of these domains could be proteolytically cleaved from mouse keratin IF, while retaining its morphological integrity. Since the end domains among IF are hypervariable in both size and amino acid sequence, and since their sequences appear to correlate with the expression characteristics of the subunit itself (see discussion below), it seems certain that the end domains specify the function of the IF in which they occur (Steinert et al 1980, 1983, 1985b). However, the way in which the end domains or constituent subdomains of the IF interact with other molecules in cells has yet to be elucidated.

In addition to the probable definition of IF function, several types of data have shown that the end domains are important in some way in IF assembly, at least in vitro. Helix-enriched fragments generated on extensive proteolytic digestion of IF do not assemble into ordered structures in vitro (Crewther & Dowling 1971; Steinert 1978; Steinert et al 1980). Partially cleaved desmin forms ribbon-like structures (Geisler et al 1982), or no filaments at all (Nelson & Traub 1983), but removal of just the N-terminal domain still permits IF assembly (Lu & Johnson 1983). Partially cleaved keratin subunits form subfilamentous or quasi-crystalline structures with or without lateral striations (Aebi et al 1983; Sauk et al 1984). However, it should be pointed out that the mechanism of IF assembly in vitro may be completely different from that in vivo; indeed, few data are yet available on assembly mechanisms in vivo. Some morphological data seems to suggest that subunit synthesis and assembly of IF occurs on the cytoskeletal framework (Fulton et al 1980). This is an exciting concept that is nevertheless fraught with many logistical and mechanical complexities that have not yet been explored experimentally. IF assembly in vitro or in vivo may be directed in part by a registration peptide within the sequence of the subunit itself (cf the telopeptides in collagen). We note here with interest that the universally conserved sequence at the end of segment 2B (and possibly that in segment 1A) might serve in this capacity.

CORRELATION OF EXPRESSION OF IF SUBUNITS WITH STRUCTURE AND FUNCTION OF END DOMAIN SEQUENCES

Rules of Keratin IF Subunit Expression

The keratins are by far the most complex class of IF. By use of high-resolution polyacrylamide gel electrophoresis 17 (Sun et al 1984) or 19 (Moll et al 1982) different subunits have been identified in various human epithelial tissues. In addition to this family of "soft" keratins, the number of keratins expressed in epidermal derivative tissues, the "hard" α-keratins of hair and nail, etc, is still uncertain (Gillespie 1983), although it is at least 10. In all cases, the molecular weights of keratin subunits lie within the range of 40–70 kDa. Most of these keratin IF subunits represent distinct protein entities because each can be translated from a unique mRNA species (Fuchs & Green 1979; Kim et al 1983; Magin et al 1983; Powell et al 1983; Roop et al 1983).

In the soft epithelial keratins, techniques such as two-dimensional gel electrophoresis, and immunology by "Western" blots and immunofluorescence have shown that the expression of keratin subunits in different epithelial tissues follows several simple rules (Table 1) (Moll et al 1982;

Woodcock-Mitchell et al 1982; Wu et al 1982; Eichner et al 1984; Sun et al 1984):

1. There are two subfamilies of keratin IF subunits: an acidic subfamily (pI < 5.5) and a neutral-basic subfamily (pI > 6), which consist of approximately equal numbers of subunits. As discussed above, the acidic keratins are more generally known as Type I IF subunits, and the neutral-basic keratins as Type II IF subunits.
2. At least one member of each Type is coexpressed in all epithelia. Thus both Types of subunits are required for IF assembly in vivo and in vitro (Steinert et al 1976, 1982; Franke et al 1983).
3. Any given epithelium expresses only a few subunits. In most cases specific pairs of Type I and II subunits are coexpressed in a particular epithelial cell type (Table 1). In some cases, however, different pairs of subunits may be expressed when cells are removed from their tissue environment, and are adapted to growth in a liquid medium. When restored to their tissue of origin they again express their characteristic subunits. Also, for example in certain simple epithelia, a given subunit may be coexpressed with another partner in a different epithelial tissue.
4. In the pairs of keratins coexpressed in this characteristic manner, the Type II keratin is always about 7–10 kDa larger than its coexpressed Type I partner (Table 1). This is due to the presence of the H1 and H2 subdomains (Figure 1) in Type II only.
5. There is a clear correlation between the complexity of an epithelium (simple versus stratified squamous versus specialized terminally differentiated) and the size of the keratin subunits expressed (Table 1). The keratins of simple embryonic or ductal epithelia, for example, are usually small, while those of the terminally differentiating epidermis are the largest.
6. When an epithelium changes from a simple to a more complex one, during embryonic development or cellular differentiation, the pattern of keratin subunits expressed also changes in conformity with Rules 3 and 5.
7. When an epithelial cell type is transformed the sets of keratin subunits characterisitc of the normal cell type are usually retained, but occasionally smaller keratins more typical of embryonic or mesothelial epithelia may be expressed.
8. The IF protein vimentin, not usually found in epithelia, is also expressed in mesothelial tissues and lens epithelia, and in certain transformed cells maintained in cell culture, e.g. HeLa and PtK1 cells.

The rules that govern "hard" keratin IF subunit expression have not yet been rigorously explored, although there is evidence that Rules 1–4 also apply (Gillespie 1983; Woods & Inglis 1984) (Table 1). Notably, the keratins expressed in these terminally differentiated tissues are smaller than their epidermal counterparts.

In summary, the expression of the different keratin subunits reflects cell type, degree of histological differentiation, growth environment, and disease state.

Keratin IF Subunit End Domains Correlate with Expression

All of the above rules clearly imply that the keratin IF expressed in different epithelia serve different functions in those epithelia. From the emerging body of amino acid sequence data, trends are appearing that permit correlation of the end domain sequences of keratin subunits with their expression characteristics and possible functions. The key observation in support of this concept is that coexpressed Type I and II subunits have generally similar V1 or V2 subdomain sequences, or both (Table 1) (Steinert et al 1984b, 1985b). Two examples are available. Terminally differentiating epidermal cells express two specific keratin subunits that are among the largest of all keratins. Their V1 and V2 subdomains are each more than 100 residues long, and contain about 60% glycines interspersed with occasional serines, tyrosines, and phenylalanines. Such unusual sequences are rather hydrophobic, which explains why these large subunits are the least soluble of all keratins. The structure(s) adopted by these V subdomains in the intact IF is unknown, but it is probably very flexible and convoluted. It is also thought that these unusual sequences may interact with an interfilamentous matrix protein of the epidermis, filaggrin, to form an insoluble yet compliant structure (Steinert et al 1984a, 1985b). Thus, the properties of these V subdomains have probably been evolutionarily adapted to maintain the insoluble yet flexible nature of the protective barrier provided by the outer layer of the skin, the epidermis.

The end domains of the "hard" wool α-keratins contain cysteine-rich sequences of a highly folded, convoluted structure. These sequences are thought to form extensive disulfide bond cross-links with cysteine-rich (high-sulfur) matrix proteins, thereby contributing to the insoluble, hard, rigid properties (Steinert et al 1984a). In contrast, the keratin subunits expressed in less differentiated inner, living epithelial cell layers are less insoluble. Their V1 subdomains contain fewer glycines, and their V2 subdomains are generally enriched in serines instead. The V1 and V2 subdomains of the two Type I and Type II keratins expressed in stratified squamous epithelia (e.g. basal epidermal cells) have very similar sequences, which supports the idea that such sequences cooperate in defining the function of the IF in these cells.

As can be seen from Table 1, the sizes of the keratins of each type, in effect the sizes of their V subdomains, increase in discrete 2–5 kDa steps in progression with the complexity of the epithelium in which they are expressed. It is difficult to escape the conclusion that the development of progressively more complex and insoluble epithelial tissues has occurred by addition of extra sequences to the V subdomains of these keratins (Sun et al 1984). We expect that further sequencing data will confirm this trend, and thus provide important clues to the functions of these subunits in the IF of diverse epithelia. Following the examples of epidermal filaggrin, and the high-sulfur proteins in hard keratins, these sequence studies may also offer clues on IFAP that may be coexpressed with the keratin IF as mediators of their functions in cells.

Other IF

The neurofilament NF-L, NF-M, and NF-H triplet proteins are characteristic products of neuronal tissues. Although all three proteins are synthesized in the neuronal cell body (Czosneck et al 1980), they are not equally distributed in the various parts of the cell. The axons, including the giant axon of *Myxicola infundibulum*, appear to contain mostly the larger NF-M and NF-H subunits (Gilbert et al 1975; Hill et al 1984). In contrast, whole squid brains contain mostly the NF-L and NF-H subunits (Zackroff & Goldman 1980). The emerging sequence data from these subunits show that the NF-M and NF-H subunits have long glutamate-rich C-terminal domain sequences (Figure 1) (Weber & Geisler 1984). Since immunoelectron microscopy data suggest that large portions of these highly charged sequences protrude from the neurofilament periphery (Willard & Simon 1981), perhaps they are involved in axonal transport of macromolecules (Hoffman & Lasek 1975), or even the transmission of electrical impulses along the axon. Complete characterizations of these neurofilament subunits is likely to provide important insights into their functions in neurons.

In contrast, vimentin is widely distributed in cells of mesenchymal origin, and is present in many established cell lines maintained in culture. Its expression in these less-differentiated types of cells seems to indicate a more general role, such as centering the nucleus, and cell motility (Zackroff et al 1981; Steinert et al 1984a). However, in some embryonic or primitive differentiating cells, such as myotubes, the expression of vimentin is closely coordinated with that of desmin, and together these two IF proteins play an important role in myofibril assembly during myogenesis (Lazarides 1980, 1982). Conversely, in certain transformed epithelial cells that coexpress vimentin and keratin, these two subclasses of IF form distinct and separate networks (Aynardi et al 1984). The rules governing the expression or coexpression of vimentin with another IF type in these cases have not yet been experimentally addressed.

GENE STRUCTURE

Complexity

Clearly, the next step in understanding the complex rules of IF gene expression is the detailed analysis of the genes themselves. Partial or complete information is now available on the genes coding for several IF subunits. It has been reported that hamster (Quax et al 1983, 1984) and avian vimentin (Zehner & Paterson 1983), hamster desmin (Quax et al 1984), mouse glial fibrillary acidic protein (Lewis et al 1984), human 50 kDa (Type I) (Marchuk et al 1984), and mouse 59 kDa (Type I) (Krieg et al 1985) keratin genes exist in only one copy per haploid genome of their species. The human 67 kDa (Type II) keratin gene may be present in more than one copy (Johnson et al 1985). These studies firmly establish that IF diversity originates at the gene level, and is not generated by unusual rearrangement events. However, the avian vimentin gene can express at least three mRNA species that differ only in the length of their 3'-noncoding regions by virtue of alternate termination at one of four possible polyadenylation sites (Zehner & Paterson 1983). Further studies are required to determine whether or not this phenomenon has regulatory significance.

Intron-Exon Structure

The complete genomic structures are now known for four IF genes of three different subunit Types (Quax et al 1983; Marchuk et al 1984; Johnson et al 1985; Krieg et al 1985). Comparison of their organizations reveals that they each contain several introns, mostly located in regions that encode the amino acid sequences predicted to form the coiled-coil rod domains (Figure 4). Interestingly, three introns occur in identical positions in all four genes, and the location of the fourth intron near the end of segment 2B is conserved. Each Type also possesses one or two unique introns in the rod domain. Further, each contains one or two other introns in regions that encode the C-terminal but not the N-terminal domain. However, the introns do not appear to delineate the exact structural boundaries in either the coiled-coil or end domains (Figure 4). Furthermore, the introns in the coiled-coil domain usually occur at points that correspond to the beginning of heptads of the coiled coils.

These data all indicate that the genes for these three Types of IF subunits arose from a common ancestor, which may have been assembled from smaller units containing multiple heptad repeats (Krieg et al 1985).

Presumably, subsequent duplication events have formed the known Types, and each of its various members. This hypothesis suggests that the genes encoding the subunits of each Type are very similar. In support of this, the two Type I keratin genes each contain six introns in the rod domain in identical or nearly identical locations (Figure 4). Some less definitive R-loop studies of several Type II bovine epidermal keratin genes (Lehnert et al 1984) showed that each has seven introns in the rod domain in positions generally similar to those of the human 67 kDa keratin (Figure 4). Characterization of more IF genes, including those of Type IV neurofilaments, is obviously desirable in confirming this trend. The interruption by introns of sequences encoding the coiled-coil regions is not a general feature of genes for other α-type fibrous proteins such as fibrinogen (Crabtree et al 1985) or myosin (Karn et al 1983). Thus the conservation of the location of introns in IF genes reflects the stringent evolutionary forces that have preserved the secondary structure of the coiled-coil rod domain in all IF subunits.

To date, the most variable intron locations appear to occur in the C-terminal domain (Figure 4), but the structural or functional significance of this observation is unclear. Accordingly, a great deal has yet to be learned about how the putative ancestral gene was assembled, how the predicted duplication events occurred to generate the different types of coiled-coil rod domains, and how the highly variable end domains evolved and were coupled with the rod domains. Moreover, the regulatory sequences and mechanisms that direct the expression of the genes during cellular

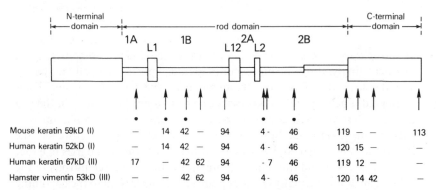

	1A	L1	1B		2A L12 L2		2B			
Mouse keratin 59kD (I)	—	14	42	—	94	4 -	46	119 — —		113
Human keratin 52kD (I)	—	14	42	—	94	4 -	46	120 15 —		—
Human keratin 67kD (II)	17	—	42 62		94	- 7	46	119 12 —		—
Hamster vimentin 53kD (III)	—	—	42 62		94	4 -	46	120 14 42		—

Figure 4   Location of introns in three types of IF genes. Using the universal IF subunit model of Figure 1, arrows indicate the positions of introns of four genes. The numbers refer to the amino acid number along the subdomain. The *arrows* with *closed circles* mark those introns that occur after position 7 of the characteristic heptad repeat in the coiled-coil segments. No introns is indicated by "—". See text for references. [Redrawn with permission from Krieg et al (1985).]

differentiation are not yet known. Indeed, the chromosomal location of IF genes is still unknown.

CONCLUDING REMARKS

From the perspective of the end of 1984, it is apparent that IF are still almost as much an enigma now as when they were first appreciated as a distinct component of the cytoskeleton. Although a little more is presently known about the chemical heterogeneity and structure of IF, different sets of questions should be addressed in the future. For example, work must be directed toward understanding how the end domains of IF subunits interact with other molecules or cytoskeletal components, if we are to better understand the functions of IF in cells. It seems to us that many of the functions of IF, alluded to here and discussed in more detail by others (Lazarides 1982; Wiche et al 1982; Steinert et al 1984a; Yang et al 1985), are mediated by IFAP, and indeed, a handful of these have now been identified. Likewise, many studies employing light and electron microscopy have shown that IF extend from the cell periphery to or into the nuclear surface (e.g. Fey et al 1984; Jones et al 1983, 1985; Ward et al 1984). How do such interactions occur? What do these observations tell us about the role of IF in the lives of cells? Further, the initial reports of the gene structures of IF subunits, summarized here, must be extended so we may better understand the rules that govern IF gene expression and direct their functions during development and differentiation. We therefore await with enthusiasm the further advances in this exciting area of cell biology.

ACKNOWLEDGMENT

We thank Drs. R. D. B. Fraser, R. D. Goldman, D. R. Roop, A. C. Steven, and T.-T. Sun for their critical comments on the manuscript.

Literature Cited

Aebi, U., Fowler, W. E., Rew, P., Sun, T.-T. 1983. *J. Cell Biol.* 97:1131–43

Astbury, W. T., Street, A. 1931. *Phil. Trans. R. Soc. London* 230A:75

Aynardi, M. W., Steinert, P. M., Goldman, R. D. 1984. *J. Cell Biol.* 98:1407–21

Carden, M. J., Eagles, P. A. M. 1983. *Biochem. J.* 215:227–37

Chou, F. Y., Fasman, G. D. 1978. *Adv. Enzmol. Relat. Subj. Biochem.* 47:45–96

Crick, F. H. C. 1953. *Acta Crystallogr.* 6:685–88

Crabtree, G. R., Comeau, C. M., Fowlkes, D.

M., Fornace, A. J., Malley, J. D., Kant, J. A. 1985. *J. Mol. Biol.* In press

Crewther, W. G., Harrap, B. S. 1967. *J. Biol. Chem.* 242:4310–19

Crewther, W. G., Dowling, L. M. 1971. *Appl. Polym. Symp.* 18:1–20

Crewther, W. G., Gough, K. H., Inglis, A. S., McKern, N. M. 1978. *Text. Res. J.* 48:160–73

Crewther, W. G., Dowling, L. M., Steinert, P. M., Parry, D. A. D. 1983. *Int. J. Biol. Macromol.* 5:267–74

Czosnek, H., Soifer, D., Wisniewski, H. M. 1980. *J. Cell Biol.* 85:726–34

Dowling, L. M., Parry, D. A. D., Sparrow, L. G. 1983. *Biosci. Rep.* 3:73–78

Eichner, R., Bonitz, P., Sun, T.-T. 1984. *J. Cell Biol.* 99:1388–96

Fey, E. G., Capco, D. G., Krochmalnic, G., Penman, S. 1984. *J. Cell Biol.* 99:203s–8s

Franke, W. W., Schiller, D. L., Hatzfeld, M., Winter, S. 1983. *Proc. Natl. Acad. Sci. USA* 80:7113–17

Fraser, R. D. B., MacRae, T. P. 1973. *Polymer* 14:61–67

Fraser, R. D. B., MacRae, T. P. 1983. *Biosci. Rep.* 3:517–25

Fraser, R. D. B., MacRae, T. P., Suzuki, E. 1976. *J. Mol. Biol.* 108:435–52

Fraser, R. D. B., MacRae, T. P., Suzuki, E., Parry, D. A. D. 1985. *Int. J. Biol. Macromol.* In press

Fuchs, E., Green, H. 1979. *Cell* 17:573–81

Fuchs, E., Coppock, S., Green, H., Cleveland, D. 1981. *Cell* 27:75–84

Fuchs, E., Marchuk, D. 1983. *Proc. Natl. Acad. Sci. USA* 80:5857–61

Fulton, A. B., Wan, K. M., Penman, S. 1980. *Cell* 20:849–57

Garnier, J., Osguthorpe, D. J., Robson, B. 1978. *J. Mol. Biol.* 120:97–120

Geisler, N., Weber, K. 1982. *EMBO J.* 1:1649–56

Geisler, N., Kaufmann, E., Weber, K. 1982. *Cell* 30:277–86

Geisler, N., Kaufmann, E., Fischer, S., Plessmann, U., Weber, K. 1983. *EMBO J.* 2:1295–1300

Geisler, N., Fischer, S., Vandekerckhove, J., Plessmann, U., Weber, K. 1984. *EMBO J.* 3:2701–6

Gilbert, D. S., Newby, B. J., Anderton, B. 1975. *Nature* 256:586–89

Gillespie, J. M. 1983. In *Biochemistry and Physiology of the Skin*, ed. L. A. Goldsmith, pp. 475–510. Oxford: Oxford Univ. Press

Gough, K. H., Inglis, A. S., Crewther, W. G. 1978. *Biochem. J.* 173:373–85

Gruen, L. C., Woods, E. F. 1983. *Biochem. J.* 209:587–98

Hanukoglu, I., Fuchs, E. 1982. *Cell* 31:243–52

Hanukoglu, I., Fuchs, E. 1983. *Cell* 33:915–24

Henderson, D., Geisler, N., Weber, K. 1982. *J. Mol. Biol.* 155:173–76

Hill, D. E., Zackroff, R. V., Goldman, R. D. 1984. *J. Cell Biol.* 99:322a

Hoffman, R., Lasek, R. J. 1975. *J. Cell Biol.* 66:351–66

Johnson, L. D., Idler, W. W., Zhou, X.-M., Roop, D. R., Steinert, P. M. 1985. *Proc. Natl. Acad. Sci. USA* 82:1886–90

Jones, J. C. R., Goldman, A. E., Steinert, P. M., Yuspa, S. H., Goldman, R. D. 1983. *Cell Motility* 2:197–213

Jones, J. C. R., Goldman, A. E., Yang, H.-Y.,

Goldman, R. D. 1985. *J. Cell Biol.* 100:93–102

Karn, J., Brenner, S., Barnett, L. 1983. *Proc. Natl. Acad. Sci. USA* 80:716–20

Kim, K. H., Rheinwald, J. G., Fuchs, E. 1983. *Mol. Cell Biol.* 3:495–502

Kim, K. H., Marchuk, D., Fuchs, E. 1984. *J. Cell Biol.* 99:1872–77

Koury, S. T., Eckert, B. S. 1984. *J. Cell Biol.* 99:320a

Krieg, T. M., Schafer, M. P., Cheng, C. K., Filpula, D., Flaherty, P., et al. 1985. *J. Biol. Chem.* 260:5867–70

Krishnan, N., Kaiserman-Abramof, I. R., Lasek, R. J. 1979. *J. Cell Biol.* 82:323–35

Lazarides, E. 1980. *Nature* 283:249–56

Lazarides, E. 1982. *Ann. Rev. Biochem.* 51:219–50

Lehnert, M. E., Jorcano, J. L., Hanswalter, Z., Blessing, M., Franz, J. K., Franke, W. W. 1984. *EMBO J.* 3:3279–87

Lewis, S. A., Balcarek, J. M., Krek, V., Shelanski, M., Cowan, N. J. 1984. *Proc. Natl. Acad. Sci. USA* 81:2743–46

Lu, Y.-J., Johnson, P. 1983. *Int. J. Biol. Macromol.* 5:347–50

McLachlan, A. D., Stewart, M. 1982. *J. Mol. Biol.* 162:693–98

Magin, T. M., Jorcano, J. L., Franke, W. W. 1983. *EMBO J.* 2:1387–94

Marchuk, D., McCrohon, S., Fuchs, E. 1984. *Cell* 39:491–98

Milam, L., Erickson, H. P. 1982. *J. Cell Biol.* 94:592–96

Moll, R., Franke, W. W., Schiller, D. L., Geiger, B., Krepler, R. 1982. *Cell* 31:11–21

Nelson, W. J., Traub, P. 1983. *Mol. Cell Biol.* 3:1146–56

Parry, D. A. D., Crewther, W. G., Fraser, R. D. B., MacRae, T. P. 1977. *J. Mol. Biol.* 113:449–54

Parry, D. A. D., Steven, A. C., Steinert, P. M. 1985. *Biochem. Biophys. Res. Commun.* 127:1012–18

Parry, D. A. D., Fraser, R. D. B. 1985. *Int. J. Biol. Macromol.* In press

Pruss, R. M., Mirsky, R., Raff, M. C., Thorpe, R., Dowdling, A. J., Anderton, B. H. 1981. *Cell* 27:419–28

Powell, B. C., Sleigh, M. J., Ward, K. A., Rogers, G. E. 1983. *Nucleic Acid Res.* 11:5327–46

Quax, W., Egberts, W. V., Hendricks, W., Quax-Jeuken, Y., Bloemendal, H. 1983. *Cell* 35:215–23

Quax, W., van den Heuvel, R., Egberts, W. V., Quax-Jeuken, Y., Bloemendal, H. 1984. *Proc. Natl. Acad. Sci. USA* 81:5970–74

Quinlan, R. A., Franke, W. W. 1982. *Proc. Natl. Acad. Sci. USA* 79:3452–56

Quinlan, R. A., Franke, W. W. 1983. *Eur. J. Biochem.* 132:477–84

Quinlan, R. A., Cohlberg, J. A., Schiller, D. L.,

Hatzfeld, M., Franke, W. W. 1984. *J. Mol. Biol.* 178:365–88

Renner, W., Franke, W. W., Schmid, E., Geisler, N., Weber, K., Mandelkow, E. 1981. *J. Mol. Biol.* 149:285–306

Roop, D. R., Hawley-Nelson, P., Cheng, C. K., Yuspa, S. H. 1983. *Proc. Natl. Acad. Sci. USA* 80:716–20

Roop, D. R., Cheng, C. K., Titterington, L., Meyers, C. A., Stanley, J. R., et al. 1984. *J. Biol. Chem.* 259:8037–40

Sauk, J. J., Krumweide, M., Cocking-Johnson, D., White, J. G. 1984. *J. Cell Biol.* 99:1590–97

Skerrow, D. 1974. *Biochem. Biophys. Res. Commun.* 59:1311–16

Skerrow, D., Matoltsy, M. N., Matoltsy, A. G. 1973. *J. Biol. Chem.* 248:4820–26

Steinert, P. M. 1978. *J. Mol. Biol.* 123:49–70

Steinert, P. M. 1981. In *Electron Microscopy of Proteins*, Vol. 1, ed. J. R. Harris, pp. 125–66. London: Academic. 352 pp.

Steinert, P. M., Idler, W. W., Zimmerman, S. B. 1976. *J. Mol. Biol.* 108:547–67

Steinert, P. M., Idler, W. W., Goldman, R. D. 1980. *Proc. Natl. Acad. Sci. USA* 77:4534–38

Steinert, P. M., Idler, W. W., Aynardi-Whitman, M., Zackroff, R. V., Goldman, R. D. 1982. *Cold Spring Harbor Symp. Quant. Biol.* 46:465–73

Steinert, P. M., Rice, R. H., Roop, D. R., Truss, B. L., Steven, A. C. 1983. *Nature* 302:794–800

Steinert, P. M., Jones, J. C. R., Goldman, R. D. 1984a. *J. Cell Biol.* 99:22s–27s

Steinert, P. M., Parry, D. A. D., Racoosin, E. L., Idler, W. W., Steven, A. C., et al. 1984b. *Proc. Natl. Acad. Sci. USA* 81:5709–13

Steinert, P. M., Steven, A. C., Roop, D. R. 1985a. *Cell.* In press

Steinert, P. M., Parry, D. A. D., Idler, W. W., Johnson, L. D., Steven, A. C., Roop, D. R. 1985b. *J. Biol. Chem.* In press

Steinert, P. M., Idler, W. W., Zhou, X.-M., Johnson, L. D., Parry, D. A. D., et al. 1985c. *Ann. NY Acad. Sci.* In press

Steven, A. C., Wall, J. S., Hainfeld, J. T., Steinert, P. M. 1982. *Proc. Natl. Acad. Sci. USA* 79:3101–5

Steven, A. C., Hainfeld, J. T., Trus, B. L., Wall, J. S., Steinert, P. M. 1983a. *J. Biol. Chem.* 258:8323–29

Steven, A. C., Hainfeld, J. T., Trus, B. L., Wall, J. S., Steinert, P. M. 1983b. *J. Cell Biol.* 97:1939–44

Steven, A. C., Trus, B. L., Wall, J. S., Hainfeld, J. T., Steinert, P. M. 1985. *Ann. NY Acad. Sci.* In press

Sun, T.-T., Eichner, R., Nelson, W. G., Tseng, S. C. G., Weiss, R. A., et al. 1983. *J. Invest. Dermatol.* 81:109s–17s

Sun, T.-T., Eichner, R., Schermer, A., Cooper, D., Nelson, W. G., Weiss, R. A. 1984. In *Cancer Cells 1, The Transformed Phenotype*, pp. 169–76. Cold Spring Harbor, NY: Cold Spring Harbor Lab.

Walter, M. F., Biessmann, H. 1984. *J. Cell Biol.* 99:1468–77

Ward, W. S., Schmidt, W. N., Schmidt, C. A., Hnilica, L. S. 1984. *Proc. Natl. Acad. Sci. USA* 81:419–23

Weber, K., Geisler, N. 1984. In *Cancer Cells 1, The Transformed Phenotype*, pp. 153–59. Cold Spring Harbor, NY: Cold Spring Harbor Lab.

Wiche, G., Herrmann, H., Leichtfried, F., Pytela, R. 1982. *Cold Spring Harbor Symp. Quant. Biol.* 46:475–82

Willard, M., Simon, C. 1981. *J. Cell Biol.* 89:198–205

Woodcock-Mitchell, J., Eichner, R., Nelson, W. G., Sun, T.-T. 1982. *J. Cell Biol.* 95:580–88

Woods, E. F., Inglis, A. S. 1984. *Int. J. Biol. Macromol.* 6:277–83

Wu, Y.-J., Parker, L. M., Binder, N. E., Beckert, M. A., Sinard, J. H., et al. 1982. *Cell* 31:693–703

Yang, H.-Y., Goldman, R. D. 1985. *Ann. Rev. Physiol.* In press

Yang, H.-Y., Lieska, N., Goldman, A. E., Goldman, R. D. 1985. *J. Cell Biol.* 100:620–31

Zackroff, R. V., Goldman, R. D. 1980. *Science* 208:1152–55

Zackroff, R. V., Steinert, P. M., Aynardi-Whitman, M., Goldman, R. D. 1981. In *Cell Surface Reviews*, Vol. 7, ed. G. Poste, G. Nicolson, pp. 55–97. New York: North-Holland

Zackroff, R. V., Goldman, A. E., Jones, J. C. R., Steinert, P. M., Goldman, R. D. 1984. *J. Cell Biol.* 98:1231–37

Zehner, Z. E., Paterson, B. M. 1983. *Proc. Natl. Acad. Sci. USA* 80:911–15

Ann. Rev. Cell Biol. 1985. 1:67–90

MOLECULAR BIOLOGY OF FIBRONECTIN

Richard Hynes

Center for Cancer Research and Department of Biology, Massachusetts Institute of Technology, Cambridge, Massachusetts 02139

CONTENTS

INTRODUCTION

The interactions of cells with one another and with extracellular materials (matrices, solid surfaces, etc) are of vital importance for cell function. These interactions have major effects on the proliferation, differentiation, and organization of cells. Our understanding of these interactions of cells has advanced considerably in recent years, and it is now clear that they are often mediated by a class of high molecular weight glycoproteins that are involved both in these interactions and in the actual structure of extracellular matrices. The most intensively studied of these glycoproteins is *fibronectin* (FN), but there is a set of proteins with analogous properties (laminin, von Willebrand protein, thrombospondin, vitronectin, etc), the analysis of which is also progressing apace. I do not review here the extensive literature on the distributions and functions of these proteins, which has been covered in several recent reviews (Hynes & Yamada 1982; Furcht 1983; Yamada 1983; Mosher 1984; Yamada & Akiyama 1984). Instead, I concentrate on recent work on the primary structure and molecular genetics of these

67

0743–4634/85/1115–0067$02.00

proteins, in particular FN, and discuss the insights this work gives into their structure-function relationships, and the way in which molecular biological approaches open the way to more precise cell biological analysis of the functions of these glycoproteins.

Three years ago, the general outlines of FN structure were clear (Hynes & Yamada 1982). The protein consists of a dimer of two subunits, each about 250 kilodaltons (kDa), which are similar but not necessarily identical. Each subunit is folded into an elongated and flexible arm ~ 60 nm long, and the two subunits are joined by disulfide bonds very near their C-termini. Within each subunit there is a series of tightly folded globular domains, each specialized for binding to other molecules or to cells. This general structure is shown in Figure 1, and the evidence for it is discussed in the reviews cited earlier. This modular structure of FN is well suited to its role as a ligand between cells and extracellular materials. FN performs this ligand function in a wide variety of biologically important phenomena (see reviews).

As indicated in Figure 1, there was some evidence by 1982 of a difference between FN subunits in the region of the C-terminal heparin-binding domain (Domain VII in Figure 1). There were also indications that the FNs from different sources (e.g. fibroblasts, plasma) were not identical (see later section for references). However, it was unclear whether the differences

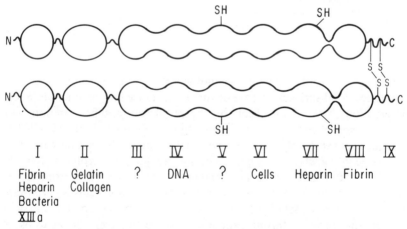

Figure 1 Domain structure of FN. The two subunits of the dimer are attached via a pair of disulfide bonds near the C-terminus. Each subunit is about 250 kDa, and consists of a series of globular domains connected by flexible polypeptide segments, some of which are very readily cleaved by proteases (*thin lines*). Individual domains are specialized for specific binding to other moieties, as shown. Different subunits have different structures in the region of Domain VII (see also Figure 2).

between subunits were due to primary sequence differences, or to carbohydrates or other posttranslational modifications. It was obvious that further structure-function analyses would depend on more detailed structural information concerning FN. The existence of subunit heterogeneity also raised questions concerning the numbers of different forms of FN and their origins, i.e. whether they arose from one gene or many. In the last three years a good deal of information has been obtained by protein sequencing and by the application of recombinant DNA methods. This information provides answers to many of the questions raised by the earlier work, and in turn raises new ones, opening the way to further analysis of the functions of FN.

PRIMARY SEQUENCE

Primary sequence information is now available for the FNs of three species: human, bovine, and rat. This information comes from two sources. Plasma FN can be purified in large amounts, and a large part of the sequence of bovine plasma FN has been determined at the protein level (Skorstengaard et al 1982, 1984; Petersen et al 1983). Significant stretches of the sequence of human plasma FN have been determined in the same way (Pierschbacher et al 1982; Pande & Shively 1982; Pande et al 1985; Gold et al 1983; Garcia-Pardo et al 1983, 1984, 1985). Extensive amino acid sequencing of cellular FN is not a feasible proposition; information on the primary sequence of cellular FN, and elucidation of the origin of the differences between FN subunits have come from the sequencing of cDNA and genomic clones for FN (Kornblihtt et al 1983, 1984a,b, 1985; Schwarzbauer et al 1983; Tamkun et al 1984; Odermatt et al 1985; Bernard et al 1985). The recombinant DNA analyses reveal that, although different subunits do differ in parts of their primary sequence, they all arise from a single gene, and are identical over much of their sequence. I discuss the FN gene and the origin of the different subunits later, but I first describe the striking results on the primary sequence of FN.

As first shown by Petersen, Skorstengaard, and their colleagues (Skorstengaard et al 1982, 1984; Petersen et al 1983), FN consists of a series of homologous repeats. These are of three types: Type I and Type II homologies are disulfide-bonded loops, each 45–50 amino acids long; Type III homologies are about 90 amino acids long, and do not contain disulfide bonds. Greater than 90% of the sequence of FN is made up of repeats of these three homologies (Figure 2).

There are twelve Type I homologies, which are 25–50% identical with each other. All show the same pattern of cystine residues, except that the twelfth repeat contains two extra cysteines. The disulfide-bonding pattern has been determined for four of the repeats (Petersen et al 1983). In each case, the disulfides are formed between cys_1 and cys_3 and between cys_2 and cys_4. Thus, a central cys_2-x-cys_3 sequence is flanked by two disulfide-bonded loops. It is assumed that all Type I homologies show the same disulfide-bonding pattern, with one extra disulfide in the twelfth repeat.

These twelve repeats of Type I homology are concentrated in three of the functional domains of FN. The N-terminal fibrin-binding 29 kDa domain (I in Figure 1) consists of five such repeats. The sequence of this region of bovine plasma FN (Skorstengaard et al 1982, 1984; Petersen et al 1983) is identical at 250/259 positions with the sequence of human FN (Garcia-Pardo et al 1983; Kornblihtt et al 1985). The C-terminal fibrin-binding domain (VIII in Figure 1) also consists entirely of Type I homologies, in this case three of them, which have been sequenced for bovine and human plasma FNs at the protein level (Skorstengaard et al 1982; Petersen et al 1983; Garcia-Pardo et al 1985), and for bovine, human, and rat FNs at the cDNA level (Kornblihtt et al 1983, 1984a,b; Schwarzbauer et al 1983).

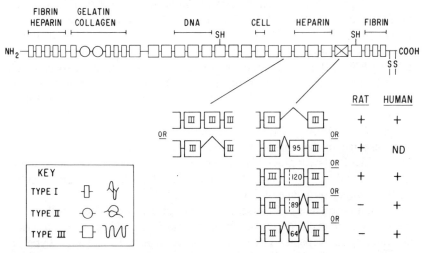

Figure 2   FN structure and variants. Schematic of the primary sequence of FN, which is largely composed of repeats of three types of homology, as diagrammed by the three symbols which represent short homologous segments of polypeptide. The locations of the various functional domains on this sequence are marked. One of the Type III homologies, between the cell- and heparin-binding domains, can be present or absent. A second region of variation falls between the last two Type III homologies, where segments of 0–120 amino acids are found. Three variants have been detected at this position in rat FNs, and four have been identified in human FNs. One insert has not yet been detected (ND) in humans, but probably does occur.

Again, there is a very high degree of sequence conservation among species.

Finally, the gelatin-binding domain contains four more Type I homologies, as well as the only two repeats of Type II homology (Skorstengaard et al 1984; Kornblihtt et al 1985). The arrangement of these homologies is I_1 II_2 I_3, and these six homology segments account for the entire gelatin-binding domain. The disulfide-bonding pattern of the Type II homologies is again cys_1-cys_3 and cys_2-cys_4, but in the Type II homologies the four cystine residues are fairly evenly spaced. The two Type II homologies are about 60% identical within species and each is 98% identical between bovine and human FNs.

A short segment of about 50 amino acids follows the final Type I repeat of the C-terminal fibrin-binding site (IX in Figure 1). This segment is apparently flexible and very easily cleaved by proteases. It contains a stretch of 10–15 amino acids, which varies among species (Kornblihtt et al 1983; Schwarzbauer et al 1983; Bernard et al 1985), followed by the interchain disulfide-bonding region, which is highly conserved among human, bovine, and rat FNs. Therefore, both ends of each FN subunit are highly enriched in cystine residues: 22 disulfide bonds in the N-terminal 70 kDa, and 8 in the C-terminal 25 kDa. In contrast, the central 150–170 kDa of each subunit is free of disulfide bonds, although there are two sulfhydryl residues per subunit which fall in this region (Figure 2) (Wagner & Hynes 1980; Smith et al 1982). This central region of FN consists almost exclusively of Type III homologies. Parts of this Type III region of FN have been sequenced from bovine (Vibe-Pedersen et al 1982; Petersen et al 1983) and human plasma FNs (Pierschbacher et al 1982; Pande & Shively 1982; Pande et al 1985; Gold et al 1983), and long continuous sequences have been obtained from cDNA encoding rat (Schwarzbauer et al 1983; Odermatt et al 1985) and human FNs (Oldberg et al 1983; Kornblihtt et al 1984a,b). Most recently, Kornblihtt et al (1985) determined the sequences of a series of overlapping cDNA's that together encode protein sequences extending from the C-terminus to the N-terminus of human FN. Since variations exist between different cDNA's and among different FN subunits (see below), it is not yet certain that the sequence deduced by Kornblihtt et al represents a "complete" sequence, but for the present we will assume that it does.

The sequences derived from the recombinant cDNA clones show that the central region of FN consists of a series of Type III homologies (15 or 16 of them). The first 40 amino acids of each Type III repeat show strong homology between repeats, as do the last 25–30 amino acids. In contrast, the central sections of each Type III repeat show much less homology between repeats (Schwarzbauer et al 1983; Kornblihtt et al 1985). The sequence of Type III homologies is interrupted at two positions by segments of polypeptide that do not correspond with any of the three types of homology. These interruptions occur (*a*) between the first and second Type III homologies, where a segment of 18 amino acids is predicted by one cDNA clone (Kornblihtt et al 1985); and (*b*) between the final and the penultimate Type III homologies, where a segment of variable length (up to 120 amino acids) is inserted (Schwarzbauer et al 1983; Tamkun et al 1984; Kornblihtt et al 1984a,b; Bernard et al 1985). The first and last Type III homologies diverge more from the Type III consensus pattern than do any of the central 14. Although these Type III homologies appear very similar in structure on the basis of primary sequence, some are relatively susceptible to proteolytic digestion, while other are extremely stable. This has allowed the isolation from this region of separate domains that bind to cells, heparin, and DNA (see Figure 2 and reviews cited earlier). Therefore, despite the homologous repeating structure of this region, different repeats apparently have specialized functions.

In summary, FN has a remarkable primary sequence consisting mostly of repeating homologies of three types (12 Type I, 2 Type II, 15–16 Type III). As mentioned in passing, there are variations in this sequence, which will be discussed in the next section.

SUBUNIT VARIATION

It has been known for years that plasma FN contains subunits of two different mobilities on SDS-polyacrylamide gels, and that FN from fibroblasts shows a different subunit pattern (Yamada & Kennedy 1979; Tamkun & Hynes 1983; Paul & Hynes 1984). Because FNs are glycoproteins, it was unclear whether this might be due to differential glycosylation, which is known to occur (Fukuda et al 1982; Paul & Hynes 1984). However, FNs can be synthesized in the presence of tunicamycin, which blocks addition of asparagine-linked carbohydrates, the major type on FN. Even such carbohydrate-free FN shows multiple subunits on 2D gels, and differences between FNs from different sources (Paul & Hynes 1984; Price & Hynes 1985). Furthermore, proteolytic fragments thought to be carbohydrate-free show differences among different FNs (Hayashi & Yamada 1981, 1983; Sekiguchi et al 1981, 1985; Sekiguchi & Hakomori 1983a,b), and a monoclonal antibody was isolated that shows higher binding to fibroblast than to plasma FN (Atherton & Hynes 1981). All of these results suggested the possibility of structural differences between subunits, but it was still possible that these arose by posttranslational modifications. It was only when cDNA clones were isolated that it could be shown conclusively that there exist FN subunits with different primary sequences (Schwarzbauer et al 1983; Kornblihtt et al 1984a,b), and the results were rather suprising.

When cDNA clones were isolated from rat liver, three classes were found, which differed in sequence at a point between the penultimate and final Type III repeats in the central region of FN (Schwarzbauer et al 1983). In clones of one class, these two Type III repeats are contiguous, whereas clones of the other two classes have inserts of 285 and 360 bases that encode 95 or 120 amino acids, respectively. The 120 amino acid segment consists of the 95 residue segment plus an extra 25 amino acids N-terminal to it. S1 nuclease analyses showed that all three clones represent genuine mRNA's found in rat liver and in other cell types. The 95 and 25 amino acid segments are completely different from any of the three types of repeating homology; they are extremely proline-rich. Outside this difference region all the clones are identical in nucleotide sequence.

The first human cDNA clone covering this region showed a further variation: It contained an insert encoding 89 amino acids homologous with the 25 amino acid segment plus the first 64 residues of the 95 amino acid segment (Kornblihtt et al 1984a,b). Another human cDNA clone from a different cell source encoded the entire 120 amino acid segment (Bernard et al 1985), and other clones have now been isolated containing no insert (K. Sekiguchi et al, personal communication), or an insert encoding only the 64 amino acid segment (A. R. Kornblihtt et al, personal communication). These variants are all diagrammed in Figure 2. It appears that 3 variants are possible in rat and 5 in human FN (see below).

A second region of variation was first detected in human cDNA's by Kornblihtt et al (1984a,b), who isolated two classes of FN cDNA clones from a human cell line. The clones differ as to the presence or absence of a segment of 270 bases encoding an entire Type III repeat. S1 nuclease analysis confirmed that these both represent genuine mRNAs. While the extra Type III repeat is not encoded by any of the cDNA clones from rat liver (Schwarzbauer et al 1983), it is present in the rat gene, and is expressed in other cell types (Odermatt et al 1985; see below). Therefore, there are (at least) two positions of subunit variation at which protein-encoding segments can be present or absent. If all combinations are possible (which is not yet certain) these variations could give rise to 6 (2×3) different variants in rat and 10 (2×5) in human.

Do these variants account for the known subunit variations? To examine this point, cDNA sequences encoding the 95 amino acid difference segment of rat FN were subcloned into the λgt11 expression vector to produce a β-galactosidase FN fusion protein. Antibodies to this fusion protein are specific for the difference region, and recognize only the larger subunits of plasma and cellular FN (Schwarzbauer et al 1985). Therefore, the variations occurring at this position account for the two size classes of plasma FN subunits; the larger ones contain the 95 or 120 amino acid segment while the smaller ones do not. This is true in rats and hamsters, but human plasma FN has not yet been analyzed carefully, and the situation may be a little more complex, although it seems likely to be comparable (see Figure 2). Each size class of FN subunit is known to comprise at least two isoelectric variants (Tamkun & Hynes 1983; Paul & Hynes 1984; Schwarzbauer et al 1985). It seems likely that this is due to the presence or absence of the 25 amino acid segment (which has a net negative charge), but this remains to be proven. For cellular FN, most subunits contain the 95 amino acid segment (Schwarzbauer et al 1985).

The second region of variation has been similarly analyzed. Antibodies raised to this segment recognize fibroblast FN, in particular the larger subunits not seen in plasma FN (Paul & Hynes 1984), but do not recognize plasma FN (J. I. Paul et al, unpublished observations). The extra Type III repeat therefore appears to be a characteristic of cellular FN. This was suggested earlier on the basis of S1 nuclease analyses (Kornblihtt et al 1984a,b), which showed that this segment was detectable in the mRNAs of a variety of cell lines but not in the mRNAs of liver. Since liver, and particularly hepatocytes, are known to be the major or sole source of plasma FN (Owens & Cimino 1982; Tamkun & Hynes 1983), these S1 nuclease data are consistent with the conclusion that inclusion of this segment in mRNAs is characteristic only of cells synthesizing cellular FN.

Therefore, these two regions of variation appear to account for most of the known subunits of FNs. The extra Type III repeat is characteristic of cellular FN, and accounts for the generally larger size of cellular FN subunits, while the non-Type III difference segment accounts for the size difference between the subunits of plasma FN. These conclusions fit well with earlier data mapping apparent size differences between cellular and plasma FNs (Hayashi & Yamada 1981; Atherton & Hynes 1981), and between plasma FN subunits (Sekiguchi et al 1981, 1985; Richter et al 1981; Richter & Hormann 1982; Ehrismann et al 1982; Sekiguchi & Hakomori 1983a,b; Hayashi & Yamada 1983).

Could there be other regions of variation? There are indirect, inconclusive indications of variations in the region just C-terminal to the collagen-binding site (Hayashi & Yamada 1981; Vartio et al 1983), and it is intriguing that a second segment of nonhomologous sequence is similarly located (Kornblihtt et al 1985). It will be interesting to determine whether any variation occurs in that region. It will also be important to use the cDNA and immunological reagents to analyze quantitatively the expression of the different subunit variants in different cell types.

GENE NUMBER, GENE STRUCTURE, AND ALTERNATIVE SPLICING

The existence of multiple subunits and mRNAs immediately leads to questions as to their origin. Do they arise from the same or different genes?

Analysis of recombinant DNA clones has shown conclusively that all the known variants are encoded by a single FN gene. Genomic Southern blots probed with cDNA clones of human and rat FNs (Kornblihtt et al 1983; Tamkun et al 1984) are consistent with the existence of a single gene. Even under low stringency conditions related genes were not detected. These results do not rule out the existence of more distantly related genes, but do indicate that all the cDNA clones, which are closely related, represent mRNAs encoded by a single gene. The question then becomes, how do mRNAs that differ by the presence or absence of internal segments arise from the transcript of a single gene?

The difference must arise from alternative splicing of the primary transcript. Analyses of genomic clones covering the two regions of variation have shown that two different forms of alternative splicing operate. Sequencing of a rat genomic clone covering the non-Type III difference region reveals that the entire 360 base pair (bp) segment encoding the difference region is encoded in a single exon, together with 107 bp encoding the first half of the final Type III repeat (Tamkun et al 1984). Therefore, mRNAs that contain or lack the 285 bp or 360 bp difference segments must arise by alternative splicing of this complex exon. The DNA sequence shows that there are indeed two splice acceptor sequences within the exon, and that they fall precisely at the ends of the 75 bp and 360 bp segments that can be omitted (see Figure 3c). Therefore, subdivision of this complex exon can give rise to three different mRNAs.

As mentioned earlier, two other variants are found in this region of human FN (Figure 2). Although the structure of the human gene has not been reported, it is possible to deduce the probable origin of the variants by examination of the cDNA sequences. The two human variants not seen in rat contain 64 or 89 amino acids (192 or 267 bases) of the difference region, and lack the last 93 bases of the 360 base difference segment. A third human variant contains all 360 bases, and the sequence at the start of the 93 base segment is GTGAG (Bernard et al 1985), which fits the consensus for a 5′ splice site (Sharp 1981; Mount 1982). It therefore seems very likely that the human transcript can undergo an extra splice within the exon to delete the last 93 bases of the difference segment, thus giving rise to the shorter variants seen in human FN (see Figure 3d). This extra splice is not possible in rat FN because the corresponding sequence is ATGAG, which cannot function as a 5′ slice site. Thus, complex splicing of a single exon appears to give rise to 3 to 5 variants, depending on the species.

The second region of variation arises by a different form of alternative splicing. Sequencing of genomic clones encoding the extra Type III segment shows that this Type III homology is encoded by a single exon of 270 bp (Vibe-Pedersen et al 1984; Odermatt et al 1985). This indicates that the presence or absence of this repeat is determined by a pattern of alternative splicing in which this exon can be included or spliced out (Figure 3*b*). Vibe-Petersen et al (1984) have shown that the sequence information that will allow both forms of splicing of the transcript is contained in this region. They inserted a short genomic segment that included the extra Type III exon into a transient expression vector, and observed generation of mRNAs of both types, i.e. either including or lacking this exon.

Therefore, two forms of alternative splicing operate on the FN gene

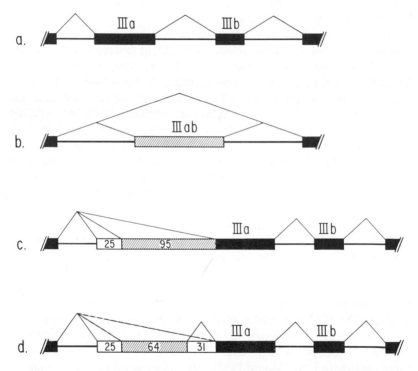

Figure 3   Intron-exon structure of Type III homologies and regions of variation. In each case exons are represented by *boxes* and introns by *lines*. Patterns of splicing are marked. (*a*) Typical Type III homology is encoded by two exons, IIIa and IIIb. (*b*) One Type III homology is encoded by a single exon (IIIab), which can be included or omitted. (*c*) The first exon (IIIa) of the last Type III repeat in the rat gene is extended for 360 base pair (bp) at its 5′ end. The 360 bp segment encodes one of the difference segments (see Figure 2), and can be spliced in three different ways, as shown, using 3′ splice sites at the start of the exon and within the coding region. Use of the internal splice sites can delete segments of the exon encoding 25 or 25 + 95 amino acids. (*d*) Probable structure of the same region of the human gene. It differs from the rat gene in that it has an internal 5′ splice site, which can allow an extra splice, thereby deleting a segment encoding 31 amino acids. Thus five variants are possible in human FN. [Based on a figure of Odermatt et al (1985).]

transcript: *Exon skipping* (Figure 3*b*) accounts for the presence or absence of the extra Type III segment (and the difference between cellular and plasma FNs), and *exon subdivision* (Figures 3*c* and 3*d*) accounts for variations in the non-Type III difference region (and the difference between plasma FN subunits). The pattern of splicing in both places is cell-type specific. Thus, the extra Type III exon is skipped in hepatocytes, leading to the synthesis of plasma FN lacking the extra Type III repeat, but is included about one-third of the time by a human cell line (Kornblihtt et al 1984a), and accounts for the largest FN subunits, which are unique to cellular FN (Paul & Hynes 1984; J. I. Paul et al, unpublished results). Similarly, the ratio of the various splices in the region of exon subdivision varies in different cell types (Schwarzbauer et al 1983), and the smallest subunits, which lack the insert (Schwarzbauer et al 1985), are prevalent in plasma and hepatocyte FN, but rare in fibroblast FN (Paul & Hynes 1984), and essentially absent in astrocyte FN (Price & Hynes 1985). Therefore, the mechanism and regulation of these two forms of alternative splicing are clearly problems of considerable interest to be tackled in the next couple of years.

Indeed, the splicing of the FN transcript is a complex problem. The chicken FN gene is reported to have at least 48 exons which are mostly 100–200 bp long (Hirano et al 1983). How does the repeating structure of the protein (Figure 2) correspond with the exon structure of the gene? All evidence to date is consistent with the assumption that the exon structures of FN genes in other species are similar, so it is possible to make reasonable assignments of several exons and tentative assignments of most of them. Although the extra Type III repeat is encoded by a single exon (Vibe-Pedersen et al 1984; Odermatt et al 1985), all other Type III repeats that have been analyzed at the genomic level are encoded by *two* exons each. This is true of the two Type III repeats on either side of the extra Type III in the human gene (Vibe-Petersen et al 1984), and of the six Type III repeats covering the cell and heparin-binding domains of rat FN (Odermatt et al 1985). Therefore, the typical Type III repeat is encoded by a pair of exons (Figure 3*a*), and this pattern has been modified at the two positions involved in alternative splicing (Figure 3). If this pattern holds for all the rest of the Type III repeats, then the 16 Type III repeats would account for 31 exons. The non-Type III sequence preceding the last Type III repeat is included in one of the Type III exons (see Figures 3*c,d*) and for the purpose of argument I will assume without evidence that the same is true of the short non-Type III segment following the first Type III repeat.

It is known that a partial sequence of exon 12 of the chicken FN gene (Hirano et al 1983) encodes part of the last Type I repeat in the collagen-binding domain (Skorstengaard et al 1984). A reasonable proposition is that exons 13–43 encode the Type III repeats and the nonhomologous segments. On this calculation, the single exon encoding the twelfth and extra Type III repeat would be exon 35, which is in fact reported to be longer than most $(246 \pm 97$ bp) (Hirano et al 1983). The exon encoding the second region of variation would be exon 42. In the rat this exon is large (467 bp) (Tamkun et al 1984), but in the chicken gene exon 42 was determined to be 200 ± 84 bp by R-looping with fibroblast mRNA, and no exons in this region of the chicken gene were reported to be any longer (Hirano et al 1983). It is possible that the non-Type III difference segment in chicken FN is shorter. This would be consistent with the smaller separation between subunits of plasma FN in chickens (Yamada & Kennedy 1979), although other reasons such as differential glycosylation could also play a role here. Resolution of this question must await further results.

If it were true that exons 13–43 encode the central Type III region of FN, then exons 2–12 could encode the N-terminal domains, and exons 44–46 the C-terminal fibrin-binding domain, with one exon for each Type I or Type II repeat. Exon 1 would then encode the 5' untranslated region and/or any pro- or presequence, exon 47 would encode the C-terminal segment, and exon 48 the 3' untranslated region. While this assignment of exons is reasonable, it needs verification. In particular, the suggestion that each Type I and Type II homology is encoded by a single exon needs examination. The answer has evolutionary implications (see later section).

The complex structure of FN is a reflection of a complex gene, which is close to 50 kilobase (kb) long with almost 50 exons. This one gene can give rise to at least 6 to 10 different FN subunits by alternative splicing of the primary transcript within the coding region. Different cell types can apparently regulate this alternative splicing to generate different populations of FN subunits. It remains unclear how or why they do so.

STRUCTURAL AND FUNCTIONAL IMPLICATIONS

Given the wealth of molecular biological information and the primary sequence data reviewed here, what can one learn about the structure and function of FN? One of the attractive features of FN is that we already know quite a bit about its functions (see Figure 1), so the primary sequence information can be compared with known functions and used to analyze them.

The sequence data show that each of the protein-binding domains (I, II, and VIII in Figure 1) is composed of a series of homologous repeats (Figure 2). It is reasonable to propose that each repeat represents a basic "binding module," and that the repeating structure confers multivalency and high affinity binding. There is some evidence that a single Type I repeat can be active in fibrin binding (Banyai et al 1983, see also next section), and it seems logical that 3 or 5 Type I's, as found in the fibrin-binding sites of FN, would be even more effective. However, this remains to be tested directly. Similarly, the relative contributions of the four Type I repeats and the two Type II repeats to the collagen-binding function of Domain II need to be investigated.

The glutamine in Domain I, which is reactive with Factor XIIIa transglutaminase, has been identified as the third amino acid (McDonagh et al 1981), and homologous sequences occur in the non-Type III difference region (Schwarzbauer et al 1983), where a second transglutaminase-reactive site has been mapped (Richter et al 1981). Thus, these reactive groups of FN have been fairly precisely located. The same is obviously true for the sulfhydryl groups (Figures 1 and 2), and the tyrosine sulfate residues of FN (Paul & Hynes 1984; Liu & Lipmann 1985) have also been mapped to the non-Type III difference region (Paul & Hynes 1984). The positions of several carbohydrate side chains in the collagen-binding domain have been precisely defined (Skorstengaard et al 1984), and another has been identified adjacent to one of the sulfhydryl groups (Vibe-Pedersen et al 1982). The positions of other carbohydrate side chains are not yet accurately defined, and may differ depending on the type of FN and the species. This also seems to be true of phosphate groups attached to serine residues near the C-terminus of FN (Ali & Hunter 1981; Ledger & Tanzer 1982). Although Skorstengaard et al (1982) identified a partially phosphorylated serine residue three residues from the carboxyl end of bovine plasma FN, Garcia-Pardo et al (1984) found the equivalent residue of human plasma FN to be nonphosphorylated. It seems likely that several serine residues can be phosphorylated (Ali & Hunter 1981; Paul & Hynes 1984) and the others need to be located.

The structural or functional significance, if any, of these various posttranslational modifications of FN remains obscure. However, the precise locations revealed by the sequencing information will make these potential functions easier to investigate.

The best examples to date of the use of primary sequence information to elucidate structure-function relationships of FN originate from the work of Pierschbacher and coworkers (1982, 1983, 1984a,b). They determined the amino acid sequence of the cell-attachment domain (VII in Figure 1), and then prepared synthetic peptides corresponding with segments of this domain. This work culminated in the demonstration that the cell-attachment activity could be mimicked by a tetrapeptide from this domain of FN (Pierschbacher & Ruoslahti 1984a,b). Short peptides containing the active sequence (L-arg-gly-asp-ser) and closely related active variants of this sequence will compete the cell attachment and spreading functions of FN

(Pierschbacher & Ruoslahti 1984a,b; Yamada & Kennedy 1984, 1985), and when such peptides are coupled to solid surfaces they will promote cell adhesion, albeit much less effectively than intact FN (Pierschbacher et al 1983; Pierschbacher & Ruoslahti 1984a,b). This tetrapeptide is predicted to be present in a hydrophilic loop between two β-strands (Pierschbacher et al 1982; Odermatt et al 1985), and appears to play a crucial role in cell-attachment activity.

Analyses of related peptides suggest that the core of the active site is arg-gly-asp, and that several variations in the fourth position are functional. One such variation may be of some interest: The sequence arg-gly-asp-val is active (Pierschbacher & Ruoslahti 1984b), and is found in the non-Type III difference segment which is alternatively spliced (Schwarzbauer et al 1983; see earlier section). Possibly, inclusion of this segment could confer a second cell-binding site on some FN subunits. Preliminary data on fusion proteins containing this difference region fused to β-galactosidase support this idea (J. W. Tamkun et al, unpublished results), but further work is necessary. It will also be important to determine whether any other sequences within FN, especially in equivalent positions in other Type III homologies, can function as cell-attachment sites. Several other proteins that contain sequences of the form arg-gly-asp-x are known to bind to various cells. These include fibrinogen, collagen, thrombin, and discoidin (Pierschbacher & Ruoslahti 1984b; Springer et al 1984), which raises the intriguing possibility that several proteins share similar active sites for interaction with cells.

Knowledge of the active site of the cell-attachment domain has also facilitated several investigations of the cellular side of this interaction. The identity of the cell surface binding sites or "receptors" for FN has been difficult to elucidate, largely because the binding of FN to cells is of relatively low affinity with dissociation constant $K_d = 3.10^{-7}$ M in platelets (Plow & Ginsberg 1981), and $K_d = 8.10^{-7}$ M in fibroblasts (Akiyama & Yamada 1985). This low affinity makes biochemical investigation of the proposed FN-receptor interaction difficult.

However, Pytela et al (1985) recently exploited the information about the FN active site for cell attachment to isolate a complex of cell surface glycoproteins. This complex binds to an affinity column of a FN cell-binding fragment, and is selectively eluted with the peptide gly-arg-gly-asp-ser-pro, but not by the closely related but inactive peptide gly-arg-gly-*glu*-ser-pro. The hapten elution provides the requisite specificity control that allows identification of this complex as a putative "FN receptor." The complex consists of 2 to 3 glycoproteins with molecular weights around 140,000, and when inserted into liposomes promotes their binding to FN-coated surfaces. It is almost certain that this complex is identical with the

140 kDa complex identified earlier by other workers (Knudsen et al 1981, 1985; Neff et al 1982; Greve & Gottlieb 1982; Giancotti et al 1985) using antibodies to cell surfaces, which interfere with cell-substratum adhesion. This 140 kDa complex has been shown to codistribute with FN and actin in immunolocalization studies (Chen et al 1985; Damsky et al 1985), and already appeared a good candidate for an FN-membrane binding site. The direct binding of this complex to FN, and its inhibition by the active site peptide (Pytela et al 1985) support this model.

Exploitation of the active site peptides has also provided insights into the nature of "FN receptors" on a different cell type. Ginsberg et al (1985) used the same set of peptides that block adhesion of fibroblastic cells to FN to inhibit the binding of FN to platelets, which have a thrombin-inducible FN receptor (Plow & Ginsberg 1981). By and large, peptides that are inhibitory in one system are inhibitory in the other, although there are some minor discrepancies. This shows that the active site in FN for binding to both these two cell types involves the sequence arg-gly-asp-ser.

Gardner & Hynes (1985) used chemical crosslinking reagents to couple the 12 kDa cell-binding domain of FN to glycoprotein IIIa at the surface of thrombin-activated platelets. This crosslinking is blocked by the active site hexapeptide, gly-arg-gly-asp-ser-pro, which confirms that the complex of 12 kDa with IIIa is involved in the specific binding of FN to platelets. Other evidence is also consistent with a glycoprotein complex consisting of two proteins, IIb and IIIa, acting as a FN binding site on platelets. These two glycoproteins are absent from the surfaces of platelets from patients with Glanzmann's thrombasthenia (Nurden & Caen 1975; Phillips & Agin 1977), whose platelets fail to bind FN when activated (Ginsberg et al 1983). Furthermore, monoclonal antibodies to the IIb/IIIa complex inhibit the binding of FN to platelets (Ginsberg et al 1984).

Other evidence suggests that there may be additional FN–cell surface interactions in both fibroblasts (McKeown-Longo & Mosher 1984, 1985; and see reviews cited earlier) and platelets (Plow et al 1984); these remain to be analyzed in similar detail. However, the results of recent work based in part on the detailed analysis of the structure and function of the cell attachment domain of FN (VI in Figure 1) strongly implicate the 140 kDa complex and the IIb/IIIa complex as "FN receptors" in fibroblasts and platelets, respectively.

A further application of synthetic peptides related to the cell-attachment site of FN is seen in the work of Boucaut et al (1984b), who injected a decapeptide containing the arg-gly-asp-ser sequence into embryos, and observed inhibition of cell migration in amphibian gastrulae and the avian neural crest. Both these embryonic systems contain FN, which is thought to be involved in promoting cell migration (Boucaut et al 1983a,b, 1984a,b;

Lee et al 1984; Mayer et al 1981; Duband & Thiery 1982; Thiery & Duband 1982; Rovasio et al 1983). The inhibition by a decapeptide from the cell-attachment region of FN but not by control peptides strongly supports this supposition.

Clearly, knowledge of the primary sequence allows one to design very precise experiments to probe the structure-function relationships of FN, and these studies have already been productive in the analysis of the cell attachment domain. Similar approaches are now possible for all the functional domains of FN. A particularly pressing series of questions concerns the *functional* significance of the variations in primary sequence among FN subunits. Since the sequences of these segments are known, and antibodies have been prepared against them, the way is now open to investigate their roles in detail. The availability of cDNA clones also makes possible the expression of defined segments of FN in bacteria or in animal cells, and the analysis of deleted or mutagenized forms of FN. In the future, such molecular genetic analyses of FN function should be extremely productive.

EVOLUTIONARY IMPLICATIONS

There are several evolutionary implications of the structure of the FN gene and protein. The first and most obvious is that the repeating homologies strongly suggest that the FN gene arose by endoduplication of several primordial minidomains or modules, corresponding with the present day homology Types I, II, and III. This supposition is supported by analyses of the intron-exon structure of the gene, which shows clearly that each Type III repeat is encoded by a basic repeating unit in the gene. Seven Type III repeats in the rat FN gene are precisely separated in each case by an intron (Odermatt et al 1985). The Type III repeat that can be alternatively spliced, the twelfth, is encoded by a single exon (Vibe-Pedersen et al 1984; Odermatt et al 1985), but all the others analyzed thus far are interrupted by an intron (Odermatt et al 1985). These introns always fall in the central nonconserved segments of each Type III repeat, while each of the two exons encodes a segment of each repeat, which has a conserved sequence and structure (Odermatt et al 1985). Therefore, it appears that the basic repeating unit is a pair of exons, which together encode a complete Type III repeat consisting of two structural minidomains. Presumably, the large exon encoding the Type III repeat whose presence is variable arose by fusion of the two exons encoding a standard Type III repeat, perhaps because this is necessary for correct alternative splicing (Figure 3). The origin of the large exon encoding the other non-Type III difference region is obscure. It could have arisen by exon fusion or by extension of the 5' exon of the last Type III into the preceding intron.

While it is not yet clear that the Type I and II repeats similarly represent exon units within the gene, this appears likely, as discussed earlier. It is certainly true that a homologous Type I repeat within tissue plasminogen activator (t-PA) *is* encoded by a single exon (Ny et al 1984). It seems very likely that the single or double exon units encoding each repeat are the modules that underwent duplication and divergence to generate the FN gene. This duplication and divergence must have occurred well before the divergence of vertebrates since, while individual repeats show only 20–60% homology *within* a species (i.e. are quite widely diverged), the homology *between* vertebrate species in a given repeat is >90%. That is, once the duplication and divergence had occurred, the gene became highly conserved. This is true not only at the level of amino acid sequence, but also at the nucleotide level, and even applies to several untranslated segments of the gene. For example, there is a highly conserved stretch of 200 bases in the 3' untranslated region immediately preceding the poly-A addition signal (Kornblihtt et al 1983; Schwarzbauer et al 1983, and unpublished results). Similarly, the 6 intron-exon boundaries that have been determined in the human gene (Vibe-Pedersen et al 1984) are very similar to those in the corresponding positions in the rat gene (Odermatt et al 1985). The strong conservation of FN sequences must reflect the importance of the protein and its gene. Since the protein plays a role in the earliest cell migration events during embryogenesis, and at least in amphibians is encoded by a maternal mRNA (Lee et al 1984; Darribere et al 1984), the protein obviously plays a crucial role in development as well as in later physiological functions. The gene is also subject to a great deal of developmental regulation, including the necessity for accurate and cell type–specific splicing of almost 50 exons. This presumably requires highly specific sequences and/or structures in the noncoding regions.

Another fascinating aspect of the evolution of the FN gene also came to light recently. Banyai et al (1983) noted that tissue plasminogen activator (t-PA) has an N-terminal Type I repeat homologous with those in FN. The related plasminogen activator, urokinase (u-PA), lacks this homology. t-PA binds to fibrin whereas u-PA does not, which is consistent with the idea that Type I homology confers affinity for fibrin, as is also suggested by the presence of 3 and 5 Type I homologies in the two fibrin-binding sites of FN. More recently, McMullen & Fujikawa (1985) determined the sequence of human Factor XIIa (activated Hagemann factor). This protease contains one copy of each of the Type I and II homologies, as well as other conserved structural domains characteristic of serum proteases. The Type II homology is identical with one or both Type II repeats of FN at more than 50% of the positions, and is also extremely similar to two Type II repeats found in a bovine seminal plasma protein (McMullen & Fujikawa 1985; Esch et al 1983). The Type I repeat of Factor XIIa is closely related to those of FN and t-PA. Thus, the two disulfide-bonded minidomains present in multiple copies in FN are also found in other proteins. They presumably moved from one gene to another by a process of exon reassortment or shuffling (Gilbert 1978). This hypothesis gains support from the observation that the Type I repeat in t-PA is encoded precisely by one exon (Ny et al 1984). Therefore, these minidomains that appear to be involved in protein-binding functions in FN may subserve similar roles in other proteins. Furthermore, Patthy et al (1984) noted that Type II homologies are related in sequence to the core of "kringles," which are protein-binding modules characteristic of proteolytic enzymes; this suggests even more distant evolutionary relationships among these various minidomains.

OTHER ADHESIVE GLYCOPROTEINS

There are several other extracellular glycoproteins with properties in common with FN (Yamada 1983; Yamada & Akiyama 1984). These glycoproteins also mediate cell adhesion in various situations, and the question of their possible analogy or homology to FN is of some interest. These proteins include von Willebrand protein (Hoyer 1981; Wagner & Marder 1984), laminin, thrombospondin, and vitronectin. Much less information is available on these proteins, although it is beginning to accumulate at an increasing rate.

The domain organization of laminin is partially established (Timpl et al 1983), cDNA clones have been isolated (Wang & Gudas 1983; Barlow et al 1984), and some primary sequence has been obtained from an α-helical coiled-coil region of laminin (Barlow et al 1984; Paulsson et al 1985). Similarly, some domain organization and short stretches of protein sequence have been determined for thrombospondin (Coligan & Slayter 1984; Dixit et al 1984, 1985; Haverstick et al 1984, 1985) and for vitronectin (Suzuki et al 1984). It seems clear that work on these other glycoproteins will progress rapidly, following to some extent the approaches used so successfully for FN. It will be of great interest to determine whether or not any of the homologous functions reflect homologous protein segments. For example, do these proteins, all of which bind heparin, contain Type III repeats homologous with those in the cell- and heparin-binding domains of FN? And do those which bind fibrinogen (thrombospondin) or collagen (laminin, thrombospondin, von Willebrand protein) contain Type I or Type II repeats like some of the blood clotting and fibrinolytic proteases?

CONCLUSION

The detailed information concerning the structure of FN that has been obtained in the past three years by protein sequencing and molecular biological methods has revealed much of interest concerning the structure of this complex multifunctional protein. Many of the earlier questions have been answered, and several surprising features have emerged. We now know roughly how the different forms of FN arise, namely via a complex pattern of alternative splicing. We also know the structure of the functional domains; they turn out to be assemblages of smaller minidomains. The single large gene that encodes this modular structure is composed of a series of multiple small exons. Some of these appear to have found their way into other genes where they presumably carry out similar functions.

The detailed structural knowledge has allowed more precise analyses of the interaction of FN with cells, and has assisted in the recent progress on studies of the cell-surface binding sites for FN. Similar benefits are likely to accrue from the detailed knowledge of the other functional domains. Furthermore, the availability of FN cDNA and genomic clones will make possible detailed molecular genetic analyses of the structure and function of this complex molecule, and will allow more precise analyses of the expression of the gene using recombinant DNA methods thus far applied in only a few cases (Fagan et al 1981; Tyagi et al 1983). Clearly, similar approaches to analogous extracellular matrix proteins will be equally productive, and their structures will soon be determined in similar detail. This will be of interest both for the information it will provide concerning each individual protein, and for the elucidation of any general or common principles.

A good many questions remain concerning the biological functions of these proteins, but molecular biological approaches have provided more precise means to address these questions, and one can anticipate continued rapid progress in this area of research.

ACKNOWLEDGMENT

I would like to acknowledge the invaluable contributions of my colleagues Jean Schwarzbauer, John Tamkun, Jeremy Paul, Erich Odermatt, and John Gardner who were responsible for the work from our group. This work was supported by grants from the USPHS, National Cancer Institute (POICA 26712, ROI CA17007). I would also like to thank all those who sent me unpublished manuscripts and communicated their latest information, allowing this review to be reasonably current.

Literature Cited

Akiyama, S. K., Yamada, K. M. 1985. The interaction of plasma fibronectin with fibroblastic cells in suspension. *J. Biol. Chem.* In press

Ali, I. U., Hunter, T. 1981. Structural comparison of fibronectins from normal and transformed cells. *J. Biol. Chem.* 256: 7671–77

Atherton, B. T., Hynes, R. O. 1981. A difference between plasma and cellular fibronectins located with monoclonal antibodies. *Cell* 25: 133–41

Banyai, L., Varadi, A., Patthy, L. 1983. Common evolutionary origin of the fibrin-binding structures of fibronectin and tissue-type plasminogen activator. *FEBS Lett.* 163: 37–41

Barlow, D. P., Green, N. M., Kurkinen, M., Hogan, B. L. M. 1984. Sequencing of laminin B chain cDNAs reveals C-terminal regions of coiled-coil alpha helix. *EMBO J.* 3: 2355–62

Bernard, M. P., Kolbe, M., Weil, D., Chu, M. L. 1985. Human cellular fibronectin: comparison of the C-terminal portion with rat identifies primary structural domains separated by hypervariable regions. *Biochemistry.* In press

Boucaut, J. C., Darribere, T. 1983a. Presence of fibronectin during early embryogenesis in amphibian *Pleurodeles waltlii. Cell Differ.* 12: 77–83

Boucaut, J. C., Darribere, T. 1983b. Fibronectin in early amphibian embryos. Migrating mesodermal cells contact fibronectin established prior to gastrulation. *Cell Tissue Res.* 234: 135–45

Boucaut, J. C., Darribere, T., Boulekbache, H., Thiery, J. P. 1984a. Prevention of gastrulation but not neurulation by antibodies to fibronectin in amphibian embryos. *Nature* 307: 364–67

Boucaut, J. C., Darribere, T., Poole, T. J., Aoyama, M., Yamada, K. M., Thiery, J. P. 1984b. Biologically active synthetic peptides as probes of embryonic development: a competitive peptide inhibitor of fibronectin function inhibits gastrulation in amphibian embryos and neural crest cell migration in avian embryos. *J. Cell Biol.* 99: 1822–30

Chen, W. T., Hasegawa, T., Hasegawa, C., Weinstock, C., Yamada, K. M. 1985. Development of cell surface linkage complexes in cultivated fibroblasts. *J. Cell Biol.* 100: 1103–14

Coligan, J. E., Slayter, H. S. 1984. Structure of thrombospondin. *J. Cell Biol.* 259: 3944–48

Damsky, C. M., Knudsen, K. A., Bradley, D., Buck, C. A., Horwitz, A. F. 1985. Distri-

bution of the CSAT cell-matrix antigen on myogenic and fibroblastic cells in culture. *J. Cell Biol.* 100: 1528–39

Darribere, T., Boucher, D., Lacroix, J. C., Boucaut, J. C. 1984. Fibronectin synthesis during oogenesis and early development of the amphibian *Pleurodeles waltlii. Cell Differ.* 14: 171–77

Dixit, V. M., Grant, G. A., Santoro, S. A., Frazier, W. A. 1984. Isolation and characterization of a heparin-binding domain from the amino terminus of platelet thrombospondin. *J. Cell Biol.* 259: 10100–5

Dixit, V. M., Haverstick, D. M., O'Rourke, K. M., Hennessy, S. W., Grant, G. A., et al. 1985. A monoclonal antibody against human thrombospondin inhibits platelet aggregation. *Proc. Natl. Acad. Sci. USA* 82: 3472–76

Duband, J. L., Thiery, J. P. 1982. Distribution of fibronectin in the early phase of avian cephalic neural crest cell migration. *Dev. Biol.* 93: 308–23

Ehrismann, R., Roth, D. E., Eppenberger, H. M., Turner, D. C. 1982. Arrangement of attachment-promoting, self-association, and heparin-binding sites in horse serum fibronectin. *J. Biol. Chem.* 257: 7381–87

Esch, F. S., Ling, N. C., Bohlen, P., Ying, S. Y., Guillemin, R. 1983. Primary structure of PDC-109, a major protein constituent of bovine seminal plasma. *Biochem. Biophys. Res. Commun.* 113: 861–67

Fagan, J. B., Sobel, M. E., Yamada, K. M., deCrombugghe, B., Pastan, I. 1981. Effects of transformation on fibronectin gene expression using cloned fibronectin cDNA. *J. Biol. Chem.* 256: 520–25

Fukuda, M., Levery, S. B., Hakomori, S. I. 1982. Carbohydrate structure of hamster plasma fibronectin. Evidence for chemical diversity between cellular and plasma fibronectins. *J. Biol. Chem.* 257: 6856–60

Furcht, L. T. 1983. Structure and function of the adhesive glycoprotein fibronectin. *Mod. Cell Biol.* 1: 53–117

Garcia-Pardo, A., Pearlstein, E., Frangione, B. 1983. Primary structure of human plasma fibronectin. The 29,000-dalton NH2 terminal domain. *J. Biol. Chem.* 258: 12670–74

Garcia-Pardo, A., Pearlstein, E., Frangione, B. 1984. Primary structure of human plasma fibronectin—characterization of the 6,000 dalton C-terminal fragment containing the interchain disulfide bridges. *Biochem. Biophys. Res. Commun.* 120: 1015–21

Garcia-Pardo, A., Pearlstein, E., Frangione, B. 1985. Primary structure of human plasma fibronectin. Characterization of a

31,000 dalton fragment from the C-terminal region containing a free sulfhydryl group and a fibrin-binding site. *J. Biol. Chem.* In press

Gardner, J. M., Hynes, R. O. 1985. Interaction of fibronectin with its receptor on platelets. *Cell.* In press

Giancotti, F. G., Tarone, G., Knudsen, K., Damsky, C., Comoglio, P. M. 1985. Cleavage of a 135KD cell surface glycoprotein correlates with loss of fibroblast adhesion to fibronectin. *Exp. Cell Res.* 156:182–90

Gilbert, W. 1978. Why genes in pieces? *Nature* 271:501

Ginsberg, M. H., Forsyth, J., Lightsey, A., Chediak, J., Plow, E. F. 1983. Reduced surface expression and binding of fibronectin by thrombin-stimulated thrombasthenic platelets. *J. Clin. Invest.* 71:619–24

Ginsberg, M. H., Wolff, R., Marguerie, G., Coller, B., McEver, R., Plow, E. F. 1984. Thrombospondin binding to thrombin-stimulated platelets: evidence for a common adhesive protein binding mechanism. *Clin. Res.* 32:308A

Ginsberg, M. H., Pierschbacher, M. D., Ruoslahti, E., Marguerie, G., Plow, E. 1985. Inhibition of fibronectin binding to platelets by proteolytic fragments and synthetic peptides which support fibroblast adhesion. *J. Biol. Chem.* 260:3931–36

Gold, L. I., Frangione, B., Pearlstein, E. 1983. Biochemical and immunological characterization of three binding sites on human plasma fibronectin with different affinities for heparin. *Biochemistry* 22:4113–19

Greve, J. M., Gottlieb, D. L. 1982. Monoclonal antibodies which alter the morphology of cultured chick myogenic cells. *J. Cell Biochem.* 18:221–30

Haverstick, D. M., Dixit, V. M., Grant, G. A., Frazier, W. A., Santoro, S. A. 1984. Localization of the hemagglutinating activity of platelet thrombospondin to a 140,000-dalton thermolysin fragment. *Biochemistry* 23:5597–5603

Haverstick, D. M., Dixit, V. M., Grant, G. A., Frazier, W. A., Santoro, S. A. 1985. Characterization of the platelet agglutinating activity of thrombospondin. *Biochemistry.* In press.

Hayashi, M., Yamada, K. M. 1981. Differences in domain structures between plasma and cellular fibronections. *J. Biol. Chem.* 256:11292–11300

Hayashi, M., Yamada, K. M. 1983. Domain structure of the carboxyl-terminal half of human plasma fibronectin. *J. Biol. Chem.* 258:3332–40

Hirano, H., Yamada, Y., Sullivan, M., deCrombugghe, B., Pastan, I., Yamada, K.

M. 1983. Isolation of genomic DNA clones spanning the entire fibronectin gene. *Proc. Natl. Acad. Sci. USA* 80:46–50

Hoyer, L. W. 1981. The factor VII complex: structure and function. *Blood* 58:1–13

Hynes, R. O., Yamada, K. M. 1982. Fibronectins: multifunctional modular glycoproteins. *J. Cell Biol.* 95:369–77

Knudsen, K. A., Rao, P., Damsky, C. M., Buck, C. A. 1981. Membrane glycoproteins involved in cell-substratum adhesion. *Proc. Natl. Acad. Sci. USA* 78:6071–78

Knudsen, K. A., Horwitz, A. F., Buck, C. A. 1985. A monoclonal antibody identifies a glycoprotein complex involved in cell-substratum adhesion. *Exp. Cell Res.* In press

Kornblihtt, A. R., Vibe-Pedersen, K., Baralle, F. E. 1983. Isolation and characterization of cDNA clones for human and bovine fibronectins. *Proc. Natl. Acad. Sci. USA* 80:3218–22

Kornblihtt, A. R., Vibe-Pedersen, K., Baralle, F. E. 1984a. Human fibronectin: molecular cloning evidence for two mRNA species differing by an internal segment coding for a structural domain. *EMBO J.* 3:221–26

Kornblihtt, A. R., Vibe-Pedersen, K., Baralle, F. E. 1984b. Human fibronectin: cell specific alternative mRNA splicing generates polypeptide chains differing in the number of internal repeats. *Nucleic Acids Res.* 12:5853–68

Kornblihtt, A. R., Umezawa, H. K., Vibe-Pedersen, K., Baralle, F. E. 1985. Primary structure of human fibronectin: Differential splicing may generate at least 10 polypeptides from a single gene. *EMBO J.* In press

Ledger, P. W., Tanzer, M. L. 1982. The phosphate content of human fibronectin. *J. Biol. Chem.* 257:3890–95

Lee, G., Hynes, R., Kirschner, M. 1984. Temporal and spatial regulation of fibronectin in early Xenopus development. *Cell* 36:729–40

Liu, M. C., Lipmann, F. 1985. Isolation of tyrosine-O-sulfate by pronase hydrolysis from fibronectin secreted by Fujinami sarcoma virus-infected rat fibroblasts. *Proc. Natl. Acad. Sci. USA* 82:34–37

Mayer, B. W., Hay, E. D., Hynes, R. O. 1981. Immunocytochemical localization of fibronectin in embryonic chick trunk and area vasculosa. *Dev. Biol.* 82:267–86

McDonagh, R. P., McDonagh, J., Petersen, T. E., Thorgersen, H. C., Skorstengaard, K., et al. 1981. Amino acid sequence of the factor XIII$_a$ acceptor site in bovine plasma fibronection. *FEBS Lett.* 127:174–78

McKeown-Longo, P. J., Mosher, D. F. 1984.

Mechanism of formation of disulfide-bonded multimers of plasma fibronectin in cell layers of cultured human fibroblasts. *J. Biol. Chem.* 259:12210–15

McKeown-Longo, P. J., Mosher, D. F. 1985. Interaction of the 70 kilodalton amino-terminal fragment of fibronectin with the matrix-assembly receptor of fibroblasts. *J. Cell Biol.* 100:364–74

McMullen, B. A., Fujikawa, K. 1985. Amino acid sequence of the heavy chain of human α-factor XIIIa (activated Hageman factor). *J. Biol. Chem.* In press

Mosher, D. F. 1984. Physiology of fibronectin. *Ann. Rev. Med.* 35:561–75

Mount, S. M. 1982. A catalogue of splice junction sequences. *Nucleic Acids Res.* 10:459–71

Neff, N. T., Lowrey, C., Decker, C., Tover, A., Damsky, C., et al. 1982. A monoclonal antibody detaches embryonic skeletal muscle from extracellular matrices. *J. Cell Biol.* 95:654–66

Nurden, A. T., Caen, J. P. 1975. Specific roles for platelet surface glycoproteins in platelet function. *Nature* 255:720–22

Ny, T., Elgh, F., Lund, B. 1984. The structure of the human tissue-type plasminogen activator gene: correlation of intron and exon structures to functional and structural domains. *Proc. Natl. Acad. Sci. USA* 81:5355–59

Odermatt, E., Tamkun, J. W., Hynes, R. O. 1985. The repeating modular structure of the fibronectin gene: relationship to protein structure and subunit variation. *Proc. Natl. Acad. Sci. USA* In press

Oldberg, A., Linney, E., Ruoslahti, E. 1983. Molecular cloning and nucleotide sequence of a cDNA clone coding for the cell attachment domain in human fibronectin. *J. Biol. Chem.* 258:10193–96

Owens, M. R., Cimino, C. D. 1982. Synthesis of fibronectin by the isolated perfused rat liver. *Blood* 59:1305–9

Pande, H., Calaycay, J., Hawke, D., Ben-Avram, C. M., Shively, J. E. 1985. Primary structure of a glycosylated DNA-binding domain in human fibronectin. *J. Biol. Chem.* In press

Pande, H., Shively, J. E. 1982. NH₂-terminal sequences of DNA-, heparin-, and gelatin-binding tryptic fragments from human plasma fibronectin. *Arch. Biochem. Biophys.* 213:258–65

Patthy, L., Trexler, M., Vali, Z., Banyai, L., Varadi, A. 1984. Kringles: modules specialized for protein binding. Homology of the gelatin-binding region of fibronectin with the kringle structure of proteases. *FEBS Lett.* 171:131–36

Paul, J. I., Hynes, R. O. 1984. Multiple

fibronectin subunits and their post-translational modifications. *J. Biol. Chem.* 259:13477–88

Paulsson, M., Deutzmann, R., Timpl, R., Dalzoppo, D., Odermatt, E., Engel, J. 1985. Evidence for coiled-coil α-helical regions in the long arm of laminin. *EMBO J.* In press

Petersen, T. E., Thogersen, H. C., Skorstengaard, K., Vibe-Pedersen, K., Sottrup-Jensen, L., Magnusson, S. 1983. Partial primary structure of bovine plasma fibronectin, three types of internal homology. *Proc. Natl. Acad. Sci. USA* 80:137–41

Phillips, D. R., Agin, P. P. 1977. Platelet membrane defects in Glanzmann's thrombasthemia. Evidence for decreased amounts of two major glycoproteins. *J. Clin. Invest.* 60:535–45

Pierschbacher, M. D., Ruoslahti, E. 1984a. Cell attachment activity of fibronectin can be duplicated by small synthetic fragments of the molecule. *Nature* 309:30–33

Pierschbacher, M. D., Ruoslahti, E. 1984b. Variants of the cell recognition site of fibronectin that retain attachment-promoting activity. *Proc. Natl. Acad. Sci. USA* 81:5985–88

Pierschbacher, M. D., Ruoslahti, E., Sundelin, J., Lind, P., Peterson, P. A. 1982. The cell attachment domain of fibronectin. Determination of the primary structure. *J. Biol. Chem.* 257:9593–97

Pierschbacher, M. D., Hayman, E. G., Ruoslahti, E. 1983. Synthetic peptide with cell attachment activity of fibronectin. *Proc. Natl. Acad. Sci. USA* 80:1224–27

Plow, E. F., Ginsberg, M. H. 1981. Specific and saturable binding of plasma fibronectin to thrombin-stimulated human platelets. *J. Biol. Chem.* 256:9477–82

Plow, E. F., Srouji, A. H., Meyer, D., Marguerie, G., Ginsberg, M. H. 1984. Evidence that three adhesive proteins interact with a common recognition site on activated platelets. *J. Biol. Chem.* 259:5388–91

Price, J., Hynes, R. O. 1985. Astrocytes in culture synthesize and secrete a variant form of fibronectin. *J. Neurosci.* In press

Pytela, R., Pierschbacher, M. D., Ruoslahti, E. 1985. Identification and isolation of a 140 kd cell surface glycoprotein with properties expected of a fibronectin receptor. *Cell* 40:191–98

Richter, H., Hormann, H. 1982. Early and late cathepsin D-derived fragments of fibronectin containing the C-terminal interchain disulfide cross-link. *Hoppe Seylers Z. Physiol. Chem.* 363:351–64

Richter, H., Seidl, M., Hormann, H. 1981.

Location of heparin-binding sites of fibronectin. Detection of a hitherto unrecognized transamidase sensitive site. *Hoppe Seylers Z. Physiol. Chem.* 362: 399–408

Rovasio, R. A., Delouvee, A., Yamada, K. M., Timpl, R., Thiery, J. P. 1983. Neural crest cell migration: requirements for exogenous fibronectin and high cell density. *J. Cell Biol.* 96: 462–73

Ruoslahti, E., Hayman, E. G., Engvall, E., Cothran, W. C., Butler, W. T. 1981. Alignment of biologically active domains in the fibronectin molecule. *J. Biol. Chem.* 256: 7277–81

Schwarzbauer, J. E., Tamkun, J. W., Lemischka, I. R., Hynes, R. O. 1983. Three different fibronectin mRNAs arise by alternative splicing within the coding region. *Cell* 35: 421–31

Schwarzbauer, J. E., Paul, J. I., Hynes, R. O. 1985. On the origin of species of fibronectin. *Proc. Natl. Acad. Sci. USA* 81: 1424–28

Sekiguchi, K., Hakomori, S. 1983a. Topological arrangements of four functionally distinct domains in hamster plasma fibronection: a study with combination of S-cyanylation and limited proteolysis. *Biochemistry* 22: 1415–22

Sekiguchi, K., Hakomori, S. 1983b. Domain structure of human plasma fibronectin. Differences and similarities between human and hamster fibronectins. *J. Biol. Chem.* 258: 3967–73

Sekiguchi, K., Fukuda, M., Hakomori, S. 1981. Domain structure of hamster plasma fibronectin. Isolation and characterization of four functionally distinct domains and their unequal distribution between two subunit polypeptides. *J. Biol. Chem.* 256: 6452–62

Sekiguchi, K., Siri, A., Zardi, L., Hakomori, S. I. 1985. Differences in domain structure between human fibronectins isolated from plasma and from culture supernatants of normal and transformed fibroblasts. *J. Biol. Chem.* 260: 5105–14

Sharp, P. A. 1981. Speculations on RNA Splicing. *Cell* 23: 643–46

Skorstengaard, K., Thogersen, H. C., Vibe-Pedersen, K., Petersen, T. E., Magnusson, S. 1982. Purification of twelve cyanogen bromide fragments from bovine plasma fibronectin and the amino acid sequence of eight of them. *Eur. J. Biochem.* 128: 605–23

Skorstengaard, J., Thogersen, H. C., Petersen, T. E. 1984. Complete primary structure of the collagen-binding domain of bovine fibronectin. *Eur. J. Biochem.* 140: 235–43

Smith, D. E., Mosher, D. F., Johnson, R. B., Furcht, L. T. 1982. Immunological identification of two sulfhydryl-containing fragments of human plasma fibronectin. *J. Biol. Chem.* 257: 5831–38

Springer, W. R., Cooper, D. N. W., Barondes, S. H. 1984. Discoidin I is implicated in cell-substratum attachment and ordered cell migration of *Dictyostelium discoideum* and resembles fibronectin. *Cell* 39: 557–64

Suzuki, S., Pierschbacher, M., Hayman, E. G., Nguyen, K., Ohgren, Y., Ruoslahti, E. 1984. Domain structure of vitronectin. Alignment of active sites. *J. Biol. Chem.* 259: 15307–14

Tamkun, J. W., Hynes, R. O. 1983. Plasma fibronectin is synthesized and secreted by hepatocytes. *J. Biol. Chem.* 258: 4641–47

Tamkun, J. W., Schwarzbauer, J. E., Hynes, R. O. 1984. A single rat fibronectin gene generates three different mRNAs by alternative splicing of a complex exon. *Proc. Natl. Acad. Sci. USA* 81: 5140–44

Thiery, J. P., Duband, J. L. 1982. Pathways and mechanisms of avian trunk neural crest cell migration and localization. *Dev. Biol.* 93: 324–43

Timpl, R., Engel, J., Martin, G. R. 1983. Laminin—a multifunctional protein of basement membranes. *Trends Biochem. Sci.* 8: 207–9

Tyagi, J. S., Hirano, H., Merlino, G. T., Pastan, I. 1983. Transcriptional control of the fibronectin gene in chick embryo fibroblasts transformed by Rous sarcoma virus. *J. Biol. Chem.* 258: 5787–93

Vartio, T., Barlati, S., de Petro, G., Miggiano, V., Stahli, C., et al. 1983. Evidence for preferential proteolytic cleavage of one of the two fibronectin subunits and for immunological localization of a site distinguishing them. *Eur. J. Biochem.* 135: 203–7

Vibe-Pedersen, K., Kornblihtt, A. R., Petersen, T. E. 1984. Expression of a human α-globin/fibronectin gene hybrid generates two mRNAs by alternative splicing. *EMBO J.* 3: 2511–16

Vibe-Pedersen, K., Sahl, P., Skorstengaard, K., Petersen, T. E. 1982. Amino acid sequence of a peptide from bovine plasma fibronectin containing a free sulfhydryl group (cysteine). *FEBS Lett.* 142: 27–30

Wagner, D. D., Hynes, R. O. 1980. Topological arrangement of the major structural features of fibronectin. *J. Biol. Chem.* 255: 4304–12

Wagner, D. D., Marder, V. J. 1984. Biosynthesis of von Willebrand protein by human endothelial cells: processing steps and their intracellular location. *J. Cell Biol.* 99: 2123–30

Wang, S. Y., Gudas, L. J. 1983. Isolation of cDNA clones specific for collagen IV and laminin from mouse teratocarcinoma cells. *Proc. Natl. Acad. Sci. USA* 80:5880–84

Yamada, K. M. 1983. Cell surface interactions with extracellular materials. *Ann. Rev. Biochem.* 52:761–99

Yamada, K. M., Akiyama, S. K. 1984. *The Interactions of Cells with Extracellular Matrix Components in Cell Membranes*, Vol. 2, ed. L. Glaser, W. Frazier, pp. 77–148. New York: Plenum

Yamada, K. M., Kennedy, D. W. 1979. Fibroblast cellular and plasma fibronec-tins are similar but not identical. *J. Cell Biol.* 80:492–98

Yamada, K. M., Kennedy, D. W. 1984. Dualistic nature of adhesive protein function: fibronectin and its biologically active peptide fragments can autoinhibit fibronectin function. *J. Cell Biol.* 99:29–36

Yamada, K. M., Kennedy, D. W. 1985. Amino acid sequence specificities of an adhesive recognition signal. *J. Cell Biochem.* In press

Ann. Rev. Cell Biol. 1985. 1 : 91–113

CELL MIGRATION IN THE VERTEBRATE EMBRYO: Role of Cell Adhesion and Tissue Environment in Pattern Formation

Jean Paul Thiery, Jean Loup Duband, and Gordon C. Tucker

Institut d'Embryologie, Centre National de la Recherche Scientifique, Collège de France, 49 Bis, avenue de la Belle Gabrielle, 94130 Nogent-sur-Marne, France

CONTENTS

INTRODUCTION

The development of an embryo results from a series of highly interconnected and complex processes. The detailed program used by the ferti-

91

0743–4634/85/1115–0091$02.00

lized egg differs considerably in different species, however, some general principles of development contribute to the elaboration of the body plan.

One of the most general features resides in the considerable discrepancy between the fate map of the blastoderm and the definitive location of the different tissues. This drastic reorganization of the presumptive territories in the young embryo results primarily from extensive morphogenetic movements, including individual cell migration and cell sheet spreading, during gastrulation. These various movements allow the transformation of the blastula into a three-layered embryo, and also contribute to the induction and shaping of the nervous system.

In this chapter we examine the migration of cells during gastrulation, and the subsequent emigration of some cells from the neural primordium to give rise to the peripheral nervous system. Our analysis focuses on the mechanisms that provide directionality for the migrating cells, as well as on the adhesive interactions that control cell behavior. The ability of other cell types to respond to different cues is discussed to help define the relative contributions of the mechanisms that are currently known to guide cells to their final destination.

GASTRULATION

Basic Features

A "typical" gastrulation cannot be described in detail because the morphology, the position of some presumptive territories, and the structure of the site of ingression vary across different species (4, 37, 74, 79).

In amphibians, after the segmentation phase the blastula appears as a hollow sphere containing a blastocoelic cavity. The first sign of gastrulation is a slit, called the blastopore, that appears on the future dorsal side of the embryo, corresponding to the site of invagination of the mesodermal and endodermal cells. Scanning electron microscopy, time-lapse microcinematography, and cultures of tissue explants have shown that gastrulation involves the epiboly of the superficial layer, the formation of bottle cells, and the active movement of mesodermal cells along the blastocoelic roof (12, 38, 39, 46).

Although spreading during epiboly must involve dynamic cytoskeletal reorganization, very little is known about the mechanisms that trigger these propagated movements in cell sheets. The development of pseudopodia and filopodia in the mesodermal cells, which progressively lose their epithelial arrangement, is also a very general mechanism that is repeatedly observed during embryogenesis but is not yet understood [for a review see Thiery 1984 (64)].

Role of Fibronectin

In amphibians, the surface of the blastocoelic roof, to which mesodermal cells adhere, is covered by a network of fibrils. This extracellular matrix (ECM) probably has a composition similar to that described in other embryos (32, 81, see also Hynes, this volume). However, to date, biochemical studies and immunocytochemistry have revealed only the presence of fibronectin (FN) (6).

FN is synthesized at a low rate from maternally derived mRNA during oogenesis; translation increases rapidly during the late blastula and early gastrula stages (14, 44). FN assembles specifically along the ectoderm lining the blastocoel (Figure 1a–c), even though most cells are able to synthesize it. Immunolabeling with fluorescent or gold-coupled antibodies has clearly shown that the fibrils described by scanning electron microscopy (47) contain FN (15).

The role of FN in the adhesion and migration of mesodermal cells has been assessed by three types of perturbation experiments, which led to the following conclusions: (a) When part of the blastocoelic roof is inverted, mesodermal cells avoid the area that now lacks an extracellular matrix. (b) Microinjection of monovalent anti-fibronectin antibodies into the blastocoelic cavity of late blastulae or early gastrulae blocks gastrulation (Figure 1d, e) (7). (c) Similarly, when Arg-Gly-Asp-Ser-containing peptides are injected, gastrulation is arrested (8). This peptide sequence, which is contained in the cell binding site of FN, has been shown to be directly involved in the mechanism of binding of fibroblasts to FN (54, 82). Competitive inhibition of the receptors and a steric hindrance effect, which prevents interaction between the cell surface and FN, both interfere with the movement of mesodermal cells during gastrulation (see Figure 5a).

In contrast to its requirement for the formation of the mesoderm, FN is probably not involved in the mechanism of neural induction. In vitro, mesodermal cells associated with the apical surface of the ectoderm devoid of matrix ECM can induce the appearance of neural elements (20). In vivo, when either the antibodies or the peptides are introduced during or at the end of gastrulation, a partial or a complete neural plate forms (7, 8).

NEURULATION

Extensive morphogenetic movements have been described throughout the period of neural plate and neural tube formation. In the urodele species *Taricha torosa*, time-lapse microcinematography and histology allowed detailed description of the trajectory of cells of the superficial layer. In addition, a precise map of the changes in the area occupied by the apical surface of each cell was constructed. These data were further analyzed by computer simulation; an intrinsic program of constriction of epithelial cells, and the rearrangement of cells of the notoplate (i.e. cells of the notochord and the neural plate remaining in close contact) are sufficient to account for the transformation from a flat disc to a keyhole-shaped neural plate (33). In this species, cells of the notoplate behave like a fluid, in

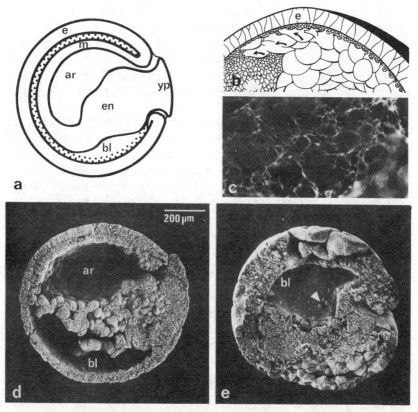

Figure 1 Fibronectin in *Pleurodeles waltlii* gastrula. (*a*) During gastrulation, mesodermal (*m*) and endodermal cells (*en*) invaginate through the blastoporal lip. The blastocoelic cavity (*bl*) is progressively displaced towards the vegetal pole, during archenteron (*ar*) formation. Mesodermal cells migrate along the basal surface of the ectoderm (*e*) in a FN meshwork (*dots*) (*yp* = yolk plug). (*b*) Schematic representation of the pioneer mesodermal cells adhering to FN and locomoting under the blastocoelic roof. (*c*) Whole-mount immunofluorescence labeling for FN of part of the blastocoelic roof. (*d*) Sagittal section of a mid-gastrula embryo, see (*a*). (*e*) Same stage embryo that was injected at the blastula stage with Arg-Gly-Asp-Ser-Pro-Ala-Ser-Ser-Lys-Pro. The embryo failed to gastrulate. Only a few cells reach the blastocoel but detach from its surface (*arrow*).

contrast to the surrounding cells of the neural plate, which move without changing their neighbors. These observations point to the existence of at least two domains of adhesion (34).

Recently, the neural cell adhesion molecules (N-CAM) and the liver cell adhesion molecules (L-CAM) were localized in the chick blastoderm (22). During gastrulation, both L-CAM and N-CAM greatly diminish at the surface of ingressing cells. The newly formed neural plate cells also progressively lose L-CAM, while retaining N-CAM at their surface.

Distinct domains of adhesion may be linked to the relative amounts and efficacy of binding mediated by L-CAM and N-CAM, particularly at the level of the notoplate in birds, where N-CAM was found to be very abundant (61). Furthermore, the progressive restriction of the calcium-dependent adhesion molecule, L-CAM, to the epidermis may create a discontinuous pattern of binding strength between cells of both compartments, which in turn can contribute to the shaping of the nervous system (63, 65). Interestingly, cells that occupy the boundary between the presumptive epidermis and the nervous system are the neural crest precursors.

THE NEURAL CREST

Definition

The neural crest is a transient embryonic structure that occupies the dorsal border of the entire neural axis. In the chick embryo, crest cells detach from the neural tube during and after its closure and migrate to various sites, where they undergo differentiation. Crest cells give rise to a large number of diverse cell types, including pigment cells; (most) peripheral neurons and glia; and cells that form muscles, cartilage, skeleton, and connective tissues in the head [for reviews see Le Douarin 1982 (42), and Noden 1984 (53)]. A variety of experiments involving transplantation of neural tubes into host embryos (43, 53, 80) have suggested that the morphology of the embryo plays a major role in the pattern of the neural crest migration and fate. In addition, the behavior of crest cells must depend on the control of adhesion, between the cells and to the ECM. From this point of view, the natural history of crest cells can be described as a series of phases involving different adhesive systems.

Separation from the Neural Tube

Prior to their migration, the presumptive crest cells are integrated in the neuroectodermal epithelium. Their release from the neural epithelium can be compared to gastrulation: It involves local disruptions of the basal lamina, and the disappearance of intercellular junctions. Gap junctions are lost among the premigratory crest cells, as shown by the absence of electrical coupling (56). As viewed by transmission and scanning electron microscopy, presumptive neural crest cells are irregular in shape, do not show tight junctions, and are frequently separated by acellular spaces (50, 66, 67). The basal lamina overlying the neural tube is interrupted and then completely disappears (19, 50, 66, 67). Thereafter, crest cells send projections out of the epithelium and are progressively surrounded by FN (17, 62, 66, 67).

The factors that trigger the disruption of the neural epithelium have not yet been identified. However, it has been shown that the disappearance of the basal lamina induces epithelial cell destabilization (60), whereas direct access to a three-dimensional ECM can promote emigration (30). The numerous morphogenetic movements that occur in the head (36), accompanied by intense cell proliferation, could generate mechanical forces responsible for damaging the basal lamina. Alternatively, plasmin and collagenases produced locally could digest the lamina components. Interestingly, crest cells synthesize large amounts of plasminogen activator (76), and plasminogen stored in the yolk can diffuse in the embryo before blood circulation is established (75). The number of L-CAM, and subsequently, N-CAM adhesive molecules diminishes at the surface of newly formed crest cells (61, 63, 65). In addition, other molecules inserted in the plasma membrane may alter the binding properties of the residual N-CAM (see section on Control of Cell Adhesion).

Patterns of Migration

The migration of the neural crest has been followed by means of various markers for crest cells themselves, such as the quail nucleolar marker (42, 43), acetylcholinesterase (11), and more recently the monoclonal antibody, NC-1 (77, 78); migration pathways have been characterized by electron microscopy and immunohistology for FN and laminin (LN) (17, 19, 62, 66, 67). These studies have shown that morphogenesis of tissues being populated by crest cells is a major element conditioning their pattern of migration. Structures adjacent to the neural tube are metamerized; the somitomeres in the head (2) and the somites in the trunk (62) provide several distinct pathways.

In the trunk, one can distinguish a transient pathway available to the crest cells between two consecutive somites, and a second path between the somite and the neural tube. The first pathway leads the crest cells to the aortic and mesonephric area, where an autonomic differentiation occurs. Cells within the other pathway give rise to the dorsal root ganglia, and to the Schwann cells that line motor and sensory nerves. The metameric pattern in the trunk is rapidly modified, because two adjacent half-somites form a vertebra. This extensive reorganization of the mesoderm surrounding the neural tube contributes to the formation of new pathways, but it primarily creates physical barriers that completely separate different crest cell populations.

In the head, ectodermal thickenings (i.e. placodes) and the neural tube itself put major constraints upon the dispersal of crest cells (17, 52).

The case of the gut is also particularly revealing, since the development of this structure conditions the progression and final distribution of crest cells (Figure 2). The gut is still open wide when the first crest cells reach the endoderm. Closure of the gut to form the umbilicus operates from each end of the embryo in opposite directions. Furthermore, at each axial level, delamination of the mesoderm surrounding the endoderm allows a ventral migration. Crest cells encounter a progressively more differentiated mesoderm; past the umbilicus, crest cells continue their migration caudally on both sides of a newly formed smooth muscle layer. Thus, colonization of the gut, and the development of one or two enteric plexuses result from the combination of active crest cell migration, local and long-range morphogenetic movements of the gut endoderm, and a complex pattern of remodeling and differentiation of the associated mesoderm (1, 72).

The distribution of crest cells can be governed in a few cases by factors inherent to the crest cells themselves. The great heterogeneity in the well-defined pattern of the pigment cells (melanophores, xanthophores, and iridiophores) in the different species and subspecies of the Californian newt, *Taricha*, allows such a hypothesis to be made (Erickson, personnal communication). Likewise, in the axolotl, the banded pattern of the pigments results from the discontinuous distribution of predetermined xanthophores along the neural axis (24). Finally, the pigment pattern produced by quail melanocytes in a chick host is very similar to that of the quail (40).

Substrate of Migration

The pathways contain an ECM limited by one or two laminin-rich basal lamina, which provide defined channels. FN, Type I and III collagens (Figure 3), hyaluronate, and small amounts of chondroitin sulfate are constituents of the three-dimensional network of fibers (16–19, 61). The ECM is deposited prior to crest migration. However, so far there is no direct evidence for its role in triggering neural crest emigration from the neural tube. Furthermore, potential pathways are not always used by crest cells. For example, the presence of high levels of chondroitin sulfate around the notochord (16), and of an undefined factor in the white axolotl skin (59),

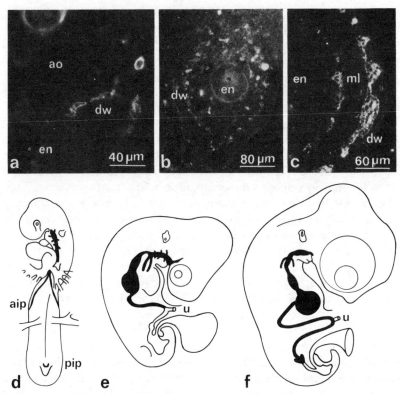

Figure 2 Colonization of the gut by vagal crest cells. (*a–c*) Labeling with FITC-coupled NC-1, a monoclonal antibody recognizing crest cells, on polyethyleneglycol sections of chick embryos. (*a*) Stage 15 of Hamburger & Hamilton (24 somites). Vagal crest cells lining the basement membrane of the digestive wall (*dw*) start a dorsoventral colonization as soon as the gut epithelia (*dw* and *en*) separate. (*b*) Stage 20 (42 somites, 3 days of incubation) at the post-bronchial level. Mesenchymal cells fill the space between the two epithelia, but most crest cells still remain at the periphery. However, some cells are already emigrating from the basement membrane of the epithelium in the dorsal part. Brightly fluorescent cells are autofluorescent erythrocytes. (*c*) Stage 41 (15 days) in the post-umbilical and pre-caecal region. Neural crest cells are aggregated on both sides of the circular muscle layer (*ml*) into two plexuses interconnected with nerve fibers. (*d–f*) Successive stages of enteric colonization. Black areas correspond to levels reached by migrating neural crest cells (pioneer cells are not necessarily distributed all around the mesenchyme). Heavy black lines delimit the area of the open gut. (*d*) Stage 16 (30 somites). Crest cells are located mainly in the dorsal and dorsolateral parts of the gut at the level of the developing branchial pouches. (*e*) Stage 26 (5 days of incubation). Vagal enteric precursors reach the umbilicus. The closure of the post-umbilical part of the gut occurs caudocephalad. Therefore, active migration is needed for crest cells to reach the distal end of the digestive tube. (*f*) Stage 28 (6 days of incubation) vagal crest cells colonize the caecal appendages, whereas the post-caecal area is not yet occupied by the few lumbosacral precursors of the myenteric or submucosal plexuses. (*aip* = anterior intestinal portal; *ao* = aorta; *en* = endoderm; *pip* = posterior intestinal portal; *u* = umbilicus.)

seem to prevent crest cell migration. Therefore, in addition to defined pathways, crest cell migration may be further restricted according to the chemical composition of the matrix.

FN alone or associated with other ECM components greatly promotes the attachment, spreading, and motility of crest cells (Figure 4d–f). In contrast, serum proteins, collagens, hyaluronate, chondroitin sulfate, and LN are very poor substrates for crest cell attachment and movement, and frequently induce their aggregation (26, 51, 58). While collagens may provide a scaffold for the organization of the ECM, hyaluronate is thought to expand the cell-free space and indirectly enhances the speed of locomotion (72).

The essential role of FN in crest migration has been confirmed by in vivo and in vitro perturbation experiments using either monovalent antibodies directed against FN, or a decapeptide that competes for the cell binding sequence of FN. Both agents reversibly block the migration of crest cells (8, 58) (Figure 5).

Behavior of Crest Cells

The dispersal of crest cells is also the result of their specific motile behavior, which differs strikingly from that of other embryonic cells (25, 28). In contrast to somitic and notochordal fibroblasts which are polarized, crest cells on a two-dimensional substrate are stellate with numerous filopodia, and do not exhibit typical contact inhibition of movement (58).

Crest cells contain very little organized cytoskeleton, and exert a weak tractional force on their substratum (73). The cell surface glycoprotein complex that is probably the receptor for FN (10, 54a) is uniformly distributed on the neural crest cell surface, in contrast to other mesenchymal cells where it is concentrated in the cell-to-substratum contact sites (Duband, Rocher, Yamada & Thiery, unpublished). In addition, most crest cells lack the ability to synthesize and deposit FN as a matrix in their immediate environment (49) (Figure 4a–c). In contrast, they synthesize large amounts of hyaluronate (29), a property that may favor their displacement (71). Similar behavior was observed in mesenchymal cells emigrating from embryonic heart explants on exogenous FN (13); these cells become stationary when they begin secreting and assembling FN fibrils at their surface. These observations prompt the hypothesis that the ability to migrate is directly linked to a particular distribution of FN receptors, and to the inability to synthesize and organize FN fibrils at the cell surface.

On a suitable substrate, isolated crest cells move very actively but randomly; their effective displacement is very small (48, 58). However, within a dense cell population, crest cells acquire a persistence in their

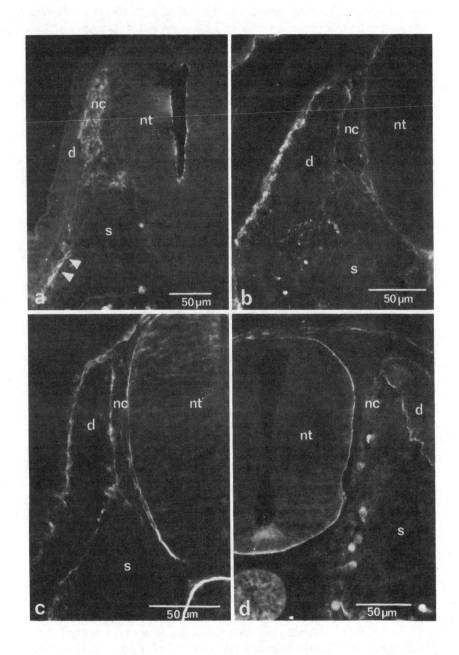

direction of movement. Contact inhibition may be involved in the mechanism controlling directional migration, although such a phenomenon is difficult to observe in vivo (28, 51). Crest cells proliferate actively in the narrow pathways of migration, thus creating a population pressure responsible for unidirectional migration, at least in vitro (58).

Final Localization and Aggregation

The loss of motility in the crest cells may result from a sudden modification of the cytoplasmic machinery, such as the formation of microfilament bundles. Cells can bind more tightly to the ECM, and particularly to FN, if FN receptors are organized in clusters or in strands. Conversely, the loss of FN receptors can induce rounding up of cells, and subsequently enhance cell-cell adhesion. The expression of new adhesive properties involved in the aggregation of crest cells can be triggered by a chemical modification in the ECM. As crest cells migrate, their local environment is progressively transformed: epithelia dissociate into mesenchymes, which expand and consequently obstruct the pathways. In some cases the ECM itself is modified. This is particularly true of the area where the spinal ganglia form: the amount of chondroitin sulfate increases; FN, hyaluronate, and Type I and III collagens disappear; and finally, laminin (LN) appears among crest cells (16, 19, 61) (Figure 6). In vitro, LN can induce crest aggregation, and when crest cells are cultured for a long period they develop a higher binding affinity for LN than for FN (58). In vivo, LN appears transiently within the dorsal root ganglion rudiments.

However, modifications of the environment are not solely responsible for crest cells' arrest; autonomic ganglia form in regions where FN is still very abundant and LN is totally absent (Figure 7a, b). In this case, the arrest of crest cells is correlated with a precocious differentiation of still dividing cells (57). Crest cells form clusters very early with N-CAM on their surface (Figure 7c). These observations suggest that the arrest of crest cells into autonomic ganglia is induced by a change of their cell surface binding properties. In addition, FN receptors acquire a polar distribution (Duband

---

Figure 3 Distribution of some ECM components at the time of thoracic neural crest cell migration. (*a–d*) Immunofluorescent staining for NC-1, FN, Type I collagen, and laminin (LN), respectively. The patterns of distribution of FN and Type I collagen appear to be very similar, both components are found in the basement membranes of epithelia, neural tube (*nt*), and dermomyotome (*d*), and around the mesenchymal cells of the sclerotome (*s*). LN is restricted to the basal surfaces of epithelia. Neural crest cells (*nc*) emerging from the neural tube migrate between the neural tube, the dermomyotome, and the sclerotome, and to a lesser extent between the dermomyotome and the sclerotome (*arrow* in *a*). Neural crest cells are found in the FN and collagen networks but do not contact LN.

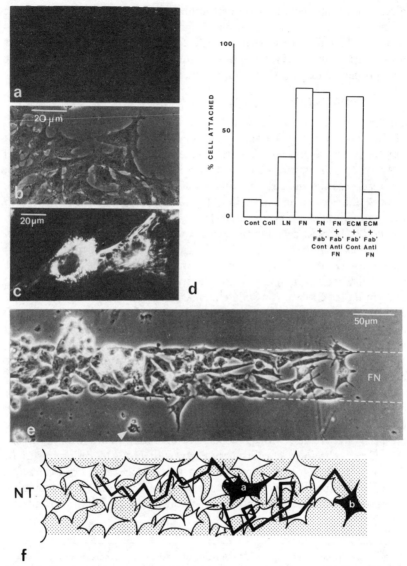

Figure 4 Quail neural crest cell adhesion and migration. (*a–c*) Immunolabeling with anti-fibronectin antibodies. (*a, b*) Trunk neural crest cells do not produce FN. (*c*) A few cranial crest cells synthesize and deposit FN in their environment. (*d*) Trunk neural crest cells adhere preferentially to pure FN or to FN-containing extracellular matrices. The binding is quantitatively inhibited by monovalent antifibronectin antibodies. (*Cont* = control; *Coll* = collagen; *LN* = laminin; *ECM* = extracellular matrix.) (*e*) Crest cells migrate almost exclusively on a FN-coated substrate. A crest cell that left the FN-rich substratum rounded up and became stationary (*arrow*). (*f*) Cells within the population (e.g. *a*) persist in their directionality in contrast to pioneer cells (cell *b*).

et al, unpublished). If they exist, environmental factors inducing this modification remain to be defined. In contrast, the crest cell precursors of spinal ganglia are maintained in a narrow environment at high density before expressing N-CAM (61, 65, 19a).

By varying the substrate of migration of crest cells in vitro, it is possible to obtain two- and three-dimensional clusters of cells that contain N-CAM

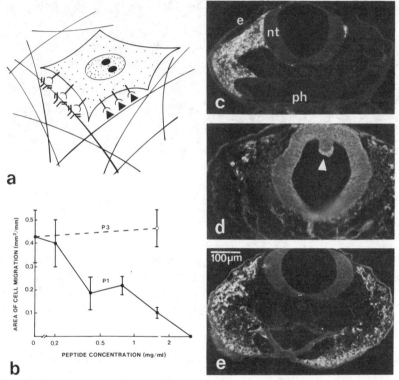

Figure 5 Inhibition of quail neural crest cell migration. (*a*) Perturbation of cell movement can be achieved by preventing FN surface receptors from interacting with the FN binding domain, either by steric hindrance (monovalent anti-fibronectin antibodies), or by competitive inhibition (cell binding site peptides). (*b*) In vitro migration is almost completely abolished in the presence of 2 mg/ml P1 peptide (Arg-Gly-Asp-Ser-Pro-Ala-Ser-Ser-Lys-Lys-Pro), whereas P3 (Cys-Gln-Asp-Ser-Glu-Thr-Arg-Phe-Tyr) taken from the collagen binding domain has no effect. (*c*) In vivo, monovalent antifibronectin antibodies injected in the rhombencephalon (*right side*) prevented crest cells from migrating at this transverse level. A normal migration pattern as revealed by NC-1 immunolabeling is found on the left side, more rostrally and more caudally on both sides. (*ph* = pharynx). (*d*) P1 peptide injected in the mesencephalon prevents most crest cells from migrating on both sides. Many crest cells remain in the dorsal neural tube (*triangle*). (*e*) At the same stage in controls, crest cells occupy the full space laterally (*c–e* from Poole, Yamada & Thiery, unpublished).

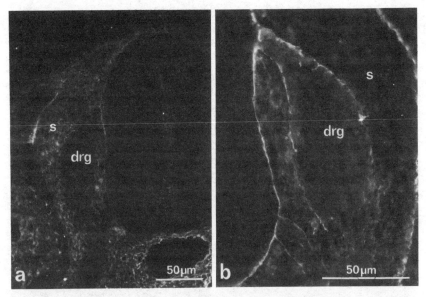

Figure 6 Aggregation of neural crest cells into spinal ganglia. (*a, b*) Immunofluorescent staining for FN and LN. The spinal or dorsal root ganglia (*drg*) forming laterally to the neural tube become surrounded by the sclerotome(s). Aggregation of crest cells is accompanied by a rapid disappearance of FN locally and the formation of a LN-rich basal lamina around the primordium of the ganglion.

(Figure 7*d, e*). An in vitro microaggregation assay revealed that crest cells, depending on their dissociation conditions, express either a calcium-dependent or a calcium-independent mechanism of aggregation. Anti-N-CAM monovalent antibodies strongly inhibit the calcium-independent aggregation of crest cells (3).

MECHANISMS FOR DIRECTIONAL MIGRATION

We have proposed that crest cells, and possibly gastrulating cells, ensure an appropriate directionality by maintaining a high density in transient pathways (58). Others have suggested that contact inhibition remains the best explanation for the dispersal of these cells (28, 51). Galvanotropism has also been invoked as a possible mechanism of directionality, because some cells can orient and migrate when subjected to electrical fields; however, such a mechanism has not been evidenced in vivo (27). The fibrillar network does not seem to be involved in a contact guidance mechanism; nor has a gradient of adhesivity (haptotaxis) been detected in the pathways (28).

Among the many other cell types that migrate at defined periods of

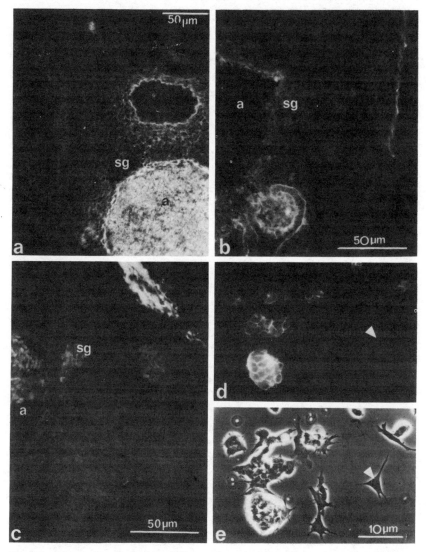

Figure 7   Aggregation of neural crest cells into sympathetic ganglia. (*a–c*) Immunofluorescent
staining for FN, LN, and N-CAM, respectively. Sympathetic ganglia (*sg*) form in a FN-rich
area along the aorta (*a*). Both FN and LN are absent from the primordium of the ganglion
which exhibits an intense staining for N-CAM. (*d, e*) Immunofluorescent staining for N-CAM,
and phase contrast of cultured neural crest cells. The two- and three-dimensional aggregates
are brightly stained for N-CAM while the migrating individual cells remain unstained (*arrow*).

development one can find good evidence for mechanisms such as contact guidance and chemotaxis. In the central nervous system, postmitotic neurons have been shown to migrate along other cells. In the pons, neurons migrate along axons already aligned along the surface of this structure. The radial glial fibers found in many areas of the brain guide the migration of postmitotic neurons. In the cerebellum, both axons and radial glia serve as substrates for the migration of external granular neurons (55). There is ample evidence that these mechanisms require specific adhesion between neurons or between neurons and glia. The two cell adhesion molecules, N-CAM and Ng-CAM, shown to mediate these interactions in several model systems are likely to play a major role in the migration of neurons (21).

Navigation of growth cones in the periphery is influenced by undefined local cues they encounter on their way to the targets (68). They may also respond to chemotactic factor secreted by the target (45).

Chemotaxis was also shown to be responsible for the colonization of the thymus by lymphoid precursors. In the quail, three waves of colonization separated by refractory phases have been described thus far (35). Secretion of a chemotactic factor in the thymus was proposed as the method of recruiting precursors, which once installed in the thymus turn off the synthesis of the factor. Maturation of these precursors into T lymphocytes then permits a new phase of attractivity (41). Using an in vitro model system developed by Zigmond (83), it was possible to demonstrate that precursor cells will respond to a soluble gradient of a low–molecular weight factor secreted only by thymuses taken during an attractive period. A second factor, with a much higher molecular weight, which is responsible for the increase in the speed of locomotion of the precursors, was produced both in attractive and refractory thymuses (5) (Figure 8). Both factors are very probably produced by the thymic epithelium (9).

CONTROL OF CELL ADHESION

Intercellular adhesive forces as well as adhesive interactions with the extracellular matrix are of primordial importance in metazoans.

CAMs, receptors for ECM molecules, and components associated with defined junctions, such as desmosomes, are the best candidates for mediators of these interactions. The elucidation of the structure and binding mechanisms, as well as of the regulatory mechanisms controlling the expression of these molecules, is a prerequisite to the understanding of morphogenesis. Different forms of modulation of adhesion must be involved during epithelium-mesenchyme interconversion, during cell migration, and in the remodeling of tissues.

So far, at least for CAMs, four types of local cell surface modulation, namely prevalence, polar distribution, cis interactions between CAM, and chemical modification of the molecules, have been shown to considerably alter the binding strength between cells (21, 31). However, during early embryogenesis changes in the type of adhesive interactions occur with time constants much smaller than that of the half-life of the CAMs, suggesting additional mechanisms of modulation.

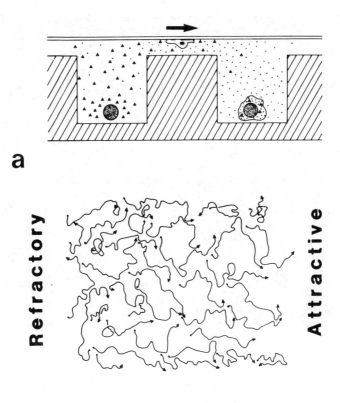

a

b

Figure 8   Chemotaxis by quail T lymphocyte precursors. (*a*) Transverse section of a Zigmond chemotactic chamber. Hemopoietic precursors migrating between the two compartments respond to a soluble gradient of a low–molecular weight component produced by an attractive thymus. In this particular experiment the attractive thymus is sealed in a dialysis tubing retaining the chemokinetic factor (*triangle*), while the low–molecular weight chemotactic factor (*dots*) is released. The chemokinetic factor is provided by a refractory thymus placed in the other compartment. (*b*) Tracks of hemopoietic precursors confronted with the two thymuses as described above. Most cells migrate towards the attractive thymus with a speed of locomotion similar to that measured when the attractive thymus is not sealed in a dialysis tubing.

Recently, the intriguing observation was made that the monoclonal antibody NC-1 recognizes a carbohydrate epitope on the surface of a great variety of cell types involved in tissue remodeling or in migration (69, 70). In the avian embryo, the epitope is present on gastrulating cells (Figure 9a). The epitope is then found on early emigrating neural crest cells (Figure 9b), and is conserved on neurectodermal derivatives. The central nervous

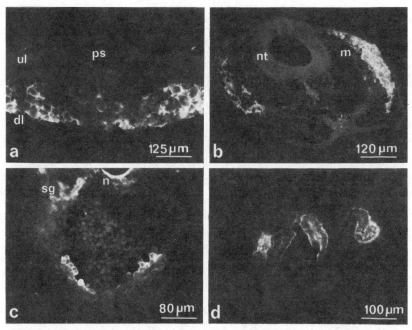

Figure 9 Immunostaining with the NC-1 and HNK-1 antibodies on sections of avian embryos at different developmental stages. The same staining was obtained using either HNK-1 or NC-1 directly coupled to fluorescein. (*a*) Gastrulating chick embryo (Stage 6 of Vakaet) at the level of the primitive streak (*ps*). Positive cells detach from the upper layer (*ul*) and correspond to invaginating cells migrating between the upper layer and the deep layer (*dl*), thus forming the mesodermal cells. Some of these cells also replace the lower layer and generate the definitive endoderm. Note that the deep layer is also stained. (*b*) Neural crest cell emigration from the neural tube (*nt*) of a 12-somite stage chick embryo at the level of the mesencephalon. Crest cells are labeled soon after their individualization, and they follow a subectodermal pathway. They migrate confluently on top of mesenchymal cells (*m*) that do not present the antigenic determinant. (*c*) Transverse section through the hematopoietic aortic foci of a 37-somite stage quail embryo. Hemopoietic precursor cells are stained only when they are detaching from the lateroventral aortic endothelium. Note also the presence of neural crest cells aggregated to form the primordium of the sympathetic ganglion (*sg*), and the diffuse staining around the notochord (*n*). (*d*) Section through the kidney tubules of a 27-somite stage (Hamburger and Hamilton) chick embryo. Cells arranged in an epithelium are not stained, whereas the growing ends react strongly.

system and neurogenic placodes acquire this antigen. Endoderm-derived structures, such as the pancreatic liver primordia or the caecal appendage, are also recognized by NC-1. Some mesodermally derived tissues, including the notochord, the aortic hematopoietic foci (Figure 9c), the pre-cartilage, the heart, the Wolffian duct, and the kidney tubules (Figure 9d), react positively. All these cell types express the antigen when they segregate from their neighbors during a conversion from an epithelium to a mesenchyme, or a similar process, or when they reorganize into a new structure.

Gangliogenesis and stratification in the central nervous system exemplify such reorganization. In each case intense tissue remodeling occurs, and new adhesive properties are involved. Biochemical analyses show that the epitope is shared by several glycoproteins and one glycolipid, at least in the central nervous system. In the adult, the antigenicity is mostly restricted to the myelin-associated glycoprotein (MAG) in the myelinated fibers (45a). A glycoprotein with a molecular weight identical to that of MAG is found in unmyelinated tissues, such as the retina, and possibly in lymphocytes (M. Cooper, personal communication). Interestingly, the same high–molecular weight proteins are detected in different tissues that briefly express the epitope. Those results suggest that the same molecules could play an identical role in these tissues. We hypothesize that these components may modulate intercellular adhesion by preventing close juxtaposition of plasma membranes, and consequently preventing interactions mediated by CAM, thus promoting plasticity and favoring migration or remodeling (70). The carbohydrate moieties could assume such a role through adequate sialylation and fucosylation. Further studies should shed light on the structural and functional relationships between the adhesion molecules N-CAM and the high–molecular weight glycoproteins detected by NC-1.

CONCLUDING REMARKS

Most, if not all, cells of the embryo are involved in active or passive movements during defined periods of development. Cues that provide directionality are diverse: contact guidance, contact inhibition of movement, chemotaxis, population pressure, as well as other mechanisms (64) must operate in the embryo. Some cell types may not utilize the same mechanism at different times in their differentiation program, or in different species. At any given time the surrounding tissues provide limited pathways of migration, whereas ECM contains a permissive substrate. Adhesive interactions are of primordial importance in controlling the assembly of cells originating in different territories into organs that continue to undergo internal reorganization. It is only over the last decade that the identification of cell adhesion molecules (CAM), and ECM adhesion molecules such as

FN, has permitted a molecular approach to one of the most exquisite problems in embryology. Our present knowledge of the structure of these molecules, as well as the results of a series of functional assays and perturbations experiments in vitro and in vivo, already indicates that they must be essential in development. A new avenue of research in molecular embryology is emerging to specifically address the problem of the regulation of these adhesive processes. As proposed recently (23), the regulatory genes involved must operate independently from those dealing with the program of differentiation.

ACKNOWLEDGMENT

Research by the authors is supported by grants from INSERM (CRL 824018), CNRS (ATP 3701), MRT (84C1312), the Ligue Nationale Française contre le Cancer, and the Fondation pour la Recherche Médicale. GCT is a fellow at the Ecole Polytechnique. Excellent technical assistance was provided by Monique Denoyelle and Louis Addade. The authors thank Dr. Julien Smith for his critical reading of the manuscript; Prof. Gerald Edelman, Dr. Kenneth Yamada, and Prof. Jean Claude Boucaut for their enthusiastic collaboration; Lydie Obert for her typing; and Sophie Tissot and Stephane Ozounoff for their help in preparing the illustrations.

Literature Cited

1. Allan, I. J., Newgreen, D. F. 1980. The origin and differentiation of enteric neurons of the intestine of the fowl embryo. *Am. J. Anat.* 157:137–54
2. Anderson, C. B., Meier, S. 1981. The influence of the metameric pattern in the mesoderm on migration of cranial neural crest cells in the chick embryo. *Dev. Biol.* 85:385–402
3. Aoyama, H., Delouvée, A., Thiery, J. P. 1985. Cell adhesion mechanisms in gangliogenesis studied in avian embryo and in a model system. *Cell Diff.* In press
4. Ballard, W. W. 1982. Morphogenetic movements and fate maps of the Cypriniform teleost, *Catostomus commersoni. J. Exp. Zool.* 219:301–21
5. Ben Slimane, S., Houllier, F., Tucker, G., Thiery, J. P. 1983. In vitro migration of avian hemopoietic cells to the thymus: Preliminary characterization of a chemotactic mechanism. *Cell Diff.* 13:1–24
6. Boucaut, J. C., Darribère, T. 1983. Fibronectin in early amphibian embryos. *Cell Tissue Res.* 234:135–45
7. Boucaut, J. C., Darribère, T., Boulekbache, H., Thiery, J. P. 1984a. Antibodies to fibronectin prevent gastrulation but do not perturb neurulation in gastrulated amphibian embryos. *Nature* 307:364–67
8. Boucaut, J. C., Darribère, T., Poole, T. J., Aoyama, H., Yamada, K. M., Thiery, J. P. 1984. Biologically active synthetic peptides as probes of embryonic development: a competitive peptide inhibitor of fibronectin function inhibits gastrulation in amphibian embryos and neural crest cell migration in avian embryo. *J. Cell Biol.* 99:1822–30
9. Champion, S., Savagner, P., Thiery, J. P. 1985. Chemotactic mechanism governing the colonization of the thymus epithelium by lymphoid precursors. Submitted for publication
10. Chen, W. T., Hasegawa, E., Hasegawa, T., Weinstock, C., Yamada, K. M. 1985. Development of membrane-fibronectin linkage complexes in cultured fibroblasts. *J. Cell Biol.* 100:1103–4
11. Cochard, P., Coltey, P. 1983. Cholinergic

traits in the neural crest: acetylcholinesterase in crest cells of the chick embryo. *Dev. Biol.* 98:211–38

12. Cooke, J. 1975. Local autonomy of gastrulation movements after dorsal lip removal in two anuran amphibians. *J. Embryol. Exp. Morphol.* 33:147–57

13. Couchman, J. R., Rees, D. A., Green, M. R., Smith, C. G. 1982. Fibronectin has a dual role in locomotion and anchorage of primary chick fibroblasts and can promote entry into the division cycle. *J. Cell Biol.* 93:402–10

14. Darribère, T., Boucher, D., Lacroix, J. C., Boucaut, J. C. 1984. Fibronectin synthesis during oogenesis and early development of the amphibian *Pleurodeles waltlii*. *Cell Diff.* 14:171–77

15. Darribère, T., Boulekbache, H., De Li, S., Boucaut, J. C. 1985. Immuno-electron-microscopic study of fibronectin in gastrulating amphibian embryos. *Cell Tissue Res.* 239:75–80

16. Derby, M. A. 1978. Analysis of glycosaminoglycans within the extracellular environments encountered by migrating neural crest cells. *Dev. Biol.* 66:321–36

17. Duband, J. L., Thiery, J. P. 1982. Distribution of fibronectin in the early phase of avian cephalic neural crest cell migration. *Dev. Biol.* 93:308–23

18. Duband, J. L., Thiery, J. P. 1985. Distribution of type I and III collagens during chick neural crest cell migration. *Cell Diff.* Submitted for publication

19. Duband, J. L., Thiery, J. P. 1985. Tissue remodeling in the early chick embryo: Distribution and fate of laminin and cell adhesion molecules during epithelium-mesenchyme interconversion. *Dev. Biol.* Submitted for publication

19a. Duband, J. L., Tucker, G. C., Poole, T. J., Vincent, M., Aoyama, H., Thiery, J. P. 1985. How do the migratory and adhesive properties of the neural crest govern ganglia formation in the avian peripheral nervous system? *J. Cell. Biochem.* 27:189–203

20. Duprat, A. M., Gualandris, L. 1984. Extracellular matrix and neural determination during amphibian gastrulation. *Cell Diff.* 14:105–12

21. Edelman, G. M. 1983. Cell adhesion molecules. *Science* 219:450–57

22. Edelman, G. M., Gallin, W. J., Delouvée, A., Cunningham, B. A., Thiery, J. P. 1983. Early epochal maps of two different cell adhesion molecules. *Proc. Natl. Acad. Sci. USA* 80:4334–88

23. Edelman, G. M. 1984. Cell adhesion and morphogenesis: the regulator hypothesis. *Proc. Natl. Acad. Sci.* 81:1460–64

24. Epperlein, H. H., Löfberg, J. 1984. Xanthophores in chromatophore groups of the premigratory neural crest initiate the pigment pattern of the axolotl larva. *Wilhelm Roux' Arch. Entwicklungsmech. Org.* 193:357–69

25. Erickson, C. A., Tosney, K. W., Weston, J. A. 1980. Analysis of migratory behavior of neural crest and fibroblastic cells in embryonic tissues. *Dev. Biol.* 77:142–56

26. Erickson, C. A., Turley, E. A. 1983. Substrata formed by combinations of extracellular matrix components alter neural crest cell motility in vitro. *J. Cell Sci.* 61:299–323

27. Erickson, C. A., Nuccitelli, R. 1984. Embryonic fibroblast motility and orientation can be influenced by physiological electric fields. *J. Cell Biol.* 98:296–307

28. Erickson, C. A. 1985. Control of neural crest cell dispersion in the trunk of the avian embryo. *Dev. Biol.* In press

29. Greenberg, J. H., Pratt, R. M. 1977. Glycosaminoglycan and glycoprotein synthesis by cranial neural crest cells in vitro. *Cell Diff.* 6:119–32

30. Greenberg, G., Hay, E. D. 1982. Epithelia suspended in collagen gels can lose polarity and express characteristics of migrating mesenchymal cells. *J. Cell Biol.* 95:333–39

31. Grumet, M., Hoffman, S., Chuong, C. M., Edelman, G. M. 1984. Polypeptide components and binding functions of neuron-glia cell adhesion molecules. *Proc. Natl. Acad. Sci. USA* 81:7989–93

32. Hay, E. D. 1981. *Cell Biology of the Extracellular Matrix.* New York: Plenum. 417 pp.

33. Jacobson, A. G. 1981. Morphogenesis of the neural plate and tube. In *Morphogenesis and Pattern Formation*, ed. T. G. Connelly, pp. 233–63. New York: Raven. 563 pp.

34. Jacobson, A. G. 1985. Adhesion and movements of cells may be coupled to produce neurulation. In *The Cell in Contact: Adhesions and Junctions as Morphogenetic Determinants*, ed. G. M. Edelman, W. E. Gall, J. P. Thiery. New York: Wiley. In press

35. Jotereau, F. V., Le Douarin, N. M. 1982. Demonstration of cyclic renewal of the lymphocyte precursor cells in the quail thymus during embryonic and perinatal life. *J. Immunol.* 129:1869–77

36. Karfunkel, P. 1974. The mechanism of neural tube formation. *Int. Rev. Cytol.* 38:245–71

37. Keller, R. E. 1975. Vital dye mapping of the gastrula and neurula of *Xenopus*

laevis. I. Prospective areas and morphogenetic movements of the superficial layer. *Dev. Biol.* 42:222–41

38. Keller, R. E. 1978. Time-lapse cinematographic analysis of superficial cell behaviour during and prior to gastrulation in *Xenopus laevis*. *J. Morphol.* 157:223–48

39. Keller, R. E. 1980. The cellular basis of epiboly: an SEM study of deep-cell rearrangement during gastrulation in *Xenopus laevis*. *J. Embryol. Exp. Morph.* 60:201–34

40. Kinutani, M., Le Douarin, N. M. 1985. Avian spinal cord chimaeras. I. Hatching ability and post-hatching survival in homo and heterospecific chimaeras. *Dev. Biol.* In press

41. Le Douarin, N. M. 1978. Ontogeny of hematopoietic organs studied in avian embryo interspecific chimeras. In *Differentiation of Normal and Neoplastic Hematopoietic Cells*, ed. B. Clarkson, P. A. Marks, J. E. Till, pp. 5–31. Cold Spring Harbor, NY: Cold Spring Harbor Conf. Cell Proliferation

42. Le Douarin, N. M. 1982. *The Neural Crest*. Cambridge, Eng.: Cambridge University Press. 259 pp.

43. Le Douarin, N. M., Teillet, M. A., Fontaine-Perus, J. 1984. Chimeras in the study of the peripheral nervous system of birds. In *Chimeras in Developmental Biology*, ed. N. M. Le Douarin, A. McLaren, pp. 313–52. London: Academic. 456 pp.

44. Lee, G., Hynes, R., Kirschner, M. 1984. Temporal and spatial regulation of fibronectin in early Xenopus development. *Cell* 36:729–40

45. Lumsden, A. G. S., Davies, A. M. 1983. Earliest sensory nerve fibers are guided to peripheral targets by attractants other than nerve growth factor. *Nature* 306:786–88

45a. McGarry, R. C., Helfand, S. L., Quarles, R. H., Roder, J. C. 1983. Recognition of myelin-associated glycoprotein by the monoclonal antibody HNK-1. *Nature* 306:376–78

46. Nakatsuji, N. 1975. Studies on the gastrulation of amphibian embryos: light and electron microscopic observation of an urodele *Cynops pyrrhogaster*. *J. Embryol. Exp. Morphol.* 34:669–85

47. Nakatsuji, N., Johnson, K. E. 1983. Comparative study of extracellular fibrils of the ectodermal layer in gastrulae of five amphibian species. *J. Cell Sci.* 59:61–70

48. Newgreen, D. F., Ritterman, M., Peters, E. A. 1979. Morphology and behaviour of neural crest cells of chick embryo *in vitro*. *Cell Tissue Res.* 203:115–40

49. Newgreen, D., Thiery, J. P. 1980. Fibronectin in early avian embryos: Synthesis and distribution along the migration pathways of neural crest cells. *Cell Tissue Res.* 211:269–91

50. Newgreen, D. F., Gibbins, I. L. 1982. Factors controlling the time of onset of the migration of neural crest cells in the fowl embryo. *Cell Tissue Res.* 224:145–60

51. Newgreen, D. F., Gibbins, I. L., Sauter, J., Wallenfels, B., Wütz, R. 1982. Ultrastructural and tissue-culture studies on the role of fibronectin, collagen and glycosaminoglycans in the migration of neural crest cells in the fowl embryo. *Cell Tissue Res.* 221:521–49

52. Noden, D. M. 1978. The control of avian cephalic crest cell cytodifferentiation. *Dev. Biol.* 67:296–329

53. Noden, D. M. 1984. The use of chimeras in analyses of cranio facial development. See Ref. 43, pp. 241–80

54. Pierschbacher, M. D., Ruoslahti, E. 1984. Variants of the cell recognition site of fibronectin that retain attachment-promoting activity. *Proc. Natl. Acad. Sci. USA* 81:5985–88

54a. Pyleta, R., Pierschbacher, M. D., Ruoslahti, E. 1985. Identification and isolation of a 140kd cell surface glycoprotein with properties expected of a fibronectin receptor. *Cell* 40:191–98

55. Rakic, P. 1985. Mechanisms of neuronal migration. In *Handbook of Physiology and Developmental Neurobiology*, ed. M. W. Cowan. In press

56. Revel, J. P., Brown, S. S. 1975. Cell junctions in development with particular reference to the neural tube. *Cold Spring Harbor Symp. Quant. Biol.* 40:433–55

57. Rothman, T., Gershon, M. D., Holtzer, M. 1978. The relationship of cell division to the acquisition of adrenergic characteristics by developing sympathetic ganglion cell precursors. *Dev. Biol.* 65:322–41

58. Rovasio, R. A., Delouvée, A., Yamada, K. M., Timpl, R., Thiery, J. P. 1983. Neural crest cell migration: Requirement for exogenous fibronectin and high cell density. *J. Cell Biol.* 96:462–73

59. Spieth, J., Keller, R. E. 1984. Neural crest cell behavior in white and dark larvae of *Ambystoma mexicanum*: Differences in cell morphology, arrangement and extracellular matrix as related to migration. *J. Exp. Zool.* 229:91–107

60. Sugrue, S. P., Hay, E. D. 1981. Response of basal epithelial cell surface and cytoskeleton to solubilized extracellular matrix molecules. *J. Cell Biol.* 91:45–54

61. Thiery, J. P., Duband, J. L., Rutishauser, U., Edelman, G. M. 1982. Cell adhesion molecules in early chicken embryogenesis. *Proc. Natl. Acad. Sci. USA* 79: 6737–41

62. Thiery, J. P., Duband, J. L., Delouvée, A. 1982. Pathways and mechanism of avian trunk neural crest cell migration and localization. *Dev. Biol.* 93: 324–43

63. Thiery, J. P., Delouvée, A., Gallin, W. J., Cunningham, B. A., Edelman, G. M. 1984. Ontogenetic expression of cell adhesion molecules: L-CAM is found in epithelia derived from the three primary germ layers. *Dev. Biol.* 102: 61–78

64. Thiery, J. P. 1984. Mechanisms of cell migration in the vertebrate embryo. *Cell Diff.* 15: 1–15

65. Thiery, J. P., Duband, J. L., Delouvée, A. 1985. Role of cell adhesion in morphogenetic movements during early embryogenesis. See Ref. 34

66. Tosney, K. W. 1978. The early migration of neural crest cells in the trunk region of the avian embryo. An electron microscopic study. *Dev. Biol.* 62: 317–33

67. Tosney, K. W. 1982. The segregation and early migration of cranial neural crest cells in the avian embryo. *Dev. Biol.* 89: 13–24

68. Tosney, K. W., Landmesser, L. T. 1984. Pattern and specificity of axonal outgrowth following varying degrees of chick limb bud ablation. *J. Neurosci.* 4: 2518–27

69. Tucker, G. C., Aoyama, H., Lipinski, M., Tursz, T., Thiery, J. P. 1984. Identical reactivity of monoclonal antibodies HNK-1 and NC-1: Conservation in vertebrates on cells derived from the neural primordium and on some leukocytes. *Cell Diff.* 14: 223–30

70. Tucker, G. C., Thiery, J. P. 1984. Surface molecules associated with cell migration and tissue remodelling during embryogenesis. In *Cellular and Pathological Aspects of Glycoconjugate Metabolism*, ed. H. Dreyfus. Paris: Inst. Natl. Santé Rech. Med. In press

71. Tucker, R. P., Erickson, C. A. 1984. Morphology and behavior of quail neural crest cells in artificial three-dimensional extracellular matrices. *Dev. Biol.* 104: 390–405

72. Tucker, G. C., Thiery, J. P. 1985. Mechanisms of neural crest cell migration in the developing gut. *Dev. Biol.* Submitted for publication

73. Tucker, R. P., Edwards, B. F., Erickson, C. A. 1985. Tension in the culture dish: microfilament organization and migratory behavior of quail neural crest cells. Submitted for publication

74. Vakaet, L. 1984. Early development of birds. See Ref. 43, pp. 71–88

75. Valinsky, J. E., Reich, E. 1981. Plasminogen in the chick embryo. Transport and biosynthesis. *J. Biol. Chem.* 256: 12470–75

76. Valinsky, J. E., Le Douarin, N. M. 1985. Production of plasminogen activator by migrating neural crest cells. *EMBO J.* In press

77. Vincent, M., Duband, J. L., Thiery, J. P. 1983. A cell determinant expressed early on migrating avian neural crest cells. *Dev. Brain Res.* 9: 235–38

78. Vincent, M., Thiery, J. P. 1984. A cell surface marker for neural crest and placodal cells: further evolution in peripheral and central nervous system. *Dev. Biol.* 103: 468–81

79. Vogt, W. 1929. Gestaltunganalyse am Amphibienkeim mit örtlicher Vitalförbung. II. Teil Gastrulation und Mesodermbildung bei Urodelen und Anuren. *Wilhelm Roux' Arch. Entwicklungsmech. Org.* 120: 384–706

80. Weston, J. A. 1963. A radioautographic analysis of the migration and localization of trunk crest cells in the chick. *Dev. Biol.* 6: 279–310

81. Yamada, K. M. 1983. Cell surface interactions with extracellular materials. *Ann. Rev. Biochem.* 52: 761–99

82. Yamada, K. M., Kennedy, D. W. 1984. Dualistic nature of adhesive protein function: fibronectin and its biologically active peptide fragments can autoinhibit fibronectin function. *J. Cell Biol.* 99: 29–36

83. Zigmond, S. H. 1978. Chemotaxis by polymorphonuclear leukocytes. *J. Cell Biol.* 77: 269–87

Ann. Rev. Cell Biol. 1985. 1 : 115–43

PROTEIN LOCALIZATION AND MEMBRANE TRAFFIC IN YEAST

Randy Schekman

Department of Biochemistry, University of California, Berkeley, California 94720

CONTENTS

INTRODUCTION

The study of intracellular protein transport has developed into a major discipline in cell biology. Progress in this area has, to a large extent, been facilitated by the development of model systems that allow greater ease of manipulation and a wider range of techniques than are possible with animal tissues. Because of the facile application of genetics, molecular biology, and biochemistry, the yeast *Saccharomyces cerevisiae* has attracted the attention of many investigators as an organism for the study of basic issues in cell biology. This chapter highlights those aspects of membrane protein assembly in yeast that are common to all eukaryotic cells. Some of this material was reviewed recently (Schekman & Novick 1982, Hay et al 1984, Cabib et al 1982, Jones 1984). I have attempted to relate subjects that are usually dealt with separately in the yeast literature.

115

0743–4634/85/1115–0115$02.00

I.   SYNTHESIS OF NONCYTOPLASMIC PROTEINS

Three major targets of protein transport in yeast are the secretory system (endoplasmic reticulum, Golgi body, vacuole, plasma membrane, and periplasm), the mitochondrion, and the nucleus. In addition, peroxisomes, which in animal cells appear to derive internal proteins from the cytoplasm rather than via a secretory organelle, have been well documented in certain yeasts, but not yet unequivocally in *S. cerevisiae*.

A principal distinction between the synthesis of secretory proteins and other localized proteins is the preferential site of synthesis on ribosomes bound to the endoplasmic reticulum (ER). Evidence for this comes from studies with reticulocyte protein synthesis reactions in which mRNAs for proteins that are localized by the secretory process (invertase, carboxypeptidase Y, killer toxin precursor, α-factor precursor) are transported during synthesis into dog pancreas microsomal vesicles. In contrast, the same reticulocyte reactions programmed with mRNA for mitochondrial proteins will produce precursors that can be transported posttranslationally into mitochondria. Secretory, mitochondrial, and perhaps even nuclear proteins have special signals, usually near the N-terminus, that contain information for initiation of transport into a specified organelle. Aside from this superficial similarity, the mechanisms used for membrane penetration are likely to be different in each case.

A.   Transport into the Endoplasmic Reticulum

TRANSLOCATION OF PROTEINS   In addition to proteins designed to remain in the ER, proteins that are secreted into the medium and periplasm (space between plasma membrane and cell wall), assembled into the plasma membrane, transported to the vacuole, and probably those that make up the Golgi membrane, all originate in the ER. The biogenesis of a number of proteins with different destinations has been investigated (secreted into medium, α-factor and killer toxin; secreted into periplasm, invertase and acid phosphatase; plasma membrane, permeases and surface labeled proteins; vacuole, carboxypeptidase Y). Two methods have been used to demonstrate initial assembly in the ER: translation of mRNA in the presence of dog pancreas microsomal vesicles, and the influence of pleiotropic secretion mutants that block translocation into or transport from the ER.

The criteria for translocation into microsomal vesicles include: signal peptide cleavage, core glycosylation, and protection against protease digestion. One or more of these conditions has been documented for invertase (Perlman & Halvorson 1981), carboxypeptidase Y (Müller & Müller 1981), α-factor precursor (Julius et al 1984a), and killer toxin precursor (Bostian et al 1983). In the case of carboxypeptidase Y and α-factor precursor, signal peptide cleavage does not accompany translocation. Unfortunately, these analyses have relied entirely on heterologous synthesis and translocation reactions programmed with yeast mRNAs. It has not yet been possible to couple the yeast in vitro protein synthesis system (Tuite et al 1980) with yeast microsomal vesicles, although there is one report of a polypeptide chain completion and ER translocation reaction in an extract from yeast (Chu & Maley 1980). In the absence of a

Table 1 Covalent modification associated with protein localization in yeast

Location and Modification	Compartment	Reference
Mitochondria		
Leader peptide cleavage	Matrix	Böhni et al 1983
		McAda & Douglas 1982
Heme attachment	Intermembrane space	Korb & Neupert 1978
		Ohashi et al 1982
Membrane anchor cleavage	Intermembrane space	Gasser et al 1982b
		Kaput et al 1982
Endoplasmic reticulum		
Signal peptide cleavage	Lumen	Perlman & Halvorson 1981
N-linked oligosaccharide attachment	Lumen	Byrd et al 1982
		Esmon et al 1984
O-linked mannose attachment	Lumen	Haselbeck & Tanner 1983
Cleavage of glucoses from N-glycosylated proteins	Lumen	Byrd et al 1982
		Esmon et al 1984
Cleavage of an α-1,2-mannose	Lumen	Byrd et al 1982
Fatty acid acylation	?	Wen & Schlesinger 1984
Golgi body		
Outer chain carbohydrate addition	Early cisternae?	Esmon et al 1984
		Julius et al 1984a
Elongation of O-linked oligosaccharides	Early cisternae?	Haselbeck & Tanner 1983
Lys Arg dibasic endoproteolysis	Late cisternae?	Julius et al 1984a,b
Dipeptidyl aminopeptidase cleavage	Late cisternae?	Julius et al 1983, 1984a
Vacuole		
Propeptide cleavage	Lumen	Hasilik & Tanner 1978b
		Hemmings et al 1981
		Stevens et al 1982
Plasma membrane		
Activation of chitin synthetase	Cytoplasmic surface	Cabib et al 1983
Activation of glucan synthetase	Cytoplasmic surface	Shematek & Cabib 1980

completely reconstituted reaction, functional equivalents of the signal recognition particle (SRP) and docking protein have not yet been detected in yeast. Such a reaction will also be useful in identifying the biochemical lesions in mutants defective in the translocation event.

Biogenesis associated with the ER has also been identified with the aid of pleiotropic secretory mutants. Class A *sec* mutants are temperature sensitive for growth, secretion, and plasma membrane assembly (Novick & Schekman 1979, Novick et al 1980). At a restrictive temperature (37°C), mutant cells accumulate secretory organelles and molecular intermediates of secretion, which in many of the mutants are discharged when cells are returned to a permissive temperature (25°C).

Secreted enzymes, such as invertase, acid phosphatase, and α-galactosidase, accumulate in their fully active, though in some cases incompletely processed, forms. Members of the class A group block secretion at one of three stages: transport from the ER (*sec12, sec13, sec16, sec17, sec18, sec19, sec20, sec21, sec22, sec23*), assembly of secretory vesicles at the Golgi body (*sec7, sec14*), and discharge of secretory vesicles (*sec1, sec2, sec3, sec4, sec5, sec6, sec8, sec9, sec10, sec15*). Analysis of haploid double *sec* mutants, in which each member of the pair represents a distinct morphological block, has shown that the typical secretory pathway of ER → Golgi body → vesicles → cell surface exists in yeast (Novick et al 1981). Thus, genesis in the ER is established for all the proteins that accumulate when transport from the ER is blocked in an appropriate *sec* mutant.

An early step in the assembly of proteins into the ER is interrupted in another group of *sec* mutants (Class B; *sec53, sec59*) (Ferro-Novick et al 1984a,b). Unlike the Class A mutants, *sec53* and *sec59* accumulate enzymatically inactive forms of invertase. Molecular precursors of invertase, carboxypeptidase Y (CPY), and α-factor that appear to have initiated but not completed penetration into the ER lumen are produced at 37°C. Three lines of evidence support this contention. First, although invertase and CPY both are soluble, the precursors that accumulate are associated with the ER membrane and resist extraction by Triton X-100. Second, invertase and the α-factor precursor both have N-glycosylation sites near the N-terminus that acquire core oligosaccharides by transfer from a dolichol-oligosaccharide on the lumenal side of the ER membrane. In *sec53*, invertase and α-factor precursors accumulate that have no N-linked core oligosaccharides. In *sec59*, forms of invertase are seen that contain 0–3 of 12 possible oligosaccharides, and forms of the α-factor precursor are found that contain 0–3 of 3 possible oligosaccharides (Julius et al 1984a). Finally, ER membranes from the mutants can be treated with trypsin under conditions that generate a fragment derived from the N-terminus of invertase. The exact orientation of the membrane associated precursors is not clear. The trypsin-sensitive portion of the invertase may protrude from the cytoplasmic surface of the ER membrane, or the membranes may be labile and lumenally oriented invertase may be degraded with a remnant left attached on the inner surface of the membrane. In either case, the translocation reaction proceeds in *sec53* to a stage beyond cleavage of the signal peptide, but not to the point of N-glycosylation. In *sec59*, translocation proceeds somewhat farther, to the point of addition of the first few oligosaccharides.

The membrane-bound forms of invertase appear to be authentic intermediates in translocation. Upon return to the permissive temperature (24°C), the invertase precursors are glycosylated, become partially active, and are secreted. Several other examples illustrate the possibility that membrane penetration and secretory polypeptide synthesis can be uncoupled. When the inducible acid phosphatase gene (*PHO5*) is introduced on a high copy number plasmid into yeast, pulse-labeling of induced cells shows a full-length unglycosylated polypeptide precursor that appears to be converted into the processed and glycosylated form during a subsequent chase period (Haguenauer-Tsapis & Hinnen 1984). This species is much less apparent in cells carrying a single copy of the *PHO5* gene, which suggests that some element of the translocation machinery may become limiting when the rate of phosphatase synthesis is increased. In bacteria, posttranslational translocation of secretory proteins has been demonstrated in two ways. Dissipation of membrane potential leads to the reversible accumulation of unprocessed cytoplasmic membrane, periplasmic, and outer membrane proteins (Date et al 1980, Enequist et al 1981, Daniels et al 1981). In unperturbed cells it has been possible to observe synthesis of a substantial length of a secretory protein prior to signal peptide cleavage and penetration to the external surface of the cytoplasmic membrane (Randall 1983, Koshland & Botstein 1982).

It seems from these examples that the requirements for initiation and completion of translocation are distinct. Assuming the SRP and docking protein mechanism operates in yeast, initiation of protein translocation will be coupled to the synthesis of a signal peptide. This is best illustrated with invertase, which is made from two alternative transcripts of the same gene: one that encodes an N-terminal signal peptide, and a shorter message whose first AUG initiates synthesis of mature length invertase (Carlson et al 1983). Transcription of the longer mRNA is regulated so that production of the secreted form of invertase is inducible, while the shorter mRNA is made constitutively, which gives rise to a constant level of cytoplasmic invertase activity. Hence, the distinction of two essentially identical forms of invertase is governed by the presence of a signal peptide on one of them.

It now appears that a signal peptide may even suffice to direct cytoplasmic proteins into the ER. Fusion of the *E. coli* β-lactamase signal peptide-coding region to the α-globin gene encodes a protein that is sequestered and processed by dog pancreas microsomal vesicles in a cell-free translation/translocation reaction (Lingappa et al 1984). Similarly, hybrids that include the N-terminal region of invertase, and are linked to *E. coli* β-galactosidase are targeted to the ER in vivo in yeast (Emr et al 1984).

The ability of a yeast signal peptide to direct mammalian secreted proteins into the ER, and subsequently through the secretory pathway, has also been shown by construction of gene fusions that contain a large N-terminal portion of the gene for the mating pheromone, α-factor (Brake et al 1984, Bitter et al 1984). Hybrid protein products resulting from these gene fusions are guided into the lumen of the ER where core glycosylation of the α-factor propeptide portion occurs. By itself, the α-factor precursor is endoproteolytically cleaved to mature pheromone at some step in transport through the Golgi body, and in favorable circumstances the heterologous protein also is clipped from the precursor.

Secretion of heterologous signal peptide–containing eukaryotic and prokaryotic proteins has met with mixed success in yeast. Leukocyte interferon is secreted with low efficiency, and is processed at two alternate sites in addition to the usual signal cleavage position (Hitzeman et al 1983). Among membrane proteins, only the hepatitis surface antigen has been produced in a membrane-assembled form in yeast (Valenzuela et al 1982). However, the expressed form accumulates in virus-like particles (Alexander particles) in the cytosol. While the surface antigen normally is glycosylated, the yeast form is not. Somehow the surface antigen recruits lipid from intracellular membranes to form a small vesicle, entirely independent of the secretory process. A final example is bacterial β-lactamase, which exhibits distinct behavior in vivo and in vitro. When expressed in yeast, the protein is not secreted and remains in the cytoplasm where the signal peptide is removed by a vacuolar protease (Hollenberg et al 1983). In a coupled translation-translocation reaction, however, β-lactamase shows SRP-dependent penetration into dog pancreas vesicles (Müller et al 1982). The apparent discrepancy could be explained if SRP is limiting in yeast cells.

In summary, a homologous signal peptide is necessary, and possibly sufficient, to allow translocation of homologous, heterologous, or even cytoplasmic proteins across the ER membrane. Heterologous signal peptides are less reliable in yeast, which suggests that the signal may be somewhat species specific.

GLYCOSYLATION AND PROTEIN ACYLATION IN THE ER   The pathway of core oligosaccharide synthesis and transfer to N-glycosidic linkage at asparaginyl residues appears to be identical in yeast and mammalian ER.

Transfer of GlcNAc from UDP-GlcNAc to dolichol-P to form dol-P-P-GlcNAc initiates the synthesis of a core. This step occurs on the cytoplasmic surface of the ER, and is inhibited by tunicamycin (Lehle & Tanner 1976, Kuo & Lampen 1974). After the addition of another GlcNAc residue in β,1-4-linkage to the first such residue, five mannose residues, all contributed by GDP-mannose, are added on the cytoplasmic surface to the growing oligosaccharide. The $Man_5GlcNAc_2$ unit then is translocated to the lumenal surface of the ER where four additional mannose units are added from dol-P-man, and three glucose units are added from dol-P-glc (Lehle 1980, Trimble et al 1980).

Several of the enzymes of core oligosaccharide synthesis have been partially purified. The role of one enzyme in translocation of mannose has been studied in a reconstituted vesicle consisting of phospholipid, dolichol-P, and GDP-Man : dol-P mannosyltransferase (Haselbeck & Tanner 1982). When such vesicles are produced so as to include GDP in the lumen, external mannose contributed from GDP-Man can be converted to intravesicular GDP-Man with no exchange of external and internal nucleotide. Hence, the enzyme serves as both a transferase and translocater. It would be interesting to examine this reaction coupled to the translocation of a $Man_5GlcNAc_2$ substrate in place of internal GDP.

A completed core oligosaccharide is transferred to an asparaginyl residue at the sequence Asn · X · Thr/Ser. In vitro the rate of this glycosylation reaction is stimulated to increase twentyfold by the presence of glucoses on the $Man_9GlcNAc_2$ core (Lehle 1980). Nevertheless, cores with as little as the $GlcNAc_2$ portion can be transferred to an acceptor peptide. At some point after polypeptide penetration into the ER lumen is complete, protein-bound core oligosaccharides are trimmed, first by the removal of the three glucoses, then by the removal of one α-1,2-linked mannose unit (Byrd et al 1982, Esmon et al 1984). This leaves core oligosaccharides of a composition $Man_8GlcNAc_2$ on glycoproteins prior to transport from the ER. Two specific oligosaccharide glucosidases have been partially purified, and the action of a mannosidase is inferred from carbohydrate structural studies (Byrd et al 1982, Kilker et al 1981, Saunier et al 1982).

Certain as yet unidentified yeast proteins undergo O-linked glycosylation, with the final product consisting of a spectrum of unbranched oligomers from Man to Man_4 linked to Ser or Thr residues. The initial mannose is added in the ER by transfer from dol-P-Man, and the subsequent mannoses are added in the Golgi body, probably by transfer from GDP-Man in the Golgi lumen (Haselbeck & Tanner 1983). O-glycosylation of a heterologous protein, *Aspergillus awamori* glucoamylase, has been demonstrated by expression of the gene in yeast (Innis et al 1985). The product is secreted into the medium, and has characteristics comparable to the form secreted by the natural host. Although O-glycosylation with artificial peptide acceptors in isolated membranes is dependent only on the presence of a Ser or Thr residue (Lehle & Bause 1984), the in vivo specificity is quite striking. Invertase, for example, has exclusively N-linked carbohydrate despite the presence of numerous Ser and Thr residues (Lehle et al 1979). It may be that N- and O-glycosylation have distinctive regulatory signals. In this regard, arrest in the G 1 cell cycle position depresses N-glycosylation specifically (Orlean et al 1984).

The physiologic significance of core glycosylation has been addressed through the use of tunicamycin. N-glycosylation is essential for progression beyond the G 1 cell cycle stage, and for production of enzymatically active forms of certain secreted enzymes, such as invertase and acid phosphatase (Arnold & Tanner 1982). Nevertheless, unglycosylated invertase made in the presence of tunicamycin is secreted, though at a reduced rate, at 25°C (Ferro-Novick et al 1984b). In this case glycosylation may facilitate, but is not absolutely required for intercompartmental transport. At an elevated temperature (37°C) unglycosylated secretory polypeptides remain irreversibly associated with the ER membrane. Invertase accumulated in this condition resembles the form retained in *sec53*; both are degraded to an N-terminal, protected fragment when ER membranes are exposed to trypsin.

A novel approach was devised recently to clone the gene encoding the tunicamycin-sensitive enzyme, UDP-GlcNAc:dol-P GlcNAc transferase (Barnes et al 1984). Yeast cells transformed with yeast genomic inserts on a high copy-number plasmid are selected on a medium supplemented with a level of tunicamycin that inhibits growth of normal cells. Transformants that survive do so by overproducing the GlcNAc transferase. The sequence of this gene reveals an extremely hybrophobic protein with two regions that may contain membrane-spanning domains.

Additional insight into the role of later steps in core oligosaccharide synthesis and processing has come from the isolation and characterization of mutants defective in these processes. The major approach here has been to screen among mutants selected by resistance to ^{3}H-mannose exposure. Mutants that fail in early steps of core oligosaccharide synthesis (*alg-asparagine-linked glycosylation*) were obtained that resisted ^{3}H-mannose suicide (Huffaker & Robbins 1982, 1983). Mutants of the *alg1* type assemble the dol-P-P-chitobiose unit, and transfer this disaccharide to protein; *alg2* mutants add only one or two mannoses to the core, and this truncated unit is transferred to protein. Both mutants are temperature sensitive lethals, although active invertase is secreted at the restrictive temperature. Mutants of the *alg4* type show no discrete block in oligosaccharide synthesis, yet invertase secretion is defective at the restrictive temperature. This complementation group is the same as *sec53*. The other *alg* mutants are not conditionally lethal. Growth of the core beyond $Man_5GlcNAc_2$ is blocked in *alg3*, and *alg5* and *alg6* are defective in glucosylation of the $Man_9GlcNAc_2$ core. Formation of dol-P-Glc is defective in *alg5*, and transfer from dol-P-Glc to the core oligosaccharide is defective in *alg6* (Runge et al 1984). One indication that glucosylation influences the rate of oligosaccharide transfer to protein is that fewer core units are found on invertase in *alg5* and *alg6* than in wild-type cells.

Glucose processing from the core oligosaccharides appears to play no role in secretion. A mutant, *gls1*-1, defective in oligosaccharide glucosidase I, grows normally and secretes invertase with normal kinetics, even though the glucoses persist on glycoproteins transported to the cell wall or vacuole (Esmon et al 1984).

A subset of the membrane-bound glycoproteins undergoes fatty acid acylation in the ER. Four minor glycoproteins are labeled with $[^3H]$-palmitate when transport from the ER is blocked in *sec18* (Wen & Schlesinger 1984). These proteins are further glycosylated in the Golgi body, but then lose the palmitate label at some stage during or after transport to the cell surface. For this reason no palmitate-labeled proteins are seen in wild-type cells. Complete assembly into the ER membrane is required for acylation because palmitate label is not incorporated when translocation is blocked in *sec53*. Unfortunately, no clear function has been associated with acylation. The possibility that it has a role in anchoring proteins to the membrane seems unlikely because most acylated proteins already contain hydrophobic membrane-spanning polypeptide domains. However, acylation could be important in association of *src* and *ras* gene products with the plasma membrane in animal cells. Clearly, a mutant defective in the enzyme that performs the acylation reaction would allow a better definition of the function of this modification.

B.   Assembly of the Mitochondrion

TARGETING TO THE MITOCHONDRIAL SURFACE   Assembly of the mitochondrion appears to be entirely independent of the secretory process. In some ways the import of proteins into the mitochondrion resembles the process observed for assembly of certain membrane proteins in *E. coli*. A large body of evidence suggests that mitochondrial precursors are made on cytoplasmic ribosomes, and the completed proteins are then inserted at the surface of the outer membrane (Neupert & Schatz 1981, Schatz & Butow 1983). By analogy, the M13 procoat and β-lactamase pass into or across the *E. coli* inner membrane posttranslationally (Koshland & Botstein 1982, Ito et al 1979, Wickner 1979).

Many different nuclear-coded mitochondrial proteins have been made in reticulocyte lysates, and without exception these precursors could be taken up by mitochondria after polypeptide synthesis was complete. While these results argue that translation and import can occur independently, some other evidence suggests that the two processes may be coupled in vivo. Thin sections of yeast cells show some enrichment of ribosomes that are apparently attached to the mitochondrial surface (Butow et al 1975). Furthermore, cytoplasmic (cycloheximide-sensitive) ribosomes found in an isolated mitochondrial fraction appear to be engaged in the synthesis of a distinct set of proteins that differ from the products made by soluble cytoplasmic ribosomes (Ades & Butow 1980). A detailed study of the enrichment of specific mRNAs has shown that some, such as the outer membrane porin protein, are nearly exclusively synthesized on soluble cytoplasmic ribosomes, while others, such as the β subunit of F_1 ATPase, show as much as a 280-fold enrichment in the mitochondrial fraction (Suissa & Schatz 1982). In spite of the impressive enrichment of certain messages, no more than 60% of a particular species was recovered in the mitochondrial fraction. Taken together the data are consistent with nonobligatory coupling of synthesis and import but do not exclude tight coupling for some proteins. Association of nascent chains with mitochondria may occur for those proteins that have a localization signal close to the N-terminus, while enrichment in the cytoplasm may be accounted for by a targeting signal closer to the C-terminus.

For many precursors a transient N-terminal "transit" peptide has been detected (Maccecchini et al 1979). In a few cases the peptide sequence has been obtained from the DNA sequence of the cloned gene (Kaput et al 1982). No rules have yet emerged that would diagnose the nature of a mitochondrial leader, although several leader sequences have a basic charge. Unlike cleavable signal peptides on secretory proteins, mitochondrial transit sequences do not have a hydrophobic core.

As for secretory signal peptides, transit peptides are necessary, and in some cases sufficient, for localization to the mitochondrion, and possibly even for transport to the correct subcompartment. Gene fusions that join as few as 140 amino acids from the N-terminus of the ATPase β-subunit onto E. coli β-galactosidase produce correctly targeted hybrid proteins (Douglas et al 1984, Geller et al 1983). In this case the minimum length necessary for transport to the mitochondrion is much longer than the 16–amino acid cleavable peptide of the β-subunit. In another case, however, a hybrid containing just the first 22 of 25 amino acids from the transient N-terminus of the cytochrome oxidase subunit IV, joined to mouse dihydrofolate reductase (a cytoplasmic enzyme), is sufficient for import of the latter into the matrix of mitochondria in vitro (Hurt et al 1984).

Conversely, deletion of an N-terminal portion of the 70 kDa outer membrane protein destroys proper assembly (Hase et al 1984). This protein contains an uncleaved N-terminus that serves for assembly and anchoring to the outer membrane. The gene has been cloned and sequenced, and has revealed an N-terminal stretch of 28 uncharged amino acids bracketed by basic amino acids (Hase et al 1983).

Identification of specific amino acids or secondary structures associated with targeting to different subdivisions within the mitochondrion will likely come from studies with gene fusions in which transit signals are exchanged, and by localized mutagenesis of regions within transit peptide sequences.

A complementary issue raised by the existence of transit sequences is the nature of receptor(s) that make contact with precursor proteins. Initially, the precursors encounter protein(s) in the cytoplasm, one of which has recently been identified as a partner in the import of the ATPase β subunit (Ohta & Schatz 1984). When pure β-subunit precursor is incubated with isolated mitochondria, import requires addition of a soluble component found either in reticulocyte or yeast lysates. This protein is small and appears not to contain an essential RNA subunit, which distinguishes it from SRP. In contrast, import of pre-ornithine transcarbamylase into rat liver mitochondria has been reported to require an RNase-sensitive component and a low molecular weight protein factor (Firgaira et al 1984). It is not known if these factors are protein-specific or general, or at what stage in the association of precursors with the mitochondrion they exert their effect. Conceivably, these factors act to modify precursor proteins to a form recognized by receptors on the outer membrane.

The existence of protein receptors on the cytoplasmic surface of the outer membrane was first inferred by destruction of import activity in intact mitochondria treated with proteases. Since then several binding sites have been examined by use of purified or specifically radiolabeled precursors. An unique receptor for apocytochrome c has been detected and partially purified from *Neurospora* mitochondria (Neupert 1983). The binding site was identified by inclusion of deuterohemin along with apocytochrome c, rendering the latter incapable of conversion to the holoenzyme (Henning & Neupert 1981). Deuterohemin-substituted protein associated with mitochondria in a saturable process that was competed by the apoprotein. Although the holoenzyme is associated with the external surface of the inner membrane, reversibly bound apoprotein is accessible to exogenous protease. When deuterohemin is displaced by protohemin the bound apoprotein can then be properly matured and localized. The receptor can be solubilized and reconstituted into liposomes, which should allow its purification.

Excess apocytochrome c does not inhibit the uptake of several other mitochondrial precursor proteins (Teintze et al 1982). Hence additional receptors must be required, though it is likely that many precursors share an import receptor. Perhaps each compartment is defined by a separate receptor. One other receptor has been detected in purified outer membrane fractions using binding of in vitro synthesized cytochrome b_2 precursor (Riezman et al 1983b). Association of this precursor with outer membranes is rapid, selective, and protein-mediated. Unfortunately, it has not yet been possible to reconstitute this receptor in liposomes.

TRANSLOCATION, PROCESSING, AND ANCHORING   Although transport into mitochondria and across the cytoplasmic membrane in bacteria share a requirement for membrane potential, proteins cross the two membranes in opposite directions with respect to the potential gradient (Date et al 1980, Schleyer et al 1982, Gasser et al 1982a). Surprisingly, a potential sufficient to drive import is developed even in respiratory deficient mitochondrial mutants. Apparently, the charge separation developed by translocater-mediated exchange of ATP from the cytoplasm for ADP in the matrix is sufficient to drive protein import. The fact that it is not possible to construct a respiratory deficient ATP/ADP translocater double-mutant strain suggests that some mitochondrial membrane potential is required for yeast cell growth.

Membrane potential seems to be required only for those proteins that pass at least partway through the mitochondrial inner membrane. Thus, outer membrane proteins and cytochrome c (located on the outer surface of the inner membrane) are assembled normally into de-energized mitochondria (Zimmerman et al 1981, Freitag et al 1982, Gasser & Schatz 1983). In contrast, at least three other proteins with active sites facing the inter-membrane space require membrane potential for assembly. Cytochrome b_2, cytochrome c_1, and cytochrome c peroxidase are each made as precursors that initially penetrate into the matrix and are proteolytically processed by a protease required for maturation of a variety of matrix and inner membrane enzymes (Gasser et al 1982b, Daum et al 1982, Reid et al 1982, Böhni et al 1983). Though initially anchored to the inner membrane, these cytochromes are then additionally processed on the external surface of the membrane by an as yet unidentified protease. This two-step processing scheme has been directly demonstrated for the c_1 and b_2 enzymes, and has been inferred from the sequence of the peroxidase gene, which contains a 23-residue long apolar segment that probably spans the inner membrane and is flanked by processing sites on both sides of the membrane (Kaput et al 1982).

The matrix protease responsible for amino-terminal processing has been partially purified from isolated mitochondria (Böhni et al 1983, McAda & Douglas 1982). Cleavage is restricted to mitochondrial precursors, and is distinguished by sensitivity to chelating agents such as EDTA, GTP, and o-

phenanthroline. Unlike processing by intact mitochondria, cleavage by the isolated protease is insensitive to membrane potential dissipaters such as CCCP. Since o-phenanthroline inhibits the protease in intact mitochondria, it has been possible to show that proteolytic processing is not required for import of proteins into energized mitochondria. In the presence of the inhibitor, precursors are taken up and are protected from the action of exogenous trypsin (Zwizinski & Neupert 1983). Reversal of the chelation block allows imported precursors to be processed.

Although the covalent modification and translocation facilitated by a membrane potential play no role in the transport of outer membrane proteins, the assembly of this membrane is nevertheless highly specific, and is probably mediated by protein receptors. In *Neurospora*, assembly of the outer membrane porin is inhibited by proteolytic treatment of intact mitochondria, however, an analogous, and perhaps identical, 29 kDa protein in *S. cerevisiae* is assembled correctly in protease-treated mitochondria (Gasser & Schatz 1983; Zwizinski et al 1984). Specificity of assembly and anchoring is clearly evident from studies on the sequence and directed-deletion analysis of a major 70 kDa outer membrane protein. This protein is anchored via a 10 kDa amino-terminal fragment, with the remaining portion facing the cytoplasm (Riezman et al 1983a). An apolar membrane-spanning peptide sequence running from residue 9 to 38 was deduced from the sequence of the gene (Hase et al 1983). Elements of this segment appear to participate both in anchoring and localization to the outer membrane. Deletion of the entire segment produces a truncated cytoplasmic form of the protein (Hase et al 1984). Surprisingly, a deletion that removes much, but not all of this region (from 12 to 106) causes $\sim 30\%$ of the truncated protein to be imported into the matrix space, with the rest remaining in the cytoplasm. Perhaps part of the anchoring segment serves as a stop transfer sequence.

As is true of translocation across the ER membrane, the actual polypeptide penetration event for mitochondrial proteins is not understood at the molecular level. Ultimately, an appreciation of the mechanism of translocation will require a reaction reconstituted in vesicles. Short of this, it has been possible to define genes that are required at some stage in the import process. Pleiotropic mutants (*mas* = mitochondrial assembly) have been obtained by screening a collection of temperature-sensitive growth mutants and selecting those that accumulate the ATPase β-subunit precursor at the nonpermissive temperature (Yaffe & Schatz 1984). Mutant alleles of two genes (*mas1* and *mas2*) were identified that cause a failure in mitochondrial assembly in respiratory competent or deficient cells. In these mutants the assembly rates of several mitochondrial proteins are reduced, though some are not as severely affected as the ATPase. This approach can be extended by screening for deficient assembly of other proteins, which could possibly reveal multiple assembly pathways. Furthermore, localization of the accumulated mitochondrial precursors, and localization of the *MAS* gene products may suggest which step(s) in the import process is affected.

C.   Transport into the Nucleus

Many of the genetic and biochemical approaches that have been useful in evaluating protein import into mitochondria and secretory organelles will probably be applied to the assembly of the nucleus. Thus far, some important insights have been gained from studies on three genes that code for nuclear proteins. Aspartate transcarbamoylase (ATCase), the product of the *URA2* gene, was identified as a nuclear protein in yeast using a histochemical stain (Denis-Duphil & Lacroute 1971; Nagy et al 1982). Surprisingly, when the *URA2* gene is present on a high copy-number plasmid, overproduction of the enzyme leads to significant build up of activity in the cytoplasm. This result could mean that some component in the import system has been overloaded. If so, import of other nuclear proteins may be reduced by competition with the *URA2* product.

Localization information on two other genes has been evaluated by joining amino-terminal segments to *E. coli* β-galactosidase. Fusions with the $\alpha2$ gene, which is involved in regulation of mating-type gene expression, produce hybrid proteins that are localized to the nucleus (Hall et al 1984). As little as 13 amino acids from the N-terminus of $\alpha2$ suffice to allow β-galactosidase to associate with the nucleus, as detected by immunofluorescent microscopy with β-galactosidase antibody. Clearer enrichment in the nucleus is seen when hybrids containing N-terminal segments of at least 67 amino acids of $\alpha2$ are examined. Similar results have been obtained with fusions containing as little as 74 amino acids from the N-terminus of the galactose regulatory protein, the product of the *GAL4* gene (Silver et al 1984). The *GAL4* and $\alpha2$ gene products have no sequence homology in the relevant N-terminal domains. Unlike the situation with *URA2*, overproduction of the *GAL4-LacZ* hybrid protein did not lead to cytoplasmic accumulation of β-galactosidase, though the relative levels of these two different proteins have not been examined quantitatively.

In general, studies on nuclear localization in other eukaryotes have shown no proteolytic maturation of precursors. Furthermore, proteins isolated from nuclei are capable of relocalization after injection into the cytoplasm of another cell (DeRobertis 1983). These features are quite distinct from mitochondrial and secretory organelle proteins. Thus, in spite of the evidence suggesting localization information on the N-terminus of many constituents of all three organelles, the mechanism for nuclear transport is likely to be entirely different.

II. INTERCOMPARTMENTAL TRANSPORT

A. Transport to the Cell Surface

SORTING IN THE ENDOPLASMIC RETICULUM   Proteins that are destined for various secretory organelles must first be sorted from those that remain in the ER. Very little is known about factors responsible for the sorting out of these two groups of proteins. One possibility is that proteins designed to be transported elsewhere have a primary signal or special structure that allows interaction with receptors in the ER membrane. These signals or structures would in turn interact on the cytoplasmic face of the membrane with proteins such as clathrin to form coated vesicles enriched in the transported proteins. Conversely, proteins designed to remain in the ER may be selectively restrained, while everything else is transported.

Two genetic approaches have been used to address the role of structural signals in intercompartmental transport of secreted enzymes. One, the evaluation of *SUC2-LacZ* gene fusions has shown that when *LacZ*, which codes for β-galactosidase, is fused to as much as 95% of the *SUC2* 5′ coding sequence, transport of the hybrid protein from the ER to the Golgi body is retarded (Emr et al 1984). It is unlikely that a transport signal resides in the C-terminal 5% of invertase that is missing in the largest hybrid. Rather, the presence of β-galactosidase on the hybrid may interfere with some essential aspect of invertase packaging in a transitional zone of the ER. Apparently this is not a problem for hybrids formed between two secreted proteins: A fusion between the precursor region of the α-factor gene and *SUC2* produces active invertase that is secreted normally (Emr et al 1983).

As an alternative to the gross conformational changes that accompany gene fusions, point mutations in the *SUC2* gene have been sought that produce invertase that is specifically defective in transport. Retention of full enzyme activity in a secretion-defective mutant protein is a criterion that may be used to distinguish a transport lesion from more general perturbations of protein structure. Two mutations that delay transport of core-glycosylated invertase, but not acid phosphatase, have been mapped in the 5′ coding region of *SUC2* (Schauer 1985). Both mutations specifically reduce the transport of invertase to a compartment, presumably in the Golgi body, where outer chain carbohydrate is added. Subsequent transport to the cell surface is not measurably delayed. One mutation (*SUC2*-s1) converts an Ala codon to Val at position −1 in the signal peptide; the other (*SUC2*-s2) changes a Thr to an Ile at position +64 in the mature protein. Mutation s1 results in a rate of invertase transport to the Golgi body approximately fifty times slower than normal, which is attributable to defective signal peptide cleavage. While peptide cleavage normally occurs at an Ala-Ser bond, the s1 mutant form is processed slowly at the adjacent Ser-Met position, giving rise to mature invertase with an N-terminal Met residue. Mutant invertase of the s2 type is transported approximately seven times more slowly than normal, with no delay in signal peptide cleavage, and no detectable abnormal physical property of the enzyme.

It is not clear why the failure to remove a signal peptide has such a dramatic effect on invertase transport from the ER. The effect has also been observed with acid phosphatase, where deletion of the signal cleavage site causes a similar delay in secretion of unprocessed active enzyme (Haguenauer-Tsapis & Hinnen 1984). In contrast, there are examples of secreted proteins, such as the precursor of α-factor, that have uncleaved signals and are nevertheless transported rapidly (Julius et al 1984a). Perhaps certain signal peptides are cleaved to prevent association with a stable component in the ER. Alternatively, certain signal peptides may be so hydrophobic that precursors remain anchored nonspecifically on the lumenal face of the membrane.

Proteins required for the sorting and transport process have not been firmly identified. However, a number of candidates can be listed. First, the *SEC* genes that are required for movement of secreted and membrane proteins from the ER play some role in this process. An understanding of the function of these genes will require experimental evidence from cell-free reactions that depend upon the relevant *SEC* proteins. Additionally, the results of cloning and structural characterization of the *SEC* gene products should aid in an evaluation of their function.

Clathrin is another protein that may play a role in this process (Pearse & Bretscher 1981). Circumstantial evidence suggests that coated vesicles mediate transport of the vesicular stomatitis G protein from the ER in mammalian cells (Rothman & Fine 1980). Coated vesicles have been isolated from *S. cerevisiae,* and clathrin was identified as a major constituent of the coat by the detection of characteristic triskelion structures recovered when the coat was removed by treatment with urea (Mueller & Branton 1984). It should be possible to introduce mutations into the gene that encodes the large subunit of clathrin to produce a protein that is conditionally defective in coated vesicle assembly. Such mutations will allow a direct test of the role(s) that clathrin plays in protein transport.

TRANSPORT THROUGH AND MODIFICATION IN THE GOLGI BODY   Convincing profiles of Golgi cisternae in wild-type *S. cerevisiae* are not common, though some have been published (Matile et al 1969). Much of what is known about the role of the yeast Golgi body in transport and processing comes from investigation of the *sec7* mutant in relation to other *sec* mutants. Mutant cells accumulate large stacks of Golgi-like cisternae when incubated at 37°C in medium that contains 0.1% glucose or a non-fermentable carbon source (Novick et al 1981). The same mutant, when incubated at 37°C in medium with 2% glucose accumulates cup and toroid-shaped organelles that are called Berkeley bodies. Many of the *sec* mutants, including *sec7*, secrete accumulated proteins when the cells are returned to a permissive temperature, even in the presence of an inhibitor of protein synthesis. Unlike the others, *sec7* is reversible only in low glucose, the condition that allows accumulation of Golgi stacks. Berkeley bodies probably represent irreversibly altered Golgi cisternae. Perhaps some glucose metabolite interferes with progressive stacking of the cisternae.

Glycosylation is a major activity associated with transport through the Golgi body. N-linked oligosaccharides are elongated to produce a structure referred to as the outer chain (Ballou 1982). The structure has an α-1,6-linked polymannose backbone that extends the α-1,6-backbone of the core oligosaccharide. In addition, mannooligomers from man_2 and man_4 are attached in α-1,3-linkage to this backbone. The man_3 and man_4 sidechains have α-1,3-mannose nonreducing ends. Some of the side chains are substituted with diester-linked mannose-phosphate or mannobiose-phosphate. α-1,3-linked mannose residues are also added to the core oligosaccharide, rendering it somewhat larger than the form made in the ER (man_{11-14} instead of man_8). Some mature N-linked oligosaccharides have as many as 150 mannose residues.

The synthesis and function of the outer chain have been investigated by isolating mutants that produce serologically altered yeast cells. Since mannoproteins are exposed on the exterior of the cell wall, the major immune response to yeast cells is antibodies directed against any of several mannose linkages of the outer chain (Ballou & Raschke 1974). Such antibodies agglutinate normal cells, but not mutant cells in which specific determinants are missing. Using this procedure, Ballou and colleagues have identified 9 genes that direct the addition of specific residues, or of whole parts of the outer chain (Ballou et al 1980). These mutants grow normally (although *mnn9* cells are clumpy), and secrete underglycosylated invertase. Even completely unglycosylated invertase, formed in the presence of tunicamycin, can be secreted at 25°C (Ferro-Novick et al 1984b). Hence, the outer chain appears to play no role in secretion, but may be important for growth of yeast cells in a more natural habitat.

In contrast to the behavior of the ER-blocked *sec* mutants, *sec7* cells accumulate mature forms of invertase and acid phosphatase that contain outer chain carbohydrate determinants (Esmon et al 1981). Mutants that accumulate secretory vesicles also produce mature glycoproteins. Double mutant analysis has shown that the *sec7* block is the earliest one in which complete glycoproteins accumulate. Hence, glycosylation probably is completed somewhere in the Golgi structure whose discharge is influenced by *sec7*. Since most of the *alg* mutations and all of the *mnn* mutations have no effect on secretion, it is unlikely that any of the *sec* mutants are directly defective in glycosylation.

Proteolysis has recently been recognized as an activity prominently associated with transport of certain proteins through the Golgi body. Both the precursor of α-factor and the precursor of killer toxin are processed to mature forms at some point in transport through the Golgi stack (Julius et al 1984a, Bussey et al 1983). In the case of α-factor, the sequence of the gene predicts four tandem repeats of pheromone sequence separated by nearly identical spacer peptides (Kurjan & Herskowitz 1982). Analysis of precursor forms arrested at various stages in secretion showed that a portion are first cleaved at the spacer peptide positions in mutant *sec7*. Processing is nearly complete by the time secretory vesicles are formed from the Golgi body. Two Golgi membrane enzymes involved in the processing have been identified by activity and by genetic analysis. The initial endoproteolytic incision is made after a pair of basic residues (Lys-Arg). An enzyme with this specificity has been detected, and is missing in a mutant strain (*kex2*) that secretes intact α-factor precursor (Julius et al 1984a). The same mutation interferes with processing of the killer toxin precursor, which is known to contain a Lys-Arg site adjacent to the C-terminal toxin sequence (Leibowitz & Wickner 1976, Bostian et al 1984). After the initial cleavage, a group of four to six amino acids (Glu-Ala-Glu-Ala or Glu-Ala-Asp-Ala) on the N-terminus of α-factor is removed by the action of a membrane-bound dipeptidyl aminopeptidase (Julius et al 1983). A mutant (*ste13*) missing this peptidase secretes a biologically inactive form of the pheromone that retains the N-terminal tetrapeptide.

It appears that outer chain carbohydrate addition occurs before the *SEC7* step, while endoproteolytic maturation of the α-factor precursor occurs primarily after the *SEC7* step (Julius et al 1984a). The yeast Golgi body may be arranged in *cis* and *trans* cisternae that are relatively enriched in mannosyl-transferases and processing proteases, respectively. Immuno-localization with specific antibodies in *sec7* mutant cells could be used to test this possibility.

CELL SURFACE ASSEMBLY   A large number of cell surface proteins fail to be exported in the *sec* mutants. In addition to those mentioned earlier, the list, documented by activity assay, includes L-asparaginase, α-galactosidase, exo- and endo-laminarinase, sulfate permease, galactose permease, arginine permease, proline-specific permease, and vanadate-sensitive Mg^{2+} ATPase (Tschopp et al 1984, W. Hansen, J. Tschopp, P. Sullivan, R. Schekman, unpublished). A more general probe of surface assembly was developed by derivatization of cell surface proteins. Modification of cell surface amino groups with trinitrobenzene sulfonate (TNBS) followed by precipitation with TNP antibody allows the analysis of newly exported proteins (Novick & Schekman 1983). In this procedure, wild-type and mutant cells are labeled with protein synthesis precursors at 37°C, and then tagged with TNBS at 0°C. Under these conditions TNBS does not penetrate into the cell. Both secreted and plasma membrane surface proteins are tagged in this procedure and can be examined separately. Secreted proteins are released when cells are converted to spheroplasts, and tagged membrane proteins are recovered in a sedimented fraction from lysed spheroplasts. Wild-type cells externalize distinct sets of membrane and secreted proteins, as revealed by SDS-gel electrophoresis of the TNP-antibody precipitates. The major proteins in both fractions are not externalized in *sec* mutant cells at 37°C, but are externalized at 24°C.

Although analysis of the *sec* mutants suggests a single linear pathway of cell surface assembly, the results are also consistent with parallel pathways in which subsets of exported proteins travel in different compartments. If plasma membrane proteins and soluble secreted proteins are transported in separate vesicles it may be possible to uncouple the processes of secretion and plasma membrane assembly. Uncoupling of this sort could explain the behavior of auxotrophic yeast cells starved for either inositol or fatty acids. During inositol deprivation yeast cells cease net cell surface growth and become dense, just as is the case for *sec* mutant cells incubated at 37°C (Henry et al 1977). In contrast to the *sec* mutants, secretion of invertase continues during inositol or fatty acid starvation, and falls off only as protein synthesis declines at the onset of cell death (Atkinson & Ramirez 1984). Unfortunately, the assembly of specific plasma membrane proteins has not yet been examined during inositol deprivation.

Even if cell surface assembly is mediated by separate classes of vesicles, it is likely that most fuse with the bud portion of the plasma membrane. Invertase and acid phosphatase are secreted into the bud, which is the major site of new surface growth during most of the cell cycle (Tkacz & Lampen 1973, Field & Schekman 1980). Polarized surface growth must be achieved by a mechanism that ensures directed vesicle transport into the bud, and selective vesicle recognition of the bud membrane. These aspects of bud growth are likely to be influenced by the 10 *sec* genes that are required at the end of the secretory pathway. Some evidence also suggests that actin could participate in vesicle transport. Immunofluorescent microscopy has shown actin filaments and cables directed toward and concentrated in the bud (Kilmartin & Adams 1984, Adams & Pringle 1984). Temperature-sensitive actin mutants show a long delay, but not an absolute block in vesicle transport to the cell surface (Novick & Botstein 1985). Microtubules, which are also directed toward the bud, do not seem to participate in secretion. Tubulin mutations do not delay secretion (P. Novick, personal communication).

Localized assembly of the cell surface is also reflected in the deposition of cell wall polysaccharides. The major structural component of the yeast cell wall is β-1,3- and β-1,6-linked glucan. Fibers of these glucans appear to be transmitted directly into the growing bud wall by plasma membrane-bound glucan synthetases (Cabib et al 1982, 1983). Somehow glucan synthetase molecules must be active in the growing bud and inactive in the mother portion of a dividing cell. Perhaps primers that stimulate glucan polymerization are provided in the bud by glucanases secreted early in the cell cycle. Alternatively, some other molecule that influences glucan synthetase activity may be available in the bud. A substantial fraction of the in vivo rate of β-1,3-glucan synthesis has been reproduced with isolated membranes incubated with either ATP or GTP, and sugar nucleotide substrate, UDPG. Cabib and colleagues (Shematek et al 1980, Shematek & Cabib 1980, Notario et al 1982) have proposed a model in which glucan synthetase is irreversibly activated by GTP, or reversibly activated by ATP and some small, enzyme-bound molecule that can be phosphorylated. This model provides an on-off regulatory feature that could explain activation of glucan synthetase in the bud and inactivation in the mother portion of the cell. This model leaves the larger issue of topological regulation unresolved.

Another equally intriguing issue of localized growth concerns assembly of the yeast division septum, which is composed principally of chitin (β-1,3-N-acetylglucosamine) (Cabib et al 1971). Chitin synthetase is an integral plasma membrane protein which, when proteolytic artifacts are controlled, can be recovered in a zymogen form (Durán et al 1975, Durán & Cabib 1978). The zymogen can be activated in vitro by trypsin, or by the trypsin-like yeast protease B. Proteolytic activation converts some site on the inner surface of the plasma membrane. Cabib has proposed that the zymogen is activated by a vesicle that somehow delivers a protease (or some other activating factor) to the nascent division septum membrane (Cabib et al 1974). This model requires that the protease be deposited with its active site exposed on the cytoplasmic surface of the membrane.

Although it is reasonable to expect that active chitin synthetase is restricted to the nascent division septum during budding growth, it is less obvious how the zymogen is distributed. Spheroplasts deposit chitin uniformly about the cell surface, and large plasma membrane fragments can be activated by trypsin to produce chitin all over the membrane surface (Durán et al 1979). In growing cells the localization of chitin deposition is disturbed by diverse mutations, including blocks at different stages in the cell cycle, and disruption of actin filaments and cables (Novick & Botstein 1985, Roberts et al 1983). The most reasonable explanation of these observations is that the zymogen is evenly distributed along the surface of the plasma membrane and is activated in unusual locations when the activating factor is redistributed.

In addition to the problem of restricting glucan and chitin synthesis to certain portions of the cell surface, prevention of intracellular synthesis of polysaccharides requires that synthetases and any activating factors be delivered separately to the cell surface. An extreme example of this is suggested by the isolation of intracellular vesicles enriched in chitin synthetase zymogen, which are believed to mediate the export of this enzyme and little else (Ruiz-Herrera et al 1977). These vesicles, called chitosomes, have not been characterized sufficiently to support the claim that no other cell surface enzymes are transported in them. Nevertheless, the possibility remains that cell surface proteins that have specific spatial and temporal regulatory requirements may be assembled in specialized vesicles. Similar specialization is believed to account for the assembly of the lumenal and basolateral plasma membrane domains of polarized epithelial cells (see Simons & Fuller, this volume).

B. Transport to the Vacuole

The yeast vacuole is analogous to a mammalian lysosome in that it contains hydrolytic enzymes (Wiemken et al 1979), many of which are derived from glycoprotein proenzyme forms. Unlike lysosomal enzymes, yeast vacuolar proteins are not detected in the culture medium or in the cell wall during logarithmic growth. Transport of several vacuolar constituents depends upon early stages in the secretory pathway. Among the vacuolar enzymes, carboxypeptidase Y (CPY) has been studied in some detail. CPY is made as a 67-kDa precursor containing four N-glycosidically-linked oligosaccharides (Hasilik & Tanner 1978a). Processing and transport to the vacuole occurs with a half-time of about six minutes in wild-type cells, and is blocked in *sec* mutant cells that are defective in transport from the ER or through the Golgi body (Hasilik & Tanner 1978b, Stevens et al 1982). Two covalent modifications of the precursor occur during transport. First, core oligosaccharides are elongated to produce a glycoprotein percursor of 69 kDa. This reaction occurs in the Golgi body, as demonstrated by the fact that one of the Golgi-blocked *sec* mutants (*sec14*) accumulates a mixture of the 67 and 69 kDa forms. Next, the amino-terminal 8 kDa propeptide portion is cleaved under the direction of at least one gene (*PEP4*) that is also required for the maturation of a variety of vacuolar hydrolases (Hemmings et al 1981). A *pep4* mutant accumulates enzymatically inactive proCPY in the vacuole, which suggests that maturation occurs after the precursor is segregated within the vacuole (Stevens et al 1982, Distel et al 1983, Zubenko et al 1983). *Sec* mutants that block transport of secretory vesicles have no effect on CPY localization. These results suggest that vacuolar and secretory proteins travel together from the ER to the Golgi body, where sorting may occur.

Although the sorting of lysosomal and secretory proteins in mammalian fibroblasts relies on a carbohydrate determinant (Kaplan et al 1977), the ultimate source of discrimination lies in an amino acid sequence or structural feature of the targeted protein (Reitman & Kornfeld 1981). Carbohydrate does not serve this role in yeast: At least two vacuolar proteins, CPY and alkaline phosphatase, are synthesized and activated normally in the absence of oligosaccharide synthesis (Schwaiger et al 1982, Onishi et al 1979).

In order to develop a genetic means of identifying the signal(s) on CPY involved in transport to the vacuole, the structural gene has been cloned. Complementation of a *prc1* (CPY structural gene) mutation is achieved with a multicopy plasmid that contains the *PRC1* gene on an insert (Stevens et al 1985). Though the level of expression of CPY is elevated only three- to fourfold by this plasmid, as much as 20% of the CPY is secreted into the periplasm and medium. Such escape from the normal sorting mechanism does not occur in cells that contain a single copy of the *PRC1* gene. An even greater fraction of the CPY is secreted when expression of the gene is enhanced with a more active promoter: 60% is secreted when the acid phosphatase promoter is substituted for the *PRC1* promoter. The principal secreted form of CPY is the 69 kDa precursor species, which suggests that the same carbohydrate modifications accompany transport to the vacuole and secretion. Unlike the normal rapid proteolytic processing event, the conversion of the secreted form to mature, enzymatically active CPY in the periplasm is slow, and does not require the *PEP4* gene. From these results it appears that CPY production is closely matched by the capacity of the sorting apparatus. An increase in the level of CPY overloads the system and excess precursor is secreted. The limiting component could be a receptor or an enzyme that converts CPY to a form recognized by a receptor.

The observation that excess CPY is secreted and slowly activated has stimulated a search for mutants that are defective in the process of sorting. Since yeast cells ordinarily do not secrete carboxypeptidases of any sort, certain amino-blocked peptides cannot be salvaged to supply an essential amino acid. This provides the basis for a selection procedure: Sorting-deficient mutant cells degrade peptides provided in the medium by virtue of external, active CPY. A large number of genes [at least 7 vacuole protein localization (*VPL*) complementation groups] have been identified thus far (T. Stevens, personal communication). Some of the mutations confer conditional lethality that is due either to a complete block in vacuole assembly, or to the presence of toxic levels of hydrolytic enzymes at the cell surface. At least one of the mutants has been shown to secrete a variety of vacuolar proteins at a restrictive growth temperature.

This selection procedure could turn up point mutations in the *PRC1* gene that disrupt the sorting signal on CPY. Though such mutations have not yet been found, another approach has implicated a region near the amino-terminus that may contain the sorting signal. Gene fusions that join the 5' regulatory and coding sequence of *PRC1* to a nearly complete piece of the *SUC* gene produce hybrid proteins that have invertase activity. As little as 40 amino acids from the amino-terminus of proCPY suffice to divert the hybrid protein into the vacuole (S. Emr, personal communication). A hybrid with 10 amino acids of proCPY is secreted. These results suggest that the sorting signal may lie within the region from position 10 to 40 of the propeptide. Assuming this reflects an interaction with a receptor, it should be possible to detect binding of proCPY to an isolated membrane fraction. CPY-invertase hybrids that contain the sorting signal should compete for binding, and any *vpl* mutation that affects receptor function should influence the binding activity.

Until recently, transport to the vacuole was assumed to be entirely by an intracellular route. It has now been shown that Lucifer yellow, a membrane impermeable fluorescent dye, is taken up by yeast cells and collects in the vacuole (Riezman 1985). Uptake is energy-dependent and not saturable, properties expected for fluid-phase endocytosis. The rate of uptake is about one-sixth that of mouse peritoneal macrophages. Surprisingly, most but not all of the *sec* mutants block uptake at 37°C. Several of the mutants blocked in translocation into the ER, and in transport from the ER to the Golgi body, are not blocked in endocytosis. All of the mutants that accumulate secretory vesicles are defective in uptake of the dye. While it is possible that a large number of gene products act directly in secretion and in endocytosis, it is equally plausible that cell surface proteins required for endocytosis must be recycled via the secretory pathway. If the latter is true, the two pathways could be coupled without extensive overlapping of direct genetic requirements.

An unresolved question is the role endocytosis plays in yeast growth or differentiation. Two possibilities may be mentioned. Response of *a* cells to α-factor and α cells to *a*-factor may require receptor-mediated internalization of the pheromone. If so, some of the sterile mutants that fail to respond to the pheromone may be defective in uptake. Aside from responding to external factors, endocytosis may play a role in turnover of the cell surface, or in recycling of cell surface components involved in the transport of other molecules.

Literature Cited

Adams, A. E. M., Pringle, J. R. 1984. Relationship of actin and tubulin distribution to bud growth in wild-type and morphogenetic-mutant *Saccharomyces cerevisiae. J. Cell Biol.* 98:934–35

Ades, I. Z., Butow, R. A. 1980. The products of mitochondria-bound cytoplasmic polysomes in yeast. *J. Biol. Chem.* 255:9918–24

Arnold, E., Tanner, W. 1982. An obligatory role of protein glycosylation in the life cycle of yeast cells. *FEBS Lett.* 148:49–53

Atkinson, K. D., Ramirez, R. M. 1984. Secretion can proceed uncoupled from net plasma membrane expansion in inositol-starved *Saccharomyces cerevisiae. J. Bacteriol.* 160:80–86

Ballou, C. 1982. Yeast cell wall and cell surface. In *The Molecular Biology of the Yeast Saccharomyces: Metabolism and Gene Expression,* ed. J. Strathern, E. Jones, J. Broach, pp. 335–60. Cold Spring Harbor, NY: Cold Spring Harbor Lab.

Ballou, L., Cohen, R. E., Ballou, C. E. 1980. *Saccharomyces cerevisiae* mutants that make mannoproteins with a truncated carbohydrate outer chain. *J. Biol. Chem.* 255:5985–91

Ballou, C. E., Raschke, W. C. 1974. Polymorphism of the somantic antigen of yeast. *Science* 184:127–34

Barnes, G., Hansen, W., Holcomb, C. L., Rine, J. 1984. Asparagine-linked glycosylation in *Saccharomyces cerevisiae:* Genetic analysis of an early step. *Mol. Cell. Biol.* 4:2381–88

Bitter, G. A., Chen, K. A., Banks, A. R., Lai, P.-H. 1984. Secretion of foreign proteins from *Saccharomyces cerevisiae* directed by α-factor gene fusions. *Proc. Natl. Acad. Sci. USA* 81:5330–34

Böhni, P., Daum, G., Schatz, G. 1983. Import of proteins into mitochondria. Partial purification of a matrix-located protease involved in cleavage of mitochondrial precursor polypeptides. *J. Biol. Chem.* 258:4937–43

Bostian, K. A., Elliot, Q., Bussey, H., Burn, V., Smith, A., et al. 1984. Sequence of the prepro-toxin dsRNA gene of Type 1 killer yeast: Multiple processing events produce a two-component toxin. *Cell* 36:741–51

Bostian, K., Jayachandran, S., Tipper, D. 1983. A glycosylated protoxin in killer yeast: Models for its structure and maturation. *Cell* 32:169–80

Brake, A. J., Merryweather, J. P., Coit, D. G., Heberlein, U. A., Masiarz, F. R., et al. 1984. α-Factor-directed synthesis and secretion of mature foreign proteins in *Saccharomyces cerevisiae. Proc. Natl. Acad. Sci. USA* 81:4642–46

Bussey, H., Saville, D., Green, D., Tipper, D. J., Bostian, K. A. 1983. Secretion of *Saccharomyces cerevisiae* killer toxin: Processing of the glycosylated precursor. *Mol. Cell. Biol.* 3:1362–70

Butow, R. A., Bennet, W. F., Finkelstein, D. B., Kellems, R. E. 1975. Nuclear-cytoplasmic interactions in the biogenesis of mitochondria in yeast. In *Membrane Biogenesis,* ed. A. Tzagoloff, pp. 155–99. New York: Plenum

Byrd, J. C., Tarentino, A. L., Maley, F., Atkinson, P. H., Trimble, R. B. 1982. Glycoprotein synthesis in yeast: Identification of $Man_8GlcNAc_2$ as an essential intermediate in oligosaccharide processing. *J. Biol. Chem.* 257:14657–66

Cabib, E., Bowers, B., Roberts, R. 1983. Vectorial synthesis of a polysaccharide by isolated plasma membranes. *Proc. Natl. Acad. Sci. USA* 80:3318–21

Cabib, E., Farkas, V., Ulane, R. E., Bowers, B. 1974. Yeast septum formation as a model system for morphogenesis. *Curr. Top. Cell. Regul.* 22:1–32

Cabib, E., Roberts, R., Bowers, B. 1982. Synthesis of the yeast cell wall and its regulation. *Ann. Rev. Biochem.* 51:763–93

Cabib, E., Ulane, R. E., Bowers, B. 1971. The control of morphogenesis. An enzymatic mechanism for the initiation of septum formation in yeast. *Proc. Natl. Acad. Sci. USA* 68:2052–57

Carlson, M., Taussig, R., Kustu, S., Bostein, D. 1983. The secreted form of invertase in *Saccharomyces cerevisiae* is synthesized from mRNA encoding a signal sequence. *Mol. Cell. Biol.* 3:439–47

Chu, F. K., Maley, F. 1980. The effect of glucose on the synthesis and glycosylation of the polypeptide moiety of yeast external invertase. *J. Biol. Chem.* 255:6392–97

Daniels, C. J., Boyle, D. G., Quay, S. C., Oxender, D. L. 1981. A role for membrane potential in the secretion of protein into the periplasm of *Escherichia coli*. *Proc. Natl. Acad. Sci. USA* 78:5396–5400

Date, T., Zwizinski, C., Ludmerer, S., Wickner, W. 1980. Mechanisms of membrane assembly: Effects of energy poisons on the conversion of soluble M13 coliphage procoat to membrane-bound coat protein. *Proc. Natl. Acad. Sci. USA* 77:827–31

Daum, G., Gasser, S., Schatz, G. 1982. Import of proteins into mitochondria. Energy-dependent two-step processing of the intermembrane space enzyme cytochrome b_2 by isolated yeast mitochondria. *J. Biol. Chem.* 257:13075–80

Denis-Duphil, M., Lacroute, F. 1971. Fine structure of the *ura2* locus in *Saccharomyces cerevisiae*. I. *In vivo* complementation studies. *Mol. Gen. Genet.* 112:354–64

DeRobertis, E. M. 1983. Nucleocytoplasmic segregation of proteins and RNAs. *Cell* 32:1021–25

Distel, B., Al, R., Tabak, H., Jones, E. W. 1983. Synthesis and maturation of the yeast vacuolar enzymes carboxypeptidase Y and aminopeptidase I. *Biochim. Biophys. Acta* 741:128–35

Douglas, M. G., Geller, B. L., Emr, S. D. 1984. Intracellular targeting and import of an F_1-ATPase β-subunit-β-galactosidase hybrid protein into yeast mitochondria. *Proc. Natl. Acad. Sci. USA* 81:3983–87

Durán, A., Bowers, B., Cabib, E. 1975. Chitin synthetase symogen is attached to yeast plasma membrane. *Proc. Natl. Acad. Sci. USA* 72:3952–57

Durán, A., Cabib, E. 1978. Solubilization and partial purification of yeast chitin synthetase. Confirmation of the zymogenic nature of the enzyme. *J. Biol. Chem.* 253:4419–25

Durán, A., Cabib, E., Bowers, B. 1979. Chitin synthetase distribution on the yeast plasma membrane. *Science* 203:363–65

Emr, S. D., Schauer, I., Hansen, W., Esmon, P., Schekman, R. 1984. Invertase-β-galactosidase hybrid proteins fail to be transported from the endoplasmic reticulum in *Saccharomyces cerevisiae*. *Mol. Cell. Biol.* 4:2347–55

Emr, S., Schekman, R., Flessel, M., Thorner, J. 1983. An MFα1-*SUC2* (α-factor-invertase) gene fusion for study of protein localization and gene expression in yeast. *Proc. Natl. Acad. Sci. USA* 80:7080–84

Enequist, H. G., Hirst, T. R., Hardy, S. J. S., Harayama, S., Randall, L. L. 1981. Energy is required for maturation of exported proteins in *Escherichia coli*. *Eur. J. Biochem.* 116:227–33

Esmon, B., Esmon, P. C., Schekman, R. 1984.

Early steps in processing of yeast glycoproteins. *J. Biol. Chem.* 259:10322–27

Esmon, B., Novick, P., Schekman, R. 1981. Compartmentalized assembly of oligosaccharides on exported glycoproteins in yeast. *Cell* 25:451–60

Ferro-Novick, S., Hansen, W., Schauer, I., Schekman, R. 1984b. Genes required for completion of import of proteins into the endoplasmic reticulum in yeast. *J. Cell Biol.* 98:44–53

Ferro-Novick, S., Novick, P., Field, C., Schekman, R. 1984a. Yeast secretory mutants that block the formation of active cell surface enzymes. *J. Cell Biol.* 98:35–43

Field, C., Schekman, R. 1980. Localized secretion of acid phosphatase reflects the pattern of cell-surface growth in *Saccharomyces cerevisiae*. *J. Biol. Chem.* 254:796–803

Firgaira, F. A., Hendrick, J. P., Kalousek, F., Kraus, J. P., Rosenberg, L. E. 1984. RNA required for import of percursor proteins into mitochondria. *Science* 226:1319–22

Freitag, H., Janes, M., Neupert, W. 1982. Biosynthesis of mitochondrial porin and insertion into the outer mitochondrial membrane of *Neurospora crassa*. *Eur. J. Biochem.* 126:197–202

Gasser, S., Daum, G., Schatz, G. 1982a. Import of proteins into mitochondria. Energy-dependent uptake of precursors by isolated mitochondria. *J. Biol. Chem.* 257:13034–41

Gasser, S., Ohashi, A., Daum, G., Böhni, P., Gibson, J., et al. 1982b. Imported mitochondrial proteins cytochrome b_2 and c_1 are processed in two steps. *Proc. Natl. Acad. Sci. USA* 79:267–71

Gasser, S., Schatz, G. 1983. Import of proteins into mitochondria. In vitro studies on the biogenesis of the outer membrane. *J. Biol. Chem.* 258:3427–30

Geller, B. L., Britten, M. L., Biggs, C. M., Douglas, M. G., Emr, S. D. 1983. Import of *ATP2-LacZ* gene fusion proteins into mitochondria. In *Mitochondria 1983—Nucleo-mitochondrial Interactions*, ed. R. J. Schweyen, K. Wolf, F. Kaudewitz, pp. 607–19. Berlin: DeGruyten

Haguenauer-Tsapis, R., Hinnen, A. 1984. A deletion that includes the signal peptidase cleavage site impairs processing, glycosylation, and secretion of cell surface yeast acid phosphatase. *J. Mol. Cell. Biol.* 4:2668–75

Hall, M. N., Hereford, L., Herskowitz, I. 1984. Targeting of *E. coli* β-galactosidase to the nucleus in yeast. *Cell* 36:1057–65

Hase, T., Muller, U., Riezman, H., Schatz, G. 1984. A 70-kd protein of the yeast mitochondrial outer membrane is targeted and anchored via its extreme amino terminus. *EMBO J.* 3:3157–64

Hase, T., Tiezman, H., Suda, K., Schatz, G. 1983. Import of proteins into mitochondria: Nucleotide sequence of the gene for a 70-kd protein of the yeast mitochondrial outer membrane. *EMBO J.* 2:2169–72

Haselbeck, A., Tanner, W. 1982. Dolichol phosphate-mediated mannosyl transfer through liposomal membranes. *Proc. Natl. Acad. Sci. USA* 79:1520–24

Haselbeck, A., Tanner, W. 1983. O-glycosylation in *Saccharomyces cerevisiae* is initiated at the endoplasmic reticulum. *FEBS Lett.* 158:335–38

Hasilik, A., Tanner, W. 1978a. Carbohydrate moiety of carboxypeptidase Y and perturbation of its biosynthesis. *Eur. J. Biochem.* 91:567–75

Hasilik, A., Tanner, W. 1978b. Biosynthesis of the vacuolar yeast glycoprotein carboxypeptidase Y. Conversion of precursor into the enzyme. *Eur. J. Biochem.* 85:599–608

Hay, R., Böhni, P., Gasser, S. 1984. How mitochondria import proteins. *Biochim. Biophys. Acta* 779:65–87

Hemmings, B. A., Zubenko, G. S., Hasilik, A., Jones, E. W. 1981. Mutant defective in processing of an enzyme located in the lysosome-like vacuole of *Saccharomyces cerevisiae. Proc. Natl. Acad. Sci. USA* 78:435–39

Henning, B., Neupert, W. 1981. Assembly of cytochrome *c*. Apocytochrome *c* is bound to specific sites on mitochondria before its conversion to holocytochrome *c*. *Eur. J. Biochem.* 121:203–11

Henry, S. A., Atkinson, K. D., Kolat, A., Culbertson, M. R. 1977. Growth and metabolism of inositol starved cells of *Saccharomyces cerevisiae. J. Bacteriol.* 130:472–84

Hitzeman, R. A., Leung, D. W., Perry, L. J., Kohr, W. J., Levine, H. L., et al. 1983. Secretion of human interferons by yeast. *Science* 219:620–25

Hollenberg, C. P., Roggenkamp, R., Erhart, E., Breunig, K., Reipen, G. 1983. The expression of bacterial *β*-lactamase and its applications to gene technology in yeast. In *Gene Expression in Yeast, Proc. Alko Yeast Symp., Helsinki 1983,* ed. M. Korhola, E. Väisänen, pp. 73–90. Helsinki: Found. Biotech. Indust. Ferment. Res.

Huffaker, T., Robbins, P. 1982. Temperature-sensitive yeast mutants deficient in asparagine-linked glycosylation. *J. Biol. Chem.* 257:3203–10

Huffaker, T., Robbins, P. 1983. Yeast mutants deficient in protein glycosylation. *Proc. Natl. Acad. Sci. USA* 80:7466–70

Hurt, E. C., Pesold-Hurt, B., Schatz, G. 1984. The cleavable prepiece of an imported mitochondrial protein is sufficient to direct cytosolic dihydrofolate reductase into the mitochondrial matrix. *FEBS Lett.* 178:306–10

Innis, M. A., Holland, M. J., McCabe, P. C., Cole, G. E., Wittman, V. P., et al. 1985. Expression, glycosylation, and secretion of an *Aspergillus* glucoamylase from *Saccharomyces cerevisiae. Science* 228:21–26

Ito, K., Mandel, G., Wickner, W. 1979. Soluble precursor of an integral membrane protein: Synthesis of procoat protein in *Escherichia coli* infected with bacteriophage M13. *Proc. Natl. Acad. Sci. USA* 76:1199–1203

Jones, E. W. 1984. The synthesis and function of proteases in *Saccharomyces*: Genetic approaches. *Ann. Rev. Genetics* 18:233–70

Julius, D., Blair, L., Brake, A., Sprague, G., Thorner, J. 1983. Yeast α-factor is processed from a larger precursor polypeptide: The essential role of a membrane-bound dipeptidyl aminopeptidase. *Cell* 32:839–52

Julius, D., Brake, A., Blair, L., Kunisawa, R., Thorner, J. 1984b. Isolation of the putative structural gene for the lysine-arginine cleaving endopeptidase required for processing of yeast prepro-α-factor. *Cell* 37:1075–89

Julius, D., Schekman, R., Thorner, J. 1984a. Glycosylation and processing of prepro-α-factor through the yeast secretory pathway. *Cell* 36:309–18

Kaplan, A., Achord, D. T., Sly, W. S. 1977. Phosphohexosyl components of a lysosomal enzyme are recognized by pinocytosis receptors on human fibroblasts. *Proc. Natl. Acad. Sci. USA* 74:2026–30

Kaput, J., Goltz, S., Blobel, G. 1982. Nucleotide sequence of the yeast nuclear gene for cytochrome *c* peroxidase precursor: Functional implications for protein transport into mitochondria. *J. Biol. Chem.* 257:15054–58

Kilker, R. D., Saunier, B., Tkacz, J. S., Herskovics, A. 1981. Partial purification from *Saccharomyces cerevisiae* of a soluble glucosidase which removes the terminal glucose from the oligosaccharide $Glc_3Man_9GlcNAc$. *J. Biol. Chem.* 256:5299–5303

Kilmartin, J. V., Adams, A. E. M. 1984. Structural rearrangements of tubulin and actin during the cell cycle of the yeast *Saccharomyces. J. Cell Biol.* 98:922–33

Korb, H., Neupert, W. 1978. Biogenesis of cytochrome *c* in *Neurospora crassa.* Synthesis of apocytochrome *c*, transfer to mitochondria and conversion to holocytochrome c. Eur. J. Biochem. 91:609–20

Koshland, D., Botstein, D. 1982. Evidence for posttranslational translocation of β-lactamase across the bacterial inner membrane. Cell 30:893–902

Kuo, S., Lampen, J. 1974. Tunicamycin, an inhibitor of yeast glycoprotein synthesis. Biochem. Biophys. Res. Commun. 58:287–95

Kurjan, J., Herskowitz, I. 1982. Structure of a yeast pheromone gene (MFα): A putative α-factor precursor contains four tandem repeats of mature α-factor. Cell 30:933–43

Lehle, L. 1980. Biosynthesis of the core region of yeast mannoproteins. Formation of a glucosylated dolichol-bound oligosaccharide precursor, its transfer to protein and subsequent modification. Eur. J. Biochem. 109:589–601

Lehle, L., Bause, E. 1984. Primary structural requirements for N- and O-glycosylation of yeast mannoproteins. Biochim. Biophys. Acta 799:246–51

Lehle, L., Cohen, R. E., Ballou, C. E. 1979. Carbohydrate structure of yeast invertase. J. Biol. Chem. 254:12209–18

Lehle, L., Tanner, W. 1976. The specific site of tunicamycin inhibition in the formation of dolichol-bound N-acetyl-glucosamine derivatives. FEBS Lett. 71:167–70

Leibowitz, M. J., Wickner, R. W. 1976. A chromosomal gene required for killer plasmid expression, mating, and sporulation in Saccharomyces cerevisiae. Proc. Natl. Acad. Sci. USA 73:2061–65

Lingappa, V. R., Chaide, J., Yost, C. S., Hedgpeth, J. 1984. Determinants for protein localization: β-lactamase signal sequence directs globin across microsomal membranes. Proc. Natl. Acad. Sci. USA 81:456–60

Maccecchini, J.-L., Rudin, Y., Blöbel, G., Schatz, G. 1979. Import of proteins into mitochondria: Precursor forms of the extra-mitochondrially made F_1-ATPase subunits in yeast. Proc. Natl. Acad. Sci. USA 76:343–47

Matile, P., Moor, H., Robinow, C. F. 1969. Yeast cytology. In The Yeasts, ed. A. H. Rose, J. S. Harrison, Vol. 1, pp. 219–301. New York: Academic

McAda, P., Douglas, M. 1982. A neutral metallo-endo-protease involved in the processing of an F_1-ATPase subunit precursor in mitochondria. J. Biol. Chem. 257:3177–82

Mueller, S., Branton, D. 1984. Identification of coated vesicles in Saccharomyces cerevisiae. J. Cell Biol. 98:341–46

Müller, M., Ibrahimi, I., Chang, C. N., Walter, P., Blobel, G. 1982. A bacterial secretory protein requires signal recognition particle for translocation across mammalian endoplasmic reticulum. J. Biol. Chem. 257:11860–63

Müller, M., Müller, H. 1981. Synthesis and processing of in vitro and in vivo precursors of the vacuolar yeast enzyme carboxypeptidase Y. J. Biol. Chem. 256:11962–65

Nagy, M., Laporte, J., Penverne, B., Hervé, G. 1982. Nuclear localization of aspartate transcarbamoylase in Saccharomyces cerevisiae. J. Cell Biol. 92:790–94

Neupert, W. 1983. Transfer of proteins into and across the membranes of mitochondria. In New Perspectives on Membrane Dynamics. Proc. CNRS-INSERM Int. Symp. NATO Workshop, Strasbourg, 1983, pp. 246–48. Strasbourg: French Biochem. Soc.

Neupert, W., Schatz, G. 1981. How proteins are transported into mitochondria. Trends Biochem. Sci. 6:1–4

Notario, V., Kawai, H., Cabib, E. 1982. Interaction between yeast β-(1 → 3)-glucan synthetase and activating phosphorylated compounds. A kinetic study. J. Biol. Chem. 257:1902–5

Novick, P., Botstein, D. 1985. Phenotypic analysis of temperature sensitive yeast actin mutants. Cell 40:405–16

Novick, P., Ferro, S., Schekman, R. 1981. Order of events in the yeast secretory pathway. Cell 25:461–69

Novick, P., Field, C., Schekman, R. 1980. Identification of 23 complementation groups required for post-translational events in the yeast secretory pathway. Cell 21:205–15

Novick, P., Schekman, R. 1979. Secretion and cell-surface growth are blocked in a temperature-sensitive mutant of Saccharomyces cerevisiae. Proc. Natl. Acad. Sci. USA 76:1858–62

Novick, P., Schekman, R. 1983. Export of major cell surface proteins is blocked in yeast secretory mutants. J. Cell Biol. 96:541–47

Ohashi, A., Gibson, J., Gregor, I., Schatz, G. 1982. Import of proteins into mitochondria. The precursor of cytochrome c_1 is processed in two steps, one of them heme-dependent. J. Biol. Chem. 257:13042–47

Ohta, S., Schatz, G. 1984. A purified precursor polypeptide requires a cytosolic protein fraction for import into mitochondria. EMBO J. 3:651–57

Onishi, H. R., Tkacz, J. S., Lampen, J. O. 1979. Glycoprotein nature of yeast alkaline phosphatase. J. Biol. Chem. 254:11943–52

Orlean, P., Schwaiger, H., Appeltauer, U., Haselbeck, A., Tanner, W. 1984. Saccharomyces cerevisiae mating pheromones specifically inhibit the synthesis of proteins destined to be N-glycosylated. *Eur. J. Biochem.* 140:183–89

Pearse, B. M. F., Bretscher, M. S. 1981. Membrane recycling by coated vesicles. *Ann. Rev. Biochem.* 50:85–101

Perlman, D., Halvorson, H. 1981. Distinct repressible mRNAs for cytoplasmic and secreted yeast invertase are encoded by a single gene. *Cell* 25:525–36

Randall, L. L. 1983. Translocation of domains of nascent periplasmic proteins across the cytoplasmic membrane is independent of elongation. *Cell* 33:231–40

Reid, G., Yonetani, T., Schatz, G. 1982. Import of proteins into mitochondria. Import and maturation of the mitochondrial intermembrane space enzymes cytochrome b_2 cytochrome c peroxidase in intact yeast cells. *J. Biol. Chem.* 257:13068–74

Reitman, M. L., Kornfeld, S. 1981. Lysosomal enzyme targeting: N-acetylglucosaminylphosphotransferase selectively phosphorylates native lysosomal enzymes. *J. Biol. Chem.* 256:11977–80

Riezman, H. 1985. Endocytosis in yeast. Several of the yeast secretory mutants are defective in endocytosis. *Cell.* In press

Riezman, H., Hase, T., van Loon, A. P. G. M., Grivell, L. A., Suda, K., Schatz, G. 1983a. Import of proteins into mitochondria: a 70 kilodalton outer membrane protein with a large carboxy-terminal deletion is still transported to the outer membrane. *EMBO J.* 2:2161–68

Riezman, H., Hay, R., Witte, C., Nelson, N., Schatz, G. 1983b. Yeast mitochondrial outer membrane specifically binds cytoplasmically-synthesized precursors of mitochondrial proteins. *EMBO J.* 2:1113–18

Roberts, R. L., Bowers, B., Slater, M. L., Cabib, E. 1983. Chitin synthesis and localization in cell division cycle mutants of *Saccharomyces cerevisiae. Mol. Cell. Biol.* 3:922–30

Rothman, J. E., Fine, R. E. 1980. Coated vesicles transport newly-synthesized membrane glycoproteins from endoplasmic reticulum to plasma membrane in two successive stages. *Proc. Natl. Acad. Sci. USA* 77:780–84

Ruiz-Herrera, J., Lopez-Romera, E., Bartnicki-Garcia, S. 1977. Properties of the chitin synthetase in isolated chitosomes from yeast cells of *Mucor rouxii. J. Biol. Chem.* 252:3338–43

Runge, K. W., Huffaker, T. C., Robbins, P. W. 1984. Two yeast mutations in glucosylation steps of the asparagine glycosylation pathway. *J. Biol. Chem.* 259:412–17

Saunier, B., Kilker, R. D., Tkacz, J. S., Quaroni, A., Herscovics, A. 1982. Inhibition of N-linked complex oligosaccharide formation by 1-deoxynojirimycin, an inhibitor of processing glucosidases. *J. Biol. Chem.* 257:14155–61

Schatz, G., Butow, R. 1983. How are proteins imported into mitochondria? *Cell* 32:316–18

Schauer, I., Emr, S., Gross, C., Schekman, R. 1985. Invertase signal and mature sequence substitutions that delay intercompartmental transport of active enzyme. *J. Cell Biol.* In press

Schekman, R., Novick, P. 1982. The secretory process and yeast cell-surface assembly. In *The Molecular Biology of the Yeast Saccharomyces: Metabolism and Gene Expression*, ed. J. Strathern, E. Jones, J. Broach, pp. 361–93. Cold Spring Harbor, NY: Cold Spring Harbor Lab.

Schleyer, M., Schmidt, B., Neupert, W. 1982. Requirement of a membrane potential for the posttranslational transfer of proteins into mitochondria. *Eur. J. Biochem.* 125:109–16

Schwaiger, H., Hasilik, A., von Figura, K., Wiemken, A., Tanner, W. 1982. Carbohydrate-free carboxypeptidase Y is transferred into the lysosome-like yeast vacuole. *Biochem. Biophys. Res. Commun.* 104:950–56

Shematek, E. M., Bratz, J. A., Cabib, E. 1980. Biosynthesis of the yeast cell wall. I. Preparation and properties of β-$(1 \to 3)$ glucan synthetase. *J. Biol. Chem.* 255:888–94

Shematek, E. M., Cabib, E. 1980. Biosynthesis of the yeast cell wall. II. Regulation of β-$(1 \to 3)$ glucan synthetase by ATP and GTP. *J. Biol. Chem.* 255:895–902

Silver, P. A., Keegan, L. P., Ptashne, M. 1984. Amino terminus of the yeast *GAL4* gene product is sufficient for nuclear localization. *Proc. Natl. Acad. Sci. USA* 81:5951–55

Stevens, T., Esmon, B., Schekman, R. 1982. Early stages of the yeast secretory pathway are required for transport of carboxypeptidase Y to the vacuole. *Cell* 30:439–48

Stevens, T., Payne, G., Rothman, J., Schekman, R. 1985. Carboxypeptidase Y is secreted when a component involved in transport to the vacuole is over-loaded. Submitted for publication

Suissa, M., Schatz, G. 1982. Import of proteins into mitochondria. Translatable mRNAs for imported mitochondrial proteins are present in free as well as mitochondria-bound cytoplasmic ribosomes. *J. Biol. Chem.* 257:13048–55

Teintze, M., Slaughter, M., Weiss, H., Neupert, W. 1982. Biogenesis of mitochondrial ubiquinol: Cytochrome c reductase (cytochrome bc_1 complex). Precursor proteins and their transfer into mitochondria. *J. Biol. Chem.* 257:10364–71

Tkacz, J. S., Lampen, J. O. 1973. Surface distribution of invertase on growing *Saccharomyces* cells. *J. Bacteriol.* 113:1073–75

Trimble, R. B., Byrd, J. C., Maley, F. 1980. Effect of glucosylation of lipid intermediates on oligosaccharide transfer in solubilized microsomes from *Saccharomyces cerevisiae*. *J. Biol. Chem.* 255:11892–95

Tschopp, J., Esmon, P. C., Schekman, R. 1984. Defective plasma membrane assembly in yeast secretory mutants. *J. Bacteriol.* 160:966–70

Tuite, M. F., Plesset, J., Moldave, K., McLaughlin, C. S. 1980. Faithful and efficient translation of homologous and heterologous mRNAs in an mRNA-dependent cell-free system from *Saccharomyces cerevisiae*. *J. Biol. Chem.* 255:8761–66

Valenzuela, P., Medina, A., Rutter, W. J., Ammerer, G., Hall, B. D. 1982. Synthesis and assembly of hepatitis B virus surface antigen particles in yeast. *Nature* 298:347–50

Wen, D., Schlesinger, M. J. 1984. Fatty acid-acylated proteins in secretory mutants of *Saccharomyces cerevisiae*. *Mol. Cell. Biol.* 4:688–94

Wickner, W. 1979. The assembly of proteins into biological membranes: The membrane trigger hypothesis. *Ann. Rev. Biochem.* 48:23–45

Wiemken, A., Schellenberg, M., Urech, K. 1979. Vacuoles: The sole compartment of digestive enzymes in yeast (*Saccharomyces cerevisiae*). *Arch. Microbiol.* 123:23–35

Yaffe, M. P., Schatz, G. 1984. Two nuclear mutations which block mitochondrial protein import in yeast. *Proc. Natl. Acad. Sci. USA* 81:4819–23

Zimmerman, R., Hennig, B., Neupert, W. 1981. Different transport pathways of individual precursor proteins into mitochondria. *Eur. J. Biochem.* 116:455–60

Zubenko, G. S., Park, F. J., Jones, E. W. 1983. Mutations in *PEP4* locus of *Saccharomyces cerevisiae* block the final step in the maturation of two vacuolar hydrolases. *Proc. Natl. Acad. Sci. USA* 80:510–14

Zwizinski, C., Neupert, W. 1983. Precursor proteins are transported into mitochondria in the absence of proteolytic cleavage of the additional sequences. *J. Biol. Chem.* 258:13340–46

Zwizinski, C., Schleyer, M., Neupert, W. 1984. Proteinaceous receptors for the import of mitochondrial precursor proteins. *J. Biol. Chem.* 259:7850–56

Ann. Rev. Cell Biol. 1985. 1 : 145–72

MICROTUBULE ORGANIZING CENTERS

B. R. Brinkley

Department of Cell Biology and Anatomy, University of Alabama in Birmingham, Birmingham, Alabama 35294

CONTENTS

1. PERSPECTIVES AND SUMMARY

Microtubules, one of three major cytoskeletal components in eukaryotic cells, are implicated in a variety of functions, including ciliary and flagellar movement, cell motility and cytoplasmic streaming, chromosome movement, maintenance of cell shape and form, intracellular and axoplasmic transport, and anchorage of cell surface receptors. It is not surprising, therefore, that microtubules are displayed in a variety of forms and arrangements inside plant and animal cells. Unlike the rigid image portrayed in electron micrographs, most microtubules are dynamic structures capable of being rapidly assembled and disassembled, and spatially arranged and rearranged to accommodate an ever-changing cytoplasm. It was obvious from early electron microscopic studies that microtubules were not arranged randomly in cells but were organized around one or more discrete foci (Porter 1966). These sites have been given a variety of names, such as microtubule initiating sites (Porter 1966), microtubule nucleating sites (Tilney & Goddard 1970), microtubule centers (Tilney 1971), microtubule generators (Wolfe 1972), and microtubule

145

0743–4634/85/1115–0145$02.00

organizing centers (Pickett-Heaps 1969). The latter term, generally expressed by the acronym MTOC, has become widely accepted in the microtubule literature and, from every indication, provides an apt description of this cellular site (see reviews by Tucker 1979, 1982, 1984). MTOCs are not only locations where microtubule assembly is believed to be initiated, but they may also be sites where free ends of microtubules are attracted or "captured" (Pickett-Heaps et al 1982). Thus, the term "organizing centers" seems appropriate because it implies a general site for gathering microtubules into arrays.

This report examines MTOCs from a variety of viewpoints, including their status as: unique structural entities with varying forms and distributions in plant and animal cells; preferred sites for the initiation, assembly, anchorage, and stabilization of microtubules and microtubule-associated molecules; structural templates for organizing microtubules with defined polarity and distribution; and endogenous determinants of cell morphology transmitted as inheritable units from one cell generation to the next.

2.   VARIATION IN FORM AND DISTRIBUTION IN CELLS

Microtubule organizing centers can have many different appearances in cells (Figures 1–7). The most easily recognizable ones are discrete foci in the cytoplasm, such as the centrosome of interphase cells (Figure 1) or the poles of the mitotic spindle (Figure 2), where large numbers of microtubules converge into a zone occupied by amorphous, electron-dense material and organelles (such as centrioles, and in some cases, smooth ER and Golgi). MTOCs themselves can be microtubule arrays, such as the pinwheel-structured basal bodies (Figure 3a,b), which elongate at their proximal ends to give rise to the $9+2$ axoneme, or form plaques at their distal ends to generate cytoplasmic microtubule arrays. MTOCs can appear as bodies of electron-dense material, which may be freely dispersed in the cytoplasm (Figure 3c) or localized around centrioles as pericentriolar material. Apparently, some cells can convert centralized, focused MTOCs into dispersed forms, and then restore them later in the cell cycle. MTOCs can appear as compact, flat, laminated plaques, such as the kinetochores (Figure 4) of many eukaryotic chromosomes or the spindle pole bodies of fungi. MTOCs can also exist as groupings of finely dispersed particles (Peterson & Berns 1980) or as fibrous rings (Russell & Burns 1984).

The best documented MTOC in eukaryotic cells is the centrosome (Mazia 1984; McIntosh 1983). Although cells vary in their display and expression of a centrosome, most animal cells and some plant cells contain one or more of these organelles. Boveri (1901) described the centrosome as "an autonomous permanent organ of the cell—the dynamic center of the cell—the true division center [whose] division creates the center of the daughter cell" (from Mazia 1984).

Today the centrosome is viewed as a major organizer of the cytoskeleton (McIntosh 1983), and one of several principle MTOCs of the cell. This view is based on electron microscopic studies, as well as observations using antitubulin antibodies or, more recently, autoantibodies specific for the centrosome (Connolly & Kalnins 1978; Brenner & Brinkley 1982; Calarco-Gilliam et al 1983). In addition, the microtubule initiating capacity of the centrosome and some of its components has been demonstrated in situ

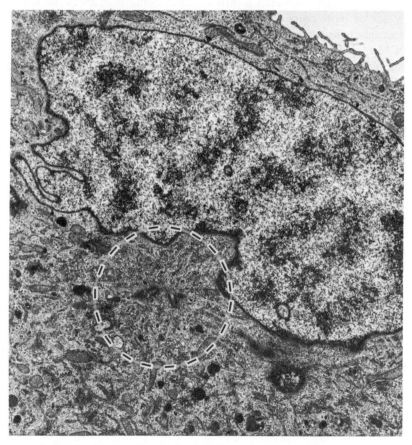

Figure 1 Survey electron micrograph of PtK$_1$ rat kangaroo cell in prophase. Note conspicuous centrosome (outlined in circle) near the nucleus.

using lysed cell models and exogenous tubulin (Brinkley et al 1981a,b; Deery & Brinkley 1983; Pepper & Brinkley 1980; Mitchison & Kirschner 1984a). Such approaches have given us important new insight into location, function, and cyclic behavior of the centrosome. Unlike the view held earlier by investigators, the centrosome is no longer thought of as a compact, centralized body stationed permanently in the center of the cytoplasm, but as a dynamic, replicating structure that can assume different forms and locations in different cells, and even within a single cell, during its life history (Mazia 1984).

As shown in Figure 1, the centrosome of animal cells is usually close to the nucleus, and may be identified by several structural features such as converging microtubule profiles, aggregations of smooth-surfaced vesicles, and even Golgi in some cells. Centrosomes in animal cells are distinguished by one or more pairs of centrioles surrounded completely or in part by electron-dense, amorphous pericentriolar material (PCM) (Figures 5, 6h). PCM may exist as a dense layer surrounding the centrioles, or as large granular masses (satellites) (De Harven 1968; Brinkley & Stubblefield 1970; Wolfe 1972), or as centriolar "appendages" dispersed around the centrioles (Figure 7d). Higher plant cells usually lack centrioles and a

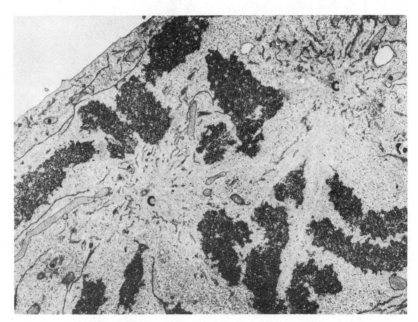

Figure 2   PtK$_1$ cell in prophase. Note the two spindle poles (C) accentuated by spindle asters, ER membranes, and fragments of nuclear envelope. (From Brinkley et al 1976.)

centrosome, but instead display large, more dispersed membranous vesicles with microtubule organizing capacity (Jackson & Doyle 1982).

Close examination of the centrosome by electron microscopy shows that microtubules rarely originate from the centrioles per se, but rather from the PCM (Gould & Borisy 1977). Thus, centrioles, which are not even present

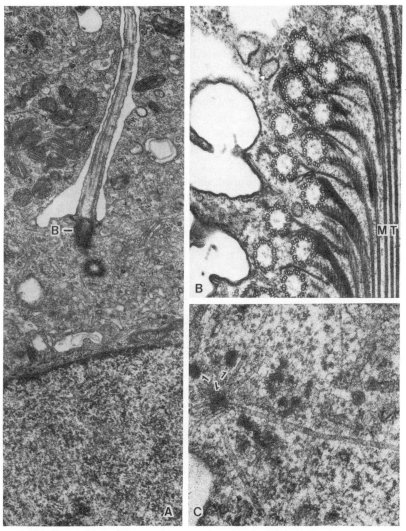

Figure 3   MTOCs from various cells. (*a*) Cilium in a Chinese hamster cell (strain Don C) *in vitro* showing basal body; (*b*) basal bodies in *Stentor* (courtesy of Dr. Bessie Huang); (*c*) microtubules extending from electron dense satellites in a Chinese hamster Don cell.

in many cells (for review see Peterson & Berns 1980), do not appear to be part of the microtubule organizing component of centrosomes. In fact, irradiation of centrioles with a laser microbeam has no apparent effect on the centrosome's function as an MTOC (Berns & Richardson 1977). Centrioles are, however, integral components of the centrosomes of many cells and, as Mazia (1984) reminds us, provide the morphological "address" and "valence" of the centrosome. As such, one pair of centrioles usually indicates a nonreplicated centrosome, two pairs, a replicated one. Therefore, the reproduction and maturation cycle of centrioles (Vorobjev & Chentsov 1982) coincides closely with that of the centrosome (Rieder & Borisy 1982).

Cells such as neuroblastoma have multiple centrioles (Sharp et al 1981; Brinkley et al 1981a). The centrioles may be confined to one large centrosome that functions as a single MTOC, or be dispersed throughout the cytoplasm where they serve as multiple MTOCs. Therefore, centrioles are convenient markers for MTOCs. Since centrioles are the precursors of basal bodies (Wolfe 1972), they are capable of becoming MTOCs independent of centrosomes during differentiation. Likewise, centrosomes can function as MTOCs independent of centrioles.

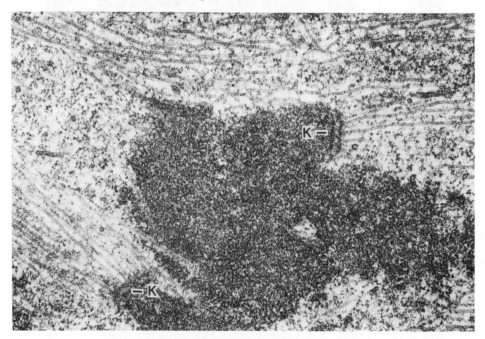

Figure 4 Kinetochores (K) with associated microtubules of PTK$_1$ chromosome at metaphase. (From Brinkley & Cartwright 1971.)

It has long been known that centrosomes undergo duplication to produce a duplex or diplosome prior to their separation to form the poles of the mitotic spindle (Figure 2). During this transition, cytoplasmic microtubules radiating from the centrosome disappear, and are replaced by large numbers of highly oriented microtubules that form the polar asters and spindle. The spindle poles, although derived from the centrosome, differ from the interphase centrosome in the shape, organization, and distribution of the microtubules that form around them. When mitosis is completed, the spindle microtubules disappear, and the poles are once again transformed into centrosomes, which give rise to an array of cytoplasmic microtubules called the cytoplasmic microtubule complex (CMTC) (Brinkley et al 1975). The amount of PCM also varies during the cell cycle; there is more present during mitosis than at interphase (Rieder & Borisy 1982). Thus, the centrosome–spindle pole cycle seen in most eukaryotic cells provides an excellent example of the structural and functional flexibility of MTOCs, and the commanding role they play in regulating the spatial distribution of microtubules throughout the life cycle of cells.

The centrosome reproduction cycle, beautifully shown by indirect immunofluorescence using anti-centrosome antibodies (Figure 6a–f), illustrates the way in which cells maintain the continuity of MTOCs from

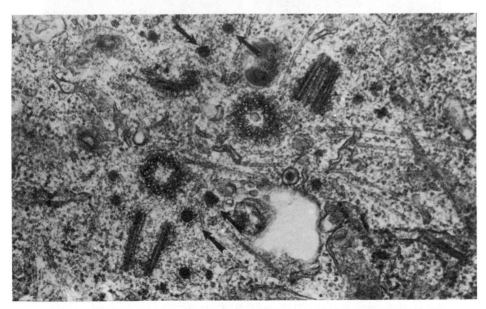

Figure 5 Higher magnification of a centrosome showing two sets of centrioles surrounded by electron dense satellites and virus-like particles (*arrows*). (From McGill et al 1976.)

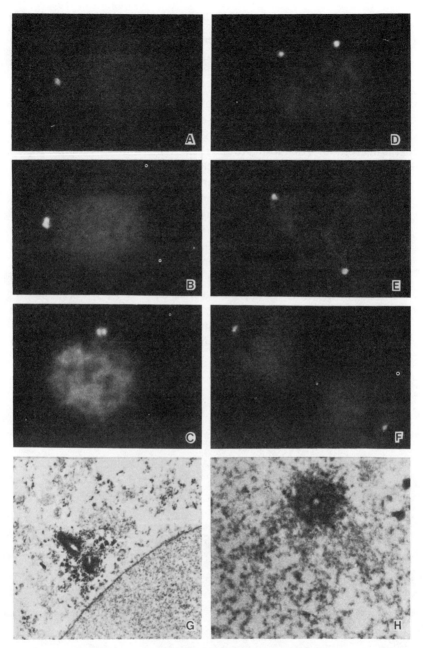

Figure 6 Localization of centrosomes in PtK$_2$ cells throughout the cell cycle (*a*). Interphase before centrosome duplication; (*b*) centrosome at the time of duplication; (*c*) post-duplication; (*d*) separation of duplicated centrosomes at prophase; (*e*) metaphase; (*f*) telophase; (*g,h*) electron micrograph of immunoperoxidase stained centrosome. (From Brenner & Brinkley 1982.)

one generation to the next. Mitosis provides each daughter cell with one major MTOC. During the subsequent interphase the centrosome is duplicated, providing the coming mitosis with two spindle poles, and the next generation of daughter cells with centrosomes.

Although recognized as a major MTOC, the centrosome appears to exert a more widespread influence on the organization of cytoplasm than just the arrangement and distribution of microtubules. For example, its position in the cytoplasm can influence intracellular movement, cell locomotion (Tucker 1984; McIntosh 1983), possibly the distribution of intermediate filaments (Goldman et al 1980; but see Eckert et al 1984), and the accumulation and positioning of membranes and organelles (Mazia 1984).

3. MTOCs ARE PREFERRED SITES FOR MICROTUBULE ASSEMBLY

Although microtubule subunits appear to be widely dispersed throughout the cytoplasm, they seldom polymerize randomly in cells. It is well established that MTOCs bind to microtubules and microtubule proteins, and assembly of microtubules is somehow preferred at these centers over other regions of the cytoplasm. This generalization was first suggested by early electron microscopic observations showing microtubules oriented toward these sites. There is now strong experimental evidence that MTOCs are indeed preferred sites for microtubule initiation and assembly.

Most evidence involves the regrowth of microtubules in cells during recovery from inhibitors such as colcemid (Brinkley et al 1967), nocodazole (De Brabander et al 1981), and cold temperatures (Rieder & Borisy 1981). For example, when exponentially growing Chinese hamster cells were treated with 0.06 $\mu g/ml$ colcemid for 2 hr and examined by electron microscopy, mitotic arrest resulted in the chromosomes becoming distributed around a single mitotic center (centrosome) (Brinkley et al 1967). Such monopolar cells displayed two centriole pairs at the cell center, with short microtubules extending radially to kinetochores that were facing the centrosome. Kinetochores not facing the centrosome were free of microtubules (see Mazia 1984 for references). When colcemid was washed from the cells, the centriole pairs separated and moved to opposite poles with bundles of microtubules forming between them. Both sets of sister kinetochores now became associated with microtubules.

Using higher doses of colcemid and examining thicker serial sections viewed with high voltage EM, Witt et al (1980) disrupted all microtubules in the cells, and observed their regrowth at kinetochores and centrosomes at brief intervals following recovery from the drug. These experiments argued against the possibility that microtubule fragments were responsible for initiation of microtubule assembly, and showed convincingly that microtubule regrowth occurred separately and independently at both kineto-chores and centrosomes. Recently, the widely held notion that kinetochores actually nucleate the assembly of microtubules has been seriously ques-tioned (Pickett-Heaps et al 1982). This issue will be discussed later in this review.

De Brabander and coworkers (1980, 1981) made important observations concerning microtubule regrowth from MTOCs in mammalian cells following recovery from nocodazole. Electron microscopic observation of ultrathin sections at brief intervals following recovery from nocodazole revealed that initially, short fragments of microtubules assembled ran-domly in a zone peripheral to the kinetochore plate. Later, the microtubule fragments elongated and became more parallel in their orientation. At this stage they appeared firmly attached to the kinetochore. De Brabander et al concluded that microtubule assembly, which is a two-step process involv-ing initiation and subsequent elongation (Bryan 1976), is spatially and temporally partitioned within the MTOC. These initial EM studies strengthened the view that MTOCs are major sites for microtubule assembly, and laid the foundation for further experiments.

The development of procedures for achieving in vitro assembly of microtubules from purified brain tubulin (Weisenberg 1972) represented a major achievement in the field of microtubule research. This technical milestone was necessary before investigators could begin to approach the more complicated issue of assembly in vivo. Given the appropriate conditions for assembly of microtubules in vitro, the capacity to initiate and maintain assembly depends upon the availability of a critical concentration of microtubule subunits (tubulin). This concentration is influenced by the presence of a number of factors, including MAPs, drugs, and nucleating seeds such as microtubule fragments. For example, tubulin purified by phosphocellulose chromatography (PC-tubulin) is generally incompetent to assemble in vitro unless very high concentrations (10 mg/ml) of the subunit are utilized. If MAPs are added to the assembly mixture in appropriate concentrations, however, polymerization proceeds at a cri-tical concentration of 0.3 mg/ml. Other factors, such as glycerol and high magnesium (Frigon & Timasheff 1975a,b), DMSO (Himes et al 1977), and the drug taxol (Schiff et al 1979), can also influence the critical concen-tration in vitro.

MTOCs, therefore, may be sites in cells where the critical concentration is lower than the surrounding cytoplasm, thereby favoring microtubule initiation and assembly. According to one model, MTOCs represent unique microenvironments in the cytoplasm where the "threshold for assembly is lowered" (De Brabander et al 1980). Thus, they may be foci that contain the appropriate assembly promoting factors, such as MAPs that maintain the essential ionic environment, and trinucleotides needed for assembly. Some features of this model can now be tested.

During the 1970s techniques were first developed for the analysis of microtubule assembly from isolated MTOCs. Weisenberg & Rosenfeld (1975a,b) first demonstrated that MTOCs could be isolated in a functional state from developing eggs of the surf clam *Spisula solidissima*. During the same period, experiments from several other laboratories demonstrated that exogenous tubulin could be assembled onto isolated organizing centers, such as sea urchin sperm and tail fragments (Burns & Starling 1974), and flagellar outer doublet microtubules (Kuriyama 1976). The assembly of brain tubulin onto flagella microtubules isolated from sea urchin sperm and *Chlamydomonas* (Allen & Borisy 1974; Binder et al 1975), and onto basal bodies of *Chlamydomonas* (Snell et al 1974; Stearns et al 1976), illustrated that the polarity of assembly in vivo was maintained in vitro. Cande et al (1974) and Inoue et al (1974) used lysed mitotic cells, and Rebhun et al (1974) used isolated spindles of *Spisula*, to demonstrate that vertebrate tubulin could be assembled onto tubules of the mitotic spindle.

In the mid-1970s several laboratories developed procedures for analyzing site-associated microtubule assembly using purified microtubule protein and extracted and partially isolated MTOCs (McGill & Brinkley 1975; Telzer et al 1975; Snyder & McIntosh 1975; Gould & Borisy 1977; Pepper & Brinkley 1979, 1980). These important technical achievements, coupled with the development of tubulin antibodies and other probes for immunofluorescent studies (Fuller et al 1975; Brinkley et al 1975; Weber et al 1975), accelerated research on microtubules and MTOCs.

Using tubulin immunofluorescence entire arrays of microtubules could be visualized in cells by light microscopy (Figure 7a–c), replacing or reducing the need for the tedious techniques of electron microscopy to visualize microtubules. Immunofluorescent images of most cells further strengthened the notion that microtubules were organized around localized MTOCs. Two microtubule arrays could be detected in proliferating cells, the microtubules of the mitotic apparatus (MA) and an elaborate tubule network of interphase cells called the cytoplasmic microtubule complex (CMTC) (Brinkley et al 1975). As cells complete mitosis, the microtubules of the MA disappear and are replaced by the CMTC. As the cells enter mitosis again, the CMTC is disassembled and replaced by a MA. The cell cycle–dependent change in microtubule arrays (i.e. MA or CMTC) typically occurred in association with one or two foci. Ostensibly, therefore, MTOCs not only serve as sites for the assembly and anchorage of microtubules, but can also influence the expression (i.e. shape and organization) of the microtubule arrays that grow about them. Indeed, experiments in recent years provide ample evidence that the number, length, polarity, distri-

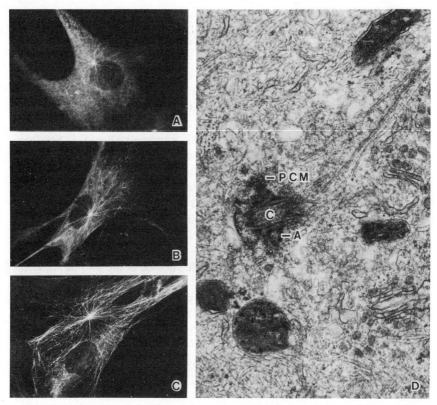

Figure 7   Microtubule reassembly in PtK$_1$ cells following recovery from exposure to 0.06 μg/ml colcemid. (*a*) 10 min after recovery; (*b*) 30 min after recovery; (*c*) 60 min after recovery. Note extensive CMTC as shown by antitubulin staining. (*d*) Electron micrograph of cell in recovery showing centriole (C) appendage (A) and pericentriolar material (PCM). (From Brinkley et al 1978.)

bution, and even lattice structure of microtubules is greatly influenced by MTOCs.

4.   MTOCs AS STRUCTURAL TEMPLATES

There is considerable evidence that the number of microtubules in cells is strictly regulated (see Tucker 1984). Thus, not only are MTOCs capable of generating microtubules, but they also appear to keep count of the number of microtubules that grow from them. Tucker (1977) has defined components of MTOCs that specify microtubule number as microtubule *nucleating elements*. According to this hypothesis, each microtubule nucleating element nucleates the assembly of one microtubule. Using tubulin immunofluorescence, Anderson et al (1983) demonstrated that isolated human polymorphonuclear leukocytes in vitro maintained a small but constant number of cytoplasmic microtubules ($\sim 32.4 \pm 4.5$ MTs per cell), whether the cells were stationary or were induced to undergo movement by chemotactic stimuli. The length distribution of microtubules extending from the MTOCs (centrosomes) was found to vary depending upon cell shape and direction of movement. Microtubules parallel to the axis of movement elongated, while those perpendicular to the axis shortened to accommodate cell elongation. The basic number of microtubules, however, remained essentially constant. Apparently, MTOC-associated microtubules can undergo simultaneous lengthening and shortening at the organizing center, a property also observed at the poles of the mitotic spindle during anaphase chromosome movement and spindle elongation. As we will discuss later, simultaneous elongation and shortening in vitro may be a unique kinetic property of MTOC-associated microtubules that can be demonstrated using isolated centrosomes (Mitchison & Kirschner 1984a).

Other examples of the constancy of microtubule number in cells include the familiar $9 + 0$ arrangement in centrioles and basal bodies, and the $9 + 2$ organization of the axonemes of cilia and flagella. The number of microtubules composing the spindle of diploid cells remains essentially constant for each species (McIntosh & Landis 1971; Brinkley & Cartwright 1971; Roos 1981). Spindles of polyploid cells have larger numbers of microtubules than their diploid counterparts (personal observations); the lengths of spindle microtubules in grasshopper spermatocytes have been shown to depend upon the number of chromosomes in the cells (Nicklas & Gordon 1985). The relationship of the latter observations to MTOCs is unknown. In *Chlamydomonas reinhardtii*, 12 cytoplasmic microtubules are arranged precisely into 4 bundles or rootlets that extend from the basal body into the cytoplasm (Tucker 1984). Probably the microtubules are maintained in a relatively constant number in most cells, and the number may increase or even double as the cells progress through the cell cycle.

Several investigators have shown experimentally that MTOCs can specify the number of microtubules grown from them. Analysis of spindle pole bodies isolated from the yeast *Saccharomyces* indicated that MTOCs initiated about the same number of microtubules as found in vivo (Hyams & Borisy 1978; Byers et al 1978). Both the number of microtubules and their positions on the MTOC are strictly controlled by isolated centrosomes of the alga *Polytomella* (Stearns & Brown 1981).

Studies of microtubule assembly in lysed cell models have also favored the notion that MTOCs exert control over the number, spatial orientation, and length distribution of microtubules grown from them. Brinkley and coworkers (1981a,b) demonstrated that MTOCs of lysed 3T3 cells retained their capacity to nucleate the assembly of PC-tubulin into microtubule arrays of a finite length, number, and distribution. Exponentially growing 3T3 cells were lysed in Triton X-100 detergent, after exposure to colcemid for 2 hr to disrupt cytoplasmic microtubules. When the lysed cells were incubated in a reassembly buffer containing bovine brain tubulin, microtubules regrew from 1 or 2 central foci (centrosomes) in a radial pattern, filling the cytoplasm with microtubules (Figure 8). The display in most cells was such that it was possible to count and measure the lengths of all microtubules as a function of time of incubation and concentration of tubulin. As shown in Figures 9a and b, both the length and number of microtubules increased with increasing concentrations of tubulin, and reached a plateau at about 1.0 mg/ml.

The leveling off of microtubule lengths, presumably at steady state, implies some type of "capping" mechanism at work in this system. Interestingly, the microtubules ceased elongation in the cell cortex near the

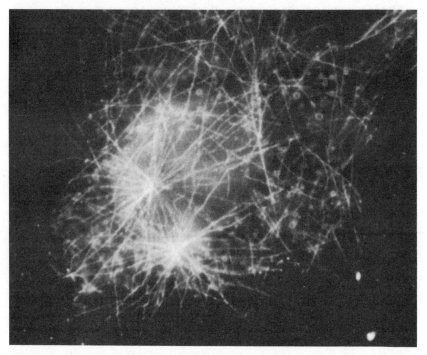

Figure 8   Microtubule pattern in lysed 3T3 cells after incubation with pure bovine brain tubulin. Bright foci represent MTOCs (centrosomes). (From Brinkley et al 1981b.)

plasma membrane, as if this region presented a morphological impass, yet the holes in the plasma membrane were sufficiently large to facilitate the continuous elongation of microtubules beyond the boundaries of the cell. If the MTOC was offset from the cell center, as was often the case, microtubules extending from the face of the MTOC furthest from the cell periphery were considerably longer than those nearest the periphery. Thus, if "capping" factors exist, they may reside in the cortical cytoplasm. Moreover, since the number of microtubules per MTOC also reached a plateau, we concluded that each organizing center contains a finite, saturable number of nucleating sites.

When the smaller transformed SV3T3 cells were lysed and exposed to the same reassembly conditions, considerably fewer and shorter microtubules reformed about the MTOCs (Brinkley et al 1981b). Interestingly, if cAMP was added to the SV3T3 cells along with the reassembly mixture a larger microtubule array approximating that of 3T3 cells was generated (Tash et al 1980). The cyclic nucleotide had no effect on the microtubule dimensions generated by 3T3 cells. These results suggest that number and length of MTOC-associated microtubules may be limited in some cells by the activity of a cyclic nucleotide-dependent kinase.

There is growing evidence that factors other than conventional MTOCs can influence the assembly and spatial distribution of cytoplasmic micro-

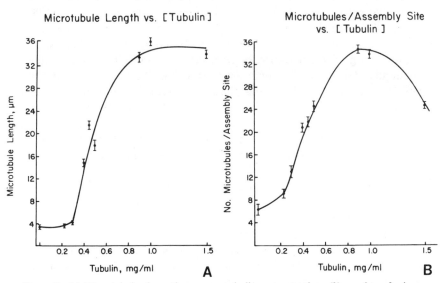

Figure 9 (*a*) Microtubule elongation versus tubulin concentration; (*b*) number of micro-tubules per MTOC versus tubulin concentration. Note that both show saturation at about 1.0 mg/ml of tubulin. (From Brinkley et al 1981b.)

tubules. In some cells, microtubules appear to assemble at dispersed sites in the cytoplasm independent of centrosomes, and in others both centrosomal and noncentrosomal microtubules coexist in the same cytoplasm (Karsenti et al 1984). Several investigators have shown that microtubule assembly can be experimentally uncoupled from the MTOC. Using a more gentle detergent, Brij-58, to lyse cells, Deery & Brinkley (1983) maintained and preserved much of the cell's endogenous tubulin. When colcemid-treated, Brij-lysed cells were incubated in reassembly buffer minus tubulin at pH 7.6, the endogenous tubulin repolymerized in a site-associated manner around the centrosome. However, if the incubation was carried out at pH 6.9, assembly occurred randomly throughout the cytoplasm. Little if any reassembly occurred at the centrosomes. A similar effect was noted earlier in living cells allowed to recover from nocodazole at acid pH (De Brabander et al 1981). Taxol also induces dispersed microtubule assembly in the cytoplasm, and suppresses the regrowth around localized MTOCs (De Brabander et al 1981; Brenner & Brinkley 1982).

Thus, in addition to the MTOC, the biochemical properties of the surrounding cytoplasm also influence the pattern of microtubule organization. The cytoplasmic ground substance or microtrabeculae (Wolosewick & Porter 1979) may play a major role in organizing the cytoskeleton. Avian erythrocytes display a marginal band of microtubules that runs along the circumference of the cell. These microtubules do not appear to be associated with a typical MTOC at either end. Miller & Solomon (1984) disrupted marginal band microtubules in avian erythrocytes, and then incubated the lysed cells in reassembly buffer containing purified bovine brain tubulin. The microtubules reformed in a marginal band pattern essentially identical to that of intact cells. In the absence of a phenotypic MTOC, it must be concluded that the spatial organization of microtubules in cells such as erythrocytes is regulated by dispersed templates in the matrix or ground substance of the cytoplasm.

The lysed cell models leave much to be desired in terms of a clean, biochemically characterized system for investigating microtubule assembly and distribution in cells. Nevertheless, these systems not only proved that MTOCs are truly preferred sites for the anchorage and assembly of microtubules, but that some "information content" exists within these sites, and in surrounding cytoplasm, that confers order and organization to the microtubule array. Since the pattern of reassembled microtubules is characteristic of cell type, and independent of the species of tubulin used in these reconstitution experiments, we must conclude that the template residues within the cytoplasm and its MTOC.

If MTOCs were removed from their surrounding cytoplasm, how much "information content" would they retain? Obviously, the answer to this question requires the successful isolation of MTOCs that are free of cytoplasmic contamination but retain the capacity to nucleate microtubule assembly. An early effort in this regard was the isolation of functional MTOCs from the alga *Polytomella agilis* (Stearns & Brown 1979). This organism contains a complex MTOC consisting of 4 basal bodies and 8 rootlets extending radially and posteriorly. Several hundred microtubules are embedded in the amorphous material along the rootlets, and extend toward the posterior just beneath the plasma membrane. If the rootlet complex is extracted to remove the amorphous material it looses its capacity to nucleate exogenous tubulin. The amorphous material, however, retains the capacity for microtubule nucleation. If the material is added back to the rootlet complex, it reassociates with the rootlets and restores the nucleation capacity to the entire complex (Stearns & Brown 1979, 1981; Stearns et al 1976). Chemical dissection of these MTOCs revealed a set of high molecular weight polypeptides, which could increase the efficiency of tubulin initiation in vitro (Stearns & Brown 1979). Furthermore, both the position and number of microtubules nucleated by these MTOCs was rigidly maintained in the isolated complex.

Spindle pole bodies isolated from *Saccharomyces* maintained the capacity to nucleate approximately the same number of microtubules as assembled in vivo (Hyams & Borisy 1978; Byers et al 1978).

Recently, Mitchison & Kirschner (1984a) succeeded in isolating functional centrosomes from cultured mammalian cells. Examination of these centrosomes by electron microscopy showed that they consisted of centriole pairs surrounded by pericentriolar material, and were essentially free of contaminating microtubule fragments and cytoplasmic debris. When the centrosomes were incubated with brain PC-tubulin, microtubules grew from the center outward in a radial pattern much like that seen in living cells or lysed cell models. The number and length of microtubules assembled per centrosome increased with tubulin concentration, starting at 2 μM and reaching a plateau at 20 μM. The shape of the curve and the existence of a plateau indicate a saturation of nucleating sites, much like that reported in lysed cell models (Brinkley et al 1981a,b) (see Figure 9). Using this system the authors defined a "steady state concentration" as a concentration of tubulin monomer that exists when polymer is assembled to steady state (Hill & Kirschner 1982). When corrected for inactive tubulin in the preparation, the steady state concentration was found to be 14 μM, considerably above the monomer concentration needed to initiate microtubule assembly from centrosomes (5 μM) in this system. It should be noted that free microtubules are unstable below 14 μM, which implies that the isolated centrosomes retained their status as preferred sites for assembly and stabilization.

Perhaps the most surprising result of this study was the discovery of a dynamic instability in microtubules when the tubulin concentration was diluted. When centrosome preparations with microtubules regrown to saturation were diluted with buffer, tubules were expected to undergo disassembly, and then reach a new steady state in which there would be fewer microtubules. Indeed, they found that the number of microtubules per centrosome decreased by 40%, but surprisingly, the average length of the remaining microtubules increased by 40%. Thus, at tubulin concentrations well below the steady state level, some centrosome-associated microtubules were shrinking while others were actually growing. This unusual effect termed a "microtubule catastrophe" (McIntosh 1984) appears to reveal some very unexpected properties of MTOCs. For example, the centrosome can remain the center of a dynamic aster at tubulin concentrations below those that lead to the rapid depolymerization of free microtubules. The mechanisms responsible for the simultaneous shrinkage and growth of microtubules around MTOCs is uncertain, as is the possibility that such unusual dynamics actually exist inside the living cell. In a companion paper, Mitchison & Kirschner (1984b) observed similar kinetics using microtubule populations assembled in vitro with PC-tubulin added to microtubule seeds or ciliary axonemes. Several mechanisms can be proposed to explain these interesting results, but a model that involves the capping of microtubule ends with unhydrolyzed GTP-tubulin, as proposed by Carlier and coworkers (1981, 1984), was favored by the investigators. Although the mechanism that regulates the lengths of microtubules, as well as those that allow simultaneous growth and shrinkage of microtubules, is presently unknown, these elegant experiments, carried out in vitro, confirm and extend the notion that MTOCs play a major role in regulating the spatial distribution of microtubules.

What other kinds of information can be imparted to microtubules by the MTOCs? It is now well established that not all microtubules are identical. Many cells contain more than one population of microtubules, based upon tubulin content (Gozes et al 1979; Krauhs et al 1981; Ponstingl et al 1981), monoclonal antibody specificity (Asai et al 1982; Thompson et al 1984; Gundersen et al 1984), and response to inhibitors and cold temperatures (Brinkley & Cartwright 1975). Not all microtubules have the same lattice structure. Examination of transverse sections of microtubules by electron microscopy, especially after tannic acid post-staining, shows that most microtubules in vivo contain 13 protofilaments in their lattice, but others may display more or less than 13 (McEwen & Adelstein 1980, Tilney et al 1973; Burton et al 1975; Peterson et al 1978; Scheele et al 1982). The binding sites for the microtubule-associated proteins, MAP-2 and Tau, on the tubule surface are highly specific, and their spatial distribution along the wall may be influenced by tubulin isoforms that make up the surface lattice (Kim et al 1979). In addition, the tubule lattice must accommodate the dynamic exchange of tubulin subunits during assembly and disassembly, and facilitate the interaction of microtubules with other cell organelles.

Recent experiments by Evans et al (1985) suggest that MTOCs do influence the lattice structure of microtubules. Utilizing isolated centrosomes from mammalian cells, and a tubulin concentration (1.2 mg/ml) that favored assembly at centrosomes as well as spontaneous assembly of free microtubules, they demonstrated that microtubules containing 13 protofilaments originated from the centrosome, while those containing 14 formed spontaneously. Since they took care to assure that no residual microtubule fragments, which might function as nucleating seeds, remained with the centrosome preparation, they concluded that the isolated centrosomes could not only support the initiation and assembly of microtubules, but could in some way influence the assembly of microtubules with a more physiological lattice structure. Whether or not MTOCs can construct a microtubule that contains multiple tubulin isoforms remains to be determined. Nevertheless, these observations support the hypothesis that MTOCs are true templates that not only control the number, length, and distribution of microtubules, but also their lattice structure.

Microtubules are polar organelles, and their functions in certain motile processes, as well as their assembly and distribution, may depend on their polar arrangement in cells (McIntosh 1981, 1983). Recent studies utilizing ingenious techniques for morphological detection of microtubule polarity showed a consistent polar arrangement of microtubules with MTOCs. Utilizing polarity-revealing "hooks" (Heidemann & McIntosh 1980; Euteneuer & McIntosh 1981), or dynein decoration of microtubules (Haimo et al 1979) it has been shown that the slow-growing ends of microtubules are embedded in the centrosome, while their plus or fast-growing ends extend away from the centrosome. Apparently, all microtubules extending from the centrosome are parallel with respect to polarity; antiparallel arrangements have not been detected. The polarity of microtubules emanating from the kinetochore appears to be the reverse of that seen at the centromere. Here the plus ends of the microtubules insert into the kinetochore plate, and the minus ends extend toward or insert into the spindle poles. This arrangement is consistent with the proposal that kinetochores capture the ends of microtubules extending from the mitotic poles, and are not sites for the initiation of microtubule assembly (Pickett-Heaps et al 1982). Other equally tenable explanations are possible however, and the significance of this finding must await further experimentation.

Nevertheless, from these limited observations it is safe to conclude that MTOCs greatly influence the polarity of microtubules emanating from them.

5. BIOCHEMISTRY OF MTOCs

The composition of centrosomes, kinetochores, and polar plaques is uncertain. It has been suggested, however, that MTOCs contain tubulin (Staprans & Dirksen 1974; Dirksen & Staprans 1975; Gould & Borisy 1977; Pepper & Brinkley 1977, 1979), other proteins (Anderson & Floyd 1980; Lin et al 1981; Turksen et al 1982; Shyamala et al 1982; Ayer & Fritzler 1984; Earnshaw et al 1984; Guldner et al 1984; Valdivia & Brinkley 1984; Vandre et al 1984), steroid hormones (Nenci 1978; Nenci & Marchetti 1978), ATPase (Hartman 1964; Anderson 1977), and RNA (Hartman et al 1974; Hartmann & Zimmerman 1974; Heidemann et al 1977; Peterson & Berns 1978; Bielek 1978; Rieder 1979).

The presence of tubulin in MTOCs is supported by the characteristic triplet microtubules in centrioles and basal bodies. Electrophoretic analysis of purified basal bodies reveals a prominent band of tubulin (Anderson & Floyd 1980). Furthermore, centriolar microtubules can elongate in a tubulin-assembly solution (Gould & Borisy 1977), and large amounts of tubulin are synthesized during ciliary multiplication (Staprans & Dirksen 1974; Dirksen & Staprans 1975; Lefebvre et al 1980). Tubulin has been immunofluorescently located in centrosomes and kinetochores using a polyclonal antibody against 6S tubulin that renders MTOCs incapable of nucleating microtubule assembly (Pepper & Brinkley 1977, 1979).

Other major proteins associated with centrioles from chicken oviduct are characterized electrophoretically as 17.4, 90, and 180 kDa (Anderson & Floyd 1980). In the ciliated protozoan, *Tetrahymena*, and chicken tracheal epithelial cells, there are additional prominent proteins of 35 and 45–47 kDa (Turksen et al 1982). A centriole-associated protein in the same species has been immunofluorescently localized with nonimmune rabbit sera. This protein corresponds to 50,000 daltons on electrophoretic immunoblots (Turksen et al 1982). An antigen that is immunocytochemically localized on centrioles and basal bodies has been found in many different cell types and species. The antigen comprises two polypeptides of 14 and 17 kDa (Lin et al 1981). When TC7 cells are infected with simian virus 40, the presence of this antigen is increased by nearly 80% (Shyamala et al 1982).

Several polypeptides have been identified in kinetochores of metaphase chromosomes using electrophoretic immunotransfers with anti-kinetochore CREST antiserum. The peptides have molecular weights of 18 and 80 (Valdivia & Brinkley 1984), 15 and 33 (Ayer & Fritzler 1984), 195

(Guldner et al 1984), 17 and 114 (Earnshaw et al 1984), 14, 20 and 34 (Cox et al 1983), and 70 (Nishikai et al 1984). Since a monoclonal antibody specific for phosphoproteins of HeLa mitotic cells immunofluorescently stains centrosomes, kinetochores, and midbodies, it seems that phosphoproteins are constituents of MTOCs (Vandre et al 1984). To date, however, the role of any reported MTOC-associated protein in binding or nucleating microtubules is unknown.

Using CREST antiserum it is possible to observe "prekinetochores" in interphase nuclei long before they become associated with spindle microtubules (Brenner & Brinkley 1982). Recently Palmer & Margolis (1985) have isolated mononucleosomes from mammalian cells, which are recognized by human CREST autoantibodies. Further studies of prekinetochores may provide valuable clues as to how these MTOCs are activated prior to attachment to microtubules.

In one report, steroid hormones appeared to be associated with centrioles and basal bodies in rats, as determined by immunofluorescence with anti-steroid antibodies (Nenci 1978; Nenci & Marchetti 1978). The role of steroids, if any, in centriole replication and ciliogenesis, and other functions of MTOCs, remains obscure.

High concentrations of the ATPase dynein have been found at the spindle poles in HeLa and sarcoma 180 cells (Hartman 1964). ATPase activity has also been measured in basal bodies isolated from chicken oviduct (Anderson 1977).

Centrioles and kinetochores may contain RNA since they specifically react with an electron dense stain for ribonucleoprotein (Bielek 1978; Rieder 1979). This staining can be specifically removed by RNAase digestion and cold perchloric acid extraction. RNAase can also dissolve the acridine orange staining of isolated basal bodies in *Tetrahymena* (Hartman et al 1974, Hartmann & Zimmerman 1974). Centrioles do not appear to contain DNA according to ^{3}H-thymidine studies (Hartman 1975). When prophase PtK$_2$ cells are treated with RNA-binding, ultraviolet light–activated psoralens, followed by ultraviolet light irradiation, the ability of centrioles to initiate microtubule assembly is abolished. This inhibition does not occur with DNA-binding psoralens (Peterson & Berns 1978). Similarly, when isolated basal bodies treated with RNAase are injected in the cytoplasm of *Xenopus* eggs, they can no longer nucleate microtubule assembly. DNAase has no such effect (Heidemann et al 1977). Therefore, it seems that RNA, not DNA, is vital in maintaining the MTOC activity of centrioles during mitotic spindle formation. Contrarily, kinetochores specifically undergo condensation when treated with DNAase I, while RNAases A and Tl have no significant effect (Pepper & Brinkley 1980). The importance of RNA in MTOCs is further obscured by the fact that enzymatic hydrolysis of RNA does not change the morphology or the microtubule initiation potential of isolated spindle pole bodies in the yeast, *Saccharomyces cerevisiae* (Hyams & Borisy 1978).

Much remains to be learned about the molecular composition of MTOCs. Until more reliable fractionation and purification schemes are worked out the chemical nature of MTOCs will remain unresolved.

6.   MTOCs AS ENDOGENOUS DETERMINANTS OF CELL FORM

The factors responsible for regulation the morphology of eukaryotic cells are vaguely defined, and a complete review of this area is beyond the scope of the present report (see Solomon 1981). Nevertheless, in view of the substantial evidence that MTOCs determine the temporal and spatial pattern of microtubules in the cytoplasm, and thereby influence cell shape, it is appropriate to briefly consider their role as an endogenous determinant of cell form.

Although the nature and source of the determinants of cell form and movement are unknown, recent experiments with cultured mouse neuroblastoma cells suggest that determinants are located, at least in part, within the cell, and more specifically within the cytoskeleton. From observing the number, positions, lengths, thicknesses, and branching patterns of neurites extending from mitotically derived daughter cells Solomon (1979a, 1981) concluded that at least 60% of the cell pairs were morphologically related. No such relatedness was noted between randomly matched, non-daughter pairs. These findings are in agreement with an earlier report by Albrecht-Buehler (1977), who found that pairs of mouse 3T3 daughter cells were related in terms of morphology and directionality of movement following division.

The role of MTOCs and microtubules in the expression of the twin morphologies was implied by experiments involving microtubule inhibitors (Solomon 1979b). Treatment of the neuroblastoma cells with nocodazole resulted in the disruption of cytoplasmic microtubules, disorganization of the cytoskeleton, and retraction of the cell processes. Removal of the drug led to the reformation of microtubules, and complete recapitulation of the detailed neurite morphologies observed before treatment.

Although these experiments fall short of actually identifying a determinant in the cytoplasm, they argue strongly for an internal organelle or apparatus that controls the pattern of neurite outgrowth from the cells. Considerable experimental evidence points to the MTOC as a likely candidate for this role (Lasek et al 1981). In conclusion, the centrosomes not only organize microtubules with defined lengths, numbers, polarities, and spatial patterns, but they may even modulate the pattern of length changes that microtubules exhibit in response to chemotactic signals (Anderson et al 1983) or localized variations in tubulin concentration (Mitchison & Kirschner 1984a). Obviously, the molecular mechanism for controlling the lengthening and shortening of microtubules, as well as their initiation, assembly, and spatial organization, is complex, and to credit the organizing center with the complete program for carying out this amazing feat is no doubt overly simplistic. Nevertheless, many facts point to MTOCs as "control centers" for the cytoskeleton, and as inheritable determinants of cell morphology.

In summary, the expression of cell form–controlling microtubule patterns appears to involve the MTOCs, and these in turn appear to be controlled by signals arising from endogenous and exogenous sources. Interestingly, Ramon y Cajal proposed many years ago that the detailed shapes of cells are controlled by both internal and external cues (see Solomon 1981), a prediction seemingly well founded by today's research. Clearly, a direction for future research is to identify the MTOC-dependent mechanism that translates internal and external signals into changes in the fabric of the cytoplasmic microtubule complex.

ACKNOWLEDGMENT

I am grateful to Albert Tousson for technical and editorial assistance, and to Pat Williams and Suzanne Saltalamacchia for secretarial assistance. Appreciation is extended to many colleagues, students, and postdoctoral fellows who have worked in my laboratory over the years, providing stimulating ideas and discussion on the topic of MTOCs. This study was supported in part by PHS grant number CA-23022, awarded by the National Cancer Institute, DHHS.

Literature Cited

Albrecht-Buehler, G. 1977. Phagokinetic tracks of 3T3 cells: Parallels between the orientation of track segments and cellular structures which contain actin or tubulin. *Cell* 12:333–39

Allen, R. D., Borisy, G. G. 1974. Structural polarity and directional growth of microtubules of *Chlamydomonas* flagella. *J. Mol. Biol.* 90:381–402

Anderson, R. G. W. 1977. Biochemical and cytochemical evidence for ATPase activity in basal bodies isolated from oviduct. *J. Cell Biol.* 74:547–60

Anderson, R. G. W., Floyd, A. K. 1980.

Electrophoretic analysis of basal body (centriole) proteins. *Biochemistry* 19: 5625–31

Anderson, D. C., Wible, L. J., Hughs, B. J., Smith, C. W., Brinkley, B. R. 1983. Cytoplasmic microtubules in polymorphonuclear leukocytes: Effects of chemotactic stimulation and colchicine. *Cell* 31:719–29

Asai, D. J., Brokaw, C. J., Thompson, W. C., Wilson, L. 1982. Two different monoclonal antibodies to α-tubulin inhibit the bending of reactivated sea urchin spermatozoa. *Cell Motil.* 2:599–614

Ayer, L. M., Fritzler, M. J. 1984. Anticentromere antibodies bind to trout testis histone 1 and a low molecular weight protein from rabbit thymus. *Mol. Immunol.* 21:761–70

Berns, M., Richardson, S. M. 1977. Continuation of mitosis after selective microbeam destruction of the centriolar region. *J. Cell Biol.* 75:977–82

Bielek, E. 1978. Structure and ribonucleoprotein staining of kinetochores of colchicine-treated HeLa cells. *Cytobiologie* 16:480–84

Binder, L., Dentler, W., Rosenbaum, J. L. 1975. Assembly of chick brain tubulin onto flagellar microtubules from *Chlamydomonas* and sea urchin sperm. *Proc. Natl. Acad. Sci. USA* 72:1122–26

Boveri, T. 1961. Zellen-Studein IV. ber Die Natur der Centrosomen. Fischer. Jena

Brenner, S. L., Brinkley, B. R. 1982. Tubulin assembly sites and the organization of microtubule arrays in mammalian cells. *Cold Spring Harbor Symp. Quant. Biol.* 46:241–54

Brinkley, B. R., Cartwright, J. Jr. 1971. Ultrastructural analysis of mitotic spindle elongation in mammalian cells *in vitro*: Direct microtubule counts. *J. Cell Biol.* 50:416–31

Brinkley, B. R., Cartwright, J. Jr. 1975. Cold labile and cold stable microtubules in the mitotic spindle of mammalian cells. *Ann. NY Acad. Sci.* 253:428–39

Brinkley, B. R., Fuller, G. M., Highfield, D. P. 1976. Tubulin antibodies as probes for microtubules in dividing and nondividing mammalian cells. In *Cell Motility*, ed. R. Goldman, T. Pollard, J. Rosenbaum, pp. 435–56. Cold Spring Harbor, NY: Cold Spring Harbor Lab. 456 pp.

Brinkley, B. R., Cox, S. M., Fistel, S. H. 1981a. Organizing centers for cell processes. *Neurosci. Res. Prog. Bull.* 19:106–24

Brinkley, B. R., Cox, S. M., Pepper, D. A., Wible, L., Brenner, S. L., Pardue, R. L. 1981b. Tubulin assembly sites and the organization of cytoplasmic microtubules in cultured mammalian cells. *J. Cell Biol.* 90:557–62

Brinkley, B. R., Miller, C. L., Fuseler, J. W., Pepper, D. A., Wible, L. J. 1978. Cytoskeletal changes in cell transformation to malignancy. In *Cell Differentiation and Neoplasia*, ed. G. F. Saunders, pp. 419–50. New York: Raven. 549 pp.

Brinkley, B. R., Stubblefield, E. 1970. Ultrastructure and interaction of the kinetochore and the centriole in mitosis and meiosis. *Adv. Cell Biol.* 1:119–85

Brinkley, B. R., Stubblefield, E., Hsu, T. C. 1967. The effects of colcemid inhibition and reversal on the fine structure of the mitotic apparatus of Chinese hamster cells

in vitro. J. Ultrastruct. Res. 19:1–18

Brinkley, B. R., Fuller, G. M., Highfield, D. P. 1975. Cytoplasmic microtubules in normal and transformed cells in culture: Analysis of tubulin immunofluorescence. *Proc. Natl. Acad. Sci. USA* 73:4981–85

Bryan, J. 1976. A quantitative analysis of microtubule elongation. *J. Cell Biol.* 10:1295–1310

Burton, P. R., Hinckley, R. E., Pierson, G. B. 1975. Tannic acid staining of microtubules with 12, 13, and 15 protofilaments. *J. Cell Biol.* 65:223–28

Burns, R. G., Starling, D. 1974. The *in vitro* assembly of tubulins from sea urchin eggs and rat brain: Use of heterologous seeds. *J. Cell Sci.* 14:411–19

Byers, B., Shriver, K., Goetsch, L. 1978. The role of spindle pole bodies and modified microtubule ends in initiation of microtubule assembly in *Saccharomyces cerevisiae. J. Cell Sci.* 30:331–52

Calarco-Gillam, P. O., Siebert, M. C., Hubble, R., Mitchison, T., Kirschner, M. 1983. Centrosome development in early mouse embryos as defined by an autoantibody against pericentriolar material. *Cell* 35:621–29

Cande, W. Z., Snyder, J., Smith, D., Summers, K., McIntosh, J. R. 1974. A functional mitotic spindle prepared from mammalian cells in culture. *Proc. Natl. Acad. Sci. USA* 71:1559–63

Carlier, M.-F., Pantaloni, D. 1981. Kinetic analysis of guanosine 5′-triphosphate hydrolysis associated with tubulin polymerization. *Biochemistry* 20:1918–24

Carlier, M.-F., Hill, T., Chen, Y.-D. 1984. Interference of GTP hydrolysis in the mechanism of microtubule assembly: An experimental study. *Proc. Natl. Acad. Sci. USA* 81:771–75

Connolly, J. A., Kalnins, V. I. 1978. Visualization of centrioles and basal bodies by fluorescent staining with nonimmune rabbit sera. *J. Cell Biol.* 79:526–32

Cox, J. V., Schenk, E. A., Olmsted, J. B. 1983. Human anticentromere antibodies: Distribution characterization of antigens and effect on microbutule organization. *Cell* 35:331–39

De Brabander, M., Geuens, G., Nuydens, R., Willebrords, R., De Mey, J. 1980. The microtubule nucleating and organizing activity of kinetochores and centrosomes in living PTK2-cells. In *Microtubule and Microtubule Inhibitors*, ed. M. De Brander, J. De Mey, pp. 255–68. Amsterdam: Elsevier/North-Holland. 576 pp.

De Brabander, M., Geuens, G., Nuydens, R., Willebrords, R., De Mey, J. 1981. Microtubule assembly in living cells after release from nocodazole block: The effect of metabolic inhibitors, taxol and pH. *Cell Biol. Int. Rep.* 5:913–20

Deery, W. J., Brinkley, B. R. 1983. Cytoplasmic microtubule assembly-disassembly from endogenous tubulin in Brij-lysed cell models. *J. Cell Biol.* 96:1631–41

De Harven, E. 1968. The centriole and the mitotic spindle. In *The Nucleus*, ed. A. J. Dalton, F. Haquenae, pp. 197–227. New York: Academic

Dirksen, E. R., Staprans, I. 1975. Tubulin synthesis during ciliogenesis in the mouse oviduct. *Dev. Biol.* 46:1–13

Earnshaw, W. C., Halligan, N., Cooke, C., Rothfield, N. 1984. The kinetochore is part of the metaphase chromosome scaffold. *J. Cell Biol.* 98:352–57

Eckert, B. S., Caputi, J. E., Brinkley, B. R. 1984. Localization of the centriole and keratin intermediate filaments in PTK1 cells by double immunofluorescence. *Cell Motil.* 4:241–47

Euteneuer, U., McIntosh, J. R. 1981. Structural polarity of kinetochore microtubules in PTK2 cells. *J. Cell Biol.* 89:338–45

Evans, L., Mitchison, T., Kirschner, M. 1985. The influence of the centrosome on the structure of the nucleated microtubule. *J. Cell Biol.* 100:1185–91

Frigon, R. P., Timasheff, S. H. 1975a. Magnesium-induced self association of calf brain tubulin. I. Stoichiometry. *Biochem.* 14:4559–66

Frigon, R. P., Timasheff, S. H. 1975b. Magnesium-induced self-association of calf brain tubulin. II. Thermodynamics. *Biochemistry* 14:4567–73

Fuller, G. M., Brinkley, B. R., Boughter, J. M. 1975. Immunofluorescence of mitotic spindles by using monospecific antibody against bovine brain tubulin. *Science* 187:948–50

Goldman, R. D., Hill, R., Steinert, P., Whitman, M. A., Zackroff, R. V. 1980. Intermediate filament-microtubule interactions. In *Microtubules and Microtubule Inhibitors*, ed. M. De Brabander, J. De Mey, pp. 91–102. Amsterdam: Elsevier North-Holland. 576 pp.

Gould, R. P., Borisy, G. G. 1977. The pericentriolar material in Chinese hamster ovary cells nucleates microtubule formation. *J. Cell Biol.* 73:601–15

Gozes, I., Saya, P., Littauer, U. Z. 1979. Tubulin heterogeneity in neuroblastoma and glioma cell lines differs from that in brain. *Brain Res.* 171:171–75

Guldner, H. H., LaKonsek, H. J., Bautz, F. A. 1984. Human anticentromere sera recognize a 19.5 Kd nonhistone chromosomal protein from HeLa cells. *Clin. Exp. Immunol.* 58:13–20

Gundersen, G. G., Kalnoski, M. H., Bulinski, J. C. 1984. Distinct populations of microtubules: Tyrosinated and nontyrosinated α-tubulin are distributed differently *in vivo*. *Cell* 38:779–89

Haimo, L. T., Telzer, B. R., Rosenbaum, J. L. 1979. Dynein binds to and cross-bridges cytoplasmic microtubules. *Proc. Natl. Acad. Sci. USA* 76:5759–63

Hartmann, J. F. 1964. Cytochemical localization of ATP in the mitotic apparatus of HeLa and sarcoma 180 tissue culture cells. *J. Cell Biol.* 23:363–70

Hartman, H., Puma, J. P., Gurney, T. 1974. Evidence for the association of RNA with the ciliary basal bodies of *Tetrahymena*. *J. Cell Sci.* 16:241–60

Hartmann, J. F., Zimmerman, A. M. 1974. The mitotic apparatus. In *The Cell Nucleus*, Vol. 2, ed. H. Busch, pp. 459–86. New York: Academic. 564 pp.

Heidemann, S. R., McIntosh, J. R. 1980. Visualization of the structural polarity of microtubules. *Nature* 286:517–19

Heidemann, S. R., Sander, G., Kirschner, M. W. 1977. Evidence for a functional role of RNA in centrioles. *Cell* 10:337–50

Hill, T. L., Kirschner, M. W. 1982. Regulation of microtubule and actin filament assembly-disassembly by associated small and large molecules. *Int. Rev. Cytol.* 84:185–234

Himes, R. H., Burton, P. R., Gaito, J. M. 1977. Dimethyl sulfoxide-induced self-assembly of tubulin lacking associated proteins. *J. Biol. Chem.* 252:6222–28

Hyams, J. S., Borisy, G. G. 1978. Nucleation of microtubules *in vitro* by isolated spindle pole bodies of the yeast *Saccharomyces cerevisiae*. *J. Cell Biol.* 78:401–14

Inoue, S., Borisy, G. G., Kiehart, D. P. 1974. Growth and lability of *Chaetopterus* oocyte mitotic spindles isolated in the presence of porcine brain tubulin. *J. Cell Biol.* 62:175–84

Jackson, W. T., Doyle, B. G. 1982. Membrane distribution in dividing endosperm cells of *Haemanthus*. *J. Cell Biol.* 94:637–43

Karsenti, E., Kobayashi, S., Mitchison, T., Kirschner, M. 1984. Role of the centrosome in organizing the interphase microtubule array: Properties of cytoblasts containing or lacking centrosomes. *J. Cell Biol.* 98:1763–76

Kim, H., Binder, L. I., Rosenbaum, J. L. 1979. The periodic association of MAPs with brain microtubules *in vitro*. *J. Cell Biol.* 80:266–76

Krauhs, E., Little, M., Kempf, T., Hofer-Warbinek, R. W., Ponstingl, H. 1981. Complete amino acid sequence of beta-tubulin from porcine brain. *Proc. Natl. Acad. Sci. USA* 78:4156–60

170 BRINKLEY

Kuriyama, R. 1976. *In vitro* polymerization of flagellar and ciliary outer fiber tubulin into microtubules. *J. Biochem.* (*Tokyo*) 80: 153–65

Lasek, R. I., Solomon, F., Brinkley, B. R. 1981. Which cytoskeletal determinants preserve ontogenetic history? *Neurosci. Res. Program Bull.* 19: 125–35

Lefebvre, P. A., Silflow, C. D., Wieben, E. D., Rosenbaum, J. L. 1980. Increased levels of mRNA for tubulin and other flagellar proteins after amputation or shortening of *Chlamydomonas* flagella. *Cell* 20: 469–97

Lin, W., Fung, B., Shyamala, M., Kasamatsu, H. 1981. Identification of antigenically related polypeptides at centrioles and basal bodies. *Proc. Natl. Acad. Sci. USA* 78: 2373–77

Mazia, D. 1984. Centrosomes and mitotic poles. *Exp. Cell Res.* 153: 1–15

McEwen, B., Adelstein, S. J. 1980. Evidence for mixed lattice in microtubules reassembled *in vitro. J. Mol. Biol.* 139: 123–45

McGill, M., Brinkley, B. R. 1975. Human chromosomes and centrioles as nucleating sites for the *in situ* assembly of microtubules from bovine brain tubulin. *J. Cell Biol.* 67: 189–99

McGill, M., Highfield, D. P., Monahan, T. M., Brinkley, B. R. 1976. Effects of nucleic acid specific dyes on centrioles of mammalian cells. *J. Ultrastruc. Res.* 57: 43–53

McIntosh, J. R. 1981. Microtubule polarity and interaction in mitotic spindle function. In *International Cell Biology, 1980–81*, ed. H. G. Schweiger, pp. 359–68. Berlin: Springer-Verlag. 1033 pp.

McIntosh, J. R. 1983. The centrosome as an organizer of the cytoskeleton. *Mod. Cell Biol.* 2: 115–42

McIntosh, J. R. 1984. Microtubule catastrophe. *Nature* 312: 196–97

McIntosh, J. R., Landis, S. C. 1971. Distribution of spindle microtubules. *J. Cell Biol.* 9: 468–97

Miller, M., Solomon, F. 1984. Kinetics and intermediates of marginal band reformation: Evidence for peripheral determinants of microtubule organization. *J. Cell Biol.* In press

Mitchison, T., Kirschner, M. 1984a. Microtubule assembly nucleated by isolated centrosomes. *Nature* 312: 232–37

Mitchison, T., Kirschner, M. 1984b. Dynamic instability of microtubule growth. *Nature* 312: 237–42

Nenci, I. 1978. Receptor and centriole pathways of steroid action in normal and neoplastic cells. *Cancer Res.* 38: 4204–11

Nenci, I., Marchetti, E. 1978. Concerning the localization of steroids in the centrioles and basal bodies by immunofluorescence. *J. Cell Biol.* 76: 255–60

Nicklas, R. B., Gordon, G. W. 1985. The total length of spindle microtubules depends on the number of chromosomes present. *J. Cell Biol.* 1001–7

Nishikai, M., Okand, O., Yamashita, H., Watanabe, M. 1984. Characterization of centromere (kinetochore) antigen reactive with sera of patients with a scleroderma variant (CREST syndrome). *Ann. Rheum. Dis.* 43: 819–24

Palmer, D. K., Margolis, R. L. 1985. Kinetochore components recognized by human autoantibodies are present on mononucleosomes. *Mol. Cell Biol.* 5: 173–86

Pepper, D. A., Brinkley, B. R. 1977. Localization of tubulin in the mitotic apparatus of mammalian cells by immunofluorescence and immunoelectron microscopy. *Chromosoma* 60: 223–37

Pepper, D. A., Brinkley, B. R. 1979. Microtubule initiation at kinetochores and centrosomes in lysed mitotic cells: Inhibition of site specific nucleation by tubulin antibody. *J. Cell Biol.* 82: 585–91

Pepper, D. A., Brinkley, B. R. 1980. Tubulin nucleation and assembly in mitotic cells: Evidence for nucleic acids in kinetochores and centrosomes. *Cell Motil.* 1–15

Peterson, S. P., Berns, M. W. 1978. Evidence for centriolar region RNA functioning in spindle formation in dividing PtK$_2$ cells. *J. Cell Sci.* 34: 289–302

Peterson, S. P., Berns, M. W. 1980. The centriolar complex. *Int. Rev. Cytol.* 64: 81–106

Peterson, G. B., Burton, P. R., Himes, R. H. 1978. Alterations in number of protofilaments in microtubules assembled by initiation sites. *J. Cell Biol.* 76: 223–28

Pickett-Heaps, J. D. 1969. The evolution of the mitotic apparatus: An attempt at comparative ultrastructural cytology in dividing plant cells. *Cytobios* 3: 257–80

Pickett-Heaps, J. D., Tippit, D. H., Porter, K. R. 1982. Rethinking mitosis. *Cell* 29: 729–44

Ponstingl, H., Krauhs, E., Little, M., Kempf, T. 1981. Complete amino acid sequence of alpha-tubulin from porcine brain. *Proc. Natl. Acad. Sci. USA* 78: 2757–61

Porter, K. R. 1966. Cytoplasmic microtubules and their function. In *Principles of Biomolecular Organization*, ed. G. E. W. Wolstenholme, M. O'Connor, pp. 308–45. London: Churchill. 491 pp.

Rebhun, L. I., Rosenbaum, J., Lefebvre, P., Smith, G. 1974. Reversible restoration of the birefringence of cold-treated, isolated mitotic apparatus of surf clam eggs with chick brain tubulin. *Nature* 249: 113–15

Rieder, C. L. 1979. Ribonucleoprotein staining of centrioles and kinetochores in newt lung cell spindles. *J. Cell Biol.* 80: 1–9

Rieder, C. L., Borisy, G. G. 1981. The attachment of kinetochores to the prometaphase spindle in PtK cells: Recovery from low temperature treatment. *Chromosoma* 82: 693–716

Rieder, C. L., Borisy, G. G. 1982. The centrosome cycle in PTK2 cells: Asymmetric distribution and structural changes in pericentriolar material. *Biol. Cell* 44: 117–32

Roos, U.-P. 1981. Quantitative structure analysis of the mitotic spindle. In *International Cell Biology, 1980–81*, ed. H. G. Schweiger, pp. 369–81. Berlin: Springer-Verlag

Russell, D. G., Burns, R. G. 1984. The polar ring of coccidian sporozoites: A unique microtubule-organizing centre. *J. Cell Sci.* 65: 193–207

Scheele, R. B., Bergin, L. F., Borisy, G. G. 1982. Control of structural fidelity of microtubules by initiating sites. *J. Cell Biol.* 154: 415–500

Schiff, P. B., Fant, J., Horowitz, S. B. 1979. Promotion of microtubule assembly *in vitro* by taxol. *Nature* 277: 665–67

Sharp, G. A., Osborn, M., Weber, K. 1981. Ultrastructure of multiple microtubule initiation sites in mouse neuroblastoma cells. *J. Cell Sci.* 47: 1–24

Shyamala, M., Atcheson, C. L., Kasamatsu, H. 1982. Stimulation of host centriolar antigen in TC7 cells by simian virus 40: Requirement for RNA and protein synthesis and an intact simian virus 40 small-t gene function. *J. Virol.* 43: 721–29

Snell, W. J., Dentler, W. L., Hamio, L. T., Binder, L. I., Rosenbaum, J. L. 1974. Assembly of chick brain tubulin onto isolated basal bodies of *Chlamydomonas reinhardii*. *Science* 185: 357–59

Snyder, J. A., McIntosh, J. R. 1975. Initiation and growth of microtubules from mitotic centers in lysed mammalian cells. *J. Cell Biol.* 67: 744–60

Solomon, F. 1979a. Detailed neurite morphologies of sister neuroblastoma cells are related. *Cell* 16: 165–69

Solomon, F. 1979b. Neuroblastoma cells recapitulate their detailed neurite morphologies after reversible microtubule disassembly. *Cell* 21: 333–38

Solomon, F. 1981. Specification of cell morphology by endogenous determinants. *J. Cell Biol.* 90: 547–53

Staprans, I., Dirksen, E. R. 1974. Microtubule protein during ciliogenesis in the mouse oviduct. *J. Cell Biol.* 62: 164–74

Stearns, M. E., Connolly, J. A., Brown, D. L. 1976. Cytoplasmic microtubule organizing centers isolated from *Polytomella agilia*. *Science* 191: 188–91

Stearns, M. E., Brown, D. L. 1979. Purification of cytoplasmic tubulin and microtubule organizing center protein functioning in microtubule initiation from the alga *Polytomella*. *Proc. Natl. Acad. Sci. USA* 76: 5745–49

Stearns, M. E., Brown, D. L. 1981. Microtubule organizing centers of the alga *Polytomella* exert spatial control over microtubule initiation *in vivo* and *in vitro*. *J. Ultrastruct. Res.* 77: 366–78

Tash, J. S., Means, A. R., Brinkley, B. R., Dedman, J. R., Cox, S. M. 1980. Cyclic nucleotide and Ca^{2+} regulation of microtubule initiation and elongation. In *Microtubules and Microtubule Inhibitors*, ed. M. De Brabander, J. De Mey, pp. 269–79. Amsterdam: Elsevier/North-Holland

Telzer, B. R., Moses, M. J., Rosenbaum, J. L. 1975. Assembly of microtubules onto kinetochores of isolated mitotic chromosomes of HeLa cells. *Proc. Natl. Acad. Sci. USA* 72: 4023–27

Thompson, W. C., Asai, D. J., Carney, D. H. 1984. Heterogeneity among microtubules of the cytoplasmic microtubule complex detected by a monoclonal antibody to an alpha tubulin. *J. Cell Biol.* 98: 1017–25

Tilney, L. G. 1971. Origin and continuity of microtubules. In *Origin and Continuity of Cell Organelles*, ed. J. Reinert, H. Ursprung, pp. 222–60. Berlin: Springer-Verlag. 342 pp.

Tilney, L. G., Bryan, J., Bush, D. J., Fujiwara, K., Mooseker, M. S., et al. 1973. Microtubules: Evidence for 13 protofilaments. *J. Cell Biol.* 59: 267–75

Tilney, L. G., Goodard, J. 1970. Nucleating sites for the assembly of microtubule in the ectodermal cells of blastula of *Arbacia punctulata*. *J. Cell Biol.* 46: 564–75

Tucker, J. B. 1979. Spatial organization of microtubules. In *Microtubules*, ed. K. Roberts, J. S. Hyams, pp. 315–57. London: Academic. 595 pp.

Tucker, J. B. 1982. Microtubule-organizing centres and microtubule-deployment in protozoa. *Acta Protozool.* 1: 45–57

Tucker, J. B. 1984. Spatial organization of microtubule-organizing centers and microtubules. *J. Cell Biol.* 99: 55s–62s

Turksen, K., Aubin, J. E., Kalnins, V. I. 1982. Identification of a centriole-associated protein by antibodies present in normal rabbit sera. *Nature* 298: 763–65

Valdivia, M. M., Brinkley, B. R. 1984. Kinetochore/centromere of mammalian chromosomes. In *Molecular Biology of the Cytoskeleton*, ed. G. G. Borisy, D. W. Cleveland, D. B. Murphy, pp. 79–86. Cold Spring Harbor, NY: Cold Spring Harbor Lab. 512 pp.

Vandre, D. D., Davis, F. M., Rao, P. N., Borisy, G. G. 1984. Phosphoproteins are components of the mitotic microtubule organizing centers. *Proc. Natl. Acad. Sci. USA* 81:4439–43

Vorobjev, I. A., Chentsov, Y. S. 1982. Centrioles in the cell cycle. *J. Cell Biol.* 93:938–49

Weber, K., Bibring, T., Osborn, M. 1975. Specific visualization of tubulin-containing structures in tissue culture cells by immunofluorescence: Cytoplasmic microtubules, vinblastine-induced paracrystals and mitotic figures. *Exp. Cell. Res.* 95:111–10

Weisenberg, R. C. 1972. Microtubule formation *in vitro* in solutions containing low calcium concentrations. *Science* 177:1104–5

Weisenberg, R. C., Rosenfeld, A. C. 1975a. Role of intermediates in microtubule assembly *in vivo* and *in vitro*. *Ann. NY Acad. Sci.* 253:78–89

Weisenberg, R. C., Rosenfeld, A. C. 1975b. *In vitro* polymerization of microtubules into asters and spindles in homogenates of surf clam eggs. *J. Cell Biol.* 64:146–58

Witt, P. L., Ris, H., Borisy, G. G. 1980. Origin of kinetochore microtubules in Chinese hamster ovary cells. *Chromosoma* 81:483–505

Wolfe, J. 1972. Basal body fine structure and chemistry. *Adv. Cell Mol. Biol.* 2:151–92

Wolosewick, J. J., Porter, K. R. 1979. Microtrabecular lattice of the cytoplasmic ground substance. Artifacts or reality. *J. Cell Biol.* 82:114–39

Ann. Rev. Cell Biol. 1985. 1 : 173–207

GENERATION OF POLARITY DURING *CAULOBACTER* CELL DIFFERENTIATION

Lucille Shapiro

Department of Molecular Biology, Albert Einstein College of Medicine, Bronx, New York 10461

CONTENTS

PERSPECTIVES AND SUMMARY

There are several questions that are fundamental to our understanding of the processes that mediate a developmental program. These include questions about how a cell knows to allow the expression of subsets of genes

0743–4634/85/1115–0173$02.00

at specific times in the cell cycle, how polarity is established, and how positional information is encoded. At the base of these questions is the problem of the generation of temporal and spatial asymmetry within a cell. Asymmetry can be generated by the selective transcription and translation of specific genetic sequences and by the programmed spatial distribution of subsets of gene products. To understand the mechanisms that dictate inherited arrangements of cell constituents in three-dimensional space we must pose questions central not only to the control of cell differentiation but to cell architecture.

In an attempt to learn how genetic information can program the timing of differential gene expression, the regulation of assembly of complex subcellular structures, and the control of spatial organization, a group of investigators has been studying a relatively simple microorganism, *Caulobacter crescentus*. Approximately 30 years ago Roger Stanier pointed out that this stalked bacterium has an elegant dimorphic life cycle that exhibits a simple and accessible cell differentiation process. Polar structures, such as the stalk and flagellum, are assembled during each cell cycle, and they provide ready landmarks of the generation of polarity. The fundamental aspects of *Caulobacter* cell physiology and behavior were established by Stanier's students, Poindexter and Schmidt (Poindexter 1964, 1981). Because this organism is amenable to complete biochemical and genetic dissection, and because a relatively small number of events mark the generation of polarity and differential gene expression, the mechanisms that participate in these fundamental aspects of cell differentiation are accessible to experimental analysis. This chapter reviews the recent developments that have allowed an extensive genetic and molecular analysis of the *Caulobacter* cell cycle. It is not meant to be a comprehensive review of all aspects of *Caulobacter*. For this the reader is referred to several recent review articles (Shapiro et al 1981; Poindexter 1981; Newton 1984; Ely & Shapiro 1984; Newton et al 1985). Our purpose here is to examine the temporal and spatial control of gene expression and the assembly of subcellular structures.

THE CELL CYCLE

The cell division cycle of *Caulobacter crescentus*, a gram negative bacterium, exhibits a clear polarity (Figure 1). Binary fission of the parent cell gives rise to daughter cells that differ structurally and metabolically. It is the establishment of polarity in the predivisional cell that allows access to one of the most fundamental aspects of development: the generation of cellular diversity. The signals that initiate differential spatial organization of structure and function in the *Caulobacter* cell are inherent and do not depend on exogeneous environmental fluctuations. Poindexter (1964) microscopically examined individual cells in microculture and found that the differentiation process was independent of cell contact or chemical interaction.

Each cell division yields a flagellum-bearing swarmer cell and a sessile stalked cell (Figure 1). The daughter swarmer cell carries several polar structures, a flagellum, pili, and receptors for several DNA phages, which were assembled earlier at one pole of the parental predivisional cell. This motile swarmer cell is unable to initiate the replication of its chromosome (Degnan & Newton 1972a). Near the end of its transient existence, the

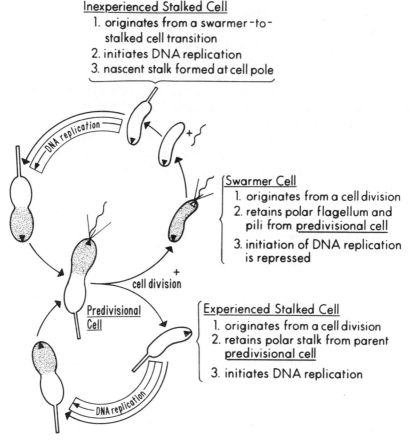

Figure 1   Generation of polarity during the cell division cycle of *Caulobacter crescentus*. As the stalked cell prepares for division, the two cell poles show increasing biochemical and structural diversity. *Shaded areas* exhibit swarmer cell characteristics. The *closed triangles* indicate the site of formation of the polar flagellum, pili, and DNA-phage receptors.

swarmer cell sheds its polar flagellum (Poindexter et al 1967; Shapiro & Maizel 1973), loses its polar pili (Shapiro & Agabian-Keshishian 1970; Smit & Agabian 1982) and polar phage receptors (Agabian-Keshishian & Shapiro 1971; Lagenaur et al 1974; Huguenel & Newton 1982; O'Neill & Bender, unpublished), and grows a stalk at the site previously occupied by the flagellum. The stalk is a slender tube formed by the localized synthesis of inner-membrane, peptidoglycan cell wall, and outer membrane (Poindexter 1964; Schmidt & Stanier 1966). This morphological transition, one of two key events in the developmental program, is accompanied by a signal to initiate the replication of the chromosome (Degnan & Newton 1972a).

The portion of the cell cycle allotted to a swarmer cell existence is a fixed fraction, one-third of the entire cell cycle, independent of the generation time (Poindexter 1964; Newton 1972; Shapiro 1976). While the newly formed stalked cell (inexperienced stalked cell) is enlarging and completing a round of DNA replication, a new set of polar appendages is being assembled at the pole opposite the stalk. In addition to the biosynthesis of the flagellum, pili, and phage receptors, the proteins involved in chemotactic signal transduction are synthesized and assembled in the incipient swarmer cell portion of the predivisional cell (Shaw et al 1983; Gomes & Shapiro 1984). Thus, prior to cell division a precisely timed program of differential gene expression is accompanied by the localized distribution of gene products to yield a predivisional cell with markedly different poles. The completion of DNA replication and the distribution of the chromosomes to the two poles of the predivisional cell is required for the subsequent cell division (Osley & Newton 1977).

The stalked cell that results from cell division retains the stalk that was present at one pole of the parent cell. The experienced stalked cell immediately initiates DNA replication (Degnan & Newton 1972a; Evinger & Agabian 1977; Iba et al 1977). It functions as a stem cell in that it continuously allows the formation of an incipient swarmer cell at the pole opposite the stalk (*shaded areas*, Figure 1). However, because the *Caulobacter* developmental program is cyclic, the differentiation pathway that produces the swarmer cell is not terminal. The mechanisms that bring about these profound cellular changes are driven by a genetic blueprint that is subject to a transient shift in interpretation, with the result that two poles of the predivisional cell express different programs.

GENETIC ANALYSIS OF CELL CYCLE EVENTS

The ability to perturb specific genetic functions and then map and engineer the genes involved in the *Caulobacter* developmental program is essential if we are to ultimately understand the generation of cellular polarity in this organism. Both classical and molecular genetic analyses are now feasible for *C. crescentus*. Mutants can be generated by treatment with a variety of agents, including UV light and nitrosoguanidine (Shapiro et al 1971; Jollick & Schervish 1972; Newton & Allebach 1975). Mutant strains generated in the latter fashion, however, have been shown to carry multiple mutations (Guerola et al 1971; Johnson & Ely 1977; Hodgson et al 1984). To circumvent the problems caused by the presence of multiple silent mutations, mutants with selectable phenotypes, such as drug resistance (Jollick & Schervish 1972; Barrett et al 1982b), altered motility (Johnson & Ely 1979; Ely et al 1984) and chemotactic behavior (Shaw et al 1983; Ely et al, in preparation), and auxotrophs (Johnson & Ely 1977), have been obtained in the absence of mutagens. Transposon mutagenesis has also been an effective method for generating mutants of *C. crescentus* because it assures the generation, by insertional inactivation, of single mutations, and simultaneously confers a specific and selectable drug resistance (Ely & Croft 1982). The transposon Tn5, which carries resistance to kanamycin and inserts randomly into the *C. crescentus* genome, has been particularly useful for the generation of mutants (Ely & Croft 1982; Ohta et al 1984), and for the cloning of insertionally inactivated genes (Purucker et al 1982).

A genetic map for *C. crescentus* has been generated (Barrett et al 1982a,b) using the combined techniques of RP4-mediated conjugation for the transfer of relatively large pieces of chromosome (Ely 1979), and transduction mediated by the bacteriophage 0Cr30 for the transfer of small pieces of chromosome (Ely & Johnson 1977). A system of low frequency genetic exchange, apparently dependent on cell–cell interaction, has been observed in several *Caulobacter* species (Shapiro et al 1971; Jollick & Schervish 1972; Newton & Allebach 1975). However, the versatility of gene transfer mediated by the broad host range plasmid, RP4, has resulted in the effective use of this heterologous system, rather than the development of the system of endogenous genetic exchange. A rapid method of gene mapping has been developed (Barrett et al 1982a) in which RP4 mediates the transfer of donor DNA, carrying a known and mapped Tn5 insertionally inactivated gene and its surrounding genetic material, into a recipient with a specific mutant phenotype. The frequency of cotransfer of the wild-type allele and the Tn5 gene, marked by virtue of its kanamycin resistance, indicates the relative position of the two genes. This method, combined with three-factor crosses using the transducing phage 0Cr30 or RP4-mediated conjugation, has allowed fine structure mapping. The resulting genetic map, represented by an open circle in Figure 2, shows the genes involved in chemotaxis and flagellar biosynthesis within the circle and those involved in other cellular functions outside the circle. The most distal markers on the *C. crescentus* genetic map have not been linked, and thus it is not known if *C. crescentus* contains a circular chromosome, as do other bacteria. The total genome size for *C. crescentus* is approximately the same as that of *Escherichia coli* (Wood et al 1976).

The genetic analysis of *C. crescentus* has been significantly aided by the fact that DNA can be efficiently transferred between *E. coli* and *C. crescentus* using both P-type and N-type plasmids (Ely 1979). For example,

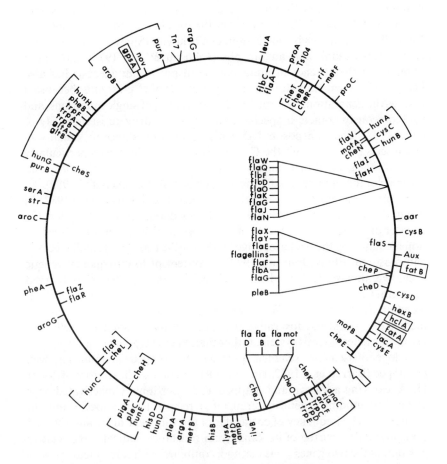

Caulobacter Crescentus CB15 Genetic Map

Figure 2 Genetic map of *C. crescentus* CB15 (Barrett et al 1982a,b). The map positions of genes involved in chemotaxis and flagellar biogenesis (Ely et al 1984) are shown on the inside of the *open circle*. The *open arrow* indicates the site where linkage has not been demonstrated for the most distal markers. The genes involved in membrane biogenesis are *boxed* (Hodgson et al 1982c).

DNA can be cloned in the plasmid RP4, transformed into *E. coli* and then transferred into *C. crescentus* by RP4-mediated conjugation (Bryan et al 1984; Schoenlein and Ely, in preparation). Complementation analysis has been successfully carried out by transferring cloned genes carried on P-type or Q-type plasmid vectors from *E. coli* to specific *C. crescentus* mutant strains (Bryan et al 1984; Ohta et al 1984; Winkler et al 1984), because many *E. coli* promoters and translation signals function in *C. crescentus*.

DNA Replication

One of the most curious characteristics of the *Caulobacter* cell cycle is the strict confinement of chromosome replication to the stalked cell (Degnan & Newton 1972a). Cell division in *Caulobacter* always yields two different daughter cells, a stalked cell and a swarmer cell, and thus only one of the progeny is able to initiate the replication of its chromosome (Degnan & Newton 1972a; Evinger & Agabian 1979; Iba et al 1977). The differential behavior of the chromosomes in the two daughter cells probably results from instructions laid down during the earlier establishment of polarity in the predivisional cells. There are several different sets of molecular events that could result in the failure of the chromosome to replicate in the swarmer cell. For example, some of the protein components required for replication may be specifically repressed in the incipient swarmer cell. It is also possible that swarmer-specific gene product(s) are synthesized that repress the initiation of replication, even in the presence of all of the proteins required for DNA synthesis. The same type of mechanism used to control the replication of stable, low copy-number plasmids may be involved in the swarmer cell (Abeles et al 1984). A third possibility is that the structure of the chromosome is altered in the incipient swarmer cell so as to prevent the formation of the initiation complex at the replication origin. In this context, Evinger & Agabian (1977, 1979) and Swoboda et al (1982) have presented evidence that the sedimentation coefficient and the protein composition of the membrane-associated nucleoid isolated from swarmer cells differs from the nucleoid isolated from the stalked cell.

Whichever mechanism is involved, selective chromosome strand segregation due to membrane attachment at a specific cell pole could differentiate the resulting two daughter cells. Although the number of chromosomes segregated upon cell division has not been resolved (Iba & Okada 1980; Newton 1984), the observation that DNA strand assortment during replication and cell division is random (Osley & Newton 1974) suggests that such a mechanism is probably not responsible for the generation of asymmetry.

As the swarmer cell continues through its cell cycle, it reaches a point at which the ability to initiate DNA replication is switched on. This switch coincides with the loss of the polar organelles (flagellum, pili, and DNA-phage receptors), and the initiation of stalk biogenesis. The sequence of events leading to that switch is being analyzed by Newton and his coworkers. Using conditional cell cycle mutants they have identified genetically defined events that are prerequisites for subsequent steps in the DNA synthesis and cell division pathway (for a review see Newton 1984). Three temperature sensitive (ts) mutants in DNA chain elongation (DNA_e), which fall into two linkage groups, and two ts mutants in DNA initiation (DNA_i) have been isolated. One of the DNA_i genes, whose expression is required for the initiation of DNA synthesis in the stalked cell, has been analyzed using reciprocal shift experiments based on the procedure described by Jarvick & Botstein (1973) and by Hereford & Hartwell (1974). These experiments show that the DNA_i gene product functions even when DNA chain elongation is blocked by agents such as hydroxyurea (Osley & Newton 1980). The function of this DNA_i gene product appears to depend upon a late stage in DNA replication during the previous cell cycle, because preventing the chromosome from completing a round of DNA replication by treatment with hydroxyurea was shown to block subsequent DNA_i action (Nathan et al 1980). These results suggest a circular pathway of dependent steps for DNA replication (Osley & Newton 1980; Nathan et al 1982).

Cell Division

The process of cell division in *C. crescentus* initiates before the replication of the chromosome is finished, and appears to result from a slow equatorial constriction, which eventually "pinches off" two daughter cells (Poindexter & Hagenzieker 1981). *C. crescentus* does not appear to form a septum or cross-wall prior to cell division. DNA replication has been shown to be a prerequisite for cell division in *C. crescentus* (Degnen & Newton 1972b; Osley & Newton 1980), as is the case in *E. coli* (Clarke 1968; Helmstetter & Pierucci 1968). The cell cycle event that allows the initiation of cell division in *C. crescentus* appears to depend on the expression of an event that occurs when about half of the DNA has been replicated (Osley & Newton 1980). Further analysis of mutants in the cell division pathway, which include those altered in the initiation of cell division (Div_i), the progression of cell constriction (Div_p), and the final steps of cell separation (CS), has led to the conclusion that these mutants reveal a dependent pathway in which each event depends on the execution of the previous step (Osley & Newton 1980; Newton 1984).

Although the gene products associated with the specific genetic lesions in the cell division process studied by Newton and his coworkers have not been identified, penicillin-binding proteins (PBPs), which are generally known to be involved in both cell wall biosynthesis and cell division, have been found in *C. crescentus* (Koyasu et al 1980, 1981, 1982, 1983). One cytoplasmic PBP with penicillinase activity was observed to be cell cycle regulated; its synthesis could only be detected in the predivisional and swarmer cell (Koyasu 1984). All other PBPs were synthesized throughout the cell cycle.

THE SWARMER-TO-STALKED CELL TRANSITION

The major structural change that occurs during the swarmer-to-stalked cell transition is the loss of the flagellum and several pili from the swarmer cell pole, and the subsequent biogenesis of a stalk at this same pole. The way in which the flagellum is released, or perhaps ejected, is an interesting question in cell mechanics. One approach to understanding this event is to define the structure of this organelle in detail, and determine how it is anchored in the cell surface. The initiation of stalk formation, which occurs at the site vacated by the flagellum (Smit & Agabian 1982b), requires localized cell wall and membrane synthesis, and may be triggered by the structural changes in the cell envelope caused by the release of the flagellum. The current state of knowledge about these surface structures is discussed in this section with a view towards understanding their protein composition, three-dimensional structure, and mechanism of assembly.

Structure of the Flagellum

The bacterial flagellum is a complex structure whose base is embedded in the cell envelope (De Pamphilis & Adler 1971b; Coulton & Murray 1978; Johnson et al 1979). A schematic representation of the *C. crescentus* flagellum is shown in Figure 3. All bacterial flagella studied thus far are composed of three subassemblies known as the filament, hook, and basal body. Both the filament and hook are outside the cell. The basal body spans the layers of the cell envelope and is composed of several rings threaded on a rod. Some of these rings are attached to specific layers of the cell envelope which, in gram-negative bacteria, include the cytoplasmic (inner) membrane, the rigid peptidoglycan cell wall, and the outer membrane. In both *E. coli* and *Salmonella typhimurium* the basal body contains four rings (De Pamphilis & Adler 1971a; Aizawa et al, submitted for publication), whereas the *C. crescentus* basal body has a fifth ring (E) (Johnson et al 1979, Stallmeyer et al 1985). In addition, the *C. crescentus* M ring has a more complex shape and a small cap- or button-like structure on its innermost surface (Stallmeyer et al 1985). SDS-polyacrylamide gel electrophoresis of purified basal body preparations has shown that in addition to a rod protein of approximately 32 kilodaltons (kDa), there are at least 5 major

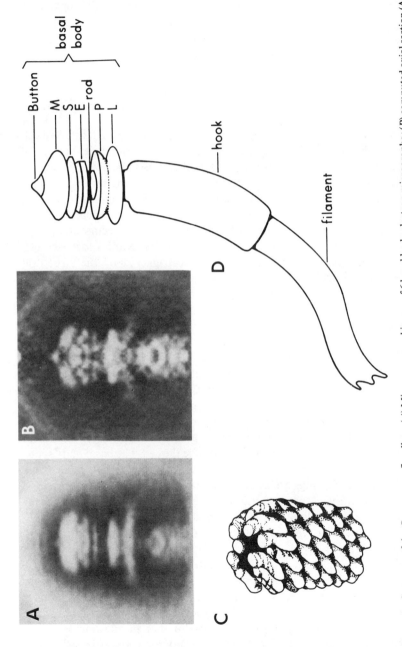

Figure 3   Structure of the *C. crescentus* flagellum. (*A*) Mirror-averaged image of 6 basal body electron micrographs; (*B*) computed axial section (Abel transform) of image (*A*) (from Stallmeyer et al 1985); (*C*) schematic of the three-dimensional model of the hook, based on the average reconstruction (from Wagenknecht et al 1981); (*D*) schematic of the flagellum basal body, hook, and filament subassemblies.

proteins and several minor protein bands (Hahnenberger & Shapiro, unpublished). Several motility mutants have been found to have altered basal body protein compositions.

The hook, a flexible structure that connects the filament to the rod of the basal body, consists of identical protein subunits (Lagenaur et al 1978; Johnson et al 1979; Sheffery & Newton 1979). Three-dimensional reconstruction of the *C. crescentus* flagellar hook has shown it to be a right-handed helical structure composed of continuous 6-start grooves that can allow the observed bending with minimum distortion (Wagenknect et al 1981). Each protein monomer appears as a finger-like subunit arranged with its long axis at 45° to the axis of the hook. The hook has 15 6-start grooves, which would yield a structure containing 295 ± 13 monomers of 70 kDa. The structure, both in *C. crescentus* and *S. typhimurium*, is notched on the end that joins to the filament, and is cone-shaped at the end that joins to the rod (Wagenknecht et al 1981, 1982). The center of the hook has a channel of approximately 25 Å, which could accommodate the passage of flagellin monomers during flagellar assembly. The bacterial flagellar filament has been shown to lengthen in vivo by the addition of new flagellin monomers to the distal tip (Iino 1969; Emerson et al 1970; Suzuki & Iino 1977). Structural data have shown an empty channel in the center of several bacterial flagellar filaments (Sleyter & Glauert 1973; Bode et al 1972; Shirakihara & Wakabayashi 1979), which could allow flagellin monomers to travel the length of the filament. It may be that the hook is similarly assembled, since the size of its central pore would allow passage of the elongated hook monomer.

Like the hook, the *C. crescentus* flagellar filament has been shown to be a right-handed helix (Koyasu & Shirakihara 1984). The filament is composed of two flagellin monomers, of 25 kDa (Fla A) and 27.5 kDa (Fla B) (Lagenaur & Agabian 1976, 1977; Marino et al 1976; Sheffery & Newton 1977; Fukuda et al 1978). Unlike phase variation in *Salmonella*, where genes for two flagellins are present but are expressed in a mutually exclusive fashion (Silverman & Simon 1977; Silverman et al 1979), each *Caulobacter* cell synthesizes and assembles both flagellins. Each filament has been shown to be composed of a relatively short length of Fla B, followed by Fla A to the end of the filament (Fukuda et al 1978; Lagenaur & Agabian 1978; Koyasu et al 1981; Weissborn et al 1982). This filament organization appears to be essential for correct assembly and function. Mutants that lack Fla B assemble a short filament composed of Fla A and the cells are immotile (Koyasu et al 1981, Fukuda et al 1981, Weissborn et al 1982, Johnson et al 1983). Several other immotile mutants that have only stub-like filaments have been shown to contain altered ratios of the flagellins, or to assemble a third flagellin of 29 kDa. The 29 kDa flagellin is normally present in the cell in very small quantities and may play a role in the assembly process (Weissborn et al 1982; Gil & Agabian 1983), but it has thus far only been detected in the filaments of mutants (Johnson et al 1983).

The mechanism that regulates the filament assembly process is assuredly multifaceted, but one component of the process must be some sort of measuring device. This process may be analogous to that proposed for the assembly of the myosin homopolymer (Davis 1981), in which the affinity constant for each additional monomer changes in a continuous manner until the filament stops growing. One would anticipate that the Fla B monomer has a higher affinity than Fla A for the nucleation site on the hook. The length of the Fla B region may be determined by the physical constraints laid down in the first interaction between the hook and Fla B monomer. The Fla A monomer may then have a higher affinity constant than the Fla B monomer for the end of the Fla B region, causing Fla A assembly to begin. One could explain the formation of stubs by immotile mutant strains using the same type of argument. If no Fla B is available, Fla A monomers form a relatively poor interaction with the hook, and soon reach a point of diminishing interaction and thus form a short (and useless) filament.

The *Caulobacter* flagellins are a family of closely related proteins that are immunologically cross-reactive and have nearly identical amino acid compositions (Lagenaur & Agabian 1976, 1978; Osley et al 1977; Fukuda et al 1978; Weissborn et al 1982). Comparable peptides isolated from the purified flagellins have been sequenced and found to be identical except for specific amino acid changes, which in most cases can be accounted for by single base changes. These results suggest that a flagellin family of at least three genes arose by gene duplication (Weissborn et al 1982; Gil & Agabian 1982), and with time acquired separate functions, each with an essential role in the formation and structure of the flagellum.

Structure of the Stalk

Stalk formation in *Caulobacter* results from the initiation of site-restricted cell wall and membrane synthesis that is distinct from lateral wall growth. Schmidt & Stanier (1966) presented evidence that the elongation of the stalk cylinder is due to a circular growth zone at the junction of the stalk and the cell body. Added evidence for this growth zone was provided by Smit & Agabian (1982b), who showed that the surface array, the outermost layer of many bacterial cells, is synthesized during stalk growth at the junction of the stalk and the cell body. Different cellular mechanisms appear to control the initiation of stalk biogenesis and its ultimate length, because fluctuations in phosphate concentration that affect stalk elongation rate do not alter the location or the timing of stalk formation (Schmidt & Stainer 1966). Also, analysis of stalk length has shown that the rate of stalk elongation during the cell cycle varies from cell to cell, unlike the site and initiation of stalk formation (Smit & Agabian 1982b).

Cross-bands have been observed at intervals perpendicular to the long axis of the stalk cylinder (Poindexter & Cohen-Bazire 1964). The bands appear to be annular discs at least partially composed of peptidoglycan (Schmidt 1973). Electron micrographs of isolated discs reveal a highly ordered and complex structure containing a series of 5–6 concentric rings (Jones & Schmidt 1973; Hahnenberger & Shapiro, unpublished). Each of the concentric rings appears to have a unique pattern of radial spokes. One of the inner concentric rings is precisely the diameter of the flagellar basal body L and P rings (23 nm) (Hahnenberger & Shapiro, unpublished). The entire disc, which is 87–88 nm in diameter, is the same size, and has the same radial organization, as a structure that resides at the base of the polar flagella in *Aquaspirillum serpens*, found by Coulton & Murray (1978). Perhaps a disc or collar, which was shown to surround the basal body of the *A. serpens* flagellum, is similarly positioned around the *C. crescentus* basal body. The released flagellum has been shown to contain only filament, hook and rod (Shapiro & Maizel 1973). Upon release of the flagellum, the initiation of wall synthesis could begin at the boundaries of the disc, which has the precise diameter of the stalk. The rings of the basal body may remain in the cell envelope and contribute to the structure of the cross-bands that ultimately appear in the stalk. It has been reported that a new cross-band appears in a stalk for each successive cell division (Staley & Jordan 1973), allowing the age of a stalked cell to be measured by counting the rings in its stalk. However, a report from Swoboda & Dow (1979) suggests that cross-band formation in their *Caulobacter* strain was not coupled to cell division. If it proves true that the appearance of each cross-band coincides with a cell division, then it may be that a polar disc is assembled at each pole of the cell upon chromosome replication, which perhaps functions as a primitive centriole. Experiments in which cell division was blocked by penicillin G have shown that a step prior to cell separation is required for subsequent stalk formation during the transition of the daughter swarmer cell to a stalked cell (Terrana & Newton 1976). During the formation of the predivisional cell, the flagellar basal body may be assembled within the disc at the incipient swarmer cell pole, and at the opposite stalked cell pole a newly formed disc may appear in the lengthening stalk to mark another cell division.

GENERATION OF POLARITY IN THE PREDIVISIONAL CELL

The biogenesis of the *Caulobacter* predivisional cell involves the synthesis of a new set of proteins, and the differential placement of these and other proteins at one cell pole (Figure 4). The visible structures assembled at the incipient swarmer cell pole include a flagellum and several pili. In addition to these structural features, receptor activity for DNA phage appears at the same pole, and the proteins involved in chemotaxis methylation reactions are sequestered in the incipient swarmer cell portion of the predivisional cell (Gomes & Shapiro 1984). This generation of polarity is followed by an equitorial cell division after which distinct programs of differential gene expression are enacted in each of the daughter cells. The developmental program for these daughter cells, which appears to be established in the predivisional cell, is most easily measured by studying specific protein synthesis and distribution. Definition of the mechanisms that alter gene expression from apparently identical sets of genetic information will

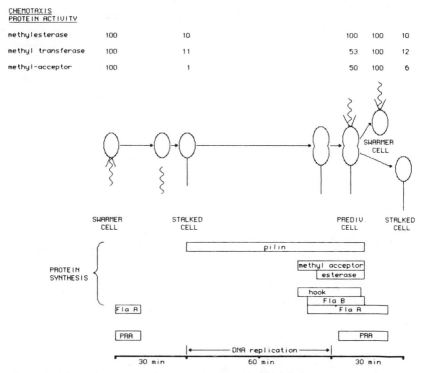

Figure 4   Schematic representation of the *C. crescentus* cell cycle. The time of synthesis of pili (Smit & Agabian 1982a), flagellar (Osley et al 1977, Agabian et al 1979), and chemotaxis proteins (Gomes & Shapiro 1984), and the time period of DNA replication (Degnen & Newton 1972) are shown below the cell cycle. The *PRA* indicates the time of appearance of DNA-phage receptor activity (Agabian-Keshishian & Shapiro 1971, Lagenaur et al 1974, Huguenel & Newton 1982, O'Neill & Bender, unpublished).

ultimately provide a window on the initial establishment of polarity in a single cell carrying newly replicated chromosomes.

Biogenesis of the Flagellum

The biosynthesis and function of the *Caulobacter* flagellum has been shown to require the expression of at least 27 flagellar (*fla*) genes and 3 motility (*mot*) genes (Figure 2) (Johnson & Ely 1979; Johnson et al 1983; Ely et al 1984). These genes must encode approximately 12 structural proteins for the filament, hook, and basal body, in addition to regulatory sequences and proteins required for flagellar assembly and localization, and for rotor function. Thus far, only the hook protein has been associated with a specific *fla* gene (Ohta et al 1984), although the physical map location of the 29 kDa flagellin (Purucker et al 1982; Milhausen et al 1982; Gil & Agabian 1983) and the 25 kDa Fla A and the 27.5 kDa Fla B flagellins (Agabian et al, unpublished) have been determined. The hook gene lies within a cluster of at least seven *fla* genes that covers approximately 17 kilobases of the chromosome (Ohta et al 1984). The hook gene transcript includes at least one additional gene, and is thus an example of an operon in *C. crescentus*. Although the sequences encoding the flagellins are grouped within the *fla* YEFG gene cluster (Figure 2), their transcriptional organization is not yet known.

The time of synthesis of the hook protein and the three flagellins has been shown to be cell cycle–regulated (Shapiro & Maizel 1973; Osley et al 1977; Laganeur & Agabian 1978; Agabian et al 1979; Osley & Newton 1980; Sheffery & Newton 1981). The initiation of synthesis of the hook protein and Fla A and Fla B occur at approximately the same time in the cell cycle (Figure 4), although the peak of hook protein synthesis precedes that of the A and B flagellins (Lagenauer & Agabian 1978; Sheffery & Newton 1981). The synthesis of the hook protein and the Fla B protein is completed prior to cell division, but the synthesis of Fla A protein (the major protein component of the filament) continues in the daughter swarmer cell (Osley et al 1977; Lagenauer & Agabian 1978). Thus, the continued synthesis of Fla A, but not Fla B, protein in the swarmer cell may be one facet of the mechanism that organizes the distribution of the two flagellin subunits in the filament.

A mutation in one of the *fla* genes, *fla* J, has been shown to result in an altered time period of flagellar protein expression. This gene appears to control the time of termination of hook protein synthesis and assembly. In *fla* J mutants a lengthened period of hook synthesis leads to the formation of a polyhook and the absence of an assembled filament (Johnson et al 1979; Sheffery & Newton 1981). Therefore, it appears that the factors contributing to the control of flagellar assembly include the time period of subunit synthesis. However, since there is considerable overlap in the time of synthesis of these proteins, other assembly control mechanisms probably contribute to the formation of the flagellum as well.

The flagellum resides at the cell surface, and thus the proteins of the flagellum must somehow pass through the inner cell membrane during flagellum assembly. The DNA sequences of the hook protein (Ohta & Newton, unpublished) and the 29 kDa flagellin (Gil & Agabian 1983) show that neither protein has a processed N-terminal signal sequence. It may be that the first proteins inserted into the polar membrane are the rings of the basal body, and that one or more of these form a pore (and perhaps specific receptors) for the passage of the rod, hook, and filament subunits.

The Chemotaxis Apparatus

In addition to the biogenesis of the flagellum, the mechanism that directs the flagellar rotor to respond to chemoattractants is under cell cycle control in *C. crescentus* (Shaw et al 1983). *Caulobacter* is able to respond to attractants such as glucose, maltose, galactose, xylose, ribose, alanine, proline, and glutamine (Frederikse & Shapiro, unpublished; Ely, unpublished). *C. crescentus*, which has a single polar flagellum, can swim a relatively long distance in a forward direction; this smooth swim is periodically interrupted by a short reversal in direction and reorientation (Frederikse, unpublished). Reversal is caused by changing the direction of flagella rotation. It is believed that in the presence of attractant the frequency of reversal is decreased, which results in a biased random path toward the attractant.

In both *E. coli* and *S. typhimurium*, signal transmission between the chemoattractant ligand and the flagellar rotor is mediated by a series of membrane-associated and cytoplasmic proteins (Springer et al 1979; Koshland 1981; Boyd & Simon 1982; Stock & Koshland 1984). In *Caulobacter crescentus* at least 8 genetic loci have been found to be involved in this process (Ely & Shapiro, unpublished). Specific membrane receptors, which are able to bind directly to exogenous ligand or to a periplasmic binding protein-ligand complex, are well characterized proteins that are involved in the early stages of signal transduction. These membrane receptors have been termed methyl-accepting chemotaxis proteins (MCPs) because adaptation to chemical stimuli can be correlated with a change in the level of MCP methylation (Springer et al 1977; Silverman & Simon 1977). MCPs have been identified in *C. crescentus* and shown to cross-react with antibodies raised against one of the MCPs purified from *Salmonella*, the TAR protein (Gomes & Shapiro 1984). In addition, two cytoplasmic proteins, which are products of the adjacent *C. crescentus che* R and *che* B genes, have been identified as the chemotaxis-specific carboxylmethyltransferase and the methylesterase, respectively (Gomes & Shapiro 1984; Alexander & Shapiro, unpublished).

These proteins have been extensively studied in both *E. coli* and *Salmonella* (Springer & Koshland 1977; Stock & Koshland 1978; Toews & Adler 1979). In all bacterial systems that have been studied, including *C. crescentus*, the methyltransferase uses S-adenosylmethionine as a methyl donor to methylate the α-carboxyl group of glutamate residues on the membrane-bound MCPs in response to ligand binding. A sensory signal is propagated through a series of as yet undefined reactions to the flagellar rotor, and the combined action of this methyltransferase and a methylesterase cancel this signal.

Analysis of chemotaxis-related functions in synchronized cultures of *C. crescentus* has shown that the MCP methyl acceptor activity and the activities of both the methyltransferase and the methylesterase are present in the predivisional cell, but that upon cell division these activities can only be detected in the daughter swarmer cell (Shaw et al 1983; Gomes & Shapiro 1984). When the swarmer cell loses its flagellum and differentiates into a stalked cell it loses all three activities (Figure 4). Using antibodies to the *Salmonella* TAR MCP (Rousso & Koshland, unpublished), and to the *Salmonella* methylesterase (Stock & Koshland 1978), the time of synthesis of these proteins during the *Caulobacter* cell cycle was determined. Both proteins were found to be synthesized coincident with the synthesis of the flagellar hook protein and the Fla B protein (Gomes & Shapiro 1984) (Figure 4). The time of synthesis of the methyltransferase has not yet been determined, but the appearance of its activity correlates with that of the other differentially synthesized chemotaxis proteins, and its synthesis is probably also under cell cycle control.

Agabian and coworkers (1979) have shown that several cytoplasmic and membrane proteins synthesized in the predivisional cell are asymmetrically distributed among the progeny swarmer and stalked cells. These results suggest that cell type–specific proteins result not only from differential synthesis but from differential distribution (Agabian et al 1979). A specific example is the synthesis and distribution of the proteins involved in the chemotaxis methylation machinery. It has been found that MCPs are synthesized in only the predivisional cell, and are sequestered to the swarmer cell upon division (Gomes & Shapiro 1984). These experiments suggest that upon synthesis the MCPs become associated with the membrane of the incipient swarmer cell portion of the predivisional cell. The appearance of these proteins in the swarmer cell therefore results from a polarity in the predivisional cell. A soluble esterase, which is specifically synthesized in the predivisional cell (Gomes & Shapiro 1984), as well as other soluble proteins (Agabian et al 1979), is also localized to the swarmer cell. Possibly, a transient membrane interaction or compartmentalization facilitates the polar distribution of soluble proteins in the predivisional cell. Alternatively, these soluble proteins may be distributed to both daughter cells, and a protease synthesized or activated in the daughter stalked cell specifically recognizes and degrades these proteins. Arguing against this hypothesis is the observation that mixing cell extracts from progeny stalked cells with extracts of swarmer cells has no effect on the activities of the chemotaxis proteins present in the swarmer cells (Gomes & Shapiro 1984).

However, specific proteolysis is suggested during the swarmer-to-stalked cell transition, when the chemotaxis protein activities are lost, along with flagellar release and the disappearance of pili and phage receptor sites (Figure 4). These changes might be due to the transient synthesis or activation of a protease that reorganizes both the flagellum-bearing pole of the cell, and a subpopulation of cytoplasmic proteins carrying a common protease-recognition sequence.

Regulation of Flagellar and Chemotaxis Gene Expression

TRANSCRIPTION REGULATION   The demonstration that de novo RNA synthesis is required for the expression of specific cell cycle events in *Caulobacter* (Newton 1972; Lagenaur et al 1976) suggested that differential protein synthesis might be controlled, at least in part, at the level of transcription. The subsequent identification of the protein components of the flagellum and the chemotaxis apparatus whose synthesis is temporally regulated, and the isolation of several of the genes encoding these proteins, has allowed experiments that provide additional evidence that transcriptional regulation is important in the generation of asymmetry during the *C. crescentus* cell cycle. The initiation of synthesis of the flagellar proteins, including the hook, Fla A, and Fla B proteins, in the predivisional cell has been shown to be inhibited by rifampicin (Osley et al 1977; Sheffery & Newton 1981). The synthesis of the Fla A protein continues in the daughter swarmer cell, where its synthesis is no longer dependent on de novo RNA synthesis (Agabian et al 1979; Osley et al 1977; Milhausen & Agabian 1983). This result predicts a relatively long half-life for the mRNA encoding the Fla A protein, as has been observed (Agabian et al 1979; Osley et al 1977). In addition, Milhausen & Agabian (1983) have reported that the Fla A mRNA, which is synthesized in the predivisional cell, segregates to the swarmer cell upon cell division. The appearance of mRNAs encoding the hook protein (Newton et al 1985) and flagellin proteins (Milhausen & Agabian 1983; Purucker et al, unpublished), is temporally regulated, although it has not been ruled out that selective mRNA degradation rather than the expected control of initiation of mRNA synthesis is involved.

Evidence has recently been obtained that genes that map in the *fla* YEFG flagellin gene cluster and near the hook protein gene cluster contain promoter regions that direct the timed expression of downstream gene products. These experiments utilized transcription fusions to a derivative of transposon Tn5 (Bellofatto et al 1984). The promoter, but not the translation start signal of the Tn5 neomycin phosphotransferase (NPT II) gene, which confers kanamycin resistance, was deleted from the transposon along with a major portion of the 5' adjacent inverted repeat (IR) sequence. Translation stop signals in all three reading frames were left intact in the retained portion of the IR sequence. This altered transposon (Tn5-VB32) was then used to obtain insertionally inactivated genes involved in motility. Those immotile insertion mutants, which were kanamycin-resistant due to the insertion of Tn5-VB32 adjacent to specific *fla* promoters, were analyzed for their ability to synthesize NPT II at specific times in the cell cycle. Using antibody raised against NPT II, it was found that a Tn5-VB32 insertion into a gene near the hook protein gene cluster caused NPT II to be synthesized only in the predivisional cell, whereas insertion of the promoter probe in the *fla* YEF flagellin gene cluster resulted in the expression of NPT II in both the predivisional and swarmer cells (Champer & Shapiro, in preparation). Other *C. crescentus* strains carrying either wild-type Tn5 or Tn5-VB32 (Bellofatto et al 1984) were found that synthesized NPT II throughout the cell cycle. These experiments provide direct evidence for transcriptional regulation of the temporal control of flagellar gene expression, which could involve either the initiation of new transcripts or some form of attenuation.

The 5' regulatory regions of the differentially transcribed genes appear to contain sequences that participate in the mechanism that controls when they are read. Specific sigma subunits of RNA polymerase participate in the selective recognition of cognate promoter sequences in subsets of sporulation genes in *Bacillus subtilis* (Haldenwang & Losick 1980; Haldenwang et al 1981; Moran et al 1981; Wiggs et al 1981; Wong & Doi 1982; Johnson et al 1983b), and a similar mechanism of gene recognition could occur during the *C. crescentus* cell cycle. Up to now, however, additional RNA polymerase sigma subunits have not been detected in *C. crescentus* (Bendis & Shapiro 1973; Amemiya et al 1977; Cheung & Newton 1977b; Purucker & Amemiya, unpublished).

CASCADE REGULATION OF SYNTHESIS AND ASSEMBLY   The biogenesis of the flagellum and other polar structures is a complex process involving more than the temporal control of specific protein components. Mechanisms must also exist to regulate the assembly of the flagellar rotor, and to control the site of concerted polar morphogenesis; the later process includes the formation of pili and DNA-phage receptors. Many of the *C. crescentus fla* mutants show disrupted flagellum assembly which results in the formation of partial structures (Johnson et al 1983a; Hahnenberger & Shapiro, unpublished). However, in no mutant examined thus far has the cellular location of the complete or partial flagellar structure been altered. These observations suggest that the spacial organization of the pole is controlled by many concerted genetic functions, and that the assembly process, like temporal regulation, is just one component.

The biogenesis of flagella in *E. coli* has been shown to result from a regulatory cascade of flagellar gene products, produced in the order of their assembly (Komeda 1982; Suzuki & Komeda 1981). This cascade has been shown to function by the action of *trans*-acting regulators that control the level of *fla* gene transcription. Control appears hierarchical, with the expression of one group of *fla* genes required for the correct level of transcription of other groups of *fla* genes. Chemotaxis (*che*) gene expression is at the bottom of this regulatory hierarchy, along with the expression of the filament components (Komeda 1982).

In several instances, products of *fla* genes in *C. crescentus* have also been shown to function in *trans* to regulate the expression of other *fla* and *che* genes (Bryan et al 1984; Ohta et al 1984). Portions of the *fla* YEFG gene cluster have been cloned (Purucker et al 1982; Milhausen et al 1982; Shoenlein & Ely, unpublished) and shown to contain genes encoding the flagellins, among other (as yet unidentified) *fla* genes (Purucker et al 1982; Milhausen et al 1982; Gil & Agabian 1983; Bryan et al 1984; Agabian et al, unpublished). Mutations in the *fla* Y gene located 3' to the flagellin genes were found to result in lowered synthesis of the flagellins and the chemotaxis MCPs, methyltransferase, and methylesterase. The genes encoding these chemotactic functions map at sites distal to the *fla* YEFG gene cluster. Mutations in *fla* Y, which produced immotile strains carrying a partial flagellar structure, were found to be effectively complemented by the cloned wild-type region carried on an extrachromosomal plasmid (Bryan et al 1984). The existence of temperature-sensitive mutants in *fla* Y suggests that it is a *fla* Y protein product that can function *in trans* to regulate the expression of flagellin and chemotaxis genes (Bryan et al 1984). These genes are comparable to the *che* and the flagellin *hag* genes at the bottom of the assembly hierarchy postulated for *E. coli* (Komeda 1982). In keeping with this observation, *fla* Y mutants do not affect hook gene expression. Another flagellar gene cluster, which includes the single structural gene for the hook protein, has been cloned (Ohta et al 1982), and mutations 3' to the hook gene have been shown to result in very low levels of expression of the hook and distal flagellin genes (Ohta et al 1984). Several of these mutations were shown to be complemented *in trans* by wild-type genes, suggesting positive regulation analogous to that seen in the flagellin gene cluster.

To determine the level at which the *trans*-acting, positive activation might be exerted, experiments were carried out with mutants generated by insertional inactivation of *fla* and *che* genes by the transcription fusion vector Tn5-VB32 (Bellofatto et al 1984). The *fla* genes carrying Tn5-VB32 were transduced into strains carrying mutations in the *fla* YE gene cluster that were known to "down-regulate" the expression of other *fla* and *che* genes (Bryan et al 1984). The expression of neomycin phosphotransferase II encoded in the Tn5-VB32 is under the control of a given *fla* gene promoter. Therefore, the effect of the mutations in the *fla* YE region on the initiation of transcription can be measured by immunoassay of NPT II synthesis and by the level of kanamycin resistance. It was found that mutations in *fla* YE caused a markedly lowered level of NPT II synthesis and kanamycin resistance (Champer & Shapiro, in preparation). These results demonstrate that *trans*-acting affectors that are products of the *fla* YE genes function at the level of transcription.

Construction of double mutants carrying Tn5-VB32 inactivated *fla* or *che* genes and other mutations in each of the *fla* and *che* genes has allowed analysis of the hierarchy for expression of the 30 *fla* genes and the 8 *che* genes in *C. crescentus* (Champer, unpublished). These experiments measure the effect of a given *fla* or *che* mutation on the level of transcription of a second *fla* gene whose promoter is driving the expression of the inserted NPT II gene. Thus far, individual *fla* mutants have been found that down-regulate the expression of different flagellar genetic loci hierarchically, but the complete order has not yet been determined (Champer, unpublished).

Accumulating evidence suggests that the mechanism controlling the hierarchy of flagellar and chemotaxis gene expression, which modulates the level of gene expression, is one component of a complex series of events that turns these genes on and off during the cell cycle (Bryan et al, in preparation). In *fla* mutants that resulted in the down-regulation of other *fla* and *che* genes it was found that the flagellin and chemotaxis genes were expressed at the correct time in the cell cycle, but in amounts too low to be useful for assembly or function. Aside from the fact that both the temporal control of gene expression and the hierarchical control of the amount of gene expression occur at the transcriptional level, little is known about how the cell regulates these processes.

CYCLIC NUCLEOTIDES Because the flagellin genes and the chemotaxis genes are scattered on the chromosome and yet are under coordinate temporal control, it is reasonable to suggest that small affector molecules may modulate the expression of these genes. It has been shown in *E. coli* that glucose represses flagella formation and that exogenous cyclic AMP relieves this repression (Adler & Templeton 1967; Dobrogosz & Hamilton 1971; Yokota & Goetz 1970). Furthermore, the expression of the *E. coli fla* I gene, which is a major regulatory gene at the top of the flagellar gene hierarchy, is under catabolite control and requires both cyclic AMP and the catabolite activator protein (CAP) for its expression (Komeda 1982).

The activities of adenylate cyclase and guanylate cyclase have been detected in *C. crescentus*, and the guanylate cyclase has been partially purified and characterized (Sun et al 1974). Although variation in the free concentrations of the cyclic nucleotides during the cell cycle was not detected, it was observed that the specific activity of the adenylate cyclase was increased two- to threefold in predivisional cells (Sun et al 1975). Binding proteins for both cyclic AMP and cyclic GMP were identified in *C. crescentus* cell extracts, and separated from each other (Sun et al 1975). It has been observed that in *C. crescentus* growing on poor carbon sources the biosynthesis of flagellar components is repressed by exogenous cyclic GMP derivatives; this repression is reversed by the addition of cyclic AMP derivatives (Kurn & Shapiro 1976). Furthermore, a diauxic lag generated by the shift from glucose to lactose can be overcome by the addition of cyclic AMP derivatives (Shapiro et al 1972; Kurn et al 1977). Mutants which no longer require exogenous cyclic AMP to abolish the diauxic lag were found to be resistant to the repression of flagellar biogenesis by cyclic GMP derivatives (Kurn & Shapiro 1976). Therefore it appears that the relative proportion of the two cyclic nucleotides can affect *C. crescentus* metabolic functions and the synthesis of flagellar components. Cyclic nucleotides might function as cell cycle messengers involved in the temporal control of flagellar biogenesis.

ROLE OF DNA REPLICATION IN TEMPORAL CONTROL   The temporal control of gene expression in *Caulobacter* is related to other cellular functions. Two categories of mutants that show disrupted temporal expression of flagellar biogenesis include those that have also been shown to have blocked DNA replication (Osley et al 1977; Sheffery & Newton 1981), and those that are also blocked in membrane biosynthesis (Shapiro et al 1982). Temperature-shift experiments using synchronized cultures of conditional mutants in DNA replication have shown that disrupting chromosome replication prior to the initiation of hook protein, flagellin A, and flagellin B synthesis blocks their synthesis. Furthermore, by causing a transient block in DNA replication in synchronized cultures, and determining the time of flagellar protein synthesis following the resumption of DNA synthesis, Newton and his coworkers have provided evidence that the replication of the chromosome can act as a timing mechanism (Osley et al 1977; Sheffery & Newton 1981), although alternative interpretations have been discussed (Newton 1984). The last time period in which DNA synthesis is required for the induction of flagellular protein synthesis has been referred to as the "execution point" (Osley & Newton 1980; Newton 1984). This execution point corresponds to the period just prior to the initiation of flagellar protein synthesis. The appearance of the polar receptor sites for DNA phage has also been shown to depend on a specific DNA replication execution point (Huguenel & Newton 1982).

An important question is whether or not the time period of specific flagellar gene replication coincides with the DNA execution points required for their expression. It was recently found that the hook gene cluster replicates at about the same time as the hook gene is expressed during the DNA replication phase of the cell cycle (Lott & Newton, unpublished). Because both the structural genes and the genes required for flagellar assembly are relatively scattered on the chromosome (Figure 2), it is reasonable to predict that only the initial steps in the hierarchy of the flagellar regulatory cascade are directly dependent on execution points. The expression of the execution point could allow the initiation of two separate regulatory controls: (1) all genes carrying a sequence in their promoter region that responds to a small affector molecule (such as a cyclic nucleotide) would be turned on to an "open position" and then (2) their level of expression would be modulated in a cascade fashion by *trans*-acting, positive regulators, which are products of genes higher in the flagellar hierarchy.

Biogenesis of Pili

The *C. crescentus* polar pili have been shown to function as receptors for RNA bacteriophage (Schmidt 1966; Bendis & Shapiro 1970; Shapiro & Agabian-Keshishian 1970), and this function has allowed the demonstration that pili biogenesis is a cell cycle–regulated event (Schmidt 1966; Shapiro & Agabian-Keshishian 1970; Shapiro et al 1971). Pili appear at the flagellum-bearing pole of the predivisional cell coincident with the appearance of the flagellum and receptor activity for DNA phage (Laganauer & Agabian 1977a). They are subsequently lost from the swarmer cell pole coincident with the release of the flagellum and the loss of DNA-phage receptor activity and the chemotaxis methylation proteins. Although the structure of the *Caulobacter* polar pili has not been determined, the major protein component, pilin, has been identified as a small protein of approximately 8 kDa (Smit et al 1981). Pilin synthesis is temporally regulated (Smit & Agabian 1982a), but both the time period of pilin synthesis and its apparent mechanism of cell cycle control differ from those of the flagellar and chemotaxis proteins. Antibody raised against purified pilin was used in radioimmune assays to demonstrate that the initiation of pilin synthesis occurs in the stalked cell soon after the swarmer-to-stalked cell transition, and continues until it is assembled into pili in the predivisional cell (Smit & Agabian 1982a). As opposed to flagellar protein synthesis and DNA-phage receptor activity, pilin synthesis does not appear to depend on a DNA execution point within the cell cycle in which its synthesis is initiated (Smit & Agabian 1982a). Although the assembly of the polar organelles in *C. crescentus* appears to occur in a coordinate fashion, these observations suggest that the temporal synthesis of their component proteins is not regulated by a common sequence of events.

MEMBRANE BIOGENESIS

The surface of the *Caulobacter* cell is the major site of structural change during the cell cycle. The biogenesis of the flagellum, the chemotaxis machinery, pili, DNA-phage receptors, and the stalk, all involve trans-membrane functions. Furthermore, the formation of these structures and functions at a specific cell pole requires some form of inherited positional information. Because portions of the cell membrane are retained from one cell cycle to another, and the division plane is the ultimate site of polar morphogenesis (see Figure 1), the membrane, in a sense, has positional memory. It is thus an attractive possibility that the structure and biosynthesis of the cell membrane contribute to the complex regulatory circuit that controls the timing and positioning of membrane and membrane-associated proteins during the cell division cycle. To understand how this might happen, the patterns of membrane protein and lipid synthesis have been examined, and mutants in membrane lipid metabolism have been analyzed with respect to their effects on the cell cycle expression of spatially regulated events.

Membrane Proteins

The membrane protein composition has been shown to change as a function of the cell cycle. Differences in membrane protein composition during the cell cycle result from asymmetric distribution of continuously synthesized proteins to the daughter cells upon division (Agabian et al 1979), and the periodic synthesis of some membrane proteins. Analysis by one- and two-dimensional SDS gel electrophoresis of total membrane protein synthesis has shown that of the detectable membrane proteins, the synthesis of ~ 6–10% appears to be differentially controlled (Cheung & Newton 1977a; Agabian et al 1979; Milhausen & Agabian 1981; Clancy & Newton 1982). The synthesis of such proteins is initiated at different stages in the cell cycle, including the swarmer cell, the swarmer-to-stalked cell transition, and the formation of the predivisional cell. In addition to the inner and outer membranes in the cell envelope, some investigators have reported the presence of a distinct internal membrane complex at the base of the stalk (Cohen-Bazire et al 1966; Schmidt & Stanier 1966). It is not known if the membrane proteins whose synthesis is under temporal control are preferentially associated with these membrane components.

Among the proteins identified as having temporally regulated synthesis are the components of the flagellum and pili. Although anchored in the cell membrane, the major portions of these structures are outside the cell; the protein components in the transmembrane anchor have not been identified in either case. Thus far, the only identified integral membrane proteins whose synthesis is temporally regulated are the methyl-accepting chemotaxis proteins (MCPs) (Gomes & Shapiro 1984). The MCPs are synthesized in the predivisional cell, and then asymmetrically sequestered in the daughter swarmer cell.

Phospholipid and Fatty Acid Metabolism

The phospholipid composition of *Caulobacter crescentus* membranes is unusual in that positively charged phosphotidylglycerol (PG) and cardiolipin (CL) are the major components. Neither phosphotidylethanolamine nor phosphotidylserine, the major lipid components of most bacterial membranes, appear to be synthesized in *C. crescentus* (Contreras et al 1978). The enzymatic pathway of PG and CL synthesis in *C. crescentus* is the same as for the comparable portion of the pathway in *E. coli* (Contreras et al 1979). Fatty acid components of the phospholipids consist of both saturated and unsaturated 14-, 16-, and 18-carbon fatty acid (Chow & Schmidt 1974; Letts et al 1982). The identification of vaccenic acid $(18:1\Delta^{11cis})$ rather than oleic acid $(18:1\Delta^{9cis})$ in *C. crescentus* phospholipids suggests that, like *E. coli* (Brock et al 1967), *C. crescentus* synthesizes fatty acids by the anaerobic pathway in which a double band is introduced on a 10-carbon chain during chain elongation (Letts et al 1982). Exogenous fatty acids were shown to repress the synthesis of new fatty acids, to induce the turnover of endogenous fatty acids, and to be efficiently incorporated, intact, into phospholipids (Letts et al 1982). These events facilitate a rapid change in the phospholipid composition in response to fatty acids in the growth environment. At a slower rate, *C. crescentus* is able to use several unsaturated fatty acids as a carbon source. The pathway of exogenous fatty acid utilization has been determined, and it has been shown that their transport and the activity of the five enzymes in the degradation pathway are repressed by glucose (O'Connell & Shapiro, unpublished).

Coordination of Membrane Synthesis and Cell Cycle Events

The membrane phospholipid composition does not change during the cell cycle, although minor phospholipids could be detected in swarmer cells of pulse-labeled cultures (Mansour et al 1980). Inexperienced stalked cells, which result from the swarmer-to-stalked cell transition, show the same rate of phospholipid synthesis as the swarmer cells and in this parameter can be distinguished from the experienced stalked cells, which result from cell division. Fatty acid synthesis has been found to occur throughout the cell cycle (Letts & Henry, unpublished). Recently, O'Neill & Bender reported that phospholipid synthesis in *C. crescentus* is periodic (O'Neill & Bender, submitted for publication).

The relationship of membrane lipid synthesis to other temporal and spatial events that occur during the cell cycle has been investigated by isolating and analyzing mutants auxotrophic for lipid precursors (Contreras et al 1979, 1980; Letts et al 1982; Shapiro et al 1982; Hodgson et al 1984a,b,c). Of the three lipid auxotrophs studied thus far, two result in cell death when growth is attempted in the absence of supplement (Contreras et al 1979, 1980; Shapiro et al 1982; Hodgson et al 1984a). Although other bacterial lipid auxotrophs have not been observed to lose viability (Mindich 1970), yeast mutants auxotrophic for fatty acids and inositol have been shown to cause cell death upon starvation, probably due to the imposition of unbalanced growth (Henry 1973; Culbertson & Henry 1975). The biogenesis of the membrane might therefore be more closely integrated with other cell cycle functions in *Caulobacter* than it is in other bacteria. Analysis of cell cycle events in *C. crescentus* strains auxotrophic for glycerol-3-PO_4 and fatty acids carrying mutations in *gps* A and *fat* A, respectively, has shown this to be the case.

Strains carrying mutations in *gps* A have tenfold lowered activities of the enzyme *sn*-glycerol-3-PO_4 dehydrogenase, and phospholipid synthesis ceases in the absence of glycerol-PO_4 (Contreras et al 1979). Accompanying phenotypes include the cessation of DNA synthesis, and a block in the synthesis of the flagellar hook protein, the flagellins, and a subset of membrane proteins shown to be under temporal contral (Contreras et al 1980; Shapiro et al 1982). These multiple phenotypes are believed to be due to a single lesion in *gps* A (for map position see Figure 2), or to mutations in coregulated genes adjacent to *gps* A, based on the analysis of revertants and transductants carrying the mutant phenotype in different genetic backgrounds (Contreras et al 1979; Hodgson et al 1984b,c). The regulatory mechanisms that account for these multiple phenotypes, including those affecting DNA synthesis, are not known. However, it is reasonable to assume that the block in flagellar protein synthesis is due to its dependence on DNA synthesis, as previously shown by Newton and his colleagues (Osley et al 1977; Sheffery & Newton 1981). An auxotroph dependent on exogenous fatty acids, altered in the *fat* B gene, has been shown to lead to a cell cycle block and the inhibition of the initiation of DNA synthesis in the absence of supplement (Hodgson et al 1984b). Although there may be a dependency relationship between membrane synthesis and the replication of the chromosome in *C. crescentus*, it has been clearly shown that a block in DNA synthesis has no effect on membrane lipid synthesis (Mansour et al 1981).

Analysis of all lipid mutants studied thus far has shown that the cessation of phospholipid synthesis does not alter the induction of the lactose transport system, or the synthesis of bulk membrane protein, but that the synthesis of a subset of temporally controlled membrane proteins is coordinated with membrane lipid biogenesis (Contreras et al 1980; Shapiro et al 1982). Recently, the synthesis of the membrane methyl-accepting chemotaxis proteins (Gomes & Shapiro, unpublished), and the appearance of polar DNA-phage receptors (O'Neill & Bender, unpublished) have been shown to depend on ongoing phospholipid synthesis. This may be a consequence of the role played by a specific membrane system that coordinates polar organization and the temporal synthesis of its constituent proteins.

SPATIAL ORGANIZATION

The generation of polarity in *Caulobacter*, as in any living cell, reflects the asymmetric distribution of specific cellular components. Although the establishment of polarity results in localized changes in cell structure and the accompanying polar distribution of a subset of protein components, other cellular asymmetries may include structural changes in one of the two newly divided chromosomes, the sequestering of mRNAs for specific proteins, or both.

Differential Protein Synthesis

The events that accompany the generation of polarity in the predivisional cell (see Figure 4) include the replication of the chromosome, and the synthesis of proteins that become part of the polar components. These are flagella, pili, the chemosensory transduction apparatus, and the as yet unidentified macromolecular receptors for several large DNA phage. Although many of the differentially allocated proteins appear to be sequestered to one portion of the predivisional cell at the time of their differential synthesis (Figure 4), the specific distribution of proteins synthesized continuously throughout the cell cycle has also been observed (Figure 5). Agabian & coworkers (1979) have shown that some soluble proteins synthesized throughout the cell cycle appear in the outer membrane of the daughter stalked cell upon cell division. Predominant among these is a protein with a molecular weight similar to the 130 kDa

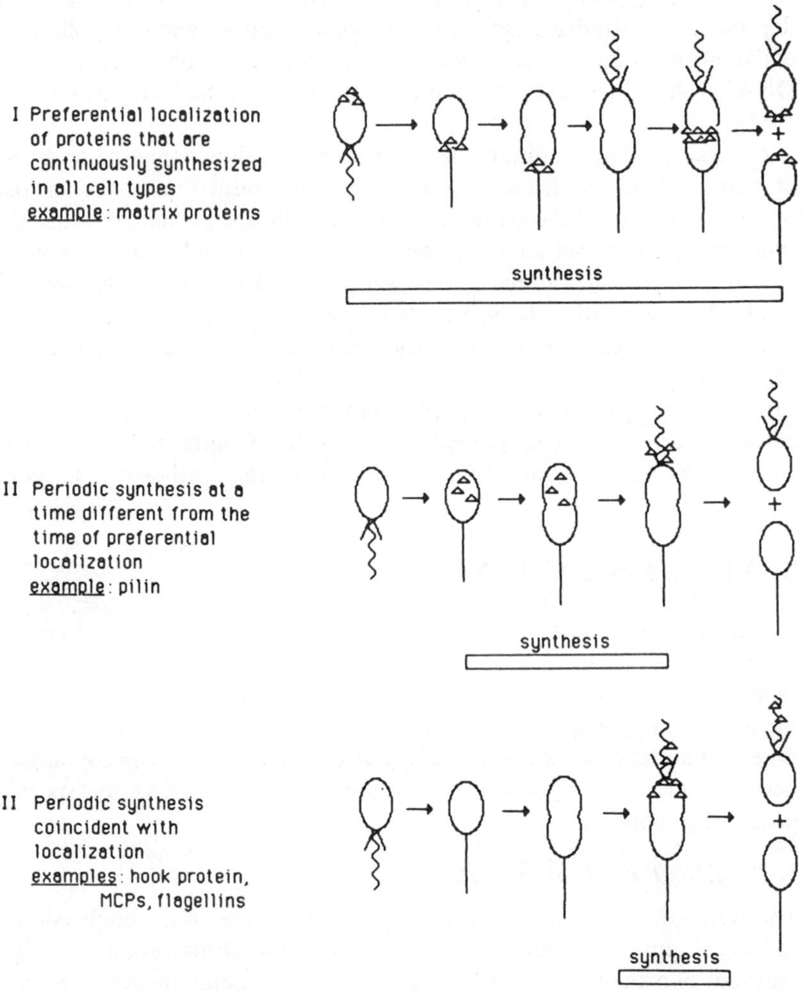

Figure 5 Different relationships observed between the time of protein synthesis and protein localization during the *Caulobacter* cell cycle. The site of matrix protein deposition was determined by Smit & Agabian (1982b). The time of synthesis and assembly of pilin (Smit & Agabian 1982a), the hook protein and flagellins (Osley et al 1977; Agabian et al 1979), and the methyl-accepting chemotaxis proteins (MCPs) (Gomes & Shapiro 1984) is represented as an open bar. The triangles represent newly synthesized proteins.

protein that is a major component of a periodic matrix found at the outermost surface of the cell (Smit et al 1981a; Smit & Agabian 1982b, 1984). This periodic surface array covers the entire surface of the *C. crescentus* cell, and is composed of three proteins that are continuously synthesized throughout the cell cycle (Agabian et al 1979; Smit et al 1981a). Immunomicroscopy, using antibody to the proteins of the surface array, showed that the location of the newly synthesized proteins is regulated such that the deposition of new matrix material occurs at the division plane upon cell division, and at the base of the stalk upon the initiation of stalk formation and during its continued elongation (Smit & Agabian 1982b). Therefore, spatial differentiation can result from a localized site of assembly for either continuously synthesized proteins, or proteins that are made periodically.

Coordinate Assembly

Examination of the proteins whose synthesis is under temporal control reveals that although both the time and site of their assembly is coordinate, the time of their synthesis during the cell cycle is not (Figure 5). For example, the pilin subunits of the polar pili are synthesized in the stalked cell long before they are assembled into pili at the predivisional cell pole (Smit et al 1982a). Conversely, the components of the flagellum and the chemotaxis machinery are synthesized fairly close to the time of their assembly (Osley et al 1977; Agabian et al 1979; Gomes & Shapiro 1984). The coordinate regulation of polar organelle assembly, but not the time of initiation of differential protein synthesis, could indicate an assembly mechanism that can function separate from the mechanism controlling the turning on of gene expression.

Genetic loci, designated *ple* A, *ple* B, and *ple* C (Figure 2), have been identified that are involved in the coordinate control of polar organelle assembly (Ely et al 1984; Fukuda et al 1981). *Ple* mutants have been identified that are able to synthesize the components of the polar organelles, but are unable to assemble flagella, pili, and functional DNA-phage receptors (Kurn et al 1974; Fukuda et al 1981). In some of these mutants, stalk assembly occurs normally in the absence of flagella, pili, and DNA-phage receptor assembly (Kurn et al 1974, Fukuda et al 1974, 1976); in others, flagella are assembled but fail to function, and the stalk as well as pili and phage receptors are not assembled (Fukuda et al 1977, 1981). In a third type of *ple* mutant stalk formation is normal but the mutant is unable to assemble the polar flagellum, pili, and DNA-phage receptors, or to synthesize the flagellar components (Fukuda et al 1981, Ely et al 1983).

These results suggest that there are several stages of coordinate polar assembly that are controlled by different gene products. It has been postulated that some kind of "organization center" (Huguenel & Newton

1982, 1984a), or a "specialized membrane complex" (Agabian et al 1979, Smit & Agabian 1982b) is put in place at the new cell poles that form at the site of cell division. Genetic evidence has been provided suggesting that a late step in the cell division pathway is required for the normal assembly of the stalk and polar structures (Huguenel & Newton 1982). Biochemical evidence from experiments in which isolated membrane vesicles bearing the polar flagellum have been separated from vesicles originating from other parts of the cell by immunoaffinity chromatography suggests that the membranes are organized in stable domains (Huguenel & Newton 1984a). Analysis of the protein composition of the polar membrane fraction showed that a unique set of proteins, including the pool of newly synthesized flagellins, becomes loosely associated with the polar membrane. The *ple* gene products might influence the formation of those polar organization centers.

Positioning of Proteins

A fundamental question in the organization of the *Caulobacter* pre-divisional cell involves the signals that direct the placement of proteins to one portion of the cell as opposed to another. The putative organizing region laid down at the new cell poles generated at the site of cell division might provide specific receptors for the initiation of assembly of the components of the flagellum, the pili, and the DNA phage receptors, as well as the basal plate. As described earlier, the basal plate might in turn later provide the spatial template for the initiation of stalk formation following flagellar release from the swarmer cell.

Receptor-ligand interaction requires that some portion of the protein ligand recognize the receptor. Recent gene fusion experiments in yeast to determine how the α-2 protein is targeted to the nucleus have suggested that a short amino acid sequence at the amino terminus of the protein is sufficient to produce accurate localization (Hall et al 1984). Similarly, a short amino acid sequence is sufficient to direct the SV40 large T antigen to the nucleus in transformed Vero cells (Kalderon et al 1984). Localization experiments with the cloned *Caulobacter* flagellar and chemotaxis protein gene fusions should be able to test whether a specific amino acid sequence can direct the correct placement of the fusion protein in the cell.

The generation of polarity in the predivisional cell appears to involve more than the correct distribution and placement of proteins. The major component of the flagellar filament, flagellin A, is synthesized in the predivisional cell as well as in the progeny swarmer cell upon cell division. The messenger RNA for this protein appears to be relatively long-lived and to be synthesized only in the predivisional cell (Osley et al 1977), which suggests that the flagellin A mRNA is segregated to the incipient swarmer cell portion of the predivisional cell, where it is retained upon cell division. Using cloned flagellin gene fragments as probes, Milhausen & Agabian (1983) have provided evidence that the flagellin mRNAs, while synthesized in the predivisional cell, are detected only in the progeny swarmer cell after division. Therefore, the localized assembly of at least the distal portion of the flagellum appears to require the correct spatial localization of mRNA, followed by either vectoral translation at the correct site of the polar membrane or posttranslational localization. There is some evidence for the latter possibility (Huguenel & Newton 1984b).

The synthesis of *Caulobacter* methyl-accepting chemotaxis proteins (MCPs) has also been studied. In this case the protein synthesis is restricted to the predivisional cell. Upon synthesis, MCPs are inserted into the membrane of the incipient swarmer cell portion of the cell (Gomes & Shapiro 1984), and they are thereby partitioned into the daughter swarmer cell upon division. The simplest possibility is that these MCPs become attached to the basal body and that MCP localization is a consequence of flagellar assembly. However, this was found not to be the case. The MCPs are not localized in the immediate vicinity of the flagellum, but are inserted in the membrane of the entire swarmer cell portion of the predivisional cell (Gomes & Shapiro 1984; Nathan & Gomes, unpublished). In addition, two soluble proteins involved in the chemotactic methylation machinery, the methyltransferase and methylesterase, are also partitioned to the swarmer cell (Gomes & Shapiro 1984). The mechanism of this localization is also unknown. One possible explanation of how the MCPs and the soluble chemotaxis proteins are localized is that the transcription of genes encoding these proteins occurs from only one of the two chromosomes present in the predivisional cell. Perhaps the interaction of the newly duplicated chromosome and the membrane at the incipient swarmer cell pole of the predivisional cell affects chromosome structure and thereby makes its chemotaxis genes available for localized transcription and translation. Evidence that the membrane-associated nucleoid differs in swarmer and stalked cells with respect to sedimentation coefficient and protein composition (Evinger & Agabian 1979; Milhausen & Agabian 1981) lends support to the idea that the structure of the daughter chromosomes prior to cell division may contribute to the generation of polarity in *Caulobacter*.

It is apparent that many cellular mechanisms, including the establishment of an organization center at the plane of division, the differential availability of genes in the daughter chromosomes for transcription and translation, specific signals in the promoter regions that dictate temporal expression and the cellular distribution of specific mRNAs, and perhaps positional signals at the amino-terminal portion of the protein, all contribute to the spatial organization of the cell. The establishment of polarity is a fundamental problem in all cells. The combined power of genetic and biochemical analyses makes the mechanisms of temporal and spatial control during cell differentiation in *Caulobacter* readily accessible for study, and a complete understanding of these processes is possible in the near future.

Literature Cited

Abeles, A. L., Snyder, K. M., Cha Horaj, D. K. 1984. *J. Mol. Biol.* 173: 307–24

Adler, J., Templeton, D. 1967. *J. Gen. Microbiol.* 46: 175–84

Agabian, N., Evinger, M., Parker, E. 1979. *J. Cell Biol.* 81: 123–36

Agabian-Keshishian, N., Shapiro, L. 1971. *Virology* 44: 46–53

Amemiya, L., Wu, C. W., Shapiro, L. 1977. *J. Biol. Chem.* 252: 4157–65

Barrett, J. T., Croft, R. H., Ferber, D. M., Gerardot, C. J., Schoenlein, P. V., Ely, B. 1982a. *J. Bacteriol.* 151: 888–98

Barrett, J. T., Rhodes, C. S., Ferber, D. M., Jenkins, B., Kuhl, S. A., Ely, B. 1982b. *J. Bacteriol.* 149: 889–96

Bellofatto, V., Shapiro, L., Hodgson, D. 1984. *Proc. Natl. Acad. Sci. USA* 81: 1035–39

Bendis, I. K., Shapiro, L. 1973. *J. Bacteriol.* 115: 848–57

Bendis, I., Shapiro, L. 1970. *J. Virol.* 6: 847–54

Blobel, G. 1980. *Proc. Natl. Acad. Sci. USA* 77: 1496–1500

Bode, W., Engel, J., Winklmair, D. 1972. *Eur. J. Biochem.* 26: 313–27

Boyd, A., Simon, M. I. 1982. *Ann. Rev. Physiol.* 44: 501–17

Brock, D. J. H., Kass, L. R., Block, K. 1967. *J. Biol. Chem.* 242: 4432–40

Bryan, R., Purucker, M., Gomes, S. L., Alexander, W., Shapiro, L. 1984. *Proc. Natl. Acad. Sci. USA* 81: 1341–45

Cheung, K. K., Newton, A. 1977a. *Dev. Biol.* 56: 417–25

Cheung, K. K., Newton, A. 1977b. *J. Biol. Chem.* 253: 2254–61

Chow, T. C., Schmidt, J. M. 1974. *J. Gen. Microbiol.* 83: 369–73

Clancy, M. J., Newton, A. 1982. *Biochim. Biophys. Acta* 686: 160–69

Clarke, D. J. 1968. *Cold Spring Harbor Symp. Quant. Biol.* 33: 823–38

Cohen-Bazire, G. R., Kunisawa, R., Poindexter, J. S. 1966. *J. Gen. Microbiol.* 42: 301–8

Contreras, I., Bender, R., Mansour, J., Henry, S., Shapiro, L. 1979. *J. Bacteriol.* 140: 612–19

Contreras, I., Shapiro, L., Henry, S. A. 1978.

J. Bacteriol. 135: 1130–36

Contreras, I., Weissborn, A., Amemiya, K., Mansour, J., Henry, S., et al. 1980. *J. Mol. Biol.* 138: 401–9

Coulton, J. W., Murray, R. G. E. 1978. *J. Bacteriol.* 136: 1037–49

Culbertson, M., Henry, S. A. 1975. *Genetics* 80: 23–40

Davis, J. S. 1981. *Biochem. J.* 197: 309–14

Degnen, S. T., Newton, A. 1972a. *J. Mol. Biol.* 64: 671–80

Degnen, S. T., Newton, A. 1972b. *J. Bacteriol.* 110: 852–56

De Pamphilis, M. L., Adler, J. 1971a. *J. Bacteriol.* 105: 384–95

De Pamphilis, M. L., Adler, J. 1971b. *J. Bacteriol.* 105: 396–407

Dobrogosz, W. J., Hamilton, P. B. 1971. *Biochem. Biophys. Res. Commun.* 42: 202–7

Ely, B. 1979. *Genetics* 91: 371–79

Ely, B., Croft, R. H. 1982. *J. Bacteriol.* 149: 620–25

Ely, B., Croft, R. H., Gerardot, C. J. 1984. *Genetics* 108: 523–32

Ely, B., Johnson, R. C. 1977. *Genetics* 87: 391–99

Ely, B., Shapiro, L. 1984. In *Microbial Development*, ed. R. Losick, L. Shapiro, pp. 1–26. New York: Cold Spring Harbor Lab. 303 pp.

Emerson, S. U., Tokuyasu, K., Simon, M. I. 1970. *Science* 169: 190–92

Evinger, M., Agabian, N. 1977. *J. Bacteriol.* 132: 294–381

Evinger, M., Agabian, N. 1979. *Proc. Natl. Acad. Sci. USA* 76: 175–78

Fukuda, A., Asada, M., Koyasu, S., Yoshida, H., Yaginuma, K., Okada, Y. 1981. *J. Bacteriol.* 145: 559–72

Fukuda, A., Iba, H., Okada, Y. 1977. *J. Bacteriol.* 131: 280–87

Fukuda, A., Koyasu, S., Okada, Y. 1978. *FEBS Lett.* 95: 70–75

Fukuda, A., Miyakawa, K., Okada, Y. 1974. *Proc. Jpn. Acad.* 50: 839–42

Fukuda, A., Miyakawa, K., Iida, H., Okada, Y. 1976. *Mol. Gen. Genet.* 149: 167–73

Fukuda, A., Okada, Y. 1977. *J. Bacteriol.* 130: 1199–1205

Gil, P. R., Agabian, N. 1982. *J. Bacteriol.* 150:925–33

Gil, P. R., Agabian, N. 1983. *J. Biol. Chem.* 258:7395–7401

Gomes, S. L., Shapiro, L. 1984. *J. Mol. Biol.* 178:551–68

Guerola, N., Ingraham, J. L., Cerda-Olmedo, E. 1971. *Nature New Biol.* 230:122–25

Haldenwang, W. G., Losick, R. 1980. *Proc. Natl. Acad. Sci. USA* 77:7000–4

Haldenwang, W. G., Lang, N., Losick, R. 1981. *Cell* 23:615–24

Hall, M. N., Hereford, L., Herskowitz, I. 1984. *Cell* 36:1057–65

Helmstetter, C. E., Pierucci, O. 1968. *J. Bacteriol.* 95:1627–33

Henry, S. A. 1973. *J. Bacteriol.* 116:1293–1303

Hereford, L. M., Hartwell, L. H. 1974. *J. Mol. Biol.* 84:445–61

Hodgson, D., Shaw, P., Letts, V., Henry, S., Shapiro, L. 1984a. *J. Bacteriol.* 158:430–40

Hodgson, D., Shaw, P., O'Connell, M., Henry, S., Shapiro, L. 1984b. *J. Bacteriol.* 158:156–62

Hodgson, D., Shaw, P., Shapiro, L. 1984c. *Genetics* 108:809–26

Huguenel, E., Newton, A. 1982. *Differentiation* 21:71–78

Huguenel, E., Newton, A. 1984a. *Proc. Natl. Acad. Sci. USA* 81:3409–13

Huguenel, E., Newton, A. 1984b. *J. Bacteriol.* 157:727–32

Iba, H., Fukuda, A., Okada, Y. 1977. *J. Bacteriol.* 129:1192–97

Iba, H., Okada, Y. 1980. *J. Mol. Biol.* 139:733–39

Iino, T. 1969. *J. Gen. Microbiol.* 56:227–39

Jarvick, J., Botstein, D. 1973. *Proc. Natl. Acad. Sci. USA* 70:2046–50

Johnson, R. C., Ely, B. 1977. *Genetics* 86:25–32

Johnson, R. C., Ely, B. 1979. *J. Bacteriol.* 137:627–34

Johnson, R. C., Ferber, D. M., Ely, B. 1983a. *J. Bacteriol.* 154:1137–44

Johnson, R. C., Walsh, J. P., Ely, B., Shapiro, L. 1979. *J. Bacteriol.* 138:984–89

Johnson, W. C., Moran, C. P., Losick, R. 1983b. *Nature* 302:800–4

Jollick, J. D., Schervish, E. M. 1972. *J. Gen. Microbiol.* 73:403–7

Jones, H. C., Schmidt, J. M. 1973. *J. Bacteriol.* 116:466–70

Kalderon, D., Roberts, B. L., Richardson, W. D., Smith, A. E. 1984. *Cell* 39:499–509

Komeda, Y. 1982. *J. Bacteriol.* 150:16–26

Koshland, D. E. Jr. 1981. *Ann. Rev. Biochem.* 50:765–82

Koyasu, A., Asada, M., Fukuda, A., Okada, Y. 1981. *J. Mol. Biol.* 153:471–75

Koyasu, S., Fukuda, A., Okada, Y. 1980. *J. Biochem.* 87:363–66

Koyasu, S., Fukuda, A., Okada, Y. 1981. *J. Gen. Microbiol.* 126:111–21

Koyasu, S., Fukuda, A., Okada, Y. 1982. *J. Gen. Microbiol.* 128:1117–24

Koyasu, S., Fukuda, A., Okada, Y. 1984. *J. Biochem.* 95:593–95

Koyasu, S., Fukuda, A., Okada, Y., Poindexter, J. S. 1983. *J. Gen. Microbiol.* 129:2789–99

Koyasu, S., Shirakihara, Y. 1984. *J. Mol. Biol.* 173:125–30

Krikos, A., Mutoh, N., Boyd, A., Simon, M. J. 1983. *Cell* 33:615–22

Kurn, N., Ammer, S., Shapiro, L. 1974. *Proc. Natl. Acad. Sci. USA* 71:3157–61

Kurn, N., Shapiro, L. 1976. *Proc. Natl. Acad. Sci. USA* 73:3303–7

Kurn, N., Shapiro, L., Agabian, N. 1977. *J. Bacteriol.* 131:951–59

Lagenaur, C., Agabian, N. 1976. *J. Bacteriol.* 128:435–44

Lagenaur, C., Agabian, N. 1977a. *J. Bacteriol.* 131:340–46

Lagenaur, C., Agabian, N. 1977b. *J. Bacteriol.* 132:731–33

Lagenaur, C., Agabian, N. 1978. *J. Bacteriol.* 135:1062–69

Lagenaur, C., De Martini, M., Agabian, N. 1978. *J. Bacteriol.* 136:795–98

Lagenaur, C., Farmer, S., Agabian, N. 1974. *Virology* 77:401–7

Letts, V., Shaw, P., Shapiro, L., Henry, S. 1982. *J. Bacteriol.* 151:1269–78

Mansour, J., Henry, S., Shapiro, L. 1980. *J. Bacteriol.* 144:262–69

Mansour, J., Henry, S., Shapiro, L. 1981. *J. Bacteriol.* 145:1404–9

Marino, W., Ammer, S., Shapiro, L. 1976. *J. Mol. Biol.* 107:115–30

Milhausen, M., Agabian, N. 1981. *J. Bacteriol.* 148:163–73

Milhausen, M., Agabian, N. 1983. *Nature* 302:630–32

Milhausen, H., Gill, P. R., Parker, G., Agabian, N. 1982. *Proc. Natl. Acad. Sci. USA* 79:6847–51

Mindich, L. 1970. *J. Mol. Biol.* 49:415–32

Moran, C. P. Jr., Lang, N., Banner, C. D. G., Haldenwang, W. G., Losick, R. 1981. *Cell* 25:783–91

Nathan, P., Osley, M. A., Newton, A. 1982. *J. Bacteriol.* 151:503–6

Newton, A. 1972. *Proc. Natl. Acad. Sci. USA* 69:447–51

Newton, A. 1984. In *The Microbial Cell Cycle*, ed. P. Nurse, E. Streiblova, pp. 51–75. Cleveland, Ohio: CRC. 285 pp.

Newton, A., Allebach, E. 1975. *Genetics* 80:1–11

Newton, A., Ohta, N., Huguenel, E., Chen,

L.-S. 1985. *Spores IX*, ed. P. Setlow, J. Hoch. Washington, D.C.: Am. Soc. Microbiol. In press

Ohta, N., Chen, L.-S., Newton, A. 1982. *Proc. Natl. Acad. Sci. USA* 79:4863–67

Ohta, N., Swanson, E., Ely, B., Newton, A. 1984. *J. Bacteriol.* 158:897–904

O'Neill, E. A., Bender, R. A. 1984. *Abstr. Ann. Meet. Am. Soc. Microbiol.* p. 138

Osley, M. A., Newton, A. 1974. *J. Mol. Biol.* 90:359–70

Osley, M. A., Newton, A. 1977. *Proc. Natl. Acad. Sci. USA* 74:124–28

Osley, M. A., Newton, A. 1978. *J. Bacteriol.* 135:10–17

Osley, M. A., Newton, A. 1980. *J. Mol. Biol.* 138:109–28

Osley, M. A., Sheffery, M., Newton, A. 1977. *Cell* 12:393–400

Poindexter, J. S. 1964. *Bacteriol. Rev.* 28:231–95

Poindexter, J. S. 1981. *Microbiol. Rev.* 45:123–79

Poindexter, J. S., Cohen-Bazire, G. 1964. *J. Cell Biol.* 23:587–607

Poindexter, J. S., Hagenzieker, J. G. 1981. *Can. J. Microbiol.* 27:704–19

Poindexter, J. S., Hornack, P. R., Armstrong, P. A. 1967. *Arch. Mikrobiol.* 59:237–46

Purucker, M., Bryan, R., Amemiya, K., Ely, B., Shapiro, L. 1982. *Proc. Natl. Acad. Sci. USA* 79:6797–6801

Schmidt, J. M. 1966. *J. Gen. Microbiol.* 45:347–53

Schmidt, J. M. 1973. *Arch. Mikrobiol.* 89:33–40

Schmidt, J. M., Stanier, R. Y. 1966. *J. Cell Biol.* 28:423–36

Shapiro, L. 1976. *Ann. Rev. Microbiol.* 30:377–407

Shapiro, L., Agabian-Keshishian, N. 1970. *Proc. Natl. Acad. Sci. USA* 67:200–3

Shapiro, L., Agabian-Keshishian, N., Bendis, I. 1971. *Science* 173:884–92

Shapiro, L., Agabian-Keshishian, N., Hirsch, A., Rosen, O. 1972. *Proc. Natl. Acad. Sci. USA* 69:1225–29

Shapiro, L., Maizel, J. V. Jr. 1973. *J. Bacteriol.* 113:478–85

Shapiro, L., Mansour, J., Shaw, P., Henry, S. 1982. *J. Mol. Biol.* 159:303–22

Shapiro, L., Nisen, P., Ely, B. 1981. In *Genetics as a Tool in Microbiology. 31st Symp. Soc. Gen. Microbiol., England*, ed. S. Glover, pp. 317–39. Cambridge, England: Cambridge Univ. Press

Shaw, P., Gomes, S. L., Sweeney, K., Ely, B., Shapiro, L. 1983. *Proc. Natl. Acad. Sci. USA* 80:5261–65

Sheffrey, M., Newton, A. 1977. *J. Bacteriol.* 132:1027–30

Sheffery, M., Newton, A. 1979. *J. Bacteriol.*
138:575–83

Sheffery, M., Newton, A. 1981. *Cell* 24:49–57

Shirakihara, Y., Wakabayashi, T. 1979. *J. Mol. Biol.* 131:485–507

Silverman, M., Simon, M. I. 1977. *Ann. Rev. Microbiol.* 31:397–419

Silverman, M., Zeig, J., Hilman, M., Simon, M. I. 1979. *Proc. Natl. Acad. Sci. USA* 76:391–95

Silverman, P. M., Simon, M. I. 1977. *Proc. Natl. Acad. Sci. USA* 74:3317–21

Sleyter, V. B., Glauert, A. M. 1973. *Nature* 241:542–43

Smit, J., Agabian, N. 1982a. *Dev. Biol.* 89:237–47

Smit, J., Agabian, N. 1982b. *J. Cell Biol.* 95:41–49

Smit, J., Agabian, N. 1984. *J. Bacteriol.* 160:1137–45

Smit, J., Grano, D. A., Glaeser, R. M., Agabian, N. 1981a. *J. Bacteriol.* 146:1135–50

Smit, J., Hermodson, M., Agabian, N. 1981b. *J. Biol. Chem.* 256:3092–97

Springer, M. S., Goy, M. F., Adler, J. 1977. *Proc. Natl. Acad. Sci. USA* 74:3312–16

Springer, M. S., Goy, M. F., Adler, J. 1979. *Nature* 280:279–84

Springer, M. S., Koshland, D. E. Jr. 1984. In *Microbial Development*, ed. R. Losick, L. Shapiro, pp. 117–32. Cold Spring Harbor, NY: Cold Spring Harbor Lab. 303 pp.

Staley, J. T., Jordan, T. L. 1973. *Nature* 246:155–56

Stallmeyer, M. J. B., De Rosier, D. J., Aizawa, S.-I., Macnab, R. M., Hahnenberger, K., Shapiro, L. 1985. *Biophys. J.* 47:48a

Stock, J. B., Koshland, D. E. Jr. 1978. *Proc. Natl. Acad. Sci. USA* 75:3659–63

Stock, J., Koshland, D. E. Jr. 1984. In *Microbial Development*, ed. R. Losick, L. Shapiro, pp. 117–31. New York: Cold Spring Harbor Lab.

Sun, I.Y.-C., Shapiro, L., Rosen, O. M. 1974. *Biochem. Biophys. Res. Commun.* 61:193–203

Sun, I. Y.-C., Shapiro, L., Rosen, O. M. 1975. *J. Biol. Chem.* 250:6181–84

Suzuki, T., Iino, T. 1977. *J. Bacteriol.* 129:527–29

Suzuki, T., Komeda, Y. 1981. *J. Bacteriol.* 145:1036–41

Swoboda, U. K., Dow, C. S. 1979. *J. Gen. Microbiol.* 112:235–39

Swoboda, U. K., Dow, C. S., Vitkovic, L. 1982. *J. Gen. Microbiol.* 128:279–89

Terrana, B., Newton, A. 1976. *J. Bacteriol.* 128:456–62

Toews, M. L., Adler, J. 1979. *J. Mol. Biol.* 254:1761–64

Tsuda, M., Oguchi, T., Iino, T. 1981. *J. Bacteriol.* 147:1008–14

Wagenknecht, T., DeRosier, D., Aizawa, S. I., Macnab, R. M. 1982. *J. Mol. Biol.* 162 : 69–87

Wagenknecht, T., DeRosier, D., Shapiro, L., Weissborn, A. 1981. *J. Mol. Biol.* 151 : 439–65

Wang, E. A., Koshland, D. E. Jr. 1980. *Proc. Natl. Acad. Sci. USA* 77 : 7157–61

Weissborn, A., Steinman, H. M., Shapiro, L. 1982. *J. Biol. Chem.* 257 : 2066–74

Wiggs, J. L., Gilman, M. Z., Chamberlin, M.

J. 1981. *Proc. Natl. Acad. Sci. USA* 78 : 2762–66

Winkler, M. E., Schoenlein, P. V., Ross, C. M., Barrett, J. T., Ely, B. 1984. *J. Bacteriol.* 160 : 279–87

Wong, S.-L., Doi, R. H. 1982. *J. Biol. Chem.* 257 : 11932–36

Wood, N. B., Rake, A. V., Shapiro, L. 1976. *J. Bacteriol.* 126 : 1305–15

Yokota, T., Goetz, J. S. 1970. *J. Bacteriol.* 103 : 513–16

Ann. Rev. Cell Biol. 1985. 1:209–41

ORGANIZATION, CHEMISTRY, AND ASSEMBLY OF THE CYTOSKELETAL APPARATUS OF THE INTESTINAL BRUSH BORDER

Mark S. Mooseker

Departments of Biology and Cell Biology, Yale University, Box 6666, Kline Biology Tower, New Haven, Connecticut 06511–8112

CONTENTS

The cytoskeletal apparatus which underlies and supports the apical, brush border (BB) surface of the intestinal epithelial cell is among the most highly ordered arrays of actin filaments and associated proteins in nature. In addition, the BB, and subfractions such as microvilli (MV) (Bretscher & Weber 1978b), can be isolated in quantities sufficient for biochemical and structural studies. For these reasons the BB has been the subject of intensive scrutiny over the past decade. There are several reviews on the BB cytoskeleton (Mooseker & Howe 1982; Bretscher 1983a; Mooseker 1983),

209

0743–4634/85/1115–0209$02.00

and a number of research summaries from many of the laboratories that have contributed to our current understanding of the BB cytoskeleton (Bretscher 1982, 1983b; Coudrier et al 1982, 1983b; Glenney & Glenney 1983b; Matsudaira 1983; Matsudaira & Burgess 1982a; Mooseker et al 1982a, 1983, 1984a,b; Weber & Glenney 1982a,b). To avoid redundancy, the main purpose of this review is to highlight several new lines of investigation, as well as give a brief overview of past contributions to this field. It is important to note at the onset of this summary that despite the wealth of information now available regarding the BB cytoskeleton, there is very little known regarding its function in vivo. Ample speculation, based on the in vitro studies described here, has been offered, and the reader is referred to the various research summaries cited above for those thoughts regarding BB function (for my own speculations see Mooseker et al 1984b).

BRUSH BORDER CYTOSKELETON: An Overview

Figure 1 (from Mooseker 1983) is a working model for the cytoskeletal apparatus underlying the BB membrane. It summarizes over a decade of work from numerous laboratories, which was primarily based on BB's isolated from the small intestines of chicken. As with any such model, there are problems. First, no attempt has been made to draw the structure to scale. Some components, such as an 80 kilodalton (kDa) polypeptide of the MV core (Bretscher 1983c), have not been "structurally" assigned as yet. More importantly, components identified here are primarily based on studies using isolated BB's. Since BB's are prepared by rather harsh procedures (for methods see Mooseker & Howe 1982), it is highly likely that this sketch lacks any number of key components. For example, immunological studies on intact chicken intestinal epithelium reveal the presence of the nonerythroid spectrin, fodrin, in the terminal web (TW) and at the basal-lateral margins of the cell, in addition to BB specific spectrin, TW 260/240, depicted in the model (see Glenney & Glenney 1983a,b). However, fodrin is lost from the BB during isolation (Mooseker, unpublished results). Similarly, several investigations (Gerke & Weber 1984; Greenberg et al 1984; Gould et al 1984) have shown that p36, one of the major substrates for the tyrosine kinase pp^{60} sarc (for review see Sefton & Hunter 1984), is present in the terminal web region of the intestinal epithelial cell, but is not retained in the isolated BB because of the removal of calcium (Gerke & Weber 1984). A third example is tropomyosin, which is present in the terminal web region of the brush border (Bretscher & Weber 1978a; Drenckhahn & Groschel-Stewart 1980). Broschat & Burgess (1984) noted that much of the tropomyosin present in the intestinal epithelial cell is lost during BB isolation as a result of chelation of divalent cations. With

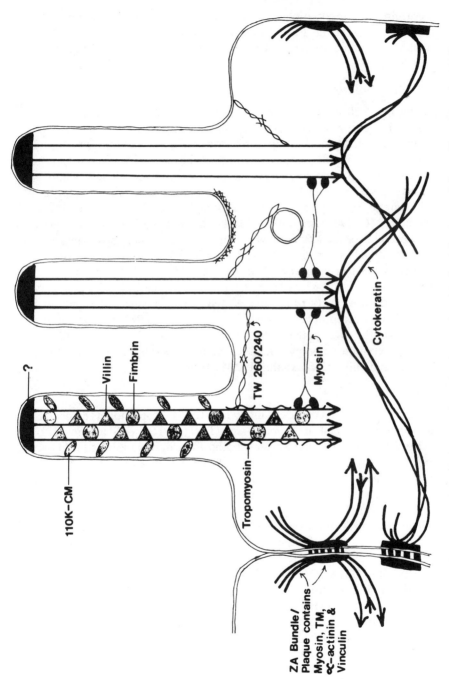

Figure 1 Working model depicting the organization and major proteins of the brush border cytoskeleton. (From Mooseker 1983.)

these important reservations in mind, I would like to briefly summarize the salient features of the BB cytoskeleton and its two structural and presumably functional domains, the microvillus (MV) core, and the terminal web (TW).

The Microvillus Core

There are ~ 20 actin filaments within each MV. These filaments are uniformly polarized (Mooseker & Tilney 1975), and have barbed (fast assembly) ends at the MV tip (for a brief review on actin assembly and structure see Pollard & Craig 1982), where they are embedded (at least in some species) in a dense plaque of unknown composition. The MV core is also attached laterally to the membrane by spirally arranged bridges (Matsudaira & Burgess 1982a) composed at least in part of a protein complex, 110-kDa-calmodulin (110-kDa-CM) (see below for references). Two proteins of the MV core with filament cross-linking properties have been identified: villin (95 kDa) and fimbrin (68 kDa) (for references see below). A third protein of 80 kDa has been identified (Bretscher 1983c) whose function is unknown. However, recent evidence indicates that it may be a substrate for the tyrosine kinase activity associated with the epidermal growth factor receptor (see below). The portion of the MV core within the TW, called the core rootlet, lacks the 110-kDa-CM (Glenney et al 1982c), but does contain tropomyosin (Drenckhahn & Groschel-Stewart 1980), fimbrin, and villin (Drenckhahn et al 1983).

The Terminal Web

The apical cytoplasm of the cell into which the MV core rootlets descend is appropriately termed the terminal web (TW) (see Hull & Staehelin 1979). The matrix of cytoskeletal elements in the TW can be subdivided into two regions. One we will call the inter-rootlet zone, which refers to the matrix of interdigitating, filamentous material between core rootlets, and the other is the cytoskeletal network associated with the junctional complex at the lateral margin of the cell.

The structural organization of the inter-rootlet zone has been best visualized using the technique of quick-freeze, deep-etch rotary replication (QFDERR), as conducted by Hirokawa and coworkers (Hirokawa & Heuser 1981; Hirokawa et al 1982, 1983a,b). The region consists of a dense meshwork of fine, non-actin filamentous material, which interdigitates between, and presumably cross-links, adjacent core rootlets. At least two chemically distinct classes of cross-linking filaments have been identified thus far; they are divided into those filaments composed of myosin (Hirokawa et al 1982), and those of nonerythroid spectrin (Hirokawa et al 1983a; Glenney et al 1983). In the chicken there is a tissue-specific form of spectrin, TW 260/240 (Glenney et al 1982a), which forms these cross-links (Glenney et al 1983; Pearl et al 1984), but in the mammalian brush border these spectrin cross-links, identified by Hirokawa et al (1983a), are probably comprised of the more widely distributed form of nonerythroid spectrin, fodrin (C. Howe, M. Mooseker & J. Morrow, unpublished results; Howe et al 1984). In the TW of the mouse brush border, Hirokawa et al (1983a) noted that the spectrin cross-links not only interconnect core rootlets, but also appear to connect core rootlets to the apical membrane, to intermediate filaments, and to vesicles embedded in the terminal web. The results of immunolocalization studies (Hirokawa et al 1983a), and selective extraction procedures (Pearl et al 1984; Mooseker et al 1984b) suggest that the myosin cross-links are absent from the apical portion of the terminal web, immediately below the plasma membrane.

The junctional complex also has an extensive cytoskeletal network associated with it. At the level of the zonula adherens (ZA) there is a circumferential bundle of actin filaments (Hull & Staehelin 1979) structurally and chemically analogous to the stress fibers of cultured cells. Like the stress fibers, the actin filaments within this bundle are of mixed polarity (Hirokawa et al 1983b); contain myosin (Bretscher & Weber 1978a; Drenckhahn & Groschel-Stewart 1980; Hirokawa et al 1983b; Burgess & Prum 1982), tropomyosin (Geiger et al 1981), and α-actinin (Craig & Pardo, 1980; Geiger et al 1979, 1981); and are associated with the membrane by an adhesion plaque-like structure that contains vinculin and α-actinin (Geiger et al 1981). Also analogous to the stress fiber, the ZA bundle is contractile (see below for references). Below the ZA is the macula adherens, which has associated with it a dense meshwork of 10 nm cytokeratin filaments, which course throughout the basal level of the TW, often making direct contact with the MV core rootlets (Franke et al 1979, 1981; Hirokawa et al 1982; Alicea 1985). These filaments are lost from avian but not mammalian BB's during isolation (Franke et al 1981; Alicea & Mooseker 1984; Alicea 1985).

CHICKEN INTESTINAL BB CYTOSKELETON VERSUS OTHER BB-CONTAINING CELL TYPES

There is ample ultrastructural data to support the generalization that the BB cytoskeletal apparatus of the vertebrate enterocytes are structurally similar regardless of species, position along the length of the small intestine, or along the villus axis (for review see Mooseker & Howe 1982; Bretscher 1983a). Recent studies in our laboratory indicate the above statement also holds for the BB's of colonic enterocytes (M. Mooseker, C. Howe, K. Barwick, and J. Morrow, unpublished observations). The main differences are matters of degree, rather than drastic differences in organization.

Common variables include length of microvilli, length of MV core rootlets, number of filaments within the MV, density of cross-linking material within the inter-rootlet zone, actual height of the TW region, and the presence or absence of obvious coated pits between microvilli (see below).

Given the universality of the structural organization, it would be surprising to find striking differences in the cytoskeletal protein composition. In fact, there is ample immunological evidence, primarily on brush borders from several mammalian species (e.g. see Figure 4, Bretscher 1982), to support the notion that the basic blueprint depicted in Figure 1 holds for mammalian BB's, and presumably for those in other vertebrate species as well. For example, the TW components myosin (Mooseker et al 1978; Drenckhahn & Groschel-Stewart 1980), nonerythroid spectrin (Hirokawa et al 1983a), tropomyosin (Drenckhahn & Groschel-Stewart 1980), filimin (Bretscher 1982), and α-actinin (Bretscher 1982); as well as the MV core components villin (Bretscher 1982), fimbrin (Bretscher 1982; Gerke & Weber 1983), the 110-kDa subunit (Bretscher 1982), and calmodulin (Bretscher 1982), have all been localized in mammalian brush borders using immunological techniques. More recently, several of the above proteins have been purified and characterized from preparations of mammalian BB's (see below).

The lumenal surface of cells lining the kidney proximal tubule also have a brush border–like array of microvilli, similar in morphology to that found in enterocytes (Maunsbach 1976; Kenny & Booth 1978). Like those present on the intestinal epithelial cell, kidney microvilli increase the effective surface area for absorption, which in the case of the kidney involves absorption of solutes and electrolytes from plasma. In addition, the kidney BB engages in extensive endocytotic uptake of intact serum proteins. This added function is apparent in the presence of prominent intermicrovillus apical invaginations, which are involved in endocytosis (Maunsbach 1976; Kerjaschki et al 1984). Recently, Rodman et al (1984) showed that these intermicrovillar invaginations have extensive clathrin coats on their cytoplasmic surface. Similar intermicrovillar invaginations are observed in the BB's of the small intestine of neonatal rats, where endocytotic uptake of maternal IgG molecules (Abrahamson & Rodewald 1981) and bulk uptake of milk protein (Knutton et al 1974; Gonnella & Neutra 1984) occur. There are other striking differences in structural organization. The MV of kidney BB's are longer and thinner than the enterocyte MV, and appear less rigid in standard thin-sectioned material. The underlying core has seven actin filaments, which in favorably fixed preparations are precisely packed in a hexagonal array (Kenny & Booth 1978). Another difference is the absence of a well-defined terminal web zone. The MV cores do not have rootlets, and there is no well-defined zone of organelle exclusion directly underneath the apical plasma membrane.

The biochemical characterization of the kidney BB cytoskeleton has lagged behind that of the intestinal BB. Immunological studies indicate the presence of fimbrin (Bretscher 1983a), villin (Bretscher et al 1981), and the 110-kDa subunit (Glenney et al 1982c) in kidney MV. Using the calmodulin-gel overlay technique (H. Alicea & M. Mooseker, unpublished observations), we noted the presence of 110- and 240-kDa subunits that bind calmodulin, which suggests the presence of 110-kDa-CM and nonerythroid spectrin (the 240-kDa subunit of spectrin is a calmodulin binding protein; see below). Despite the current paucity of information regarding the biochemical characterization of the kidney BB cytoskeleton, the system should not be neglected, as it may prove a useful model system for elucidating the role of cytoskeletal elements in endocytosis.

CHARACTERIZATION OF BB CYTOSKELETAL PROTEINS

It has been a surprising realization that a given cell probably contains dozens of proteins that interact specifically (and often competitively) with actin (for reviews see Craig & Pollard 1982; Korn 1982; Weeds 1982; Stossel, this volume). The list of actin-binding proteins is long, somewhat confusing, and ever expanding. Fortunately, most actin-binding proteins characterized thus far can be grouped into a relatively small number of functional classes. These classes include those proteins that effect the association of actin filaments with membranes, those that connect filaments together into supramolecular arrays, and those that regulate actin assembly. In addition, there is myosin, and proteins that regulate its interaction with actin to produce force. The BB cytoskeleton contains representatives of all these classes of actin-binding proteins. The characterization of these proteins has been extremely useful in elucidating structure-function relationships between actin and these various binding proteins. In the section below I briefly summarize results obtained in the characterization of the major cytoskeletal proteins of the BB cytoskeleton.

Fimbrin

One of the major proteins of the MV core is a 68-kDa polypeptide called fimbrin (Bretscher & Weber 1980a). Analysis of fimbrin purified from calcium extracts of chicken BB's (Bretscher 1981; Glenney et al 1981b) indicates that it is a mildly acidic protein that has hydrodynamic properties consistent with it being a globular monomer. The structural and chemical interaction of fimbrin with actin has been extensively characterized. Under certain ionic conditions, fimbrin will form bundles of actin filaments (Bretscher 1981; Glenney et al 1981b) that are highly ordered (Matsudaira et al 1983), and uniformly polarized (Glenney et al 1981b), closely resembling the properties of the native MV core. These results taken together suggest that fimbrin may function in vivo as the primary cross-linker of MV core filaments.

The interaction of fimbrin with actin is optimal at physiological pH, but at relatively low ionic strength (Bretscher 1981; Glenney et al 1981b). Moreover, this interaction is inhibited by high levels of Ca^{++} (Glenney et al 1981b), and by Mg^{++} concentrations greater than ~ 0.5 mM (Bretscher 1981; Glenney et al 1981b). These conditions mirror the association of fimbrin with the isolated BB cytoskeleton, because both elevated salt (Pearl et al 1984) and Ca^{++} (Glenney et al 1981b; Bretscher, 1981) readily solubilize fimbrin from the BB cytoskeleton, although the release of fimbrin by Ca^{++} is partially dependent on MV core solation by villin (see Figure 5 in Howe et al 1980). Mg^{++} does not have this effect, however. These results raise some questions regarding the exact nature of actin-fimbrin interactions in vivo, because physiological levels of salt and Mg^{++} inhibit this interaction. Perhaps other components potentiate the interaction of fimbrin with actin in vivo. A simpler explanation is that the in vitro experiments are conducted at protein concentrations roughly two orders of magnitude below that in the MV, and that at higher concentrations of fimbrin and actin the affinity of this cross-linker for actin filaments is sufficient to bind and cross-link actin into precisely ordered bundles.

Unlike the other two major components of the MV core, villin and the 110-kDa subunits, fimbrin appears to be a ubiquitous protein. Immunolocalization studies (for references, and discussion see Bretscher 1983a) indicate that fimbrin is present in a wide variety of tissues and cultured cells. This protein is enriched in membrane ruffles, microspikes, and in such structures as stereocilia, found on hair cells of the ear, but is not enriched in stress fibers. It is interesting to note that all such structures contain assemblies of uniformly polarized actin filaments, much as is found in the microvillus—perhaps a reflection of the filament cross-linking properties of fimbrin (see above). Recently, Gerke & Weber (1983) purified fimbrin from preparations of porcine intestinal microvilli. Like the avian counterpart, this protein cross-links actin filaments into bundles, and is immunologically related to the avian protein.

Villin

At Ca^{++} concentrations below $\sim 10^{-6}$ M, the MV core is a stable, rigid structure. Above this Ca^{++} threshold the filaments of the core, at least in vitro are rapidly chopped up into short fragments, as first observed by Howe et al (1980) using preparations of demembranated MV cores. This phenomenon, termed MV core solation, is mediated by the Ca^{++}-dependent severing action of villin on the actin filaments of the MV core (Mooseker et al 1980; Glenney et al 1980). Villin, a 95-kDa protein of the MV core, was initially purified and characterized by Bretscher & Weber (1980b), and by three other research groups working independently at about the same time (Mooseker et al 1980; Craig & Powell 1980; Matsudaira & Burgess 1982b). It is one member of a ubiquitous class of Ca^{++}-dependent actin-binding proteins present in cells. The properties of villin in its Ca^{++}-dependent interaction with actin has been reviewed extensively elsewhere (Bretscher 1983a; Korn 1982; Craig & Pollard 1982; Weber & Glenney 1982a,b). Here I briefly summarize those results, and focus on more recent studies on the interaction of villin with actin not found in those reviews.

Avian villin is an acidic polypeptide, which on isoelectric focusing gels can be resolved into two equimolar isoelectric variants (Bretscher & Weber 1980b). It is not known if these variants represent different gene products, or result from posttranslational modification. In an elegant series of studies, Hesterberg & Weber (1983a,b) demonstrated that avian villin has three calcium binding sites of high affinity. Two sites are readily exchangeable (apparent $K_d = 3.5$ and 7.4×10^{-6} M) and one site is not. They observed that upon Ca^{++}-binding, villin undergoes a marked conformational change. In the absence of Ca^{++}, villin is an asymmetric molecule (axial ratio 4.5 : 1); it becomes more so in the presence of Ca^{++} (axial ratio 8 : 1). The calculated lengths of the protein in the absence and presence of Ca^{++} are 8.4 and 12.3 nm, respectively. This is a gratifying result, given the dramatic changes in the behavior of this protein and its interaction with actin when it binds calcium.

Before reviewing the various in vitro effects of villin on actin, it is first necessary to briefly outline the parameters of actin assembly (for reviews see Korn 1982; Pollard & Craig 1982). The assembly of actin is a two-, or possibly three-step process, all phases of which are affected by villin. If the actin monomer concentration is above a certain level, termed the critical concentration (C_0), assembly of actin monomers into filaments can occur. First, nuclei consisting of 3–4 monomers must form (Korn 1982). This is the rate-limiting step of assembly. Next, these nuclei elongate from both of their ends, but under physiological conditions the barbed end (so called because of the arrowhead morphology of actin filaments decorated with heavy meromyosin) grows faster than the pointed end. Moreover, the barbed end has a lower C_0 (~ 0.1 μM) than the pointed end (~ 0.6–1.0 μM) (Bonder et al 1983). Elongation continues until steady state is achieved, i.e. a constant amount of polymer is in equilibrium with the concentration of monomer at the net C_0 for the two ends of the filament. Because the two C_0's are different at the two ends, treadmilling of subunits (barbed to pointed end) might occur at least under in vitro conditions in the presence of salt and Mg^{++}. Finally, actin filaments break spontaneously, particularly when exposed to even mild shear forces. There is indirect evidence that filaments can also reanneal, but no kinetic analysis of this phenomenon has been conducted. While breakage and reannealing of actin filaments is the least understood aspect of actin assembly, it is likely to be intensely scrutinized in the near future, given the recent demonstrations that bound ATP on the actin monomer is not coupled to monomer addition to the filament, but occurs sometime later (for review see Wegner 1984). This yields actin filaments with "caps" of ATP monomer at both ends, and ADP-containing monomers internally. Most importantly, filament ends exposed by breakage contain ADP monomer and are much less stable. However, the C_0's, dissociation, and association rate constants for barbed and pointed ends containing ADP monomers have not as yet been determined.

EFFECTS OF VILLIN ON ACTIN ASSEMBLY   In the absence of Ca^{++} ($< 10^{-7}$ M), villin has no detectable effect on the various phases of actin assembly outlined above (Mooseker et al 1980; Glenney et al 1981c; Bonder & Mooseker 1983; Walsh et al 1984b; Wang et al 1983). At Ca^{++} concentrations greater than $1-10\ \mu M$ the effects of villin on C_0, nucleation, and elongation phases of assembly include the following: (a) Villin accelerates the nucleation phase of assembly (Mooseker et al 1980, 1982a; Glenney et al 1981c) by rapidly forming stable, oligomeric complexes that can nucleate actin monomer addition (Glenney et al 1981c; Wang et al 1983). (b) Villin binds with very high affinity ($\sim 10^{11}\ M^{-1}$; Walsh et al 1984b) to and caps the barbed end of the filament (Glenney et al 1981c; Bonder & Mooseker 1983), thereby inhibiting the elongation rate to that of the pointed, slow-growing end (Bonder & Mooseker 1983). (c) By capping the barbed end, villin raises the C_0 to that of the slow-growing pointed end. This villin-dependent increase in C_0 saturates at low villin to actin ratios ($\sim 1:400-1000$; Wang et al 1983; Walsh et al 1984b), and transient decreases in C_0 can be induced by shearing villin-capped filaments (Walsh et al 1984b), which results in exposure of uncapped barbed ends. (d) The length filaments achieve at steady state is inversely proportional to the concentration of villin present (Craig & Powell 1980; Mooseker et al 1980), and the length distribution of filaments formed in the presence of villin is far more uniform than in its absence, even at relatively low concentrations of villin (Alicea 1985; Alicea & Mooseker 1984).

EFFECTS OF VILLIN ON F-ACTIN   At low Ca^{++} concentrations ($< 10^{-7}$), villin bundles actin filaments (Bretscher & Weber 1980b; Mooseker et al 1980;

Matsudaira & Burgess 1982b). Villin-formed actin bundles are less ordered than those formed by fimbrin (Matsudaira et al 1983). The effects of villin on F-actin in the presence of Ca^{++} include the following: (*a*) When added to F-actin, villin caps the barbed filament ends, causing a rapid depolymerization to a new steady state that reflects the higher C_0 of the pointed end (Wang et al 1983; Walsh et al 1984b). This shift in C_0, predicted for a barbed end capper, is saturable at low villin-actin ratios. (*b*) Villin inhibits the rate of filament depolymerization when those filaments are diluted below their C_0 (Walsh et al 1984b). This is consistent with the finding that the dissociation rate constant for ATP-capped polymer is greater at the barbed than at the pointed end (Pollard & Mooseker 1981; Bonder et al 1983). (*c*) Villin inhibits monomer-polymer subunit exchange (Glenney et al 1981c; Wang et al 1983). (*d*) Villin rapidly decreases filament length (Bretscher & Weber 1980b; Mooseker et al 1980; Craig & Powell 1980), (and increases the number-concentration of filaments) by direct severing of monomer-monomer interactions of the filament (Glenney et al 1981c; Bonder & Mooseker 1983). The average length of filaments formed by addition of villin to F-actin is greater, and the distribution of filament lengths more varied, than the population of filaments derived from coassembly of actin with villin (Mooseker et al 1980; Alicea & Mooseker 1984). (*e*) The severing action of villin requires higher $[Ca^{++}]$ than barbed end capping (5–10 μM versus 1–5 μM; Mooseker et al 1980; Walsh et al 1984a; Alicea & Mooseker 1984; Alicea 1985). In fact, severing activity of villin continues to be potentiated at Ca^{++} concentrations well above that required to saturate its Ca^{++}-binding sites. (For example, severing is much greater at 0.1 mM than at 10 μM Ca^{++}.) This suggests the possibility that direct effects of Ca^{++} on the actin filament, presumably through low affinity Ca^{++}-binding sites on actin, might potentiate the severing effect of villin. (*f*) Tropomyosin blocks the severing, but not barbed end–capping activity of villin (Mooseker et al 1982a; Bonder & Mooseker 1983; Walsh et al 1984a), an observation that may explain the resistance of the rootlet portion of the MV core to Ca^{++}-dependent solation (Mooseker et al 1982a; Burgess & Prum 1982).

In addition to the studies summarized above, another approach to elucidating the molecular basis for villin's interaction with actin and calcium has been to dissect the villin molecule with proteases into its various functional domains. Using V-8 protease, Glenney and coworkers (Glenney & Weber 1981; Glenney et al 1981a) cleaved the villin molecule into two fragments, with apparent molecular weights of 90,000 and 8,500. The larger fragment, called the villin core, retains Ca^{++}-dependent nucleation and severing activities, but does not bundle filaments. The smaller fragment (called the head piece) binds to actin with or without Ca^{++}, with a stoichiometry of 1:1. Although the head piece does not bundle actin filaments, it does inhibit the bundling activity of intact villin. The head piece, consisting of 76 amino acids derived from the C-terminus of the villin molecule, has been sequenced. Hesterberg and Weber (1983b) have examined the distribution of Ca^{++}-binding sites in these two fragments. The nonexchangeable Ca^{++}-binding site is on the larger core fragment, as is one of the two exchangeable sites. The head piece contains one of the exchangeable binding sites for Ca^{++}, which according to sequence data is not of the "E-F hand" class found in Ca^{++}-binding proteins such as calmodulin. This is a surprising result, since the core protein retains essentially all the Ca^{++}-dependent activity of intact villin except bundling. Perhaps the Ca^{++}-binding site on the head piece is involved in regulation of filament cross-linking by villin. However, the binding of the head piece to actin is Ca^{++}-insensitive, making this explanation problematical. It is important to note that the villin core, unlike the intact molecule, does not undergo conformational change in the presence of Ca^{++}, indicating that the head piece, and presumably its Ca^{++}-binding site, must participate in the molecular rearrangements of the villin molecule that occur in the presence of Ca^{++}. (For further discussion and references see Hesterberg & Weber 1983b.)

In recent experiments, Matsudaira & Jakes (1984) further dissected the villin molecule, using a combination of protease treatments. In brief, they have shown that a 45-kDa tryptic fragment derived from the N-terminus of the villin molecule retains severing activity. Further cleavage of this peptide with V-8 protease yielded a 13-kDa subfragment that contains a Ca^{++}-binding site, and which binds to F-actin in a Ca^{++}-dependent fashion. In addition, they used chemical cleavage methods in combination with a novel mapping technique, utilizing an antibody to the C-terminus of the molecule, to map cysteines and methionines of the villin molecule. Using this method, they have identified a Ca^{++}-sensitive actin-binding domain 30 kDa from the N-terminus. Finally, they sequenced a portion of the N-terminus, and demonstrated that this region shares homology with another filament-severing protein, gelsolin.

Recently, villin was purified from two different mammalian brush border sources, the pig and the rat. Gerke & Weber (1983) observed that purified porcine villin, like chicken villin, bundles actin filaments in the absence of Ca^{++}, and severs them in its presence. In addition, they were able to cleave porcine villin into core and head piece fragments similar in molecular weight to those observed for avian villin. In our laboratory, we have conducted an extensive characterization of rat villin, and its interaction with actin (Alicea & Mooseker 1984; Alicea 1985). Villin purified from rat brush borders via the technique of Bretscher & Weber (1980b) is slightly smaller in molecular weight (~ 93 kDa), and like chick villin consists of two isoelectric variants, both slightly more acidic than the avian protein. Using a combination of techniques, including cosedimentation, viscometry, and quantitative electron microscopy, we have determined that rat villin is virtually indistinguishable from avian villin with respect to its interaction with actin. This includes the difference in Ca^{++} threshold for nucleation capping activity versus severing activity.

The function of villin in vivo is unknown. Clearly, this protein could play any number of roles in the regulation of core filament assembly and structure (for discussion of possible functions of villin in vivo see Matsudaira 1983; Mooseker et al 1984a,b; Weber & Glenney 1982b). Recent immunological studies identifying the presence of villin in the pancreatic acinar cell (Drenckhahn & Mannherz 1983) provide additional insight into the function of this protein in vivo. In these cells Ca^{++} stimulates zymogen granule exocytosis. Perhaps villin helps mediate this event through Ca^{++}-dependent rearrangement of the cortical cytoplasm in these cells (e.g. by solating actin filaments, which might physically prevent vesicle approach and subsequent fusion with the plasma membrane). In addition to the role of villin as a regulator of actin assembly and filament length, we noted in recent experiments that villin (and presumably other Ca^{++}-dependent severing proteins) may affect the interaction of actin with myosin. Using villin to generate actin filaments of various lengths, we observed (Coleman & Mooseker 1984) that at short filament lengths ($< \sim 0.1$ μm) the interaction of actin with myosin was inhibited. Surprisingly, however, we noted that at intermediate filament lengths (~ 0.2–0.3 μm) there is a superactivation of actin-activated ATPase of myosin, suggesting that at least in vitro there is an optimal filament length for interaction with myosin.

The 110-kDa-Calmodulin Complex

The evidence that the spirally arranged bridges that attach the MV core laterally to the membrane are comprised of 110-kDa-CM is compelling, but still indirect. Immunological localization studies by Glenney et al (1982c) showed that the 110-kDa subunit of the complex, like the lateral bridge, is confined to that portion of the MV core within the membrane. However, Coudrier et al (1981) also observed some 110-kDa subunits on basal-lateral membranes and in the TW. Moreover, the 110-kDa-CM complex, like the lateral bridge, is dissociated from the MV core by ATP treatment (Matsudaira & Burgess 1979; Howe & Mooseker 1983; Verner & Bretscher 1983; Collins & Borysenko 1984). However, successful reconstitution of lateral bridges by addition of purified 110-kDa-CM to ATP-stripped MV cores has not been accomplished as yet.

Recently, several laboratories purified the 110-kDa-CM complex from

ATP extracts of isolated BB's (Howe & Mooseker 1983; Collins & Borysenko 1984, Verner & Bretscher 1984). Gel filtration analyses of purified 110-kDa-CM complex suggest that it consists of a dimer of 110 kDa (Howe & Mooseker 1983) with variable amounts of associated calmodulin (0.2–2.0 molecules) per 110-kDa subunit (Howe & Mooseker 1983; Collins & Borysenko 1984). The interaction of the 110-kDa subunit with calmodulin is unusual, because it is Ca^{++}-independent. This was first demonstrated using gel overlay procedures (Glenney & Weber 1980; Howe et al 1982). By such overlay methods it was observed that the 110 kDa subunit has an apparent higher affinity for calmodulin in the presence of Ca^{++} (Howe et al 1982). Recently Glenney & Glenney (1985) quantitated the Ca^{++}-dependent interaction of calmodulin with the 110-kDa subunit using the overlay technique, and noted a strong pH dependence of the binding of the 110-kDa subunit to calmodulin in the absence of Ca^{++}. Nevertheless, the native complex as isolated by the ATP extraction appears to be equally stable whether or not Ca^{++} is present; although no quantitative studies assessing the association of calmodulin and the 110-kDa subunit under native conditions have yet been reported.

The only indication that the complex purified from ATP extracts of BB's is a native one comes from studies of its interaction with actin. Like the lateral bridge, the 110-kDa-CM complex binds to actin in the absence but not the presence of ATP (Howe & Mooseker 1983; Collins & Borysenko 1984; Verner & Bretscher 1984). The morphology of 110-kDa-CM-actin complexes is remarkable only in noting our initial disappointment at first viewing these in the electron microscope. We observed no periodic structures, instead, we saw rather loose aggregates of filaments covered with 110-kDa-CM "fuzz", which gave a "pipe-cleaner" effect (Howe & Mooseker 1983). Given the effect of ATP on the interaction of 110-kDa-CM with actin, it is not surprising that the 110-kDa subunit is an ATP-binding protein (Mooseker et al 1984a). In our initial characterization of 110-kDa-CM, we noted that this complex had little detectable MgATPase activity with or without Ca^{++} or actin; nor did we detect any myosin light chain kinase activity associated with the complex (Howe & Mooseker 1983). Recently, however, Collins & Borysenko (1984) made the extremely exciting observation that 110-kDa-CM may be a myosin-like ATPase. Like myosin, 110-kDa-CM exhibits high levels of ATPase activity in the presence of EDTA, a condition considered a diagnostic for many myosin-like ATPases. Although these workers have as yet been unable to show any actin activation of MgATPase, this preliminary finding raises the extremely exciting possibility that the complex is a new form of membrane-associated myosin-like mechanoenzyme. Experiments to critically test this hypothesis include the demonstration that the MgATPase activity can be activated by actin under some condition. For example, does phosphorylation of the 110-kDa subunit affect its interaction with actin? [The 110-kDa subunit is phosphorylated by an endogenous kinase in the BB; Keller & Mooseker (1982).] Does the 110-kDa-CM complex have mechanochemical potential? For example, will it induce superprecipitation of actin filaments, or mediate the movement of Covaspheres when assayed by the *Nitella*-based myosin movement assay of Sheetz & Spudich (1983)?

The molecular basis for the interaction of 100-kDa-CM with the MV membrane is currently the subject of some controversy. As noted in studies discussed above, it has been shown by several laboratories that *apparently* intact 110-kDa-CM complex can be readily dissociated from the plasma membrane and the cytoskeleton by treatment with ATP, which suggests a peripheral association with the membrane. Moreover, this dissociation is potentiated by elevated salt (Howe & Mooseker 1983). These observations are consistent with the notion that the interaction of 110-kDa-CM with the membrane is, like its interaction with actin, ATP-sensitive. However, a simpler explanation, particularly given the effects of elevated salt on the dissociation of the complex, is that 110-kDa-CM is only weakly associated with the membrane, at least in vitro in isolated BB's, and that it is readily dissociated from the BB when its interaction with the MV core is broken by ATP addition. The suggestion that 110-kDa-CM binds to the membrane in a peripheral fashion to a specific binding protein in the membrane is supported by the studies of Coudrier et al (1983a), who have identified a 200-kDa glycoprotein in porcine microvilli (usually present as a 140-kDa proteolytic fragment), which binds specifically to the porcine 110-kDa subunit in gel overlays. Finally, Cowell & Danielsen (1984) have provided preliminary evidence suggesting that the 110-kDa subunit is synthesized on free rather than membrane-bound polysomes.

However, Glenney & Glenney (1984) have suggested that the 110-kDa subunit is an integral membrane protein, based on several lines of evidence that indicate it is a hydrophobic protein. The evidence is as follows: (*a*) detergent is required for solubilization of the subunit, and upon removal of detergent the 110-kDa subunit aggregates; (*b*) the protein partitions into the detergent phase of Triton X-114 solutions (Bordier 1981); (*c*) purified 110-kDa subunits are incorporated into liposomes; and (*d*) the subunit can be labeled with the hydrophobic probe phenyl isothiocyanate (PITC). Although these results provide a clear indication that the 110-kDa subunit has a hydrophobic domain(s), the 110-kDa subunit used in these studies was purified by a series of treatments, including the addition of the detergent SDS, and subsequent ammonium sulfate treatment, which might have either artificially exposed, or generated, in the case of SDS addition, hydrophobic domains not present or exposed on the native 110-kDa subunit. Moreover, the 110-kDa subunit used in these studies (with the exception of the PITC labeling) was stripped of its calmodulin by the above treatments, raising the possibility that any hydrophobic domains involved in calmodulin binding might be exposed. Recent experiments in our laboratory (K. Conzelman & M. Mooseker, unpublished observations) indicate that the above reservations are well founded. First, 110-kDa-CM complex partially purified without detergent by the method of Howe & Mooseker (1983) is not hydrophobic when assayed by the Triton-114 partition assay, but the 110-kDa subunit in the complex can be rendered so by subsequent treatment with SDS and ammonium sulfate, following the procedures of Glenney & Glenney (1984). Moreover, the 110-kDa subunit purified by the procedure of Glenney & Glenney (1984) can be partially shifted into the aqueous phase by addition of exogenous calmodulin. However, treatment of the 110-kDa-CM complex with the drug trifluoperazine to dissociate its calmodulin (Glenney et al 1980; Howe & Mooseker 1983) does not result in a shift of the 110-kDa subunit into the aqueous phase. Based on the above observations, it is clear that considerably more work is required to define the molecular basis for the interaction of this fascinating complex with the MV membrane.

The BB Specific Spectrin: TW 260/240

It is now clearly established that spectrin-like proteins are ubiquitous cytoskeletal components of both metazoan and protozoan cells (Baines 1984; Lazarides & Nelson 1982; Pollard 1984). While the extensive studies on erythroid spectrin (reviewed in Branton et al 1981) provide a rich paradigm for the function of spectrin in nonerythroid cells, it is almost certain that this paradigm (like the sliding filament model for the function of actin in nonmuscle cells) is incomplete. A good example of this is the BB-specific spectrin of the chicken intestinal epithelial cell, TW 260/240. Structural and chemical studies indicate that the main function of this spectrin is in actin filament-filament cross-linking in the terminal web region of the BB, rather than filament-membrane interaction. As noted above, TW 260/240 is present in the TW of the BB, and comprises cross-links between adjacent core rootlets.

Like other avian spectrins, TW 260/240 consists of a common alpha subunit of 240 kDa, and a beta subunit of 260 kDa, which is found only in the TW region of the intestinal epithelial cell (for review see Glenney & Glenney 1983b). (As discussed previously, mammalian BB's probably do not contain a specific spectrin, but rather use the more commonly distributed nonerythroid spectrin, fodrin.) The 240-kDa subunit binds calmodulin; the functional significance of this calmodulin interaction has not yet been determined (Glenney et al 1982b; Glenney & Glenney 1983b). At relatively low ionic strength (≤ 100 mM KCl) TW 260/240 exists as a tetrameric complex of two alpha-beta dimers associated in a head to tail fashion, as described for other spectrins (Glenney & Glenney 1983a,b; Pearl et al 1984). At salt concentrations at or above 150 mM, partial dissociation into dimers occurs. It has not yet been shown whether this represents a simple dimer-tetramer equilibrium, or the existence of two forms of tetramers, only one of which displays salt-dependent dissociation (Pearl et al 1984). The 260-kDa subunit is phosphorylated in situ by an endogenous kinase in isolated BB's (see Keller & Mooseker 1982), but no effects of phosphorylation on the various properties of TW 260/240, for example actin- or calmodulin-binding, have yet been observed (D. Fishkind & M. Mooseker, unpublished observations).

Since the main function of TW 260/240 in the BB appears to be cross-linking of actin filaments of the MV core rootlets, we have extensively investigated the interaction of this spectrin-like protein with actin (Pearl et al 1984; Fishkind et al 1983, 1985). Under optimal ionic conditions (see below) TW 260/240 is a potent cross-linker of actin filaments. High resolution electron microscopic examination of the interaction of TW 260/240 with actin filaments suggests that the TW 260/240 tetramer actually wraps around the actin filament (Fishkind et al 1985), rather than binding tangentially, as suggested by rotary-shadowed images of other spectrin-actin complexes (see Branton et al 1981). The TW 260/240 binding to and cross-linking of actin filaments is favored at physiological pH, but is optimal at relatively low ionic strength. In fact, at physiological levels of KCl, the interaction of TW 260/240 with F-actin in vitro, and with the BB cytoskeleton, is greatly inhibited (Pearl et al 1984). Recently we observed that cross-linking of filaments by TW 260/240 enhances the activation of myosin MgATPase by actin (Coleman & Mooseker 1984). A similar finding was reported for mammalian fodrin (Wagner 1984). This is potentially an important finding since TW 260/240 and myosin "cohabit" in the TW.

Using the uniformly polarized bundles of actin filaments from *Limulus* sperm (Tilney 1975) as nuclei for actin assembly, we conducted an extensive quantitative examination of the effects of TW 260/240 on actin assembly (Fishkind et al 1983, 1985). The elongations rates (see Bonder et al 1983 for methods) and C_0's for assembly at the barbed and pointed ends of the actin filament are not affected by TW 260/240. Nor does TW 260/240 affect the nucleation phase of assembly (Pearl et al 1984). This is in contrast to studies on mammalian fodrin, which apparently stimulates actin assembly (Sobue et al 1982).

We (Howe et al 1984, and submitted for publication) also conducted an extensive comparative study of the interaction of TW 260/240 and avian erythroid spectrin with membranes, using the standard inside-out vesicle (IOV) assay to examine the interaction (Branton et al 1981). In brief, chicken erythroid spectrin binds to IOV's purified from human erythrocyte membranes with an affinity indistinguishable from that of human erythroid spectrin, presumably through an ankyrin binding site on its beta subunit (Branton et al 1981; Davis & Bennett 1984). In contrast, TW 260/240, which has the same alpha subunit but a different beta subunit, binds very poorly to IOV's. Using a native gel electrophoretic assay, TW 260/240 has no detectable affinity for ankyrin, the protein that mediates the specific interaction of erythroid spectrin with the plasma membrane. These results indicate that the 260-kDa subunit lacks a high-affinity binding site for ankyrin. Note that the chicken intestinal epithelial cell also contains fodrin (Glenney & Glenney 1983a), which may be involved in actin filament-membrane interactions.

Myosin and Its Role in BB Contractility

PROPERTIES OF BB MYOSIN   The BB cytoskeleton has associated with it a large amount of myosin, which is localized exclusively in the terminal web. Studies of purified BB myosin (Mooseker et al 1978) indicate that it is an orthodox vertebrate nonmuscle myosin consisting of 2 heavy chains and 2 pairs of light chains of 19 and 16 kDa. As isolated, BB myosin has no actin-activated MgATPase activity (Mooseker et al 1978), a finding consistent with the observation that myosin in the isolated BB (Brotschat et al 1983) or purified BB myosin is in the unphosphorylated state (Keller et al 1985). However, BB myosin can be phosphorylated in situ by a cytoskeletal-associated myosin light chain kinase (MLCK), or by addition of exogenous MLCK from the gizzard to purified BB myosin (Mooseker et al 1984b; Keller et al 1983, 1985). The MgATPase activity of phosphorylated BB myosin is activated approximately tenfold by actin.

Using electrophoretic techniques (Keller et al 1985), we noted that when the 19-kDa regulatory light chain BB myosin is phosphorylated, either in situ by the endogenous MLCK(s) or by exogenously added MLCK, much of it is phosphorylated on two sites rather than on a single site, as previously assumed (see Keller & Mooseker 1982). It will be important to characterize the actin-activated MgATPase of singly versus doubly phosphorylated BB myosin, and determine the specific sites phosphorylated. As has been shown for other vertebrate nonmuscle and smooth muscle myosin, the unphosphorylated form of BB myosin exhibits the kinked-tail conformational change induced by ATP-binding (Mooseker et al 1984b), which presumably prevents filament formation.

BRUSH BORDER CONTRACTILITY   There are potentially at least two distinct sites of myosin-actin interaction in the TW, reflecting the dual localization of myosin in the inter-rootlet zone, and in the circumferential actin bundle associated with the ZA junction. Numerous studies, summarized below, have implicated ZA myosin in the ATP-dependent contraction of the ZA bundle, a phenomenon analogous to that observed for ZA contraction in pigmented retinal epithelium by Owaribe et al (1981). Conversely, the function of myosin within the inter-rootlet zone, where the bulk of BB myosin is localized, is as yet unknown.

In 1976, Rodewald et al reported an ATP-dependent contraction on the TW and BB's isolated from neonatal rats. Recently, we (and other laboratories) were prompted to further investigate this phenomenon because of the disarming discovery that my previous studies of microvillar contraction (Mooseker 1976) were actually a misinterpretation of a rather complicated artifact of Ca^{++}-dependent core solation, coupled with simultaneous TW contractility and ATP-dependent lifting of the BB membrane (as recently documented by Howe & Mooseker 1983), which made it appear that the MV cores had actually retracted into the TW region (for further discussion see Mooseker et al 1983).

Recent studies of TW contractility focused on isolated BB's interconnected by junctions (Keller & Mooseker 1982; Broschat et al 1983), or on preparations on glycerinated epithelium (Burgess 1982; Hirokawa et al 1983b) using methods first described by Owaribe et al (1981). In all cases, addition of ATP, and in the case of isolated BB's, Ca^{++} ($> 1\ \mu M$), results in contraction of the TW over a period of over 1–10 min, which results in a fanning out of the contracted BB's microvilli. Movement or shortening of MV in these preparations has not been detected. Ultrastructural analysis of these various preparations indicate that the phenomenon is due, at least in part, to a constriction of the ZA bundle.

The molecular basis for ZA contractility has been extensively investigated, although a clear picture of this phenomenon at the molecular level has yet to emerge. Not surprisingly, myosin appears to be involved: TW contractility can be blocked by addition of an antibody to BB myosin, which in vitro inhibits both myosin ATPase and filament formation (Keller et al 1983, 1985). However, most of the myosin in the BB is not involved in, or at least is not necessary for, TW contractility. In our initial studies of BB contraction, using isolated BB's (Keller & Mooseker 1982) or glycerinated intestinal epithelial cells (Hirokawa et al 1983b), we noted that much of the myosin originally associated with the BB cytoskeleton was solubilized during contraction, leaving us with serious doubts regarding its actual participation in the contractile event. Moreover, this release was not dependent on contraction. It is possible to extract up to 80% of the total BB myosin by extraction of BB's with ATP at 0°C, conditions that do not promote contraction (see Mooseker et al 1984b; Keller et al 1985). Ultrastructural examination of such ATP-treated preparations indicated that the inter-rootlet zone was greatly depleted of the cross-linkers identified as myosin by selective extraction procedures (Mooseker et al 1984b). Electron microscopic examination of myosin removed by ATP treatment revealed that it is predominantly in the form of bipolar dimers, which suggests that this is the "native" configuration of myosin in the inter-rootlet zone (Mooseker et al 1984b). Most importantly, despite the removal of up to 80% of this myosin, BB's extracted in this fashion undergo contractions indistinguishable from those in control preparations, both with respect to the extent of contractility and number of BB's in the preparation that contract (Keller et al 1983, 1985). However, the fact that myosin-depleted BBs contract normally does not rule out the involvement of ATP-labile myosin in the TW contraction of control BBs, since only a few myosin molecules may be necessary to produce constriction of the ZA bundle. That is, it is likely that the force requirements for contraction are negligible. Nevertheless, based on these observations, we have suggested that there may be two subsets of myosin in the BB: one tightly associated with the cytoskeletal apparatus in an ATP-resistant fashion, which is at least partially responsible for ZA contractility, and another set, predominantly in the inter-rootlet zone, that is organized as bipolar dimers associated with the core rootlets in an ATP-labile fashion, and whose function is unknown.

A number of studies have addressed questions regarding the regulation of TW contractility, and in particular the role of calcium and phosphorylation of BB myosin in this phenomenon. As noted above, the contraction requires ATP, and recent studies by Broschat et al (1983) indicate that neither GTP nor ITP are effective substitutes. In the contraction of isolated BB's, we have observed (Keller & Mooseker 1982) that at least in some preparations Ca^{++} concentrations above 1 μM are required to support maximum contractility. There is, however, considerable variability in this Ca^{++} sensitivity, as we have suggested that the loss of Ca^{++} dependence is due to the partial proteolysis of a BB MLCK. However, it is also a possibility that the lack of Ca^{++}-dependent sensitivity may result from variable extraction of other regulatory components, such as caldesmon (Owada et al 1984) or perhaps variable amounts of tropomyosin, which may contribute to the regulation of myosin-actin interaction in nonmuscle and smooth muscle cells (see Cote 1983). In fact, Bretscher & Lynch (1985) recently reported the presence of an immunoreactive form of caldesmon in the TW of the intestinal epithelial cell. Nevertheless, a Ca^{++}-dependent phosphorylation of the 19-kDa light chain, which occurs with kinetics that parallel BB contraction, was observed. Moreover, both contractility and the Ca^{++}-dependent phosphorylation of the myosin light chain were inhibited to levels observed in the absence of Ca^{++} by addition of the calmodulin inhibitor, trifluoperazine (Keller & Mooseker 1982). These results suggest that the regulation of BB contractility occurs at least in part through a cytoskeletal-associated calmodulin-dependent MLCK. Although this basic conclusion is at least partly correct, some of the observations presented in our original study of BB myosin and TW contractility are, in essence, irrelevant because most of the myosin that was phosphorylated in these preparations, as shown by the subsequent studies described above (Keller et al 1985), was actually not involved in TW contractility.

To clarify the role of myosin phosphorylation in TW contractility, and the involvement of ATP-dependent dissociation of myosin from the cytoskeleton, we (Keller et al 1983, 1985; Mooseker 1984b) examined the phosphorylation state of (Preparation 1) myosin dissociated from the BB by treatment with ATP at 0°C; (Preparation 2) myosin in 0°C ATP extracts, subsequently warmed in the presence of ATP to determine the presence or absence of kinase in these ATP-supernatants; and (Preparation 3) myosin that remained associated with the BB after ATP treatment at 0°C, both before and after contraction, stimulated by 37°C incubation with ATP. The myosin released from the BB by ATP treatment at 0°C (Preparation 1) was 80–100% unphosphorylated, depending on the BB preparation used. However, warming these ATP extracts (Preparation 2) resulted in quantitative phosphorylation of the 19-kDa light chain, often resulting in the "double phosphorylation" described above. These results indicate that at least some of the BB-associated MLCK is dissociated from the cytoskeleton by ATP treatment. Finally, the myosin that remained associated with the BB after ATP extraction was phosphorylated when warmed to 37°C in the presence of ATP, which suggests that phosphorylation of this cytoskeletal associated myosin does occur during TW contraction.

Results of similar studies with some important condition differences were reported by Broschat et al (1983). These workers reported that myosin in the isolated BB was unphosphorylated, but that in the presence of ATP, added at room temperature, it became phorphorylated, and was then released from the cytoskeleton where it was assembled into bipolar filaments. While we agree with the observation that ATP-treatment results in dissociation of a considerable amount of myosin from the BB cytoskeleton, we disagree with the interpretation that phosphorylation is required for this dissociation, since in our studies phosphorylation of BB myosin was not required for ATP-dependent dissociation. These authors also suggested, based on this observation, that once released, the myosin assembled into filaments and reassociated with the cytoskeleton to contribute to ZA-dependent contractility. We also disagree with this interpretation, because this ATP-dissociable myosin can be removed from the BB cytoskeleton without inhibiting contractility. This same study (Broschat et al 1983) made an additional finding that provides the most direct evidence to date for the involvement of phosphorylation in TW contractility. These workers pretreated isolated BB's with the ATP analog, ATP-γ-S, which MLCK can use as a substrate for myosin phosphorylation, but which myosin cannot use to produce force. Such pretreated BB's (but not control BB's) could then contract in the presence of ITP, even though ITP cannot be used by the kinase for phosphorylation, because it can be used as a substrate for force production by myosin. This experiment elegantly demonstrates the requirement for myosin phosphorylation in TW contractility.

ASSOCIATION OF TYROSINE KINASE SUBSTRATES WITH BB CYTOSKELETON

One of the most exciting areas of investigation in cell biology centers on the investigation of a class of kinases that phosphorylate proteins on tyrosine residues (for review and references see Sefton & Hunter 1984). Thus far, three classes of such kinases have been identified. One class includes those kinases coded for by transforming viral genes (e.g. sarc gene kinase, pp60[src]) and by the family of closely related cellular "oncogenes," whose products are implicated in the regulation of cell growth, shape, position, and metabolism, which in the case of the expression of the viral gene results in oncogenic transformation. The second class of kinases includes receptors for growth hormones, whose activity is stimulated by hormone binding. Third, several distinct tyrosine kinases have been identified in various normal tissues and cell lines. One approach to understanding the mechanism of action of tyrosine kinases has been to identify their primary substrates. Recent studies revealed that the BB has associated with it some of these substrates. Given the wealth of information available regarding cytoskeletal protein interactions in the BB, this system should provide considerable insight into the function of these tyrosine kinase substrates, and what effect, if any, their phosphorylation has on the biological activity of the various cytoskeletal components in vitro and in vivo.

Currently, the best characterized tyrosine kinase substrate associated with the BB is p36, a major substrate of the sarc gene kinase. Immunolocalization studies have shown that this protein is present in the apical portion of the intestinal epithelium, where it is restricted to the TW region of the BB (Gerke & Weber 1984; Greenberg et al 1984; Gould et al 1984). In BB prepared by conventional procedures no p36 is present, however, recently Gerke & Weber (1984, 1985) showed that if BB membranes are prepared by the Ca^{++}-precipitation methods routinely used for the preparation of transporting vesicles (Kenny & Maroux 1982) the p36 is retained. It is isolable as a complex (called Protein I), and consists of two molecules of p36 and two molecules of a 10-kDa polypeptide (Gerke & Weber 1984, 1985). Protein I binds to pig brain fodrin, and to F-actin in the presence but not absence of Ca^{++}. However, the Ca^{++} sensitivity of Protein I binding to both fodrin (Gerke & Weber 1984, 1985) and actin (Gerke & Weber 1984, 1985; Glenney & Glenney 1985) is not in the physiological range; it requires greater than $\sim 10^{-4}$ M Ca^{++} for binding. At this level of Ca^{++} the p36 undergoes a conformational change, exposing a tyrosine residue to a more aqueous environment (Gerke & Weber 1985). Glenney & Glenney (1985) examined the effect of pp60^{src} phosphorylation of p36 on the interaction of p36 and actin, and observed no effect, as assayed by cosedimentation.

Preliminary evidence has also been cited as a personal communication in Sefton & Hunter (1984) for the presence of an 80-kDa protein in the microvillus core, related to the major tyrosine kinase substrate of the epidermal–growth factor receptor. It will be important to establish whether or not this protein is identical to the 80-kDa component first characterized by Bretscher (1983c), and to determine what effects, if any, phosphorylation has on its interaction with actin and other components of the MV core.

BB CYTOSKELETON AND VITAMIN D–DEPENDENT TRANSPORT OF Ca^{++}

The intestinal epithelium absorbs Ca^{++} by a mechanism thought to involve passive movement of Ca^{++} across the BB membrane, transport of absorbed Ca^{++} through the cell cytoplasm, and finally, active expulsion of the Ca^{++} out of the cell and into the blood space by an ATP-dependent Ca^{++} pump (for review see Wasserman 1981). Vitamin D deficiency alters the metabolic state of the intestinal epithelium in a complex fashion, with direct effects, including (a) altered phospholipid composition of the BB membrane, which affects Ca^{++} permeability; (b) reduced levels of the Vitamin D–dependent Ca^{++}-binding protein (CaBP); and (c) reduced levels of a CaATPase associated with the basal lateral membrane. The role of the BB cytoskeleton in this process is unknown. However, the fact that virtually all the major components of the cytoskeleton are either directly or indirectly affected by Ca^{++} or by calmodulin strongly suggests that the cytoskeletal elements of the BB either participate in, or at least are affected by, the considerable movement of Ca^{++} ions across the BB membrane and through the cytoplasm of the intestinal epithelial cell.

Studies on cytoskeletal composition and structure in BB's isolated from rachitic chicks (Howe et al 1982) revealed no gross alterations in the structural organization or constituent proteins (including 110-kDa-CM and villin). Moreover, rachitic BB's exhibit Ca^{++}-dependent contractility and phosphorylation of BB myosin as well as villin-dependent solation of MV core filaments, which indicates that the various aspects of Ca^{++}-dependent regulation of cytoskeletal structure and contractility identified thus far are unaffected by vitamin D deficiency. Some subtle and possibly important differences have been observed, however. The 110-kDa subunit of rachitic BB's binds calmodulin, as assayed by gel overlay techniques, with greater affinity than control BB's in both the presence and absence of Ca^{++}. In addition, total levels of phosphate incorporated into BB proteins in situ by addition of ATP to isolated BB's is much higher in rachitic BB's, although no qualitative difference in the spectrum of polypeptides phosphorylated was observed (Howe et al 1982). More recently, Putkey & Norman (1983) noted that the "cytoskeletal matrix" isolated by Triton treatment of BB membrane vesicles may be more labile in rachitic samples, based on greater accessibility to iodination by lactoperoxidase.

The cytoskeletal milieu with which a Ca^{++} ion can potentially interact as it crosses the BB membrane includes a number of high-affinity Ca^{++}-binding proteins, calmodulin (present in mM concentrations within MV), villin, and CaBP. Recently, Glenney & Glenney (1985) conducted a thorough, quantitative study of the relative affinity of calmodulin, villin, and CaBP for Ca^{++} under identical conditions, and they also determined the actual amounts of calmodulin and CaBP present in the intestinal epithelium. They observed that the hierarchy of affinity for Ca^{++} of these three major Ca^{++}-binding proteins in the intestinal epithelial cell is CaBP > CM > villin. Based on these observations, they proposed a simple model for the movement of Ca^{++} through the intestinal epithelial cell. According to this model, as Ca^{++} crosses the brush border membrane it is bound in part by the CM present in high concentration in the MV, and also by any soluble CaBP present within the microvillus. Because CaBP has a higher affinity for Ca^{++} than the "stationary" calmodulin within the microvillus, CaBP in effect provides a mobile sink and carrier for Ca^{++} through the cytoplasm of the cell. Moreover, because calmodulin has a higher affinity for Ca^{++} than villin, the 110-kDa-CM complex can successfully buffer the free ion concentration below levels required for activation of villin-severing, and perhaps barbed end–capping activity, thus protecting the BB cytoskeleton from the potentially destructive effects of villin. In fact, Glenney & Glenney (1985) showed that in the presence of high concentrations of calmodulin, similar to that in the MV (1 mM), there is sufficient Ca^{++}-buffering activity to protect actin filaments from the severing activity of villin at Ca^{++} concentrations as high as 10^{-4} M.

ASSEMBLY OF THE BB CYTOSKELETON

Given the extensive information now available regarding the molecular architecture of the BB cytoskeleton, this system is particularly well suited for analysis of the molecular aspects of cytoskeletal assembly. Two distinct types of BB assembly must be considered: that which occurs during embryogenesis, and that which accompanies differentiation of epithelial cells arising from the crypt in the adult intestine. In addition, there are "steady state" rearrangements of the BB cytoskeleton that occur in the fully differentiated intestinal epithelial cell, such as reversible changes in MV length.

BB assembly during embryogenesis is a complex, gradual process. For example, in the chick (Overton & Shoup 1964; Chambers & Grey 1979; Stidwell & Burgess 1984, 1985), short, irregularly shaped MV first appear on the lumenal surface of presumptive intestinal epithelial cells in the 5–9-day-old embryo. By 11 days the MV cores are more ordered, and core rootlets begin to elongate. The cytoskeletal apparatus associated with the junctional complex is in place even in the early embryo, although an extensive network of 10 nm filaments is not yet present. The circumferential bundle of actin filaments associated with the ZA is in place, and may play an important role in the morphogenetic movements involved in pre-villus ridge formation (Burgess 1975). The inter-rootlet zone of the TW, as evidenced by a region of organelle exclusion below the apical membrane, is not formed until relatively late in embryonic development (16–20-day-old embryo; Chambers & Grey 1979; Mooseker & Conzelman, unpublished observations). Consistent with these morphological results, Glenney & Glenney (1983a) showed that TW 260/240 is not detectable in the embryonic intestine until the 15–16-day-old embryo, where it is exclusively localized in the TW region. Conversely, they observed that fodrin, the other form of spectrin present in the embryonic intestinal epithelial cell, is present much earlier in development, where it is associated with both the apical and basal-lateral membranes of the cell. Until quite late in development (day 21 embryo to 1-day posthatch chick), the MV occur at random angles to the cell surface. After hatching, until day 9 posthatch, dramatic increases in MV length occur (Chambers & Grey 1979).

Recently, Stidwell & Burgess (1984, 1985) conducted an extensive structural and biochemical analysis of these posthatch changes in MV length. They observed a number of discrete morphological and temporal changes, including an increase in total MV number at the surface, and an increase in actin filament number within the MV core, which occurred concomitantly with an increase in hexagonal packing of the core filaments. The increase in MV length occurred in two bursts: the first between the 21-

day embryo and the 1-day posthatch chick, and the second between day 2 and day 8 posthatch (resulting in average length changes from 0.7–2.5 μm). These workers also showed, using the DNAase-I inhibition assay, that the total levels of cellular actin and ratios of G-: F-actin changed during the periods of MV length change. They observed a shift in the G : F ratio from 3 : 7 to 1 : 1 immediately before MV elongation, and a return to the 3 : 7 ratio after elongation was complete, which suggests that increases in monomeric actin may drive MV elongation. Moreover, the total actin present in the cell increased concomitantly with the number of MV and core filaments.

Gradual and synchronous (within each cell) elongation of MV also occurs on intestinal epithelial cells of the adult as the cells migrate up from the crypt to the villus tip. Morphological studies indicate that crypt epithelial cells have reasonably well-formed BB's, although with very short MV. However, studies on the differentiation of the BB membrane indicate that levels of membrane-bound hydrolases found in the differentiated cells are much reduced on crypt cells (for review see Danielsen et al 1984).

New insights into the mechanism of BB assembly during differentiation have come from recent studies using a tissue-culture model for differentiation. Pinto et al (1982) made the remarkable observation that a human cell line, HT-29, derived from an adenocarcinoma of the colon, could be induced to differentiate into enterocyte-like cells with fully assembled BB surfaces, by simply substituting galactose for glucose in the culture medium. Coudrier et al (1983c) examined the rates of synthesis, and total levels of protein, for villin and the 140-kDa glycoprotein (thought to be involved in binding of 110-kDa-CM to the membrane) in both undifferentiated and differentiated HT-29 cells. Surprisingly, they observed that both villin and the 140-kDa glycoprotein were present in undifferentiated cells. However, the levels of both proteins were approximately tenfold higher in the differentiated cells. In recent experiments, Louvard et al (1984) showed that the HT-29 cell line is essentially a stem cell, which upon induced differentiation gives rise to several· populations of cells with distinct phenotypes that resemble a number of the cell types present in the epithelium, including small and large intestinal enterocytes, and goblet cells. Even if this model system for differentiation turns out not to closely resemble the differentiation of epithelium in vivo, this cell line should prove invaluable for investigating basic questions regarding molecular aspects of cytoskeletal assembly.

As noted above, once the BB cytoskeleton is assembled it is not a static structure. Most notable are changes in MV length that occur in vivo, including the elongation of MV as cells migrate up the villus axis, and the transient shortening of the MV that accompanies fasting (Misch et al 1980), treatment (in organ culture) with cycloheximide (LeCount & Grey 1972), or colchicine (Buschmann 1983). These observations suggest that the synthesis and turnover of BB cytoskeletal proteins is involved in these length changes, and that the metabolic state of the organ (i.e. eating versus fasting) regulates the synthesis and turnover of cytoskeletal components.

In vivo pulse-chase experiments done in the chicken by Stidwell et al (1984) indicate that steady state turnover of actin and other BB cytoskeletal components occurs in vivo, albeit at relatively slow rates. Using much shorter pulse-labeling times (30 min), Cowell & Danielsen (1984, 1985) analyzed the synthesis rates of villin and the 110-kDa subunit in culture explants of pig intestinal mucosa. These workers observed rapid rates of synthesis of these two proteins, 5–6 times faster than those observed for BB membrane proteins. Based on these observations, the authors suggest that villin and the 110-kDa subunit are synthesized on cytosolic rather than membrane-bound polysomes. Even with very short pulse times (2 min) both villin and the 110-kDa subunit were rapidly incorporated into the MV compartment, and no newly synthesized protein was detectable in the cytosolic (non-cytoskeletal-associated) compartment.

Although synthesis and turnover of actin and other cytoskeletal components is clearly a key aspect of MV length changes in vivo, it is also important to consider the role of actin assembly, such as the case discussed above of MV elongation in the posthatch chick (Stidwell & Burgess 1984, 1985). The classic studies of Tilney & Cardell (1970) demonstrated that the complete reassembly of the BB cytoskeleton after hydrostatic pressure–induced disruption could occur in brief time periods (< 30 min) presumably from preexisting pools of cytoskeletal components. Based on ultrastructural examination of early stages of MV reformation, these authors suggested that MV assembly might occur by nucleated polymerization of core filaments from the dense plaque material that ultimately resides at the tip of the "mature" MV. If actin assembly and disassembly is involved in MV length changes in vivo, the question arises as to which end(s) of the MV core filaments are the sites of monomer addition and loss. Although the barbed, preferred-assembly ends of the core filaments are attached to the membrane at the MV tips, we have shown that in isolated BB's, monomer addition resulting in MV elongation in vitro can occur at the barbed ends of the MV without disrupting the apparent attachment of those filaments to the MV membrane (Mooseker et al 1982b). Based on these observations we have suggested that the barbed ends of the MV core filaments may not be capped by their association with the plasma membrane. A similar argument for the elongation of actin filaments by monomer addition at their membrane-associated ends has been proposed by Tilney et al (1981) for the assembly of the acrosomal process during spermeogenesis in *Limulus*. (For further discussion on this topic, and the possible role of villin in the regulation of actin assembly associated with MV length see Mooseker et al 1984b.)

ACKNOWLEDGMENT

I would like to thank Drs. David Burgess, John Glenney, and Klaus Weber for sending manuscripts and preprints of their unpublished work. Thanks also to Drs. Jimmy Collins, Tony Bretscher, and Daniel Louvard for discussions of ongoing work in their laboratories. Special thanks to Kristine Hall Mooseker for her help in preparing this manuscript.

Literature Cited

Abrahamson, D. R., Rodewald, R. 1981. Evidence for the sorting of endocytotic vesicle contents during the receptor-mediated transport of IgG across the newborn rat intestine. *J. Cell Biol.* 91:270–80

Alicea, H. A. 1985. *Morphological and biochemical comparison of the brush borders of the intestinal epithelial cells of chickens and mammals, with special emphasis on the actin-binding protein villin.* PhD thesis. Yale Univ., New Haven, Conn. 180 pp.

Alicea, H. A., Mooseker, M. S. 1984. Comparative analysis of the actin-binding proteins of the chicken and mammalian intestinal brush border. *J. Cell Biol.* 99:306a

Baines, A. J. 1984. A spectrum of spectrins. *Nature* 312:310–11.

Bonder, E. M., Fishkind, D. J., Mooseker, M. S. 1983. Direct measurement of critical concentrations and assembly rate constants at the two ends of an actin filament. *Cell* 34:491–501

Bonder, E., Mooseker, M. 1983. Direct electron microscopic visualization of barbed end capping and filament cutting by intestinal microvillar 95-kdalton protein (villin). A new actin assembly assay using the *Limulus* acrosomal process. *J. Cell Biol.* 96:1097–1107.

Bordier, C. 1981. Phase separation of integral membrane proteins in Triton X-114 solution. *J. Biol. Chem.* 256:1604–7.

Branton, D., Cohen, C., Tyler, J. 1981. Interaction of cytoskeletal proteins on the human erythrocyte membrane. *Cell* 24:24–32

Bretscher, A. 1981. Fimbrin is a cytoskeletal protein that cross-links F-actin in vitro. *Proc. Natl. Acad. Sci. USA* 78:6849–53

Bretscher, A. 1982. Characterization and ultrastructural role of the major components of the intestinal microvillus cyto-skeleton. *Cold Spring Harbor Symp. Quant. Biol.* 46:871–79

Bretscher, A. 1983a. Microfilament organization in the cytoskeleton of the intestinal brush border. In *Cell and Muscle Motility*, Vol. IV, pp. 239–68. New York: Plenum. 333 pp.

Bretscher, A. 1983b. Molecular architecture of the microvillus cytoskeleton. *Brush Border Membranes, Ciba Found. Symp.* 95:164–79

Bretscher, A. 1983c. Purification of an 80K protein that is a component of the isolated microvillus cytoskeleton, and its localization in nonmuscle cells. *J. Cell Biol.* 97:425–32.

Bretscher, A., Lynch, A. 1985. Identification and localization of immunoreactive forms of caldesmon in smooth and nonmuscle cells: A comparison with the distributions of tropomyosin and alpha actinin. *J. Cell Biol.* 100:1656–63

Bretscher, A., Osborn, A., Wehland, J., Weber, K. 1981. Villin associates with specific microfilamentous structures as seen by immunofluorescence microscopy on tissue sections and cells microinjected with villin. *Exp. Cell Res.* 135:213

Bretscher, A., Weber, K. 1978a. Localization of actin and microfilament associated proteins in the microvilli and terminal web of the intestinal brush border by immuno-fluorescent microscopy. *J. Cell Biol.* 79:839–45

Bretscher A., Weber, K. 1978b. Purification of microvilli and an analysis of the protein components of the microfilament core bundle. *Exp. Cell Res.* 116:397–407

Bretscher, A., Weber, K. 1980a. Fimbrin, a new microfilament-associated protein present in microvilli and other cell surface structures. *J. Cell Biol.* 86:335

Bretscher, A., Weber, K. 1980b. Villin is

a major protein of the microvillus cytoskeleton which binds both G and F-actin in a calcium-dependent manner. *Cell* 20: 839–47

Broschat, K. O., Burgess, D. R. 1984. Regulatory roles of tropomyosin in brush border structure and motility. *J. Cell Biol.* 99: 5a

Broschat, K. O., Stidwell, R. P., Burgess, D. R. 1983. Phosphorylation controls brush border motility by regulating myosin structure and association with the cytoskeleton. *Cell* 35: 561–71

Burgess, D. R. 1975. Morphogenesis of intestinal villi. II. Mechanism of formation of previllous ridges. *J. Embryol. Exp. Morphol.* 34: 723–40

Burgess, D. R. 1982. Reactivation of intestinal epithelial cell brush border motility. ATP-dependent contraction via a terminal web contractile ring. *J. Cell Biol.* 95: 853–66

Burgess, D. R., Prum, B. F. 1982. A reevaluation of brush border motility: calcium induces core filament solation and microvillar vesiculation. *J. Cell Biol.* 94: 97–107

Buschmann, R. J. 1983. Morphometry of the small intestinal enterocytes of the fasted rat and the effects of colchicine. *Cell Tissue Res.* 231: 298–99

Chambers, C., Grey, R. D. 1979. Development of the structural components of the brush border in the absorptive cells of the chick intestine. *Cell Tissue Res.* 204: 387–405

Coleman, T., Mooseker, M. S. 1984. Effects of actin-filament crosslinking and filament length on actin-myosin interaction. *J. Cell Biol.* 99: 302a

Collins, J. H., Borysenko, C. W. 1984. The 110,000-dalton actin- and calmodulin-binding protein from intestinal brush border is a myosin-like ATPase. *J. Biol. Chem.* 259: 14128–35

Cote, G. P. 1983. Structural and functional properties of the non-muscle tropomyosins. *Cell Mol. Biol.* 57: 127–46

Coudrier, E., Reggio, H., Louvard, D. 1981. Immunolocalization of the 110,000 molecular weight cytoskeletal protein of intestinal microvilli. *J. Mol. Biol.* 152: 49–66

Coudrier, E., Reggio, H., Louvard, D. 1982. The cytoskeleton of intestinal microvilli contains two polypeptides immunologically related to proteins of striated muscle. *Cold Spring Harbor Symp. Quant. Biol.* 46: 881–92

Coudrier, E., Reggio, H., Louvard, D. 1983a. Characterization of an integral membrane glycoprotein associated with the microfilaments of pig intestinal microvilli. *EMBO J.*

2: 469–75

Coudrier, E., Reggio, H., Louvard, D. 1983b. Characterization of membrane glycoproteins involved in attachment of microfilaments to the microvillar membrane. *Brush Border Membranes, Ciba Found. Symp.* 95: 216–32

Coudrier, E., Robine, S., Huet, C., Arpin, M., Sahuwquillo, C., Louvard, D. 1983c. Expression of two structural markers of brushborder intestinal mucosa: villin and a membrane glycoprotein (140Kd), in a human colon carcinoma cell line HT29. *J. Submicrosc. Cytol.* 16: 159–60

Cowell, G. M., Danielsen, E. M. 1984. Biosynthesis of intestinal microvillar proteins: Rapid expression of cytoskeletal components in microvilli of pig small intestinal mucosal explants. *FEBS Lett.* 172: 309–14

Cowell, G. M., Danielsen, E. M. 1985. Membrane insertion of the microvillar 110-kDa protein of the enterocyte. *Biochem. J.* 225: 275–76

Craig, S. W., Pardo, J. V. 1979. Alpha actinin localization in the junctional complex of intestinal epithelial cells. *J. Cell Biol.* 80: 203–10

Craig, S. W., Pollard, T. D. 1982. Actin binding proteins. *Trends Biochem. Sci.* 7: 88–91

Craig, S. W., Powell, L. D. 1980. Regulation of actin polymerization by villin, a 95,000 dalton cytoskeletal component of intestinal brush borders. *Cell* 22: 739–46

Danielsen, E. M., Cowell, G. M., Noren, O., Sjostrom, H. 1984. Biosynthesis of microvillar proteins. *Biochem. J.* 221: 1–14

Davis, J. Q., Bennett, V. 1984. Brain ankyrin: A membrane-associated protein with binding sites for spectrin, tubulin, and the cytoplasmic domain of the erythrocyte anion channel. *J. Biol. Chem.* 259: 13550–59

Drenckhahn, D., Groschel-Stewart, U. 1980. Localization of myosin, actin, and tropomyosin in rat intestinal epithelium: immunohistochemical studies at the light and electron microscope levels. *J. Cell Biol.* 86: 475–82

Drenckhahn, D., Hofmann, H.-D., Mannberg, H. 1983. Villin is associated with core filaments and rootlets of intestinal epithelial microvilli. *Cell Tissue Res.* 278: 409–14

Drenckhahn, D., Mannherz, H. G. 1983. Distribution of actin and the actin-associated proteins myosin, tropomyosin, alpha-actinin, vinculin, and villin in rat and bovine exocrine glands. *Eur. J. Cell Biol.* 30: 167–76

Fishkind, D. J., Mooseker, M. S., Bonder, E.

M. 1985. Actin assembly and filament cross-linking in the presence of TW 260/240, the tissue specific spectrin of the chicken intestinal brush border. *Cell Motil.* In press

Fishkind, D., Mooseker, M., Bonder, E., Keene, D., Morrow, J., Pelligrino, L. 1983. Brush border spectrin (TW 260/240) and its interaction with actin and membranes. *J. Cell Biol.* 97:285a

Franke, W., Appelhans, B., Schmid, E., Freudenstein, C., Osborn, M., Weber, K. 1979. The organization of cytokeratin filaments in the intestinal epithelium. *Eur. J. Cell Biol.* 19:255

Franke, W., Winter, S., Grund, C., Schmid, E., Schiller, D., Jarasch, E. 1981. Isolation and characterization of desmosome-associated tonofilaments from rat intestinal brush border. *J. Cell Biol.* 90:116–27

Geiger, B., Tokuyasu, K. T., Singer, S. J. 1979. Immunocytochemical localization of α-actinin in intestinal epithelial cells. *Proc. Natl. Acad. Sci. USA* 76:2833–37

Geiger, B., Dutton, A., Tokuyasu, T., Singer, S. 1981. Immunoelectron microscope studies of membrane-microfilament interactions: distributions of α-actinin, tropomyosin, and vinculin in intestinal epithelial brush border and chicken gizzard smooth muscle cells. *J. Cell Biol.* 91:614–28

Gerke, V., Weber, K. 1983. Isolation and characterization of mammalian villin and fimbrin, the two bundling proteins of the intestinal microvilli. *Eur. J. Cell Biol.* 31:249–55

Gerke, V., Weber, K. 1984. Identity of p36K phosphorylated upon Rous sarcoma virus transformation with a protein purified from brush borders: calcium-dependent binding to nonerythroid spectrin and F-actin. *EMBO J.* 3:227–33

Gerke, V., Weber, K. 1985. Calcium-dependent conformational changes in the 36-kDa subunit of intestinal protein I related to the cellular 36-kDa target of Rous sarcoma virus tyrosine kinase. *J. Biol. Chem.* 260:1668–95

Glenney, J., Bretscher, A., Weber, K. 1980. Calcium control of the intestinal microvillus cytoskeleton: Its implications for the regulation of microfilament organization. *Proc. Natl. Acad. Sci. USA* 77:6458–62

Glenney, J. R., Geisler, P., Kaulfus, P., Weber, K. 1981a. Demonstration of at least two different actin-binding sites in villin, a calcium-regulated modulator of F-actin organization. *J. Biol. Chem.* 256:8156–61

Glenney, J. R., Glenney, P. 1983a. Fodrin is the general spectrin-like protein found in most cells whereas spectrin and the TW protein have a restricted distribution. *Cell* 34:503–12

Glenney, J. R., Glenney, P. 1983b. Spectrin, fodrin, and TW 260/240: A family of related proteins lining the plasma membrane. *Cell Motil.* 3:671–82

Glenney, J. R., Glenney, P. 1984. The microvillus 110 K cytoskeletal protein is an integral membrane protein. *Cell* 37:743–51

Glenney, J. R., Glenney, P. 1985. Comparison of Ca^{++} regulated events in the intestinal brush border. *J. Cell Biol.* 100:754–63

Glenney, J. R., Glenney, P., Osborn, M., Weber, K. 1982a. An F-actin and calmodulin-binding protein from isolated intestinal brush borders has a morphology related to spectrin. *Cell* 28:843–54

Glenney, J. R., Glenney, P., Weber, K. 1982b. Erythroid spectrin, brain fodrin, and intestinal brush border proteins (TW 260/240) are related molecules containing a common calmodulin-binding subunit bound to a variant cell type specific subunit. *Proc. Natl. Acad. Sci. USA* 79:4002–5

Glenney, J. R., Glenney, P., Weber, K. 1983. The spectrin-related molecule, TW 260/240, cross-links actin bundles of the microvillus rootlets in the brush borders of intestinal epithelial cells. *J. Cell Biol.* 96:1491–96

Glenney, J. R., Kaulfus, P., Matsudaira, P., Weber, K. 1981b. F-actin binding and bundling properties of fimbrin, a major cytoskeletal protein of the microvillus core filaments. *J. Biol. Chem.* 256:9283–88

Glenney, J. R., Kaulfus, P., Weber, K. 1981c. F-actin assembly modulated by villin: Ca^{++}-dependent nucleation and capping of the barbed end. *Cell* 24:471–80

Glenney, J. R., Osborn, P., Weber, K. 1982c. The intracellular localization of the microvillus 110K protein, a component considered to be involved in side-on membrane attachment of F-actin. *Exp. Cell Res.* 138:199–205

Glenney, J. R., Weber, K. 1980. Calmodulin binding proteins of the microfilaments present in isolated brush borders and microvilli of intestinal epithelial cells. *J. Biol. Chem.* 255:10551–54

Glenney, J. R., Weber, K. 1981. Calcium control of microfilaments: Uncoupling of the F-actin-severing and -bundling activity of villin by limited proteolysis *in vitro. Proc. Natl. Acad. Sci. USA* 78:2810–14

Gonnella, P. A., Neutra, M. R. 1984. Membrane-bound and fluid-phase macromolecules enter separate prelysosomal compartments in absorptive cells of suckling rat ilium. *J. Cell Biol.* 99:909–17

Gould, K. L., Cooper, J. A., Hunter, T. 1984. The 46,000-dalton tyrosine protein kinase substrate is widespread, whereas the 36,000-dalton substrate is only expressed at high levels in certain rodent tissues. *J. Cell Biol.* 98:487–97

Greenberg, M. E., Brackenbury, R., Edelman, G. M. 1984. Changes in the distribution of the 34-kdalton tyrosine kinase substrate during differentiation and maturation of chicken tissues. *J. Cell Biol.* 98:473–86

Hesterberg, L., Weber, K. 1983a. Ligand-induced conformational changes in villin, a calcium-controlled actin-modulating protein. *J. Biol. Chem.* 258:359–64

Hesterberg, L., Weber, K. 1983b. Demonstration of three distinct calcium-binding sites in villin, a modulator of actin assembly *J. Biol. Chem.* 258:365–69

Hirokawa, N., Cheney, R. E., Willard, M. 1983a. Location of a protein of the fodrin-spectrin-TW 260/240 family in the mouse intestinal brush border. *Cell* 32:953–65

Hirokawa, N., Heuser, J. 1981. Quick-freeze, deep-etch visualization of the cytoskeleton beneath surface differentiations of intestinal epithelial cells. *J. Cell Biol.* 91:399–409

Hirokawa, N., Keller, T. C. S., Chasen, R., Mooseker, M. S. 1983b. Mechanism of brush border contractility studied by the quick-freeze-deep-etch method. *J. Cell Biol.* 96:1325–36

Hirokawa, N., Tilney, L. G., Fujiwara, K., Heuser, J. E. 1982. The organization of actin, myosin, and intermediate filaments in the brush border of intestinal epithelial cells. *J. Cell Biol.* 94:425–43

Howe, C. L., Keller, T. C. S., Mooseker, M. S., Wasserman, R. H. 1982. Analysis of cytoskeletal proteins and Ca^{++}-dependent regulation of structure in intestinal brush borders from rachitic chicks. *Proc. Natl. Acad. Sci. USA* 79:1134–38

Howe, C. L., Mooseker, M. S. 1983. Characterization of the 110-kdalton actin-, calmodulin-, and membrane-binding protein from microvilli of intestinal epithelial cells. *J. Cell Biol.* 97:974–85

Howe, C. L., Mooseker, M. S., Graves, T. A. 1980. Brush border calmodulin. A major component of the isolated microvillus core. *J. Cell Biol.* 85:916–23

Howe, C. L., Sacramone, L. M., Mooseker, M. S., Morrow, J. S. 1984. Modulation of membrane affinity in spectrin isoforms. *J. Cell Biol.* 99:302a

Hull, B. E., Staehelin, L. A. 1979. The terminal web. A re-evaluation of its structure and function. *J. Cell Biol.* 81:67–82

Keller, T. C. S., Conzelman, K. A., Chasan, R., Mooseker, M. S. 1983. Terminal web

contraction in isolated brush borders requires little of the brush border myosin. *J. Cell Biol.* 97:295a

Keller, T. C. S., Conzelman, K. A., Chasan, R., Mooseker, M. S. 1985. The role of myosin in terminal web contraction in isolated intestinal epithelial brush borders. *J. Cell Biol.* 100:1647–55

Keller, T. C. S., Mooseker, M. S. 1982. Ca^{++}-calmodulin-dependent phosphorylation of myosin, and its role in brush border contraction in vitro. *J. Cell Biol.* 95:943–59

Kenny, A. J., Booth, A. G. 1978. Microvilli: their ultrastructure, enzymology and molecular organization. *Essays Biochem.* 14:1–43.

Kenny, A. J., Maroux, S. 1982. Topology of microvillar membrane hydrolases of kidney and intestine. *Physiol. Rev.* 62:91–128

Kerjaschki, D., Noronha-Blob, L., Sacktor, B., Farquhar, M. G. 1984. Microdomains of distinctive glycoprotein composition in the kidney proximal tubule brush border. *J. Cell Biol.* 98:1505–13

Knutton, S., Limbrick, A. R., Robertson, J. D. 1974. Regular structures in membranes: membranes in the endocytic complex of ileal epithelial cells. *J. Cell Biol.* 62:679–94

Korn, E. D. 1982. Actin polymerization and its regulation by proteins from nonmuscle cells. *Physiol. Rev.* 62:672–737

Lazarides, E., Nelson, W. J. 1982. Expression of spectrin in nonerythroid cells. *Cell* 31:505–8

LeCount, T. S., Grey, R. D. 1972. Transient shortening of microvilli induced by cycloheximide in the duodenal epithelium of chicken. *J. Cell Biol.* 53:601–5

Louvard, D., Arpin, M., Coudrier, E., Huet, C., Pringault, E., et al. 1984. Experimental manipulation of intestinal cell differentiation using a human adenocarcinoma cell-line (HT-29). *J. Cell Biol.* 99:6a

Matsudaira, P. Y. 1983. Structural and functional relationship between the membrane glycoproteins involved in attachment of microfilaments to the microvillar membrane. *Brush Border Membranes, Ciba Found. Symp.* 95:233–44

Matsudaira, P. T., Burgess, D. R. 1979. Identification and organization of the components in the isolated microvillus cytoskeleton. *J. Cell Biol.* 83:667–73

Matsudaira, P., Burgess, D. 1982a. Structure and function of the brush border cytoskeleton. *Cold Spring Harbor Symp. Quant. Biol.* 46:845–54

Matsudaira, P. T., Burgess, D. R. 1982b. Partial reconstruction of the microvillus core bundle: characterization of villin as a Ca^{++}-dependent, actin bundling/

depolymerizing protein. *J. Cell Biol.* 92: 657–64

Matsudaira, P., Jakes, R. 1984. Mapping the functions of villin structural domains. *J. Cell Biol.* 99:307a

Matsudaira, P., Mandelkow, E., Renner, W., Hesterberg, L. K., Weber, K. 1983. Role of fimbrin and villin in determining the interfilament distances of actin bundles. *Nature* 301:209–13

Maunsbach, A. E. 1976. Cellular mechanism of tubular protein transport. *Int. Rev. Physiol.* 2:145–67

Misch, P., Giebel, P., Faust, R. 1980. Intestinal microvilli responses to feeding and fasting. *Eur. J. Cell Biol.* 21:264–79

Mooseker, M. S. 1976. Brush border motility. Microvillar contraction in Triton-treated brush borders isolated from intestinal epithelium. *J. Cell Biol.* 71:417–32

Mooseker, M. S. 1983. Actin binding proteins of the brush border. *Cell* 35:11–13

Mooseker, M. S., Bonder, E. M., Conzelman, K. A., Fishkind, D. J., Howe, C. L., Keller, T. C. S. 1984a. The cytoskeletal apparatus of the intestinal brush border. In *Mechanisms of Intestinal Electrolyte Transport and Regulation by Calcium, Kroc Found. Ser.*, ed. M. Donowitz, G. Sharp, Vol. 17, pp. 287–307. New York: Allen Liss

Mooseker, M. S., Bonder, E. M., Conzelman, K. A., Fishkind, D. J., Howe, C. L., Keller, T. C. S. 1984b. The brush border cytoskeleton and integration of cellular functions. *J. Cell Biol.* 99:104s–12s

Mooseker, M. S., Bonder, E. M., Grimwade, B. G., Howe, C. L., Keller, T. C. S., et al. 1982a. Regulation of contractility, cytoskeletal structure, and filament assembly in the brush border of intestinal epithelial cells. *Cold Spring Harbor Symp. Quant. Biol.* 46:855–70

Mooseker, M. S., Graves, T. A., Wharton, K. A., Falco, N., Howe, C. L. 1980. Regulation of microvillus structure: calcium-dependent solation and cross-linking of actin filaments in the microvilli of intestinal epithelial cells. *J. Cell Biol.* 87:809–22

Mooseker, M. S., Howe, C. L. 1982. The brush border of intestinal epithelium: A model system for analysis of cell-surface architecture and motility. In *Methods in Cell Biology*, ed. L. Wilson, Vol. 25. pp. 144–75. New York: Academic, 426 pp.

Mooseker, M. S., Keller, T. C. S., Hirokawa, N. 1983. Regulation of contractility and cytoskeletal structure in the brush border. In *Brush Border Membranes, Ciba Found. Symp.* 95:195–215

Mooseker, M. S., Pollard, T. D., Fujiwara, K. 1978. Characterization and localization of myosin in the brush border of intestinal

epithelial cells. *J. Cell Biol.* 79:444–53

Mooseker, M. S., Pollard, T. D., Wharton, K. A. 1982b. Nucleated polymerization of actin from the membrane-associated ends of microvillar filaments in the intestinal brush border. *J. Cell Biol.* 95:222–33.

Mooseker, M. S., Tilney, L. G. 1975. The organization of an actin filament-membrane complex: filament polarity and membrane attachment in the microvilli of intestinal epithelial cells. *J. Cell Biol.* 67:725–43

Overton, J., Shoup, J. 1964. Fine structure of cell surface specializations in the maturing duodenal mucosa of the chick. *J. Cell Biol.* 21:75–85

Owada, M. K., Hakura, A., Iida, K., Yahara, I., Sobue, K., Kakiuchi, S. 1984. Occurrence of caldesmon (a calmodulin-binding protein) in cultured cells: comparison of normal and transformed cells. *Proc. Natl. Acad. Sci. USA* 81:3133–37

Owaribe, K., Kodama, R., Eguchi, G. 1981. Demonstration of contractility of circumferential actin bundles and its morphogenetic significance in pigmented epithelium in vitro and in vivo. *J. Cell Biol.* 90:507–14

Pearl, M., Fishkind, D., Mooseker, M. S., Keene, D., Keller, T. C. S. 1984. Studies on the spectrin-like protein from the intestinal brush border, TW 260/240, and characterization of its interaction with the cytoskeleton and actin. *J. Cell Biol.* 98:66–78

Pinto, M., Appay, M-D., Simon-Assman, P., Chevalier, G., Dracopoli, N., et al. 1982. Enterocytic differentiation of cultured human colon cancer cells by replacement of glucose by galactose in the medium. *Biol. Cell* 44:193–96

Pollard, T. D. 1984. Purification of a high molecular weight actin filament gelation protein from *Acanthamoeba* that shares antigenic determinants with vertebrate spectrins. *J. Cell Biol.* 99:1970–80

Pollard, T. D., Craig, S. W. 1982. Mechanism of actin polymerization. *Trends Biochem. Sci.* 7:55–58

Pollard, T. D., Mooseker, M. S. 1981. Direct measurement of actin polymerization rate constants by electron microscopy of actin filaments nucleated by isolated microvillus cores. *J. Cell Biol.* 88:654–59

Putkey, J. A., Norman, A. W. 1983. Vitamin D: Its effect on the protein composition and core material structure of the chick intestinal brush border membrane. *J. Biol. Chem.* 258:8971–78.

Rodewald, R., Newman, S. B., Karnovskyu, M. J. 1976. Contraction of isolated brush borders from the intestinal epithelium. *J. Cell Biol.* 70:541–54

Rodman, J. S., Kerjaschki, D., Merisko, E., Farquhar, M. G. 1984. Presence of an extensive clathrin coat on the apical plasmalemma of the rat kidney proximal tubule cell. *J. Cell Biol.* 98:1630–36

Sefton, B. A., Hunter, T. 1984. Tyrosine protein kinases. *Adv. Cyclic Nucleotide Protein Phosphorylation Res.* 18:195–226

Sheetz, M. P., Spudich, J. A. 1983. Movement of myosin-coated fluorescent beads on actin cables in vitro. *Nature* 303:31–35

Sobue, K., Kanda, K., Innui, M., Morimoto, K., Kakiuchi, S. 1982. Actin polymerization induced by calspectin, a calmodulin binding spectrin-like protein. *FEBS Lett.* 148:221–25

Stidwell, R. P., Burgess, D. R. 1984. A burst of elongation of intestinal microvilli during embryogenesis coincides with dramatic changes in the G:F-actin ratios. *J. Cell Biol.* 99:30a

Stidwell, R. P., Burgess, D. R. 1985. Changes in actin state and content may drive brush border microvillus growth during embryogenesis. Submitted for publication

Stidwell, R. P., Wysolmerski, T., Burgess, D. R. 1984. The brush border cytoskeleton is not static: in vivo turnover of proteins. *J. Cell Biol.* 98:641–45

Tilney, L. G. 1975. Actin filaments in the acrosomal reaction of *Limulus* sperm: motion generated by alterations in the packing of filaments. *J. Cell Biol.* 64:289–310

Tilney, L. G., Bonder, E. M., DeRosier, D. J. 1981. Actin filaments elongate from their membrane associated ends. *J. Cell Biol.* 90:485–94

Tilney, L. G., Cardell, R. R. Jr. 1970. Factors controlling the reassembly of the microvillus border of the small intestine of the salamander. *J. Cell Biol.* 47:408–22

Verner, K., Bretscher, A. 1983. Induced morphological changes in isolated microvilli: Regulation of membrane topography in vitro by submembranous microfilaments. *Eur. J. Cell Biol.* 29:187–92

Verner, K., Bretscher, A. 1984. Microvillus 110K-CM: Nucleoside triphosphate effects in isolated cytoskeletons and a purified F-actin/110K-CM system. *J. Cell Biol.* 99:351a

Wagner, P. D. 1984. Calcium-sensitive modulation of the actomyosin ATPase by fodrin. *J. Biol. Chem.* 259:6306–10

Walsh, T. P., Weber, A., Davis, K., Bonder, E., Mooseker, M. S. 1984a. Calcium dependence of villin-induced actin depolymerization. *Biochemistry* 23:6099–102

Walsh, T. P., Weber, A., Higgins, J., Bonder, E. M., Mooseker, M. S. 1984b. Effect of villin on the kinetics of actin polymerization. *Biochemistry* 23:2613–21.

Wang, Y., Bonder, E. M., Mooseker, M. S., Taylor, D. L. 1983. Effects of villin on the polymerization and subunit exchange of actin. *Cell Motil.* 3:151–65

Wasserman, R. H. 1981. Intestinal absorption of calcium and phosphorus. *Fed. Proc.* 40:68–72.

Weber, K., Glenney, J. R. 1982a. Calcium-modulated multifunctional proteins regulating F-actin organization. *Cold Spring Harbor Symp. Quant. Biol.* 46:541–52

Weber, K., Glenney, J. R. 1982b. Microfilament-membrane interaction: The brush border of intestinal epithelial cells. *Philos. Trans. R. Soc. London B* 299:207–14

Weeds, A. 1982. Actin-binding proteins—regulators of cell architecture and motility. *Nature* 296:811–16

Wegner, A. 1984. Subtleties of actin assembly. *Nature* 313:97–98

Ann. Rev. Cell Biol. 1985. 1 : 243–88

CELL SURFACE POLARITY IN EPITHELIA

Kai Simons and Stephen D. Fuller

European Molecular Biology Laboratory, Postfach 10.2209, 6900 Heidelberg, Federal Republic of Germany

CONTENTS

INTRODUCTION

Epithelial cells are organized into sheets that separate compartments of the organism (Berridge & Oschman 1972). The cells in the epithelium are linked through junctional complexes so that they form a selective permeability barrier. Epithelial cells are specialized to perform a wide variety of vectorial

243

0743–4634/85/1115–0243$02.00

functions. Transporting epithelia such as those of the renal tubule, absorptive epithelia such as those of the intestine, and secretory epithelial cells such as the hepatocytes are typical examples of epithelia that create and maintain concentration gradients between the compartments they separate. These vectorial functions reflect the polar organization of the epithelial cell. Epithelial cells accomplish these functions by localizing distinct sets of cell surface components to separate plasma membrane domains. The mechanisms by which cell surface polarity is generated and maintained in epithelia form the subject of this review.

Although the morphology and other characteristics of different epithelial cells are extremely varied, their cell surfaces are similarly organized (Figure 1). The plasma membrane of each cell is divided into two domains: an apical domain, often covered with microvilli, which faces the external milieu of the organism (or an equivalent thereof), whereas the basolateral domain is in contact with the internal milieu, facing the mesenchymal space and blood supply. The domains are separated by tight junctions that encircle the apex of the cell, and seal neighboring cells to each other. Intermediate junctions and desmosomes connect the lateral membranes of adjacent cells. A set of epithelial intermediate filaments, cytokeratins, which are specific for epithelial cells, attach to the cytoplasmic side of the desmosomes. The basal aspect of the epithelial cell layer usually rests on an extracellular matrix, which is often organized into a basal lamina.

An underlying assumption in our treatment of cell polarity is that the mechanisms responsible for regulating traffic to the cell surface are identical in all epithelial cells. We employ a consistent nomenclature for the plasma membrane domains of epithelial cells to emphasize their similarity in design. The basal surface of the basolateral membrane is that which faces the blood supply on the serosal side, while the apical surface faces the luminal side (Figure 1). Although many epithelia are stratified and multilayered, we limit our discussion to simple epithelia, for which more information is available concerning cell surface polarity. We expect that similar principles guide the establishment and maintenance of polarity in all epithelia.

EXPERIMENTAL SYSTEMS TO STUDY CELL SURFACE BIOGENESIS

Tissues

Many of the early investigations of the biogenesis of secretory and cell surface proteins were performed in tissues with epithelial organization, such as the liver or the pancreas (Jamieson 1972, Palade 1975, Evans 1980). While these studies did not focus on the biogenesis of the polarized cell

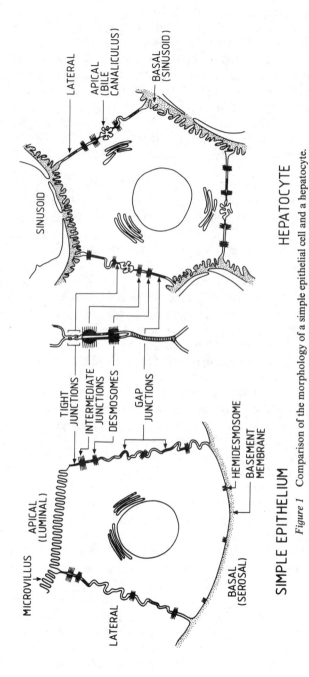

Figure 1 Comparison of the morphology of a simple epithelial cell and a hepatocyte.

surface per se, they did delineate the pathways of newly synthesized proteins from the rough endoplasmic reticulum (ER) over the Golgi complex to the cell surface. Studies of the biogenesis of specific apical or basolateral proteins were performed in intact tissues such as liver or small intestine using pulse-chase labeling, subcellular fractionation, and antibodies against specific surface proteins. These investigations provided important information on the biogenesis of apical and basolateral proteins. The use of whole tissues as the experimental system has the advantage that epithelial organization is intact. The disadvantages are the difficulty of experimentally manipulating the system, and the presence of many cell types in a single tissue.

Cultured Epithelial Cells

Some of the technical difficulties involved in the manipulation of intact tissues are avoided by the use of cultured epithelial cells. Such systems provide a homogeneous cell population and greater ease of handling. The main drawback is the difficulty of maintaining epithelial organization. Subcellular fractionation is also more difficult to use than in the case of intact tissues like liver, for which the methods are well established. Primary cell cultures can be obtained by selection for epithelial cell growth from explants using the appropriate serum-free medium (Taub & Sato 1979, Barnes & Sato 1980), and through microdissection or fractionation of epithelial cells from tissues (Bulger & Dobyan 1982, Burg et al 1982, Wilson & Horster 1983). Established cell lines are more useful for studies of surface polarity. An increasing number of polarized epithelial cell lines are becoming available. These are derived from both normal and malignant tissues, and include cells that originate from the kidneys (Leighton et al 1970, McRoberts et al 1981, Perkins & Handler 1981), mammary gland (Lippman et al 1980, Danielson et al 1984), urinary bladder (Handler et al 1979), thyroid (Ambesi-Impiombato et al 1980), and intestine (Pinto et al 1982 and 1983, Louvard et al 1985).

Several factors are important for optimal expression of the epithelial phenotype in vitro. A primary consideration is the polarity of nutrient uptake. In vivo, many nutrients reach the epithelial sheet from the basolateral side, which faces the blood supply. However, when epithelial cells are cultured on glass or plastic, they are forced to feed from the apical surface, which faces the culture medium. Hence, the basolateral surface becomes isolated from the growth medium as the monolayer is sealed by the formation of tight junctions. To grow properly the epithelial sheet must remain somewhat leaky, or expose basolateral proteins responsible for uptake of nutrients and binding of growth factors on the apical side (Balcarova-Stander et al 1984, Handler et al 1984, Fuller et al 1984). These problems can be overcome simply by growing the epithelial cells on permeable supports, such as nitrocellulose filters (Misfeldt et al 1976), or collagen-coated nylon discs (Cereijido et al 1978). Epithelial cells form monolayers with a higher degree of differentiation when the basolateral surface is directly accessible to the growth medium (Valentich 1982, Handler et al 1984, Fuller et al 1984). This is evident from the morphology of the cells, their increased responsiveness to hormones, and the exclusion of basolateral proteins from their apical surfaces.

Other factors that may be important for the expression of the epithelial phenotype in vitro are proper growth factors (Barnes & Sato 1980), and a suitable extracellular matrix (Bernfield & Banerjee 1978, Gospodarowicz & Tauber 1980, Kleinman et al 1981). Several serum-free media have been devised that selectively promote the growth of different polarized epithelial cells over nonepithelial cells (Barnes & Sato 1980, Taub & Livingstone 1981). In some cases the biochemical and cytological differentiation of the epithelial cells can be improved by seeding the cells on collagen, in collagen gels, or on more complete extracellular matrices (Sugrue & Hay 1981, Wicha et al 1982, Gospodarowicz et al 1984).

The best characterized epithelial cell line is the Madin-Darby canine kidney (MDCK) cell, which contains all the elements of a polarized cell surface depicted in Figure 1. This line has been extensively used by cell physiologists to study transepithelial transport of ions (Misfeldt et al 1976, Cereijido et al 1981a, Rindler et al 1982, Simmons et al 1984). Two strains with different characteristics have been defined (Richardson et al 1981, Fuller et al 1984, Balcarova-Ständer et al 1984), and are used by cell biologists for studies of cell surface polarity. These studies have shown that use of the appropriate cell line and optimal growth conditions yields epithelia in vitro which mimic the epithelial organization found in vivo.

Viruses and Polarity

Rodriguez-Boulan & Sabatini (1978) first made the observation that enveloped viruses bud in a polarized manner from infected MDCK cells. This finding paved the way for the use of viral glycoproteins as probes to study the biogenesis and transport of apical and basolateral proteins in MDCK cells (Rodriguez-Boulan 1983, Sabatini et al 1983). The G protein of vesicular stomatitis virus (VSV) was found to be inserted mainly into the basolateral plasma membrane, whereas the spike glycoproteins of influenza virus behaved as apical plasmalemmal proteins (Rodriguez-Boulan & Pendergast 1980, Roth et al 1983).

Enveloped viruses were used previously to study the biosynthesis and transport of proteins to the cell surface in nonepithelial cells grown in culture. Following infection by these viruses, host protein synthesis is suppressed and large quantities of viral surface glycoproteins are synthesized. This amplification facilitates the studies of plasma membrane biogenesis. The use of viral probes complemented by studies on endogenous plasma membrane proteins has contributed significantly to our understanding of plasma membrane biogenesis in nonpolarized cells (Sabatini et al 1982, Rothman & Lenard 1984, Simons & Warren 1984), and holds a similar potential for analyzing cell surface polarity.

The cytopathic effects that accompany infection are the major limitation of viral probes. In MDCK cells grown on plastic or glass and infected with VSV, the tight junctions of many cells begin to open by the time virus budding starts (Rodriguez-Boulan & Pendergast 1980). Immunocyto-chemical studies demonstrate significant amounts of the G protein on the apical plasma membrane even early in infection (Rindler et al 1984, 1985). Strikingly different results are obtained with filter-grown MDCK cells (Fuller et al 1984, Pfeiffer et al 1985). VSV infects these cells only when applied to the basolateral surface through the pores of the filter because the VSV receptors are located exclusively on the basolateral side of the cells (Fuller et al 1984, 1985b). When VSV is applied to MDCK cells growing on plastic, the virus only has access to the apical surface, and hence will preferentially infect those cells that are less polar and express the normally basolateral receptor apically (Fuller et al 1984). Filter-grown MDCK cells also appear more resistant to the cytopathic effects of viral infection, so that transport of viral glycoproteins to the cell surface is polar under these conditions. These features coupled with access to both sides of the monolayer have made filter-grown MDCK cells an attractive model system for the study of cell surface polarity.

MOLECULAR ORGANIZATION OF THE EPITHELIAL CELL SURFACE

Lipids

Although little is known about the lipid compositions of apical and basolateral membranes, the available evidence suggests that they differ. The only complete analyses were carried out using brush border (apical) and basolateral membranes of the epithelial cells isolated from rat and mouse small intestine (Forstner et al 1968, Douglas et al 1972, Kawai et al 1974, Brasitus & Schachter 1980). The apical membrane was enriched in glycolipids (about twofold) and cholesterol compared to the basolateral membrane, and had half as much phosphatidylcholine. The amount of the second major phospholipid, phosphatidylethanolamine, was almost equal in the two membranes. Similar differences were demonstrated in the lipid composition of viruses budding from the apical and basolateral membranes of two polarized cell lines, MDBK, Madin-Darby bovine kidney (Klenk & Choppin 1970a,b, Rothman et al 1976), and MDCK (van Meer & Simons 1982), although the analyses were not as complete as those with intestinal cells. These viruses are known to incorporate into their envelopes the lipids present in the plasma membrane from which they bud (Patzer et al 1979). The same viruses (VSV, influenza) produced in nonpolarized baby hamster kidney (BHK) cells had identical lipid compositions (van Meer et al 1985). These results can be best explained by postulating that the lipids in the two membrane domains of the epithelial cell surface cannot freely intermix.

About one third of the lipid content of the apical membrane of intestinal cells is glycolipid. Since glycolipids usually reside exclusively in the external half of the lipid bilayer (Op den Kamp 1979), this leaflet of the apical membrane may be almost solely composed of glycolipids. Calculations show that simply maintaining the polarized and asymmetric organization of glycolipids would result in the observed polarity of their total lipid compositions (van Meer et al 1985).

Whether or not lipid polarity is a general phenomenon remains unclear. Lipid analyses of epithelial cells from the rat kidney have not revealed such clear differences in composition between the two surface domains (Bode et al 1976, Hise et al 1984). The apical (biliary) and the basal (sinusoidal) membranes isolated from rat hepatocytes show smaller differences (Meier et al 1984) than those seen in intestinal cells. More lipid analyses of different epithelial cell types are needed to find out how widely the lipid composition of the cell surface domains can vary.

Proteins

The distribution of individual proteins over the apical and basolateral plasma membranes has been intensively studied in different epithelia. Several methods have been used based on enzyme cytochemistry, immunocytochemistry, and subcellular fractionation of epithelial tissues and, more recently, ligand binding, surface radioimmunoassay, and uptake studies with filter-grown epithelial cells. Table 1 shows a selected list of proteins that have been used as markers for apical and for basolateral membranes. [For reviews on the apical and basolateral localization of transport and enzyme activities see Murer & Kinne (1980), Kenny & Maroux (1982), Rodriguez-Boulan (1983) and Almers & Stirling (1984).]

Two trends are apparent in Table 1. First, a specific protein is usually localized to a single surface domain in each epithelial cell. Second, a protein that has been assigned to one surface domain in one epithelial cell type is usually found in the same domain of other epithelial cells. There are exceptions, however. Alkaline phosphatase is localized to the apical membrane in several epithelial cell types (Table 1). However, it was found

Table 1 Typical polarity markers found in two or more epithelia

Protein	Domain[a]	Tissue[b]	References
Adenylate cyclase	B	Enterocyte	Lodja (1974)
	B	Hepatocyte	Reik et al (1970)
	B	Renal tubule	Schwartz et al (1974)
Alkaline phosphatase	A	Enterocyte	Lodja (1974)
	A		Borgers (1973)
	B		Mircheff et al (1975)
	B	Hepatocyte	Wisher and Evans (1975)
	B		Meier et al (1984)
	A	Placenta	Carlson et al (1976)
	A	Pig kidney line	Rabito et al (1984)
	A	Renal tubule	Gomori (1941)
	A		Heidrich et al (1972)
Amiloride sensitive sodium channel	A	Colon	Stoner (1979)
	A		Turnheim et al (1978)
	A	Frog skin	Cuthbert & Schum (1974)
	A	Renal tubule	O'Neil & Boulpaep (1979)
	A	Salivary gland	Schneyer (1970)
	A	Urinary bladder	Lewis et al (1976)
ATP-dependent calcium uptake	B	Enterocyte	Hildeman et al (1979)
	B	Renal tubule	Gmaj et al (1979)
Dipeptidyl peptidase	A	Enterocyte	Barth et al (1974)
	A	Renal tubule	Hopsu-Havu et al (1968)
Furosemide-sensitive sodium/potassium chloride cotransport	B	Cornea	Ludens et al (1980)
	B	MDCK	Aiton et al (1982)
	A	Renal tubule	Koenig et al (1983)
	A	Shark rectal gland	Hannafin et al (1983)
γ-glutamyl transferase	A	Enterocyte	Marathe et al (1979)
	A	Hepatocyte	Inoue et al (1983)
	A	Pig kidney line	Rabito et al (1984)
	A	Renal tubule	Marathe et al (1979)
H-2 antigens	B	Enterocyte	Parr & Kirby (1979)
	B	Gall bladder	Parr & Kirby (1979)
	B	Hepatocyte	Parr (1979)
	B	Trachea	Parr & Kirby (1979)
	B	Uterus	Parr & Kirby (1979)
Insulin receptor	B	Hepatocyte	Bergeron et al (1979)
	B	Renal tubule	Taylor et al (1982)
Leucine aminopeptidase (aminopeptidase N)	A	Enterocyte	Semenza (1976)
	A	Hepatocyte	Roman & Hubbard (1984a,b)
	A	MDCK	Louvard (1980)
	A	Renal tubule	Desnuelle (1979)
	A	Thyroid follicle	Feracci et al (1981)
Maltase	A	Enterocyte	Semenza (1976)
	A	Renal tubule	Kerjaschki et al (1984)
Magnesium ATPase	A	Hepatocyte	Meier et al (1984)
	A		Wisher & Evans (1975)
	A	Renal tubule	George & Kenny (1973)

Table 1 (*continued*)

Protein	Domain[a]	Tissue[b]	References
Na[+]/ K[+] ATPase	B	Avian salt gland	Ernst (1972)
	B	Choriod plexus	Milhorat et al (1975)
	A		Quinton et al (1973)
	B	Enterocyte	Stirling (1972)
	B	Frog skin	Mills et al (1977)
	B	Gallbladder	Mills & DiBona (1978)
	B	Hepatocyte	Meier et al (1984)
	B		Blitzer and Boyer (1978)
	A		Takemura et al (1984)
	B	MDCK	Louvard (1980)
	B		Lamb et al (1981)
	B	Pancreas	Bundgaard et al (1981)
	B	Renal tubule	Shaver & Stirling (1978)
	B		Kyte (1976a,b)
	B	Salivary gland	Bundgaard et al (1977)
	B	Shark rectal gland	Karnaky et al (1976)
	B	Sweat gland	Quinton & Tormey (1973)
	B	Teleost gill	Hootman & Philpott (1979)
	B	Urinary bladder	Mills & Ernst (1975)
Neutral Endopeptidase	A	Enterocyte	Danielsen et al (1980)
	A	Renal tubule	Kerr & Kenny (1974)
5' Nucleotidase	A	Enterocyte	Colas & Maroux (1980)
	A	Hepatocyte	Inoue et al (1983)
	A		Meier et al (1984)
	A	Renal tubule	George & Kenny (1973)

[a] A = apical; B = basolateral.
[b] Organism is mammalian unless indicated.

that in rat duodenal cells the enzyme was localized to a basolateral membrane fraction, which had been highly purified both by density gradient centrifugation and free-flow electrophoresis (Mircheff et al 1979). Also in the hepatocyte the enzyme has been localized to the basolateral domain. Whether the enzymatic activity reflects the same protein as in the apical membrane of other cells is not known. Another possible exception is Na$^+$K$^+$-ATPase, which is basolateral in almost all epithelial cells. However, its position in two cell types is ambiguous. In choroid plexus cells, ^{3}H-oubain autoradiography located the protein on the apical surface (Milhorat et al 1975), while enzyme cytochemistry yielded the opposite result (Quinton et al 1973b). In hepatocytes the results also vary. A recent immunocytochemical study of dog liver (Takemura et al 1984) using antibodies against the Na$^+$K$^+$-ATPase revealed a higher density of label on the apical than on the basolateral surface. In contrast, no enzyme activity was detected in a bile canalicular fraction isolated by subcellular fractionation from rat liver (Meier et al 1984), although a Mg-ATPase was localized to the apical membrane. The immunological localization could reflect cross-reactivity since the catalytic subunits of several ATPases are related (Kyte 1981).

In the case of the furosemide-sensitive NaKCl cotransport system, different research groups agree that an identical functional activity is localized to either the apical or the basolateral side of different epithelial cells (Table 1) (Almers & Stirling 1984). Since the protein responsible for this transport activity has not been isolated, it is not known whether or not this activity is the result of the same protein in all cells. Similar functions may, of course, be specified by different genes that code for the functional activity and for the signals necessary for transport of the protein to the correct surface domain. Another possibility is that the protein has an oligomeric structure in which the functional sites reside on one homologous subunit, which is associated with another subunit that varies in different cell types. This latter subunit would carry the signals for apical or basolateral sorting. The elucidation of the molecular structure of the proteins responsible for sorting in two cells with opposite localization could give important clues to the sorting signals.

Data on the distribution of specific proteins between the two surface domains are necessary to define the efficiency of the sorting of apical and basolateral proteins. It would be useful to know not only the total amounts of a protein in each domain but also its relative surface density. This requires an estimate of the surface areas of the apical and basolateral membrane domains by morphometric or other means. Such studies are only beginning to appear. Table 2 shows data for hepatocytes and MDCK. From these very limited data it appears that the basolateral proteins and functions are exclusively localized to their correct domain (better than 97% of these proteins), whereas the two apical proteins are less stringently polarized, with significant amounts in the basolateral domain. This trend disappears when the values are corrected for surface area.

Data are now needed for more proteins and from different epithelial cells, both in tissues and in culture, to elucidate how exclusively apical and basolateral proteins are localized to their respective surface domains. The levels of missorted proteins may simply reflect the difference between apical and basolateral surface areas. Alternatively, there may be a fundamental difference in the sorting mechanisms for apical and basolateral proteins that leads to a greater amount of missorted protein in the basolateral membrane. Whether or not proteins exist that can be distributed without polarity to the two surface domains in polarized epithelial cells also remains to be seen. The IgA receptor is transported first to the basolateral surface, and then to the apical surface, where it is released into the medium (Kühn & Kraehenbuhl 1982). This system should also reveal interesting facets of epithelial sorting.

Table 2 Polarity marker and surface area distribution in hepatocytes and MDCK cells

Markers	Cell type	Total marker[a] $B:A$	Surface density[b] $[B]:[A]$	Reference; method
Asialoglycoprotein receptor	rat hepatocyte	85:1	14:1	Matsuura et al (1982); ferritin immunoelectron microscopy
Asialoglycoprotein receptor	rat hepatocyte	51:1	7.5:1	Hubbard et al (1983); gold-ligand binding
5'-nucleotidase	rat hepatocyte	1.3:1	1:4.5	Matsuura et al (1984); ferritin immunoelectron microscopy
VSV G protein	MDCK	28:1	3.9:1	Pfeiffer et al (1985); surface radioimmune assay
Influenza HA protein	MDCK	1:7	1:53	Pfeiffer et al (1985); surface radioimmune assay
Methionine uptake	MDCK	>40:1	>5.5:1	Balcarova-Ständer et al (1984); ratio of methionine incorporation into protein
Transferrin receptors	MDCK	>300:1	>41:1	Fuller & Simons (1985); ratio of transferrin-mediated iron uptake

Cell type	Surface area[c] $Ba:Lat:A$ (μm²)	Surface area[d] $\%B:\%A$	Reference
rat hepatocyte	1756:785:407	86:14	Matsuura et al (1982)
rat hepatocyte	3796:780:676	87:13	Weibel (1976)
MDCK	1404:173:218	87.9:12.1	von Bonsdorff et al (1985)

[a] Ratio of the total amount of marker on the basal plus lateral surfaces (B) versus the total amount on the apical (A).
[b] Ratio of the surface density of marker on the basal plus lateral surfaces ($[B]$) versus the surface density on the apical ($[A]$).
[c] Morphometric data used to generate surface densities of markers from total amounts. The surface areas are reported for the basal or sinusoidal (Ba), the lateral (Lat) and the apical or canalicular (A) in μm².
[d] The percentage of the total surface area contained in the basal plus lateral surface ($\%B$) and the apical ($\%A$).

The basolateral membrane can be divided into two parts: the lateral, which attaches to adjacent cells in the epithelial sheet through cellular junctions, and the basal, which usually rests on a basal lamina. These parts have many components in common, and whether or not differences exist has not yet been carefully studied. In hepatocytes, quantitative differences have been reported. The density of the asialoglycoprotein receptors in rat liver was found to be somewhat higher in the basal membrane than in the lateral membrane (Matsuura et al 1982, Hubbard et al 1983).

Most of the proteins listed in Table 1 are integral membrane proteins. However, a number of other proteins also associate with one or the other cell surface domain. For example, villin (Bretscher & Weber 1979) and fimbrin (Bretscher & Weber 1980) associate specifically with actin filament cores in the brush border membrane of small intestine. Two proteins of 140 (Coudrier et al 1983) and 110 kilodaltons (kDa) (Coudrier et al 1981, Glenney & Glenney 1984) localized to the apical membrane of the enterocyte may serve as attachment sites of the actin filaments to the membrane. Less is known about the proteins associated with the basolateral membrane. Several proteins are known to attach to junctional elements in this membrane domain (see below).

Extracellular Matrix

The basal surface of an epithelial sheet usually rests on a basal lamina. This structure represents a special type of extracellular matrix found under epithelial and endothelial cells (Kleinman et al 1981, Timpl & Martin 1982), and it contains mainly Type IV collagen, laminin, and proteoglycans, particularly heparan sulfate proteoglycans. The molecular details of the attachment of the basolateral membrane to the basal lamina are not known. Hemidesmosomes, specialized adhesion junctions, are present in the basal part of the basolateral membrane of some epithelial cells, and these adhere to the basal lamina (Gipson et al 1983). Specific receptors for laminin and for collagen have been localized on the basolateral membrane; they may function in attaching epithelial cells to the extracellular matrix (Sugrue & Hay 1981, von der Mark et al 1984, Salas et al 1985).

A fully developed basal lamina recognizable by electron microscopy is not essential for epithelial organization. For instance, no basal lamina is found beneath the basolateral membrane in the liver of most mammals (Jones & Schmucker 1977), although collagen bundles and proteoglycans are present. Many epithelial cells adapted to growth in culture can form epithelial sheets with no apparent basal lamina. The MDCK I cells form a very polarized monolayer with a transepithelial electrical resistance of more than $1000 \ \mathrm{ohm \cdot cm^2}$ on nitrocellulose filters without collagen coating in serum-free growth medium (Fuller et al 1984). No basal lamina is found by electron microscopy. However, the cells may be secreting extracellular matrix components, thereby forming a rudimentary matrix on the filter (Takeuchi et al 1977, Valentich 1981). The basal surface remains fairly flat over the pores of the filter, even when they are as large as 3 μm in diameter.

Cues for defining the apical-basal axis in an epithelial sheet seem to be mediated in some instances by interactions with the extracellular substratum. Chambard et al (1981) showed that when inverted follicles of thyroid cells (with their apical surface outward) were embedded in collagen gels reversal of cell polarity was induced. They proposed that binding sites for collagen were recruited to the apical surface, and this interaction of the apical membrane with the extracellular matrix induced the formation of a basal pole, which led to reorganization of the epithelial structure.

Other epithelial cells respond in different ways when the apical pole of the cell is embedded in collagen. Anterior lens epithelium undergoes a transformation during which some epithelial cells give rise to mesenchymal cells that migrate out of the epithelium into the collagen gel (Greenburg & Hay 1982, Hay 1983). Similar cellular transformations are known to play an important role during embryonic development (Ekblom 1984, Stern 1984).

Cellular Junctions

Epithelial cells are connected by a junctional complex encircling the apex of each cell (Farquhar & Palade 1963). The tight junction, *zonula occludens*, is the most apical member of the complex (Figure 1). It is found at the intersection of the apical and the lateral plasma membranes and joins each cell to its neighbors, limiting the diffusion of molecules between the luminal and serosal compartments (Diamond 1977). Immediately basal to the tight junctions is the intermediate junction, *zonula adhaerens*, or belt desmosomes (Volk & Geiger 1984). The other, more basal junctional elements are desmosomes, *maculae adherentes*, and gap junctions, which attach the lateral membranes of adjacent cells to each other (Grinnell 1978, Loewenstein 1981).

The tight junctions were first thought to form impervious seals between epithelial cells, but electrophysiological studies soon demonstrated that they act as gates, selectively regulating the passage of ions between the cells (Diamond 1977). Some epithelia, such as that of the urinary bladder, are very tight to ions and have transepithelial electrical resistances over 100,000 ohm $\cdot$ cm^2, whereas other epithelia, for instance those of proximal tubules, are "leaky," with resistances of only a few ohm $\cdot$ cm^2 (Frömter & Diamond 1972). Similar behavior is seen in epithelial cells grown in culture (McRoberts et al 1981). MDCK I cells form tight monolayers with resistances over 3000 ohm $\cdot$ cm^2 on filters, whereas filter-grown MDCK II cells are more leaky, with resistances of about 100 ohm · cm^2 (Richardson et al 1981, Balcarova-Ständer et al 1984). The tight junctions of MDCK cell monolayers are more permeable to Na$^+$ than to Cl$^-$, as are those of kidney tubules (Misfeldt et al 1976, Cereijido et al 1981b). Both tight and leaky epithelia are impermeable to larger molecules such as lanthanum hydroxide, which has been used as an electron-dense marker to assess the permeability of tight junctions (Friend & Gilula 1972).

The tight junctions may also function as a fence to limit the diffusion of membrane constituents between the apical and the lateral membranes (Diamond 1977). Both cytochemical and autoradiographic methods suggest a very steep gradient of molecules near the tight junction. For instance, antibody labeling that localizes H-2 antigens to the basolateral surface stops exactly at the level of the junction (Parr & Kirby 1979), as does ^{3}H-oubain-binding used to identify basolateral Na$^+$K$^+$-ATPase (Almers & Stirling 1984). Antibody labeling of apical aminopeptidase (Kinne & Kinne-Saffran 1978) and sucrase-isomaltase (Fransen et al 1985) ceases on the opposite side of the junction. Lipids are probably also restricted in their movement from one side to the other (Pinto da Silva & Kachar 1982). There is presently no evidence that suggests differences in the polarity of apical and basolateral components between tight and leaky epithelia, but more studies are needed on this point.

Little is known about the structure and molecular composition of tight junctions. Freeze fracture electron microscopy has shown that they are composed of an anastomosing network of strands (Goodenough & Revel 1970, Pinto da Silva & Kachar 1982). The tightness of the junctions in restricting the passage of ions between the cells seems to be logarithmically correlated with the number of strands in the network (Claude 1978, Easter et al 1983, Marcial et al 1984). The peculiarities of the strand structure and the membrane fusion images seen by electron microscopists led Kachar and coworkers to postulate that the strands of the tight junctions represent cylindrical, inverted lipid micelles in which the exoplasmic (outer) leaflets of the plasma membrane bilayer of adjacent cells have fused (Kachar & Reese 1982, Pinto da Silva & Kachar 1982). The fusion site may be stabilized by transmembrane proteins. This model implies that lipid molecules in the outer leaflet can move by lateral diffusion from one cell to another, but not to the other plasma membrane domain. The lipid in the cytoplasmic leaflet can freely diffuse only between the apical and lateral membranes of the same cell. Studies using fluorescent lipid probes suggest that movement from the apical to the basolateral side does take place (Dragsten et al 1982). However, the use of water-soluble lipids in this study leaves open the possibility that movement occurred via the cytoplasmic aqueous phase.

MDCK cells require protein synthesis to redevelop tight junctions after disruption by trypsin treatment (Griepp et al 1983). This suggests that proteins are necessary components for the formation of tight junctions (see also Stevenson & Goodenough 1984). An assay to identify proteins involved in the gating function of tight junctions was recently devised with filter-grown MDCK I cells (B. Gumbiner & K. Simons, submitted). This assay is based on the use of transepithelial electrical resistance to monitor the reversible opening of tight junctions upon removal of extracellular calcium (Martinez-Palomo et al 1980). Using this resistance assay a monoclonal antibody has been identified that blocks the reformation of the tight junctions, and reacts with a polypeptide of about 120,000 daltons.

The diffusion barrier for membrane proteins is lost after tight junctions are opened by removal of calcium (Galli et al 1976, U et al 1979). Glycoproteins restricted to the apical surface of urinary bladder epithelium become uniformly distributed over the whole cell surface about 80 min after junction disruption (Pisam & Ripoche 1976). H-2 antigens confined to the basolateral membrane in intestinal cells undergo similar redistribution to the apical side (Parr & Kirby 1979). These reports do not prove that tight junctions act as barriers to restrict the diffusion of apical and basolateral proteins because other cellular functions may be disrupted by removing calcium. However, the simplest interpretation of the data is that the tight junction functions as a fence between the two surface domains, and is essential for the maintenance of cell surface polarity.

The other encircling member of the junctional complex, the intermediate junction (*zonula adhaerens*), is composed of three parts: (*a*) an integral membrane site containing a 135,000 dalton protein (Volk & Geiger 1984) and uvomorulin (Boller et al 1985); (*b*) a membrane-bound cytoplasmic plaque composed of vinculin, and possibly additional peripheral proteins (Geiger et al 1981); and (*c*) a contractile cytoplasmic actin filament bundle that forms a belt around the cell and connects the junctional complex to the actin network in the cell. The best example of this actin network is seen in the terminal web and the microfilament cores of epithelial cells with brush borders (Hull & Staehelin 1979, Tilney 1983), although less developed networks are found in other cells.

One further important characteristic of epithelial cell organization is the presence of spot desmosomes along the lateral membranes. The desmosomes seem to act as rivets between the lateral membranes (Grinnell 1978). These junctions also consist of a membrane domain made of three antigenically distinct glycoprotein families (Cohen et al 1983), and of a cytoplasmic plaque (Geiger et al 1983) that contains the nonglycosylated desmoplakins and other peripheral proteins (Franke et al 1981, Mueller &

Franke 1983, Gorbsky et al 1985). This plaque is anchored to tonofilaments of the cytokeratin type (Franke et al 1978, Lazarides 1980), which also attach to hemidesmosomes in the basal surface (Gipson et al 1983).

Cytoskeletal elements play an important role in epithelial cell organization. Most of the lipid and 65% of the total cell protein can be extracted from MDCK cell monolayers by detergent treatment, leaving a skeletal framework that retains cell nuclei, apical microvilli, and intermediate filaments connected via desmosomes throughout the epithelial sheet. In contrast, the corresponding skeletal framework of fibroblasts shows little regularity and lacks junctional complexes between the cells (Fey et al 1984). Cytokeratins are known to be specific to epithelial cells (Franke et al 1978, Lazarides 1980, Schmid et al 1983). Their function is not known, but they may stabilize the organization of epithelial sheets. Some epithelial cells also have intermediate filament proteins in common with those of nonepithelial cells (Kartenbeck et al 1984).

Gap junctions are also found in the basal and lateral membranes of epithelial cells, as they are in nonepithelial cell membranes. These junctions are assembled from a 27,000 molecular weight protein that forms modulatable channels for the passage of small molecules between adjoining cells (Hertzberg & Gilula 1979, Loewenstein 1981, Hertzberg & Skibbens 1984). Electron microscopic and X-ray analyses have revealed the structure of the gap junction, and suggested a plausible model for its function (Unwin & Zampighi 1980, Unwin & Ennis 1984). Many epithelial layers have been shown to be electrically and metabolically coupled (Loewenstein 1981). These include the trophoblast cells of the mammalian embryo, which express gap junctions at the stage when they first take on an epithelial character (Ducibella 1977). These observations suggest that gap junctions link the epithelial sheet into a communicating layer so that it functions as a unit. Recently, however, MDCK cells and pig kidney cells were shown not to be electrically or metabolically coupled after reaching confluence on glass (Cereijido et al 1984, Sepulveda & Pearson 1984). This difference may reflect the physiological state of the cell, and it deserves further study.

BIOGENESIS OF APICAL AND BASOLATERAL MEMBRANES

Lipids

Studies on the biogenesis of the lipids of the animal cell surface are still in their infancy. Most lipids are known to be synthesized in the endoplasmic reticulum (ER), but how they are moved from there has not yet been elucidated (Bell et al 1981, Pagano & Langmuir 1983). The newly synthesized lipids could move with the proteins by vesicular traffic from the ER over the Golgi complex to the cell surface (Dower et al 1982, Pfenninger & Johnson 1983, Mills et al 1984), or by transfer through the aqueous, cytoplasmic phase, possibly by binding to lipid exchange proteins (Wirtz 1974, De Grella & Simoni 1982, Yaffe & Kennedy 1983, Sleight & Pagano 1983). No studies have yet been reported on lipid traffic in epithelial cells. It would be interesting to follow the intracellular transport of lipids, especially of glycolipids, to find out whether the mechanisms responsible for the different lipid compositions of the apical and basolateral membranes are intracellular, or in the plasma membrane itself.

Proteins

Most surface proteins of epithelial cells are glycosylated, as are surface proteins in other cells (Bretscher & Raff 1975). Both membrane glycoproteins and secretory proteins are translated on ribosomes bound to the rough ER (Blobel & Dobberstein 1975, Walter et al 1984). N-linked core glycosylation takes place during translocation through the ER membrane (Rothman & Lodish 1977, Sabatini et al 1982), and from this compartment the newly synthesized proteins are transported to the Golgi complex for further processing and extension of N-linked, and addition of O-linked, carbohydrate side chains (Hanover & Lennarz 1981, Farquhar & Palade 1981, Rothman 1981, Staneloni & LeLoir 1982, Tartakoff 1983). The ER-Golgi pathway is common to all surface and secretory proteins in all cell types studied (for epithelial cells see Rodriguez-Boulan 1983, Danielsen et al 1984).

Apical and basolateral proteins seem to remain together during transport through the Golgi complex. Rindler et al (1984, 1985) infected MDCK cells grown on plastic with both VSV and influenza virus, and using immunoelectron microscopy they localized influenza hemagglutinin and VSV G protein in all Golgi cisternae. They concluded that the pathways of these two proteins could not diverge before passage through the Golgi stack. This interpretation should be tempered by their observation that the cells were somewhat apolar with respect to appearance of G protein at the surface, as a result of viral cytopathic effects.

Another assay based on the enzymatic activity of the influenza neuraminidase showed that the VSV G protein and influenza neuraminidase remained together through the *trans* compartment of the Golgi complex in doubly infected cells (Fuller et al 1985a). The basolaterally directed G protein was found to lose most of its sialic acid as a result of cleavage of its glycans by the apically directed neuraminidase. The sensitivity of the assay allowed analysis early in double infection, before loss of cell polarity had begun. The cleavage indicates that apical proteins must still be in physical contact with basolateral proteins during the terminal (post-sialylation) steps of processing in the Golgi complex.

Given that the proteins traverse the Golgi together, at least three alternatives have been proposed to describe where sorting of newly synthesized apical and basolateral proteins occurs (Quaroni et al 1979a, Hauri et al 1979, Evans 1980, Rodriguez-Boulan et al 1984, Danielsen et al 1984, Rindler et al 1984): (a) *intracellularly* during or after exit from the Golgi complex; (b) *after transport* of the glycoproteins in a random fashion *to both surface domains,* either by selective passage through the tight junctions or by subsequent vesicular transport to the correct domain; (c) *after transport* of both apical and basolateral proteins *to the same surface domain,* either by diffusion through the junction or by vesicular transport.

The first results obtained using pulse-chase labeling and subcellular fractionation of intestinal cells suggested that the apical proteins (Quaroni et al 1979a), including the sucrase-isomaltase enzyme (Hauri et al 1979), passed over the basolateral membrane en route to their brush border destination. In contrast, similar studies with intestinal aminopeptidase yielded no evidence for a basolateral intermediate on the Golgi–apical membrane route (Ahnen et al 1982). These authors suggested that the apparent incorporation of apical enzymes into the basolateral membrane in the previous studies could have been due to contamination of the basolateral membrane fraction with Golgi membranes.

Detailed studies on the transport of apical proteins to the cell surface were performed using influenza hemagglutinin as a probe in MDCK cells grown on filters (Matlin & Simons 1984, Misek et al 1984). These studies indicated that the newly synthesized hemagglutinin is delivered directly from an intracellular site to the apical membrane in amounts that account for its final polarized distribution. Although influenza virus–infected MDCK cells do express a small fraction of the hemagglutinin on the basolateral side, quantitative assays based on extracellular trypsin cleavage or on antibody accessibility showed that the hemagglutinin molecule does not have to pass the basolateral membrane on the route to the apical membrane.

Less work has been done on the delivery of basolateral proteins. Sztul et al (1985a,b) studied the transport of the IgA receptor to the sinusoidal membrane in rat liver, and found by subcellular fractionation that the protein reached the sinusoidal surface about 30 min after synthesis, i.e. within the time range observed for apical protein transport to the cell surface in other cells. Quantitative studies with newly synthesized VSV G protein in filter-grown MDCK cells revealed direct delivery from an intracellular site to the basolateral cell surface (Pfeiffer et al 1985).

Based upon the studies on the viral glycoproteins, we conclude that sorting of apical and basolateral proteins is an intracellular event. More work is obviously needed on endogenous surface proteins in both cultured cells and tissues to generalize this conclusion.

The precise site at which the pathways to the surface diverge is not yet known but the *trans*-most part of the Golgi complex is emerging as a good candidate. The transport of newly synthesized influenza hemagglutinin (Matlin & Simons 1983) and of VSV G protein (Fuller et al 1985a) to the cell surface is inhibited at 20°C in MDCK cells. The proteins pass into a late Golgi compartment, as judged by their endoglycosidase H resistance and by their sialylation. Thin frozen-section immunolabeling showed that the proteins accumulate in a *trans*-Golgi location. At 20°C a tubular network of membranes is formed, perhaps due to the cessation of membrane transport to the cell surface from the Golgi complex. In nonepithelial BHK cells this *trans*-Golgi compartment shows positive reaction for both thiamine pyrophosphatase and acid phosphatase. Exit of newly synthesized surface proteins bypassed the endosomes en route to the plasma membrane (Griffiths et al 1985). Further work on this *trans*-Golgi compartment is necessary to define its role in intracellular transport in epithelial and nonepithelial cells.

Less is known of the biosynthesis and transport of peripheral proteins that interact with the cytoplasmic face of apical and basolateral membranes. These are probably translated on ribosomes in the cytoplasm (Lodish et al 1981). The correct membrane localization of peripheral proteins may be determined by their binding affinity for the cytoplasmic domains of spanning membrane glycoproteins (Danielsen et al 1984, Lazarides & Moon 1984). Alternatively, a peripheral protein may become localized to a specific membrane domain solely by interacting with the lipids in the membrane, although no such example is known. Lipid tails may also be added to cytoplasmic proteins. This is the case with the p60-src protein (Sefton et al 1982), which becomes amphiphilic through the attachment of myristic acid to the protein. This leads to specific binding to the plasma membrane (Cross et al 1984). How much of the protein associates with intracellular membranes remains to be clarified.

ENDOCYTOSIS AND TRANSCYTOSIS

Cell surface components of animal cells are continuously internalized through endocytic vesicles (Steinman et al 1983). Studies with BHK cells have shown that most of these vesicles are bounded by a coat (Marsh & Helenius 1980). After internalization the vesicles usually fuse with endosomes (Helenius et al 1983), and the membrane must be recycled to the plasma membrane to compensate for lost surface area. The magnitude of this membrane recycling is considerable: In macrophages and L cells the equivalent of the entire cell surface has been estimated to recycle every 0.5–2.0 hr (Steinman et al 1976).

The apical and the basolateral membranes of epithelial cells also undergo continuous endocytosis. A number of protein ligands are taken up by receptor-mediated endocytosis from either the apical or the basolateral side of different epithelial cells (Anderson & Kaplan 1983, Steinman et al 1983). Four different pathways have been described for the internalized ligands and their receptors. Most ligands are dissociated from their receptors in the low pH (<5.5) of the endosomes (Helenius et al 1983), which are also called receptosomes by Pastan & Willingham (1983), and CURL (the compartment for uncoupling of receptors from ligands) by Geuze et al (1983). The ligands are then routed to the lysosomes for degradation while the receptors return to the cell surface. In other cases both the receptor and the ligand are delivered to the lysosomes to be degraded (Kasuga et al 1981, Mellman 1982, Beguinot et al 1984). Transferrin uptake represents a third case: The receptor-bound transferrin releases its iron in the endosomes, and is returned to the cell surface (Dautry-Varsat et al 1983, Klausner et al 1983), where it is released from the transferrin receptor at the neutral pH of the extracellular medium. The fourth case, characteristic of epithelial cells, is represented by several protein ligands that are transported from the endosomes across the cell. The ligands (IgG, thyroglobulin, IgA) bind to their receptors either on the apical or on the basolateral side, the complexes are endocytosed, and transported to endosomes via coated vesicles (Renston et al 1980; Abrahamson & Rodewald 1981; Mullock & Hinton 1981; Herzog 1983, 1984; Geuze et al 1984b). In the endosomes these complexes are segregated from other ligand-receptor complexes, and are transported to the opposite side of the cell (Geuze et al 1984b), in as yet uncharacterized membrane vesicles.

Only one study using fluid phase markers has been performed to measure the volume of fluid taken up from both the apical and the basolateral sides, and to determine the fraction passed across the cell (von Bonsdorff et al 1985). In MDCK I cells grown on filters the amounts of apical and basolateral endocytosis corresponded to a fluid uptake of 1×10^{-8} and 20×10^{-8} nl min^{-1} per cell, respectively. In addition, the measured transcellular passage of horseradish peroxidase and fluoresceinated dextran corresponded to a fluid volume of about 3×10^{-8} nl min^{-1} per cell in both directions. When the surface areas of the apical and basolateral membranes are taken into account, the total uptake of fluid (the sum of the endocytic and transcellular values) corresponded to 0.019×10^{-8} nl min^{-1} μm^{-2} cell surface area from the apical side and 0.015×10^{-8} nl min^{-1}

μm^{-2} from the basolateral side, which shows that per unit surface area, the fluid uptake was similar from both sides.

This uptake of fluid is in the same range as that observed for BHK cells (Marsh & Helenius 1980), but smaller than that measured for L cells (fibroblasts) and macrophages (Steinman et al 1976). In BHK cells with a total cell surface area of 3400 μm^2 (Griffiths et al 1984) the fluid uptake corresponded to the formation of 2300 coated vesicles per min per cell (Marsh & Helenius 1980). Assuming that similar 90 nm vesicles carried the estimated 3×10^{-8} nl across the MDCK cell, about 100 such vesicles would traverse the cell per min in each direction. A transcellular flow of membrane of this magnitude implies that sorting must take place during transcytosis. Otherwise, rapid intermixing of apical and basolateral components would occur. How this intermixing is avoided is not known. However, it has recently been shown that epithelial cells have mechanisms for routing misplaced surface proteins by transcytosis. If the VSV G protein is implanted into the apical membranes of MDCK cells a considerable fraction of the protein is transported through the endosomes to the basolateral side (Matlin et al 1983; Pesonen & Simons 1983). Biochemical and morphological studies suggested that the transcellular route is the same as that used by receptor-bound ligands undergoing transcytosis: lysosomes and the Golgi complex are bypassed (Pesonen et al 1984a,b).

The selection mechanisms acting during transcytosis probably operate both in the coated pits on the cell surface and in the endosomes. Coated pits are known to act as molecular filters during endocytosis (Bretscher & Pearse 1984). This is not sufficient to account for sorting during transcytosis because receptors destined for different locations co-localize in coated pits and coated vesicles (Geuze et al 1984b). Sorting also has to take place in the endosomal compartment to direct membrane traffic into at least three directions: (a) back to the same side of the epithelium, e.g. to recycle receptors; (b) across the cell to the opposite side; and (c) to the lysosomes. In addition, there may exist a fourth pathway connecting the endosomes with the Golgi complex (Farquhar 1982).

The amount of ligand passed into the transcellular pathway during receptor-mediated endocytosis has been estimated in perfused liver. Thomas & Summers (1978) and Schiff et al (1984) measured the passage of asialoglycoprotein from the blood side of the liver into the bile, and found that, although the bulk is degraded in the lysosomes, 1–4% of the injected dose passes intact across the hepatocyte. In a similar study, about 4% of the epidermal growth factor injected into the blood side passed intact into the bile (St. Hilaire et al 1983, Burwen et al 1984). It is interesting to note the difference in the handling of receptor-bound transferrin by filter-grown MDCK I cells (S. Fuller & K. Simons, unpublished). Less than 0.3% of the transferrin endocytosed from the basolateral side is missorted to the apical side. The low error frequency might reflect the association of apotransferrin and its receptor during passage through the endosome (Klausner et al 1983).

In many respects the endosomes are acquiring a role in directing membrane traffic during endocytosis analogous to that of the Golgi complex in biosynthetic traffic (Rothman & Lenard 1984). Little is known about the organization of the endosomal compartment in epithelial cells. Are the endosomes underlying the apical membrane continuous with those underlying the basolateral membrane? Or are they structurally and functionally distinct? Preliminary evidence suggests that differences may exist in exocrine acinar cells and in filter-grown MDCK cells. For both of these cell types the endosomes containing fluid-phase markers endocytosed from either the apical or the basolateral side differ in distribution and appearance (Oliver 1982, von Bonsdorff et al 1985).

SORTING OF EPITHELIAL SURFACE PROTEINS

Newly synthesized apical and basolateral proteins must be separated from each other during their transport to the cell surface. This sorting occurs intracellularly, at least for viral membrane glycoproteins and secreted proteins. It may occur in the *trans*-Golgi compartment. This process involves recognition of a sorting signal, segregation of the proteins into separate vesicles, transport, and fusion of the carrier vesicle with the correct cell surface domain.

Sorting Signals

Our discussion of the distribution of surface components showed that a particular protein is usually directed to the same plasma membrane domain in different epithelial cells. This suggests that the signal, which is interpreted by the sorting machinery of the cell, is intrinsic to the protein, residing in its three-dimensional structure.

No specific signal responsible for directing proteins to a particular plasma membrane domain has been identified. Other signals involved in intracellular transport have been characterized. These include specific, apparently quite short, sequences that are involved in directing proteins to the mitochondria (Schatz & Butow 1983, Hurt et al 1984) and the nucleus (Kalderon et al 1984, Lanford & Butel 1984, Dingwall 1985), as well as the signal sequence necessary for the translocation of a nascent polypeptide across the membrane of the endoplasmic reticulum (Blobel 1980, Walter et al 1984). These signals can be divided into two classes: those that must be interpreted only once, and those that must be read many times. The signal sequence for translocation of nascent polypeptides, and those involved in transport into mitochondria are examples of the first type. Transport into these compartments is unidirectional; the signal is needed only during transport of the newly synthesized proteins to their correct location, and is indeed often removed during translocation.

The recently characterized signals that direct proteins from the cytoplasm to the nucleus should be of the second sort. Mature nuclear proteins can be injected into oocytes and become localized in the nucleus (Bonner 1975, Dingwall 1985). The nucleus breaks down during mitosis, and must be reassembled from cytoplasmically dispersed components. These must bear signals that can be reread, and thus direct their relocalization. The signals needed for localization to a plasma membrane domain are also probably of the second type. Epithelial plasma membrane proteins are continuously recycled, hence their signals must be continually rechecked. Further, direct evidence of the ability of the cell to redirect a mislocalized protein to the proper domain is provided by the apical implantation and redistribution experiments with VSV G protein described above (Pesonen et al 1984a,b). However, it is not yet known whether the signals operating during exocytic and endocytic sorting of epithelial surface proteins are the same.

The three-dimensional nature of the signals described above is probably quite important. Accessibility, at least, appears to be essential. The translocation signals are predominantly N-terminal, and the final folding of the protein, which could block access to the signal, is delayed until translocation is complete (Blobel 1980, Walter et al 1984, Hurt et al 1984). The nuclear signals have common sequence elements that are characteristic of the surface rather than the interior of the protein (Lanford & Butel 1984). The mannose-6-phosphate signal (described below) also fulfills this criterion.

These considerations have motivated a search of the primary structures of apical and basolateral proteins to identify possible sorting signals. No clear sequence homology has been found among proteins localized to the same plasma membrane domain, although the structural trends depicted in Table 3 suggested several possible signals. For example, apical proteins tend to have their amino-termini on the cytoplasmic side of the membrane, while basolateral proteins have the opposite orientation. The apical hemagglutinin and the basolateral asialoglycoprotein and transferrin receptors are exceptions, however. Similarly, basolateral proteins tend to have cytoplasmic tails longer than 24 amino acids, while apical ones have tails shorter than 11 amino acids. Again, two counter examples are known,

Table 3 Features of apical (*A*) and basolateral (*B*) surface proteins

Protein	Domain	Cytoplasmic tail[a]	Glycosylation[b]	References
Asialoglycoprotein receptor	B	N-terminal 24 aa chicken 38 aa rat	Complex, sialic acid	Table 1; Chiacchia & Drickamer (1984); Drickamer (1981); Drickamer et al (1984); Schwartz & Rup (1983); Paulson et al (1977)
Epidermal growth factor receptor	B	C-terminal 542 aa	Complex, high mannose, sialic acid	Dunn & Hubbard (1984); Ullrich et al (1984); Mayes & Waterfield (1984)
E2 protein Semliki Forest virus	B	C-terminal 31 aa	Complex, high mannose, sialic acid	Fuller et al (1985); Simons & Warren (1984)
F protein Sendai virus	A	C-terminal 42 aa	Complex, no sialic acid	Rodriguez-Boulan & Pendergast (1980); Scheid & Choppin (1977); Hsu & & Choppin (1984); Kohama et al (1979)
γ-glutamyl transferase	A	N-terminal	Complex, high mannose, sialic acid	Table 1; Finidori et al (1984); Yamashita et al (1983)
H-2 antigen	B	C-terminal 31–40 aa	Complex, sialic acid	Table 1; Coligan et al (1981); Kvist et al (1981); Bregégère et al (1981); Dobberstein et al (1977); Nathenson & Cullen (1974)
G protein vesicular stomatitis virus	B	C-terminal 28–29 aa	Complex, sialic acid	Rodriguez-Boulan & Pendergast (1980); Rose & Gallione (1981);· Gallione & Rose (1983); Reading et al (1978)

Hemagglutinin influenza virus	A	C-terminal 10–11 aa	Complex, high mannose, no sialic acid	Rodriguez-Boulan & Pendergast (1980); Air (1981); Porter et al (1979); Min-Jou et al (1980); Keil et al (1984)
Insulin receptor	B	C-terminal 403 aa	Complex, high mannose, sialic acid	Table 1; Ullrich et al (1985); Hedo et al (1983)
Leucine aminopeptidase	A	N-terminal 7 aa	Complex, sialic acid	Table 1; Feracci et al (1982); Danielsen (1982)
Neuraminidase influenza virus	A	N-terminal 6 aa	Complex, no sialic acid	Fuller et al (1985a); Fields et al (1981); Ward et al (1982); Bos et al (1984); Ward et al (1983)
p15[env] murine retroviruses	B	N-terminal 32 aa	Non-glycosylated, but attached to p70, which contains complex & high mannose glycans and sialic acid	Roth et al (1983); Dickson et al (1982); Dickson et al (1982)
Rhodopsin	A	C-terminal 37 aa	Unusual, N-linked	Nir et al (1984); Ovchinnikov (1982); Liang et al (1979)
Transferrin receptor	B	N-terminal 63 aa	Complex, no sialic acid	Fuller & Simons (1985); Schneider et al (1984); Schneider et al (1982)

[a] The orientation and the length in amino acids (aa) of the cytoplasmic tail of the protein.
[b] Description of the carbohydrate in the mature protein in terms of terminal glycosylation (complex or high mannose), and sialic acid content.

the F protein of the Sendai virus and rhodopsin, which are apical proteins with long cytoplasmic tails.

It remains unclear whether the signal that is recognized is within the cytoplasmic, membrane-spanning or luminal regions of the protein. A recombinant DNA approach to this problem seems a natural one since the behavior of hybrid proteins containing domains from apical and basolateral parents should yield an unambiguous answer. Extensive work has yielded few results so far because the expression systems that have been developed for nonpolar cells work poorly in polar ones. Roth et al (1983) utilized epithelial African green monkey cells to show that the influenza hemagglutinin is properly localized when expressed from cDNA in the absence of other influenza proteins. This confirms that the signal for localization is carried in the glycoprotein, and not by the other proteins of the virus. However, this system is difficult to manipulate, and the quantitation necessary for characterizing polarity must await the development of other expression systems.

A posttranslational modification could also be involved in the sorting of apical and basolateral proteins. The signal that is responsible for the localization of lysosomal hydrolases to the lysosome is an added carbohydrate, mannose-6-phosphate (Kornfeld et al 1982), which mediates recognition of the enzyme by means of the mannose-6-phosphate receptor, both intracellularly (Brown & Farquhar 1984, Geuze et al 1984a) and at the cell surface (Neufeld et al 1977). The role of glycosylation in the routing of apical and basolateral proteins to the correct surface domain has been studied in several laboratories (Roth et al 1979, Green et al 1981, Danielsen et al 1984). Too few structures of the oligosaccharide chains of epithelial surface proteins have been elucidated to allow a meaningful comparison of apical and basolateral proteins. Both N-linked and O-linked side chains are known to be present (Table 3) (Herscovics et al 1980, Danielsen et al 1984).

Roth et al (1979) and Green et al (1981) have studied viral protein expression in MDCK cells in the presence of tunicamycin, an inhibitor of N-linked glycosylation, and saw no marked effect on polarity. Similar results were obtained with mutant MDCK cells that were defective in glycosylation (Green et al 1981, Meiss et al 1982). The proper localization of viral glycoproteins after such dramatic changes in N-linked glycosylation argues against its importance in the sorting of these apical and basolateral proteins. Whether carbohydrates can act as sorting signals for other apical and basolateral proteins remains to be seen. A transient carbohydrate modification that may be relevant for epithelial sorting has been reported for thyroid cell proteins. Parodi et al (1983) found that proteins from extracts of thyroid glands were transiently glucosylated during N-linked carbohydrate processing, and they suggested that this modification might be responsible for routing glycoproteins to specific cellular locations.

Membrane Sorting

Membrane traffic in the animal cell achieves the selective localization of membrane components in distinct compartments of the cell. The specific transport of proteins to plasma membrane domains in a polarized cell is just one example of the phenomenon. This traffic is believed to be mediated by membrane vesicles that bud from one cellular compartment with a selected fraction of its components, and fuse selectively with the next compartment of the transport pathway (Palade 1975, 1982).

To understand the selective formation of membrane vesicles we need to know: first, how the proteins to be transported to the next organelle are selectively included in the budding vesicle; second, how membrane proteins that should remain in the organelle are selectively excluded from the forming vesicle; and third, what generates the forces that distort the membrane of a budding vesicle (McCloskey & Poo 1984). Although detailed answers are not available in any system, the best characterized example of membrane sorting, that of the enveloped animal viruses, provides a useful paradigm for considering these questions (Simons & Warren 1984).

Enveloped animals viruses form at the surface of the cell by an interaction between the transmembrane viral glycoproteins and the cytoplasmic nucleocapsid (Simons & Garoff 1980). The interaction wraps the glycoprotein-containing membrane around the nucleocapsid, distorting the membrane into a vesicle-like bud, which is released from the cell. The nucleocapsid probably acts as a cytoplasmic template for the budding process, while the localization of the viral glycoproteins ensures that the budding occurs only from the plasma membrane. Budding seems to require a critical concentration of glycoproteins, and this is achieved only at the cell surface (Simons & Warren 1984). The protein composition of the virus is strictly controlled, since practically all host membrane proteins are excluded. Usually greater than 99% of the protein in the viral membrane is virally encoded (Holland & Kiehn 1970, Strauss 1978, Calafat et al 1983). This selectivity has two complementary facets: the inclusion of viral glycoproteins and the exclusion of host proteins.

For alpha-viruses the inclusion of viral proteins occurs via their interaction with the nucleocapsid (Simons & Warren 1984). We envisage that the nucleocapsid interacts with the cytoplasmic tails of a patch of viral glycoproteins, initiating the budding event by bending the membrane. The bud would grow by the collection of more viral glycoproteins from the surface, hence concentrating them (Garoff & Simons 1974). Finally, the budding virus would seal by filling the remaining sites on the capsid template, closing the bud. The mechanism of exclusion of host proteins might simply be that of steric hindrance. The density of spike proteins in an alpha-virus is so high, about 30,000 spikes μm^{-2}, that little room remains for host proteins (Quinn et al 1984). This steric hindrance is mediated by the extramembranous portions of the protein because the transmembranous domain is quite slender; it occupies only 20% of the lipid bilayer area (Simons & Warren 1984).

The variations of this budding scheme that are exhibited by influenza virus or M-protein viruses, such as vesicular stomatitis, show its applicability to the sorting of several proteins in a single budding event. The interaction of the viral glycoproteins and the nucleocapsid that is responsible for the budding is mediated by another cytoplasmic viral protein, the M protein (Simons & Garoff 1980, Dubois-Dalcq et al 1984). In exchange for this extra level of complexity in their assembly, these viruses gain flexibility in their composition. The ratio of the M protein to the nucleocapsid protein in the budded virus remains approximately constant for a virus such as VSV, while the ratio of the glycoprotein G to the M protein can vary severalfold (Lodish & Porter 1980).

Further evidence of flexibility in VSV envelope formation is provided by the pseudotype phenomenon (Zavada 1982). Double infections of a cell with VSV and another enveloped virus often result in the production of hybrid virions that contain the M proteins and nucleocapsid of VSV, and the envelope proteins of both. Witte & Baltimore (1977) showed that mixed virus envelope formation requires a critical amount of G protein. (For an alternative view, see Weiss & Bennett 1980.) Studies in MDCK cells by Roth & Compans (1981) support the critical role of G protein in the budding process. Doubly infected MDCK cells that expressed both influenza and VSV did not yield VSV particles containing influenza glycoproteins until the tight junctions had broken down and the glycoproteins were allowed to mix. Expression of influenza glycoprotein on the apical surface, even at high concentration in the presence of cytoplasmic VSV M protein and nucleocapsid, was not sufficient; some VSV G protein had to be present in the same membrane.

The scheme of VSV formation that emerges from these considerations suggests that budding is initiated when a critical concentration of G protein is reached in the plasma membrane. The interaction of M protein with the cytoplasmic tail of G protein provides the link between the membrane and the nucleocapsid. The interaction with M protein organizes the nucleocapsid, which can then serve as the template for the formation of a budding particle. Foreign viral membrane proteins are included in the budding virion either by interacting directly with M protein through their cytoplasmic tails, or indirectly via lateral interactions of their extracellular domains with those of G. As budding proceeds, more G and foreign spike proteins are collected from the plasma membrane. The combined effect of many weak individual interactions between foreign viral proteins and the G protein, which is held by an M protein lattice, would provide the force to control the composition of the envelope.

This scheme for virus budding can be adapted to a description of vesicular traffic within the cell. The selectivity of the transport vesicles must also be the result of inclusion of specific proteins and the exclusion of all others. This inclusion could be mediated by direct interaction with cytoplasmic molecules analogous to M proteins. Lateral interactions with the luminal portions of membrane-spanning proteins that have been organized by cytoplasmic interactions with M-like proteins are also conceivable. These lateral interactions might be favored by the pH and ionic composition of the compartment. Exclusion might be accomplished sterically as sites in the underlying lattice are progressively filled. Vesicle budding might be initiated by the formation of a cytoplasmic scaffold that recognizes the lattice on the cytoplasmic surface, and closes the vesicle when the lattice is complete. Only the direction of budding differs from the viral system. In the virus the nucleocapsid scaffold forces the membrane into a convex shape so that it is enclosed; in the intracellular transport vesicle the clathrin (or other type of vesicle coat) forces the membrane into a concave shape enclosed by the coat (Harrison & Kirchhausen 1983).

After formation, the intracellular transport vesicle moves to its specific site of delivery. The involvement of cytoskeleton has been postulated (Quaroni et al 1979b, Danielsen et al 1983, Rindler et al 1984), but little is known about this process in molecular terms. The final event in vesicular transport is specific fusion with the target membrane. The viral glycoproteins might provide a paradigm for this process as well. Fusion of two membranes requires bringing them into close apposition, and a transient disruption of the bilayer structure. The viral glycoproteins of several enveloped viruses are responsible for receptor binding at the cell surface, and for the fusion event that allows entry of the genome into the cell (White et al 1983). The combination of multivalent receptor binding and fusion activity in multiple copies of the same protein ensures both efficiency and specificity in the process. Such a combination of features could also be characteristic of putative cellular fusion proteins that guide the targeting of intracellular vesicles.

DEVELOPMENT AND REDEVELOPMENT OF SURFACE POLARITY

A fundamental problem in development is how cell surface polarity is generated during the differentiation of epithelial cells. We describe the development of polarity in two systems that provide complementary views of the process. The differentiation of the trophoblast epithelium in the mammalian embryo illustrates the assembly of the elements that organize the polarity of the cell as they arise. In contrast, the reestablishment of polarity in the model epithelium MDCK provides a system that has expressed all the components necessary for polarity, and assembles them rapidly in response to external interactions.

Formation of the Trophoblast Epithelium

The development of the mammalian embryo, as in the mouse or rabbit, proceeds by several concerted cleavages, which generate an 8-cell stage (Ducibella 1977). The cells of the embryo do not display any obvious polarity through the early 8-cell stage. With the formation of the morula, dramatic changes in the structure of the embryo occur in membrane transport systems, surface glycoproteins and antigens, intercellular junctions, and cell morphology. At this stage the outer cells of the embryo differentiate into a well-ordered transporting epithelium, the trophoblast, which seals the inner cells from the outer medium, and changes the composition of the blastocoel fluid by active transport.

The trophoblast cells display all the properties of a polar epithelium, which we have described. The outward apical domain and the inward basolateral domain of the plasma membrane each display distinct morphology, and are separated by zonular tight junctions. These junctions form a seal, of up to 6 junctional strands, between the outer cells, as confirmed by freeze fracture observation (Hastings & Enders 1975, Ducibella et al 1975), and evince transepithelial resistances of greater than 2000 ohm $\cdot$ cm^2 in the rabbit blastocyst (Cross 1973). The lateral membranes of the outer cells are connected by desmosome-like structures associated with tufts of intermediate filaments of the cytokeratin type (Jackson et al 1980).

The formation of the trophoblast layer is accomplished through compaction of the outer cells of the 8-cell embryo. Compaction begins with polarization of the cells, as revealed by markers such as surface ligand binding (Handyside 1980), redistribution of microvilli (Ducibella & Anderson 1975), and organization of microtubules and microfilaments (Ducibella & Anderson 1975, Lehtonen & Badley 1980). The process is completed by the flattening of cells upon one another to maximize contact, and the linking of the cells through tight and gap junctions.

The polarization of the early 8-cell stage cells is dependent on cell-cell or cell-substrate interaction. Ziomek & Johnson (1980, 1981) studied the effect of cell surface interactions by following the polarity of 8-cell stage cells with fluoresceinated concavalin A. The patches of lectin-binding material moved to one pole of the cell if dissociated cells were brought in contact with other cells or an adherent substratum. In the absence of such contact, the lectin-binding sites formed patches randomly over the surface of the cell. This first step of the polarization process can be characterized as the formation of a bipolar structure. The cell contact surface lies at one pole, and the free surface marked by the lectin-binding sites forms the other. There is little evidence of an actual concentration of surface components in this bipolar stage, rather, the polarity appears to reflect the reorganization of the cytoskeleton, and the formation of microvilli in the free pole.

The flattening of the cells and the establishment of zonular junctions occur in concert, and are dependent on cytoskeletal reorganization. As described by Ducibella (1977), arrays of parallel microtubules are seen under the surface of the opposed membranes, and macular junctions are formed between them. As compaction continues, these junctions move apically, coalescing, and eventually forming a zonular tight junction at the top of the cell. The establishment of this zonular junction is followed by stabilization of cellular polarity and physical strengthening of intercellular contacts through formation of desmosomes. Gap junctions are also formed between the trophoblast cells, as well as with the inner cells. This process can be reversibly disrupted by the removal of calcium (Ducibella & Anderson 1975, 1979). The arrays of microtubules that underlie opposed membranes are still formed, but the formation of macular, and eventually zonular, junctions is inhibited.

During compaction the outer cells of the embryo are thus first induced to form a bipolar structure, and this structure probably provides the base for the formation of zonular junctions, which in turn leads to the development of distinct cell surface domains.

Reestablishment of Polarity in MDCK Cells

The development of polarity in freshly seeded MDCK cells has many features in common with the development of the trophoblast epithelium. In suspension, single cells display no obvious polarity as assayed by virus budding. Clusters of cells in which contact defines a free (apical) and a contacting (basal) surface do display such polarity (Rodriguez-Boulan et al 1983). When seeded onto an appropriate substrate in the presence of calcium, some of the cells of the population display polarity of virus budding, even when they have not formed tight junctions with cells on all sides (Rodriguez-Boulan et al 1983). Polarity of virus budding is not seen in the absence of calcium. It is possible that the contact of the cell with the substrate results in the formation of hemi-tight junctions analogous to the formation of hemidesmosomes. This structure could act as a fence, i.e. as a diffusion barrier for membrane components between the apical and basolateral surfaces, but not as an intercellular gate. Following the formation of cell-cell contacts the degree of polarity of the cell surface can be shown to increase quantitatively. The polarity of the cell surface, as monitored by monoclonal antibodies or unilateral methionine uptake, increases sharply about the time that transepithelial resistance can first be detected, and continues to increase thereafter, in parallel with the resistance (Balcarova-Ständer et al 1984, Herzlinger & Ojakian 1984). The need for the fence function of the tight junction in the development of polarity appears evident in this system, as it was in the embryonic system.

Epithelial Cells in Mitosis

The behavior of epithelial cells during mitosis provides an intriguing example of the interaction of the cytoskeleton and the tight junctions to preserve the polarity of the cell. Epithelial cells usually cleave perpendicular to the plane of the epithelium so that both daughters remain within the layer, and the apical and basolateral membranes are divided equally between them (Zeligs & Wollman 1979a,b,c). The process of cell division is accompanied by a rounding of the cell, but the zonular tight junctions with neighboring cells remains intact through the end of telophase. During cytokinesis there may be a local breakdown of the tight junction at the point where it meets the cleavage furrow, however it remains intact elsewhere (Zeligs & Wollman 1979c). This scheme minimizes the disturbance to the polarity of the cell. Since the majority of the fence remains intact throughout the process, diffusion is not rapid enough to significantly mix the plasma membrane components over an area as large as the plasma membrane surface (von Bonsdorff et al 1985).

During differentiation mitosis does not always occur perpendicular to the plane of the epithelial sheet. Unequal mitosis is a process through which epithelial cells give rise to epithelial and nonepithelial daughters. This is thought to occur by the establishment of a cleavage furrow oblique to the plane of the epithelium, so that only one daughter retains tight junctions, and the other cell becomes mesenchymal (Watterson 1966, Franke et al 1982). Nonepithelial cells can also differentiate into epithelia during development (Ekblom 1984, Stern 1984). Conversion between epithelial and nonepithelial cells is a normal feature of organ formation. It is through such interconversions that all three primary germ layers give rise to the epithelia of the adult organism.

CONCLUDING REMARKS

From our discussion it is clear that several elements of epithelial cell structure interact to generate and maintain cell surface polarity. These include sorting devices responsible for the selective formation of membrane vesicles for newly synthesized and recycling surface proteins; mechanisms for the delivery of these transport vesicles to their correct surface domains; and tight junctions that function both as gates between extracellular compartments, and as barriers to diffusion between the apical and basolateral membranes. The cytoskeleton and the extracellular matrix clearly play an important but still imprecisely defined role. For more than a century, cell biologists have ascribed organization of the polarity of the cell to that little understood organelle, the centrosome (Wilson 1896, McIntosh 1983). It is a measure of the progress and the limitations of the field that although this organelle remains mysterious we now understand enough about the elements of the epithelial cell to begin exploring more precisely how they are organized. It is possible to test the hypothesis that the centrosome directs the bipolar organization of the cell, which leads to the establishment of polarity. The next few years promise to bring us close to an understanding of the basic interactions that lie at the heart of cell polarity.

ACKNOWLEDGMENT

We would like to thank Annie Steiner, Marianne Remy, and Joyce de Bruyn for excellent secretarial assistance, and Barry Gumbiner, Werner Franke, Kathryn Howell, Gerrit van Meer, Lennart Philipson, Laurie Roman, John Tooze, and Graham Warren for critically reading the manuscript.

Literature Cited

Abrahamson, D. R., Rodewald, R. 1981. Evidence for the sorting of endocytic vesicle contents during the receptor-mediated transport of IgG across the newborn rat intestine. *J. Cell Biol.* 91:270–80

Ahnen, D. J., Santiago, N. A., Cezard, J.-P., Gray, G. M. 1982. Intestinal amino oligopeptidase. In vivo synthesis on intracellular membranes of rat jejunum. *J. Biol.* *Chem.* 257:12129–35

Air, G. M. 1981. Sequence relationships among the haemagglutinin genes of twelve types of influenza A virus. *Proc. Natl. Acad. Sci. USA* 78:7639–43

Aiton, J. F., Brown, C. D. A., Ogden, P., Simmons, N. L. 1982. K^+ transport in tight epithelial monolayers of MDCK cells. *J. Membrane Biol.* 65:99–109

Almers, W., Stirling, C. 1984. Distribution of transport proteins over animal cell membranes. *J. Membr. Biol.* 77:169–86

Ambesi-Impiombato, F. S., Parks, L. A. M., Coon, H. G. 1980. Culture of hormone-dependent functional epithelial cells from rat thyroids. *Proc. Natl. Acad. Sci. USA* 77:3455–59

Anderson, R. G. W., Kaplan, J. 1983. Receptor-mediated endocytosis. In *Modern Cell Biology*, ed. B. Satir, Vol. 1, pp. 1–52. New York: Alan Liss

Balcarova-Ständer, J., Pfeiffer, S. E., Fuller, S. D., Simons, K. 1984. Development of cell surface polarity in the epithelial Madin-Darby canine kidney (MDCK) cell line. *EMBO J.* 3:2687–94

Barnes, D., Sato, G. 1980. Methods for growth of cultured cells in serum-free medium. *Anal. Biochem.* 102:255–70

Barth, A., Schulz, H., Neubert, K. 1974. Untersuchungen zur Reinigung und Characterisierung der Dipeptidylaminopeptidase IV. *Acta Biol. Med. Ger.* 32:157–74

Beguinot, L., Lyall, R. M., Willingham, M. C., Pastan, I. 1984. Down regulation of the EGF receptor in KB cells is due to receptor internalization and subsequent degradation in lysosomes. *Proc. Natl. Acad. Sci. USA* 81:2384–88

Bell, R. M., Ballas, L. M., Coleman, R. A. 1981. Lipid topogenesis. *J. Lipid Res.* 22:391–403

Bergeron, J. J. M., Sikstrom, A. R., Hand, A. R., Posner, B. I. 1979. Binding and uptake of ¹²⁵I-insulin into rat hepatocytes and endothelium. *J. Cell Biol.* 80:427–33

Bernfield, M. R., Banerjee, S. D. 1978. The basal lamina in epithelial-mesenchymal morphogenetic interactions. In *Biology and Chemistry of Basement Membranes*, ed. N. Kefalides, pp. 137–48. New York: Academic

Berridge, M. J., Oschman, J. L. 1972. *Transporting Epithelia*. New York: Academic. 91 pp.

Blitzer, B. L., Boyer, J. L. 1978. Cytochemical localization of Na^+, K^+, ATPase in the rat hepatocyte. *J. Clin. Invest.* 62:1104–8

Blobel, G. 1980. Intracellular protein topogenesis. *Proc. Natl. Acad. Sci. USA* 177:1496–1500

Blobel, G., Dobberstein, B. 1975. Transfer of proteins across membranes 1. Presence of proteolytically processed and unprocessed nascent immuno-globulin light chains on membrane-bound ribosomes of murine myeloma. *J. Cell Biol.* 67:835–51

Bode, F., Baumann, K., Kinne, R. 1976. Analysis of the pinocytic process in rat kidney. II. Biochemical composition of pinocytic vesicles compared to brush border microvilli, lysosomes and basolateral plasma membranes. *Biochim. Biophys. Acta* 433:294–310

Boller, K., Vestweber, D., Kemler, R. 1985. Cell-adhesion molecule uvomorulin is localized in the intermediate junctions of adult intestinal epithelial cells. *J. Cell Biol.* 100:327–32

Bonner, W. M. 1975. Protein migration into nuclei. I. Frog oocyte nuclei *in vivo* accumulate microinjected histones, allow entry to small proteins and exclude large proteins. *J. Cell Biol.* 64:421–30

Borgers, M. 1973. The cytochemical application of new potent inhibitors of alkaline phosphatases. *J. Histochem. Cytochem.* 21:812–24

Bos, T. J., Davis, A. R., Nayak, D. P. 1984. NH_2-terminal hydrophobic region of influenza virus neuraminidase provides the signal function in translocation. *Proc. Natl. Acad. Sci. USA* 81:2327–31

Brasitus, T. A., Schachter, D. 1980. Lipid dynamics and lipid-protein interactions in rat enterocyte basolateral and microvillus membranes. *Biochemistry* 19:2763–69

Brégégère, F., Abastado, J. P., Kvist, S. M., Rusk, L., Lalanne, J. L., et al. 1981. Structure of the C-terminal half of two H-2 antigens from cloned mRNA. *Nature* 292:78–81

Bretscher, M. S., Pearse, B. M. F. 1984. Coated pits in action. *Cell* 38:3–4

Bretscher, M. S., Raff, M. C. 1975. Mammalian plasma membranes. *Nature* 258:43–49

Bretscher, A., Weber, K. 1979. Villin: The major microfilament-associated protein of the intestinal microvillus. *Proc. Natl. Acad. Sci. USA* 76:2321–25

Bretscher, A., Weber, K. 1980. Fimbrin, a new microfilament-associated protein present in microvilli and other cell surface structures. *J. Cell Biol.* 86:335–40

Brown, W. J. T., Farquhar, M. G. 1984. The mannose-6-phosphate receptor for lysosomal enzymes is concentrated in *cis* Golgi cistenae. *Cell* 36:295–307

Bulger, R. E., Dobyan, D. C. 1982. Recent advances in renal morphology. *Ann. Rev. Physiol.* 44:147–79

Bundgaard, M., Möller, M., Hedemark-Poulsen, J. 1977. Localization of sodium pump sites in cat salivary gland. *J. Physiol.* (London) 273:339–53

Bundgaard, M., Möller, M., Hedemark-Poulsen, J. 1981. Localization of sodium pump sites in cat pancreas. *J. Physiol. London* 313:405–14

Burg, M., Green, N., Sohraby, S., Steele, R., Handler, J. 1982. Differentiated epithelia derived from thick ascending limbs clones. *Am. J. Physiol.* 242:C1229–33

Burwen, S. J., Barker, M. E., Goldman, I. S., Hradek, G. T., Raper, S. E., et al. 1984. Transport of epidermal growth factor by rat liver: evidence for a nonlysosomal pathway. *J. Cell Biol.* 79:1259–65

Calafat, J., Janssen, H., Démant, P., Hilgers, J., Závada, J. 1983. Specific selection of host cell glycoproteins during assembly of murine leukaemia virus and vesicular stomatitis virus: presence of thy-1 glycoprotein and absence of H-2, Pgp-1 and T-200 glycoproteins on the envelopes of these virus particles. *J. Gen. Virol.* 64: 1241–53

Carlson, R. W., Wada, H. G., Sussman, H. H. 1976. The plasma membrane of human placenta; isolation of microvillus membrane and characterization of protein and glycoprotein subunits. *J. Biol. Chem.* 251:4139–46

Cereijido, M., Ehrenfeld, J., Fernández-Castelo, S., Meza, J. 1981a. Fluxes, junctions and blisters in cultured monolayers of epitheliod cells (MDCK). *Ann. NY Acad. Sci.* 372:422–41

Cereijido, M., Meza, I., Martínez-Palomo, A. 1981b. Occluding junctions in cultured epithelial monolayers. *Am. J. Physiol.* 240:C96–C102

Cereijido, M., Robbins, E. S., Dolan, W. J., Rotunno, C. A., Sabatini, D. D. 1978. Polarized monolayers formed by epithelial cells on a permeable and translucent support. *J. Cell Biol.* 77:853–80

Cereijido, M., Robbins, E., Sabatini, D. D., Stefani, E. 1984. Cell to cell communication in monolayers of epithelial cells (MDCK) as a function of the age of the monolayers. *J. Membr. Biol.* 81:41–48

Chambard, M., Gabrion, J., Mauchamp, J. 1981. Influence of collagen gel in the orientation of epithelial cell polarity: follicle formation from isolated thyroid cells and from preformed monolayers. *J. Cell Biol.* 91:157–67

Chiacchia, K. B., Drickamer, K. 1984. Direct evidence for the transmembrane orientation of the hepatic glycoprotein receptors. *J. Biol. Chem.* 259:15440–46

Claude, P. 1978. Morphological factors influencing transepithelial permeability: a model for the resistance of the zonula occludens. *J. Membr. Biol.* 39:219–32

Cohen, S. M., Gorbsky, G., Steinberg, M. S. 1983. Immunochemical characterization of related families of glycoproteins in desmosomes. *J. Biol. Chem.* 258:2621–27

Colas, B., Maroux, S. 1980. Simultaneous isolation of brush border and basolateral membrane from rabbit enterocytes: Presence of brush border hydrolases in the basolateral membrane of rabbit enterocytes. *Biochim. Biophys. Acta* 600:406–20

Coligan, J. E., Kindt, T. J., Uehara, H., Martinko, J., Nathenson, S. G. 1981. Primary structure of a murine transplantation antigen. *Nature* 291:35–39

Coudrier, E., Reggio, H., Louvard, D. 1981. Immunolocalization of the 110,000 molecular weight cytoskeletal protein of intestinal microvilli. *J. Mol. Biol.* 152:49–66

Coudrier, E., Reggio, H., Louvard, D. 1983. Characterization of an integral membrane glycoprotein associated with the microfilaments of pig intestinal microvilli. *EMBO J.* 2:469–75

Cross, M. 1973. Active sodium and chloride transport across the rabbit blastocell wall. *Biol. Reprod.* 8:556–75

Cross, F. K., Garber, E. A., Pellman, D., Hanafusa, H. 1984. A short sequence in the p60 src N-terminus is required for p60 src myristylation and membrane association and for cell transformation. *Mol. Cell Biol.* 4:1834–42

Cuthbert, A. W., Schum, W. K. 1974. Amiloride and the sodium channel. *Naunyn Schmiedebergs Arch. Pharmakol.* 281:261–69

Danielsen, E. M. 1982. Biosynthesis of microvillar proteins pulse-chase labelling studies on amino-peptidase N and sucraseisomaltase. *Biochem. J.* 204:639–45

Danielsen, E. M., Cowell, G. M., Norén, O., Sjöström, H. 1984. Biosynthesis of microvillar proteins. *Biochem. J.* 221:1–14

Danielsen, E. M., Cowell, G. M., Poulsen, S. S. 1983. Biosynthesis of intestinal microvillar proteins. Role of the Golgi complex and microtubules. *Biochem. J.* 216:37–42

Danielsen, E. M., Vyas, J. P., Kenny, A. J. 1980. A neutral endopeptidase in the microvillar membrane of pig intestine. *Biochem. J.* 191:645–48

Danielson, K. G., Osborn, C. J., Durban, B. M., Butel, J. S., Medina, D. 1984. Epithelial mouse mammary cell line exhibiting normal morphogenesis *in vivo* and functional differentiation *in vitro*. *Proc. Natl. Acad. Sci. USA* 81:3756–60

Dautry-Varsat, A., Ciechanover, A., Lodish, H. F. 1983. pH and the recycling of transferrin during receptor-mediated endocytosis. *Proc. Natl. Acad. Sci. USA* 80:2258–62

De Grella, R. F., Simoni, R. D. 1982. Intracellular transport of cholesterol to the plasma membrane. *J. Biol. Chem.* 257:14256–62

Desneulle, P. 1979. Intestinal and renal aminopeptidase: A model of a transmembrane protein. *Eur. J. Biochem.* 101:1–11

Diamond, J. M. 1977. The epithelial junction: Bridge, gate and fence. *Physiologist* 20:10–18

Dickson, C., Eisenman, R., Fan, H., Hunter,

E., Teich, N. 1982. Protein biosynthesis and assembly in RNA tumor viruses. *Molecular Biology of Tumor Viruses*, ed. R. Weiss, N. Teich, H. Varmus, J. Coffin, pp. 513–648. New York: Cold Spring Harbor Lab.

Dingwall, C. 1985. The accumulation of proteins in the nucleus. *Trends Biochem. Sci.* 10:64–66

Dobberstein, B., Garoff, H., Warren, G., Robinson, P. J. 1979. Cell free synthesis and membrane insertion of mouse H-2D$_d$ histocompatibility antigen and β_2 microglobulin. *Cell* 17:759–69

Douglas, A. P., Kerley, R., Isselbacher, K. J. 1972. Preparation and characterization of the lateral and the basal plasma membranes of the rat intestinal epithelial cell. *Biochem. J.* 128:1329–38

Dower, S., Miller-Podraza, H., Fishman, P. H. 1982. Translocation of newly synthesized gangliosides to the cell surface. *Biochemistry* 21:3265–70

Dragsten, P. R., Handler, J. S., Blumenthal, R. 1982. Fluorescent membrane probes and the mechanism of maintenance of cellular asymmetry in epithelia. *Fed. Proc.* 41:48–53

Drickamer, K. 1981. Complete amino acid sequence of a membrane receptor for glycoproteins: Sequence of the chicken hepatic lectin. *J. Biol. Chem.* 259:5827–33

Drickamer, K., Mamon, J. F., Binns, G., Leung, J. O. 1984. Primary structure of the rat liver asialoglycoprotein receptor. Structural evidence for multiple protein species. *J. Biol. Chem.* 259:770–78

Dubois-Dalcq, M., Holmes, K., Rentier, B. 1984. Assembly of enveloped RNA viruses. Vienna: Springer. 235 pp.

Ducibella, T. 1977. Surface changes in the developing trophoblast cell. In *Development in Mammals*, Vol. 1, ed. M. H. Johnson, pp. 5–32. Amsterdam: Elsevier/North Holland

Ducibella, T., Albertini, D. F., Anderson, E., Biggers, J. D. 1975. The preimplantation embryo: characterization of intercellular junctions and their appearance during development. *Dev. Biol.* 45:231–50

Ducibella, T., Anderson, E. 1975. Cell shape and membrane changes in the eight-cell mouse embryo: Prerequisites for morphogenesis of the blastocyst. *Dev. Biol.* 47:45–58

Ducibella, T., Anderson, E. 1979. The effects of calcium deficiency on the formation of the zonula occludens and blastocoel in the mouse embryo. *Dev. Biol.* 73:46–58

Dunn, W. A., Hubbard, A. L. 1984. Receptor mediated endocytosis of epidermal growth factor (EGF) by hepatocytes: receptor pathway. *J. Cell Biol.* 99:371a

Easter, D. W., Wade, J. B., Boyer, J. L. 1983. Structural integrity of hepatocyte tight junctions. *J. Cell Biol.* 96:745–49

Ekblom, P. 1984. Basement membrane proteins and growth factors in kidney differentiation. In *The Role of Extracellular Matrix in Development*, ed. R. L. Trelstad, pp. 173–206. New York: Alan Liss

Ernst, S. A. 1972. Transport adenosine triphosphate cytochemistry. II. Cytochemical localization. *J. Histochem. Cytochem.* 20:23–38

Evans, W. H. 1980. A biochemical dissection of the functional polarity of the plasma membrane of the hepatocyte. *Biochim. Biophys. Acta* 604:27–64

Farquhar, M. G. 1982. Membrane recycling in secretory cells: pathway to the Golgi complex. *Ciba Found. Symp.* 92:157–83

Farquhar, M. G., Palade, G. E. 1963. Junctional complexes in various epithelia. *J. Cell Biol.* 17:375–412

Farquhar, M. G., Palade, G. E. 1981. The Golgi apparatus (complex)—(1954–1981)—from artifact to center stage. *J. Cell Biol.* 91:77s–103s

Feracci, H., Bernadac, A., Hovsépian, S., Fayet, G., Maroux, S. 1981. Aminopeptidase N is a marker for the apical pole of porcine thyroid epithelial cells in vivo and in culture. *Cell Tissue Res.* 221:137–46

Feracci, H., Maroux, S., Bonicel, J., Desnuelle, P. 1982. The amino acid sequence of the hydrophobic anchor of rabbit intestinal brush border aminopeptidase N. *Biochem. Biophys. Acta* 684:133–36

Fey, E. G., Wan, K. M., Penman, S. 1984. Epithelial cytoskeletal framework and nuclear matrix-intermediate filament scaffold: three-dimensional organization and protein composition. *J. Cell Biol.* 98:1973–84

Fields, S., Winter, G., Brownlee, G. G. 1981. The structure of the neuraminidase gene in human influenza virus A/PR/8/34. *Nature* 290:213–17

Finidori, J., Laperche, Y., Haguenauer-Tsapis, R., Barouki, R., Guellaen, G., et al. 1984. *In vitro* biosynthesis and membrane insertion of γ-glutamyl transpeptidase. *J. Biol. Chem.* 259:4695–98

Forstner, G. G., Tanaka, K., Isselbacher, K. J. 1968. Lipid composition of the isolated rat intestinal microvillus membrane. *Biochem. J.* 109:51–59

Franke, W. W., Grund, C., Kuhn, C., Jackson, B. W., Illmensee, K. 1982. Formation of cytoskeletal elements during mouse embryogenesis. III. Primary mesenchymal cells and the first appearance of vimentin filaments. *Differentiation* 23:43–59

Franke, W. W., Schmid, E., Grund, C., Müller, H., Engelbrecht, I., et al. 1981. Antibodies to high molecular weight polypeptides of desmosomes: specific localization of a class of junctional proteins in cells and tissues. *Differentiation* 20:217–41

Franke, W. W., Weber, K., Osborn, M., Schmid, E., Freudenstein, C. 1978. Antibody to prekeratin: decoration of tonofilament-like arrays in various cells of epithelial character. *Exp. Cell Res.* 116:429–45

Fransen, J. A. M., Jinsel, L. A., Hauri, H.-P., Sterchi, E., Blok, J. 1985. Immunoelectron microscopic localization of a microvillus membrane oligosaccharidase in the human small intestinal epithelium with monoclonal antibodies. *Eur. J. Cell Biol.* In press

Friend, D. S., Gilula, N. B. 1972. Variations in tight and gap-junctions in mammalian tissues. *J. Cell Biol.* 53:758–76

Frömter, E., Diamond, J. 1972. Route of passive ion permeation in epithelia. *Nature New Biol.* 235:9–13

Fuller, S. D., Bravo, R., Simons, K. 1985a. An enzymatic assay reveals that proteins destined for the apical in basolateral domains of an epithelial cell line share the same late Golgi compartments. *EMBO J.* 4:297–307

Fuller, S. D., von Bonsdorff, C.-H., Simons, K. 1985b. Cell surface haemagglutinin can mediate the infection of other enveloped viruses. Submitted for publication

Fuller, S. D., von Bonsdorff, C.-H., Simons, K. 1984. Vesicular stomatitis virus infects and matures only through the basolateral surface of the polarized epithelial cell line, MDCK. *Cell* 38:65–77

Galli, P., Brenna, A., deCamilli, P., Meldolesi, J. 1976. Extracellular calcium and organization of tight junctions in pancreatic acinar cells. *Exp. Cell Res.* 99:178–83

Gallione, C. J., Rose, J. K. 1983. Nucleotide sequence of a cDNA clone encoding the entire glycoprotein from the New Jersey serotype of Vesicular stomatitis virus. *J. Virol.* 46:162–69

Garoff, H., Simons, K. 1974. Location of the spike glycoproteins in the Semliki Forest virus membrane. *Proc. Natl. Acad. Sci. USA* 71:3988–92

Geiger, B., Dutton, A. H., Tokuyasu, K. T., Singer, S. J. 1981. Immunoelectron microscope studies of membrane-microfilament interactions: Distributions of α-actinin, tropomyosin and vinculin in brush border and chicken gizzard smooth muscle cells. *J. Cell Biol.* 91:614–28

Geiger, B., Schmid, E., Franke, W. W. 1983. Spatial distribution of proteins specific for desmosomes and adhaerens junctions in epithelial cells demonstrated by double immunofluorescence microscopy. *Differentiation* 23:189–205

George, S. G., Kenny, A. J. 1973. Studies on the enzymology of purified preparations of brush border from rabbit kidney cortex. *Exp. Cell Res.* 51:123–40

Geuze, H. J., Slot, J. W., Strous, G. J. A. M., Lodish, H. F., Schwartz, A. L. 1983. Intracellular site of asialoglycoprotein receptor-ligand uncoupling: double-labeling immunoelectron microscopy during receptor-mediated endocytosis. *Cell* 32:277–87

Geuze, H. J., Slot, J. W., Strous, G. J. A. M., Hasilik, A., von Figura, K. 1984a. Ultrastructural localization of the mannose-6-phosphate receptors in rat liver. *J. Cell Biol.* 98:2047–54

Geuze, H. J., Slot, J. W., Strous, G. J. A. M., Peppard, J., von Figura, K., et al. 1984b. Intracellular receptor sorting during endocytosis: comparative immunoelectron microscopy of multiple receptors in rat liver. *Cell* 37:195–204

Gipson, I. K., Grill, S. M., Spurr, S. J., Brennan, S. J. 1983. Hemidesmosome formation in vitro. *J. Cell Biol.* 97:849–57

Glenney, J. R. Jr., Glenney, P. 1984. The microvillus 110K cytoskeletal protein is an integral membrane protein. *Cell* 37:743–51

Gmaj, P., Murer, H., Kinne, K. 1979. Calcium ion transport across plasma membranes isolated from rat kidney cortex. *Biochem. J.* 178:253–56

Gomori, G. 1941. The distribution of phosphatase in normal organs and tissues. *J. Cell Comp. Physiol.* 17:71–83

Goodenough, D. A., Revel, J. P. 1970. A fine structural analysis of the intercellular junctions in the mouse liver. *J. Cell Biol.* 45:272–90

Gorbsky, G., Cohen, S. M., Shida, M., Gincide, G. J., Steinberg, M. S. 1985. Isolation of non-glycosylated proteins of desmosomes and immunolocalization of a third plaque protein: desmoplakin III. *Proc. Natl. Acad. Sci. USA* 82:810–14

Gospodarowicz, D., Lepine, J., Massoglia, S., Wood, I. 1984. Comparison of the ability of basement membranes produced by corneal endothelial and mouse-derived endodermal PF-HR-9 cells to support the proliferation of bovine kidney tubule epithelial cells in vitro. *J. Cell Biol.* 99:947–61

Gospodarowicz, D., Tauber, J.-P. 1980. Growth factors and the extracellular matrix. *Endocrinol. Rev.* 1:201–27

Green, R., Meiss, H., Rodriguez-Boulan, E. J. 1981. Glycosylation does not determine segregation of viral envelope proteins in the plasma membrane of epithelial cells. *J. Cell Biol.* 89:230–39

Greenburg, G., Hay, E. D. 1982. Epithelia suspended in collagen gels can lose polarity and express characteristics of migrating mesenchymal cells. *J. Cell Biol.* 95:333–39

Griepp, E. B., Dolan, W. J., Robbins, E. S., Sabatini, D. D. 1983. Participation of plasma membrane proteins in the formation of tight junctions by cultured epithelial cells. *J. Cell Biol.* 96:693–702

Griffiths, G., Pfeiffer, S., Simons, K., Matlin, K. 1985. Exit of newly synthesized membrane proteins from the *trans* cisterna of the Golgi complex to the plasma membrane. *J. Cell Biol.* In press

Griffiths, G., Warren, G., Quinn, P., Mathieu-Costello, D., Hoppeler, H. 1984. Density of newly synthesized plasma membrane proteins in intracellular membranes I. Stereological studies. *J. Cell Biol.* 98:2133–41

Grinnell, F. 1978. Cellular adhesiveness and extracellular substrata. *Int. Rev. Cytol.* 53:65–144

Handler, J. S., Preston, A. S., Steele, R. E. 1984. Factors affecting the differentiation of epithelial transport and responsiveness to hormones. *Fed. Proc.* 43:2221–24

Handler, J. S., Steele, R. E., Sahib, M. K., Wade, J. B., Preston, S. A., et al. 1979. Toad urinary bladder epithelial cell in culture: maintenance of epithelial cell structure sodium transport and response to hormones. *Proc. Natl. Acad. Sci. USA* 76:4151–55

Handyside, A. H. 1980. Distribution of antibody- and lectin-binding sites on dissociated blastomeres from mouse morulae: evidence for polarization at compaction. *J. Embryol. Exp. Morphol.* 60:99–116

Hannafin, J., Kinne-Saffran, E., Friedman, D., Kinne, R. 1983. Presence of a sodium-potassium chloride co-transport system in the rectal gland of squalus acanthias. *J. Membr. Biol.* 75:73–83

Hanover, J. A., Lennarz, W. J. 1981. Transmembrane assembly of membrane and secretory proteins. *Arch. Biochem. Biophys.* 211:1–19

Harrison, S. C., Kirchhausen, T. 1983. Clathrin, cages and coated vesicles. *Cell* 33:650–52

Hasting, R., Enders, A. D. 1975. Junctional complexes in the preimplantation rabbit embryo. *Anat. Rec.* 181:17–34

Hauri, H.-P., Quaroni, A., Isselbacher, K. J. 1979. Biosynthesis of intestinal plasma membrane: Post-translational route and cleavage of sucrose-isomaltase. *Proc. Natl. Acad. Sci. USA* 76:5183–86

Hay, E. D. 1983. Cell and extracellular matrix: their organization and mutual dependence. *Modern Cell Biology 2*, pp. 509–48. New York: Alan Liss

Hedo, J. A., Kahn, C. R., Hayashi, M., Yamada, K. H., Kasuya, M. 1983. Biosynthesis and glycosylation of the insulin receptor: Evidence for a single polypeptide precursor of the two major subunits. *J. Biol. Chem.* 258:10020–26

Heidrich, H.-G., Kinne, R., Kinne-Saffran, E., Hannig, K. 1972. The polarity of the proximal tubule cell in rat kidney: Different surface charges for the brush border microvilli and plasma membranes from the basal infoldings. *J. Cell Biol.* 54:232–45

Helenius, A., Mellman, I., Wall, D., Hubbard, A. 1983. Endosomes. *Trends Biochem. Sci.* 8:245–50

Herscovics, A., Bugge, B., Quaroni, A., Kirsch, K. 1980. Characterization of glycopeptides labelled from D-[2-³H] mannose and L-[6-³H] fucose in intestinal epithelial cell membranes during differentiation. *Biochem. J.* 192:145–53

Hertzberg, E. L., Gilula, N. B. 1979. Isolation and characterization of gap junctions from rat liver. *J. Biol. Chem.* 254:2138–47

Hertzberg, E. L., Skibbens, R. V. 1984. A protein homologous to the 27,000 dalton liver gap junction protein is present in a wide variety of species and tissues. *Cell* 39:61–69

Herzlinger, D. A., Ojakian, G. K. 1984. Studies on the development and maintenance of epithelial cell surface polarity with monoclonal antibodies. *J. Cell Biol.* 98:1777–87

Herzog, V. 1983. Transcytosis in thyroid follicle cells. *J. Cell Biol.* 97:607–71

Herzog, V. 1984. Pathways of endocytosis in thyroid follicle cells. *Int. Rev. Cytol.* 91:107–39

Hildemann, B., Schmidt, A., Murer, H. 1979. Ca⁺⁺ transport in basal-lateral plasma membranes isolated from rat small intestinal epithelial cells. *Pflügers Arch.* 382:R23

Hise, M. K., Mantulin, W. W., Weinman, E. J. 1984. Fluidity and composition of brush border and basolateral membranes from rat kidney. *Am. J. Physiol.* 247:F434–39

Holland, J. J., Kiehn, E. D. 1970. Influenza virus effects on cell membrane proteins. *Science* 167:202–5

Hootman, S. R., Philpott, C. W. 1979. Ultracytochemical localization of Na⁺K⁺ activated ATPase in chloride cells from gills of a teleost. *Anat. Rec.* 193:99–129

Hopsu-Havu, V. K., Rintala, P., Glenner, G. G. 1968. A frog kidney amino-peptidase cleaving N-terminal dipeptides. Partial purification and characteristics. *Acta Chem. Scand.* 22:299–308

Hsu, M.-C., Choppin, P. W. 1984. Analysis of Sendai virus mRNA's with cDNA clones of viral genes and sequences of biologically important regions of the fusion protein. *Proc. Natl. Acad. USA* 81:7732–36

Hubbard, A. L., Wall, D. A., Ma, A. 1983. Isolation of rat hepatocyte plasma membranes. I. Presence of three major domains. *J. Cell Biol.* 96:217–29

Hull, B. E., Staehelin, L. A. 1979. The terminal web. A reevaluation of its structure and function. *J. Cell Biol.* 81:67–82

Hurt, E. D., Pesold-Hurt, B., Schatz, G. 1984. The amino-terminal region of an imported mitochondrial precursor polypeptide can direct cytoplasmic dihydrofolate reductase into the mitochondrial matrix. *EMBO J.* 3:3149–56

Inoue, M., Kinne, R., Tran, T., Blempica, L., Arias, I. M. 1983. Rat liver canalicular membrane vesicles: isolation and topological characterization. *J. Biol. Chem.* 258:5183–88

Jackson, B. W., Grund, C., Schmid, E., Bürki, K., Franke, W. W., et al. 1980. Formation of cytoskeletal elements during mouse embryogenesis. Intermediate filaments of cytokeratin type and desmosomes in preimplantation embryos. *Differentiation* 17:161–79

Jamieson, J. D. 1972. Transport and discharge of exportable proteins in pancreatic acinar cells: In vitro studies. *Curr. Top. Membr. Transp.* 3:273–338

Johnson, M. H., Ziomek, C. A. 1981. Induction of polarity in mouse 8-cell blastomeres: specificity, geometry and stability. *J. Cell Biol.* 91:303–8

Jones, A. L., Schmucker, D. L. 1977. Current concepts of liver structure as related to junction. *Gastroenterology* 73:833–51

Kachar, B., Reese, T. S. 1982. Evidence for the lipidic nature of tight junction strands. *Nature* 296:464–66

Kalderon, D., Richardson, W. D., Markham, A. F., Smith, A. E. 1984. Sequence requirements for nuclear location of simian virus 40 large T antigen. *Nature* 311:33–38

Karnaky, K. J. Jr., Kinter, L. B., Kinter, W. B., Stirling, C. E. 1976. Telost chloride cell II. Autographic localization of gill NaK ATPase in killfish *Fundulus heteroclitus* adapted to low and high salinity environments. *J. Cell Biol.* 70:157–77

Kartenbeck, J., Schwechheimer, K., Moll, R., Franke, W. W. 1984. Attachment of vimentin filaments to desmosomal plaques in human meningeomal cells and arachnoidal tissue. *J. Cell Biol.* 98:1072–81

Kasuga, M., Kahn, R. C., Hedo, J. A., Obbergen, E. V., Yamada, K. M. 1981.

Insulin-induced receptors loss in cultured human lymphocytes is due to accelerated receptors degradation. *Proc. Natl. Acad. Sci. USA* 78:6917–21

Kawai, K., Fujita, M., Nakao, M. 1974. Lipid components of two different regions of an intestinal epithelial cell membrane of mouse. *Biochim. Biophys. Acta* 369:222–33

Keil, W., Niemann, H., Schwartz, R. T., Klenk, H. D. 1984. Carbohydrates of influenza virus. V. Oligosaccharides attached to individual glycosylation sites on fowl plaque virus. *Virology* 133:77–91

Kenny, A. J., Maroux, S. 1982. Topology of microvillar membrane hydrolases of kidney and intestine. *Physiol. Rev.* 62:91–128

Kerjaschki, D., Noronha-Blob, L., Sacktor, B., Farquhar, M. G. 1984. Microdomains of distinctive glycoprotein composition in the kidney proximal tubule brush border. *J. Cell Biol.* 98:1505–13

Kerr, M. A., Kenny, A. J. 1974. The purification and specificity of a neutral endopeptidase from rabbit kidney brush border. *Biochem. J.* 137:477–88

Kinne, R., Kinne-Saffran, E. 1978. Differentiation of cell faces in epithelia. In *Molecular Specialization and Symmetry in Membrane Function*, ed. A. K. Salomon, M. Karnovsky, pp. 272–93. Cambridge, Mass.: Harvard Univ. Press

Klausner, R. D., Van Renswoude, J., Ashwell, G., Kempf, C., Schechter, A. N., et al. 1983. Receptor-mediated endocytosis of transferrin in K562 cells. *J. Biol. Chem.* 258:4715–24

Kleinman, H. K., Klebe, R. J., Martin, R. G. 1981. Role of collagenous matrices in the adhesion and growth of cells. *J. Cell Biol.* 88:473–85

Klenk, H.-D., Choppin, P. W. 1970a. Plasma membrane lipids and parainfluenza virus assembly. *Virology* 40:939–47

Klenk, H.-D., Choppin, P. W. 1970b. Glycosphigolipids of plasma membranes of cultured cells and an enveloped virus (SV5) grown in these cells. *Proc. Natl. Acad. Sci. USA* 66:57–64

Keonig, B., Ricapito, S., Kinne, R. 1983. Chloride transport in thick ascending limb of Henle's loop: potassium dependence and stoichiometry of the NaCl cotransport system in plasma membrane vesicles. *Pflügers Arch.* 399:173–79

Kohama, T., Shimizu, K., Ishida, N. 1979. Carbohydrate composition of the envelope glycoproteins of Sendai virus. *Virology* 90:226–34

Kornfeld, S., Reitman, M. L., Varki, A., Goldberg, D., Gabel, C. A. 1982. Steps in the phosphorylation of the high mannose oligosaccharides. *Ciba Found. Symp.* 92: 138–56

Kühn, L. C., Kraehenbuhl, J.-P. 1982. The sacrificial receptor-translocation of polymeric IgA. *Trends Biochem. Sci.* 7:299–302

Kvist, S., Brégégère, F., Rask, L., Cami, B., Garoff, H., Daniel, F., et al. 1981. cDNA clone coding for part of a mouse H-2^d major histocompatibility antigen. *Proc. Natl. Acad. Sci. USA* 78:2772–76

Kyte, J. A. 1976a. Immunoferritin determination of the distribution of Na$^+$/K$^+$ ATPase over the plasma membranes of renal convoluted tubules. I. Distal segment. *J. Cell Biol.* 68:287–303

Kyte, J. 1976b. Immunoferritin determination of Na$^+$/K$^+$ATPase over the plasma membrane of renal convoluted tubules. II. Proximal segment. *J. Cell Biol.* 68:304–18

Kyte, J. A. 1981. Molecular considerations relevant to the mechanism of active transport. *Nature* 292:201–4

Lamb, J. F., Ogden, P., Simmons, N. L. 1981. Autoradiographic localization of ^{3}H-Oubain bound to cultured epithelial cell monolayers of MDCK cells. *Biochim. Biophys. Acta* 644:333–40

Lanford, R. E., Butel, J. S. 1984. Construction and characterization of an SV40 mutant defective in nuclear transport of T antigen. *Cell* 37:801–13

Lazarides, E. 1980. Intermediate filaments as mechanical integrators of cellular space. *Nature* 283:249–56

Lazarides, E., Moon, R. T. 1984. Assembly and topogenesis of the spectrin-based membrane skeleton in erythroid development. *Cell* 37:354–56

Lehtonen, E., Badley, R. A. 1980. Localization of cytoskeletal proteins in preimplantation mouse embryos. *J. Embryol. Exp. Morphol.* 48:211–25

Leighton, E., Estes, L. W., Mansukhani, S., Brada, Z. 1970. A cell line derived from normal dog kidney (MDCK) exhibiting qualities of papillary adenocarcinoma and renal tubular epithelium. *Cancer Brussels* 26:1022–28

Lewis, S. A., Eaton, D. C., Diamond, J. M. 1976. The mechanism of Na$^+$ transport by rabbit urinary bladder. *J. Membr. Biol.* 28:41–50

Liang, C.-J., Yamushita, K., Muellenberg, C. C., Schichi, H., Kobata, A. 1979. Structure of carbohydrate moieties of bovine rhodopsin. *J. Biol. Chem.* 254:6414–18

Lippman, M. E., Strobl, J., Allegro, J. C. 1980. Effects of hormones on human breast cancer cells in tissue culture. In *Cell Biology of Breast Cancer*, ed. L. McGrath, M. Brennan, M. Rich, pp. 265–75. New York: Academic

Lodish, H. F., Braell, W. A., Schwartz, A. L., Strous, G. J. A. M., Zilberstein, A. 1981. Synthesis and assembly of membrane and organelle proteins. *Int. Rev. Cytol. Suppl.* 12:247–307

Lodish, H. F., Porter, M. 1980. Heterogeneity of vesicular stomatitis virus particles: Implications for virion assembly. *J. Virol.* 33:52–58

Lodja, Z. 1974. Cytochemistry of enterocytes and of other cells in the mucus membrane of the small intestine. *Biomembranes* 4A: 43–122

Loewenstein, W. R. 1981. Junctional intercellular communication: the cell-to-cell membrane channel. *Physiol. Rev.* 61:829–913

Louvard, D. 1980. Apical membrane aminopeptidase appears at sites of cell-cell contact in cultured epithelial cells. *Proc. Natl. Acad. Sci. USA* 77:4132–36

Louvard, D., Godefroy, O., Huet, C., Sahnquillo-Merino, C., Robine, S., et al. 1985. Basolateral membrane proteins are expressed at the surface of immature intestinal cells whereas transport of apical proteins is abortive. In *Transport and Secretion of Proteins*, ed. M. J. Gething, J. Sambrook. Cold Spring Harbor, NY: Cold Spring Harbor Pub. In press

Ludens, J. H., Zimmerman, W. B., Schieders, J. R. 1980. Nature of the inhibition of Cl$^-$ transport by furosemide: evidence for a direct effect on active transport in the toad cornea. *Life Sci.* 27:2453–58

Maratje, G. V., Nash, B., Haschemeyer, R. H., Tate, S. S. 1979. Ultrastructural localization of γ-glutamyl transferase in rat kidney and jejunum. *FEBS Lett.* 107:436–40

Marcial, M. A., Carlson, S. L., Madara, J. L. 1984. Partitioning of paracellular conductance along the ileal crypt-villus axis: a hypothesis based on structural analysis with detailed consideration of tight junction structure-function relationships. *J. Membr. Biol.* 80:59–70

Marsh, M., Helenius, A. 1980. Adsorptive endocytosis of Semliki Forest virus. *J. Mol. Biol.* 142:439–54

Martinez-Palomo, A., Meza, I., Beaty, G., Cereijido, M. 1980. Experimental modulation of occluding junctions in a cultured transporting epithelium. *J. Cell Biol.* 87: 736–45

Matlin, K. S., Bainton, D. F., Pesonen, M., Louvard, D., Genty, N., et al. 1983. Transepithelial transport of a viral membrane glycoprotein implanted into the apical plasma membrane of MDCK cells. I. Morphological evidence. *J. Cell Biol.* 97:627–37

Matlin, K. S., Simons, K. 1983. Reduced

temperature prevents transfer of a membrane glycoprotein to the cell surface but does not prevent terminal glycosylation. *Cell* 34:233–43

Matlin, K., Simons, K. 1984. Sorting of a plasma membrane glycoprotein occurs before it reaches the cell surface in cultured epithelial cells. *J. Cell Biol.* 99:2131–39

Matsuura, S., Eto, S., Kato, K., Tashino, Y. 1984. Ferritin immunoelectron microscopic localization of 5'-nucleotidase on rat liver cell surface. *J. Cell Biol.* 99:166–73

Matsuura, S., Nakada, H., Sawamura, T., Tashiro, Y. 1982. Distribution of an asialoglycoprotein receptor on rat hepatocyte cell surface. *J. Cell Biol.* 95:864–75

Mayes, E. L., Waterfield, M. D. 1984. Biosynthesis of the epidermal growth factor receptor in A4331 cells. *EMBO J.* 3:531–37

McCloskey, M., Poo, M. 1984. Protein diffusion in cell membranes: some biological implications. *Int. Rev. Cytol.* 87:19–81

McIntosh, J. R. 1983. The centrosome as an organizer of the cytoskeleton. *Modern Cell Biology 2*, pp. 115–42. New York: Alan Liss

McRoberts, J. A., Taub, M., Saier, M. H. Jr. 1981. The Madin-Darby canine kidney (MDCK) cell line. In *Functionally Differentiated Cell Lines*, ed. G. Sato, pp. 117–39. New York: Alan Liss

Meier, P. J., Sztul, E. S., Reuben, A., Boyer, J. L. 1984. Structural and functional polarity of canalicular and basolateral plasma membrane vesicles isolated in high yield from rat liver. *J. Cell Biol.* 98:991–1000

Meiss, H. K., Green, R. F., Rodriguez-Boulan, E. J. 1982. Lectin-resistant mutant of polarized epithelial cells. *Mol. Cell Biol.* 2:1287–94

Mellman, I. S. 1982. Endocytosis, membrane recycling and Fc receptor function. *Ciba Found. Symp.* 92:35–51

Milhorat, T. H., Davis, D. A., Hammock, M. K. 1975. Localization of oubain-sensitive Na$^+$-K$^+$-ATPase in frog, rabbit and rat choroid plexus. *Brain Res.* 99:170–74

Mills, J. T., Furlong, S. T., Dawidowicz, E. A. 1984. Plasma membrane biogenesis in eukaryotic cells: translocation of newly synthesized lipid. *Proc. Natl. Acad. Sci. USA* 81:1385–88

Mills, J. W., Ernst, S. A. 1975. Localization of sodium pump sites in frog urinary bladder. *Biochim. Biophys. Acta* 375:268–73

Mills, J. W., Ernst, S. A., Di Bona, D. R. 1977. Location of Na$^+$ pump sites in frog skin. *J. Cell Biol.* 73:88–110

Min-Jou, W., Verhoeyen, M., Devos, R., Saman, E., Fang, R., et al. 1980. Complete structure of the haemagglutinin gene from the human influenza A/Victoria/3/75 DNA. *Cell* 19:683–96

Mircheff, A. K., Sachs, G., Hanna, S. D., Labiner, C. S., Rabon, E., et al. 1979. Highly purified basal lateral plasma membranes from rat duodenum. Physical criteria for purity. *J. Membr. Biol.* 50:343–63

Misek, D. E., Vard, E., Rodrigeuz-Boulan, E. 1984. Biogenesis of epithelial cell polarity: intracellular sorting and vectorial exocytosis of an apical plasma membrane glycoprotein. *Cell* 39:537–46

Misfeldt, D. S., Hamamoto, S. T., Pitelka, D. R. 1976. Transepithelial transport in cell culture. *Proc. Natl. Acad. Sci. USA* 73:1212–16

Mueller, H., Franke, W. W. 1983. Biochemical and immunological characterization of desmoplakins I and II. The major polypeptides of the desmosomal plaque. *J. Mol. Biol.* 163:647–71

Mullock, B. M., Hinton, R. H. 1981. Transport of proteins from blood to bile. *Trends Biochem. Sci.* 6:188–91

Murer, H., Kinne, R. 1980. The use of isolated membrane vesicles to study epithelial transport processes. *J. Membr. Biol.* 55:81–95

Nathenson, S. G., Cullen, S. E. 1974. Biochemical properties and immunochemical-genetic relationship of mouse H-2 alloantigens. *Biochim. Biophys. Acta* 344:1–25

Neufeld, E. F., Sando, G. N., Garvin, A. J., Rome, L. H. 1977. The transport of lysosomal enzymes. *J. Supramol. Struct.* 6:95–101

Nir, I., Cohen, D., Papermaster, D. S. 1984. Immunochemical localization of opsin in the cell membrane of developing rat retinal photoreceptors. *J. Cell Biol.* 98:1788–95

Oliver, C. 1982. Endocytic pathway at the lateral and basal cell surfaces of exocrine acinar cells. *J. Cell Biol.* 95:154–61

O'Neil, R. G., Boulpaep, E. L. 1979. Effect of amiloride on the apical cell membrane cation channels of a sodium-absorbing potassium-secreting renal epithelium. *J. Membr. Biol.* 50:365–87

Op den Kamp, J. A. F. 1979. Lipid asymmetry in membranes. *Ann. Rev. Biochem.* 48:47–71

Ovchinnikov, Y. A. 1982. Rhodopsin and bacteriorhodopsin: structure-function relationships. *FEBS Lett.* 148:179–89

Owens, R. 1976. Selective cultivation of mammalian epithelial cells. In *Methods in Cell Biology*, ed. D. M. Prescott, Vol. XIV, pp. 341–55. New York: Academic

Palade, G. 1975. Intracellular aspects of the process of protein secretion. *Science* 189:347–58

Palade, G. E. 1982. Problems in intracellular

membrane traffic. *Ciba Found. Symp.* 92:1–14

Parodi, A. J., Mendelzon, D. H., Leder-kremer, G. Z. 1983. Transient glucosylation of protein-bound $Man_9GlcNaC_2$, $Man_8GlcNaC_2$ and $Man_7GlcNaC_2$ in calf thyroid cells. *J. Biol. Chem.* 258:8260–65

Parr, E. L. 1979. Diversity in the expression of H-2 antigens on liver cells. *Transplantation* 27:45

Parr, E. L., Kirby, W. N. 1979. An immunoferritin labeling study of H-2 antigens on dissociated epithelial cells. *J. Histochem. Cytochem.* 27:1327–36

Pastan, I., Willingham, M. C. 1983. Receptor-mediated endocytosis: coated pits, receptosomes and the Golgi. *Trends Biochem. Sci.* 8:250–54

Patzer, E. J., Wagner, R. R., Dubovi, E. J. 1979. Viral membranes: model systems for studying biological membranes. *CRC Crit. Rev. Biochem.* 6:165–217

Paulson, J. C., Hill, R. L., Tanabe, T., Ashwell, G. 1977. Reactivation of asialo-rabbit liver binding protein by resialation with β-D-galactoside α-2-6-sialotransferase. *J. Biol. Chem.* 252:8624–28

Perkins, F. M., Handler, J. S. 1981. Transport properties of toad kidney epithelial in culture. *Am. J. Physiol.* 241:C154–59

Pesonen, M., Ansorge, W., Simons, K. 1984a. Transcytosis of the G protein of vesicular stomatitis virus after implantation into the apical plasma membrane of Madin-Darby canine kidney cells. I. Involvement of endosomes and lysosomes. *J. Cell Biol.* 99:796–802

Pesonen, M., Bravo, R., Simons, K. 1984b. Transcytosis of the G protein of vesicular stomatitis virus after plantation into the apical membrane of MDCK cells. II. Involvement of the Golgi complex. *J. Cell Biol.* 99:803–9

Pesonen, M., Simons, K. 1983. Trans-epithelial transport of a viral membrane glycoprotein implanted into the apical plasma membranes of MDCK cells. II. Immunological quantitation. *J. Cell Biol.* 97:638–43

Pfeiffer, S., Fuller, S. D., Simons, K. 1985. Intracellular sorting and basolateral appearance of the G protein of vesicular stomatitis virus in MDCK cells. *J. Cell Biol.* In press

Pfenninger, K. H., Johnson, M. P. 1983. Membrane biogenesis in the sprouting neuron. I. Selective transfer of newly synthesized phospholipid into the growing neurite. *J. Cell Biol.* 97:1038–42

Pinto, M., Appay, M.-D., Simon-Assmann, P., Chevalier, G., Dracopoli, J., et al. 1982. Enterocyte differentiation of cultured human colon cancer cells by replacement of glucose by galactose in the medium. *Biol. Cell.* 44:193–96

Pinto, M., Robine-Leon, S., Appay, M. D., Kedinger, M., Triadon, N., et al. 1983. Enterocyte-like differentiation and polarization of the human colon carcinoma cell line caco-2 in culture. *Biol. Cell* 47:323–30

Pinto da Silva, P., Kachar, B. 1982. On tight-junction structure. *Cell* 28:441–50

Pisam, M., Ripoche, P. 1976. Redistribution of surface macromolecules in dissociated epithelial cells. *J. Cell Biol.* 71:907–20

Porter, A. A., Barber, C., Carey, N. H., Hallewell, R. A., Threlfall, G., et al. 1979. Complete nucleotide sequence of an influenza virus haemagglutinin gene from cloned DNA. *Nature* 282:471–77

Quaroni, A., Kirsch, K., Weiser, M. M. 1979a. Synthesis of membrane glycoproteins in rat small-intestinal villus cells. Redistribution of L-$[1,5,6-^3H]$-glucose-labeled membrane glycoproteins among Golgi, lateral, basal, and microvillus membranes *in vivo*. *Biochem. J.* 182:203–12

Quaroni, A., Kirsch, K., Weiser, M. M. 1979b. Synthesis of membrane glycoproteins in rat small-intestinal villus cells. Effect of colchicine. *Biochem. J.* 182:213–21

Quinn, P., Griffiths, G., Warren, G. 1984. Density of newly synthesized plasma membrane proteins in intracellular membranes. II. Biochemical studies. *J. Cell Biol.* 98:2142–47

Quinton, P. M., Tormey, J. M. 1973. Localization of Na/K ATPase sites in the secretory and reabsorptive epithelia of perfused exocrine sweat glands: A question as to the role of the enzyme in secretion. *J. Membr. Biol.* 29:383–99

Quinton, P. M., Wright, E. M., Tormey, J. M. 1973. Localization of sodium pumps in the choroid plexus epithelium. *J. Cell Biol.* 58:724–30

Rabito, C. A., Kreiberger, J. I., Wight, D. 1984. Alkaline phosphatase and glutamyl transpeptidase as polarization markers during the organization of LLC-PK_1 cells into an epithelial membrane. *J. Biol. Chem.* 259:574–82

Reading, C. L., Renhoet, E. E., Ballon, C. E. 1978. Carbohydrate structure of vesicular stomatitis virus. *J. Biol. Chem.* 253:5600–12

Reik, L., Petzold, G. L., Higgins, J. A., Green-gard, P., Bannett, R. J. 1970. Hormone-sensitive adenyl cyclase: cytochemical localization in rat liver. *Science* 168:382–86

Renston, R. H., Maloney, D. G., Jones, A. L., Hradek, G. T., Wong, K. Y., Goldfine, I. D. 1980. Bile secretory apparatus: evidence for a vesicular transport mechanism for proteins in the rat using horseradish peroxidase and insulin. *Gastroenterology* 78:1373–88

Richardson, J. C. W., Scalera, V., Simmons, N. L. 1981. Identification of two strains of MDCK cells which resemble separate nephron tubule segments. *Biochim. Biophys. Acta* 673:26–36

Rindler, M. J., Ivanov, I. E., Plesken, H., Rodriguez-Boulan, E., Sabatini, D. D. 1984. Viral glycoproteins destined for apical or basolateral plasma membrane domains traverse the same Golgi apparatus during their intracellular transport in double infected Madin-Darby canine kidney cells. *J. Cell Biol.* 98:1304–19

Rindler, M. J., Ivanov, I. E., Plesken, H., Sabatini, D. D. 1985. Polarized delivery of viral glycoproteins to the apical and basolateral plasma membranes of MDCK cells infected with temperature-sensitive viruses. *J. Cell Biol.* 100:136–51

Rindler, M. J., McRoberts, J. A., Saier, M. H. 1982. (Na,K) cotransport in the MDCK cell line. *J. Biol. Chem.* 257:2254–59

Rodriguez-Boulan, E. 1983. Membrane biogenesis, enveloped RNA viruses, and epithelial polarity. In *Modern Cell Biology*, Vol. 1, pp. 119–70, ed. B. H. Satin. New York: Alan Liss

Rodriguez-Boulan, E., Paskiet, K. T., Sabatini, D. D. 1983. Assembly of enveloped viruses in MDCK cells: polarized budding from single attached cells and from clusters of cells in suspension. *J. Cell Biol.* 96:866–74

Rodriguez-Boulan, E., Paskiet, K. T., Salas, P. J. I., Bard, E. 1984. Intracellular transport of influenza virus hemagglutinin to the apical surface of MDCK cells. *J. Cell Biol.* 98:308–19

Rodriguez-Boulan, E. J., Pendergast, M. 1980. Polarized distribution of viral envelope glycoprotein in the plasma membrane of infected epithelial cells. *Cell* 20:45–54

Rodriguez-Boulan, E. J., Sabatini, D. D. 1978. Asymmetric budding of viruses in epithelial monolayers: a model system for the study of epithelial cell polarity. *Proc. Natl. Acad. Sci. USA* 75:5071–75

Roman, L. M., Hubbard, A. L. 1983. A domain-specific marker for the hepatocyte plasma membrane: localization of leucine aminopeptidase to the bile canalicular domain. *J. Cell Biol.* 96:1548–58

Roman, L. M., Hubbard, A. L. 1984a. A domain specific marker for the hepatocyte plasma membrane. II. Ultrastructural localization of leucine aminopeptidase to the bile canalicular membrane of isolated rat liver plasma membranes. *J. Cell Biol.* 98:1488–96

Roman, L. M., Hubbard, A. L. 1984b. A domain-specific marker for the hepatocyte plasma membrane. III. Isolation of bile canalicular membrane by immunoadsorption. *J. Cell Biol.* 98:1497–1504

Rose, J. K., Gallione, C. J. 1981. Nucleotide sequences of the mRNA's encoding the vesicular stomatitis virus G and M proteins determined from cDNA clones containing the complete coding regions. *J. Virol.* 39:519–28

Roth, M. G., Compans, R. W. 1981. Delayed appearance of pseudotypes between vesicular stomatitis and influenza viruses. *J. Virol.* 40:848–60

Roth, M. G., Compans, R. W., Giusti, L., Davis, A. R., Nayak, D. P., et al. 1983. Influenza virus hemagglutinin expression is polarized in cells infected with recombinant SV40 viruses carrying cloned hemagglutinin DNA. *Cell* 33:435–43

Roth, M. G., Fitzpatrick, J. P., Compans, R. W. 1979. Polarity of influenza and vesicular stomatitis virus maturation in MDCK cells: lack of a requirement for glycosylation of viral glycoproteins. *Proc. Natl. Acad. Sci. USA* 76:6430–34

Roth, M. G., Srinivas, R. V., Compans, R. W. 1983. Basolateral maturation of retroviruses in polarized epithelial cells. *J. Virol.* 45:1065–73

Rothman, J. E. 1981. The Golgi apparatus: two organelles in tandem. *Science* 213:1212–19

Rothman, J. E., Lenard, J. 1984. Membrane traffic in animal cells. *Trends Biochem. Sci.* 9:176–78

Rothman, J. E., Lodish, H. F. 1977. Synchronized transmembrane insertion and glycosylation of a nascent membrane protein. *Nature* 269:775–80

Rothman, J. E., Tsai, D. K., Dawidowicz, E. A., Lenard, J. 1976. Transbilayer phospholipid asymmetry and its maintenance in the membrane of influenza virus. *Biochemistry* 15:2361–70

Sabatini, D. D., Griepp, E. B., Rodriguez-Boulan, E. J., Dolan, W. J., Robbins, E. S., et al. 1983. Biogenesis of epithelial surface polarity. In *Modern Cell Biology 2*, p. 419–50. New York: Alan Liss

Sabatini, D. D., Kreibich, G., Morimoto, T., Adesnik, M. 1982. Mechanisms for the incorporation of proteins in membranes and organelles. *J. Cell Biol.* 92:1–22

Salas, P. J. I., Vega-Salas, D., Misek, D., Rodriguez-Boulan, E. 1985. Intracellular routes of apical and basolateral plasma membrane proteins to the surface of epithelial cells. *Pflügers Arch.* In press

Schatz, G., Butow, R. A. 1983. How are proteins incorporated into mitochondria? *Cell* 32:316–18

Scheid, A., Choppin, P. W. 1977. Two disulfide-linked polypeptide chains constitute the active F protein of paramyxoviruses. *Virology* 80:54–66

Schiff, J. M., Fisher, M. M., Underdown, B. J. 1984. Receptor-mediated biliary transport of IgA and asialoglycoprotein: sorting and missorting of ligands revealed by two radiolabeling methods. *J. Cell Biol.* 98:79–89

Schmid, E., Schiller, D. L., Grund, C., Stadler, J., Franke, W. W. 1983. Tissue type-specific expression of intermediate filament proteins in a cultured epithelial cell line from bovine mammary gland. *J. Cell Biol.* 96:37–50

Schneider, C., Owen, M. J., Banville, D., Williams, J. G. 1984. Primary structure of the human transferrin receptor deduced from the mRNA sequence. *Nature* 311:675–78

Schneider, C., Sutherland, R., Newman, R., Greaves, M. 1982. Structural feature of the cell surface receptor for transferrin that is recognized by the monoclonal antibody OKT 9. *J. Biol. Chem.* 257:8516–22

Schneyer, L. H. 1970. Amiloride inhibition of ion transport in perfused excretory duct or rat submaxillary gland. *Am. J. Physiol.* 219:1050–55

Schwartz, A. L., Rup, D. 1983. Biosynthesis of the human asialoglycoprotein receptor. *J. Biol. Chem.* 258:11249–55

Schwartz, I. L., Shlatz, L. J., Kinne-Saffran, E., Kinne, R. 1974. Target cell polarity and membrane phosphorylation in relation to the mechanism of action of antidiuretic hormone. *Proc. Natl. Acad. Sci. USA* 71:2595–99

Sefton, B. M., Trowbridge, I. S., Cooper, J. A., Scolnick, E. M. 1982. The transforming proteins of Rous sarcoma virus, Harvey sarcoma virus and Abelson virus contain tightly bound lipid. *Cell* 31:465–74

Semenza, G. 1976. Small intestinal disaccharidases: their properties and role as sugar translocators across natural and artificial membranes. In *Membrane Transport, Enzymes of Biological Membrane*, Vol. 3, pp. 349–82. New York: Plenum

Sepulveda, F. V., Pearson, J. D. 1984. Deficiency in intercellular communication in two established renal epithelial cell lines (LLC-PK$_1$ and MDCK). *J. Cell Sci.* 66:81–93

Shaver, J. F., Stirling, C. 1978. Ouabain binding to renal tubules of the rabbit. *J. Cell Biol.* 76:278–92

Simmons, N. L., Brown, C. D. A., Rugg, E. L. 1984. The action of epinephrine on Madin-Darby canine kidney cells. *Fed. Proc.* 43:2225–29

Simons, K., Garoff, H. 1980. The budding mechanisms of enveloped animal viruses. *J. Gen. Virol.* 50:1–21

Simons, K., Warren, G. 1984. Semliki Forest virus: a probe for membrane traffic in the animal cell. *Adv. Protein Chem.* 36:79–132

Sleight, R. G., Pagano, R. E. 1983. Rapid appearance of newly synthesized phosphatidylethanolamine at the plasma membrane. *J. Biol. Chem.* 258:9050–58

Staneloni, R. J., LeLoir, L. F. 1982. The biosynthetic pathway of the asparagine-linked oligosaccharides of glycoproteins. *CRC Crit. Rev. Biochem.* 12:289–326

Steinman, R. M., Brodie, S. E., Cohn, Z. A. 1976. Membrane flow during pinocytosis. A stereologic analysis. *J. Cell Biol.* 68:665–87

Steinman, R. M., Mellman, I. S., Muller, W. A., Cohn, Z. A. 1983. Endocytosis and recycling of plasma membrane. *J. Cell Biol.* 96:1–27

Stern, C. D. 1984. A simple model for early morphogenesis. *J. Theor. Biol.* 107:229–42

Stevenson, B. R., Goodenough, D. A. 1984. Zonulae occludentes in junctional complex-enriched fractions from mouse liver: Preliminary morphological and biochemical characterization. *J. Cell Biol.* 98:1209–21

St. Hilaire, R. J., Hradek, G. T., Jones, A. L. 1983. Hepatic sequestration and biliary secretion of epidermal growth factor: evidence for a high-capacity uptake system. *Proc. Natl. Acad. Sci. USA* 80:3797–3801

Stirling, C. E. 1972. Radioautographic localization of sodium pumps in rabbit intestine. *J. Cell Biol.* 53:704–14

Stoner, L. C. 1979. Studies with amiloride on isolated distal nephron segments. In *Amiloride and Epithelial Sodium Transport*, ed. A. W. Cuthbert, G. M. Fanelli, A. Scriabine, pp. 51–60. Baltimore, MD: Urband & Schwarzenberg

Strauss, E. G. 1978. Mutants of Sindbis virus. III. Host polypeptides present in purified HR and ts 103 virus peptides. *J. Virol.* 28:466–74

Sugrue, S. P., Hay, E. D. 1981. Response of basal epithelial cell surface and cytoskeleton to solubilized extracellular matrix molecules. *J. Cell Biol.* 91:45–54

Sztul, E. S., Howell, K. E., Palade, G. E. 1985a. Biogenesis of the polymeric IgA receptor in rat hepatocytes. I. Kinetic studies of its intracellular forms. *J. Cell Biol.* 100:1248–54

Sztul, E. S., Howell, K. E., Palade, G. E. 1985b. Biogenesis for the polymeric IgA receptor in rat hepatocytes. II. Localization of its intracellular forms by cell fractionation. *J. Cell Biol.* 100:1255–61

Takemura, S., Omiro, K., Tanaka, K.,

Omori, K., Matsuura, S., et al. 1984. Quantitative immunoferritin localization of Na$^+$/K$^+$ ATPase on canine hepatocyte cell surface. *J. Cell Biol.* 99:1502–10

Takeuchi, J., Sobue, M., Shamoto, M., Yoshida, M., Sato, E., et al. 1977. Cell surface glycosaminoglyans of cell line MDCK derived from canine kidney. *Cancer Res.* 37:1507–12

Tartakoff, A. M. 1983. The confined function model of the Golgi complex: center for ordered processing of biosynthetic products of the rough endoplasmic reticulum. *Int. Rev. Cytol.* 85:221–52

Taub, M., Livingstone, M. 1981. The development of serum-free hormone supplemental media for primary kidney cultures and their use in examining renal functions. *Ann. NY Acad. Sci.* 372:406–20

Taub, M., Sato, G. H. 1979. Growth of kidney epithelial cells in hormone-supplemented serum-free medium. *J. Supramolec. Res.* 11:207–16

Taylor, Z., Emmanoeul, D. S., Katz, A. I. 1982. Insulin binding and degradation by luminal and basolateral tubular membranes from rabbit kidney. *J. Clin. Invest.* 69:1136–46

Thomas, P., Summers, J. 1978. The biliary secretion of circulating asialoglycoproteins in the rat. *Biochem. Biophys. Res. Comm.* 80:335–39

Tilney, L. 1983. Interactions between actin filaments and membranes give spatial organization to cells. *Modern Cell Biology*, Vol. 2, pp. 163–99. New York: Alan Liss

Timpl, R., Martin, G. R. 1982. Components of basement membranes. *Immunochemistry of the Extracellular Matrix*, ed. H. Furthmayr, Vol. II, pp. 119–50. Boca Raton, Fla: CRC

Turnheim, K., Frizzell, R. A., Schultz, S. G. 1978. Interaction between cell sodium and the amiloride-sensitive sodium entry step in rabbit colon. *J. Membr. Biol.* 39:233–56

U, H.-S., Saier, M. H., Ellisman, M. H. 1979. Tight junction formation is closely linked to polar redistribution of intramembranous particles in aggregating MDCK cells. *Exp. Cell Res.* 122:384–91

Ullrich, A., Bell, J. R., Chen, E. Y., Herrera, R., Petruzelli, L. M., et al. 1985. Human insulin receptor and its relationship to the tyrosine kinase family of oncogenes. *Nature* 313:756–61

Ullrich, A., Coussens, L., Hayflick, J. S., Dull, T. J., Gray, A., et al. 1984. Human epidermal growth factor receptor cDNA sequence and aberrant expression of the amplified gene in A431 epidermoid carcinoma cells. *Nature* 309:418–25

Unwin, P. N. T., Ennis, P. D. 1984. Two conformations of a channel forming membrane protein. *Nature* 307:609–13

Unwin, P. N. T., Zampighi, G. 1980. Structure of the junction between communicating cells. *Nature* 283:545–49

Valentich, J. D. 1981. Morphological similarities between the dog kidney cell line MDCK and mammalian cortical collecting tubule. *Ann. NY Acad. Sci.* 372:384–405

Valentich, J. D. 1982. Basal-lamina assembly by the dog kidney epithelial cell line MDCK. *Cold Spring Harbor Conf. Cell Proliferation* 9:567–79

van Meer, G., Fuller, S. D., Simons, K. 1985. Sorting of (glyco-?) lipids in epithelial cells: possible implications for tight junction structure. In *Transport and Secretion of Proteins*, ed. M.-J. Gething, J. Sambrook. Cold Spring Harbor, NY: Cold Spring Harbor Pub. In press

van Meer, G., Simons, K. 1982. Viruses budding from either the apical or the basolateral plasma membrane domain of MDCK cells have unique phospholipid compositions. *EMBO J.* 1:847–52

Volk, T., Geiger, B. 1984. A 135-Kd membrane protein of intercellular adherens junctions. *EMBO J.* 3:2249–60

von Bonsdorff, C.-H., Fuller, S., Simons, K. 1985. Apical and basolateral endocytosis in MDCK cells grown on nitrocellulose filters. Submitted for publication

von der Mark, R., Mollenhauer, J., Kühl, K., Bee, J., Lesot, H. 1984. Anchorins: a new class of membrane proteins, involved in cell-matrix interactions. In *The Role of Extracellular Matrix in Development*, ed. R. L. Trelstad, pp. 67–87. New York: Alan Liss

Walter, P., Gilmore, R., Blobel, G. 1984. Protein translocation across the endoplasmic reticulum. *Cell* 38:5–8

Ward, C. W., Elleman, T. C., Azad, A. A. 1982. Amino acid sequence of the pronase-released heads of neuraminidase subtype N2 from asian strain A/Tokyo/3/67 of influenza virus. *Biochem. J.* 207:207:91–95

Ward, C. W., Murray, J. M., Roxburgh, C. M., Jackson, D. C. 1983. Chemical and antigenic characterization of the carbohydrate side chains of an Asian (N2) influenza virus neuraminidase. *Virology* 126:370–75

Watterson, R. L. 1966. Structure and mitotic behavior of the early neural tube. In *Organogenesis*, ed. R. L. Haan, H. Ursprung, pp. 129–59. New York: Holt, Rinehart & Winston

Weibel, E. R. 1976. Stereological approach to the study of cell surface morphology. *6th Eur. Congr. Electron Microsc., Jerusalem*, pp. 6–9

Weiss, R. A., Bennett, P. L. P. 1980. Assembly of membrane glycoproteins studies by phenotypic mixing between mutants of vesicular stomatitis virus and retroviruses. *Virology* 100:252–74

White, J., Kielian, M., Helenius, A. 1983. Membrane fusion proteins of enveloped animal viruses. *Q. Rev. Biophys.* 16:151–95

Wicha, M. S., Lowrie, G., Kohn, E., Bagavandoss, P., Mahn, T. 1982. Extracellular matrix promotes mammary epithelial growth and differentiation in vitro. *Proc. Natl. Acad. Sci. USA* 79:3213–17

Wilson, E. B. 1896 (1966). The cell in development and inheritance. New York: Johnson Reprint 371 pp. (Reprint)

Wilson, P. D., Horster, M. F. 1983. Differential response to hormones of defined nephron and epithelia culture. *Am. J. Physiol.* 244:C166–74

Wirtz, K. W. A. 1974. Transfer of phospholipids between membranes. *Biochim. Biophys. Acta* 344:95–117

Wisher, M. H., Evans, W. H. 1975. Functional polarity of rat hepatocyte surface membrane: isolation and characterization of plasma membrane subfractions from the blood-sinusoidal, bile canalicular and contiguous surfaces of the hepatocyte. *Biochem. J.* 146:375–88

Witte, D. N., Baltimore, D. 1977. Mechanism of formation of pseudotypes between vesicular stomatitis virus and murine leukemia virus. *Cell* 11:505–11

Yaffe, M. P., Kennedy, E. P. 1983. Intracellular phospholipid and the role of phospholipid transfer protein in animal cells. *Biochemistry* 22:1497–1507

Yamashita, K., Hitoi, A., Matsuda, Y., Tsuji, A., Katunuma, N., et al. 1983. Structural studies of the carbohydrate moieties of rat kidney γ-glutamyltranspeptidase. An extremely heterogeneous pattern enriched with non-reducing terminal N-terminal acetylglucosamine residues. *J. Biol. Chem.* 258:1098–1107

Zavada, J. 1982. The pseudotypic paradox. *J. Gen. Virol.* 63:15–24

Zeligs, J. D., Wollman, S. H. 1979a. Mitosis in rat thyroid cells *in vivo*. I. Ultrastructural changes in cytoplasmic organelles during the mitotic cycle. *J. Ultrastruct. Res.* 66:53–77

Zeligs, J. D., Wollman, S. H. 1979b. Mitosis in rat thyroid epithelial cells in vivo. II. Centrioles and pericentriolar material. *J. Ultrastruct. Res.* 66:97–108

Zeligs, J. D., Wollman, S. H. 1979c. Mitosis in thyroid follicular cells in vivo. III. Cytokinesis. *J. Ultrastruct. Res.* 66:288–303

Ziomek, C. A., Johnson, M. H. 1980. Cell surface interaction induces polarization of mouse 8-cell blastomeres at compaction. *Cell* 21:935–42

Ann. Rev. Cell Biol. 1985. 1: 289–315

CHROMOSOME SEGREGATION IN MITOSIS AND MEIOSIS

Andrew W. Murray and Jack W. Szostak

Department of Molecular Biology, Massachusetts General Hospital, Boston, Massachusetts 02114

CONTENTS

INTRODUCTION

The faithful inheritance of genetic information depends on the orderly segregation of chromosomes in mitosis and meiosis. Mitotic segregation produces genetically identical daughter cells, while meiotic segregation produces cells that contain only one member of each chromosome pair that

289

0743–4634/85/1115–0341$02.00

was present in the parental cell. Chromosome segregation is an extremely accurate process: Chromosome loss in mitosis in yeast occurs once in every 10^5 divisions (1), while aberrations in meiotic chromosome segregation occur once in every 10^4 meioses in both yeast (2) and *Drosophila* (reviewed in 3).

Recent reviews have discussed mitosis with reference to mechanisms of chromosome movement (4–6), the role of membranes in the behavior of the mitotic spindle (7), and the structure and behavior of kinetochores (8). The genetic control of meiosis has been reviewed for yeast (9), *Drosophila* (3), and other organisms (10). Ultrastructural aspects of meiosis have also been reviewed (11, 12). We do not attempt to review the voluminous literature on all aspects of mitosis and meiosis here, but concentrate on the fundamental problem of chromosome segregation: the nature of the mechanisms that ensure that sister chromatids (in mitosis and the second meiotic division) or homologous chromosomes (in meiosis I) segregate to opposite poles of the spindle. We emphasize studies on genetically tractable organisms, especially baker's yeast (*Saccharomyces cerevisiae*). We use the term kinetochore to describe the structural components that are assembled on the centromeric DNA, reserving the term centromere for the DNA itself. We use the term spindle pole body (SPB) to refer to the general concept of spindle pole organizing activity, although the morphological SPB is found only in fungi. In animal cells the centrosome is the microtubule organizing center.

Mitosis and meiosis have been studied in many organisms using a wide variety of techniques. We have tried to integrate our discussion by considering the experimental evidence in the context of a specific model for chromosome segregation. The model rests on a synthesis of ideas that have all been put forward at least once by other workers. Six central concepts underlie the model:

1. Normal chromosome segregation is dependent on the integrity of microtubules (reviewed in 4 and 5).
2. Separate mechanisms exist that can generate force on chromosomes either towards or away from the spindle pole to which they are attached (reviewed in 4).
3. The stable attachment of a chromosome pair to the spindle requires the generation of forces that pull the two kinetochores of that chromosome pair towards opposite poles of the spindle, creating tension on a flexible linkage between the two chromosomal units to be segregated (13, 14).
4. The replication of DNA molecules can generate linkage of the daughter DNA molecules in a chromosome by intertwining (catenation) of the DNA duplexes (17). Catenation can be destroyed during anaphase by the activation of type II DNA topoisomerases, which catalyze the passage of one DNA duplex through another (reviewed in 18).

5. In most organisms normal segregation of homologs at meiosis I is dependent on genetic recombination (3, 15).
6. The difference between the behavior of sister chromatids in meiosis I from that in mitosis and meiosis II is the result of differences in the organization of the chromosomes rather than that of the spindle (16).

A MODEL FOR CHROMOSOME BEHAVIOR DURING MITOSIS AND MEIOSIS

The following model is highly speculative but has the advantages of providing a unified interpretation of a wide variety of data and of making testable predictions.

The Mitotic Cell Cycle

INTERPHASE   Our model for the mitotic cell cycle is shown in Figure 1. At the beginning of the cell cycle each chromosome has a single kinetochore. We follow Sundin & Varshavsky in proposing that the recognition that a given pair of kinetochores are sisters is a result of the flexible linkage between them produced by the intercatenation of sister chromatids (19). Intercatenation arises when duplex DNA is replicated in the absence of topoisomerase activity. Under these conditions each double-helical turn of the parental duplex is converted into one intertwining of the daughter molecules (Figure 2). Sundin & Varshavsky proposed that linkage arises at the points where replication forks meet as a result of the steric exclusion of DNA topoisomerase by the replication complexes (17).

If intercatenation is stable from S to mitosis then each chromatid should be catenated to its sister at each of the many points where replication forks met. Alternatively, most intercatenation could be resolved during G2. In this case the catenation that we propose links sister chromatids together at metaphase could be generated by the late replication of the centromeric DNA (19). Tschumper & Carbon proposed that the yeast centromere acts as a block to the passage of replication forks (20). Release of this block at prophase would allow replication of the centromeric DNA with retention of sister chromatid linkage as a local catenation of the DNA at the centromere.

METAPHASE   Our proposal for the mechanics of chromosome movement during mitosis is based on a recent review by Pickett-Heaps et al (4). Microtubules in the two halves of the spindle have opposite polarities, and all those within one half have the same polarity (21, 22). Because of the observed bipolarity of the spindle there are no fundamental difficulties in inventing schemes that will supply the vectorial chromosome movement required at anaphase.

Pickett-Heaps et al (4) suggested that the kinetochores can interact with microtubules in two orientations. In the polar (P) orientation the kineto-chore is pulled towards the spindle pole from which the microtubule originates. We suggest that attachment in the P orientation acts as a molecular ratchet: As long as the P force is opposed by a force acting in the opposite direction attachment is stable. If the kinetochores of two sister chromatids are attached in P orientation to microtubules from opposite poles of the spindle, they will move towards the poles to which they are

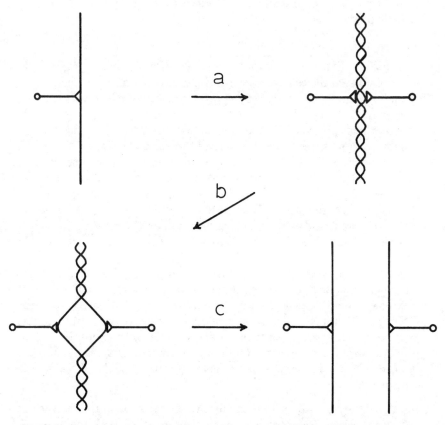

Figure 1 A model for mitotic chromosome segregation. (*a*) DNA replication leads to two sister chromatids attached to each other by catenation (each sister chromatid is represented by a single line). (*b*) Metaphase: stable attachment to the spindle is achieved when the kinetochores are attached to opposite poles. (*c*) Anaphase: DNA topoisomerase activity decatenates the sister chromatids and allows them to move to opposite poles of the spindle. Symbols: *triangle* = kinetochore; *circle* = spindle pole.

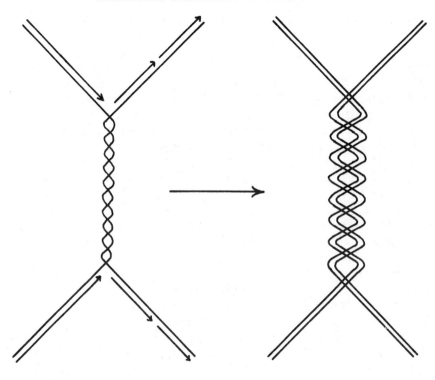

Figure 2 Generation of catenation. When two replication forks meet the last 10–20 turns of the DNA helix are converted into intertwinings of the daughter DNA duplexes.

attached until the linkage between them is stretched so that the P force on one kinetochore is opposed by that acting in the opposite direction on its sister. Attachment of a kinetochore to a microtubule in the opposite polarity yields the antipolar (AP) orientation. In this orientation the kinetochore generates force that moves it away from the pole from which the microtubule originates. Because all attachments at metaphase are in the P orientation, attachment in the AP orientation must be unstable.

ANAPHASE We propose that during metaphase the separation of sister chromatids is prevented by their catenation. We suggest that a feedback mechanism exists that activates DNA topoisomerase II only after all the kinetochores are attached to microtubules in the P orientation. The activation of topoisomerase marks the transition from metaphase to anaphase, as it is the destruction of the catenation between sister chromatids that allows them to separate and move polewards under the influence of the P force.

Meiosis

We argue that the meiotic cell cycle represents a mitotic cell cycle modified by the addition of a second duplication of the SPB without an intervening S phase, allowing two rounds of chromosome segregation to occur in a single cell cycle. Specifically, we propose that kinetochore duplication and centromere replication is delayed until the completion of meiosis I. During meiosis II the centromere DNA is replicated, leaving the kinetochores attached to each other by intercatenation, exactly as they are in mitotic metaphase.

We propose that the proper disjunction of homologs in meiosis I also requires intercatenation between segregating chromosomes, and that this linkage arises as the result of recombination. If the sister chromatids that make up a homolog are catenated at many points as a result of the collision of replication forks in meiotic S phase, then any reciprocal cross-over (but not gene conversion events) between nonsister chromatids will generate linkage of the homologs (Figure 3). After recombination each recombinant chromatid will be catenated to its sister centromere proximal to the point of recombination, and to its nonsister distal to the site of cross-over. (We define the sister of a recombinant chromatid as being the chromatid with which it shares a kinetochore in meiosis I.)

If homologs are linked by intercatenation, this linkage must be destroyed by the activation of a type II topoisomerase at the metaphase-anaphase transition of meiosis I. The topoisomerase activation will also destroy the intercatenation of sister chromatids. We propose that the linkage is regenerated at meiosis II as the result of replication of the centromeric DNA after the completion of meiosis I.

The delayed replication of the kinetochore's structural elements and the centromeric DNA means that at meiosis I the homologs each have only one kinetochore and no separation of sister chromatids can occur (Figure 3). In this scheme kinetochore duplication commits the cell to a mitotic division.

Two modifications are required to allow meiosis to evolve from mitosis: (a) introduction of a precocious SPB duplication that allows meiosis I to occur before the mitosis-like meiosis II; and (b) the suppression of kinetochore duplication until after the completion of meiosis I. If the regulation of SPB duplication and spindle assembly in meiosis I is different from that in meiosis II and mitosis we expect to find mutations that block the assembly of the meiosis I spindle, leading to meiotic divisions in which recombination is normal but meiosis I fails to occur. Such mutants are in fact known (83).

MEIOSIS I

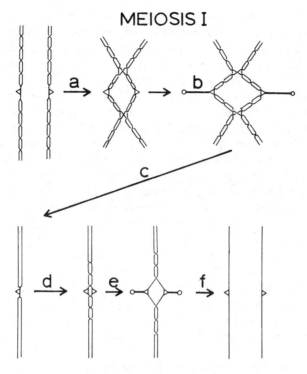

MEIOSIS II

Figure 3   A model for chromosome segregation in meiosis. In meiosis I two homologous pairs of sister chromatids are shown. Each sister chromatid pair has a single kinetochore and the centromeric DNA has not been replicated. (*a*) Recombination leads to the catenation of sister and nonsister chromatids, providing a flexible linkage between the two pairs of sister chromatids. (*b*) Metaphase I: stable attachment to the spindle occurs when the kinetochores are attached to opposite poles of the spindle. (*c*) Anaphase I: DNA topoisomerase destroys the catenation, allowing the sister chromatid pairs to segregate to opposite poles of the spindle. (*d*) Replication of the centromeric DNA generates a pair of catenated sister chromatids (only one of the two products of meiosis I is shown). (*e*) Metaphase II; (*f*) Anaphase II: DNA topoisomerase decatenates the sister chromatids allowing them to segregate to opposite poles of the spindle. Symbols: *triangle* = kinetochore; *circle* = spindle pole.

EXPERIMENTAL EVIDENCE

The Mitotic Cell Cycle

TIME OF KINETOCHORE DUPLICATION   We suggest that structural components of the kinetochore, but not the centromeric DNA, are duplicated during G1, although the timing of kinetochore duplication is not essential to the integrity of the model. There is no direct assay available for kinetochore duplication. Immunofluorescence studies on mammalian cells using antikinetochore sera from patients suffering from autoimmune diseases show that the kinetochores appear as single structures in most interphase cells (25, 26). In a small fraction of G2 cells the kinetochores appear as pairs of closely apposed dots (26). There are approximately the same number of paired dots as there are chromosomes in the cell line, suggesting that the staining reveals pairs of sister kinetochores.

In the fungus *Saproglenia*, kinetochores are visible throughout the cell cycle and are always associated with microtubules (27). Electron microscopy reveals that the morphological duplication of the kinetochore occurs immediately before mitosis (27). However, it is possible that the structural components of the kinetochore actually duplicate much earlier in the cell cycle, but that the morphological duplication is not visible until the centromeric DNA replicates during G2. Because the kinetochores of *S. cerevisiae* are not visible, even in electron micrographs, we do not know if they remain associated with microtubules throughout metaphase, although nuclear microtubules are present in this period (reviewed in 11). Immunofluorescence of yeast cells with antitubulin antisera does not show extensive microtubule arrays in the interphase nucleus (28), suggesting that in interphase the microtubules may extend only a short distance from the SPB.

There is no compelling reason to argue against the duplication of the structural components of the kinetochore in S or in early G2, although it is interesting to note that during mitosis of cells in which endoreduplication has occurred the sister chromatids that arise from the extra round of DNA replication fail to segregate from each other (29). This suggests that this pair of DNA duplexes shares a single kinetochore, and thus that kinetochore duplication does not occur during S phase. The yeast cell cycle mutant *cdc31* may also be defective in kinetochore duplication. At the nonpermissive temperature the duplication of the spindle pole body, which normally occurs in G1, is blocked (30). However, the SPB increases in size, and the processes of DNA replication and the initial stages of spindle formation occur as normal (30). The cells arrest in nuclear division, and electron microscopy shows that a normal spindle forms that has a SPB only at one end (30). This raises obvious questions about the role of the SPB in spindle formation. When *cdc31* cells are grown at a semipermissive temperature their ploidy increases (31). This suggests that under these conditions sister chromatids fail to separate, so that all the chromosomes go to a single pole. We suggest that the *CDC31* gene product is required for kinetochore as well as SPB duplication, consistent with a simultaneous duplication of the SPB and kinetochores in G1. One unresolved question is whether it is the mother or daughter cell that receives the chromosomes when a *cdc31* strain divides at the semipermissive temperature.

Another yeast gene product that may be required for kinetochore duplication is that of the *NDC1* gene. In cold-sensitive *ndc1* mutants the ploidy of the cells increases when they are grown at semipermissive temperatures. At the nonpermissive temperature pairs of cells accumulate in which one of the progeny inherits all of the DNA, as judged by staining with the DNA-specific dye DAPI, although the spindle appears normal by antitubulin immunofluorescence. The cell that inherits the DNA can be either the mother or the daughter cell (J. Thomas & D. Botstein, personal communication). The behavior of this mutant is compatible with the possibility that lack of the functional *NDC1* gene product blocks kineto-chore duplication so that sister chromatids fail to separate.

Kinetochores may remain attached to one of the SPBs throughout interphase. This would provide an attractive explanation for the cosegregation of chromatids whose oldest strand is of the same age, which has been observed in yeast (32), *Neurospora* (33), and some mammalian cells (34). If kinetochore duplication was conservative such that the older kinetochores remained attached both to the oldest DNA strands and to one of the two SPBs, then all the old kinetochores would segregate to the same spindle pole, carrying with them the chromatids that contain the oldest DNA strands. In yeast the mother and daughter cells were equally likely to inherit the old set of chromatids (32), which suggests that the SPB attached to the old kinetochores is equally likely to be inherited by either cell; this is consistent with our explanation of the behavior of the *ndc1* mutant.

CENTROMERIC DNA REPLICATION   The time at which different chromosomal regions replicate has been investigated in plant and animal cells. In general the centromeres and telomeres are late replicating (reviewed in 35). However, since these regions are also the sites of satellite DNA sequences (36), which are often late replicating (36), it is not possible to tell if the functional centromeres and telomeres are themselves late replicating.

DNA REPLICATION AND CATENATION   Sundin & Varshavsky (17) demonstrated that the replication of the animal virus SV40 produced catenated dimers. They suggested that sister chromatids could also become catenated as a result of DNA replication, and that this might help direct mitotic chromosome segregation (19). Topoisomerase II was shown to be required for the separation of bacterial chromosomes in vivo, and catenated chromosomes could be induced to separate in vitro by the action of topoisomerase (38). Would the same result be obtained with eucaryotic chromosomes isolated from metaphase cells?

In the yeast *top2* mutant, which carries a temperature-sensitive DNA topoisomerase II, cells arrest during nuclear division at the nonpermissive temperature (39). In addition, the 2 μm circle becomes highly intercatenated (39). No intercatenation of the circle is seen at the permissive temperature (39), which suggests that the half-life of catenated forms is short, and therefore that decatenation occurs prior to anaphase. If sister chromatid intercatenation, generated during S phase, survives from the end of S through metaphase, the removal of catenation by topoisomerase activity must occur much more slowly for long linear DNA molecules than it does for short circular ones.

By performing shift-up and shift-down experiments on strains carrying a temperature-sensitive *top2* mutation C. Holm & D. Botstein have shown that the activity of the *TOP2* gene product is required at mitosis (personal communication). These experiments do not reveal whether topoisomerase II is also active at other points in the cell cycle. In the absence of topoisomerase activity the cells arrest at mitosis, and cell death rapidly ensues. The death of these mitotic cells can be prevented by adding nocodazole (a drug that inhibits tubulin polymerization) before shifting cells to the nonpermissive temperature (C. Holm & D. Botstein, personal communication). This suggests that death is due to chromosome breakage or aneuploidy that occurs when cells attempt to enter anaphase with catenated sister chromatids.

When diploid yeast cells are treated for prolonged periods with the microtubule depolymerizing drug, methyl benzimadazol carbamate (MBC), a high fraction of aneuploid cells are seen in cells rescued on drug-free medium (40). One interpretation of this finding is that aneuploidy arises because the catenation between sister chromatids slowly decays in cells that are arrested in mitosis, allowing a defective anaphase separation process to occur. If this idea is correct, *top2* mutants should show a much lower frequency of aneuploidy when they are arrested in mitosis by MBC at the nonpermissive than at the permissive temperature.

At metaphase in diatoms the sister kinetochores lie close to their respective poles, separated from each other by about 5–10 μm (41). Thus, the proximity of the kinetochores cannot be what allows them to be recognized as sisters. The hypothesis that sister chromatid intercatenation normally holds daughter DNA molecules together at metaphase provides exactly the sort of flexible, long range connection this observation requires.

PROMETAPHASE CHROMOSOME MOVEMENT   Pickett-Heaps and his collaborators have made extensive studies of prometaphase chromosome movement in diatoms, which led to the concepts of P and AP force (reviewed in 4). From experiments using colchicine and drugs that prevent ATP generation they drew the following conclusions:

1. Attachment to the kinetochores in the P orientation leads to the generation of P force, and the movement of the kinetochores towards the pole to which the microtubule is attached. P movement does not require ATP and can occur in the absence of microtubules (42, 43).
2. Attachment of the kinetochores to microtubules in the AP orientation leads to the generation of AP force, and the movement of the kinetochore away from the spindle pole to which the microtubule is attached. AP movement requires ATP and intact microtubules (42, 43).
3. In prometaphase the initial movement of the kinetochores is always polewards. This suggests that if the kinetochores remain attached to microtubules during interphase, they do so in the P orientation (41).

Attachment of both kinetochores in the P orientation to opposite poles of the spindle does not necessarily specify the position of the chromosome on the spindle. The chromosome will come to rest only when the forces applied from opposite directions are equal. Ostergren (44) proposed that if the force on a kinetochore were proportional to its distance from the spindle pole then chromosomes would lie precisely between the poles at metaphase. This idea was tested by examining the position of trivalents (formed in spermatocytes from grasshoppers which had been treated with X-rays) at metaphase of meiosis I (45). At equilibrium the two kinetochores that face one pole must collectively generate no more force than the single kinetochore that faces the other pole. The kinetochore-to-pole distance for the single kinetochore was twice that for the two kinetochores that faced the opposite pole, as predicted by Ostergren (44). It is difficult to see how force generated at either the kinetochore or the spindle pole could be affected by the distance between them. This suggests either that P force is generated along the length of the microtubules, or alternatively that it represents some elastic component to which the kinetochore is attached, with the microtubule acting as the guide rather than the agent that communicates force to the kinetochore. The energy independence and the microtubule independence of P movement make the latter possibility attractive.

An example of an elastic protein is the contractile protein spasmin found in certain rotifers. The rotifers are carried on contractile stalks composed largely of spasmin, which changes from a highly extended to a globular conformation in response to an increase in the calcium concentration. This conformational change produces a rapid contraction of the stalk (46). Possibly a similar elastic protein generates the P force in the spindle.

MICROMANIPULATION OF METAPHASE CHROMOSOMES   The ability to micromanipulate chromosomes has made it possible to directly test the effect of stretching the interkinetochore linkage on chromosome orientation and segregation. So far these experiments have been possible only in meiotic cells. In an elegant series of experiments Nicklas and his colleagues micromanipulated the chromosomes of grasshopper primary spermatocytes in metaphase of meiosis I (13, 14, 47). The formation of chiasmata links the kinetochores of a pair of homologs, with the chromosome arms forming the linkage. By detaching one of the kinetochores of a pair of homologs from the spindle and moving it into the opposite half of the spindle, both kinetochores were attached to the same spindle pole (unipolar orientation). After this operation both kinetochores moved towards the pole to which they were attached, until one kinetochore detached from the spindle. The detached kinetochore would reattach at random to fibers from either spindle pole. Reattachment that maintained the unipolar orientation that resulted from the original manipulation was unstable. Only when the kinetochore reattached to the pole from which it had originally been detached did the two kinetochores move towards opposite poles and the chromosome orientation become stable. Such reorientation events were no more likely to occur near the spindle poles than at the equator (14).

The reorientation could be prevented if the needle used for micromanipulation was inserted between the homologs, and used to apply force towards the spindle pole that was not attached to the kinetochores (15). The simplest interpretation of this result is that spindle attachment becomes stable only when the linkage between the kinetochores is stretched. The high frequency of reorientation seen in spermatocytes recovering from cold treatment could also be prevented by stressing the linkage between the kinetochores (47). In addition, univalents (either sex chromosomes in the heterogametic sex, or autosomes that have no chiasmata) have been observed to reorient continually during the first meiotic division (48).

The transition from metaphase to anaphase failed to occur in cells in which one chromosome had been detached from the spindle (13). This suggests that the cell is able to monitor the attachment of kinetochores to the spindle. As one example, the kinetochores could have a protein kinase activity that phosphorylates and thereby inactivates topoisomerase II. This activity might be expressed only in mitosis and be inhibited when the kinetochore is attached to the spindle in the P orientation. When all the kinetochores become attached in the P orientation, the inhibition of topoisomerase activity would be abolished and decatenation of sister chromatids would accompany the transition from metaphase to anaphase.

ANAPHASE   In our model, anaphase is initiated by the decatenation of the sister chromatids, which allows each chromatid to move polewards as a result of P force. Indirect arguments suggest that the magnitude of the poleward forces increase during anaphase. Dicentric chromosomes are broken in anaphase if the two kinetochores are attached to opposite poles of the spindle (49). In contrast, ring chromosomes that appear to be catenated in metaphase are not broken, which suggests that the magnitude of the polewards forces is smaller than in anaphase. At anaphase these ring chromosomes appear to pass through each other without breakage, an observation that strongly suggests the action of DNA topoisomerases (50). The fate of dicentrics suggests that chromosomes linked by catenation would be broken if cells made the transition to anaphase without decatenating sister chromatids. The force pulling sister chromatids apart could increase as a result of an increase in the magnitude of the P forces transmitted by the kinetochore-to-pole microtubules, or the elongation of the spindle mediated by interactions between the pole-to-pole microtubules, or both.

Force on the kinetochores does not seem to be required for the initial stages of sister kinetochore separation at anaphase. Sister chromatid separation occurs in cells treated with colchicine (51), in acentric fragments (51), and in monopolar spindles (52). These observations strongly suggest that some linkage between sister chromatids is destroyed as cells progress through mitosis. Our model postulates that the separation of sister chromatids is due to the activation of type II topoisomerase at the beginning of anaphase.

What mediates the transition between metaphase and anaphase, and the activation of topoisomerase? One attractive candidate is destruction of maturation promoting factor (MPF), an activity that was first identified in Xenopus by the ability of cytoplasm from hormonally activated oocytes to induce non-hormone-treated oocytes to pass through meiosis I and yield mature eggs (reviewed in 53). Since then MPF activity has been found in a wide variety of mitotic and meiotic cells. When injected into cells of early Xenopus embryos, MPF causes chromosome condensation, nuclear envelope breakdown, and spindle formation (54). Cells remain in metaphase until the level of MPF decays (55). Either MPF or some component required for its activation is destroyed at the end of mitosis, since new protein synthesis is required for the generation of MPF activity in the next cell cycle (56). The oscillation in MPF levels during the cell cycle probably represents a component of the cell cycle clock that is activated by fertilization in Xenopus eggs (57).

One way in which MPF could prolong metaphase is by acting either directly or indirectly to inhibit topoisomerase II activity. Perhaps attachment of all the kinetochores to the spindle in the P orientation activates a system that destroys MPF activity.

ARTIFICIAL CHROMOSOMES   The availability of cloned replicators (58), telomeres (59, 60), and centromeres (61, 62) in yeast has opened up a new approach to chromosome structure and behavior: the construction in vitro of molecules that contain all the elements known to be required for normal chromosome function. These artificial chromosomes are then introduced into yeast cells to examine their segregation in mitosis and meiosis (63, 64).

The first linear artificial chromosomes were 10–15 kilobases (kb) in length (63, 64). Surprisingly, they were present in many copies per cell, and were much less mitotically stable than circular centromeric plasmids, which were present at one copy per cell and were lost about once in every hundred mitoses (61, 62, 64). The behavior of these linear plasmids was consistent with random segregation at mitosis, rather than the ordered segregation of sister molecules characteristic of circular centromeric plasmids or natural chromosomes. The defect in segregation is not due to mutations in the centromere (64), interference between the centromeric and telomeric chromatin structures (64), or the failure of the centromeres to attach to the mitotic spindle. The last point was demonstrated by constructing linear plasmids that contained two centromeres. Like dicentric circular plasmids (65), these linear molecules are structurally unstable, and they yield rearrangements that delete one of the two centromeres (Murray & Szostak, unpublished). Such deletions are the result of mechanical breakage that occurs after the two kinetochores of a dicentric molecule have attached to opposite spindle poles (49). Since one of the centromeres was only 2.5 kb from the end of the linear dicentric plasmid, it seems unlikely that there is a minimum telomere-centromere separation that must be exceeded before spindle attachment can occur.

The short linear centromeric plasmids are less than 5% of the length of the smallest yeast chromosome [chromosome I is 300 kb (66)]. Larger artificial chromosomes (55 kb) show evidence of normal sister chromatid segregation: their copy number is one, and their mitotic loss frequency is about 10^{-2} (64). A further increase in length to 104 kb increases the stability by a factor of five, but such molecules are still at least one hundred–fold less stable than natural chromosomes (Murray & Szostak, unpublished). The 55 kb artificial chromosomes pair and segregate from each other normally in meiosis, although again the fidelity of segregation is substantially less than that of natural chromosomes (64).

DERIVATIVES OF NATURAL CHROMOSOMES   The ability to create directed alterations in the structure of the yeast genome (67, 68) has made it possible to test the effect of various structural changes on chromosome behavior.

The first type of alteration is created by integrating a linear telomere-bearing plasmid into a yeast chromosome (Figure 4a) (Murray & Szostak, unpublished). Plasmid integration is directed by cleavage with a restriction enzyme that cuts in a region of the plasmid that is homologous to the

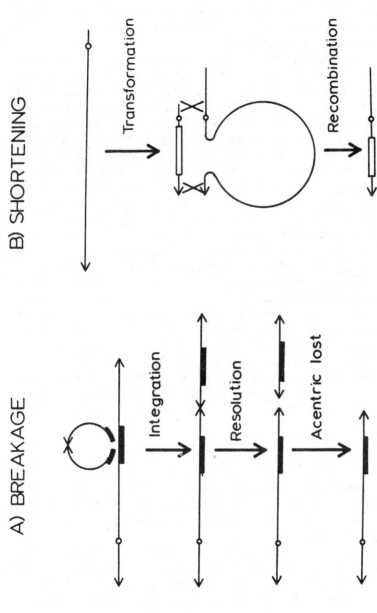

Figure 4   Manipulation of chromosome structure. (*a*) A chromosome can be broken and a new telomere generated by the directed introduction of a plasmid carrying an inverted repeat of telomeric sequences. *Dark bars* represent homologous sequences on the plasmid and chromosome; *arrows* telomeric DNA sequences. (*b*) A chromosome arm can be shortened by transformation with a linear DNA fragment whose ends are homologous to distant sites on the chromosome. *Open bars* represent fragment sequences that are not homologous to the chromosome.

desired site of integration. The chromosome is broken into two fragments at the site of integration. The fragment that carries the centromere is mitotically stable and is retained, while the acentric fragment is rapidly lost. The experiments are conducted in diploid strains so that the loss of a chromosome fragment is not lethal.

To test the effect of length on chromosome behavior we introduced telomeres at various points on yeast chromosome III. Centromeric chromosome fragments of 42 and 70 kb are lost mitotically at frequencies of 10^{-2} and 10^{-3} respectively, in comparison to chromosome III (530 kb) (66), which is lost at a frequency of 10^{-5}. Similar results have been obtained by Tye & Surosky (personal communication). For the smaller fragment, pedigree analysis has shown that 70% of the loss events occur by nondisjunction [the sister of the cell that lacks the chromosome fragment carries two copies (Murray & Szostak, unpublished)]. This is in marked contrast to the behavior of circular centromeric plasmids, in which only 25% of the events are due to nondisjunction (69, 70). The remaining events have only one copy in the sister cell that inherited the plasmid, which suggests that the plasmid failed to replicate in the cell cycle preceding loss, or that one copy of the plasmid was destroyed after replication. Increasing the size of circular centromeric plasmids from 5 to 100 kb causes a tenfold decrease in mitotic loss frequency from 10^{-2} to 10^{-3} (69). Unlike linear centromeric plasmids, the copy number of the circular centromeric molecules is one, irrespective of their size. A circular version of chromosome III is lost at a frequency of 10^{-3} (69, 71), and many of the events are due to sister chromatid exchange that creates a dicentric dimeric chromosome. Since sister chromatid exchange does not affect the structure of linear chromosomes, this may have been one of the selective forces favoring the evolution of linear chromosomes as genome size increased during evolution.

A number of results suggest that the length of DNA molecules is one of the main determinants of mitotic stability:

1. The mitotic loss frequencies of chromosome fragments and artificial chromosomes of similar sizes are within a factor of 5 of each other, despite the fact that most of the DNA on the artificial chromosomes is procaryotic (Murray & Szostak, unpublished).

2. Fragments of chromosome III of decreasing size show gradually decreasing stability, which demonstrates that there is no single element (apart from the centromere) whose presence is required for a high level of stability (R. Surosky & B.-K. Tye, personal communication; Murray & Szostak, unpublished).

3. The yeast mutant *chl1* increases the spontaneous frequency of

chromosome loss (72). In this mutant the frequency of chromosome loss decreases with increasing chromosome size (72).

4. In yeast zygotes where nuclear fusion is prevented by the *kar1* mutation, chromosomes are transferred between the two nuclei. The frequency of chromosome transfer decreases with increasing chromosome size (73).

The second type of chromosome manipulation is the creation of large internal deletions, by transforming with a DNA fragment one of whose ends is homologous to a site on the chromosome, and the other end of which is homologous to some distant chromosomal site (Figure 4*b*) (R. Surosky & B.-K. Tye, personal communication; Murray & Szostak, unpublished). A recombination event at both ends of the fragment replaces all the chromosomal information that lies between the two sites of homology with the sequences of the fragment. This technique has been used to produce an almost full length version of chromosome III whose centromere is only 3.5 kb from the nearest telomere. The telocentric chromosome is ten times less stable than the normal metacentric chromosome III (Murray & Szostak, unpublished). The telocentric chromosome is still one thousand times more stable than the 10–15 kb linear centromeric plasmids; this demonstrates that the primary cause of the plasmids' instability is not the closeness of their centromeres to their telomeres.

The results of studies on both artificial chromosomes and derivatives of natural yeast chromosomes suggest that the fundamental determinant of mitotic stability is the length of linear DNA molecules. This is consistent with the idea that catenation is an essential prerequisite for proper chromosome segregation. If the products of DNA replication are catenated circular molecules they are topologically linked to each other. In contrast, short linear molecules are not linked: The intertwinings can be resolved simply by twisting one molecule about the other, which explains why they, but not circular centromeric plasmids, segregate randomly at mitosis. To account for the ordered segregation of long artificial chromosomes we suggest that they can become linked to each other during replication. Such linkage requires the existence of topological barriers that constrain the catenation generated by replication. These barriers could be either attachment of the chromosome to the nuclear matrix, or the formation of chromosomal loops by DNA-binding proteins, which would noncovalently cross-link the DNA. Such domains of supercoiling have been seen in eucaryotic chromosomes, and it is intriguing that they are about 50 kb long (74), the size of the smallest artificial chromosome that shows ordered segregation at mitosis.

The length of DNA molecules may also affect their ability to become catenated due to the collision of replication forks. On a circular molecule a single origin of replication suffices to give catenation because the two forks arising from this origin will meet on the opposite side of the circle. For a linear molecule the forks will meet only if the plasmid has at least two replication origins, and both origins fire before either has been replicated by the fork emanating from the other origin. The chance of fulfilling the second criterion increases as the distance between the origins increases. In particular, linear molecules with a single replication origin should be incapable of generating intercatenation, and should show a very high frequency of nondisjunction. Such molecules have not yet been constructed.

Meiosis

PAIRING AND DISJUNCTION   All the available evidence suggests that homolog disjunction is the result of some form of chromosome pairing (reviewed in 3, 9, 12). In most organisms pairing proceeds through the formation of homologous pairs of chromosomes joined by the synaptonemal complex (SC). Pairing is initiated in zygotene when the SC begins to form. This process has been studied in yeast (75), insects (12), mammals (12), and plants (76) by examining nuclei reconstructed from serial section electron micrographs. It appears that SC formation is initiated at the telomeres, which are in contact with the nuclear envelope, and then proceeds towards the centromeres. However, the presence of telomeres is not essential for SC formation and recombination since diploids containing two circular copies of chromosome III show normal levels of recombination for markers on chromosome III (71). By the beginning of pachytene the SC covers the full length of each bivalent.

Two lines of evidence suggest that SC formation is an essential prerequisite for recombination in yeast: (a) In the cdc7 mutant, cells arrested at the nonpermissive temperature fail to form SC or commit to recombination, while both processes occur after shifting to the permissive temperature (77); and (b) arrest of wild-type yeast cells in pachytene by exposure to 36°C leads to marked increases in recombination (78). SC formation is not sufficient for recombination, since female silkmoths (Bombyx) form apparently normal SC but fail to undergo recombination (12). The genetic analysis of meiotic recombination and the different models invoked to explain it have recently been reviewed elsewhere (79).

In pachytene, small electron-dense structures are found associated with the SC, and these have been named recombination nodules (80) because (a) their number and distribution is roughly equal to the number and distribution of recombination events (75, 80); and (b) nodules are absent in organisms like female Bombyx that lack recombination, even if they have morphologically normal SC (12).

Although pairing appears to precede recombination, and recombination is dispensable in certain organisms [male *Drosophila* (reviewed in 3), female *Bombyx*], a number of lines of evidence suggest that recombination is normally involved in ensuring homolog segregation. Recombination results in the formation of chiasmata that act as points of linkage between homologs. As a result, the kinetochores of a pair of homologs are physically linked but can be widely separated unless the chiasmata lie close to the centromere. In both *Drosophila* (81) and yeast (E. Lambie & S. Roeder, personal communication) meiotic recombination near the centromere is suppressed. In yeast, moving the centromere of chromosome III from its normal location near the *LEU2* gene to the *HIS4* gene increases the frequency of recombination near *LEU2* and decreases it near *HIS4*; this demonstrates that the centromere itself is responsible for suppressing recombination (E. Lambie & S. Roeder, personal communication).

One interpretation of this suppression of recombination is that it ensures that the kinetochores are not too tightly linked, and makes it easy for the two kinetochores of a pair of homologs to capture microtubules from opposite poles of the meiotic spindle. To achieve stable bipolar orientation an unattached kinetochore must find a microtubule with a polarity opposite the one to which its sister is attached. A flexible interkinetochore linkage will allow the unattached kinetochore to scan a larger nuclear volume than if it were rigidly attached to its sister.

While it seems clear that chiasmata link homologous chromosomes that have recombined, the nature of the linkage is unknown. In maize there is an exact correlation between the number of chiasmata and the number of reciprocal exchanges (reviewed in 82). This suggests that recombination events that are resolved as cross-overs give rise to chiasmata, while those that are resolved as gene conversions do not. This observation implies that the sister chromatids of homologs are attached to each other along their length. Thus, a cross-over will leave the recombinant chromatids catenated to their sister's centromere proximal to the point of exchange, and to their nonsister's centromere distally (Figure 3). Gene conversion, which merely replaces the information on one chromatid with a short stretch from its nonsister, would not lead to this linkage.

As the homologs separate the chiasmata move toward the telomeres. The linkage of nonsister chromatids by catenation explains this phenomenon. Topoisomerase-mediated removal of intertwining at the chiasmata allows the homologs to move apart under the influence of the polewards force exerted at the kinetochores. However, removal of intertwinings at sites other than the chiasmata fails to lead to separation because the homologs remain linked at the chiasmata. Thus the chiasmata move progressively towards the ends of the chromosomes as if topoisomerase acted preferentially at the chiasmata.

YEAST MEIOTIC MUTANTS   The mutation *spo13* blocks the first meiotic division, and diploid spores are produced (83). These spores show normal levels of recombination, which demonstrates that this mutant has the expected phenotype for a failure of the "extra" SPB duplication invoked in the model for the meiotic cell cycle described previously. The duplication of SPBs in the mitotic cell cycle is dependent on the *CDC31* gene product. To explain the behavior of the *spo13* mutant we postulate that:

1. The *CDC31* gene product can induce both SPB and kinetochore duplication, while the *SPO13* gene product can induce SPB but *not* kinetochore duplication.

2. The normal mitotic cell cycle incorporates a feedback mechanism that prevents the SPBs from being duplicated twice between rounds of chromosome segregation.

3. The activity of the *SPO13* and *CDC31* gene products are regulated in a mutually exclusive fashion: *CDC31* is active in mitotically grown cells and in meiosis II, while *SPO13* is active only during meiosis I.

4. During meiosis I the continued activity of the *SPO13* gene product is dependent on maintenance of the conditions required to induce sporulation.

The expression of the *SPO13* gene product is known to be induced by sporulation medium (R. E. Esposito, personal communication).

According to our scenario, cells starting the cell cycle in sporulation media will have high levels of *SPO13* gene product, and will duplicate their SPBs, but not their kinetochores, which will lead to meiosis I. The completion of meiosis I will allow the *CDC31* function to become active, kinetochore and SPB duplication will occur, and meiosis II (which is mechanistically identical to mitosis) will follow. If cells are returned to growth conditions before meiosis I occurs the *CDC31* function will become active, the kinetochores will be duplicated, and meiosis I will be replaced by a mitotic division. The diploid products of this division will have suffered meiotic levels of recombination. Such cells are observed when sporulating cultures are returned to vegetative growth (24). Thus our model explains why cells cannot enter the meiotic cell cycle after SPB duplication, and how they can escape from the meiotic cell cycle before meiosis I.

The effects of the *cdc31* and *ndc1* mutants on meiosis have been examined. When strains mutant for *either* gene are put through meiosis at the nonpermissive temperature two diploid spores are recovered, and genetic analysis indicates that homologous chromosomes have separated at meiosis I, but that meiosis II and the segregation of sister chromatids has failed to occur (30, J. Thomas & D. Botstein, personal communication). This behavior strongly suggests that meiosis II is functionally equivalent to mitosis, and is consistent with the idea that these functions are required for kinetochore duplication, and that kinetochore duplication in meiosis does not occur until after the completion of meiosis I.

SPO13 (wild-type) cells that contain the recombination deficient mutations spo11, rad50, or rad52 produce inviable spores (reviewed in 9, 15). Electron microscopic studies show that the rad50 mutant lacks synaptonemal complex, while spo11 possesses it (B. Byers & R. E. Esposito, personal communication), which demonstrates that the formation of SC is not sufficient to ensure homolog disjunction in yeast. In the case of spo11 a small number of viable spores can be recovered, and these have not undergone meiotic recombination (84). The spo11, spo13 double mutant produces normal levels of viable diploid spores that have not undergone recombination (84). The presence of spo11 increases spore viability in spo13 diploids, which suggests that meiotic recombination events can interfere with chromosome segregation in the second (mitosis-like) division (84). Perhaps the linkage of homologs interferes with their correct orientation on the metaphase plate, or with the segregation of sister chromatids at anaphase. The spo13, rad50 double mutant also produces viable but unrecombined spores (15). This suggests that both RAD50 and SPO11 gene products are required to initiate meiotic recombination, and that the failure to do so leads to the failure of homolog disjunction at meiosis I, and thus to inviable spores.

DROSOPHILA MEIOTIC MUTANTS   In Drosophila a number of meiotic mutants that produce inviable and aneuploid gametes have been shown to be deficient for meiotic recombination (reviewed in 3). In female Drosophila those homologs (or unpaired chromosomes) that have not recombined enter a second pairing system, the distributive pairing pool, in which chromosomes disjoin from each other on the basis of size (reviewed in 85). The distributive pairing mechanism is apparently sufficient to deal with about four unpaired chromosomes. Higher numbers lead to its breakdown, and very high frequencies of nondisjunction (3). In male Drosophila, which lack homologous pairing, recombination, and distributive pairing, the homologs appear to pair via specific pairing sites in the heterochromatin (reviewed in 3).

The Drosophila meiotic mutant pal leads to chromosome loss in meiosis, and chromosome loss in the early divisions of the zygote (86). pal affects meiosis only in males, and the chromosomes that are lost from the zygote nucleus and its progeny are always the paternal chromosomes (86). In one stock the X chromosome was much more resistant to pal-induced loss. In crosses this resistance to loss maps to the centromeric heterochromatin of the X (86). We suggest that a wild-type product of the pal gene acts to suppress kinetochore assembly before meiosis I in males. The mutant alleles allow kinetochore duplication to occur before meiosis I; therefore, the newly assembled kinetochores are defective because some substance or condition required for kinetochore assembly is absent at this stage of meiosis. In meiosis II this would lead to chromosomes with one normal and one defective kinetochore, and a high frequency of nondisjunction.

Simultaneous loss of both chromosomes X and 4 is more frequent than expected from the product of the frequencies of exceptional segregation for either one alone (86). The behavior of the two chromosomes is correlated. In the case of early mitotic loss both chromosomes are almost always lost from the same pole in division (86). This behavior is reminiscent of the yeast *ndc1* mutant in that the chromosomes that nondisjoin all arrive at the same pole in mitosis and meiosis. However, in the *ndc1* mutant either all or none of the chromosomes in a cell nondisjoin and segregate to a single pole, while in the *pal* mutant some of the chromosomes disjoin and others segregate normally. The correlated behavior of chromosomes could be explained if the defective kinetochores assembled in the meiotic cell cycle always face the same pole in subsequent divisions. One way in which such coorientation can be achieved is to attach all the kinetochores of the same age to the same pole. If our interpretation is correct, the behavior of these mutants argues that kinetochores of the same age are all attached to one pole in meiosis II as well as in mitosis.

Goldstein (87) has shown that in wild-type males at prometaphase of meiosis I the kinetochore of each homolog is a single hemispherical structure. By late metaphase the kinetochore has metamorphosed into a double structure that consists of two discs lying side by side (87). While these two discs could be functionally independent they have the same orientation, and thus will attach to microtubules of the same polarity, so there will be no tendency of sister chromatids to separate. We predict that electron microscopy of meiosis in *pal* males would disclose the occurrence of kinetochore duplication before meiosis I.

Are there meiosis-specific genes that prevent sister kinetochore separation before metaphase of meiosis II? Defects in such genes would be expected to give rise to precocious separation of sister chromatids, which would be scored as nondisjunction at either meiosis I or meiosis II, depending on when separation occurred. Two mutants with this property have been described in *Drosophila*, *mei-S322* (88) and *ord* (89). They are unique among meiotic mutants in that they act in both males and females, and both have been cytologically observed to cause the precocious separation of sister chromatids in male meiosis (90). The fraction of nondisjunction events that occurs at meiosis II is 95% and 30% for *mei-S322* and *ord*, respectively. For *mei-S322* the frequency of gametes that carry no copies of a given chromosome is much greater than that of gametes that carry two copies of the same chromosome, which suggests that some chromosomes are not included in the nuclei that form after the completion of meiosis.

MEIOSIS I VERSUS MEIOSIS II   Sister chromatids separate from each other in mitosis and meiosis II, but not in meiosis I. Does this reflect differences in the structure of the meiosis I and II spindles, or is it a reflection of the organization of the chromosomes themselves? One test between these alternatives is provided by an elegant experiment of Nicklas (16). He induced primary and secondary spermatocytes to fuse with each other to produce a single cell that had both meiosis I and meiosis II spindles. He then transferred either a bivalent from the meiosis I to the meiosis II spindle, or a pair of sister chromatids from the meiosis II to the meiosis I spindle. In both cases the transferred entities behaved as they would have done on the spindle they originally belonged to: The bivalent separated into two homologs, while the sister chromatids from meiosis II separated from each other (16). This result is entirely consistent with our argument that kinetochore duplication does not occur until the interval between meiosis I and II. Homologs transferred from the meiosis I spindle would have a single kinetochore, while those transferred from the meiosis II to meiosis I spindle would have duplicated kinetochores, allowing the sister chromatids to disjoin from each other.

EVOLUTION OF MEIOSIS

Our model provides a unified view of the mitotic and meiotic cell cycles, and suggests how meiosis evolved from mitosis. In the primitive forms of meiosis, disjunction was probably ensured by synaptonemal complex formation without recombination. Although in mitosis the chromosomes were flexibly attached to each other by intercatenation, the rigid association of homologs in meiosis mediated by the SC probably led to a high frequency of nondisjunction. The introduction of recombination to create a flexible linkage between homologs would have increased the fidelity of chromosome segregation. Thus the initial selection for the evolution of meiotic recombination may have been for a decrease in nondisjunction, rather than an increase in the potential for genetic diversity in the succeeding generation.

ACKNOWLEDGMENT

Because of space limitations we were unable to quote every reference to a given topic. To those whose names are not mentioned we offer our apologies. We are extremely grateful to all those who provided information before publication. We thank D. Dawson, P. Szauter, and B. Alberts for their helpful comments on the manuscript, and T. Claus for photography.

Literature Cited

1. Hartwell, L. H., Dutcher, S. K., Wood, J. S., Garvik, B. 1982. The fidelity of mitotic chromosome reproduction in *S. cerevisiae. Rec. Adv. Yeast Mol. Biol.* 1:28–38

2. Sora, S., Luchini, G., Magni, G. E. 1982. Meiotic diploid progeny and meiotic non-disjunction in Saccharomyces cerevisiae. *Genetics* 101:17–33

3. Baker, B. S., Hall, J. C. 1976. Meiotic mutants: genic control of meiotic recombination and chromosome segregation. In: *Genetics and Biology of Drosophila,* Vol. I, ed. E. Novitski, M. Ashburner, pp. 351–429. New York: Academic

4. Pickett-Heaps, J. D., Tippit, D. H., Porter, K. R. 1982. Rethinking Mitosis. *Cell* 29:729–44

5. Inoue, S. 1981. Cell division and the mitotic spindle. *J. Cell Biol.* 91:131s–47s

6. Cabral, F. 1984. Genetic dissection of the assembly of microtubules and their role in mitosis. *Cell. Muscle Motil.* 5:313–40

7. Hepler, P. K., Wolniak, S. M. 1984. Membranes in the mitotic apparatus: their structure and function. *Int. Rev. Cytol.* 90:169–238

8. Rieder, C. L. 1982. The formation, structure and composition of the mammalian kinetochore and kinetochore fiber. *Int. Rev. Cytol.* 79:1–58

9. Esposito, R. E., Klapholz, S. 1981. Meiosis and ascospore development. In *Molecular Biology of the Yeast Saccharomyces, Life Cycle and Inheritance,* ed. J. Strathern, E. W. Jones, J. R. Broach, pp. 211–88. Cold Spring Harbor, New York: Cold Spring Harbor Lab.

10. Baker, B. S., Carpenter, A. T. C., Esposito, M. S., Esposito, R. E., Sandler, L. 1976. The genetic control of meiosis. *Ann. Rev. Genet.* 10:53–134

11. Byers, B. 1981. Cytology of the yeast life cycle. See Ref. 9, pp. 59–96

12. Rasmussen, S. W., Holm, P. B. 1980. Mechanics of meiosis. *Heriditas* 93:187–216

13. Nicklas, R. B. 1967. Chromosome micromanipulation II. Induced reorientation and the experimental control of segregation in meiosis. *Chromosoma* 21:17–57

14. Nicklas, R. B., Koch, C. A. 1969. Chromosome micromanipulation III.

Spindle fiber tension and the reorientation of maloriented chromosomes. *J. Cell Biol.* 43:40–50

15. Malone, R. E., Easton-Esposito, R. 1981. Recombinationless meiosis in Saccharomyces cerevisiae. *Mol. Cell. Biol.* 1:891–901

16. Nicklas, R. B. 1977. Chromosome distribution: experiments on cell hybrids and in vitro. *Philos. Trans. R. Soc. London B* 277:267–76

17. Sundin, O., Varshavsky, A. 1980. Terminal stages of SV40 DNA replication proceed via multiply intertwined catenated dimers. *Cell* 21:103–14

18. Gellert, M. 1981. DNA topoisomerases. *Ann. Rev. Biochem.* 50:879–910

19. Sundin, O., Varshavsky, A. 1981. Arrest of segregation leads to accumulation of highly intercatenated dimers: dissection of the final stages of SV40 DNA replication. *Cell* 25:659–69

20. Tschumper, G., Carbon, J. 1983. Copy number control by a yeast centromere. *Gene* 23:221–32

21. Telzer, B. R., Haimo, L. T. 1981. Decoration of spindle microtubules with dynein: evidence for uniform polarity. *J. Cell Biol.* 89:373–78

22. Euteneur, U., McIntosh, J. R. 1981. Structural polarity of kinetochore microtubules in PtK1 cells. *J. Cell Biol.* 89:338–45

23. Deleted in proof

24. Sherman, F., Roman, H. 1963. Evidence for two types of allelic recombination in yeast. *Genetics* 48:255–61

25. Moroi, Y., Hartman, A. L., Nakane, P. K., Tan, E. M. 1981. Distribution of kinetochore (centromere) antigen in mammalian cell nuclei. *J. Cell Biol.* 90:254–59

26. Brenner, S., Pepper, D., Berns, M. W., Tan, E., Brinkley, B. R. 1982. Kinetochore structure duplication and distribution in mammalian cells: Analysis by autoantibodies from scleroderma patients. *J. Cell Biol.* 91:95–102

27. Heath, I. B. 1980. Behavior of the kinetochores in the fungus Saproglenia ferax. *J. Cell Biol.* 84:531

28. Kilmartin, J. V., Adams, A. E. M. 1984. Structural rearrangements of actin and

tubulin during the cell cycle of the yeast *Saccharomyces. J. Cell Biol.* 98:922–33

29. Brenner, S., Branch, A., Meredith, S., Berns, M. W. 1977. The absence of centrioles from spindle poles of rat kangaroo (PtK2) cells undergoing meiotic like reductional division in vitro. *J. Cell Biol.* 72:368–79

30. Byers, B. 1981. Multiple roles of the spindle pole bodies in the life cycle of *Saccharomyces cerevisiae.* In *Molecular Genetics in Yeast. Alfred Benzon Symp. 16,* ed. D. von Wettstein et al, pp. 119–34. Copenhagen: Munksgaard

31. Schild, D., Ananthaswarmy, H. N., Mortimer, R. K. 1981. An endomitotic effect of a cell cycle mutation of *Saccharomyces cerevisiae. Genetics* 97:551–62

32. Williamson, D. H., Fennell, D. J. 1981. Non-random assortment of sister chromatids in yeast mitosis. See Ref. 30, pp. 89–102

33. Rosenburger, R. F., Kessel, M. 1968. Non-random sister chromatid segregation and nuclear migration in hyphae of *Aspergillus nidulans. J. Bacteriol.* 96:1208–13

34. Lark, K. G., Consigli, R. A., Minocha, H. C. 1966. Segregation of sister chromatids in mammalian cells. *Science* 154:1202–4

35. Back, F. 1976. The variable condition of euchromatin and heterochromatin. *Int. Rev. Cytol.* 45:25–64

36. Appels, R., Peacock, W. J. 1978. The arrangement and evolution of highly repeated (satellite) DNA sequences with special reference to *Drosophila. Int. Rev. Cytol. Suppl.* 8:70–127

37. Deleted in proof

38. Steck, F. R., Drilca, K. 1984. Bacterial chromosome segregation: evidence for DNA gyrase involvement in decatenation. *Cell* 36:1081–88

39. DiNardo, S., Voelkel, K. A., Sternaglanz, R. L. 1984. DNA topoisomerase mutant of S. cerevisiae: topoisomerase II is required for segregation of daughter molecules at the termination of DNA replication. *Proc. Natl. Acad. Sci. USA* 81:2616–20

40. Wood, J. S., Hartwell, L. H. 1982. A dependent pathway of gene functions leading to chromosome segregation in *Saccharomyces cerevisiae. J. Cell Biol.* 94:718–26

41. Tippet, D. H., Pickett-Heaps, J. D., Leslie, R. L. 1980. Cell division in two large pennate diatoms *Hantzchia* and *Nitzchia* III. A new proposal for kinetochore function during prometaphase. *J. Cell Biol.* 86:402–16

42. Pickett-Heaps, J. D., Spurck, T. P. 1982.

Studies on kinetochore function in mitosis. II. The effects of metabolic inhibitors on mitosis in the diatom *Hantzchia amphioxus. Eur. J. Cell Biol.* 28:83–91

43. Pickett-Heaps, J. D., Spurck, T. P. 1982. Studies on kinetochore function in mitosis. I. The effects of cytochalasin and colchicine on mitosis in the diatom *Hantzchia amphioxus. Eur. J. Cell Biol.* 28:77–82

44. Ostergren, G. 1950. Considerations on some elementary features of mitosis. *Heriditas* 36:1–19

45. Hays, T. S., Wise, D., Salmon, E. D. 1982. Traction force on a kinetochore fiber acts as a linear function of kinetochore fiber length. *J. Cell Biol.* 93:374–82

46. Amos, W. B., Routledge, L. L., Yew, F. F. 1975. Calcium binding proteins in a vorticellid contractile organelle. *J. Cell Sci.* 19:203–13

47. Henderson, S. A., Nicklas, R. B., Koch, C. A. 1970. Temperature induced orientation instability during meiosis: an experimental analysis. *J. Cell Sci.* 6:323–50

48. Nicklas, R. B. 1961. Recurrent pole to pole movements of the sex chromosome of *Melanoplus differentialis* in spermatocytes. *Chromosoma* 12:97–115

49. McClintock, B. 1938. The production of homozygous deficient tissues with mutant characteristics by means of the aberrant mitotic behavior of ring shaped chromosomes. *Genetics* 23:315–76

50. Bajer, A., Ostergren, G. 1963. Observations on transverse movement within the phragmoblast. *Hereditas* 50:179–95

51. Bajer, A., Mole-Bajer, J. 1963. Cine analysis of some aspects of mitosis in endosperm. In *Cinematography in Cell Biology,* ed. G. Rose, pp. 357–409. New York: Academic

52. Bajer, A. S. 1982. Functional anatomy of monopolar spindle and evidence for oscillatory movement in mitosis. *J. Cell Biol.* 93:33–48

53. Matsui, Y., Clarke, H. J. 1979. Oocyte maturation. *Int. Rev. Cytol.* 57:185–282

54. Miake-Lye, R., Newport, J. W., Kirschner, M. 1983. Maturation promoting factor induces nuclear envelope breakdown in cycloheximide arrested embryos of *Xenopus laevis. J. Cell Biol.* 97:81–91

55. Newport, J. W., Kirschner, M. 1984. Regulation of the cell cycle during early *Xenopus* development. *Cell* 37:731–42

56. Gerhart, J., Wu, M., Kirschner, M. 1984. Cell cycle dynamics of an M-phase specific cytoplasmic factor in Xenopus laevis oocytes and eggs. *J. Cell Biol.* 98:1247–55

57. Hara, K., Tydeman, P., Kirschner, M. 1980. A cytoplasmic clock with the same period as the division cycle in *Xenopus* eggs. *Proc. Natl. Acad. Sci. USA* 77:462–66

58. Struhl, K., Stinchcomb, D. T., Scherer, S., Davis, R. W. 1979. High frequency transformation of yeast: Autonomous replication of hybrid DNA molecules. *Proc. Natl. Acad. Sci. USA* 76:1035–39

59. Szostak, J. W., Blackburn, E. H. Cloning yeast telomeres on linear plasmid vectors. *Cell* 29:245–55

60. Szostak, J. W. 1982. Structural requirements for telomere resolution. *Cold Spring Harbor Symp. Quant. Biol.* 47:1187–94

61. Clarke, L., Carbon, J. 1980. Isolation of a yeast centromere and construction of functional small circular chromosomes. *Nature* 287:504–9

62. Stinchcomb, D. T., Mann, C., Davis, R. W. 1982. Centromeric DNA from *Saccharomyces cerevisiae. J. Mol. Biol.* 158:157–79

63. Dani, G. M., Zakian, V. A. 1983. Mitotic and meiotic stability of linear plasmids in yeast. *Proc. Natl. Acad. Sci. USA* 80:3406–10

64. Murray, A. W., Szostak, J. W. 1983. Construction of artificial chromosomes in yeast. *Nature* 305:189–93

65. Mann, C., Davis, R. W. 1983. Instability of dicentric plasmids in yeast. *Proc. Natl. Acad. Sci. USA* 80:228–32

66. Schwartz, D. C., Cantor, C. R. 1984. Separation of chromosome sized DNAs by pulsed field gradient gel electrophoresis. *Cell* 37:67–95

67. Orr-Weaver, T. L., Szostak, J. W., Rothstein, R. J. 1983. Genetic applications of yeast transformation with linear and gapped plasmids. *Methods Enzymol.* 101:228–45

68. Rothstein, R. J. 1983. One step gene disruption in yeast. *Methods Enzymol.* 101:202–11

69. Hieter, P., Mann, C., Snyder, M., Davis, R. W. 1985. Mitotic stability of yeast chromosomes: a colony assay that measures non-disjunction and chromosome loss. *Cell* 40:381–92

70. Koshland, D., Kent, J. C., Hartwell, L. H. 1985. Genetic analysis of the mitotic transmission of mini-chromosomes. *Cell* 40:393–403

71. Haber, J. E., Thornburn, P. C., Rogers, D. 1984. Meiotic and mitotic behaviour of dicentric chromosomes in *Saccharomyces cerevisiae. Genetics* 106:185–205

72. Liras, P., McCusker, J., Masciola, S., Haber, J. E. 1978. Characterization of a mutation in yeast causing non-random chromosome loss during mitosis. *Genetics* 88:650–71

73. Dutcher, S. K. 1981. Internuclear transfer of genetic information in kar1/KAR1 heterokaryons in *Saccharomyces cerevisiae. Mol. Cell. Biol.* 1:245–53

74. Banyajati, C., Worcel, A. 1976. Isolation, characterization and structure of the folded interphase chromosome of *Drosophila melanogaster. Cell* 9:393–407

75. Byers, B., Goetsch, L. 1975. Electron microscopic observations on the karyotype of diploid and tetraploid *Saccharomyces cerevisiae. Proc. Natl. Acad. Sci. USA* 72:5056–60

76. Gillies, C. B. 1984. The synaptonemal complex in higher plants. *Crit. Rev. Plant. Sci.* 2:81–116

77. Schild, D., Byers, B. 1978. Meiotic effects of DNA-defective cell division cycle mutants of *Saccharomyces cerevisiae. Chromosoma* 70:109–30

78. Davidow, L. S., Byers, B. 1984. Enhanced gene conversion and post-meiotic segregation in pachytene arrested *Saccharomyces cerevisiae. Genetics* 106:165–83

79. Orr-Weaver, T. L., Szostak, J. W. 1985. Fungal recombination. *Microbiol. Rev.* 49:33–58

80. Carpenter, A.T.C. 1975. Electron microscopy of meiosis in *Drosophila melanogaster* females. II. The recombination nodule—a recombination associated structure of pachytene. *Proc. Natl. Acad. Sci. USA* 72:3186–90

81. Carpenter, A. T. C., Baker, B. S. 1982. The control of the distribution of meiotic exchange in *Drosophila melanogaster. Genetics* 101:81–90

82. John, B., Lewis, K. R. 1976. *The Meiotic Mechanism* (Oxford Biology Readers), ed. J. J. Head. Oxford: Oxford Univ. Press

83. Klapholz, S., Esposito, R. E. 1980. Recombination and chromosome segregation during the single division meiosis in *spo12-1* and *spo13-1* diploids. *Genetics* 96:589–611

84. Wagstaff, J. E., Klapholz, S., Esposito, R. E. 1982. Meiosis in haploid yeast. *Proc. Natl. Acad. Sci. USA* 72:2986–90

85. Grell, R. F. 1976. Distributive pairing. In *Genetics and Molecular Biology of Drosophila*, Vol. 1, pp. 435–85, ed. E. Novitski, M. Ashburner. New York: Academic

86. Baker, B. S. 1975. Paternal loss (*pal*): a meiotic mutant in *Drosophila melanogas-*

ter causing loss of paternal chromosomes. *Genetics* 80:267–96

87. Goldstein, L. S. B. 1981. Kinetochore structure and its role in chromosome orientation during the first meiotic division in male *Drosophila melanogaster*. *Cell* 25:591–602

88. Davis, B. K. 1971. Genetic analysis of a meiotic mutant resulting in precocious sister centromere separation in *Drosophila melanogaster*. *Mol. Gen. Genet.* 113:251–72

89. Mason, J. M. 1976. Orientation disruptor (*ord*): a recombination defective and disjunction defective meiotic mutant in *Drosophila melanogaster*. *Genetics* 84:545–72

90. Goldstein, L. S. B. 1980. Mechanisms of chromosome segregation revealed by two meiotic mutants of *Drosophila melanogaster*. *Chromosoma* 78:79–111

Ann. Rev. Cell Biol. 1985. 1:317–51
Copyright © 1985 by Annual Reviews Inc. All rights reserved

ACETYLCHOLINE RECEPTOR STRUCTURE, FUNCTION, AND EVOLUTION

Robert M. Stroud and Janet Finer-Moore

Department of Biochemistry and Biophysics, University of California, San Francisco, California 94143

CONTENTS

1. INTRODUCTION

The acetylcholine receptor (Achr) is the best understood cell surface receptor involved in cell-to-cell signaling, principally due to two factors that have aided in its characterization. The first is the rich source of Achr present in species of electric fish, *Torpedo marmorata* and *Torpedo californica* (marine elasmobranch fish), and in *Electrophorus electricus* (a fresh-water teleost, the electric eel). Nicotinic Achr's are components of all

317

0743–4634/85/1115–0317$02.00

vertebrate skeletal neuromuscular junctions, and provide one means of signaling between nerve and muscle cells. The embryonic electric fish contains similar neuromuscular junctions on muscle cells, some of which change through the process of development into electrocytes, or specially evolved cells designed to generate the electric potential that these species use to stun or kill their prey. The second factor that has contributed to the characterization of the Achr is the presence of neurotoxins in snake venoms that bind specifically to Achr. Toxins and cholinergic ligands provide the means for assaying receptor binding and for affinity purification of Achr molecules.

The Achr structure has been characterized in terms of ultrastructural organization in the cell, divergence within one species at different sites in the neuron or the neuromuscular junction, and postsynaptic membrane organization, as well as in terms of sequence and structure within the membranes (reviewed by Popot & Changeux 1984; Stroud 1983; Lindstrom 1983; Conti-Tronconi & Raftery 1982; Karlin 1980, 1983). The minimal functional molecule is a pentamer of five similar subunits of stoichiometry $\alpha_2\beta\gamma\delta$. Achr function has been characterized electrophysiologically (Raftery et al 1983; Hess et al 1983; Spivak & Albuquerque 1982); and its conductivity has been measured as a function of transmembrane potential and the nature, size, and charge of the conducted ions. The amino acid sequences of all four gene products have been determined. These sequences have provided insight into the structure and evolution of Achr that could not have been forseen a few years ago. The impact of these discoveries on the field of neurobiology is enormous. Transcripts of cDNA clones of all chains have been coexpressed, and functional assembly of all chains has been established. Many site-specific mutagenesis experiments have been done to elucidate which elements of the sequence are important to the function of the complex, or its correct assembly. Some aspects of the topology of chains as they fold through and across the membrane have been determined, and the three-dimensional structure is forthcoming at increasing resolution from X-ray and electron crystallography.

2.   Ach RELEASE, LOCATION, AND FUNCTION OF RECEPTORS

Acetylcholine (Ach) is released at electron-dense "active zones" of the presynaptic plasma membrane of a nerve terminal by fusion with the cell membrane of nearby acetylcholine-containing vesicles (Reichardt & Kelly 1983). Presynaptic vesicles, which can be seen in the electron microscope, contain Ach (Whittaker et al 1964), a finding which supports the vesicle hypothesis (Katz 1962), and provides a possible explanation for the quantal release of Ach. Fusion is triggered by an influx of Ca^{2+} ions through voltage-sensitive calcium channels. The latter is in turn triggered by the potential change relayed by the wave of depolarization down the axon, mediated by voltage-dependent sodium channels in the axon. However, alternate views raise the question of whether or not the vesicular mechanism is the only one the cells use (Tauc 1982; Cooper & Meyer 1984; Dunant & Israel 1985), and proof of the vesicle hypothesis has been difficult to establish; it relies mostly on its ability to explain a rapid and large release of Ach, and the lack of viable alternatives.

Upon release, the concentration of acetylcholine reaches about 1 mM in the synaptic space, where it triggers the opening of a cation-conducting channel within responding Ach receptors on the target muscle cell, or electrocyte plasma membrane. There are morphogenic and functional similarities between Achr in the neuromuscular junctions and those in the membranes of the electrocytes in *Torpedo marmorata* or *Electrophorus electricus*. In *Torpedo*, the embryonic fish contains muscle cells that develop into the oriented stack of polar cells that generate the electric organs. Thus, on the molecular level they consist of cells that are the counterparts of muscle cells. Only one side of the electrocyte plasmalemma is endowed with acetylcholine receptors. The endplate components in these cells have been extensively characterized biochemically, though these cells are much less suitable for electrophysiological studies than vertebrate neuromuscular junctions. Their Ach receptors, however, are structurally, functionally, and evolutionarily related to those of the neuromuscular junctions in higher mammals. For this reason biochemical and structural analyses have progressed furthest with the electrocyte receptors, while electrophysiological characterization has proceeded furthest with receptors in vertebrate neuromuscular junctions. Analysis of cDNA and amino acid sequences has proceeded from *Torpedo* and *Electrophorus* to higher vertebrates (cow, mouse, chick), including humans, through studies at the genetic level. Based on sequence homology, most aspects of the $\alpha_2\beta\gamma\delta$ structure and its corresponding amino acid sequences are shared between *Torpedo*, *Electrophorus*, and the higher vertebrates.

3. MECHANISMS OF THE Ach RECEPTOR

After binding acetylcholine, the Achr responds by an extensive change of conformation that affects all subunits (Witzemann & Raftery 1978; Lindstrom et al 1979; Rübsamen et al 1978; Kistler et al 1982), and leads to opening of an ion-conducting channel across the plasma membrane. This channel conducts all small cations, but predominantly the sodium ions that are continuously pumped into the synaptic space by Na^+/K^+-ATPase. Each Ach receptor molecule conducts about 10^4 ions per millisecond (ms). This rate is within a factor of 10 of the diffusion-controlled rate found for an open pore of the size defined by the largest organic cations that can be conducted through the channel. Dimethyldiethylammonium, one of these cations, measures about 7 Å in diameter, close to that of a hydrated sodium ion. Thus, the cell membrane, which is normally insulating to a potential gradient of 100,000 V cm^{-1}, is transformed within about 50 microseconds (μs) to a highly conducting state.

There is no cooperativity in ligand binding; however, there is cooperativity in channel opening, which shows a Hill coefficient of 1.97 ± 0.06 (Neubig & Cohen 1980). At least two Ach molecules per Achr bound are required to effect a channel-opening event (Cash & Hess 1980; Sine & Taylor 1980). The common interpretation is that the two α-chains, the chains that bind neurotoxins, also bind Ach and contain the binding sites for the activating agonist. These sites normally have low affinity ($K_d \approx$ 100 μM to 1 mM) for agonists (Raftery et al 1983). In constant presence of agonist, for example as required in equilibrium binding studies, Ach is bound about 10^2–10^3 times more tightly to high affinity sites ($K_d \approx 1 \mu$M). It has been shown that the conformation of Achr changes on the same time scale as channel opening, independent of occupancy of high affinity sites on the α-chains. This led to the proposal that perhaps as many as five channel-opening Ach sites of low affinity are located on the similar subunits, and that the high affinity sites on the α subunits may be different from these (Dunn & Raftery 1982; Dunn et al 1983).

Each Ach receptor closes within 0.1–10 ms (Colquhoun & Sakmann 1981). Individual single-channel open times are described by two Poisson distributions, characteristic of stochastic processes. Acetylcholine diffuses away, and is hydrolyzed in the synaptic space by the enzyme acetylcholine esterase, localized on the basal lamina. One challenge is to understand the mechanism for conducting cations by the Achr, for the selectivity of the ion channel that conducts cations but not anions, and for the high rates of conductance via Achr across the membrane in the open state (Horn & Stevens 1980). The mechanism of gating the channel, and the factors involved in the many types of regulation of Achr activity are current foci of attention.

In the continued presence of agonists the Achr becomes desensitized and no longer responds to agonist. This state is analogous to pharmacological desensitization of receptor molecules. Its physiological relevance is undetermined, however, there are at least two rate constants involved in reaching this state. The overall kinetic scheme is represented by the minimal model (Hess et al 1982, 1983):

$$R + L \rightleftharpoons RL \overset{+L}{\rightleftharpoons} RL_2 \rightleftharpoons R^*L_2 \rightleftharpoons \text{conduction}$$

with the lower branch:

$$R'L \rightleftharpoons R'L_2 \rightleftharpoons R''L_2$$

(Open)

In this model R is the resting state of the receptor, R^* represents the open-channel state, and R' and R'' are desensitized states.

In addition to the physiological effects of acetylcholine itself, several other physiological and nonphysiological agents modulate activity. The former include Ca^{2+}, phosphorylation of cytoplasmic sites, N- and O-glycosylation, and bound fatty acids. The latter include snake neurotoxins, curare, agents described and used clinically as muscle relaxants, such as suberyldicholine, and numerous compounds described variously as non-competitive blockers or local anesthetics. The primary functional site of many local anesthetics is the cytoplasmic side of voltage-activated sodium channels in neurons. The effects of this class of extrinsic reagents on the Ach receptors suggest modes of attachment to membrane proteins in general and to Achr, which are being characterized in terms of the sequences and structure of their sites of attachment.

4. STRUCTURE

A. The Ach Receptor is a Pentameric Complex of Five Similar Transmembrane Subunits

Subcellular fractionation of electric organs led to the first detailed characterization of the minimum components responsible for Achr function, initially defined in terms of binding of neurotoxins, such as α-bungarotoxin or curare (D-tubocurarine), and in terms of ability to conduct ions across membranes in response to binding cholinergic agonists. More recently, assays have involved the response of individual molecules under patch-clamp conditions. Four minimal essential protein components were identified by establishing reconstitution of molecules that were purified, first by detergent solubilization and then by affinity chromatography on agonist-like, or neurotoxin-containing columns. Initial deductions of an $\alpha_2\beta\gamma\delta$ stoichiometry (Reynolds & Karlin 1978; Lindstrom et al 1979) were verified by quantitative amino acid sequencing of the affinity-purified molecules (Raftery et al 1980). This pentameric structure is associated with other proteins, all about 43 kDa, that can be removed by treatment with a base at pH 10–11 (Neubig et al 1979). It has subsequently been shown that these 43,000 dalton proteins, sometimes called ν proteins, are localized on the cytoplasmic side of the membrane. They are intracellular proteins that probably form a cytoskeletal matrix that is associated with acetylcholine receptors on the surface (Cartaud et al 1981; Rousselet et al 1982; Sealock et al 1984). At least one of these components seems to contain phosphatase activity (Gordon et al 1983). They may or may not include some small amount of actin, and one has actin-binding ability (Walker et al 1984). Although these proteins have been copurified with Achr, they have as yet no clearly identified function in regulating Achr activity. They may be involved in clustering of the receptors concentrated on the crests of the synaptic folds in between invaginations, or in the cytostructural integrity of the membrane.

B.   The Structure is a Pentamer of Quasi-Fivefold Symmetry

Acetylcholine receptors in the plasma membrane were initially identified by immunoelectronmicroscopy using anti-Achr antibodies (Klymkowsky & Stroud 1979). The pentameric complex of five similar subunits extends into the synapse as a funnel-shaped molecule. In projection onto the membrane surface, the density distribution around the central well is asymmetric, but has strongly fivefold symmetric characteristics (Ross et al 1977; Zingsheim et al 1980; Kistler & Stroud 1981; Bon et al 1984; Brisson & Unwin 1984). In some images this structure appears to be pentameric, although the mass distribution of subunits estimated from the sequences predicts approximately 33% variation in their apparent projected size when glycosylation of each subunit is taken into account. Subunits surround the ion channel like staves. The angle between the two α-subunits has been measured precisely using monoclonal antibody Fab fragments as $144 \pm 4°$; the results indicate that the fivefold symmetry is well preserved (Fairclough et al 1983).

The transmembrane electron density profile, determined from X-ray diffraction measurements, shows that the complex of five similar subunits extends about 60 Å into the synaptic space (Ross et al 1977). The entrance to the central channel begins at about 40 Å and narrows to about 7 Å in diameter at the level of the membrane surface (Klymkowsky & Stroud 1979).

Traversing the lipid bilayer is a central channel, measured by structural analysis as about 7 Å in diameter (Kistler et al 1982), close to the maximum diameter derived from electrophysiological measurements of conductivity of organic cations of various sizes (Furukawa & Furukawa 1959; Maeno et al 1977; Huang et al 1978; Dwyer et al 1980). In two- or three-dimensional electron-image reconstructions, which are currently limited to about 30 Å resolution, the synaptic entrance of the channel is well represented. However, the narrowest region of the channel, located in the bilayer, is not.

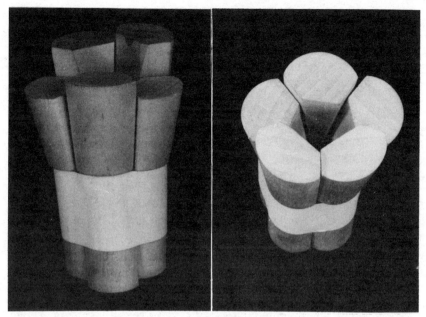

Figure 1   A model of the Ach receptor, which displays the mass distribution expected on the basis of sequence homology, and the topology with respect to the membrane. While there is quasi-fivefold symmetry in the structure, the differences in subunit masses are apparent on the larger synaptic side, and also, on the cytoplasmic surface (lower part of model, *left*) where there are larger differences in chain length, though the largest difference is due to polysaccharides on the extracellular side.

At this level its dimensions were determined by quantitative densitometry of negatively stained electron micrographs of Achr, in 2.4 × 2.4 Å pixels. These measurements clearly show that the central well is about 7 Å in diameter and can hold uranyl (UO_2^{2+}) ions throughout 90–100% of the vertical dimension of the Achr molecule (Kistler et al 1982).

A 10 Å equatorial X-ray spacing shows that Achr contains between 12 and 30 α-helices, oriented perpendicular to the membrane plane. The structural model derived from electronmicroscopic and X-ray diffraction analysis is shown in Figure 1, which illustrates the mass distribution implied by the molecular weights of the glycosylated chains.

C. *Arrangement of Subunits*

Each subunit is accessible to proteolytic cleavage by trypsin from either side of the membrane. This finding establishes the transmembrane orientation of each subunit (Klymkowsky et al 1980; Strader & Raftery 1980;

Lindstrom et al 1980; Wennogle & Changeux 1980), implied for the whole complex by X-ray diffraction (Stroud & Agard 1979; Ross et al 1977). A proteolytic cleavage point located 16,000 daltons from the carboxyl terminus of the δ chain occurs on the cytoplasmic side of the membrane (Wennogle et al 1981). The profound homology of all subunit types seems to imply that each one of them is in a similar environment, which is represented by their packing around the central ion channel like five $\sim 72°$ segments.

The arrangement of subunits and their order around the channel have been studied by analysis of inter-subunit cross-links between monomers through either β- or δ-chains in conjunction with avidin-biotinyl toxin receptor complexes (Holtzman et al 1982; Karlin et al 1983a,b), and by using Fab fragments of monoclonal antibodies (Kistler et al 1983; Fairclough et al 1983). The angle between equivalent epitopes recognized by antibody fragments on the synaptic surface of the α subunits ($144 \pm 4°$, or two 72° segments) demonstrates that α subunits are separated by one other subunit and so are in intrinsically different environments within the assembled complex. The avidin-labeled toxin molecules, which attach to the α subunits, are located at an angle of $114 \pm 29°$. Both these findings seem to imply that the α subunits are separated by another subunit. Direct visualization by overlapped images suggests a similar angular disposition (Zingsheim et al 1982).

D. Amino Acid Sequences

Purification of the four gene products that constitute the polypeptide chains of the minimal acetylcholine receptor from *T. californica* led to generation of antibodies and to sequencing of parts of the polypeptide chains (Devillers-Thiery et al 1979). Comparison of the amino termini of the four gene products established their homology, as well as the $\alpha_2\beta\gamma\delta$ stoichiometry of the protein (Raftery et al 1980). The homology between chains initially identified at the amino acid level (Raftery et al 1980) was dramatically corroborated by sequencing of all chains at the cDNA level (Sumikawa et al 1982; Noda et al 1982, 1983b,c; Claudio et al 1983; Devillers-Thiery et al 1983). These gene products are homologous from their amino- to C-terminal residues, with various insertions and deletions in one chain with respect to others, as is often characteristic of evolutionarily related protein sequences. All four sequences are characterized by four hydrophobic stretches of about the same length as the membrane-spanning helices in bacteriorhodopsin (Ovchinnikov et al 1979), though they are even more hydrophobic. They traverse the membrane as α-helices, and thus would contribute to the oriented helix-helix packing reflection seen in X-ray patterns from oriented membranes. Three of the hydrophobic segments occur in the middle of each sequence and are separated from each other only by short linking regions, while the fourth hydrophobic segment is close to the subunit carboxyl terminus.

The amino terminal ends of Achr subunits from *Electrophorus* electric organ (Conti-Tronconi et al 1982a) and from calf muscle (Conti-Tronconi et al 1982b) were sequenced, and they showed substantial homology with the *Torpedo* Achr sequences. The similarity of Achr receptors from different organisms was demonstrated at the DNA level when a probe from *Torpedo* α subunit detected corresponding sequences in a chicken genomic library (Ballivet et al 1983). Sequences have since been determined for calf and human α-chains (Noda et al 1983a); the calf β-chain (Tanabe et al 1984); calf (Takai et al 1984), chick (Nef et al 1984), and human (Shibahara et al 1985) γ-chains; and chick (Nef et al 1984) and mouse δ-chains (LaPolla et al 1984). When aligned as in Figure 2, 55–81% of the residues in these sequences are identical to residues in the corresponding *Torpedo* sequences.[1] All have the four characteristic stretches of hydrophobic amino acids.

5.   CORRELATION OF STRUCTURE AND FUNCTION IN Achr FROM *TORPEDO*

The cloning and sequencing of cDNA for all four subunits from *Torpedo* Achr (Ballivet et al 1982; Sumikawa et al 1982; Noda et al 1982, 1983b,c; Giraudat et al 1982; Claudio et al 1983; Devillers-Thiery et al 1983), and the expression of functional receptor from the cDNA clones (Mishina et al 1984; White et al 1985) have made it possible to probe the relationship between structure and function using site-directed mutagenesis (Mishina et al 1985; White 1985). High (atomic) resolution information on the protein's three-dimensional structure is necessary to best plan alterations of specific residues and interpret the results of these changes. In the absence of a high resolution crystal structure, investigators have relied on experimental and model-building methods that map the sequence into the available low resolution structure. Detailed model building is helped by the fact that the receptor is divided into extracellular, membranous, and cytoplasmic domains, and that the partitioning of the sequence into these domains may be inferred from patterns of hydrophobicity of amino acids in the sequences, and established by testing which side of the membrane certain sequences lie on.

---

[1] The numbering of amino acids in this review follows the *Torpedo* consensus alignment of homologous residues of Finer-Moore & Stroud (1984) and Fairclough et al (1983).

```
                    -24 |              | -1  1   |            | 23 |

T. californica α    MILCSYWHVGLVLLLFSCCGLVLG        SEHETRLVANLL  EN  YNKVIRPVEHHTHFVDITVGLQLIQLINVDEV
                           S  +        SS                       -+-  +      +- +++++-+-+-              --
T. marmorata α      MILCSYWHVGLVLLLFSCCGLVLG        SEHETRLVANLL  EN  YNKVIRPVEHHTHFVDITVGLQLIQLINVDEV
                           S  +        SS                       -+-  +      +- +++++-+-+-              --
Calf α              MEPRPLLLLLGLCSAGLVLG            SEHETRLVAKLF  ED  YNSVVRPVEDHRQAVEVTVGLQLIQLINVDEV
                           S +                                  -+-         +   +  +   +               --
Human α             MEPWPLLLLLFSLCSAGLVLG           SEHETRLVAKLF  KD  YSSVVRPVEDHRQVVEVTVGLQLIQLINVDEV
                           S +                                  -+-         +   +  +   +               --
T. californica β    MENVRRMALGLVVMMALALSGVGA        SVMEDTLLSVLF  ET  YNPKVRPAQTVGDKVTVRVGLTLTNLLILNEK
                             ++                                 - +-        +                         -+
Calf β              MITPGALLLLLLGVLGAHLAPGARG       SEABGRLREKLF  SG  YDSTVRPAREVGDRVWVSIGLTLAQLISLNEK
                           --  --                               - - -       -   +                    -+
T. californica γ    MVLTLLLIICLALEVRS               ENEBGRLIEKLL  GD  YDKRIIPAKTLDHIIDVTLKLTLTNLISLNEK
                            S. -+                              - - -+- +    +-  ++++-+-+  - - - -      -+
Chicken γ           RNQEEKLLQDLM  TN  YNRHLRPALRGDQVIDVTLKLTLTNLISLNER
                                                               + +- -++      ++- + + -  -  -  -  +- +
Calf γ              MCGGQRPLFLLPLLAVCLGAKG          RNQEERLLGDLM  QG  YNPHLRPAEHDSDVVNVSLKLTLTNLISLNER
                             S    S                             + +- -++      ++ + - - - -  -  -  -    -+
Human γ             MHGGQGPLLLLLLLAVCLGAQG          RNQEERLLADLM  QN  YDPNLRPAERDSDVVNVSLKLTLTNLISLNER
                              +   S                             + +- +++  *   ++  + - - - -  -  -  -    -+
T. californica δ    MGNIHFVYLLISCLYYSGCSG           VNEEERLINDLL  IVNKVRPVKHNNEVVNIALSLTLSNLISLKET
                          S                                    - -+-+       + ++ +  ++  - - -  -      -+
Chicken δ           VNQEERLIHHLFEERGYNKEVRPVASADEVVDVYLALTLSNLISLKEV
                                                              - -+ -+       ++ + + -  - -  -  -       +-
Mouse δ             MAGPVLTLGLLAALVVCALPGSWG        LNEEGRLIQHLFNEKGYDKDLRPVARKEDKDVALSLTLSNLISLKEV
                          S                                    -- + +    * -+  ++-++-+ -               +-
```

```
  | 55               | 81              | 100           | 120        | 130          | *

NQIVETNVRLRQQWIDVRLRWNPADYGGIKKIRLPSDDVWLPDLVLYNNADGDFAIVHMTKLLLDYTGKIMWTPPAIFKSYCEIIVTHFPFDQQN
S  +       ++                                                      +                  S- -    +
NQIVETNVRLRQQWIDVRLRWNPADYGGIKKIRLPSDDVWLPDLVLYNNADGDFAIVHMTKLLLDYTGKIMWTPPAIFKSYCEIIVTHFPFDQQN
S  +       ++                                                      +                  S- -    +
NQIVTTNVRLKQQWVDYNLKWNPDDYGGVKKIHIPSEKIWRPDLVLYNNADGDFAIVKFTKVLLQYTGHITWTPPAIFKSYCEIIVTHFPFDEQN
S  +       ++                                                      +                  S -     +
NQIVTTNVRLKQQWVDYNLKWNPDDYGGVKKIHIPSEKIWRPDLVLYNNADGDFAIVKFPTKVLLQYTGHITWTPPAIFKSYCEIIVTHFPFDEQN
S  +       ++                                                      +                  S -     +
IEEMTTNVFLNLAWIDYRLQWDPAAYEGIKDLRIPSSDVWQPDIVLMNNNDGSFEITLHVNVLVQHTGAVSWQPSAIYRSSCTIKVMYFPFDWQN
           +  -                                                    +                       -    +
DEEMSTKVYLDLEWIDYRLSWDPEEHEGIDSLRISARSVWLPDVVLLNNNDGNFDVALDINVVVSSDGSMRWQPPGIYRSSCSIQVTYFPFDWQN
           +  -                                                    +                            +
EEALTTNVWIEIQWNDYRLSWNTSEYEGIDLVRIPSELLWLPDVVLENNVDGQFEVAYYANVLYRYTCCPEVAYIRSTCPIAVTYFPFDWQN
           +  +                                                    +                    S       +
EETLTTNVWIEMQWSDYRLRWDPKDYEGIQQLRVPSAMVWLPDIVLENNVDGTFEITLYTNVLVYPDGSIYWLPPAIFRSSCSIHVTYFPFDWQN
           +  +                                                    +                    S-      +
EEALTTNVWIEMQWCDYRLRWDPRDYGGLWVLRVPSTMVWRPDIVLENNVDGVFEVALYCNVLVSPDGCVYWLPPAIFRSSCPVSVTFPFDWQN
           +  +                                                    +                  S- +     +
EEALTTNVWIEMQWCDYRLRWDPRDYEGLWVLRVPSTMVWRPDIVLENNVDGVFEVALYCNVLVSPDGCIYWLPPAIFRSACSISVTYFPFDWQN
           +  +                                                    +                  S  +     +
DETLTSNVVWMDHAWYDHRLTWNASEYSDISILLRLPPELWIPDIVLQNNNDGQYHVAYFCNVLVRPNGYVTWLPPAIFRSSCPINVLYFPFDWQN
              -       +                                            +                           - *
DETLTTNVWVEQSWIDYRLQWNTSEFGGVDVLRLLPEMLWLPEIVLENNNDGLFEVAYYSNVLVNYGYVVWLPPAIFRSACPINVNFFPFDWQN
              +       +                                            +                           -  +
EETLTTNVWIEMQWNDYRLQWDANDFGNITVLRLPPDMVWLPEIVLENNNDGSFQISYACNVLVYDSGYVTWLPPAIFRSSCPISVTYFPFDWQN
              -       +                                 *          +                    S       *
```

```
144      | 160       |       | 185E G W        | 208      |      | 228|

CTMKLGIWTYDGTKVSIS       PES DRP    DLSTFMESGEWVMKDYRGWKH  WWYYTCCPD  TPYLD   ITYHFIMQRIPLYFVV
S  +                        -+-                       SS -+                             +
CTMKLGIWTYDGTKVSIS       PES DRP    DLSTFMESGEWVMKDYRGWKH  WWYYTCCPD  TPYLD   ITYHFIMQRIPLYFVV
S  +                        -+-                       SS -+                             +
CSMKLGTWTYDGSVVVIN       PES DQP    DLSNFMESGEWVIKESRGWKH  WFYACCPS   TPYLD   ITYHFVMQRLPLYFIV
S  +                        -+-                       - -+                              +
CSMKLGTWTYDGSVVAIN       PES DQP    DLSNFMESGEWVIKESRGWKH  SVTYSCCPD  TPYLD   ITYHFVMQRLPLYFIV
S  +                        -+-                       - -+                              +
CTMVFKSYTYDTSEVTLQ       HALDAKG    EREVKEIVINKDAFTENGQWS IEH  KPSRKNW RSD DP  S YED   VTFYLIIQRKPLFIV
           ++                                           +-                              +
CTMVFSYSYDSSEVSLQ        TGLSPEG    QER QEVYIHEGTFIENGQWE III HK KPSRLIQP SV DPRGGG EGRREEVTFYLIIRRKPLFLV
           ++ -                                         ++                              +++
CSLVFRSQTYNAHEVNLQLSAEEGE        AVEWIHIDPEDFTENGWAIFH RPAKKNYNWQLTKDD TDFQE   IIFFLIIQRKPLFYII
   S +          --                                      +  +                            +++
CTMVFQSQTYSANEINLLLTVEEGQ        TIEWIFIDPEAFTENGWAIKH RPARKIINSGRFTPDDIQYQQ   VIFYLIIQRKPLFYII
   S +                                                  ++                              +++
CSLIFQSQTYSTNEINLQLSQEDGQ        TIEWIFIDPEAFTENGWAIRH RPAKMLLDEAAPAEE AGHQK   VVFYLLIQRKPLFYVI
   S +                                                  ++                              +++
CSLIFQSQTYSTNEIDLQLSQEDGQ        TIEWIFIDPEAFTENGWAIQH RPAKMLLDPAAPAQE AGHQK   VVFYLIIQRKPLFYVI
   S +                                                  ++                              +++
CSLKFTALNYDANEITMDLMTDTIDGK      DYPIEWIIIDPEAFTENGEWEIIH KPAKKN IYPDKFPNGTNYQD   VTFYLIIRRKPLFYVI
   S +                                                  +  ++                           +++
CTLKFSSLAYNAQEINMHLKEESDPTEKNYRVEWIIIDPEGFTENGWEIIH RPARKN IHPSYPTESSEHQD   IIFYLIIKRKPLFYVI
   S +                                                  +                               +++
CSLKFSSLKYTAKEITLSLKQEEENNR      SYPIEWIIIDPEGFTENGBWEIVH RAAKLN VDPSVPMDSTNHQD   VTFYLIIRRKPLFYII
   S +  +    +-                                         ++                              +++
```

```
     | 253      260       |          | 283       290     |              | 316 R|

NVIIPCLLFSFLTVLVFYLPTDSG  EKMTLSISVLLSLTVFLLVIVELIPSTSSAVPLIGKYMLFTMIFVISSIIVTVVVINTHHRSPSTHMPQ
S                                                                                 +++
NVIIPCLLFSFLTVLVFYLPTDSG  EKMTLSISVLLSLTVFLLVIVELIPSTSSAVPLIGKYMLFTMIFVISSIIVTVVVINTHHRSPSTHMPQ
S                                                                                 +++
NVIIPCLLLFSFLTGLVFYLPTDSG EKMTLSISVLLSLTVFLLVIVELIPSTSSAVPLIGKYMLFTMVFVIASIIITVIVINTHHRSPSTHVMPE
S                                                                                 +++
NVIIPCLLFSFLTGLVFYLPTDSG  EKMTLSISVLLSLTVFLLVIVELIPSTSSAVPLIGKYMLFTMVFVIASIIITVIVINTHHRSPSTHVMPN
S                                                                                 +++
YTIIPCILISILAILVFYLPPDAG  EKMSLSISALLAVTVFLLLLADKVPETSLSVPIIIRYLMFIMILVAFSVILSVVVLNLHHRSPNTHTMPN
S                                                                                 +++
NVIAPCILITLLAIFVFYLPPDAG  EKMGLSIFALLTLTVFLLLLADKVPETSLNVPIIIKYLMFTMVLVTFSVILSVVVLNLHHRSPHTHQMPL
S                                                                                 +++
NIIAPCVLISSLVVLVYFLPAQAGQQKCTLSISVLLAQTIFLFLIAQKVPETSLNVSLIGKYLIFVMFVSMLIVMNCVILNVSLRTPNTHSLSE
   S                                                                              +++
NIIVPCVLISSMAVLVYFLPAKAGGQKCTVSINVLLAQTVFLFLIAQKVPETSAVPLIGKYLTFLMVVTVVIVVNAVILNVSLRTPNTHSMSQ
   S                              -+S                                              +++
NIIAPCVLISSVAILIYFLPAKAGGQKCTVAINVLLAQTVFLFLVAKKVPETSQAVPLISKYLTFLLVVTILIVVNAVVVLNVSLRSPHTHSMAR
   S                              -+S                                              +++
NIIAPCVLISSVAILIHFLPAKAGGQKCTVAINVLLAQTVFLFLVAKKVPETSQAVPLISKYLTFLLVVTILIVVNAVVVLNVSLRSPHTHSMAR
   S                              -+S                                              +++
NFITPCVLISFLASLAFYLPAESG  EKMSTAISVLLAQAVFLLLTSQRLPETALAVPLIGKYLMFIMSLVTGVIVNCGIVLNFHFRTPSTHVLSE
S                                                                                 +++
NILVPCVLISFMINLVFYLPGDCG  EKMTLVISVLLAQSVFLLLVSQRLPATSMAIPLVGKFLLFGMVLVTMVVVICIVLNIHFRTPSTHVLSE
S                                S    -+                                           + +
```

```
 330          |          |          |          |          |          |          400          |          |

WVRKIFINTIPNLMFFS                                              TMKRASKEK   QE NKIFADDIDISDISGKQVTGEV
   ++                                                            ++  ++-+   -  -  - -- -     -    +    -
WVRKIFINTIPNLMFFS                                              TMKRASKEK   QE NKIFADDIDISDISGKQVTGEV
   ++                                                            ++  ++-+   -  -  - -- -     -    +    -
WVRKVFIDTIPNLMFFS                                              TMKRPSREK   QD KKIFTEDIDISDISGKFGPPPM
   ++ -                                                           ++  ++-+   -  -+ - -- -     -    -    +
WVRKVFIDTIPNLMFFS                                              TMKRPSREK   QD KKIFTEDIDISDISGKFGPPPM
   ++                                                             ++  ++-+   -  -+ - -- - + -  -++ -  -  +
WIRQIFIETLPPFLWIQRPVTTPSPD         SKPTIISRANDEYFIRKPAGDFVCPVDNAVAVQP ERLFS EMKWH              LN
  -+                               +    +    +   +  +    ++-     -  -  +  -+ ++
WVRQIFIHKLPLYLGLKRPKPERDQMQEPPSIAPRDSPGSGWGRGTDEYFIRKPPNDFLFPKPN RF  QP ELSAPDLRRF    IDG
  +   ++   ++ ++ -+-         + -     +        ++     S -+  +  - -+   +
KIKHLFLGFLPKYLGMQLEPSEETPEKPQ PR RRSSFGIMI KA EEYILLKKPRSELMFEEQKDRHGLKRYVNKMTS DIDI  GTT       VD
  +  +   -   +       -    -++  -   +++++    +   ++ +  ++  + -    +  -   - ++
RVRQWLHLLPRYLGMHM PEEAPGPFQQATR RRSSLGLMV KA DEYMLWKARTELLFEKQKERDGLMKTVLEKI GRGLESNCA        QD
  + + +  +         +   +    +  ++++  +  +  +   +  ++ +  - -     +  -   - - +
GVRKVFLRLLPQLLRMHVRPLAPVAVQDAHPRLQNGSSSGWPITAGEEVALCLPRSELLF RQRQRNGLVRAALEKL EKGPESGQS      PE
  +   +++       + +++        -  +   +    +     ++  ++  -+  -        + +  -     -  +
GVRKVFLRLLPQLLRMHVRPLAPAAVQDTQSRLQNGSS GWSITTGEEVALCLPRSELLF QQWQRQGLVAAALEKL EKGPELGLS       Q
  +++       + +++            +  +   •       -- S -   +   +      +  - -     -++
RVKQIFLEKLPRILHMSRADESEQPDWQNDLKLRRSSSVGYISKAQ EYFNIKSRSELMFEKQSERHGLVP RVTP RIGFG NN        NE
  ++  -  -+ -+  +  -      - ++  +       -- +  - ++-  -  -++    +  -  + + +
WVRGYFLEILPRLLHMSHPAES  PAGAPCIRRCSSAGYIAKAE EYYSVKSRSELMFEKQSERHGLAS RVTP ARFAP AA          SE
  ++  -  -+ +  + +  -+ -    -   + -     S  ++-  -  - ++-   + + -   ++
GVKKPFLETLPKLLHMSRPAEE  DPGPRALIRRSSSLGYICKAE EYFSLKSRSDLMFEKQSERHGLAR RLTT ARRPP AS         SE
  ++-    + + + --         +   ++     S+- + + -   -+ -++  +  +   +  ++
```

```
 426  431  436   |   443   450        462          |          |          |          |          |   517

IFQTPLIKNPDVKSAIEGVKYIAEHMKSDEESSNAAEEMKYVAMVIDHILLCVFMLICIIGTVCVFAGRLIELSQEG
   + -   -   + -  +    -+ + ---   -   S    S   S    -   +    +     +
IFQTPLIKNPDVKSAIEGVKYIAEHMKSDEESSNAAEEMKYVAMVIDHILLCVFMLICIIGTVCVFAGRLIELSQEG
   + -   -   + -  +    -+ + ---   -   S    S   S    -   +    +     +
GFHSPLIKHPEVKSAIEGIKYIAETMKSDQESNNAAEBMKYVAMVMDHILLAVFMLVCIIGTLAVFAGRLIELNQQG
   +  ++   -   - -  +    -+ + --- -  +  S    -+  +    +  -    +     +
GFHSPLIKHPEVKSAIEGIKYIAETMKSDQESNNAAEBMKYVAMVMDHILLGVFMLVCIIGTLAVFAGRLIELNQQG
   +  ++   -   - -  +    -+ + --- -  +  S    -+  +    +  -    +     +
GLTQPVTLPQDLKEAVEAIKYIAEQLESASEFDDLKKDMQYVAMVADRLFLYVFFVICSIGTFSIFLDASHNVPPDNPFA
  -  -+   -        - +  +    -  S -     +     + +  +   -   +   +
PNRAVGLPPELREVVSSISYIARQLQEQEDHDVLKEDMQFVAMVDRLFLWIFIIFTSVGTLVIFLDATYHLPPADPFP
    +-   - -  +   - + -        -+    +       -++   +     + -
LYKDLANFAPEIKSCVEACNFIAKSTKEQNDSGSENENMWLIGKVIDKACFWIALLLFSIGTLAIFLTGHFNQVPEFPFFGDPRKYVP
  +-   -   + S- S    + - +-     -     +     -+    +      +    +
FCQSLEEASPEIRACVEACNHIANATREQNDFSSENEEWILVGRVIDRVCFFIMASLFVCGTIGIFLMAHFNQAPALPFPGDPKTYLPP
  S    -  + -   -  +   + +     -  - S   +  +       +    +
WCGSLKQAAPAIQACVEACNLIARARHQQTHFDSGNKEMFLVGRVLDRVCFLAMLSLFVCGTAGIFLMAHYNRVPALPFPGDPFSYLPSSD
  S   -  +   - +    + +++  +     -+ S     +  +    +       +  -
FCGSLKQAAPAIQACVEACNLIACARHQQSHFDNGNEEWFLVGRVLDRVCFLAMLSFLICGTAGIFLMAHYNRVPALPFPGDPRPYLPSPD
  S +  +   - +    + ++  +     -   S      +  +     +       +  -
NIAASDQLHDEIKSGIDSTNYIVKQIKEKNAYDEEVGMWNLVGQTIDRLSMFIITPVMVLGTIFIFVMGNFNHPPAKPFEGDPFDYSSDHPRCA
  -  -    +    - -  + ++      -+  S     - +    ++      +      +++
ISEBQLYDHLKPTLDEANFIVKHMREKNSYNEEKDNMNRVARTLDRLCLFLITPMLVVGTLWIFLMGIYNHPPPLPFSGDPFDYREENKRYI
  -   - +  + -   -  ++ ++-     +  -  S     - +    +  -      +    -+ +S
Q  VQQELFNEMKPAVDGANFIVNHMRDQNSYNEEKDNMNQVARTVDRLCLFVVTPVMVVGTAWIFLQGVYNQPPLQPFGDPFSYSEQDKRFI
  -  - +        + +-      -- -+-    +  -+ S         - +  -     - -++
```

Figure 2 Sequences of the Achr subunits from *Torpedo* electric organ and from the neuromuscular junctions of chicken and mammalian species. Charged residues (+, −), possible glycosylation sites (*), and introns (*arrows*) are indicated below each sequence. Sequence numbering is that of Finer-Moore & Stroud (1984), which optimizes alignment of *Torpedo* sequences. Numbers listed correspond to the topological diagram shown in Figure 3.

A. Topology

The low resolution structure of Achr restricts proposals for the arrangement of the sequence with respect to the lipid bilayer. A valid model for the partitioning of the sequence between the inside and outside of the membrane must preserve the distribution of electron density observed in X-ray diffraction and electron microscopy (Ross et al 1977; Kistler et al 1982). Furthermore, the homology between the sequences of the four subunit types and the observed quasi-fivefold symmetry of the assembled protein dictates that tertiary structure models for the different subunits be similar.

Threading the four obvious hydrophobic stretches in each sequence into the membrane as helices distributes the protein mass between extracellular and cytoplasmic regions in approximately the correct ratio (Claudio et al 1983; Finer-Moore & Stroud 1984). However, close-packed hydrophobic helices are not likely to form a water-filled pore like that deduced from electrophysiological measurements (Lewis & Stevens 1983). A fifth amphipathic helix has been proposed to form the interface between this pore and the hydrophobic membrane domains (Finer-Moore & Stroud 1984; Stroud 1983; Guy 1984), as shown in Figure 3. The five-crossing scheme, depicted in Figure 3, is reasonably consistent with the observed distribution of mass between the two sides of the membrane (Finer-Moore & Stroud 1984), and the cross-sectional area of the receptor in the membrane region is approximately equal to the area occupied by 5 close-packed helices per subunit (Brisson & Unwin, personal communication). The density of protein on the cytoplasmic side of the membrane as observed by X-ray diffraction or electron microscopy is less than predicted from either the four-crossing or five-crossing models. This is presumed to result from disordered arrangement of the sequences between regions M3 and A, which differ considerably between the four subunits.

The scheme shown in Figure 3 for partitioning the sequence into synaptic, membranous, and cytoplasmic regions is consistent with other experimental observations. Proteolysis experiments have identified cytoplasmic and extracellular domains of the protein. Extensive trypsinization from the cytoplasmic side of either Achr in vesicles (Wennogle & Changeux 1980), or of in vitro transcripts in microsomal membranes (Anderson & Blobel 1981), reduced the molecular weight of the subunits to 35 kDa (α), 37 kDa (β), 45 kDa (γ), and 44 kDa (δ). These fragments contained the N-termini (proven for the α and δ subunits), and most if not all of the glycosylation sites of the chains (Wennogle & Changeux 1980; Anderson & Blobel 1981). The 44 kDa glycosylated domain of δ could be cleaved to give a soluble 26 kDa glycosylated extracellular domain, which would imply that as much as 18 kDa of the protein is embedded in the membrane (Anderson et al 1983). Limited trypsinization of the δ and β subunits from their cytoplasmic sides gave, in addition to the 44 and 37 kDa glycosylated fragments, 12 and 8 kDa fragments, respectively. These fragments were only partially soluble even at pH 10 (Anderson et al 1983).

If the sizes of the primary translation products of the four chains are compared with the sizes of the mature subunits, and loss of the signal peptide during translocation is taken into account, the polysaccharide molecular weights can be calculated (Anderson & Blobel 1981), as shown in Table 1. The number of possible N-glycosylation sites (S, T-X-N) in the *Torpedo* α, β, and δ sequences (one, one, and three, respectively) matches the number of intermediates seen during glycosylation (Anderson & Blobel 1981; Anderson et al 1983). For the γ-chain, the apparent polysaccharide weight is larger than for δ, and either two or three intermediate bands are seen in glycosylation (Figure 3a, "lane 3" of Anderson & Blobel 1981). Three

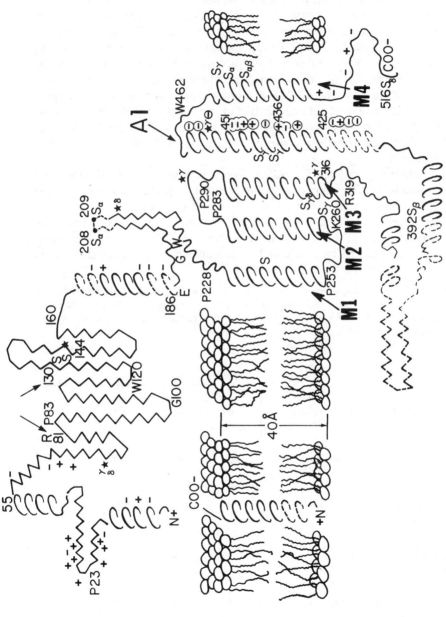

Figure 3   Model for the secondary structure and topography of the acetylcholine receptor. Hydrophobic membrane-spanning helices are marked *M1*, *M2*, *M3*, and *M4*. An amphipathic membrane-spanning helix is marked *A*. Possible N-glycosylation sites are marked with stars. The scheme is for each of the five subunits; dotted lines refer to differences in subunit lengths.

consensus glycosylation sites occur in the first 45 kDa of the chain, all on the extracellular side of the topological diagram, Figure 3. A fourth sequence is embedded in the membrane near the cytoplasmic side of M3, and is unlikely to be glycosylated. A fifth consensus sequence is located between A and M4, on the extracellular side of the topological map. Thus, in the five-crossing scheme four sites are potentially accessible for glycosylation on the extracellular side of the γ-chain. The estimated weights of the oligosaccharides on each chain allow a more accurate calculation of the mass distribution of each subunit relative to the bilayer surfaces, based on the proposed topology (Table 2). These mass distributions are the relevant numbers for comparison of the model with the results of protein structural studies.

Additional evidence for the topological scheme in Figure 3 comes from antibody labeling experiments that mapped specific residues into the cytoplasmic domain. Antibodies were produced against three synthetic peptides, one comprised of residues 350–358 of the β-chain, another of residues 372–389 of the γ-chain, and the third containing the carboxyl-terminal residues of the δ-chain. The antibodies were used to label receptors

Table 1 Polysaccharide mass in kilodaltons and number of glycosylation sites for the four subunits of Achr from *T. californica* as deduced from the differences in the mobilities of glycosylated and unglycosylated peptide chains on SDS-PAGE (Anderson & Blobel 1981, Anderson et al 1982)

	α	β	γ	δ
translation products	38	50	49	59
mature chains	40	50	60	64
signal sequences	3	3	2	3
polysaccharide	5	3	13	8
number glycosylation sites	1	1	3 or 4	3

Table 2 Mass distributions in kDa for the four subunits of Achr from *T. californica* as deduced from the known amino acid sequences, the polysaccharide masses described in Table 1, and the topology shown in Figure 3*

	α	β	γ	δ
synaptic side	29.6	29.9	38.1	37.1
	(31.8)	(29.9)	(40.6)	(36.1)
membrane imbedded	14.3	14.8	14.1	14.4
cytoplasmic	8.64	11.5	14.0	14.4
total	52.5	56.2	66.2	65.9
	(54.2)	(56.2)	(68.7)	(64.9)

* Values in brackets are based on polysaccharide mass from Table 1, and unbracketed numbers are based on an estimated 3 kDa per oligosaccharide.

in *Torpedo* electroplax, and the labeled sites were visualized in the electron microscope by means of colloidal gold-conjugated second antibodies (Young et al 1985; LaRochelle et al 1985). Antibodies to C-terminal peptides of the α, β, and γ subunits effectively labeled acetylcholine receptors in membrane vesicles only when the vesicles were permeabilized with detergent or lithium diiodosalicylate. This suggests that the C-termini of these subunits were also on the cytoplasmic side of the membrane in the assembled protein (Lindstrom et al 1984; Ratnam & Lindstrom 1984). The cytoplasmic orientation of the subunit carboxy-termini implies an odd number of transmembrane crossings in each subunit.

B. Detailed Modeling of Functional Regions

LIGAND BINDING SITES *Structure of the N-terminal domains* The various agonists and antagonists that interact with the acetylcholine receptor bind to the synaptic side of the protein. The number and location of the binding sites for each type of ligand is still debated. The three-dimensional structure of the entire extracellular domain is therefore of considerable interest. Examination of the N-terminal regions of the subunit sequences suggests that this domain may be composed primarily of β structures. Common turn-forming residues (e.g. gly or pro) or clusters of hydrophilic residues, which would tend to form exterior loops, occur every 10–15 residues, often bracketing stretches of amino acids that commonly adopt the β conformation in soluble proteins (Garnier et al 1978).

Protein domains containing primarily β structure are usually composed of antiparallel β-sheet in a β "barrel" or "sandwich" arrangement (Richardson 1981). In these structures each β-sheet generally uses one surface to form part of the hydrophobic protein core, and its other surface to interface with solvent or with other protein domains. The amino acids in the β-strands are therefore frequently alternately hydrophobic and hydrophilic, giving the β-sheet an amphipathic character. This periodicity may be detected most conveniently by Fourier transformation of the hydrophobicities of the amino acids (Finer-Moore & Stroud 1984; Eisenberg et al 1984). Fourier transforms of every stretch of 25 contiguous residues in the N-terminal halves of the *T. californica* receptor sequences have been averaged over the different subunits. They are plotted in Figure 4a as a two-

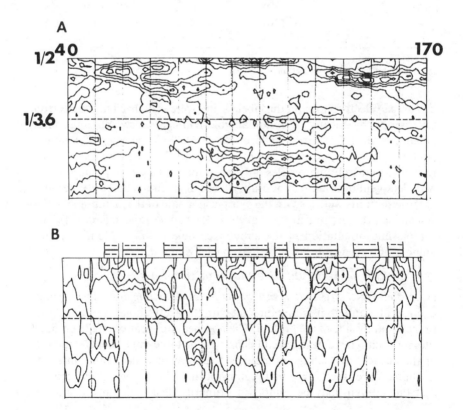

A

1/2 **40** 0   **170**

1/3.6

B

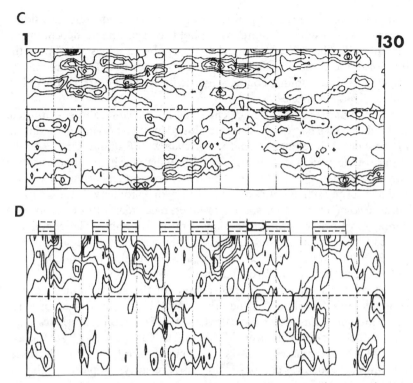

Figure 4   Amphipathic Fourier transforms between frequencies 0 and $\frac{1}{2}$ residues^{-1} for the acetylcholine receptor sequences (*a* and *b*) and from concanavalin A (*c* and *d*). The transforms were calculated with a moving window of 25 residues (*a* and *c*) and 11 residues (*b* and *d*). The central residue number is plotted on the horizontal axes and vertical lines are drawn every 10 residues. The frequency is plotted on the vertical axes. The locations of predicted (in *b*) or observed (in *d*) β-strands are indicated by a solid line between two dashed lines. An α-helix in concanavalin A is indicated in *d* by a cylinder.

dimensional contour plot, with frequency as the abscissa and the residue number of the central residue in the window as the ordinate. The plot shows a strong signal between residues 60 and 180 in the frequency range $(1/2)$–$(1/2.5)$ residue^{-1}, which is characteristic of amphipathic β-sheet. The dramatic length of the continuous signal, and its deviation from a frequency of $(1/2)$ residues^{-1} results from the large window size (25 residues) used in the calculation. When 11 contiguous residues are used in the Fourier transform, a plot that is more sensitive to short regions of periodicity is obtained (Figure 4*b*), from which locations of potential surface-forming β-strands can be identified by discrete maxima at the $(1/2)$ residues^{-1} frequency. Inspection of the *Torpedo* receptor sequences shows that in many of the predicted amphipathic β-strands, strictly hydrophobic residues alternate with residues that are hydrophobic or hydrophilic, depending on the subunit. The more conserved hydrophobic residues may form the interior of the domain, while the alternate, less-conserved residues may be involved in intersubunit contacts, particularly if they are conserved in widely divergent species.

For comparison, the same calculations for concanavalin A, a protein with a flattened β barrel structure, are plotted in Figure 4c and d. Maxima in Figure 4d at $(1/2)$ residues^{-1} tend to correlate with β-strands that lie on the surface of the protein with one side contributing to the hydrophobic protein core (Reeke et al 1975). The correlation between surface-forming β-strands and a strict periodicity of two residues in hydrophobicity is not complete because some of the outer surface residues, particularly those involved in contacts with other protein domains, are hydrophobic. The completely buried β-strands and the strands that do not interface with the protein core, or which have only one or two buried side-chains, do not generally show regular periodicity in the hydrophobicities of their amino acids.

The proposal that the synaptic regions of the acetylcholine receptor subunits are domains of β structure is corroborated by Raman spectroscopy measurements, which indicate that 34% of the receptor residues are in antiparallel β-sheets (Chang et al 1983), and by CD spectroscopy (Moore et al 1974), which estimates the β-sheet content of Achr to be 27%. The three-dimensional structures of β barrel domains are fairly complex, and there is still considerable ambiguity in the assignment of secondary structure to the Achr sequences. Therefore, a detailed model of the extracellular domain based on a β barrel motif has not yet been constructed.

Structure of the high affinity acetylcholine binding sites   More tractable than the challenge of deriving the structures of the extracellular domains of the subunits is the problem of modeling a specific ligand-binding site on the acetylcholine receptor. The best-characterized ligand-binding sites on the receptor are the two high-affinity agonist binding sites, one on each of the α subunits (reviewed in Conti-Tronconi & Raftery 1982; Popot & Changeux 1984; McCarthy et al 1986). While these binding sites do not behave identically, they probably involve the same residues of each α subunit (Conti-Tronconi et al 1984). The differences in their behavior have been attributed to different environments, and possibly to different states of glycosylation of the α subunits in the functional receptor (Conti-Tronconi et al 1984). Agonist binding to these sites induces conformational changes that lead to desensitization or to channel opening. The high-affinity sites may play a role in both processes (Hess et al 1979; Neubig & Cohen 1980). However, their involvement in the latter process is uncertain since the conformational changes that have been experimentally observed are slower than the channel-opening event (Raftery et al 1983).

Snake neurotoxins block agonist binding to the high-affinity sites. Bound α-bungarotoxin, one of these toxins, has been localized by electron microscopy (Klymkowsky & Stroud 1979; Holtzman et al 1982; Zingsheim et al 1982) and X-ray diffraction (Fairclough et al 1983) to the top surface of the synaptic crest of the Achr. Under reducing conditions, affinity reagents also block agonist binding to the sites, covalently modifying the α subunit at a cysteine residue(s) (Damle & Karlin 1978). The affinity labeling experiment, as well as a variety of less direct evidence (to be reviewed in McCarthy et al 1986), indicates the existence of a disulfide bond within about 10 Å of each site. The requirement that the high-affinity sites lie on the surface of the synaptic domain near a disulfide bond led to an early proposal that they involved residues 130–144 in the α sequence (Noda et al 1982). Cys 130 and Cys 144 were postulated to form a disulfide bond, and the residues between them to form two strands of antiparallel β-sheet, separated by a hydrophilic turn, and including a possible glycosylation site. Residues 130–144 are highly conserved in all known Achr sequences, and are indeed predicted by sequence analysis to form amphipathic (surface-forming) β-strands (Noda et al 1982; Finer-Moore & Stroud 1984). A glycosylation site near the binding site (residue 143) is consistent with the proposal that agonist binding may be sensitive to the glycosylation state of the α subunit (Conti-Tronconi et al 1984; Mishina et al 1985). Three-dimensional modeling of the two β-strands shows that they may provide a surface with an electrostatic distribution that nicely complements that of the acetylcholine molecule (Noda et al 1982; Smart et al 1984; White 1985).

In two separate experiments the cysteines, which are covalently modified by the affinity label 4-(N-maleimido)benzyltrimethyl ammonium iodide (MBTA) (Karlin 1969), have been identified as cys 208 and/or 209 (Kao et al 1984), which are unique to the α subunit, and as cys 144 (Cahill & Schmidt 1984). These seemingly contradictory results may simply indicate that cysteines 130, 144, 208, and 209 are all close to one another in the three-dimensional structure of the α subunit. Genetically altering any one of these cysteines to serine greatly impairs the response to Ach, which suggests that they may all be in the vicinity of the binding site (Mishina et al 1985). The specific effects of the alterations also suggest, but do not prove, that disulfide bonds exist between cys 130 and 144, and between cys 208 and 209. A disulfide bond between two adjacent cysteines is feasible (Mitra & Chandrasekaran 1984), and has been seen in small peptides (Capasso et al 1977), but is unprecedented in protein structures. Peptide mapping experiments could definitively identify the disulfide bonds in the α subunit, and provide a firmer basis for detailed modeling of the high affinity sites.

ION CHANNEL   The ion channel is the functional region of the acetylcholine receptor most amenable to model building because information is available about the properties and structure of the channel from electrophysiological experiments and low-resolution structural studies (Horn & Stevens 1980). The residues involved specifically in channel function have been localized to the membrane domain of the topological map (Figure 3) by studying the effects on Achr function of deletions throughout the carboxyl-terminal half of the α subunit from *T. californica* (Mishina et al 1985). It is reasonably certain that the membrane domain is composed of close-packed helices. This structural motif, supported by X-ray diffraction, has now been seen in bacteriorhodopsin (Henderson & Unwin 1975), and in other membrane proteins. Recognition of this motif in Achr drastically reduces the degrees of freedom available for ion channel modeling.

The simplest channel model consistent with available information is one in which the amphipathic helix of each subunit contributes its hydrophilic face to the wall of the channel (Finer-Moore & Stroud 1984; Stroud 1983; Guy 1984). A 20° tilt of the five helices allows them to form the best close-packed side chain arrangement about a pore with a diameter of 6.5–7.0 Å at the narrowest point. A stereoview of this model is shown in Figure 5. Ions passing through the ion channel are thought to experience an environment similar to that of free water (Lewis & Stevens 1983), and thus cations do not appear to interact directly with the side-chains of the amino acids that line the channel. The effect of the side-chain conformations on ion conduction should therefore be small, provided that they are at an energy minimum and do not block the channel.

The free energy an ion would experience at any point in a channel, such as that shown in Figure 5, may be estimated using statistical mechanics methods (Bash & Stroud, manuscript in preparation). The results of such calculations are plotted in Figure 6 as a function of distance through the channel. As a result of the even distribution of positive and negative charge along the channel there are no local fluctuations of more than 2–3 kcal in the calculated free energy profile. This channel model is consistent with electrophysiological data in that it does not have tight ion binding sites or prohibitively large barriers that would limit the passage of cations.

Because the side-chains lining the channel wall do not appear to interact directly with cations, conservative changes in these side-chains are likely to alter, but not abolish, the conductance properties of the channel. Calculations that estimate ion conduction by a model channel are being developed (P. Bash, personal communication). They will allow us to predict the results of site-specific mutations of the postulated channel-forming residues. The correspondence between prediction and results may validate the model, or may suggest substantial revisions. In either case, the methods

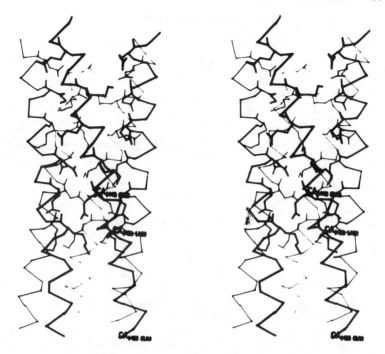

Figure 5   Stereo view of an atomic model of the ion channel. The proposed structure places alternating rings of positive and negative charge in the narrowest region of the channel as seen here. The high charge density may help to preserve the water-filled nature of the channel where purely hydrophobic or polar residues would permit collapse of the channel. This region has a net neutral charge probably important for ion conduction since a negatively charged channel would tend to capture cations, and a positively charged channel would repel them. The entrance to the channel is negatively charged and can account for specificity.

for estimating conductance properties of a given channel model will be a valuable tool for understanding the relationship between structure and function in the acetylcholine receptor.

WHERE IS THE GATE IN Achr?   Considering the narrow size of the channel across the bilayer where the gate must act, the gate can only be visualized directly in a high resolution structure determined by X-ray crystallographic analysis. However, there is one pertinent X-ray study carried out at a Rayleigh resolution of 6.25 Å, using one-dimensional terms to a resolution of 12.5 Å, that indirectly places the most likely location of the gate 15 Å in from the plane of the phosphatidyl head groups on the cytoplasmic surface. In this experiment, Tb^{3+} ions bound to the resting state of Achr were

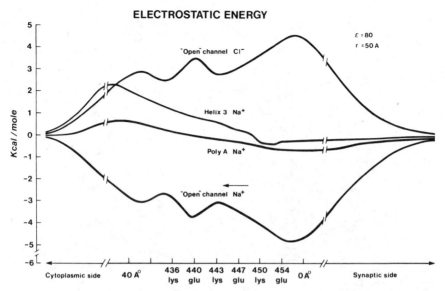

Figure 6 Energy profiles for a Na$^+$ and a Cl$^-$ ion as a function of distance from the extracellular side of the membrane are shown; they display cation selectivity. Calculations are for the model shown in Figure 5, and for similar models constructed with hydrophobic helices *M3* or for polyalanine helices. In this case conductivity would also be high, though the channel could easily collapse without need for solvation.

located by the unusually large anomalous dispersion of X-rays by Tb^{3+} or other lanthanides with unfilled d-orbitals (Fairclough et al 1985). This experiment was extremely well controlled; the differences in X-ray scattering due to the presence of terbium were recorded on single samples of terbium-treated Achr by precisely tuning the incident X-ray beam across the range of 7505–7525 electron volts. Only the scattering of Tb^{3+} changes under these conditions, so the positions of Tb^{3+} relative to the bilayer can be accurately determined. Three Tb^{3+} binding sites were located in the bilayer region, and therefore must be located within the central channel. These sites are about 11 Å apart from each other, which is very close to the distance between rings of negative charges in the ion channel model. The lowest density of Tb^{3+} electron content is located 15 Å from the cytoplasmic side of the bilayer. This is therefore the most likely place for the gate. The position of narrowest occlusion in closed channels can be measured (Coronado & Miller 1982). However, until dynamic structures at high resolution are determined, the stereochemistry of the gate will not be visualized. Site-specific mutagenesis in concert with structural and electrophysiological analysis can best focus on this critical and fascinating issue.

BIOSYNTHESIS AND ASSEMBLY OF Achr IN THE PLASMA MEMBRANES   The process of folding and assembling Achr requires several steps. The protein synthesis, cotranslational insertion, signal cleavage, and glycosylation in the endoplasmic reticulum take only about 1 minute or 1% of the time required for appearance of assembled, functional Ach receptors on the cell surface (Anderson & Blobel 1983; Merlie 1984). Monoclonal antibodies made against either α- or β-chains were used to follow conformational maturation, a process that within 15–30 min yields isolated 5S α-chains that only after this step can bind α-bungarotoxin. α- and β-chains are apparently over-produced by a factor of 3–4 over those eventually incorporated into Achr (Merlie et al 1982). After maturation, assembly of the subunits proceeds, possibly through formation of homopolymer aggregates (Anderson & Blobel 1983), over a period of 30–90 min, during passage through the Golgi to the cell surface (Merlie & Lindstrom 1983). Unglycosylated chains, and chains that are not assembled into a complex, are degraded within about 30 min.

The structural model of Achr derived from sequence analysis suggests the following method of assembly, which includes a testable hypothesis for the formation of the water-filled ion channel within a hydrophobic environment. The N-terminal region of each subunit folds into a β-sheet domain independent of the rest of the chain and before assembly. In the α subunit, this process would create the α-bungarotoxin binding site which appears within 15–30 min of protein synthesis. The β-sheet domain probably has little requirement for other determinants after residue 228, where the chain enters the bilayer; very few residues beyond this point protrude into the luminal side of the endoplasmic reticulum. The amphipathic helix A and the last hydrophobic helix would be unstable in the bilayer, since A contains many charged residues; it would cost about 2.3 kcals/mole to insert this sequence into a hydrophobic environment. Thus helix A would not easily enter the bilayer unless compensated by other counterionic stretches. One notion is that assembly may occur by the process indicated in Figure 7. Subunits come together initially through their synaptic components. The ionic amphipathic sequences only enter the membrane together as the subunits assemble.

6.   SEQUENCES FROM OTHER GENERA

A.   Sequence Homology and Evolution

The homology of the amino terminal ends of the four subunits of Achr from *T. californica* suggests that the subunits arose from a fourfold duplication of a single ancestral gene (Raftery et al 1980). The proposal that the subunits all derive from a common origin was borne out by sequencing of their

AcCh ASSEMBLY

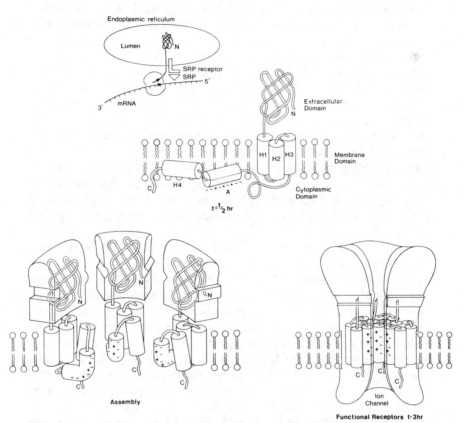

Figure 7 One proposed model for assembly of Achr. In the early phase subunits are assembled, but the hydrophilicity of *A* is too great to permit entry into the bilayer. The ribosomes are released from the membrane after *M3*, and *A* and *M4* float in the membrane until other subunits are correctly recruited. (In this figure *M* is represented by *H*.) Thus subunits appear after 0.5 hours, but fully assembled Achr appears after infolding of the *A* helices with counter-ions after about 3 hours.

cDNA clones: homology was evident throughout the chains (Noda et al 1983b). The amino acid differences per residue between all pairs of the four sequences indicated that a single gene duplication probably gave rise to two protogenes, each of which was in turn duplicated to give rise to the α and β, and the γ and δ lineages (Noda et al 1983b). The β sequence is more similar to the sequences of the γ and δ subunits than is the α sequence, indicating different rates of change for the α and β lineages (Noda et al 1983b).

Genomic or cDNA encoding components of Achr from the neuromuscular junctions of chicken and mammal species have been sequenced (Noda et al 1983a; Takai et al 1984; Tanabe et al 1984; Nef et al 1984; LaPolla et al 1984; Shibahara et al 1985). Each sequence shows greatest homology to the *Torpedo* subunit of comparable molecular weight (Table 3). Sequence homology decreases with an increase in the evolutionary distance between species (Table 4). These results were anticipated from inspection of the N-terminal sequences of the Achr subunits from *E. electricus* electric organ, and calf muscle (Conti-Tronconi et al 1982a,b). They suggest that Achr from all sources studied to date are directly descended from the products of the original fourfold gene duplication, and that the four genes developed specific functions which have been preserved through evolutionary history (Conti-Tronconi et al 1982a).

Table 4 indicates that provided additional gene duplications have not occurred the α subunit is changing through evolution at a much slower rate than the other subunits. The α subunit sequences from organisms as diverse as humans and electric fish show 80% homology (Noda et al 1983a). The evolutionary rate for a particular class of proteins is believed to be related to the tolerance for change of that protein and to its dispensability (reviewed in Wilson et al 1977). Based on this notion, the unusually high homology between the α subunits emphasizes that it is the most important component of the Achr complex.

The proposal that each subunit of the Achr has a unique role has been validated by experiments using artificial expression systems. Such systems allow large quantities of functional protein to be expressed from the four cDNA clones of *T. californica* Achr (Mishina et al 1984; White et al 1985). When any one of the cDNAs is omitted, the amount of functional receptor detected is, at most, 3% of the control (Mishina et al 1984; White et al 1985).

Table 3 Homology of chains from calf, human, and chicken Achr to chains from *T. californica*

| | | α | | β | | γ | | δ | |
		calf	human	calf	chick	calf	human	chick	mouse
	M.W.*	49897	49694	55055	56484	55936	55724	57215	57104
α	50116	114†	112	61	55	53	53	59	58
β	53683	66	66	84	65	63	63	66	62
γ	56279	55	55	64	93	82	81	76	72
δ	57565	57	58	61	73	65	65	92	89

* Molecular weight calculated from sequence.

† Average of the top 400 scores obtained by matching all possible 25 residue segments in one sequence with all possible 25 residue segments in the second (George et al 1984). Scores for a pair of segments are computed by assigning every residue pair a number based on their tendency to replace one another in evolutionarily related sequences (Schwartz & Dayhoff 1979). The score average is related to the percent of homology, but is independent of sequence alignment.

Table 4 Homology between sequences from different organisms

α

	T. marmorata	calf	human
T. californica	131*	114	112
T. marmorata	—	114	112
calf	—	—	127

γ

	chick	calf	human
T. californica	93*	82	81
chick	—	95	93
calf	—	—	120

δ

	chick	mouse
T. californica	92*	89
chick	—	100

* Scores computed as for Table 3.

However, when cDNA for the δ subunit of *Torpedo* is replaced with cDNA encoding the δ subunit of the mouse Achr, protein is obtained that responds to Ach with a fourfold greater conduction than normal *Torpedo* Achr (White et al 1985). Combinations of *Torpedo* α, β, and δ cDNAs with the mouse δ cDNA gave no detectable functional protein when γ cDNA was omitted.

These results are surprising because the sequence homology between mouse and *Torpedo* δ chains (59%) is not much greater than the homology between the *Torpedo* γ and mouse δ chains (50%) (LaPolla et al 1984). Based on sequence homology alone, the mouse δ subunit might be expected to substitute equally well for *Torpedo* γ or δ subunits. Inspection of the sequences showed no regions that were exceptionally homologous in both species of δ chains and that might therefore encode a specific function for this subunit (LaPolla et al 1984). It was proposed that the residues that give the δ subunit its unique character are those involved in subunit-subunit contacts. These residues would lie on the surface of secondary structure elements, and would therefore be detected as stretches of sequence in which residues unique to and homologous in the δ subunits occur in a periodic fashion (White et al 1985).

B.   Homology of the Functional Regions of the Sequences

The arguments used to justify different evolutionary rates for different classes of proteins have also been used to explain why different regions of related protein sequences are more highly conserved than others (Wilson et al 1977). Regions of exceptionally high homology in a set of related sequences are presumably regions that are crucial for protein structure and function. Stretches of low homology indicate regions, such as exterior loops, that are of little significance during protein folding, and whose role in protein function is general enough that frequent sequence modification can be tolerated.

In general, the functional regions of Achr identified by deletion mapping and oligonucleotide mutagenesis (Mishina et al 1985) show less variation among organisms for all subunit types than do the sequences as a whole (Table 5). The exception is the proposed channel-forming helix, which for the β, γ, and δ subunits lies in one of the regions of lowest sequence homology. However, when conservative substitutions (Dayhoff et al 1978) are counted as matches (Table 6) a substantial degree of structural homology among these putative helices is evident. The proposed role for these helices is to provide the hydrophilic wall of a water-filled channel, and not to interact directly with ions. The primary requirement for this function is amphipathicity, and this quality is preserved in all sequences.

Table 5 Percent of identical residues among all sequences of each subunit type

	α	β	γ	δ
mature chain	76.6	56.3	48.1	52.5
synaptic domain	74.4	56.2	55.3	53.9
residues 125-150	88.5	84.6	61.5	61.5
M1	88.5	73.1	73.1	65.4
M2	100.0	84.6	76.9	69.2
M3	84.6	84.6	53.9	61.5
cytoplasmic domain	72.7	38.8	25.0	51.5
A	77.8	44.4	40.7	22.2
M4	80.8	57.7	46.1	50.0

C. Intron-Exon Boundaries

Genomic DNA sequences have been determined for chicken DNA encoding the γ, δ, and part of the α Achr subunits (Nef et al 1984; Ballivet et al 1983), and for human genomic DNA encoding the α and γ subunits (Noda et al 1983a; Shibahara et al 1985). The intron-exon structures of these genes are very similar (Figure 2). For the γ and δ subunits of both organisms, introns interrupt the protein coding regions in the same relative positions. Twelve protein regions, including the prepeptide, a postulated Ach-binding sequence, and each of the membrane-spanning regions, are encoded by separate exons. In the α gene the introns are similarly placed, but some are missing, namely those immediately preceding the first hydrophobic stretch, those immediately preceding the amphipathic helix, and the intron between hydrophobic stretches two and three.

The intron-exon structures are consistent with the hypothesis that exons frequently encode functional or structural units in a protein (Gilbert 1978; Blake 1978). The tendency for membrane-spanning regions to be encoded

Table 6   Percent of similar residues among all sequences of each subunit type

	α	β	γ	δ
mature chain	90.2	70.4	68.7	74.1
synaptic domain	85.9	72.6	77.4	75.7
residues 125-150	96.2	88.5	92.3	92.3
M1	96.2	84.6	88.5	80.8
M2	100.0	96.2	92.3	76.9
M3	100.0	96.2	88.5	80.8
cytoplasmic domain	90.9	51.5	43.3	65.7
A	88.9	66.7	63.0	66.7
M4	92.3	73.1	61.5	84.6

by separate exons has been observed in at least two other membrane proteins (Nathans & Hogness 1983; Liscum et al 1985). The intron positions are also consistent with the observation that intron-exon splice sites, whether or not they form the boundary of a structural or functional unit, generally occur at protein surfaces (Craik et al 1982). One explanation for this phenomenon is that in the processes of combining exons to form a modified protein and of randomly incorporating intron sequences into a gene a greater-than-average risk of mutation exists at the splice sites. These mutations would be best tolerated at the residues not intimately involved in the packing interactions of the tertiary structure (Craik et al 1982). As predicted, almost all of the intron-exon splice sites in the Achr genes occur at turns or loops in the synaptic region of the protein, or at the ends of membrane-spanning helices. The only exception is an intron in each of the γ and δ genes that lies in the middle of the predicted amphipathic, surface-forming β strand just following cys 144.

CONCLUSION

The Achr presents a uniquely advantageous format for concerted resolution of some of the most challenging problems of neurobiology, membrane biogenesis, membrane protein structure, transmembrane signaling, and genetic regulation.

ACKNOWLEDGMENT

Related research was supported by NIH-GM 24485, and by NSF, PCM 83-16401 to R.M.S. We thank Mel Jones for constructing the model of Figure 1, Paul Bash and Mei Lie Wong for preparation of Figure 5, and Dr. Hugo Martinez for use of his sequence alignment programs. We also thank Kathy Mixter-Mayne and Drs. Michael White and Henry Lester for discussing the results of their work before publication.

Literature Cited

Anderson, D. J., Blobel, G. 1981. In vitro synthesis, glycosylation and membrane insertion of the four subunits of *Torpedo* acetylcholine receptor. *Proc. Natl. Acad. Sci. USA* 78:5598–5602

Anderson, D. J., Blobel, G. 1983. Identification of homo-oligomers as potential intermediates in acetylcholine receptor subunit assembly. *Proc. Natl. Acad. Sci. USA* 80:4359–63

Anderson, D. J., Blobel, G., Tzartos, S., Gullick, W., Lindstrom, J. 1983. Transmembrane orientation of an early biosynthetic form of acetylcholine receptor δ subunit determined by proteolytic dissection in conjunction with monoclonal antibodies. *J. Neurosci.* 3:1773–84

Anderson, D. J., Walter, P., Blobel, G. 1982. Signal recognition protein is required for the integration of acetylcholine receptor δ subunit, a transmembrane glycoprotein, into the endoplasmic reticulum membrane. *J. Cell Biol.* 93:501–6

Ballivet, M., Nef, P., Stadler, R., Fulpius, B. 1983. Genomic sequences encoding the α-subunit of acetylcholine receptor are conserved in evolution. *Cold Spring Harbor Symp. Quant. Biol.* 48:83–87

Ballivet, M., Patrick, J., Lee, J., Heinemann, S. 1982. Molecular cloning of cDNA coding for the γ subunit of *Torpedo* acetylcholine receptor in its membrane environment. *Proc. Natl. Acad. Sci. USA* 74:4460–70

Blake, C. C. F. 1978. Do genes in pieces imply proteins in pieces? *Nature* 273:267–68

Bon, F., Lebrun, E., Gomel, J., van Rapenbusch, R., Cartaud, J., et al. 1984. Image analysis of the heavy form of the acetylcholine receptor from *Torpedo marmorata*. *J. Mol. Biol.* 176:205–37

Brisson, A., Unwin, P. N. T. 1984. Tubular crystals of acetylcholine receptor. *J. Cell Biol.* 99:1202–11

Cahill, S., Schmidt, J. 1984. An immunological approach to the identification of the MBTA binding site of the nicotinic acetylcholine receptor of *Torpedo californica*. *Biochem. Biophys. Res. Commun.* 122:602–8

Capasso, S., Mattia, C., Mazzarella, L., Puliti, R. 1977. Structure of a cis-peptide unit: Molecular conformation of the cyclic disulphide L-cysteinyl-L-cysteine. *Acta Crystallogr.* B33:2080–83

Cartaud, J., Sobel, A., Rousselet, A., Devaux, P. F., Changeux, J.-P. 1981. Consequences of alkaline treatment for the ultrastructure of the acetylcholine receptor-rich membranes from *Torpedo marmorata* electric organ. *J. Cell Biol.* 90:418–26

Cash, D. J., Hess, G. P. 1980. Molecular mechanisms of acetylcholine receptor-controlled ion translocation across cell membranes. *Proc. Natl. Acad. Sci. USA* 77:842–46

Chang, E. L., Yager, P., Williams, R. W., Dalziel, A. W. 1983. The secondary structure of reconstituted acetylcholine receptor as determined by Raman spectroscopy. *Biophys. J.* 41:65a

Claudio, T., Ballivet, M., Patrick, J.,

Heinemann, S. 1983. Nucleotide and deduced amino acid sequences of *Torpedo californica* acetylcholine receptor α subunit. *Proc. Natl. Acad. Sci. USA* 80:1111–15

Colquhoun, D., Sakmann, B. 1981. Fluctuations in the microsecond time range of the current through single acetylcholine receptor ion channels. *Nature* 294:464–66

Conti-Tronconi, B. M., Gotti, C. M., Hunkapiller, M. W., Raftery, M. A. 1982b. Mammalian muscle acetylcholine receptor: a supramolecular structure formed by four related proteins. *Science* 218:1227–29

Conti-Tronconi, B. M., Hunkapiller, M. W., Lindstrom, J. M., Raftery, M. A. 1982a. Subunit structure of the acetylcholine receptor from *Electrophorus electricus*. *Proc. Natl. Acad. Sci. USA* 79:6489–93

Conti-Tronconi, B. M., Hunkapiller, M. W., Raftery, M. A. 1984. Molecular weight and structural nonequivalence of the mature α subunits of *Torpedo californica* acetylcholine receptor. *Proc. Natl. Acad. Sci. USA* 81:2631–34

Conti-Tronconi, B. M., Raftery, M. A. 1982. The nicotinic cholinergic receptor: Correlation of molecular structure with functional properties. *Ann. Rev. Biochem.* 51:491–530

Cooper, J. R., Meyer, E. M. 1984. Possible mechanisms involved in the release and modulation of release of neuroactive agents. *Neurochem. Int.* 6:419–33

Coronado, R., Miller, C. 1982. Conduction and block by organic cations in a K^+ selective channel from sarcoplasmic reticulum incorporated into planar phospholipid bilayer membranes. *J. Gen. Physiol.* 79:529–47

Craik, C. S., Sprang, S., Fletterick, R., Rutter, W. J. 1982. Intron-exon splice junctions map at protein surfaces. *Nature* 299:180–82

Damle, V. N., Karlin, A. 1978. Affinity labeling of one of two α-neurotoxin binding sites in acetylcholine receptor from *Torpedo californica*. *Biochemistry* 17:2039–45

Dayhoff, M. O., Schwartz, R. M., Orcutt, B. C. 1978. A model of evolutionary change in proteins in *Atlas of Protein Sequence and Structure*, Vol. 5, Sup. 3, ed. M. O. Dayhoff, pp. 345–52. Washington D.C.: Natl. Biomed. Res. Found.

Devillers-Thiery, A., Changeux, J.-P., Parotaud, P., Strosberg, A. D. 1979. The amino terminal sequence of the 40 K subunit of the acetylcholine receptor protein from *Torpedo marmorata*. *FEBS Lett.* 104:99–105

Devillers-Thiery, A., Giraudat, J., Bentaboulet, M., Changeux, J.-P. 1983. Complete mRNA coding sequence of the acetylcholine binding α-subunit from *Torpedo marmorata*: A model for the transmembrane organization of the polypeptide chain. *Proc. Natl. Acad. Sci. USA* 80:2067–71

Dunant, Y., Israel, M. 1985. The release of acetylcholine. *Scientific American* 252:58–83

Dunn, S. M. J., Conti-Tronconi, B. M., Raftery, M. A. 1983. Separate sites of low and high affinity for agonists on *Torpedo californica* acetylcholine receptor. *Biochemistry* 22:6264–72

Dunn, S. M. J., Raftery, M. A. 1982. Multiple binding sites for agonists on *Torpedo californica* acetylcholine receptor. *Biochemistry* 21:6264–72

Dwyer, T. M., Adams, D. J., Hille, B. 1980. The permeability of the endplate channel to organic cations in frog muscle. *J. Gen. Physiol.* 75:469–92

Eisenberg, D., Weiss, R. M., Terwilliger, T. C. 1984. The hydrophobic moment detects periodicity in protein hydrophobicity. *Proc. Natl. Acad. Sci. USA* 81:140–44

Fairclough, R. H., Finer-Moore, J., Love, R. A., Kristofferson, D., Desmeules, P. J., Stroud, R. M. 1983. Subunit organization and structure of an acetylcholine receptor. *Cold Spring Harbor Symp. Quant. Biol.* 48:9–20

Fairclough, R. H., Miake-Lye, R. C., Stroud, R. M., Hodgson, K. O., Doniach, S. 1985. Location of Tb (III) binding sites on acetylcholine receptor enriched membranes. In press

Finer-Moore, J., Stroud, R. M. 1984. Amphipathic analysis and possible formation of the ion channel in an acetylcholine receptor. *Proc. Natl. Acad. Sci. USA* 81:155–59

Furukawa, T., Furukawa, A. 1959. Effects of methyl- and ethyl-derivatives of NH_4^+ on the neuromuscular junction. *Jpn. J. Physiol.* 9:130–42

Garnier, J., Osguthorpe, D. J., Robson, B. 1978. Analysis of the accuracy and implications of simple methods for predicting secondary structure of globular proteins. *J. Mol. Biol.* 120:97–120

George, D. G., Orcutt, B. C., Dayhoff, M. O., Barker, W. C. 1984. Program for detecting distant relationships. *PIR Rep. REL-0484*. Washington D.C.: Natl. Biomed. Res. Found.

Gilbert, W. 1978. Why genes in pieces? *Nature* 271:501

Giraudat, J., Devillers-Thiery, A., Auffrey, C., Rougeon, F., Changeux, J.-P. 1982. Identification of a cDNA clone coding for the acetylcholine binding subunit of *Torpedo marmorata* acetylcholine receptor. *EMBO J.* 1:713–17

Gordon, A. S., Milfoy, D., Diamond, I. 1983. Identification of a molecular weight 43,000 protein kinase in acetylcholine receptor-enriched membranes. *Proc. Natl. Acad. Sci. USA* 80:5862–65

Guy, H. R. 1984. A structural model of the acetylcholine receptor channel based on partition energy and helix packing calculations. *Biophys. J.* 45:249–61

Henderson, R., Unwin, P. N. T. 1975. Three-dimensional model of purple membrane obtained by electron microscopy. *Nature* 257:28–32

Hess, G. P., Cash, D. J., Aoshima. H. 1979. Acetylcholine receptor-controlled ion fluxes in membrane vesicles investigated by fast reaction techniques. *Nature* 282:329–31

Hess, G. P., Cash, D. J. Aoshima, H. 1983. Acetylcholine receptor-controlled ion translocation: Chemical kinetic investigations of the mechanism. *Ann. Rev. Biophys. Bioeng.* 12:443–73

Hess, G. P., Pasquale, E. B., Walker, J. W., McNamee, M. G. 1982. Comparison of acetylcholine receptor-controlled cation flux in membrane vesicles from *Torpedo californica* and *Electrophorus electricus*: Chemical kinetic measurements in the millisecond region. *Proc. Natl. Acad. Sci. USA* 79:963–67

Holtzman, E., Wise, D., Wall, J., Karlin, A. 1982. Electron microscopy of complexes of isolated acetylcholine receptor, biotinyltoxin, and avidin. *Proc. Natl. Acad. Sci. USA* 79:310–14

Horn, R., Stevens, C. F. 1980. Relations between structure and function of ion channels. *Comments Mol. Cell. Biophys.* 1:57–68

Huang, L.-Y. M., Catterall, W. A., Ehrenstein, G. 1978. Selectivity of cations and nonelectrolytes for acetylcholine-activated channels in cultured muscle cells. *J. Gen. Physiol.* 71:397–410

Kao, P. N., Dwork, A. J., Kaldany, R.-R. J., Silver, M. L., Wideman, J., et al. 1984. Identification of the α-subunit half-cystine specifically labeled by an affinity reagent for the acetylcholine receptor binding site. *J. Biol. Chem.* 259:11662–65

Karlin, A. 1969. Chemical modification of the active site of the acetylcholine receptor. *J. Gen. Physiol.* 54:245s–64s

Karlin, A. 1980. Molecular properties of nicotinic acetylcholine receptors. *Cell Surf. Rev.* 6:191

Karlin, A. 1983. The anatomy of a receptor. *Neurosci. Comm.* 1:111–23

Karlin, A., Cox, R., Kaldany, R. R., Lobel, P., Holtzman, E. 1983a. The arrangements and functions of the chains of the acetylcholine receptor of *Torpedo* electric tissue.

Cold Spring Harbor Symp. Quant. Biol. 48:1–8

Karlin, A., Holtzman, E., Yodh, N., Lobel, P., Wall, J., Hainfeld, J. 1983b. The arrangement of the subunits of the acetylcholine receptor of *Torpedo californica. J. Biol. Chem.* 258:6678–81

Katz, B. 1962. The transmission of impulses from nerves to muscle and the subcellular unit of synaptic action. *Proc. R. Soc. London B* 155:455–77

Kistler, J., Stroud, R. M. 1981. Crystalline arrays of membrane-bound acetylcholine receptor. *Proc. Natl. Acad. Sci. USA* 78:3678–82

Kistler, J., Stroud, R. M., Klymkowsky, M. W., Lalancette, R. A., Fairclough, R. H. 1982. Structure and function of an acetylcholine receptor. *Biophys. J.* 37:371–83

Klymkowsky, M. W., Heuser, J. E., Stroud, R. M. 1980. Protease effects on the structure of acetylcholine receptor membranes from *Torpedo californica. J. Cell Biol.* 85:823–38

Klymkowsky, M. W., Stroud, R. M. 1979. Immunospecific identification and three-dimensional structure of a membrane-bound acetylcholine receptor *Torpedo californica. J. Mol. Biol.* 128:319–34

LaPolla, R. J., Mayne, K. M., Davidson, N. 1984. Isolation and characterization of cDNA clone for the complete protein coding region of the δ subunit of the mouse acetylcholine receptor. *Proc. Natl. Acad. Sci. USA* 81:7970–74

LaRochelle, W. J., Wray, B. E., Sealock, R., Froehner, S. C. 1985. Immunochemical demonstration that amino acids 360–377 of the acetylcholine receptor gamma-subunit are cytoplasmic. *J. Cell Biol.* 100:684–91

Lewis, C. A., Stevens, C. F. 1983. Acetylcholine receptor channel ionic selectivity: Ions experience an aqueous environment. *Proc. Natl. Acad. Sci. USA* 80:6110–13

Lindstrom, J. 1983. Using monoclonal antibodies to study acetylcholine receptors and *myasthenia gravis. Neurosci. Comm.* 1:139–56

Lindstrom, J., Criado, M., Hochschwender, S., Fox, J. L., Sarin, V. 1984. Immunochemical tests of acetylcholine receptor subunit models. *Nature* 311:573–75

Lindstrom, J., Einarson, B., Francy, M. 1977. Acetylcholine receptors and myasthenia gravis: the effect of antibodies to eel acetylcholine receptors on eel electric organ cells. In *Cellular Neurobiology*, ed. Z. Hall, R. Kelly, C. F. Fox, pp. 119–30. New York: Liss

Lindstrom, J., Gullick, W., Conti-Tronconi, B., Ellisman, M. 1980. Proteolytic nicking of the acetylcholine receptor. *Biochemistry* 19:4791–95

Lindstrom, J., Merlie, J., Yogeeswaran, G. 1979. Biochemical properties of acetylcholine receptor subunits from *Torpedo californica. Biochemistry* 18:4465–70

Liscum, L., Finer-Moore, J., Stroud, R. M., Luskey, K. L., Brown, M. S., Goldstein, J. L. 1985. Domain structure of 3-hydroxy-3-methylglutaryl coenzyme A reductase, a glycoprotein of the endoplasmic reticulum. *J. Biol. Chem.* 260:522–30

Maeno, T., Edwards, C., Anraku, M. 1977. Permeability of the end-plate membrane activated by acetylcholine to some organic cations. *J. Neurobiol.* 8:173–84

McCarthy, M. P., Earnest, J. P., Young, E. F., Choe, S., Stroud, R. M. 1986. The molecular neurobiology of the acetylcholine receptor. *Ann. Rev. Neurosci.* In press

Merlie, J. P. 1984. Biogenesis of the acetylcholine receptor, a multisubunit integral membrane protein. *Cell* 36:573–75

Merlie, J. P., Lindstrom, J. 1983. Assembly in vivo of mouse muscle acetylcholine receptor: identification of an α subunit species that may be an assembly intermediate. *Cell* 34:747–57

Merlie, J. P., Sebbane, R., Tzartos, S., Lindstrom, J. 1982. Inhibition of glycosylation with tunicamycin blocks assembly of newly synthesized acetylcholine receptor subunits in muscle cells. *J. Biol. Chem.* 257:2694–2701

Mishina, M., Kurosaki, T., Tobimatsu, T., Marimoto, Y., Noda, M. et al. 1984. Expression of functional acetylcholine receptor from cloned cDNAs. *Nature* 307:604–8

Mishina, M., Tobimatsu, T., Imoto, K., Tanaka, K., Fujita, Y. et al. 1985. Location of functional regions of acetylcholine receptor α-subunit by site-directed mutagenesis. *Nature* 313:364–69

Mitra, A. K., Chandrasekaran, R. 1984. Conformational flexibilities in malformin A. *Biopolymers* 23:2513–24

Moore, W. M., Holliday, L. A., Puett, D., Brady, R. N. 1974. On the conformation of the acetylcholine receptor protein from *Torpedo nobiliana. FEBS Lett.* 45:145–49

Nathans, J., Hogness, D. S. 1983. Isolation, sequence analysis, and intron/exon arrangement of the gene encoding bovine rhodopsin. *Cell* 34:807–14

Nef, P., Mauron, A., Stalder, R., Alliod, C., Ballivet, M. 1984. Structure, linkage and sequence of the two genes encoding the δ and γ subunits of the nicotinic acetylcholine receptor. *Proc. Natl. Acad. Sci. USA* 81:7975–79

Neubig, R. R., Cohen, J. B. 1980. Permea-

bility control by cholinergic receptors in *Torpedo* post-synaptic membranes: Agonist dose-response relations measured at second and millisecond times. *Biochemistry* 19:2770–79

Neubig, R. R., Krodel, E. K., Boyd, N. D., Cohen, J. B. 1979. Acetylcholine and local anesthetic binding to *Torpedo* nicotinic post-synaptic membranes after removal of non-receptor peptides. *Proc. Natl. Acad. Sci. USA* 76:690–94

Noda, M., Furutani, Y., Takahashi, H., Toyosato, M., Tanabe, T., et al. 1983a. Cloning and sequence analysis of calf cDNA and human genomic DNA encoding α-subunit precursor of *Torpedo californica* acetylcholine receptor deduced from cDNA sequence. *Nature* 299:793–97

Noda, M., Takahashi, H., Tanabe, T., Toyosato, M., Furutani, Y., et al. 1982. Primary structure of α-subunit precursor of *Torpedo californica* acetylcholine receptor deduced from a cDNA sequence. *Nature* 299:793–97

Noda, M., Takahashi, H., Tanabe, T., Toyosato, M., Kikyotani, S., et al. 1983b. Structural homology of *Torpedo californica* acetylcholine receptor subunits. *Nature* 302:528–32

Noda, M., Takahashi, H., Tanabe, T., Toyosa, M., Kikyotani, S., et al. 1983c. Primary structures of β- and δ-subunit precursors of *Torpedo californica* acetylcholine receptor deduced from cDNA sequences. *Nature* 301:251–55

Ovchinnikov, Yu. A., Abdulaev, N. G., Feigina, M. Yu., Kiselev, A. V., Lobanov, N. A. 1979. The structural basis of the functioning of bacteriorhodopsin: an overview. *FEBS Lett.* 100:219–24

Popot, J.-L., Changeux, J.-P. 1984. Nicotinic receptor of acetylcholine: Structure of an oligomeric integral membrane protein. *Physiol. Rev.* 64:1162–1239

Raftery, M. A., Dunn, S. M. J., Conti-Tronconi, B. M., Middlemas, D. S., Crawford, R. D. 1983. The nicotinic acetylcholine receptor: Subunit structure, functional binding sites, and ion transport properties. *Cold Spring Harbor Symp. Quant. Biol.* 48:21–23

Raftery, M. A., Hunkapiller, M. W., Strader, C. D., Hood, L. E. 1980. Acetylcholine receptor: Complex of homologous subunits. *Science* 208:1454–57

Ratnam, M., Lindstrom, J. 1984. Structural features of the nicotinic acetylcholine receptor revealed by antibodies to synthetic peptides. *Biochem. Biophys. Res. Comm.* 122:1225–33

Reeke, G. N. Jr., Becker, J. W., Edelman, G. M. 1975. The covalent and three-dimensional structure of concanavalin A.

IV. Atomic coordinates, hydrogen bonding, and quarternary structure. *J. Biol. Chem.* 250:1525–47

Reichardt, L. F., Kelly, R. B. 1983. A molecular description of nerve terminal function. *Ann. Rev. Biochem.* 52:871–926

Reynolds, J., Karlin, A. 1978. Molecular weight in detergent solution of acetylcholine receptor from *Torpedo californica*. *Biochemistry* 17:2035–38

Richardson, J. 1981. The anatomy and taxonomy of protein structure. *Adv. Prot. Chem.* 34:167–339

Ross, M. J., Klymkowsky, M. W., Agard, D. A., Stroud, R. M. 1977. Structural studies of a membrane-bound acetylcholine receptor from *Torpedo californica*. *J. Mol. Biol.* 116:635–59

Rousselet, A., Cartaud, J., Devaux, P. F. Changeux, J.-P. 1982. The rotational diffusion of the acetylcholine receptor in *Torpedo marmorata* membrane fragments studied with a spin-labeled α-toxin: Importance of the 43,000 protein(s). *EMBO J.* 1:439–45

Rübsamen, H., Eldefrawi, A. T., Eldefrawi, M. E., Hess, G. P. 1978. Characterization of the calcium-binding sites of the purified acetylcholine receptor and identification of the calcium-binding subunit. *Biochemistry* 17:3818–25

Schwartz, R. M., Dayhoff, M. O. 1979. Matrices for detecting distant relationships. In *Atlas of Protein Sequence and Structure*, Vol. 5, Sup. 3, ed. M. O. Dayhoff, pp. 353–58. Washington D.C.: Natl. Biomed. Res. Found.

Sealock, R., Wray, B. E., Froehner, S. C. 1984. Ultrastructural localization of the M_r 43,000 protein and the acetylcholine receptor in *Torpedo* postsynaptic membranes using monoclonal antibodies. *J. Cell Biol.* 98:2239–44

Shibahara, S., Kubo, T., Perski, H. J., Takahashi, H., Noda, M., Numa, S. 1985. Cloning and sequence analysis of human genomic DNA encoding γ subunit precursor of muscle acetylcholine receptor. *Eur. J. Biochem.* 146:15–22

Sine, S. M., Taylor, P. 1980. The relationship between occupation and the permeability response of the cholinergic receptor revealed by bound cobra α-toxin. *J. Biol. Chem.* 255:10144–56

Smart, L., Hans-Wilhelm, M., Hilgenfeld, R., Saenger, W., Maelicke, A. 1984. A structural model for the ligand-binding sites at the nicotinic acetylcholine receptor. *FEBS Lett.* 178:64–68

Spivak, C. E., Albuquerque, E. X. 1982. Dynamic properties of the nicotinic acetylcholine receptor ionic channel complex: Activation and blockade. In *Progress in*

Cholinergic Biology, Vol. 2, ed. I. Hanin, A. M. Goldberg, pp. 323–57. New York: Raven

Strader, C. D., Raftery, M. A. 1980. Topographic studies of *Torpedo* acetylcholine receptor subunits as a transmembrane complex. *Proc. Natl. Acad. Sci. USA* 77:5807–11

Stroud, R. M. 1983. Acetylcholine receptor structure. *Neurosci. Comm.* 1:124–38

Stroud, R. M., Agard, D. A. 1979. Structure determination of asymmetric membrane profiles using an iterative Fourier method. *Biophys. J.* 25:495–512

Sumikawa, K., Houghton, J., Smith, J. G., Bell, L., Richards, B. M., Barnard, E. A. 1982. The molecular cloning and characterization of cDNA coding for the α subunit of the acetylcholine receptor. *Nucleic Acids Res.* 10:5809–22

Takai, T., Noda, M., Furutani, Y., Takahashi, H., Notake, M., et al. 1984. Primary structure of γ subunit precursor of calf-muscle acetylcholine receptor deduced from the cDNA sequence. *Eur. J. Biochem.* 143:109–15

Tanabe, T., Noda, M., Furutani, Y., Takai, T., Takahashi, H., et al. 1984. Primary structure of β subunit precursor of calf muscle acetylcholine receptor deduced from cDNA sequence. *Eur. J. Biochem.* 144:11–17

Tauc, L. 1982. Nonvesicular release of neurotransmitter. *Physiol. Rev.* 62:857–93

Walker, J. H., Baustead, C. M., Witzemann, V. 1984. The 43-K protein, v_1, associated with acetylcholine receptor containing membrane fragments is an actin-binding protein. *EMBO J.* 3:2287–90

Wennogle, L. P., Changeux, J.-P. 1980. Transmembrane orientation of proteins present in acetylcholine receptor rich membranes from *Torpedo marmorata* studied by selective proteolysis. *Eur. J. Biochem.* 106:381–93

Wennogle, L. P., Oswald, R., Saitoh, T., Changeux, J.-P. 1981. Dissection of the 66,000 dalton subunit of the acetylcholine receptor. *Biochemistry* 20:2492–97

White, M. M. 1985. Designer channels: site-directed mutagenesis as a probe for structural features of channels and receptors. *Trends Neurosci.* In press

White, M. M., Mayne, K. M., Lester, H. A., Davidson, N. 1985. Mouse-torpedo hybrid acetylcholine receptors: functional homology does not equal sequence homology. *Proc. Natl. Acad. Sci. USA.* In press

Whittaker, V. P., Michaelson, I. A., Kirkland, R. J. A. 1964. The separation of synaptic vesicles from nerve-ending particles ("synaptosomes"). *Biochem. J.* 128:833–46

Wilson, A. C., Carlson, S. S., White, T. J.

1977. Biochemical evolution. *Ann. Rev. Biochem.* 46:573–639

Witzemann, V., Raftery, M., 1978. Ligand binding sites and subunit interactions of *Torpedo californica* acetylcholine receptor. *Biochemistry* 17:3593–604

Young, E. F., Ralston, E., Blake, J., Ramachandran, J., Hall, Z. W., Stroud, R. M. 1985. Topological mapping of acetylcholine receptor: Evidence for a model with five transmembrane segments and a cytoplasmic COOH-terminal peptide.

Proc. Natl. Acad. Sci. USA 82:626–30

Zingsheim, H. P., Barrantes, F. J., Frank, J., Hanicke, W., Neugebauer, D.-C. 1982. Direct structural localization of two toxin-recognition sites on an acetylcholine receptor protein. *Nature* 299:81–84

Zingsheim, H. P., Neugebauer, D.-C., Barrantes, F. J., Frank, J. 1980. Structural details of membrane-bound acetylcholine receptor from *Torpedo marmorata*. *Proc. Natl. Acad. Sci. USA* 77:952–56

Ann. Rev. Cell Biol. 1985. 1 : 353–402

NONMUSCLE ACTIN-BINDING PROTEINS[1]

T. P. Stossel, C. Chaponnier, R. M. Ezzell,
J. H. Hartwig, P. A. Janmey, D. J. Kwiatkowski,
S. E. Lind, D. B. Smith, F. S. Southwick, H. L. Yin,
K. S. Zaner

Hematology-Oncology Unit, Massachusetts General Hospital,
Department of Medicine, Harvard Medical School, Boston,
Massachusetts 02114

CONTENTS

1. INTRODUCTION

Actin is one of the most abundant and highly conserved proteins in nature, and is a principal and often the major protein component of eukaryotic cells. Actin molecules self-assemble to form a large variety of three-dimensional structures that influence cell form and function, and this assembly of actin is regulated in a remarkably precise manner; it occurs at particular times and in discrete places within cells. It is now evident that this

[1] Supported by grants from the USPHS (HL19429, HL29113, HL00912, GM09523, GM09976, AI18294), the Council for Tobacco Research, and the Edwin S. Webster Foundation.

0743–4634/85/1115–0353$02.00

control is conferred by proteins collectively called "actin-binding proteins," which are the objects of this review. Actin and actin-binding proteins are both intrinsically interesting and generally relevant.

By the mid 1970s, after a quarter century of muscle research, the list of actin-binding proteins of known and unknown function was short. It included myosin, tropomyosin, and troponin, which are integral to the force-generating mechanism and its regulation in muscle (Szent-Gyorgyi 1947; Huxley 1963; Ebashi 1972). Four other myofibrillar constituents, named alpha-, beta-, gamma-, and eu-actinins, were reported to bind to actin and affect its assembly, although their role in the function of muscle is still unclear (Ebashi et al 1964; Ebashi & Ebashi 1965; Maruyama & Ebashi 1965; Kuroda & Maruyama 1976).

Actin was purified from nonmuscle cells in the 1960s (Hatano & Oosawa 1966; Adelman & Taylor 1969), and was recognized as a major structural component of the peripheral cytoplasm (Ishikawa et al 1969), a region long recognized as important for cell structure and function (Just 1924). Many of the initial efforts to understand actin's function in nonmuscle cells understandably focused on a search for other known muscle proteins. By 1975, myosin, tropomyosin, and a troponin-like protein (later identified as calmodulin) had been isolated from such cells. Thereafter, previously unknown "actin-binding" proteins began to be identified as constituents of nonmuscle cells, and in many cases were subsequently found in muscle. The list of such proteins has increased at an average rate of about six per year.

On reflection, this large number of actin-binding proteins is not surprising. The investigations of Oosawa (Oosawa & Asakura 1975), established fundamental principles of the assembly of actin molecules into filaments. These studies, now extensively confirmed and embellished, indicated that pure actin has a relatively limited repertoire of three-dimensional configurations under conditions thought to be "physiological" for most cells. Therefore, the many supramolecular forms actin creates in cells seem to depend on the action of accessory proteins, and the extent of the variety of forms predicts the existence of many such proteins. Although a single actin molecule can only present a limited and fixed area of surface contour to bind other proteins, the intricate structure of the polymer helix theoretically affords a much larger surface area, and a considerably greater number of conformations are possible among actin monomers in adjacent positions within the filament.

Because the workings of actin-binding proteins depend upon the characteristics of actin itself, the first section of this article summarizes current knowledge of actin assembly. The ensuing review of actin-binding proteins classifies these proteins according to their functions, as has been done previously (Schliwa 1981; Weeds 1982; Craig & Pollard 1983; Stossel

1984). There are actin-binding proteins that bind predominantly to actin monomers and inhibit their self-assembly into filaments. Other proteins bind to one or more actin monomers when they are in the form of filaments, either to the ends of the filaments, or in some instances to monomers within the filament, causing the filament to break. They then remain bound to one end of the fractured filament. Still other proteins bind simultaneously to actin monomers in different filaments, which causes the actin filaments to be cross-linked into particular conformations relative to one another. Another class of proteins binds along the sides of actin filaments without cross-linking them. Finally, there are proteins that bind actin filaments to other structures, and some that do not fit into any of the above classes. The word "predominant" is very important here, because as work progresses on particular actin-binding proteins, their functions tend to become less "pure" and more complex. This review emphasizes biochemical data concerning the structural and functional properties of actin-binding proteins and their interactions with actin in vitro. Some effort is made, however, to speculate on how these interactions might operate in vivo.

In this review we have tried to provide a balanced approach, and to give appropriate credit to the many scientists who have contributed to this field. We apologize to those who might find that our biases were inadequately curbed and our diligence inadequate in citing important contributions.

2. ASSEMBLY OF ACTIN

Slightly elongated globular subunits (G-actin) of molecular weight 42,000 assemble to form linear polymers (F-actin) of a size distribution up to several micrometers in length. The polymerization reaction is reversible, and the extent of polymerization in the absence of other proteins is determined by the temperature, and by the ionic character of the solution. Both hydrophobic and electrostatic forces contribute to the noncovalent bonds holding the subunits together within the polymer, and the conformation of the monomers is altered following incorporation into polymers. The thermodynamic characteristics of the reaction are significantly different for muscle actins derived from different sources (Swezey & Somero 1982). The polymerization reaction can be initiated by addition of greater than 10 mM concentrations of monovalent salts, or by lower concentrations of divalent cations (Maruyama & Tsukagoshi 1984), which bind to specific sites on the G-actin subunits (Mornet & Ue 1984). Mg^{2+} binds with higher affinity than Ca^{2+}, and is particularly effective in supporting polymerization (Selden et al 1983). The resulting linear polymers consist of two staggered, parallel rows, twisted in a right-handed helix (Hanson & Lowy 1963), which can also be described as a single-start left-handed helix

(Korn 1982). Although they appear rod-like in electron micrographs, actin filaments have a torsional flexibility that allows 5–10 degrees of fluctuation (Egelman et al 1982; Yoshimura et al 1984), and there is some evidence that they may also have significant longitudinal flexibility (Fujime & Ishiwata 1971; Fujime et al 1984). The flexibility of F-actin increases following binding of Ca^{2+} (Miki et al 1982). Under some conditions, probably nonphysiological, actin filaments can assemble into a variety of supramolecular structures, including bundles (Hanson 1973), tubes (Curmi et al 1984), and sheets (dos Remedios & Dickens 1978).

Although the extent of polymerization of actin is determined by the thermodynamic properties of the system, the average filament length can also be strongly influenced by the kinetics of polymerization. Under ionic conditions promoting polymerization, actin dimers are thermodynamically unstable and generally dissociate before larger aggregates form (Wegner & Engel 1975). However, when aggregates containing a critical number of monomers assemble, monomer addition becomes more probable than oligomer dissociation, and successive addition of monomers to these actin "nuclei" is rapid (Wegner & Engel 1975; Tobacman & Korn 1983; Cooper et al 1983; Frieden 1983). Therefore, nucleus formation is generally the rate-limiting step in the polymerization reaction, so the average filament length, which is inversely proportional to the number of filaments, is determined by the number of actin nuclei formed. In addition, a delay called the "lag phase," which precedes detectable polymerization of an actin solution to which neutral salt has been added, is ascribable to the time required for nucleation.

Actin monomers bind the adenine nucleotides, ATP and ADP, and actin polymerization is associated with the hydrolysis of actin-bound ATP to ADP. This hydrolysis is not kinetically coupled to the addition of monomers to the polymer (Pardee & Spudich 1982; Carlier et al 1984a; Pollard & Weeds 1984; Brenner & Korn 1984), but both the rate and extent of polymerization vary greatly depending on whether the nucleotide bound to the actin monomers is ATP or ADP (Carlier et al 1984a,b; Pollard 1984b). Whenever actin polymerization is faster than the ATP hydrolysis rate, the actin subunits in the interior of the filament contain ADP, but the first few monomers at the ends of the filaments have ATP. ATP monomers dissociate more slowly from filaments than ADP monomers (Lal et al 1984a; Pollard 1984; Carlier et al 1984a), so ATP monomers at the ends of actin filaments act as functional blocks to retard monomer loss from filament ends (Carlier et al 1984a). Therefore, when actin is assembling, the effect of delayed ATP hydrolysis is to promote actin addition. Conversely, when depolymerization takes precedence, monomer loss is facilitated by the exposure of ADP monomers formerly in the interior of a filament (reviewed by Wegner 1985). After dissociation of ADP monomers from filaments, ADP is exchanged for ATP, provided sufficient ATP is in the medium. From a thermodynamic standpoint, actin filaments in the presence of an excess of ambient ATP are not in equilibrium with free monomers, but exist in an energy-consuming steady state.

Actin filaments are bipolar and have "barbed" and "pointed" ends, as determined by the appearance of heavy meromyosin-labeled filaments (Huxley 1963). Actin monomers add more rapidly to the barbed, designated $(+)$, filament end than to the pointed, or $(-)$, end (Woodrum et al 1975). The concentration of free monomers required to maintain the steady state, known as the critical monomer concentration, is different for the two ends of the filaments (Pollard & Mooseker 1981). These critical concentrations are a function of the ionic properties of the solution, and under solvent conditions designed to be approximately physiological (100 mM KCl, 2 mM Mg^{2+}, 0.5 mM ATP, pH 7.5, 37°C) are 0.1 μM at the $(+)$ end and 1.5 μM at the $(-)$ end (Wegner & Isenberg 1983).

This difference in the critical concentration dictated by the opposite ends of actin filaments is consistent with the idea that actin monomers might cycle through the filament from the $(+)$ end to the $(-)$ end by the process called "head-to-tail polymerization" or "treadmilling" (Neuhaus et al 1983). Whether such a process can play a physiological role, however, depends not only on the critical concentrations but also on the rates of monomer exchange at both ends of the filament. Pardee et al (1982) have demonstrated that only limited actin subunit exchange occurs at the ends of actin filaments under physiologic ionic conditions. The rates of monomer addition and dissociation at the $(+)$ end, which are the dominant processes occurring during actin polymerization in vitro, have been determined in several studies (Pollard & Mooseker 1981; Bonder et al 1983; Lal et al 1984b) and the results suggest rapid exchange at this end. Although measurements of these rates at the $(-)$ end have led to widely varying results (Pollard & Mooseker 1981; Bonder et al 1983; Coluccio & Tilney 1983), the evidence clearly shows that the rates of filament growth and disassembly, as well as the extent of polymerization, depend on which end of the filament is free to exchange with actin monomers in solution.

In addition to considerations of polymerization rate and extent, the length of actin filaments can also be influenced by their susceptibility to breakage. Measurements of linear viscoelasticity have shown that actin filaments are relatively resistant to breakage under small shear stresses (Zaner & Stossel 1982), but a number of experiments have shown that breakage of actin filaments does occur at a low rate. Analysis of the polymerization kinetics can yield an estimate of the number of nuclei formed during the course of polymerization, and thus the expected average filament length. Such calculations suggest that actin filaments should be over 1 mm long, but such lengths have never been observed (Wegner & Savko 1982; Frieden & Goddette 1983). In addition, it has long been known that actin solutions are thixotropic; that is, their viscosity is drastically decreased following shear (Kasai et al 1960), and this viscosity slowly increases with time (Jen et al 1982). Filament shortening has been observed following sonication (Asakura et al 1963; Carlier et al 1984b; Pantaloni et al 1984). Such a process can be explained if filaments are broken during flow or when subject to sonic forces, and slowly elongate if left unperturbed. Although it has been proposed that fragmented actin filaments can elongate by annealing, this has not been proven, and the growth kinetics of sonically disrupted actin suggest that redistribution of monomers from the shortest fibers onto longer ones is the predominant mechanism (Carlier et al 1984b).

Because solutions containing long polymers generally have much higher viscosity than those of unaggregated subunits, the most dramatic change that takes place in a solution of actin following polymerization is an increase in its resistance to flow. This mechanical aspect of actin assembly is clearly the most important property as far as actin's function in cells. The ability or inability of actin to resist deformation, and to transmit forces across the cell, affects cell movement and shape. The viscosity is a function of the concentration, average length, and structure of actin polymers. Under some conditions subtle changes in any of these properties can cause abrupt changes in the viscoelastic properties of the solutions. This is particularly the case for solutions, termed semidilute, in which the solute molecules are too long to undergo rotational diffusion without interference from other solute molecules (reviewed by DeGennes & Leger 1982). The concentration at which polymer solutions become semidilute depends strongly on the length of the polymer (Doi & Edwards 1978; Edwards & Evans 1982). Because of the great average length of pure actin filaments, the semidilute state can be attained at very low protein concentration (Janmey et al 1985b). Interfilament bonds can further immobilize the polymer strands, and thus increase the resistance to deformation, but even in the absence of specific interfilament interactions semidilute polymer solutions can have properties similar to those of viscoelastic solids.

The simplest method for altering the consistency of an actin-containing solution, such as peripheral cytoplasm, would therefore be to vary the number and length of actin filaments. This might be accomplished by regulating their assembly such that transformations can occur between solutions that are free to flow and those in which deformation is restricted, like sol-gel transformations observed in many polymeric systems. The situation in cells, however, seems to be more complex. First, in the few cases where measurements are available, cytoplasmic actin filaments that are not arrayed in bundles may be relatively short, and present at a concentration that would promote only borderline semidilute behavior (Hartwig & Stossel 1985). Second, except at extremely high concentrations or very long lengths, the elastic behavior of actin fibers is transient in the presence of applied stress (Kasai et al 1960; Zaner & Stossel 1982). Therefore, to produce sustained, strong rigidity in a system of relatively short and dilute actin filaments, added bonds between fibers are necessary, and such bonds appear to be conferred by actin filament cross-linking proteins. As will be discussed, the mechanical consequences of such cross-linking are influenced by the filament number and length. From the foregoing it is evident that the regulation of actin filament assembly and length is extremely important in determining actin's mechanical properties.

3. PROTEINS THAT PREDOMINANTLY SEQUESTER ACTIN MONOMERS

This class of actin-binding proteins binds to actin monomers, but does not form stable complexes with actin filaments. The major effect of such proteins on the polymerization of actin in vitro is to retard the incorporation of actin monomers into filaments. There are a number of ways in which such an interaction may occur. To polymerize, each actin subunit must have at least two, and perhaps four, binding sites for other actin monomers (Oosawa & Kasai 1971; Korn 1982), and the actin subunit binding site(s) for the ($+$) filament end must differ from that for the ($-$) end. A protein that binds actin monomers may link to the same site at which the monomer ordinarily binds to the filament end, and thereby prevent its association with actin polymers. Such a protein would be expected to be unable to bind to actin subunits already bound to the end of a filament.

An actin monomer–sequestering protein might also bind to actin monomers at a site different from that at which the actin monomer binds to the polymer. Such a sequestering protein might therefore bind to the ends or the side of an actin filament as well as to monomeric actin. After binding of such a protein to the end of a filament, elongation of the filament at this end may be prevented until dissociation of the sequestering protein-actin complex from the filament end. Profilin and DNAase I appear to have such binding properties, since they bind to both monomeric actin and filament ends, and retard the polymerization of actin. Other actin monomer–sequestering proteins, such as depactin, appear to extract an actin monomer from the interior of a filament and cause breakage of the filament.

An important feature of actin monomer–binding proteins is their affinity for actin relative to the affinity of actin filament ends for actin monomers. In

Table 1 Actin monomer sequestering proteins

Protein	Molecular wt. (kDa)	Binding affinity $(K_a)(M^{-1})$	Isoelectric point	Source
Profilin	12–15	0.05–2×10^7	9.1	Eukaryotic cells
Vitamin D–binding protein (Gc-globulin)	57	1.8×10^7	4.9	Blood plasma
Deoxyribonuclease I	31	5×10^8	4.7	Bovine pancreas
Depactin	19	2×10^6	6.1	Starfish oocytes
19 kDa protein	19	$? \times 10^6$	6	Porcine brain

the presence of Mg^{2+}, KCl, and ATP the apparent dissociation constant for actin monomers from the $(+)$ end of actin filaments is 0.1 μM, and from the $(-)$ end is 1.5 μM (Wegner & Isenberg 1983). Proteins that bind to actin monomers but not polymers with an affinity greater than either end of the filament will be able to depolymerize a stoichiometric amount of actin at steady state. The vitamin D–binding protein and DNAase I can depolymerize actin in this manner. In contrast, monomer-binding proteins that have an affinity intermediate between that of the $(+)$ and $(-)$ ends may be unable to depolymerize an equimolar amount of F-actin, but may have dramatic effects on the kinetics of filament assembly, and may interact differently with $(+)$ end-blocked filaments than with unblocked filaments. The profilins characterized thus far have affinities near this intermediate range, and have a more complex interaction with actin than do DNAase I or the vitamin D–binding protein. The characteristics of the various actin monomer sequestering proteins are summarized in Table 1.

3.1 Vitamin D Binding Protein (Gc-Globulin)

Although there is some evidence that vitamin D–binding protein is associated with the membrane of lymphocytes, where it may act as a link between cytoplasmic actin and membrane-associated immunoglobulin (Petrini et al 1983), vitamin D–binding protein is mainly a mammalian serum protein (Van Baelen et al 1980; Haddad 1982). The mechanism of action of this 57 kilodalton (kDa) protein is similar to that of profilin (Vandekerckhove & Sandoval 1982; Coué et al 1983), but differs in that the binding affinity (K_a) for actin is higher $(K_a = 1.2 \times 10^7 \text{ M}^{-1})$. Therefore, vitamin D–binding protein can completely depolymerize an equimolar amount of F-actin. The rate at which this depolymerization occurs does not depend on the concentration of vitamin D–binding protein (Lees et al 1984), and is equal to the rate of dissociation of actin monomers from the filament end (Janmey et al 1985a), which suggests that vitamin D–binding protein does not interact directly with the filaments to depolymerize them. The concentration of vitamin D–binding protein in mammalian plasma is 6–10 μM, making it a potent actin depolymerizing factor in the blood.

3.2 *Deoxyribonuclease I (DNAase I)*

One of the first nonmuscle actin-binding proteins to be identified was pancreatic deoxyribonuclease I (DNAase I) (Lazarides & Lindberg 1974). DNAase I forms a tight complex with monomeric actin ($K_a = 5 \times 10^8 \, \text{M}^{-1}$) (Mannherz et al 1980), and can therefore depolymerize a stoichiometric amount of actin. The binding of actin inhibits the enzymatic activity of DNAase I (Mannherz et al 1975; Hitchcock et al 1976; Lacks 1981). This inhibitory property of actin is the basis of a method for the quantification of monomeric actin (Blikstad et al 1978). Unlike vitamin D–binding protein, DNAase I can also bind with high affinity to the (+) end of F-actin (Mannherz et al 1980; Pinder & Gratzer 1982). DNAase I bound to the end of an actin filament retains its enzymatic activity (Pinder & Gratzer 1982).

Another consequence of the binding of DNAase I to actin is the retardation of the exchange of nucleotide bound to the actin subunit (Mannherz et al 1980; Hitchcock 1980). Moreover, the extent of this retardation depends on whether DNAase I is added to F-actin or G-actin (Hitchcock 1980), which provides additional evidence that DNAase I interacts directly with actin filaments. DNAase I and vitamin D–binding protein bind to different sites on the actin subunit since complexes of actin and vitamin D–binding protein bind to DNAase I (Haddad 1982). DNAase I differs from the filament end–binding proteins discussed in the next section in that it is not calcium sensitive, does not sever filaments, and does not act as a stable end-blocking protein. It also must bind to a different site on actin from these proteins, since several of them do not block the binding of DNAase I to actin.

3.3 *Profilin*

Profilin is defined as a low molecular weight protein that predominantly binds a single actin monomer, and thereby impairs its ability to self-assemble into filaments. The first profilin was purified from spleen extracts (Carlsson et al 1976), and has subsequently been isolated from platelets (Markey et al 1978; Harris & Weeds 1978), lymphocytes, calf thymus (Carlsson et al 1976; Blikstad et al 1980), calf brain (Blikstad et al 1980), porcine brain (Nishida et al 1984), thyroid (Fattoum et al 1980), *Acanthamoeba* (Reichstein & Korn 1979; Tseng & Pollard 1982), *Physarum* (Ozaki & Hatano 1984), and macrophages (DiNubile & Southwick 1985).

This is a sufficient number of cell types to suggest that this protein has a wide distribution among eukaryotic cells. The molecular weights of all the mammalian forms are approximately 15,000; the *Acanthamoeba* and *Physarum* profilins have a molecular weight of 12,000 (Reichstein & Korn 1979; Tseng et al 1984; Ozaki et al 1984). The relatively low M_r of profilin has facilitated determination of its complete amino acid sequence in a few cases. Comparison of the amino acid sequences of *Acanthamoeba* and calf spleen profilin reveal homology at the amino-terminal region, which suggests that this is the region of the profilin molecule that participates in actin-binding activity (Ampe et al 1985; Nyström et al 1979).

All profilins purified to date appear to function similarly. Experimental findings suggest that profilin reversibly binds actin monomers. The apparent dissociation constants (K_d) of the *Acanthamoeba* profilin/*Acanthamoeba* actin, and macrophage profilin/skeletal muscle actin complexes are 4–9 μM and 2–4 μM, respectively (Tobacman & Korn 1982; Tseng & Pollard 1982; Tobacman et al 1983; DiNubile & Southwick 1985). *Acanthamoeba* profilin complexed with skeletal muscle actin has a much higher apparent K_d of 50 μM (Tobacman & Korn 1982; Tseng & Pollard 1982).

In solutions that contain mixtures of purified actin and profilin the fraction of actin bound by profilin, as determined by the assembly competence of actin, depends strictly on the total profilin concentration and the critical actin monomer concentration. This relationship can be described by the equation:

$$PA = P \times C_0/(C_0 + K_d), \hspace{2cm} 1.$$

where PA is the profilin/actin complex concentration, P is the total profilin concentration, C_0 is the critical actin monomer concentration, and K_d the dissociation constant of the profilin/actin complex (Tobacman & Korn 1982). In most studies the K_d of *Acanthamoeba* and rabbit lung macrophage profilin/actin complexes remained constant under a wide variety of ionic conditions (Tobacman & Korn 1982; Tobacman et al 1983; Pollard & Cooper 1984; DiNubile & Southwick 1985). However, one group of investigators reported a marked reduction in *Acanthamoeba* profilin's affinity for actin in $MgCl_2$ solutions (Tseng & Pollard 1982). There is some evidence that the interaction of profilin with actin may be regulated by additional factors. A partially purified cytoplasmic extract has been found to stabilize the profilin-actin complex (Malm et al 1983), and Lassing & Lindberg (1985) have shown that phosphatidylinositol 4,5-biphosphate can bind to profilin-actin complexes and promote their dissociation. This latter interaction is potentially important in that phosphatidylinositol 4,5-bisphosphate can be formed during cell activation.

Acanthamoeba and macrophage profilins decrease the steady-state F-actin concentration in solutions of various ionic composition favoring polymerization, and this decrease is directly proportional to the amount of profilin present (Tobacman & Korn 1982; Tseng & Pollard 1982; DiNubile & Southwick 1985). For a given concentration of free profilin and actin, the magnitude of the reduction in F-actin concentration is strictly dependent on the critical monomer concentration: the higher the critical concentration, the greater the decrease in F-actin concentration induced by the same concentration of profilin. These observations are consistent with the conclusion that profilin interacts only with monomeric actin, and that profilin amplifies the small changes in critical concentration induced by changes in the ionic conditions. These effects of profilin also explain how this protein is able partially to depolymerize F-actin, to prolong the lag phase of polymerization, and to slow the elongation rate of filament assembly.

High concentrations of profilin slow the actin assembly rate to an extent not easily explained simply by monomer sequestration. One interpretation of this finding is that profilin may interact weakly with the ($+$) end of actin filaments. The estimated K_d of profilin from actin filaments is 50–100 μM (Tilney et al 1983; Pollard & Cooper 1984).

Most of the profilins markedly prolong the lag phase of polymerization, and reduce the rate and extent of actin polymerization. Unlike *Acanthamoeba* and macrophage profilin, which reproducibly lower the steady-state F-actin concentration, the effects of other profilins on F-actin concentration are variable. These functional differences may be explained by differences in purification techniques. *Acanthamoeba* and macrophage profilin have been purified from a population of apparently free profilin molecules. Many of the other forms of profilin have been purified from a fraction of profilin molecules tightly complexed to actin. To obtain free profilin this complex must be treated with 2 M urea, which denatures the actin and may alter the profilin. It is also possible that profilin is functionally heterogeneous. However, different profilin fractions isolated from *Acanthamoeba* were found to be functionally identical (Tseng et al 1984).

3.4 *Monomer Sequestering and Filament Severing Proteins*

Another type of actin monomer sequestering protein is similar to profilin in molecular weight and affinity for monomeric actin. However, these proteins may also bind to actin subunits in the interior of filaments, causing filament breakage as well as sequestration of monomers. Such proteins, rather than slowing the rate of actin polymerization, can increase the rate of assembly during reactions in which filament nucleation is the rate-determining step.

This is because as the filaments grow the monomer-binding proteins can sever them, exposing new ends from which further addition can occur. The proteins that act in this manner include depactin, isolated from starfish oocytes (Mabuchi 1983), a 19 kDa protein from porcine brain (Nishida et al 1984; Bamberg et al 1980), and a protein isolated from *Acanthamoeba* called actophorin (Cooper et al 1984), as well as the end-blocking and severing proteins discussed below. Although all of these proteins are reported to sever actin filaments, they differ from the severing proteins discussed in the following section in that they have a lower affinity for actin filaments (and therefore are much less potent in severing), they are not affected by calcium, and they do not remain bound to the filament end after severing.

3.5 Binding Sites on the Actin Subunit

The sites on the actin subunit to which monomer sequestering proteins bind are beginning to be elucidated. DNAase I binds near residues 50–69, as determined by chemical cross-linking studies (Sutoh 1984); this is also supported by the finding that DNAase I binding protects actin residues 62–69 from proteolytic cleavage. The binding of DNAase I to actin may involve one or two lysine residues, possibly Lys 68 (Ng & Burtnick 1982). This region is near the apparent nucleotide binding site (Mornet & Ue 1984), which is consistent with the effect of DNAase I binding on actin nucleotide exchange. The binding of profilin to actin can be inhibited by removal of Phe 375 (Malm et al 1983), or modification of Cys 374 (Malm 1984), which suggests that the carboxyl terminus of actin is involved with binding to profilin. Cross-linking studies have shown that the binding of depactin to actin involves actin residues 1–11 and 358–372 (Sutoh & Mabuchi 1984). These two regions may have particular importance since a region near the amino terminus of actin is involved in actin-actin binding (Sutoh 1984; Elzinga & Phelan 1984), and in the binding of actin to divalent cations (Mornet & Ue 1984) and to myosin subfragment S_1 (Sutoh 1982). The observation that profilin and depactin bind to similar sites on the actin subunit that are distinct from the site of DNAase I binding is also in agreement with the finding that complexes of actin and profilin or depactin, as well as other actin-binding proteins discussed below, can bind to immobilized DNAase I.

4. PROTEINS THAT PREDOMINANTLY BLOCK THE ENDS OF ACTIN FILAMENTS

End-blocking proteins (also called "capping" proteins) bind to one end of actin filaments and prevent the exchange of monomers with that end. As described above, the two ends of actin filaments have different critical concentrations in the presence of KCl, Mg^{++}, and ATP at steady state. Therefore, blocking one end of the filament will reduce the contribution of that end to the overall monomer concentration. For example, if the (+) end of the filament is blocked, the monomer concentration at steady state with the actin polymers increases to reflect the higher critical concentration for the (−) end. Conversely, the critical concentration should theoretically fall when the (−) ends are blocked. The increase in critical concentration due to blocking of the (+) ends requires that nearly all of the filament ends be blocked (Wanger & Wegner 1985). With some end-blocking proteins of sufficiently high affinity, saturation of this effect is observed at low molar ratios of these proteins to actin (Harris & Schwartz 1981; Kurth et al 1983; Janmey et al 1985). Other end-blocking proteins depolymerize more actin than can be accounted for by differences in critical concentrations of the filament ends, but require higher, though still substoichiometric, concentrations to do so (Yamamoto et al 1982; Wang & Spudich 1984).

The majority of the end-blocking proteins identified to date bind to the (+) end of actin filaments (see Table 2). The stage for elucidation of how such end-blocking proteins function was set by earlier work on the fungal metabolites, the cytochalasins, which were shown to shorten actin filaments and to inhibit actin assembly by blocking their (+) ends (Brenner & Korn 1979b; Hartwig & Stossel 1979; Flanagan & Lin 1980; Brown & Spudich 1979; MacLean-Fletcher & Pollard 1980). However, the (+) end–blocking proteins do not necessarily inhibit cytochalasin binding, which suggests that different actin-binding sites are involved. Most of the proteins of this class have multiple functions. They can bind to one or two actin monomers, and they accelerate the nucleation phase of actin assembly, presumably by forming stable oligomeric complexes with actin that nucleate polymerization. In the case of the (+) end–blocking proteins, nucleation leads to elongation provided the actin monomer concentration is sufficiently greater than the C_0 to drive growth from the (−) end. The rate of elongation in this setting, however, is slow in comparison with the growth of unblocked filaments. Also, the average filament length achieved at steady state is shorter than that of control actin because a larger number of filaments are formed. The (+) end–blocking proteins should also be able to shorten preformed actin filaments by promoting dissociation of actin monomers from the (−) ends, which will subsequently assemble from spontaneously formed actin nuclei or actin and blocking protein complexes. However, such a redistributive mechanism is too slow to account for the rapid shortening observed when some of the (+) end–blocking proteins are added to preformed F-actin. In these cases, clearly the actin filaments are shortened primarily by severing along their length. Based on the ex-

Table 2 Actin filament end blocking proteins

Protein	Source	M_r (kDa)	Calcium Interaction			F-actin Binding		G-actin Binding[a]			Nucleation	Severing
			Effect	Capacity	K_a	Site	K_a	EGTA-Reversed	Capacity	K_a		
Gelsolin	Mammalian cells Plasma (ADF, Brevin)	90 93	Yes	2	10^6	+ end	$>10^9$	Partial	2	$>10^8$	Yes	Yes, inhibited by tropomyosin
Villin	Amphibian, eggs Avian epithelium Mammalian epithelium	95	Yes	2+1	3.5	+ end	$>10^{11}$ 7.4×10^{-6}	Yes	2	—	Yes	Yes, inhibited by tropomyosin
Fragmin/severin	Physarum Dictyostelium Sea urchin eggs	42 40 40–45	Yes	—	—	+ end	—	No	1	$>10^{10}$	Yes	Yes
Capping protein	Acanthamoeba	31+28	No	—	—	+ end	—	—	—	—	Yes	No
Cap 42a + b	Physarum	42a + 42b	Yes	—	—	+ end	—	—	—	—	—	—
Acumentin	Leukocytes	65	No	—	—	− end	—	—	—	—	Yes	No
β-actinin	Kidney, skeletal muscle	37+35	No	—	—	− end	—	—	—	—	Yes	No

[a] Inferred from ability to accelerate actin polymerization.

tent of filament shortening observed in relation to the blocking protein concentration, fragmentation is likely to involve a stoichiometric interaction between the proteins to break actin-actin bonds within the filament.

The potent effects of these blocking proteins on actin filament length suggest that they may locally regulate the rheological properties of the cortical cytoplasm (gel-sol transformations). As amplified in the following discussion of actin filament cross-linking, it is clear that the effectiveness of gelation of actin filaments by cross-linking proteins is inversely proportional to the filament length. It has been shown for some actin-severing proteins that the critical cross-linker concentrations for actin gelation are entirely consistent with those predicted from the degree of shortening according to Flory's theory (1953). The fact that these severing proteins can fragment actin filaments rapidly suggests that they may also have a role in the remodeling of the actin cytoskeleton. The ability of these proteins to block the ($+$) end suggests they may anchor actin filaments to various cell structures, such as membranes. Their ability to promote the nucleation of actin filaments suggests that they may locally control the number of actin filaments in the cytoplasm.

Central to any understanding of the physiological role of the end-blocking proteins is the question of how they are regulated in vivo. Ca^{2+} influences the activity of all of the end blocking proteins that also sever actin filaments (Table 2). Other blocking proteins are apparently Ca^{2+} insensitive, and factors regulating them have not yet been identified. In one case, phosphorylation has also been shown to modulate the activity of a Ca^{2+}-sensitive actin end-blocking protein that does not sever actin filaments. The role of Ca^{2+} in the activation/deactivation of the Ca^{2+}-sensitive end-blocking proteins is complex. Their binding affinity to actin increases when they complex Ca^{2+}, but the different actin modulating functions are expressed at different threshold levels of Ca^{2+}, which suggests that multiple Ca^{2+}-binding sites may be present. Moreover, with one exception, the blocking proteins studied thus far that require Ca^{2+} to bind and sever actin cannot be dissociated from the actin monomers to which they are bound in vitro by merely lowering the free Ca^{2+} concentration. The interaction of end-blocking proteins with actin may also be modulated by other actin-binding proteins. Tropomyosin and myosin subfragment S_1 interfere with filament severing but not end blocking in some cases.

4.1 Gelsolin

Gelsolin is a Ca^{2+}-sensitive, ($+$) end–blocking protein that promotes actin nucleation and also severs actin filaments. It was first identified in rabbit lung macrophages (Yin & Stossel 1979, 1980; Yin et al 1981), and subsequently was found in many mammalian cells, including muscles (Lind et al 1982; Wang & Bryan 1981; Yin et al 1981; Tellam & Frieden 1982; Petrucci et al 1983; Rouayrenc et al 1984; Hinssen et al 1984; Verkhovsky et al 1984; Kobayashi et al 1983; Olomucki et al 1984). Gelsolin binds 2 moles of Ca^{2+} per mole with high affinity ($K_a = 1 \times 10^6$ M^{-1}) (Yin & Stossel 1980; Bryan & Kurth 1984), and can bind to both actin monomers and filaments. When added to G-actin in the presence of micromolar Ca^{2+} it forms a complex with two actins (2:1 actin-gelsolin complex), which promotes the nucleated assembly of actin filaments. When added to preformed actin filaments it rapidly shortens them, primarily by directly severing the actin-actin bonds along the filaments, but also by blocking the (+) end, and thereby raising the critical concentration. Shortening of actin filaments raises the critical cross-linker concentration required for gelation of actin filaments; and it has been shown that localized, Ca^{2+}-induced solation of actin gels results in directed movements in the presence of myosin and ATP (Stendahl & Stossel 1980).

The expression of the three effects of gelsolin on actin, severing, nucleation, and end blocking, depend on the Ca^{2+} concentration (Bryan & Kurth 1984; Coluccio et al 1984; Janmey et al 1985a). Severing has the most stringent requirement for Ca^{2+}, and does not occur at submicromolar Ca^{2+} concentrations. Blocking of filament ends by gelsolin takes place to a small extent in the presence of submicromolar Ca^{2+}, although it is considerably enhanced at higher Ca^{2+} concentrations.

Although binding of gelsolin to actin is highly Ca^{2+} dependent, it is only partially reversible on subsequent removal of free Ca^{2+} with EGTA. This phenomenon was studied extensively by Kurth & Bryan (Kurth et al 1983; Kurth & Bryan 1984), who found that unless platelets were rigorously prevented from being activated to avoid an increase their cytoplasmic Ca^{2+} concentration, the gelsolin was isolated as a 1:1 complex with actin, even in the presence of EGTA.

Although gelsolin forms a complex with two actin monomers in the presence but not absence of micromolar Ca^{2+}, only one of the actins can subsequently be dissociated by chelating calcium with EGTA; the 1:1 actin-gelsolin complex retains a Ca^{2+} ion that resists elution by EGTA. Unlike free gelsolin, this complex does not fragment actin filaments, and it blocks actin filaments with high affinity even at submicromolar Ca^{2+} concentrations (Kurth et al 1983; Janmey et al 1985a). Therefore, once gelsolin blocks the (+) end of a filament in vitro it can no longer dissociate from actin by a simple on-off mechanism. Nevertheless, filaments shortened by gelsolin-Ca^{2+} can slowly elongate when the calcium concentration is lowered. In the presence of EGTA, actin monomers dissociate one by one from the (−) ends of filaments blocked at their (+) ends. Since the 1:1 gelsolin-actin complex neither nucleates actin assembly effectively at submicromolar Ca^{2+} concentrations, nor fragments actin filaments under any circumstances, the new filaments formed from spontaneously formed actin nuclei (no gelsolin) are long. The net result is a relative separation of gelsolin-actin complexes from newly formed actin filaments. This redistribution process, however, is slow because of the apparently low dissociation rate constant of the ($-$) end. If this were the sole mechanism for switching off gelsolin in cells, the recovery from Ca^{2+}-induced actin filament shortening would be slow, and free gelsolin could not be regenerated except by new protein synthesis. Since motile cells are exposed to episodic increases in cytoplasmic Ca^{2+} concentration, it follows that most of their gelsolin would ultimately be complexed with actin and could no longer sever actin filaments. However, this does not appear to be true of a number of highly motile cells, from which a large proportion of the total gelsolin has been purified as a free molecule, provided that the gelsolin has not been exposed to greater than micromolar Ca^{2+} concentrations in the presence of actin during purification. To explain these cases one would have to postulate a cell mechanism that dissociates the gelsolin-actin complex, or one that selectively degrades the complex and replenishes free gelsolin rapidly from de novo synthesis. There is evidence that while gelsolin is synthesized constitutively, its rate of synthesis is enhanced by dexamethasone treatment (Lanks & Kasambalides 1983).

A variant of gelsolin exists in blood plasma in micromolar concentrations (Smith et al 1984), and is secreted by a variety of cells (Yin et al 1984). It has also been called actin-depolymerizing factor (Chaponnier et al 1979; Norberg et al 1979; Harris et al 1980) or brevin (Harris & Schwartz 1981), and is functionally similar to cytoplasmic gelsolin (Thorstensson et al 1982; Harris & Weeds 1983; Doi & Frieden 1984; Lees et al 1984; Chaponnier et al 1985). Human plasma gelsolin is slightly larger than cytoplasmic gelsolin (93 versus 90 kDa), and contains a 25 amino acid extension on its NH_2-terminus (Yin et al 1984). The two proteins are otherwise structurally very similar (Markey et al 1982; Yin et al 1984). Plasma gelsolin could be involved in the clearance of actin filaments from the circulation, and may therefore operate in conjunction with another plasma protein, the vitamin D–binding protein, which sequesters actin monomers, to reduce and maintain actin in the depolymerized state. Alternatively, plasma gelsolin may have some as yet unidentified functions, and may interact with proteins other than actin. For instance, it has been shown that gelsolin binds to fibronectin with high affinity (Lind & Janmey 1984).

4.2 Villin

Villin was first isolated from the brush border microvilli of chicken intestinal epithelial cells (Bretscher & Weber 1980a; Glenney et al 1980; Mooseker et al 1980), and subsequently shown to be present in other vertebrate cells containing microvilli (Corwin & Hartwig 1983; Alicea & Mooseker 1984). In the presence of micromolar concentrations of Ca^{2+} it functions essentially like gelsolin: It severs, nucleates, and blocks the (+) end of actin filaments (Craig & Powell 1980; Glenney & Weber 1981; Glenney et al 1981; Nunnally et al 1981; Matsudaira & Burgess 1982; Bonder & Mooseker 1983; Wang et al 1983; Walsh et al 1984). Titration experiments show that severing is activated between 5 and 10 μM Ca^{2+}, and is saturated at millimolar concentrations of Ca^{2+}, while end blocking and nucleation are maximally expressed at 1–2 μM Ca^{2+} (Walsh et al 1984). Although these functional studies suggest that multiple Ca^{2+}-binding sites with large differences in affinities may be involved, direct Ca^{2+}-binding experiments reveal only high affinity Ca^{2+}-binding sites (Hesterberg & Weber 1983a). It is therefore not clear if the higher Ca^{2+} requirement for severing reflects saturation of undetected lower affinity binding sites on villin itself or on actin.

Villin is different from gelsolin, fragmin, and severin in that its binding to G-actin and actin filament ends is rapidly and completely reversed by EGTA (Bretscher & Weber 1980a; Walsh et al 1984). Furthermore, in EGTA at sufficiently high villin concentration, villin can cross-link actin filaments into bundles (Bretscher & Weber 1980; Glenney et al 1980; Mooseker et al 1980; Matsudaira et al 1983). The ability of villin to bundle actin filaments is unique among the blocking proteins, and together with its exclusive localization in microvilli suggests that villin may have evolved specifically for maintaining the organization of actin filament bundles in these structures.

The EGTA-promoted actin cross-linking function of villin is associated with a 8.5 kDa peptide at the COOH-terminal of villin, which has been designated the villin headpiece (Glenney & Weber 1981). The headpiece can be cleaved from the 87 kDa core of the protein which contains the Ca^{2+}-dependent blocking, nucleating, and severing functions. The headpiece binds one mole of Ca^{2+} with K_d 7.4×10^{-6} M, and binds F-actin independent of Ca^{2+}. The villin core has an exchangeable Ca^{2+}-binding site (K_d 3.5×10^{-6} M), and a very slowly exchangeable high affinity Ca^{2+}-binding site. Villin undergoes a marked conformational change when it binds Ca^{2+}, whereas the villin core does not. Therefore, the villin headpiece is likely to be involved in altering the hydrodynamic properties of intact villin in response to changes in Ca^{2+} concentration (Hesterberg & Weber 1983b). Based on secondary structural predictions and a comparison of the amino acid sequence of the villin headpiece with other known Ca^{2+}-binding proteins, a region suggestive of a Ca^{2+}-binding site has been identified, although the headpiece does not exhibit a classical EF-hand Ca^{2+}-binding structure.

4.3   Fragmin/Severin

Fragmin was first isolated from the slime mold *Physarum polycephalum* and was estimated to have a molecular weight of 42,000 (Hasegawa et al 1980; Hinssen 1981a,b; Sugino & Hatano 1982). A functionally similar protein named severin was subsequently purified from another slime mold, *Dictyostelium discoideum*, and was shown to differ from fragmin in its slightly lower M_r (40,000), and its different amino acid composition (Brown et al 1982; Yamamoto et al 1982; Giffard et al 1984). Subsequently, other functionally similar proteins of comparable molecular weights have also been identified in sea urchin eggs (Hosoya & Mabuchi 1984; Wang & Spudich 1984; Sedlar & Bryan 1984). Ca^{2+} induces a conformational change in these proteins, and promotes actin severing, nucleating, and ($+$) end–blocking activities that resemble those of gelsolin. These proteins are half the size of gelsolin and villin, and bind one molecule of actin only. The actin-fragmin complex contains a trapped Ca^{2+}, and cannot be dissociated by EGTA (Sedlar & Bryan 1984; Hosoya & Mabuchi 1984).

Since some of the blocking proteins, especially the Ca^{2+}-sensitive species, are functionally similar, an obvious question is whether they are structurally related as well. Most published studies have emphasized their structural divergence based on differences in molecular weights, tissue and species distributions, and immunological cross-reactivity. However, recent evidence suggests that parts of the molecules (possibly the active sites) may be homologous. Actin-binding domains located at the NH_2-termini of gelsolin and villin have striking homology in certain regions (Matsudaira & Jakes 1984; Kwiatkowski et al 1985). Fragmin, which is half the size of villin and gelsolin, but has the same molecular weight as actin, is immunologically and functionally similar to another *Physarum* end-blocking protein designated Cap 42 a, which in turn may be a nonpolymerizable variant of actin (Maruta et al 1984). The homology between Ca^{2+}-dependent actin capping proteins, as well as their homology to actin, raises the intriguing possibility that the actin end-blocking proteins may block filament ends by binding to actin much in the same way as actin binds to itself, and that they may have evolved by duplication and modification of ancestral actin genes (Maruta et al 1984).

4.4   *Physarum polycephalum* Cap 42 (a+b)

A protein called Cap 42 (a + b) that is antigenically closely related to fragmin and actin has been purified from *Physarum polycephalum* (Maruta et al 1983; Maruta & Isenberg 1983, 1984; Maruta et al 1984). This protein consists of two 42 kDa peptides, designated a and b. The Cap 42 b polypeptide is phosphorylated in vivo. In vitro, this same protein can be phosphorylated by purified 42 Cap kinase; phosphorylation is inhibited at high calcium concentration when actin or DNAase I are also present (Maruta & Isenberg 1983, 1984). When Cap 42 b is phosphorylated it binds to and blocks actin monomer exchange at the (+) end of actin filaments, but only in the presence of micromolar calcium concentrations (Maruta & Isenberg 1984). When Cap 42 b is dephosphorylated, the ability of Cap 42 (a + b) to bind actin becomes Ca^{2+}-independent (Maruta & Isenberg 1983, 1984). The Cap 42 (a + b) can also nucleate actin filament assembly and induce rapid shortening of actin filaments, as assessed by viscosity (Maruta & Isenberg 1983). However, this complex does not sever preformed actin filaments.

4.5   Capping Protein

Acanthamoeba capping protein is a (+) end–blocking protein that nucleates actin assembly but does not sever actin filaments. The mechanism by which capping protein accelerates actin polymerization has been analyzed (Cooper & Pollard 1985). It has a native molecular weight of 74,000, and is a heterodimer of two immunologically distinct subunits of 31 and 28 kDa. Its blocking of actin filaments is not affected by changes in Ca^{2+} concentrations (Isenberg et al 1980; Cooper et al 1984). It is localized in the amoeba cortex and copurifies with plasma membrane fractions (Cooper et al 1984). Similar proteins have also been purified from brain, *Dictyostelium*, and muscle.

4.6   Beta Actinin

Beta actinin was first purified from rabbit skeletal muscle, and more recently from rat kidney cells. The skeletal form is a heterodimer containing 34 and 37 kDa subunits (Maruyama 1971; Maruyama et al 1977), while the kidney cell form is a monomer of approximately 80 kDa (Maruyama & Sakai 1981). Both protein forms are functionally similar. They promote the nucleation of actin filament assembly, inhibit the elongation of fragmented actin filaments, and they shorten the actin filament length distribution at steady state. This protein's ability to shorten actin filaments is enhanced by the presence of millimolar magnesium chloride concentrations (Kamiya et al 1972). Electron micrographs of heavy meromyosin-decorated actin filaments suggest that skeletal muscle β-actinin inhibits actin monomer exchange at the (−) end of actin filaments (Maruyama et al 1977).

4.7   Acumentin

This 65 kDa monomeric protein has been purified from human granulocytes (Southwick & Stossel 1981) and rabbit alveolar macrophages (Southwick et al 1982). Acumentin is present in high concentrations in cytoplasmic extracts from these two cells, possibly representing up to 6% of the total protein. Acumentin binds to actin filaments with relatively low affinity, and blocks the addition of actin monomers at the (−) ends of heavy meromyosin-labeled actin filaments whose (+) ends are blocked by gelsolin (Southwick & Hartwig 1982).

Like many other end-blocking proteins acumentin can nucleate actin filament assembly. This function, in combination with its ability to block monomer exchange at one filament end, can explain how acumentin reduces the viscosity of actin solutions. Nucleation produces a greater number of short filaments than results from the same concentrations of actin polymerized in buffer. Blocking one end prevents filament reassociation and redistribution produced by nucleation. When acumentin is added to preformed actin filaments it causes only small reductions in viscosity, which indicates that this protein can not sever actin filaments.

Blocking monomer exchange at the (−) end in the presence of acumentin would be expected to lower the critical monomer concentration. However, the critical concentration of unblocked actin filaments is not a simple average of the critical concentrations at the two ends, but depends strongly on the rates of exchange at each of the filament ends. Since the bulk of the exchange of monomers with unblocked actin filaments appears to occur at the (+) end, the critical concentration of unblocked filaments is much closer to that of the (+) end than the (−) end (Wegner 1982b). Therefore, proteins that block the (−) ends of actin filaments would be expected to have a weaker effect on actin critical concentration than those that block the (+) ends. Consistent with this expectation, acumentin has been found to produce little if any change in actin critical concentration (Southwick et al 1982; DiNubile & Southwick 1985).

This end-blocking protein is equally active under physiologic salt conditions in the presence or absence of micromolar calcium concentrations. Acumentin's activity, however, is inhibited by KCl concentrations above 0.1 M, and is completely inactivated at a KCl concentration of 0.3 M.

5. REGULATION OF ACTIN ASSEMBLY AND DISASSEMBLY BY THE INTERACTION OF END BLOCKING AND MONOMER SEQUESTERING PROTEINS

Actin monomer sequestering proteins, filament severing proteins, and end-blocking proteins all decrease the amount of polymerized actin and the average filament length. They do so by different mechanisms that can act in combination to regulate each other's effects; in some cases they interact competitively, and in others they have amplifying effects. Comprehension of the interaction of these proteins in vivo will require knowledge, currently being accumulated, of their relative affinities for actin, their concentration and location in the cell, and their regulation by ions and metabolites.

Monomer sequestering proteins have a relatively low affinity for actin, are not known to be functionally regulated by any other factors, and appear to be present at relatively high concentration in the cytoplasm, at nearly equimolar amounts to actin. In contrast, proteins with both filament-severing and (+) end–blocking activities have a very high affinity for actin, are generally potentiated by Ca^{2+}, and are present in cells at low molar ratios to actin, approximately 1 : 100. The functional regulation of monomer sequestering proteins in vivo may occur through their combined effect on actin with the severing and blocking proteins. For example, Equation 1 predicts that the amount of actin bound to profilin increases as the critical actin monomer concentration is raised (Tobacman et al 1983). Such an effect has been demonstrated by varying the critical concentration either by changing the ionic environment (Tobacman & Korn 1983) or by blocking the (+) filament ends (DiNubile & Southwick 1985; Nishida et al 1984b). The extent of this increase in profilin-actin complex concentration is greatest when the affinity of profilin for actin is close to that of the filament (−) end, but much less than that of the (+) end. However, the changes in profilin-actin complex concentration induced by end-blocking proteins may be greater than predicted by this model, which suggests that in addition to the interaction of blocking proteins with the ends of actin filaments, and the binding of profilin to G-actin, other reactions may be taking place (DiNubile & Southwick 1985).

As a first approximation, the major function of actin monomer sequestering proteins seems to be to prevent unregulated nucleation. The prime function of actin filament end-blocking proteins (+ or −) could be to regulate the time and place of actin assembly; severing proteins may mainly act to effect rapid actin filament shortening, and thereby control gel-to-sol phase changes.

In addition to their effects on actin polymerization at steady state, monomer-sequestering and end-blocking proteins also affect the kinetics of actin filament polymerization and depolymerization. The major effect of profilins in vitro is to retard the formation of actin filament nuclei and, to a lesser extent, addition of monomers to filaments. However, these effects are seen with purified proteins at low concentration relative to their concentrations in cytoplasm, and there may be situations in which the presence of profilin and profilin-actin complexes may accelerate actin filament assembly and disassembly. For example, changes in the structure of the actin network in the cell are thought to involve simultaneous assembly of actin filaments in one part of the cell and disassembly in another part. If the amount of nonpolymerized actin were small compared to the total amount of polymerized actin, then the rate of polymerization of assembling filaments could be no greater than the rate of monomer release from disassembling filaments. Since this reordering of filament structure can be initiated by activation of severing and (+) end–blocking proteins, the disassembly of filaments is likely to occur from the more slowly exchanging (−) ends. However, if a high concentration of nonpolymerized actin, either free or complexed to profilin, exists throughout the cytoplasm, then filaments can assemble rapidly by addition of monomers from this pool. In this way the kinetics of actin polymerization in one region of a cell can be uncoupled from the depolymerization at another.

Although the net effect of monomer sequestering proteins on F-actin in vitro, especially for (+) end–blocked filaments, is to decrease the amount of polymerized actin, the affinity of the profilin-actin complex is low enough that theoretically monomer addition can occur at either end of a growing filament. The binding affinity of the profilin-actin complex ($K_d = 2$–$4\ \mu M$) is less than that of the (−) filament end, so assembly at this end, possibly stimulated by formation of gelsolin-actin complexes, for example, could cause at least a transient increase in the amount of polymerized actin, and a concomitant decrease in profilin-actin complex. Another element to be factored into the regulation of actin filament assembly by actin-binding proteins is the state of adenine nucleotides bound to actin. For example, it is possible that exposure of (+) filament ends by dissociation of severing and blocking proteins could promote either depolymerization, if the exposed actin monomers at the ends of the filaments bound ADP, or polymerization, if they bound ATP.

6.   ACTIN FILAMENT CROSS-LINKING PROTEINS

A number of proteins have been purified that can join actin filaments together because they have at least two sets of binding sites for actin filaments (Table 3). These proteins differ in mass and subunit structure, in sensitivity of function to calcium concentration, and in their cross-linking potency.

Cross-linking of actin filaments results in the formation of two general types of structures: actin networks of varying isotropy, and anisotropic arrays or filament bundles. These structures can be evaluated by electron microscopy and by their physical properties.

Actin filaments, like other polymers, can form gels when cross-linked,

Table 3 Actin cross-linking proteins

Protein	Source	Subunit M_r (kDa)	Sediment coefficient $S_{20,w}$ ($\times 10^{-13}$)	Stokes radius (nm)	Frictional ratio (f/fo)	Molecular dimensions	V_c^a (nM)	Calcium sensitive	Other modifications
Isotropic Gelation Proteins									
1. Actin-binding protein	Macrophages Platelets Xenopus oocytes HeLa cells	2 × 270	9.4	13.5	2.56	160 nm dimer	40	No	Phosphorylated in vivo and in vitro. Phosphorylation reported to increase cross-linking ability
Filamin	Smooth muscle Avian gizzard Myocytes	2 × 250	8.86	12.5	2.32	160 nm dimer	90	No	—
2. Spectrin	Erythrocytes	2 × 240 2 × 220	11.0	21.4	2.9	196 nm tetramer	200	Maybe. Binds calmodulin and 4.1 polypeptide	Phosphorylated in vitro and in vivo
(Fodrin)	Brain Avian lens	2 × 240 2 × 235	—	—	—	200 nm tetramer	—	—	—
(TW 260/240)	Intestinal Epithelium	2 × 260 2 × 240	—	—	—	260 nm tetramer	—	—	—
	Acanthamoeba	? × 260	—	—	—	80–230 nm	720–1150	No	—
3. Other HMWPs	Physarum	2 × 230	11.5	9.0	—	60 nm strands with 2 globular regions on each (dumb-bell shaped)	70	No	—
4. 120 kDa	Dictyostelium	2 × 120	7.4	8.6	2.1	4 × 35 nm rod	350	None	—

Anisotropic Filament Bundling Proteins

	Source	Subunit				Dimensions		Calcium sensitivity	
1. Alpha-actinin	Striated and smooth muscle	2 × 95	6.2	—	—	4 × 40 nm rod	?	None	—
	Non-muscle: *Vertebrate* Platelets Lymphocytes Ascites cells	2 × 105	—	6.4–6.7	—	4 × 40 nm rod	—	Yes. Inhibited by µM calcium	—
	Invertebrate Sea urchin eggs	2 × 95	7.1	7.7	—	4 × 50 rod	1000	Yes. Inhibited by µM calcium	—
	Dictyostelium	2 × 95	—	7.3	1.9	4 × 40 nm rods	300–3000	Yes, it is 100% inhibited by 0.1 µM calcium	—
	Acanthamoeba	? × 85	—	8.5	—		1000 (if dimer)	Yes. Only 100 µM calcium tested	—
(Actinogelin)	Elrlich ascites cells Liver	2 × 115	—	—	—		—	Yes. Only 100 µM calcium tested	—
2. Fascin	Sea urchin eggs	1 × 57	—	—	—		—	None	—
	Porcine brain	1 × 53	—	—	—		7000	—	—
	Starfish sperm	1 × 57	—	—	—		7000	—	—
3. Fimbrin	Intestinal epithelium	? × 68	—	—	—		—	None	—
4. Villin	Intestinal epithelium Toad oocytes	1 × 95	—	—	—		—	Yes, becomes an actin filament fragmenter	—
5. 30 kDa	Dictyostelium	1 × 30	—	3.0	1.3		600	Yes, it is 100% inhibited by 0.1 µM calcium	—
6. Aldolase									—
7. Nerve growth factor		1 × 27							—

Table 3 *continued*

Protein	Source	Subunit M_r (kDa)	Sediment coefficient $S_{20,w}$ ($\times 10^{-13}$)	Stokes radius (nm)	Frictional ratio (f/fo)	Molecular dimensions	V_c[a] (nM)	Calcium sensitive	Other modifications
Miscellaneous Proteins									
1. Gelactins	*Acanthamoeba*								
I		? × 23	—	—	—	—	150	—	—
II		? × 28	—	—	—	—	225	—	—
III		? × 32	—	—	—	—	100	—	—
IV		? × 38	—	—	—	—	4[b]	—	—
2. MAPs (MAP-2 and Tau)	Brain	? × 300; ? × 50–68	—	—	—	—	400[c]	—	Phosphorylated in vitro. Reported to decrease gelation activity
3. 220 kDa	Sea urchin eggs	? × 220	—	—	—	—	—	Does not bind to actin but binds to fascin. Crosslinks fascin-actin bundles	—

[a] Standardized for 23.8 μM f-actin, assuming actin filaments of equivalent lengths in the various experiments.
[b] Reported value is below the theoretical V_c unless filaments are >10 μm.
[c] Low phosphate–containing form.

and the quantitative relationship between cross-linking and gelation is one property that distinguishes isotropic from anisotropic networks. On theoretical grounds, gelation occurs abruptly when the added cross-linking protein links a sufficient number of the actin filaments (Flory 1953). For a heterodispersed polymer length distribution, such as exists when pure actin assembles in vitro, the theoretical minimum molar concentration (V_c) of a bivalent cross-linking protein required to gel a solution of actin filaments is described by the following expression:

$$V_c = c/M_w, \qquad\qquad 2.$$

where c is the concentration of actin in a solution and M_w is the weight-average length of the filaments (Flory 1953). Provided that the binding affinity of an actin cross-linking protein is known, V_c can be easily defined experimentally by determining the added cross-linker concentration at which the actin filament solution abruptly changes from a liquid to a gel (Stossel 1984). Similar values for minimum concentrations of the same actin cross-linking proteins that result in actin gelling have been experimentally obtained by measurements of apparent static rigidity (Brotschi et al 1978; Hartwig & Stossel 1979) or apparent viscosity of actin solutions, determined at relatively low (Yin et al 1980; Zaner & Stossel 1982, 1983; Corwin & Hartwig 1983) or high shear rates (Nunnally et al 1981). With low shear viscosity measurements, the shear rate dependence of apparent viscosity must be normalized for the actin concentration, because failure to do so (as in Rockwell et al 1984) causes variations in actin concentration to affect anomalously the minimum gelling concentration. Gel point analysis that employs a defined cross-linking protein is also useful in determining the weight-average actin filament length produced by actin-binding proteins that influence the steady-state length of F-actin.

Table 3 lists minimum gelling concentrations of actin cross-linking proteins, standardized to an equivalent actin concentration. However, since the minimum gelling concentration is also dependent on the actin filament length, a parameter that has rarely been measured, few of the values in Table 3 are strictly comparable. The theoretical minimum actin gelling concentration expression (Equation 2) assumes that each added cross-linker molecule can recruit two polymers into a network. Therefore, the experimentally determined minimum actin gelling concentration for actin filaments of a given length deviates upwards from the theoretical if the affinity of the added cross-linker for actin is low, or if the geometry of the polymers prevents recruitment of filaments into the network. All of the cross-linking proteins that have been characterized have two actin-binding sites, and similar actin-binding affinities. Therefore, the higher minimal actin gelling concentrations are best explained by more than one molecule of cross-linking protein being required to join individual filaments. This redundant cross-linking is favored by conditions in which filaments align in parallel. Therefore, proteins that promote or stabilize isotropic branching of actin fibers are more potent agents in gelling actin solutions than are actin-bundling proteins.

6.1   Actin Cross-Linking Binding Proteins That Promote Isotropic Networks

The relatively rigid, rod-like nature of actin filaments dictates that for actin-binding proteins to create isotropic actin gels with a low volume fraction of polymer mass they must, at a minimum, stabilize the isotropy inherent in solutions of overlapping, semiflexible actin filaments (Kasai et al 1960; Zaner & Stossel 1983; Niederman et al 1983). Another way to produce isotropic gels is for the cross-linking species to cause filaments to branch at angles approaching 90°, thereby counteracting the tendency of long rod-like fibers to align in parallel (Flory 1956). The large size of actin polymers and their complex helical structure suggest that for actin-binding proteins to promote such isotropic gelation they must be sufficiently large and flexible to accommodate the specific binding sites in overlapping actin fibers (Hartwig & Stossel 1981). This prediction has been fulfilled by several actin-binding proteins of high molecular weight. The interactions of these proteins with actin have been studied using binding analyses, by determining minimum gelling concentrations, and by using quantitative morphometry of electron micrographs of cross-linked actin networks prepared so as to retain three-dimensional structure.

6.1.1   ACTIN-BINDING PROTEIN AND FILAMIN   Actin-binding protein has been purified from rabbit and human leukocytes (Hartwig & Stossel 1975; Stossel & Hartwig 1975, 1976; Boxer & Stossel 1976), human platelets (Lucas et al 1976; Schollmeyer et al 1978; Rosenberg et al 1981b), HeLa cells (Weihing 1983), BHK-21 cells (Schloss and Goldman 1979), and toad oocytes (Corwin & Hartwig 1983). A very similar protein called filamin has been purified from chicken gizzard tissue (Shizuta et al 1976; Wang et al 1975; Wang 1977), and guinea pig vas deferens (Wallach et al 1978). Antibodies against the latter cross-react with a variety of cell and tissue types as well as leukocytes (Wallach et al 1978). Although actin-binding protein and filamin are structurally similar, they differ functionally. Actin-binding protein and filamin are homodimers composed of 270 kDa subunits. When rotary-shadowed molecules are viewed in the electron microscope, individual molecules are flexible strands of 3–5 nm in diameter and 160 nm in length (Tyler et al 1980; Hartwig & Stossel 1981). They have binding sites for actin located near their ends (Hartwig & Stossel 1981).

The minimum concentration of actin-binding protein to produce actin gelling approaches the theoretical value (Table 3). Actin-binding protein has a minimum actin gelling concentration that is 3–5 times lower than the smooth muscle counterpart (Brotschi et al 1978). Failure to recognize this difference has led to unwarranted criticism of the applicability of minimum gelling concentration measurements for assessing filament length (Rockwell et al 1984). The effectiveness of actin-binding protein in actin gelation relative to filamin appears to derive from its ability to promote perpendicular branching of actin filaments (Hartwig et al 1980; Niederman et al 1983). This perpendicular branching of actin by actin-binding protein is more obvious when actin is polymerized in the presence of actin-binding protein than when actin-binding protein is added to preformed actin filaments. This difference was initially ascribed to nucleation of actin assembly by actin-binding protein (Hartwig et al 1980), but recent evidence suggests the actin-binding protein does not nucleate actin assembly (P. A. Janmey, unpublished results). A better explanation is that actin-binding protein molecules can orient the branching of short actin fibers, which they bind to during actin assembly, more easily than they can orient extensively entangled, long preformed filaments. Filamin has not been reported to cause perpendicular actin branching. The structural basis of the different functional properties of actin-binding protein and filamin are unclear, although filamin has a somewhat more compact configuration in solution than actin-binding protein (Shizuta et al 1976; Wang 1977), and has been reported to aggregate into higher oligomers (Davies et al 1980). Until it can be shown that differences in purification, speciation, or other modifications can account for the different functional properties of actin-binding protein and filamin, it is reasonable to consider them related but distinct proteins designed to cross-link actin into different configurations in nonmuscle and smooth muscle cells, respectively.

Actin-binding protein accounts for 1–2% of the total cell protein of leukocytes and tissue culture cells, and as much as 7% of the total cellular protein of platelets. The actin cross-linking activity of actin-binding protein has been demonstrated to account for the majority of actin crosslinking activity in several cell types (Brotschi et al 1978; Mimura & Asano 1979; Corwin & Hartwig 1983). By immunofluorescence microscopy, anti-actin–binding protein and filamin antibodies stain the cortical cytoplasm of leukocytes, and the cortical cytoplasm and actin-containing stress fiber bundles of cultured cells.

Thus far, actin-binding protein is the only protein that has been shown to localize at junctions between actin filaments both in vitro in purified protein assembles, and in situ in a cortical cytoplasmic actin network. Macrophages have cortical actin filaments arranged in an orthogonal network that qualitatively resembles the actin assemblies that form in the presence of actin-binding protein. As shown by immuno-gold electron microscopy, actin-binding protein locates in the network selectively at points where filaments intersect (Hartwig & Stossel 1983). These findings indicate that actin cross-linking by actin-binding protein accounts in an important way for the architecture of actin in the cortex of macrophages. It can be inferred that this protein has a similar role in the other cell types in which it has been identified. Since each end of the actin-binding protein molecule has an actin-binding site, a template is present at each end of the molecule to orient filaments in three-dimensional space such that an orthogonal array results.

6.1.2 ERYTHROCYTE SPECTRIN   Spectrin, a 460 kDa protein, connects actin oligomers into a two-dimensional, submembranous network that laminates the cytoplasmic face of the red cell membrane. Erythrocyte spectrin is a heterodimer of 240 (alpha) and 220 kDa (beta) chains. These dimers appear to be more rigid in structure than the subunits of actin-binding protein and filamin, in that the two spectrin subunit chains are generally observed in electron micrographs arranged in parallel and interwoven along their length. Spectrin dimers contain only one binding site for actin (Brenner & Korn 1979a), which is located near one end of the molecule. Spectrin dimers also have binding sites for other red cell proteins, specifically ankyrin, band 4.1 protein, and calmodulin. By binding to ankyrin, which is itself bound to the integral membrane protein band 3, the spectrin-actin network is fastened to the membrane (Bennett & Stenbuck 1980; Hargreaves et al 1980). Band 4.1 protein increases ten- to one hundredfold the affinity of spectrin for actin (Cohen & Korsgren 1980), and aids in membrane attachment of spectrin, because band 4.1 protein also binds to the integral membrane protein glycophorin (Anderson & Lovrien 1984). The role of calmodulin associated with spectrin is unknown. At the end of the spectrin dimer opposite the actin-binding site is a self-association site. In physiological salt solutions, spectrin dimers reversibly associate into elongated tetramers, 200 nm long (Ungewickell et al 1979; Shotton et al 1979). The spectrin molecules must be in the form of tetramers to cross-link actin filaments (Brenner & Korn 1979a; Fowler & Taylor 1980). The minimum actin gelling concentration of spectrin tetramers is severalfold higher than that of actin-binding protein or filamin (Schanus et al 1985); presumably the lower flexibility of spectrin molecules accounts for this difference.

Although spectrin tetramers can cross-link actin filaments to form a three-dimensional gel, a fact that supports our concept of the mechanism of isotropic actin gelation, the actin-spectrin network of the erythrocyte is not considered a three-dimensional gel of actin filaments cross-linked by spectrin molecules. Instead, it consists of a two-dimensional sheet of elongated spectrin molecules joined by very short actin oligomers containing fewer than 12 subunits. How this two-dimensional lamina of spectrin and actin assembles in the red cell is not clear, but it seems likely that two–dimensionality in vivo results from the low molar ratio of actin to spectrin in the red cell, and to the tight adherence of spectrin at the membrane.

6.1.3 NONERYTHROID SPECTRINS   Spectrin-related molecules are present in diverse cell and tissue types including brain (Levine & Willard 1981, 1983; Bennett et al 1982; Burns et al 1983; Davis & Bennett 1983; Glenney et al 1982a,b; Burridge et al 1982; Goodman et al 1981; Lazarides & Nelson 1982, 1983; Nelson & Lazarides 1983; Smino-Oka et al 1983), intestinal epithelial cells (Glenney et al 1982c,d; Pearl et al 1984), avian lens (Nelson et al 1983), sea urchin coelomocytes (Edds & Venute-Henderson 1983), and amoebae (Pollard 1984a). The brain and intestinal spectrin-like proteins have also been called fodrin (Levine & Willard 1981) and TW260/240 (Glenney et al 1982c). Like the erythrocyte protein, these molecules are asymmetric heterodimers 100–130 nm long. The larger subunit has been designated alpha, and the smaller chain has been termed either beta or gamma, depending on whether or not it is antigenically related to the beta chain of vertebrate erythroid cells. Cell- or tissue-specific alpha chains vary from 235 to 260 kDa, and the beta chains are 230–235 kDa. The beta chains cross–react with IgG's prepared against the erythrocyte beta chains (Glenney et al 1982c; Burridge et al 1982; Lazarides & Nelson 1983). Chicken nonerythroid cells are the exception to this rule: They have alpha chains that are homologous to their erythroid alpha chains, and a variable nonhomologous beta chain (Lazarides & Nelson 1983).

Like erythrocyte spectrin, the nonerythroid spectrin analogues must be in the form of tetramers to cross-link actin filaments. In electron micrographs, the tetramers are asymmetrical, rigid rods 200–260 nm long, depending on the tissue source. The spectrins bind to actin filaments at their ends. These proteins also bind calmodulin but, as with erythroid spectrin, the functional significance of this binding has not been determined. Since the gelation of actin by the TW260/240 intestinal brush border cell variant of this protein is reportedly diminished by micromolar calcium concentrations (Glenney et al 1982c), calmodulin may have a role in this effect.

It has been suggested, based on its structural similarity to the red cell protein, that nonerythrocyte spectrin attaches actin filaments to membranes. This idea finds support in the diffuse, mesh-like localization of spectrin in cells, as determined by labeling cells with immunofluorescent antibodies (Burridge et al 1982; Lazarides & Nelson 1983; Levine & Willard 1981; Lehto & Virtanen 1983). However, TW260/240 localizes only in the rootlets of intestinal epithelial cell microvilli, where it appears to connect the actin filament bundles that compose the cores of these surface specializations (Glenney et al 1982c).

6.1.4 PHYSARUM 230-kDa PROTEIN   A protein has been purified from *Physarum polycephalum* that shares physical properties with both actin-binding protein and with spectrins (Sutoh et al 1984). In physiological solvents this protein exists as a dimer composed of apparently identical 230 kDa subunits. In the electron microscope, the subunit chains appear to be connected head-to-head like actin-binding protein to form bipolar structures 60 nm in contour length. The bipolar molecules are much more condensed than actin-binding protein, because each subunit strand has a globular region near its middle. Therefore, the *Physarum*-derived molecules have a dumbbell-like appearance which is consistent with the hydro-dynamic properties of the protein. Its low minimum actin gelling concentration suggests that it forms isotropic actin networks (Table 3).

6.1.5 DICTYOSTELIUM 120 kDa   A dimeric protein of 120 kDa subunits has been purified from *Dictyostelium discoideum*, and is reported to form branched actin networks (Condeelis et al 1984). Molecules of this 120 kDa protein appear as short rods, 35 nm long, in the electron microscope (Condeelis et al 1984). The 120 kDa protein has a high minimum gelling concentration (Table 3), which is consistent with the idea that only large asymmetrical actin cross-linking molecules are effective in forming iso-tropic actin networks. This protein has been localized by immunofluor-escence in the cell periphery. Its ability to cause actin gelation is not affected by calcium.

Recently, a chicken gizzard protein of similar subunit size, called caldesmon, has been shown to cross-link actin filaments (Bretscher 1984). This protein binds calmodulin, which results in inhibition of filament cross-linking in the presence of calcium. Caldesmon isoforms with different M_r's have been purified from brain tissue (Sobue et al 1982).

6.1.6 MYOSINS   Although myosin is usually considered in terms of its mechanico-chemical activity in generating contractile forces and move-ments with actin, it is important to note that this large asymmetrical molecule is a relatively potent actin cross-linking agent (Brotschi et al 1978). From the potentially high tensile strength of the shaft of bipolar myosin filaments (Huxley 1963) it follows that a matrix of actin fibers cross-linked by such filaments would have a high rigidity. This prediction has been experimentally verified (Abe & Maruyama 1974). It is possible, therefore, that myosin contributes to the consistency of the cytoplasmic actin matrix.

6.2   *Actin Filament Bundling Proteins*

Proteins of this group stabilize the alignment of actin filaments into side-by-side aggregates. Once actin filaments align in parallel, homologous regions on adjacent fibers can take on fixed relationships, thus relatively short-range actin filament cross-linkers are capable of stabilizing bundles. Since individual filaments in the bundles are connected by multiple cross-linking molecules, they are predicted to have high minimum gelling concentrations relative to cross-linkers that promote more isotropic structures. As shown in Table 3, this is what occurs experimentally. The interaction of these proteins and actin have been documented by a number of techniques. The alignment of actin fibers alters their hydrodynamic properties, and these changes can be detected by increases in solution viscosity, flow birefringence, and protein sedimentation rate. Electron microscopy provides direct visualization of bundles formed, and in the case of tightly packed bundles, optical diffraction of the micrographs is especially helpful in identifying the nature and spacing of cross-linking proteins on the filaments. These cross-linkers fall into two broad structural categories: rigid rods and globular proteins.

6.2.1   ALPHA-ACTININ   Alpha-actinin is an asymmetric homodimer, which when viewed in the electron microscope appears as a short rod, 4 nm in diameter and 40 nm in length. The orientation of the individual subunit chains in the rods is unknown. The binding sites for actin filaments appear to be located at the ends of these rods, because the rods are observed in electron micrographs to space filaments within the loose bundles formed by the addition of alpha-actinin to actin (Podlubnaya et al 1975; Jockusch & Isenberg 1981; Maruyama et al 1977). In 1975, antigens cross-reactive with the muscle alpha-actinin were shown to be structural components of nonmuscle cells (Lazarides & Burridge 1975), and alpha-actinin has now been isolated and characterized from a number of different cells, including various cultured mammalian cells (Burridge & Feramisco 1981; Mimura & Asano 1979; Yeltman et al 1981), platelets (Rosenberg et al 1981; Landon & Olomucki 1983), amoeba (Brier et al 1983; Condeelis & Vahey 1982; Fechheimer et al 1982; Pollard 1981), brain cells (Duhaiman & Bamburg 1984), and macrophages (Bennett et al 1984). A principal feature that differentiates muscle from nonmuscle cell alpha-actinins is that the latter are calcium-binding proteins, and this calcium binding inhibits the ability of alpha-actinin to interact with actin. As nonmuscle alpha-actinin binds calcium, which it does with high affinity, its binding affinity for actin diminishes (Duhaiman & Bamburg 1984; Bennett et al 1984). Although alpha-actinin has clearly been demonstrated to bind and cross-link actin filaments into a side-by-side configuration, these effects, at least in the case of vertebrate muscle and several cytoplasmic alpha-actinins, are maximally manifested at low temperatures. Increasing temperatures diminish alpha-. actinins' effects on actin viscosity, and on the gelation of actin, and no gelation of actin by the macrophage variant occurs when the temperature is raised to 37°C (Bennett et al 1984). These effects occur despite the finding that the binding affinity of alpha-actinins for actin increases with temperature. Nonmuscle alpha-actinin can also bind diacylglycerol and palmitic acid. Increased actin bundle formation has been reported to occur under these conditions (Burn et al 1985).

6.2.2   FASCIN   A 57-kDa protein that bundles actin filaments has been purified from eggs of the sea urchin *Tripneustes gratilla* (Bryan & Kane 1978, 1982). Antigens that cross-react with this protein have been identified by immunofluorescence in other sea urchin eggs (Otto et al 1980), and in sea urchin coelomocytes (Otto et al 1979). Actin filament bundling proteins of similar structure have recently been isolated from such diverse sources as porcine brain and starfish sperm (Maekawa et al 1982, 1983).

These proteins are monomers that have a high capacity for binding actin filaments, one molecule for every 4–5 actin monomers in the filament (Bryan & Kane 1978). Optical diffraction studies of actin bundles formed with fascin indicate that fascin exists within actin filament bundles as a regular repeat inserted once every 13 nm; this value agrees with predictions from its binding capacity (DeRosier et al 1977; Bryan & Kane 1978; DeRosier & Edds 1980). This cross-linking results in the formation of tightly packed bundles in which each filament is approximately 9 nm from its neighbor.

Fascin interacts with another sea urchin protein, which has a subunit M_r of 220,000. This protein appears to cross-link fascin molecules exposed on fascin-actin bundles, thus providing a mechanism for gelling actin bundles (Kane 1975, 1976). Since the 220-kDa protein does not bind to actin filaments with high affinity (Bryan & Kane 1978) it is presumed that it binds to the actin-fascin complex exposed on the bundle surface, although direct evidence for such binding has not been reported.

6.2.3   FIMBRIN   Fimbrin is the name given to a protein found in the actin filament bundles of microvilli (Bretscher & Weber 1980b; Bretscher 1981; Glenney et al 1981b). It is a 68 kDa polypeptide that binds to actin filaments with high capacity, similar to fascin, and forms tightly packed filament bundles in which all filaments have the same polarity. The filament to filament spacing in these bundles is approximately 9 nm (Matsudaira et al 1983).

Antibodies against the intestinal fimbrin cross-react with a component, and localize in the cortical cytoplasm of a variety of cells (Bretscher &

Weber 1980b). In most cells the domain in which the cross-reactivity occurs lacks filament bundles.

6.2.4 VILLIN   Villin, as mentioned above, is a versatile protein originally purified from intestinal epithelial cells. It cross-links actin filaments into bundles in the absence of calcium by binding actin to its "headpiece." The structure of actin bundles cross-linked by villin is different from those cross-linked by fascin- and fimbrin-actin bundles. They are more loosely packed, and have a center-to-center spacing of approximately 12 nm (Matsudaira et al 1983). The polarity of filaments in villin-actin bundles has not been reported.

6.2.5 OTHER BUNDLING PROTEINS   A 30 kDa protein that bundles actin filaments was recently purified from *Dictyostelium discoideum* (Fechheimer & Taylor 1984). Binding of this protein to actin filaments, as assessed by cosedimentation, is inhibited by 0.1 μM calcium. The protein is monomeric in physiologic solvent solution. Its amino acid composition differs from severin, another *Dictostelium* protein of similar M_r, and from a 35 kDa actin-binding protein isolated from *Physarum* by Ogihara & Tonomura (1982).

The glycolytic enzyme aldolase can induce the parallel alignment of actin filaments in vitro when present at high molar ratios to actin (Clarke & Morton 1976, 1982; Morton et al 1977). Although some bundling of filaments occurs when aldolase is added to F-actin alone, a more ordered lattice structure showing transverse bands every 38 nm forms when aldolase is added to actin filaments that contain tropomyosin and troponin. This was found to result from the binding of aldolase to the tropomyosin-troponin complex in addition to its interaction with actin (Walsh et al 1980; Stewart et al 1980). Binding studies show that aldolase possesses multiple binding sites for these proteins, with an affinity constant of 20 μM^{-1} for F-actin, and 600 μM^{-1} for the F-actin–tropomyosin–troponin complex.

Nerve growth factor (NGF) also forms well-ordered actin bundles (Castellani & O'Brien 1981). These structures, which are formed only by the biologically active form of NGF, resemble actin paracrystals induced by Mg^{2+}. While no binding studies have been done, optical diffraction patterns show that NGF cross-links actin filaments in pairs of bridges 9 nm apart, and at repeating intervals of 37 nm.

6.3   Other Actin Gelling Proteins

6.3.1 GELACTINS   Preliminary data concerning four actin gelling proteins purified from *Acanthamoeba* were reported in 1977 (Maruta & Korn 1977).

Judging from their minimum actin gelling concentrations these proteins are of a potency associated with cross-linkers that form isotropic networks. However, no further information on the nature of the structure formed by these proteins and actin has subsequently appeared.

6.3.2 MICROTUBULE-ASSOCIATED PROTEINS Another group of proteins that have been reported to gel solutions of actin are microtubule-associated proteins (MAP's), a term used to encompass several proteins that bind to microtubules (Griffith & Pollard 1978, 1982b; Nishida et al 1981; Selden & Pollard 1983). In terms of cross-linking actin filaments, fractions enriched in MAP-2 and the tau proteins have been shown to gel actin at relatively high concentrations, which suggests either parallel alignment of actin or a low affinity interaction. Actin filaments incubated with high concentrations of MAP's form bundles visible in the electron microscope (Griffith & Pollard 1982b; Sattilaro et al 1981). The binding affinity of these protein(s) for actin has not been determined. It is interesting that MAP-2 has been reported to be antigenically related to spectrin, which suggests that it may be another specialized spectrin variant that shares common actin-binding subunits.

Phosphorylation of MAP's decreases their ability to interact with actin filaments, as evidenced by an increase in the minimum actin gelling concentration (Selden & Pollard 1983). Since MAP's localize by immunofluorescence to microtubules, and not to cortical networks or bundles of actin in cells, the functional significance of actin cross-linking by MAP's is unclear. However, since MAP's increase the viscosity of mixtures of actin filaments and microtubules, it is speculated that they may mediate some interaction between these two cytoplasmic polymer types (Griffith & Pollard 1978).

7. PROTEINS THAT PREDOMINANTLY BIND THE SIDES OF ACTIN FILAMENTS

The high molecular weight actin cross-linking proteins, actin-binding protein and filamin, bind to several monomers within the actin filament helix (Zeece et al 1979; Hartwig & Stossel 1981). Proteins that sever actin filaments must transiently bind to the sides of actin filaments. The proteins tropomyosin and cofilin are noteworthy in that they bind to the sides of actin filaments, but do not appear to have direct cross-linking or fragmenting functions.

7.1 Tropomyosin

Tropomyosin, together with troponin, confers calcium sensitivity upon the interaction of myosin and actin in the sarcomeres of striated muscle (Smillie

Table 4   Nonmuscle cells from which tropomyosins have been isolated

Cell type	References
Blood platelets	Cohen & Cohen 1972; Cote & Smillie 1981; Fine & Blitz 1975; der Terrossian et al 1981
Brain	Fine et al 1973; Fine & Blitz 1975
Sea urchin eggs	Ishimodo-Takagi 1978
Dictyostelium discoideum	Kato & Tonomura 1975
Macrophages	Fattoum et al 1983
Fibroblasts and epithelial cells	Fine & Blitz 1975; Matsumura et al 1983a,b; Schloss & Goldman 1980; Lin et al 1985; Giometti & Anderson 1981

1979). Tropomyosins have also been isolated from smooth muscles and from nonmuscle cells (Table 4). Native tropomyosin is an asymmetric dimer; its elongated structure allows it to rest within the groove of the actin filament helix. The molecular weights of the protein subunits vary slightly from cell to cell, and even within the same cells, in the range of 29–39 kDa. The binding of tropomyosin to F-actin exhibits positive cooperativity (Wegner 1979), and the K_d for this binding is in the micromolar range. The various tropomyosin isoforms have somewhat different binding affinities for actin. Interestingly, malignant transformation of cultured cells results in the expression of lower affinity isoforms (Lin et al 1985).

The only known function of tropomyosin, other than its modulating effects on actin-myosin interaction defined by studies in vitro is its ability to inhibit incompletely the spontaneous fragmentation of actin filaments (Kawamura & Maruyama 1970; Wegner 1982), or the rupture of actin filaments by actin-fragmenting proteins (Bernstein & Bamburg 1982; Fattoum et al 1983).

7.2   Cofilin

Cofilin is a 19 kDa protein purified from bovine brain (Nishida et al 1984a). When present with monomeric actin undergoing assembly it slows the rate of polymerization, and can be isolated together with actin as an equimolar complex. In this respect the protein resembles profilin. When added to actin polymers, however, the protein clings to the sides of the filaments, as evidenced by the thickened appearance of negatively stained actin filaments in the electron microscope, by the cosedimentation of cofilin with actin, and by the quenching of pyrene fluorescence of F-actin by the addition of cofilin. The ability of actin filaments to interact with skeletal muscle myosin is also inhibited by the binding of cofilin.

8. OTHER PROTEINS REPORTED TO BIND TO ACTIN

Proteins not normally thought to be cytoskeletal proteins have been reported to interact with actin. Some of these (aldolase, other glycolytic enzymes, and nerve growth factor) cross-link actin filaments into bundles and are listed in Table 3; others are shown in Table 5.

Current thought about the cytoskeleton focuses on its putative determination of the mechanics of cell division, shape, and movement. Consideration of the proteins listed in Table 5 may be useful insofar as it may stimulate thought about other possible roles for actin. Adsorbtion of various enzymes such as aldolase and other glycolytic enzymes to actin

Table 5 Other proteins reported to interact with actin

Cellular proteins	
protamine	Magri et al 1978
aldolase	Stewart et al 1980
adenylosuccinate synthetase	Ogawa et al 1978
lysozyme	Griffith & Pollard 1982a
tubulin	Verkhovsky et al 1981
Nuclear constituents	
histone	Magri et al 1978
poly(A) polymerase-endoribonuclease IV	Schroder et al 1982
ribonuclease A	Griffith & Pollard 1982a
DNAase I	Lazarides & Lindberg 1974
Extracellular proteins	
immunoglobulin	Fechheimer et al 1979
C1q	Nishioka et al 1982
fibronectin	Keski-Oja et al 1980
plasma gelsolin (brevin, ADF)	Chaponnier et al 1979; Norberg et al 1979; Harris & Schwartz 1981; Janmey et al 1985a
vitamin D-binding protein (G_c-globin)	Van Baelen et al 1980; Haddad 1982
lysozyme	Griffith & Pollard 1982a
fibrin	Laki & Muszbek 1974; Janmey et al 1985b
connectin	Brown et al 1983
crystallin	Delvecchio et al 1984
Membrane linkage proteins	
Dictyostelium 24 kDa	Stratford & Brown 1985
brush border 110 kDa	Glenney et al 1982d

filaments may be important in functionally compartmentalizing the cytoplasm. Alternatively, the association of various enzymes with actin filaments may play a physiological role in regulating their activity, such as has been shown for poly(A) polymerase-endoribonuclease IV. The presence of nuclear actin (Bremer et al 1981) suggests that the binding of actin to histone may prove to be of physiological significance.

The reported association of actin with extracellular proteins, such as plasma gelsolin, C1q, IgG, IgM, connectin, the vitamin D-binding protein, and fibronectin, suggests that actin may be found outside of cells. Since many tissue proteins pass through the plasma transiently as a result of cell turnover, defense mechanisms may exist to deal with actin that passes through the extracellular space as a result of cell death.

Actin is the major protein of many mammalian cells, and as such constitutes a principle commodity to be managed by the body's scavenging mechanisms. Furthermore, much of the actin present in the body exists in a filamentous form. Such filaments, often micrometers long might have significant effects upon the rheology of blood in the microcirculation should they circulate intact. It is possible that plasma proteins interact with actin to facilitate the clearance of actin from the circulation, thereby minimizing the deleterious effects that this long biopolymer might cause.

Literature Cited

Abe, S. I., Maruyama, K. 1974. Dynamic viscoelastic study of acto-heavy meromyosin in solution. *Biochim. Biophys. Acta* 160:160–74

Adelman, M. R., Taylor, E. W. 1969. Isolation of an actomyosin-like protein from slime mold plasmodium and the separation of the complex into actin-myosin-like fractions. *Biochemistry* 12:4964–75

Alicea, H. A., Mooseker, M. S. 1984. Comparative analysis of actin-binding-proteins of the chicken and mammalian intestinal brush borders. *J. Cell Biol.* 99:306a

Ampe, C., Vandekerckhove, J., Brenner, S. L., Tobacman, L., Korn, E. D. 1985. The amino acid sequence of acanthamoeba profilin. *J. Biol. Chem.* 260:834–40

Anderson, R. A., Lovrien, R. E. 1984. Glycophorin is linked by band 4.1 protein to the human erythrocyte membrane skeleton. *Nature* 307:655–58

Asakura, S., Taniguchi, M., Oosawa, F. 1963. Mechano-chemical behavior of F-actin. *J. Mol. Biol.* 7:55–69

Bamburg, J. R., Harris, H. E., Weeds, A. G. 1980. Partial purification and characterization of an actin depolymerizing factor from brain. *FEBS Lett.* 121:178–82

Bennett, J. P., Zaner, K. S., Stossel, T. P. 1984. Isolation and some properties of macrophage a-actinin: evidence that it is not an actin gelling protein. *Biochemistry* 23:5081–86

Bennett, V., Davis, J., Fowler, V. E. 1982. Brain spectrin, a membrane-associated protein related in structure and function to erythrocyte spectrin. *Nature* 299:126–31

Bennett, V., Stenbuck, P. J. 1980. Association between ankyrin and the cytoplasmic domains of band 3 isolated from the human erythrocyte membrane. *J. Biol. Chem.* 255:6424–32

Bernstein, B. W., Bamburg, J. R. 1982. Tropomyosin binding to F-actin protects the F-actin from disassembly by actin depolymerizing factor (ADF). *Cell Motil.* 2:1–8

Blikstad, I., Markey, F., Carlsson, L., Persson, T., Lindberg, U. 1978. Selective assay of monomeric and filamentous actin in cell extracts, using inhibition of deoxyribonuclease I. *Cell* 15:935–43

Blikstad, I., Sundkvist, I., Eriksson, S. 1980. Isolation and characterization of profilactin and profilin from calf thymus and brain. *Eur. J. Biochem.* 105:425–33

Bonder, E., Mooseker, M. S. 1983. Direct electron microscopic visualization of barbed end capping and filament cutting by intestinal microvillar 95-kdalton protein (villin). A new actin assembly assay using the limulus acrosomal process. *J. Cell Biol.* 96:1097–1107

Bonder, E. M., Fishkind, D. J., Mooseker, M. S. 1983. Direct measurement of critical concentrations and assembly rate constants at the two ends of an actin filament. *Cell* 34:491–501

Boxer, L. A., Stossel, T. P. 1976. Isolation and properties of actin, myosin, and a new actin-binding protein of chronic myelogenous leukemia leukocytes. *J. Clin. Invest.* 57:5696–5705

Bremer, J. W., Busch, H., Yeoman, L. C. 1981. Evidence for a species of nuclear actin distinct from cytoplasmic and muscle actins. *Biochemistry* 20:2013–17

Brenner, S. L., Korn, E. D. 1979a. Spectrin-actin interaction. Phosphorylated and dephosphorylated spectrin tetramer crosslink f-actin. *J. Biol. Chem.* 254:8620–27

Brenner, S. L., Korn, E. D. 1979b. Substoichiometric concentrations of cytochalasin and inhibit actin polymerization. Additional evidence for an f-actin treadmill. *J. Biol. Chem.* 254:9982–85

Brenner, S. L., Korn, E. D. 1984. Evidence that F-actin can hydrolyze ATP independent of monomer-polymer end interactions. *J. Biol. Chem.* 259:1441–46

Bretscher, A. 1981. Fimbrin is a cytoskeletal protein that crosslinks F-actin in vitro. *Proc. Natl. Acad. Sci. USA* 78:6849–53

Bretscher, A. 1984. Smooth muscle caldesmon. Rapid purification and f-actin cross-linking properties. *J. Biol. Chem.* 259:12873–88

Bretscher, A., Weber, K. 1980a. Villin is a major protein of microvillus cytoskeleton which binds both G and F actin in a calcium-dependent manner. *Cell* 20:839–47

Bretscher, A., Weber, K. 1980b. Fimbrin, a new microfilament-associated protein present in microvilli and other cell surface structures. *J. Cell Biol.* 86:335–40

Brier, J., Fechheimer, M., Swanson, J., Taylor, D. L. 1983. Abundance, relative gelation activity, and distribution of the 95,000-dalton actin-binding protein from *Dictyostelium discoideum*. *J. Cell Biol.* 97:178–85

Brotschi, E. A., Hartwig, J. H., Stossel, T. P. 1978. The gelation of actin by actin-binding protein. *J. Biol. Chem.* 253:8988–93

Brown, S. S., Malinoff, H. L., Wicha, M. S. 1983. Connectin: cell surface protein that binds both laminin and actin. *Proc. Natl. Acad. Sci. USA* 80:5927–30

Brown, S. S., Spudich, J. A. 1979. Cytochalasin inhibits the rate of elongation of actin filament fragments. *J. Cell Biol.* 83:657–62

Brown, S. S., Yamamoto, K., Spudich, J. A. 1982. A 40,000 protein from *dictyostelium discoideum* affects assembly properties of actin in a Ca^{2+} dependent manner. *J. Cell Biol.* 93:205–10

Bryan, J., Kane, R. E. 1978. Separation and interaction of the major components of sea urchin actin gel. *J. Mol. Biol.* 125:207–24

Bryan, J., Kane, R. E. 1982. Actin gelation in sea urchin egg extracts. In *Methods in Cell Biology*, ed. L. Wilson, pp. 176–99. New York: Academic

Bryan, J., Kurth, M. 1984. Actin-gelsolin interactions: evidence for two actin-binding sites. *J. Biol. Chem.* 259:7480–87

Burn, P., Rotman, A., Meyer, R. K., Burger, M. M. 1985. Diacylglycerol in large α-actinin/actin complexes and in the cytoskeleton of activated platelets. *Nature* 314:469–72

Burns, N. R., Ohanian, V., Gratzer, W. B. 1983. Properties of brain spectrin (fodrin). *FEBS Lett.* 153:165–68

Burridge, K., Feramisco, J. R. 1981. Nonmuscle a-actinins are calcium-sensitive actin-binding proteins. *Nature* 294:565–67

Burridge, K., Kelly, T., Mangeat, P. 1982. Nonerythrocyte spectrins: actin-membrane attachment proteins occurring in many cell types. *J. Cell Biol.* 95:478–86

Carlier, M. F., Pantaloni, D., Korn, E. D. 1984. Evidence for an ATP cap at the ends of actin filaments and its regulation of the F-actin steady state. *J. Biol. Chem.* 259:9983–86

Carlier, M. F., Pantaloni, D., Korn, E. D. 1984. Steady state length distribution of F-actin under controlled fragmentation and mechanism of length redistribution following fragmentation. *J. Biol. Chem.* 259:9987–91

Carlsson, L., Nystrom, L. E., Sundkvist, I., Markey, F., Lindberg, U. 1976. Profilin, a low molecular weight protein controlling actin polymerisability. In *Contractile Systems in Non-Muscle Tissues*, ed. S. V. Perry, A. Margreth, R. S. Adelstein, pp. 39–48. Amsterdam: Elsevier/North Holland

Castellani, L., O'Brien, E. J. 1981. Structure of actin paracrystals induced by nerve growth factor. *J. Mol. Biol.* 147:205–13

Chaponnier, C., Borgia, R., Rungger-Brandle, E., Weil, R., Gabbiani, G. 1979. An actin-destabilizing factor is present in human plasma. *Experientia (Basel)* 35: 1039–40

Chaponnier, C., Patebex, P., Gabbiani, G. 1985. Human plasma actin-depolymerizing factor. Purification, biological activity and localization in leucocytes and platelets. *Eur. J. Biochem.* 146: 267–76

Clarke, F. M., Morton, D. J. 1976. Aldolase binding to actin-containing filaments. Formation of paracrystals. *Biochem. J.* 159: 797–98

Clarke, F. M., Morton, D. J. 1982. Glycolytic enzyme binding in fetal brain—the role of actin. *Biochem. Biophys. Res. Comm.* 109: 388–93

Cohen, C. M., Korsgren, C. 1980. Band 4.1 causes spectrin-actin gels to become thixiotropic. *Biochem. Biophys. Res. Comm.* 97: 1429–35

Cohen, I., Cohen, C. 1972. A tropomyosin-like protein from human platelets. *J. Mol. Biol.* 68: 383–87

Coluccio, L. M., Kurth, M. C., Bryan, J. 1984. Gelsolin, a multifunctional actin associated protein. *J. Cell Biol.* 99: 307a

Coluccio, L. M., Tilney, L. G. 1983. Under physiological conditions actin disassembles slowly from the nonpreferred end of an actin filament. *J. Cell. Biol.* 97: 1629–34

Condeelis, J., Vahey, M. 1982. A calcium- and pH-regulated protein from *Dictyostelium discoideum* that cross-links actin filaments. *J. Cell Biol.* 94: 466–71

Condeelis, J., Vahey, M., Carboni, J. M., DeMey, J., Ogihara, S. 1984. Properties of the 120,000- and 95,000-dalton actin-binding proteins from *Dictyostelium discoideum* and their possible functions in assembling the cytoplasmic matrix. *J. Cell Biol.* 99: 119s–26s

Cooper, J. A., Blum, J. D., Pollard, T. D. 1984. *Acanthamoeba castellanii* capping protein: properties, mechanism of action, immunologic cross-reactivity, and localization. *J. Cell Biol.* 99: 217–25

Cooper, J. A., Pollard, T. D. 1985. Effect of capping protein on the kinetics of actin polymerization. *Biochemistry* 24: 793–99

Cooper, J. A., Buhle, E. L. Jr., Walker, S. B., Tsong, T. Y., Pollard, T. D. 1983. Kinetic evidence for a monomer activation step in actin polymerization. *Biochemistry* 22: 2193–2202

Corwin, H. L., Hartwig, J. H. 1983. Isolation of actin-binding protein and villin from toad oocytes. *Devel. Biol.* 99: 61–74

Côte, G. P., Smillie, L. B. 1981. The interaction of equine platelet tropomyosin with

skeletal muscle actin. *J. Biol. Chem.* 256: 7257–61

Coué, M., Constans, J., Viau, M., Olomucki, A. 1983. The effect of serum vitamin D–binding protein on polymerization and depolymerization of actin is similar to the effect of profilin on actin. *Biochem. Biophys. Acta* 759: 137–45

Craig, S. W., Pollard, T. D. 1982. Actin binding proteins. *Trends Biochem. Sci.* 7: 88–92

Craig, S. W., Powell, L. D. 1980. Regulation of actin polymerization by villin, a 95,000 dalton cytoskeletal component of intestinal brush borders. *Cell* 22: 739–46

Curmi, P. M. G., Barden, J. A., dos Remedios, C. G. 1984. Actin tube formation: effects of variations in commonly used solvent conditions. *J. Muscle Res. Cell Motil.* 5: 423–30

Davies, P. J. A., Wallach, D., Willingham, M., Pastan, I., Lewis, M. S. 1980. Self-association of chicken gizzard fillamin and heavy merofilamin. *Biochemistry* 19: 1366–72

Davis, J., Bennett, V. 1983. Brain spectrin. *J. Biol. Chem.* 258: 7757–66

DeGennes, P. G., Leger, L. 1982. Dynamics of entangled polymer chains. *Ann. Rev. Phys. Chem.* 33: 49–61

Delvecchio, P. J., Macelroy, K. S., Rosser, M. P., Church, R. L. 1984. Association of alpha crystallin with actin in cultured lens cells. *Curr. Eye Res.* 3: 1213–19

der Terrossian, E., Fuller, S. D., Stewart, M., Weeds, A. G. 1981. Porcine platelet tropomyosin. Purification, characterization, and paracrystal formation. *J. Mol. Biol.* 153: 147–67

DeRosier, D., Mandelkow, E., Siliman, A., Tilney, L., Kane, R. 1977. Structure of actin-containing filaments from two types of non-muscle cell. *J. Mol. Biol.* 113: 679–95

DeRosier, D. J., Edds, K. T. 1980. Evidence for fascin cross-links between the actin filaments in coelomocyte filopodia. *Exp. Cell Res.* 126: 490–94

DiNubile, M. J., Southwick, F. S. 1985. Effects of macrophage profilin on actin in the presence of acumentin and gelsolin. *J. Biol. Chem.* 260: 7402–9

Doi, M., Edwards, S. F. 1978. Dynamics of rod-like macromolecules in concentrated solution. Part 2. *J. Chem. Soc. Faraday Trans.* 74: 918–32

Doi, Y., Frieden, C. 1984. Actin polymerization: the effect of brevin on filament size and rate of polymerization. *J. Biol. Chem.* 259: 11868–75

dos Remedios, C. G., Dickens, M. J. 1978. Actin microcrystals and tubes formed in the presence of gadolinium ions. *Nature* 276:731–33

Duhaiman, A. S., Bamburg, J. R. 1984. Isolation of brain a-actinin. Its characterization and a comparison of its properties with those of muscle a-actinins. *Biochemistry* 23:1600–8

Ebashi, S. 1972. Calcium ions and muscle contraction. *Nature* 240:217–18

Ebashi, S., Ebashi, F. 1965. a-actinin, a new structural protein from striated muscle I. Preparation and action on actomyosin-ATP interaction. *J. Biochem.* 58:7–12

Ebashi, S., Ebashi, F., Maruyama, K. 1964. A new protein factor promoting contraction of actomyosin. *Nature* 203:645–46

Edds, K., Venute-Henderson, J. 1983. Coelomocyte spectrin. *Cell Motil.* 3:683–91

Edwards, S. F., Evans, K. E. 1982. Dynamics of highly entangled rod-like molecules. *J. Chem. Soc. Faraday Trans. 2* 78:113–21

Egelman, E. H., Francis, N., DeRosier, D. J. 1982. F-actin is a helix with a random variable twist. *Nature* 298:131–35

Elzinga, M., Phelan, J. J. 1984. F-actin is intermolecularly crosslinked by N,N'-p-phenylene-dimaleimide through lysine-191 and cystein 374. *Proc. Natl. Acad. Sci. USA* 81:6599–6602

Fattoum, A., Hartwig, J. H., Stossel, T. P. 1983. Isolation and some structural and functional properties of macrophage tropomyosin. *Biochemistry* 22:1187–93

Fattoum, A., Roustan, C., Feinberg, J., Prandel, L. A. 1980. Biochemical evidence for a low molecular weight protein (profilin-like protein) in hog thyroid gland and its involvements in actin polymerisation. *FEBS Lett.* 118:237–40

Fechheimer, M., Brier, J., Rockwell, M., Luna, E. J., Taylor, D. L. 1982. A calcium- and pH-regulated actin binding protein from *D. discoideum*. *Cell Motil.* 2:287–308

Fechheimer, M., Daiss, J. L., Cebra, J. J. 1979. Interaction of immunoglobulin with actin. *Mol. Immunol.* 16:881–88

Fechheimer, M., Taylor, D. L. 1984. Isolation and characterization of a 30;000-dalton calcium-sensitive actin cross-linking protein from *dictyostelium discoideum*. *J. Biol. Chem.* 259:4515–20

Fine, R. E., Blitz, A. L. 1975. A chemical comparison of tropomyosins from muscle and non-muscle tissues. *J. Mol. Biol.* 95:447–54

Fine, R. E., Blitz, A. L., Hitchcock, S. E., Kaminer, B. 1973. Tropomyosin in brain and growing neurones. *Nature New Biol.* 245:182–86

Flanagan, M. D., Lin, S. 1980. Cytochalasins block actin filament elongation by binding to high affinity sites associated with F-actin. *J. Biol. Chem.* 255:835–38

Flory, P. 1953. *Principles of Polymer Chemistry*. Ithaca, New York: Cornell Univ. Press. 342 pp.

Flory, P. 1956. Statistical thermodynamics of semiflexible chain molecules. *Proc. R. Soc. London A* 234:60–72

Fowler, V., Talyor, D. L. 1980. Spectrin plus band 4.1 cross-link actin. Regulation by micromolar calcium. *J. Cell Biol.* 85:361–76

Frieden, C. 1983. Polymerization of actin: Mechanism of the Mg^{2+}-induced process at pH 8 and 20°C. *Proc. Natl. Acad. Sci. USA* 80:6513–17

Frieden, C., Goddette, D. W. 1983. Polymerization of actin and actin-like systems: evaluation of the t time course of polymerization in relation to the mechanism. *Biochemistry* 22:5836–43

Fujime, S., Ishiwata, S. 1971. Dynamic study of F-actin by quasielastic scattering of laser light. *J. Mol. Biol.* 62:251–65

Fujime, S., Ishiwata, S., Maeda, T. 1984. Dynamic light scattering study of muscle F-actin. *Biophys. Chem.* 20:1–21

Giffard, R. G., Weeds, A. G., Spudich, J. A. 1984. Ca^{2+}-dependent binding of severin to actin: A one-to-one complex is formed. *J. Cell Biol.* 98:1796–1803

Giometti, C. S., Anderson, N. L. 1981. A varient of human nonmuscle tropomyosin found in fibroblasts by using two-dimensional electrophoresis. *J. Biol. Chem.* 256:11840–46

Glenney, J. R., Bretscher, A., Weber, K. 1980. Calcium control of the intestinal microvillins cytoskeleton: its complications for regulation of microfilament organizations. *Science* 77:6458–62

Glenney, J. R., Geislen, N., Kaulfus, P., Weber, K. 1981. Demonstration of at least two different actin binding sites in villin, a calcium regulated produlator of F-actin organization. *J. Biol. Chem.* 256:8156–66

Glenney, J. R., Glenney, P., Osborn, M., Weber, K. 1982c. A high molecular weight f-actin and calmodulin binding protein with spectrin-related morphology is a major constituent of the terminal web microfilament organization of isolated intestinal brush borders. *Cell* 28:843–54

Glenney, J. R., Glenney, P., Weber, K. 1982a. F-actin-binding and cross-linking properties of porcine brain fodrin, a spectrin-related molecule. *J. Biol. Chem.* 257:9781–87

Glenney, J. R., Glenney, P., Weber, K. 1982b. Erythroid spectrin, brain fodrin, and intestinal brush border proteins (TW-260/240) are related molecules containing a common calmodulin-binding subunit bound to a variant cell type-specific subunit. *Proc. Natl. Acad. Sci. USA* 79:4002–5

Glenney, J. R., Kaulfus, P., Matsudaira, P., Weber, K. 1981. F-actin binding and bundling properties of fimbrin, a major cytoskeletal protein of microvillus core filaments. *J. Biol. Chem.* 256:9283–88

Glenney, J. R., Osborn, M., Weber, K. 1982d. The intracellular localization of the microvillus 110K protein, a component considered to be involved in side-on membrane attachment of F-actin. *Exp. Cell Res.* 138:199–205

Glenney, J. R., Weber, K. 1981. Calcium control of microfilaments: uncoupling of the f-actin severing and bundling activity of villin by limited proteolysis in vitro. *Proc. Natl. Acad. Sci. USA* 78:2810–14

Goodman, S. R., Zagon, I. S., Kulikowski, R. R. 1981. Identification of a spectrin-like protein in nonerythroid cells. *Proc. Natl. Acad. Sci. USA* 78:7570–74

Griffith, L. M., Pollard, T. D. 1978. Evidence for actin filament-microtubule interaction mediated by microtubule-associated proteins. *J. Cell Biol.* 78:958–65

Griffith, L. M., Pollard, T. D. 1982a. Crosslinking of actin filament networks by self-association and actin-binding macromolecules. *J. Biol. Chem.* 257:9135–42

Griffith, L. M., Pollard, T. D. 1982b. The interaction of actin filaments with microtubules and microtubule-associated proteins. *J. Biol. Chem.* 257:9143–51

Haddad, J. G. 1982. Human serum binding protein for vitamin D and its metabolites (DBP): evidence that actin is the DBP binding component in human skeletal muscle. *Arch. Biochem. Biophys.* 213:538–44

Hanson, J. 1973. Evidence from electron microscope studies on actin paracrystals concerning the origin of the cross-striation in the thin filaments of vertebrate skeletal muscle. *Proc. R. Soc. London B* 183:39–58

Hanson, J., Lowy, J. 1963. The structure of F-actin and of actin filaments isolated from muscle. *J. Mol. Biol.* 6:46–60

Hargreaves, W. R., Giedd, K. N., Verkleij, A., Branton, D. 1980. Reassociation of ankyrin with band 3 in erythrocyte membranes and in lipid vesicles. *J. Biol. Chem.* 255:11965–72

Harris, D. A., Schwartz, J. H. 1981. Characterization of brevin, a serum protein that shortens actin filaments. *Proc. Natl. Acad. Sci. USA* 78:6798–6802

Harris, H. E., Bamburg, J. R., Weeds, A. G. 1980. Actin filament disassembly in blood plasma. *FEBS Lett.* 121:175–77

Harris, H. E., Weeds, A. G. 1978. Platelet actin: sub-cellular distribution and association with profilin. *FEBS Lett.* 90:84–88

Harris, H. E., Weeds, A. G. 1983. Plasma actin depolymerizing factor has both calcium dependent and calcium independent effects on actin. *Biochemistry* 22:2728–41

Hartwig, J. H., Stossel, T. P. 1975. Isolation and properties of actin, myosin, and a new actin-binding protein in rabbit alveolar macrophages. *J. Biol. Chem.* 250:5696–5705

Hartwig, J. H., Stossel, T. P. 1979. Cytochalasin B and the structure of actin gels. *J. Mol. Biol.* 134:539–54

Hartwig, J. H., Stossel, T. P. 1981. The structure of actin-binding protein molecules in solution and interacting with actin filaments. *J. Mol. Biol.* 145:563–81

Hartwig, J. H., Stossel, T. P. 1983. The architecture of actin filaments and the ultrastructural location of actin-binding protein in the periphery of lung macrophages. *J. Cell Biol.* 27:274A

Hartwig, J. H., Tyler, J., Stossel, T. P. 1980. Actin-binding protein promotes the bipolar and perpendicular branching of actin filaments. *J. Cell Biol.* 87:841–48

Hasegawa, T., Takahashi, S., Hayashi, H., Hatano, S. 1980. Fragmin: a calcium ion sensitive regulatory factor on the formation of actin filaments. *Biochemistry* 19:2677–83

Hatano, S., Oosawa, F. 1966. Isolation and characterization of plasmodium actin. *Biochem. Biophys. Acta* 127:488–98

Hesterberg, L. K., Weber, K. 1983a. Ligand-induced conformational changes in villin, a calcium controlled actin-modulating protein. *J. Biol. Chem.* 258:359–64

Hesterberg, L. K., Weber, K. 1983b. Demonstration of three distinct calcium-binding sites in villin, a modulator of actin assembly. *J. Biol. Chem.* 258:365–69

Hinssen, H. 1981a. A actin-modulating protein from *Physarum polycephalum*. I. *Eur. J. Cell Biol.* 23:225–33

Hinssen, H. 1981b. An actin-modulating protein from *Physarum polycephalum*. II. Ca^{++} dependence and other properties. *Eur. J. Cell Biol.* 23:234–40

Hinssen, H., Small, J. V., Sobiebzek, A. 1984. A Ca^{2+}-dependent actin modulator from vertebrate smooth muscle. *FEBS Lett.* 166:90–95

Hitchcock, S. E. 1980. Actin: deoxyribonuclease I interaction. Depolymerization and nucleotide exchange. *J. Biol. Chem.* 255:5668–73

Hitchcock, S. E., Carlsson, L., Lindberg, U. 1976. Depolymerization of F-actin by deoxyribonuclease I. *Cell* 7:531–42

Hosoya, H., Mabuchi, I. 1984. A 45,000-mol-wt protein-actin complex from unfertilized sea urchin egg affects assembly properties of actin. *J. Cell Biol.* 99:994–1001

Huxley, H. E. 1963. Electron microscope studies on the structure of natural and synthetic protein filaments from striate muscle. *J. Mol. Biol.* 7:281–308

Isenberg, G., Aebi, U., Pollard, T. P. 1980. An actin-binding protein from *acanthamoeba* regulator actin filament polymerization and interactions. *Nature* 288:455–59

Ishikawa, H., Bischoff, R., Holtzer, H. 1969. Formation of arrowhead complexes with heavy meromyosin in a variety of cell types. *J. Cell Biol.* 43:312–28

Ishimoda-Takagi, T. 1978. Immunological purification of sea urchin egg tropomyosin. *J. Biochem.* 83:1757–62

Janmey, P. A., Chaponnier, C., Lind, S. E., Zaner, K. S., Stossel, T. P., Yin, H. L. 1985a. Interactions of gelsolin and gelsolin actin complexes with actin. Effects of calcium on actin nucleation, filament severing and end blocking. *Biochemistry* 24:3714–23

Janmey, P. A., Lind, S. E., Yin, H. L., Stossel, T. P. 1985b. Effects of semidilute actin solutions on the mobility of fibrin protofibrils during clot formation. *Biochim. Biophys. Acta.* In press

Jen, C. J., McIntire, L. V., Bryan, J. 1982. The viscoelastic properties of actin solutions. *Arch. Biochem. Biophys.* 216:126–32

Jockusch, E. M., Isenberg, G. 1981. Interaction of a-actinin and vinculin with actin: opposite effects on filament network formation. *Proc. Natl. Acad. Sci. USA* 78:3005–9

Just, E. E. 1924. *The Biology of the Cell Surface.* Philadelphia: Blakiston. 125 pp.

Kamiya, R., Maruyama, K., Kuroda, M., Kawmura, M., Kikuchi, M. 1972. Mg-polymer of actin formed under the influence of beta-actinin. *Biochem. Biophys. Acta* 256:120–31

Kane, R. E. 1975. Preparation and purification of polymerized actin from urchin egg extracts. *J. Cell Biol.* 66:305–15

Kane, R. E. 1976. Actin polymerization and interaction with other proteins in temperature-induced gelation of sea urchin egg extracts. *J. Cell Biol.* 71:704–14

Kasai, M., Kawashima, H., Oosawa, F. 1960. Structure of f-actin solutions. *Polymer Sci.* 44:51–69

Kato, T., Tonomura, Y. 1975. *Physarum* tropomyosin-troponin complex: isolation and properties. *J. Biochem.* 78:583–88

Kawamura, M., Maruyama, K. 1970. Polymorphism of F-actin. I. Three forms of paracrystals. *J. Biochem.* 68:885–99

Keski-Oja, J., Sen, A., Todaro, G. F. 1980. Direct association of fibronectin and actin molecules in vitro. *J. Cell Biol.* 85:527–33

Kobayashi, R., Bradley, W. A., Bryan, J., Field, J. B. 1983. Identification and purification of calcium ion dependent modulators of actin polymerization from bovine thyroid. *Biochemistry* 22:2463–69

Korn, E. D. 1982. Actin polymerization and its regulation by proteins from nonmuscle cells. *Physiol. Rev.* 62:672–737

Kuroda, M., Maruyama, K. 1976. Gamma-actinin, a new regulatory protein from rabbit skeletal muscle I. Purification and characterization. *J. Biochem.* 80:315–22

Kurth, M., Bryan, J. 1984. Platelet activation induces the formation of a stable gelsolin-actin complex from monomeric gelsolin. *J. Biol. Chem.* 259:7473–79

Kurth, M. C., Wang, L. L., Dingus, J., Bryan, J. 1983. Purification and characterization of a gelsolin-actin complex from human platelets. *J. Biol. Chem.* 258:10895–903

Kwiatkowski, D. J., Janmey, P. A., Mole, J. E., Yin, H. L. 1985. Isolation and properties of two actin-binding domains in gelsolin. *J. Biol. Chem.* In press

Lacks, S. A. 1981. Deoxyribonuclease I in mammalian tissues. Specificity of inhibition by actin. *J. Biol. Chem.* 256:2644–48

Laki, K., Muszbek, L. 1974. On the interaction of F-actin with fibrin. *Biochim. Biophys. Acta* 371:519–25

Lal, A. A., Brenner, S. L., Korn, E. D. 1984a. Preparation and polymerization of skeletal muscle ADP-actin. *J. Biol. Chem.* 259:13061–65

Lal, A. A., Korn, E. D., Brenner, S. L. 1984b. Rate constants for actin polymerization in ATP determined using cross-linked actin trimers as nuclei. *J. Biol. Chem.* 259:8794–8800

Landon, F., Olomucki, A. 1983. Isolation and physio-chemical properties of blood platelet a-actinin. *Biochim. Biophys. Acta* 742:129–34

Lanks, K. L., Kasambalides, E. J. 1983. Dexamethasone induces gelsolin synthesis and altered morphology in L929 cells. *J. Cell Biol.* 96:577–81

Lassing, I., Lindberg, U. 1985. Specific interaction between phosphatidylinositol 4,5-bisphosphate and profilactin. *Nature* 314:472–74

Lazarides, E., Burridge, K. 1975. a-actinin: immunofluorescent localization of a muscle structural protein in nonmuscle cells. *Cell* 6:289–98

Lazarides, E., Lindberg, U. 1974. Actin is the naturally occurring inhibitor of deoxyribonuclease I. *Proc. Natl. Acad. Sci. USA* 71:4742–46

Lazarides, E., Nelson, W. J. 1982. Expression of spectrin in nonerythroid cells. *Cell* 31:505–8

Lazarides, E., Nelson, W. J. 1983. Erythrocyte and brain forms of spectrin in cerebellum: distinct membrane-cytoskeletal domains in neurons. *Science* 220:1295–96

Lees, A., Haddad, J. G., Lin, S. 1984. Brevin and vitamin D binding protein: comparison of the effects of two serum proteins on actin assembly and disassembly. *Biochemistry* 23:3038–47

Lehto, V. P., Virtanen, I. 1983. Immunolocalization of a novel, cytoskeleton-associated polypeptide of Mr 230,000 daltons (P230). *J. Cell Biol.* 96:703–16

Levine, J., Willard, M. 1981. Fodrin: axonally transported polypeptides associated with the internal periphery of many cells. *J. Cell Biol.* 90:631–43

Levine, J., Willard, M. 1983. Redistribution of fodrin (a component of the cortical cytoplasm) accompanying capping of cell surface molecules. *Proc. Natl. Acad. Sci. USA* 80:191–95

Lin, J. J.-C., Helfman, D. M., Hughes, S. H., Chou, C.-S. 1985. Tropomyosin isoforms in chicken embryo fibroblasts: purification, characterization, and changes in Rous sarcoma virus-transformed cells. *J. Cell Biol.* 100:692–703

Lind, S., Yin, H. L., Stossel, T. P. 1982. Human platelets contain gelsolin, a regulator of actin filament length. *J. Clin. Invest.* 69:1384–87

Lind, S. E., Janmey, P. A. 1984. Human plasma gelsolin binds to fibronectin. *J. Biol. Chem.* 259:13262–66

Lucas, R. C., Gallagher, M., Stracher, A. 1976. Actin and actin binding protein in platelets. In *Contractile Systems in Non-Muscle Tissues*, ed. S. V. Perry, pp. 133–39. Amsterdam: Elsevier/North Holland

Mabuchi, I. 1983. An actin-depolymerizing protein (depactin) from starfish oocytes: properties and interaction with actin. *J. Cell Biol.* 97:1612–21

MacLean-Fletcher, S., Pollard, T. D. 1980. Mechanism of action of cytochalasin b on actin. *Cell* 20:329–41

Maekawa, S., Endo, S., Sakai, H. 1982. A protein in starfish sperm head which bundles actin filaments in vitro: purification and characterization. *J. Biochemistry* 92:1959–72

Maekawa, S., Endo, S., Sakai, H. 1983. Purification and partial characterization of a new protein in porcine brain which bundles actin filaments. *J. Biochem.* 94:1329–37

Magri, E., Zaccarini, M., Grazi, E. 1978. The interaction of histone and protamine with actin. *Biochem. Biophys. Res. Comm.* 82:1207–10

Malm, B. 1984. Chemical modification of Cys-374 of actin interferes with the formation of the profilactin complex. *FEBS Lett.* 173:399–402

Malm, B., Larsson, H., Lindberg, U. 1983. The profilin-actin complex: further characterization of profilin and studies on the stability of the complex. *J. Muscle Res. Cell Motil.* 4:569–88

Mannherz, H. G., Goody, R. S., Konrad, M., Nowak, E. 1980. The interaction of bovine pancreatic deoxyribonuclease I and skeletal muscle actin. *Eur. J. Biochem.* 104:367–79

Mannherz, H. G., Leigh, J. B., Leberman, R., Pfrang, H. 1975. A specific 1:1 G-actin:DNAse I complex formed by the action of DNAase I on F-actin. *FEBS Lett.* 60:34–38

Markey, F., Lindberg, U., Eriksson, L. 1978. Human platelets contain profilin, a potential regulator of actin polymerizability. *FEBS Lett.* 88:75–79

Markey, F., Persson, T., Lindberg, U. 1982. A 90,000 dalton actin binding protein form platelets. Comparison with villin and plasma brevin. *Biochem. Biophys. Acta* 709:122–33

Maruta, H., Isenberg, G. 1983. Ca^{2+}-dependent actin-binding phosphoprotein in *physarum polycephalum*. *J. Biol. Chem.* 258:10151–58

Maruta, H., Isenberg, G. 1984. Ca^{2+}-dependent actin-binding phosphoprotein in *physarum polycephalum*. Subunit b is a DNase I-binding and f-actin capping protein. *J. Biol. Chem.* 259:5208–13

Maruta, H., Isenberg, G., Schreckenbach, T., Risse, G., Shibayama, T., Hesse, J. 1983. Ca^{2+}-dependent actin-binding phosphoprotein in *physarum polycephalum*: I. Ca^{2+}/actin-dependent inhibition of its phosphorylation. *J. Biol. Chem.* 258:10144–50

Maruta, H., Knoerzer, W., Hinssen, H., Isenberg, G. 1984. Regulation of actin polymerization by non-polymerizable actin-like proteins. *Nature* 312:424–27

Maruta, H., Korn, E. D. 1977. Purification from *Acanthamoeba castellanii* of proteins that induce gelation and syneresis of f-actin. *J. Biol. Chem.* 252:399–402

Maruyama, K. 1971. A study of B-actinin, myofibrillar protein from rabbit skeletal muscle. *J. Biochem.* 69:369–86

Maruyama, K., Ebashi, S. 1965. a-actinin, a new structural protein from striated muscle. II. Action on actin. *J. Biochem.* 58:13–19

Maruyama, K., Kimura, S., Ishil, T., Kuroda, M., Ohashi, K., Muramstsu, S. 1977. Beta-actinin a regulatory protein of muscle. *J. Biochem.* 81:215–32

Maruyama, K., Sakai, H. 1981. Cell B-actinin, an accelerator of actin polymerization, isolated from rat kidney cytosol. *J. Biochem.* 89:1337–40

Maruyama, K., Tsukagoshi, K. 1984. Effects of KCl, MgCl2, and CaCl2 concentrations on the monomer-polymer equilibrium of actin in the presence and absence of cytochalasin D. *J. Biochem.* 96:605–11

Matsudaira, P., Jakes, R., Walker, J. E. 1985. A gelsolin-like Ca^{2+}-dependent actin-binding domain in villin. *Nature* 315:248–50

Matsudaira, P., Mandelkow, E., Renner, W., Hesterberg, L. K., Weber, K. 1983. Role of fimbrin and villin in determining the inter-filament distances of actin bundles. *Nature* 301:209–14

Matsudaira, P. T., Burgess, D. R. 1982. Partial reconstitution of the microvillus core bundle: characterization of villin as a Ca^{2+}-dependent actin-bundling/depolymerizing protein. *J. Cell Biol.* 92:648–56

Matsumura, F., Yamashiro-Matsmura, S., Lin, J. J.-C. 1983a. Isolation and characterization of tropomyosin-containing microfilaments from cultured cells. *J. Biol. Chem.* 258:6636–44

Matsumura, F., Lin, J. J.-C., Yamashiro-Matsumura, S., Thomas, G. P., Topp, W. C. 1983b. Different expression of tropomyosin forms in the microfilaments isolated from normal and transformed rat cultured cells. *J. Biol. Chem.* 258:13954–64

Miki, M., Wahl, P., Auchet, J. C. 1982. Fluorescence anisotropy of labelled F-actin. Influence of Ca^{2+} on the flexibility of F-actin. *Biophys. Chem.* 16:165–72

Mimura, N., Asano, A. 1979. Ca^{2+}-sensitive gelation of actin filaments by a new protein factor. *Nature* 282:44–47

Mooseker, M. S., Graves, T. A., Wharton, K. A., Falco, N., Howe, C. L. 1980. Regulation of microvillins structure: calcium-dependent solation and crosslinking of actin filaments in microvilli of intestinal epithelial cells. *J. Cell Biol.* 87:809–22

Mornet, D., Ue, K. 1984. Proteolysis and structure of skeletal muscle actin. *Proc. Natl. Acad. Sci. USA* 81:3680–84

Morton, D. J., Clarke, F. M., Masters, C. J. 1977. An electron microscope study of the interaction between fructose diphosphate aldolase and actin-containing filaments. *J. Cell Biol.* 74:1016–23

Nelson, W. J., Granger, B. L., Lazarides, E. 1983. Avian lens spectrin: subunit composition compared with erythrocyte and brain spectrins. *J. Cell Biol.* 97:1271–76

Nelson, W. J., Lazarides, E. 1983. Expression of the B subunit of spectrin in nonerythroid cells. *Proc. Natl. Acad. Sci. USA* 80:363–67

Neuhaus, J. M., Wanger, M., Keisler, T., Wegner, A. 1983. Treadmilling of actin. *J. Muscle Res. Cell Motil.* 4:507–27

Ng, J. S. Y., Burtnick, L. D. 1982. Chemical modifications of actin: effects on interaction with deoxyribonuclease I. *Int. J.*

Biol. Macromol. 4:215–18

Niederman, R., Amrein, P., Hartwig, J. H. 1983. The three dimensional structure of actin filaments in solution and an actin gel made with actin-binding protein. *J. Cell Biol.* 96:1400–13

Nishida, E., Kuwaki, T., Sakai, H. 1981. Phosphorylation of microtubule-associated proteins (MAPs) and pH of the medium control interaction between MAPs and actin filaments. *J. Biochem.* 90:575–78

Nishida, E., Maekawa, S., Muneyuki, E., Sakai, H. 1984a. Action of a 19K protein from bovine brain on actin depolymerization: a new functional class of actin-binding proteins. *J. Biochem.* 95:387–98

Nishida, E., Maekawa, S., Sakai, H. 1984b. Characterization of the action of porcine brain profilin on actin polymerization. *J. Biochem.* 95:399–404

Nishioka, M., Kobayashi, K., Uchida, M., Nakamura, T. 1982. A binding activity of actin with human C1q. *Biochem. Biophys. Res. Comm.* 108:1307–12

Norberg, R., Thorstensson, R., Utter, G., Fargaeus, A. 1979. F-actin depolymerizing activity of human serum. *Eur. J. Biochem.* 100:575–83

Nunnally, M. H., Powell, L. D., Craig, S. W. 1981. Reconstitution and regulation of actin gelsol transformation with purified filamin and villin. *J. Biol. Chem.* 256:2083–86

Nystrom, L. E., Lindberg, I., Kendrick-Jones, J., Jakes, R. 1979. The amino acid sequence of profilin from calf spleen. *FEBS Lett.* 101:161–65

Ogawa, H., Shiraki, H., Matsuda, Y., Nakagawa, H. 1978. Interaction of adenylosuccinate synthestase with F-actin. *Eur. J. Biochem.* 85:331–37

Ogihara, S., Tonomura, Y. 1982. A novel 36,000-dalton actin-binding protein purified from microfilaments in *Physarum plasmodia* which aggregates actin filaments and blocks actin-myosin interaction. *J. Cell Biol.* 93:604–14

Olomucki, A., Huc, C., Lefebure, F., Coué, M. 1984. Isolation and characterization of human blood platelet gelsolin. *FEBS Lett.* 174:80–85

Oosawa, F., Asakura, S. 1975. *Thermodynamics of the Polymerization of Protein.* 162 pp.

Oosawa, F., Kasai, M. 1971. Actin. In *Subunits in Biological Systems*, ed. S. N. Timasheff, G. D. Fasman, pp. 261–322. New York: Dekker

Otto, J. J., Kane, R. E., Bryan, J. 1979. Formation of filopodia in coelomocytes: localization of fascin, a 58,000 dalton actin cross-linking protein. *Cell* 17:285–93

Otto, J. J., Kane, R. E., Bryan, J. 1980. Redistribution of actin and fascin in sea urchin eggs after fertilization. *Cell Motil.* 1:31–40

Ozaki, K., Hatano, S. 1984. Mechanism of regulation of actin polymerization by *Physarum* profilin. *J. Cell Biol.* 98:1919–25

Pantaloni, D., Carlier, M. F., Coué, M., Lal, A. A., Brenner, S. L., Korn, E. D. 1984. The critical concentration of actin in the presence of ATP increases with the number concentration of filaments and approaches the critical concentration of actin: ADP. *J. Biol. Chem.* 259:6274–83

Pardee, J. D., Simpson, P. A., Stryer, L., Spudich, J. A. 1982. Actin filaments undergo limited subunit exchange in physiological salt conditions. *J. Cell Biol.* 94:316–24

Pardee, J. D., Spudich, J. A. 1982. Mechanism of K$^+$-induced actin assembly. *J. Cell Biol.* 93:648–54

Pearl, M., Fishkind, D., Mooseker, M., Keene, D., Keller, T. 1984. Studies on the spectrin-like protein from the intestinal brush border, TW 260/240, and characterization of its interaction with the cytoskeleton and actin. *J. Cell Biol.* 98:66–78

Petrini, M., Emerson, D. L., Galbraith, R. M. 1983. Linkage between surface immunoglobulin and cytoskeleton of B lymphocytes may involve Gc protein. *Nature* 306:73–74

Petrucci, T. C., Thomas, C., Bray, D. 1983. Isolation of a Ca^{2+}-dependent actin-fragmenting protein from brain, spinal cord, and cultured neurones. *J. Neurochem.* 40:1507–16

Pinder, J. C., Gratzer, W. B. 1982. Investigation of the actin-deoxyribonuclease I interaction using a pyrene-conjugated actin derivative. *Biochemistry* 21:4886–90

Podlubnaya, Z. A., Tskhovrebova, L. A., Zaalishvili, M. M., Stefanenko, G. A. 1975. Electron microscopic study of a-actinin. *J. Mol. Biol.* 92:357–59

Pollard, T. D. 1981. Purification of a calcium-sensitive actin gelation protein from *Acanthamoeba. J. Biol. Chem.* 256:7666–70

Pollard, T. D. 1984a. Purification of a high molecular weight actin filament gelation protein from *Acanthamoeba* that shares antigenic determinants with vertebrate spectrins. *J. Cell Biol.* 99:1970–80

Pollard, T. D. 1984b. Polymerization of ADP-actin. *J. Cell Biol.* 99:769–77

Pollard, T. D., Cooper, J. A. 1984. Quantitative analysis of the effect of *Acanthamoeba* profilin on actin filament nucleation and elongation. *Biochemistry* 23:6631–41

Pollard, T. D., Mooseker, M. S. 1981. Direct measurement of actin polymerization rate constants by electron microscopy of actin filaments nucleated by isolated microvillus cores. *J. Cell Biol.* 88:654–59

Pollard, T. D., Weeds, A. G. 1984. The rate constant for ATP hydrolysis by polymerized actin. *FEBS Lett.* 170:94–98

Reichstein, E., Korn, E. D. 1979. *Acanthamoeba* profilin: a protein of low molecular weight from *Acanthamoeba castellanii* that inhibits actin nucleation. *J. Biol. Chem.* 254:6174–79

Rockwell, M. A., Fechheimer, M., Taylor, D. L. 1984. A comparison of methods used to characterize gelation of actin in vitro. *Cell Motil.* 4:197–213

Rosenberg, S., Stracher, A., Burridge, K. 1981. Isolation and characterization of a calcium-sensitive a-actinin-like protein from human platelet cytoskeletons. *J. Biol. Chem.* 256:12986–91

Rosenberg, S., Stracher, A., Lucas, R. C. 1981. Isolation and characterization of actin and actin-binding protein from human platelets. *J. Cell Biol.* 91:201–11

Rouayrenc, J. F., Fattoum, A., Gabrion, J., Audemard, E., Kassab, R. 1984. Muscle gelsolin: isolation from heart tissue and characterization as an integral myofibrillar protein. *FEBS Lett.* 167:52–58

Sattilaro, R. F., Dentler, W. L., LeCluyse, E. L. 1981. Microtubule-associated proteins (MAPs) and the organization of actin filaments in vitro. *J. Cell Biol.* 90:467–73

Schanus, E., Booth, S., Hallaway, B., Rosenberg, A. 1985. The elasticity of spectrin-actin gels at high protein concentration. *J. Biol. Chem.* 260:3724–30

Schliwa, M. 1981. Proteins associated with cytoplasmic actin. *Cell* 25:587–90

Schloss, J. A., Goldman, R. D. 1979. Isolation of a high molecular weight actin-binding protein from baby hamster kidney (BHK-21) cells. *Proc. Natl. Acad. Sci. USA* 76:4484–88

Schloss, J. A., Goldman, R. D. 1980. Microfilaments and tropomyosin of cultured mammalian cells: isolation and characterization. *J. Cell Biol.* 87:633–42

Schollmeyer, J. V., Rao, G. H. R., White, J. G. 1978. An actin-binding protein in human platelets. *Amer. J. Path.* 93:433–45

Schroder, H. C., Zahn, R. K., Muller, W. E. G. 1982. Role of actin and tubulin in the regulation of poly(A) polymerase-endoribonuclease IV complex from calf thymus. *J. Biol. Chem.* 257:2305–9

Sedlar, P. A., Bryan, J. 1984. Partial characterization of a 45 KDa actin filament severing protein from sea urchin eggs. *J. Cell Biol.* 99:308a

Selden, L. A., Estes, J. E., Gershman, L. C. 1983. The tightly bound divalent cation regulates actin polymerization. *Biochem. Biophys. Res. Commun.* 116:478–85

Selden, S. C., Pollard, T. D. 1983. Phosphorylation of microtubule-associated proteins regulates their interaction with actin filaments. *J. Biol. Chem.* 258:7064–71

Shizuta, Y., Shizuta, H., Gallo, M., Davies, P., Pastan, I. 1976. Purification and properties of filamin, an actin-binding protein from chicken gizzard. *J. Biol. Chem.* 251:6562–67

Shotton, D. M., Burke, B. E., Branton, D. 1979. The molecular structure of human erythrocyte spectrin. Biophysical and electron microscopical studies. *J. Mol. Biol.* 131:303–29

Smillie, L. B. 1979. Structure and functions of tropomyosins from muscle and nonmuscle sources. *Trends Biochem. Sci.* 4:151–55

Smino-Oka, T., Ohnishi, K., Watanabe, Y. 1983. Further characterization of a brain high molecular weight actin-binding protein (BABP): interaction with brain actin and ultrastructural studies. *J. Biochem.* 93:977–87

Smith, D. B., Lind, S. E., Yin, H. L., Stossel, T. P. 1984. Low levels of plasma gelsolin in patients with adult respiratory distress syndrome. *Clin. Res.* 32:708a

Sobue, K., Morimoto, K., Kanda, K., Maruyama, K., Kakiuchi, S. 1982. Reconstitution of Ca^{2+}-sensitive gelation of actin filaments with filamin, caldesmon, and calmodulin. *FEBS Lett.* 138:289–92

Southwick, F. S., Hartwig, J. H. 1982. Acumentin, a protein in macrophage which caps the "pointed" end of actin filaments. *Nature* 297:303–7

Southwick, F. S., Stossel, T. P. 1981. Isolation of an inhibitor of actin polymerization from human polymorphonuclear leukocytes. *J. Biol. Chem.* 256:3030–36

Southwick, F. S., Tatsumi, N., Stossel, T. P. 1982. Acumentin an actin-modulating protein of rabbit pulmonary macrophages. *Biochemistry* 21:6321–26

Stendahl, O. I., Stossel, T. P. 1980. Actin-binding protein amplifies actomyosin contraction and gelsolin confers calcium controls in the direction of contraction. *Biochem. Biophys. Res. Commun.* 92:675–81

Stewart, M., Morton, D. J., Clarke, F. M. 1980. Interaction of aldolase with actin-containing filaments. *Biochem. J.* 186:99–104

Stossel, T. P. 1984. Contribution of actin to the structure of the cytoplasmic matrix. *J. Cell. Biol.* 99:15s–21s

Stossel, T. P., Hartwig, J. H. 1975. Interactions of actin, myosin and an actin-binding protein of rabbit alveolar macrophages. Macrophage myosin Mg^{++}-adenosine triphosphatase requires a cofactor for activation by actin. *J. Biol. Chem.* 250:5706–12

Stossel, T. P., Hartwig, J. H. 1976. Interaction of actin, myosin, and a new actin-binding protein of rabbit pulmonary macrophages. II. Role in cytoplasmic movement and phagocytosis. *J. Cell Biol.* 68:602–14

Stratford, C. A., Brown, S. S. 1985. Isolation of an actin-binding protein from membrane of *Dictyostelium discoideum*. *J. Cell Biol.* 100:727–35

Straub, F. B. 1942. Actin. *Stud. Szeged.* 2:3–15

Sugino, H., Hatano, S. 1982. Effects of fragmin on actin polymerization: Evidence for enhancement of nucleation and capping of the barbed end. *Cell Motil.* 2:457–70

Sutoh, K. 1982. An actin binding site on the 20k fragment of myosin subfragment 1. *Biochem.* 21:4800–4

Sutoh, K. 1984. Actin-actin and actin-deoxyribonuclease I contact sites in the actin sequence. *Biochemistry* 23:1942–46

Sutoh, K., Iwane, M., Matsuzaki, F., Kikuchi, M., Ikai, A. 1984. Isolation and characterization of a high molecular weight actin-binding protein from *Physarum polycephalum* plasmodia. *J. Cell Biol.* 98:1611–18

Sutoh, K., Mabuchi, I. 1984. N-terminal and C-terminal segments of actin participate in binding depactin, an actin depolymerizing protein from starfish oocytes. *Biochemistry* 23:6757–61

Swezey, R. R., Somero, G. N. 1982. Polymerization thermodynamics and structural stabilities of skeletal muscle actins from vertebrates adapted to different temperatures and hydrostatic pressures. *Biochemistry* 21:4496–4503

Szent-Gyorgyi, A. 1947. *Chemistry of Muscular Contraction*, New York: Academic. 150 pp.

Tellam, R., Frieden, C. 1982. Cytochalasin D and platelet gelsolin accelerate actin polymer formation. A model for regulation of the extent of actin polymer formation in vivo. *Biochemistry* 21:3207–14

Thorstensson, R., Utter, G., Norberg, R. 1982. Further characterization of the Ca^{2+} dependent F-actin depolymerizing protein of human serum. *Eur. J. Biochem.* 126:11–16

Tilney, L. G., Bonder, E. M., Coluccio, L. M., Mooseker, M. S. 1983. Actin from Thyone sperm assembles on only one end of an actin filament: a behavior regulated by profilin. *J. Cell Biol.* 97:112–24

Tobacman, L. S., Brenner, S. L., Korn, E. D. 1983. Effect of acanthamoeba profilin on the pre-steady state kinetics of actin polymerization and on the concentration of F-actin at steady state. *J. Biol. Chem.* 258: 8806–12

Tobacman, L. S., Korn, E. D. 1982. The regulation of actin polymerization and the inhibition of monomeric actin ATPase activity by *Acanthamoeba* profilin. *J. Biol. Chem.* 257: 4166–70

Tobacman, L. S., Korn, E. D. 1983. The kinetics of actin nucleation and polymerization. *J. Biol. Chem.* 258: 3207–14

Torbet, J., Dickens, M. J. 1984. Orientation of skeletal muscle actin in strong magnetic fields. *FEBS Lett.* 173: 403–6

Tseng, P. C., Pollard, T. D. 1982. Mechanism of action of *Acanthamoeba* profilin: demonstration of actin species specificity and regulation by micromolar concentrations of MgCl2. *J. Cell Biol.* 94: 213–18

Tseng, P. C., Runge, M. S., Cooper, J. A., Williams, R. C., Pollard, T. D. 1984. Physical, immunochemical, and functional properties of *Acanthamoeba* profilin. *J.Cell Biol.* 98: 214–21

Tyler, J. M., Anderson, J. M., Branton, D. 1980. Structural comparison of several actin-binding molecules. *J. Cell Biol.* 85: 489–95

Ungewickell, E., Bennett, P. M., Calvert, R., Ohaian, V., Gratzer, W. B. 1979. In vitro formation of a complex between cytoskeletal proteins of the human erythrocyte. *Nature* 280: 811–14

van Baelen, H., Bouillon, R., de Moor, P. 1980. Vitamin D-binding protein (Gc-globulin) binds actin. *J. Biol. Chem.* 255: 2270–72

Vandekerckhove, J. S., Sandoval, I. V. 1982. Purification and characterization of a new mammalian serum protein with the ability to inhibit actin polymerization and promote depolymerization of actin filaments. *Biochemistry* 21: 3983–91

Verkhovsky, A. B., Surgucheva, I. G., Gelfand, V. I., Rosenblat, V. A. 1981. G-actin-tubulin interaction. *FEBS Lett.* 135: 290–94

Verkhovsky, A. B., Surgucheva, I. G., Gelfend, V. I. 1984. Phalloidin and tropomyosin do not prevent actin filament shortening by the 90 kD protein-actin complex from brain. *Biochem. Biophys. Res. Commun.* 123: 596–603

Wallach, D., Davies, P. J. A., Pastan, I. 1978. Cyclic AMP-dependent phosphorylation of filamin in mammalian smooth muscle. *J. Biol. Chem.* 253: 4739–45

Walsh, T. P., Weber, A., Higgins, J., Bonder, E. M., Mooseker, M. S. 1984. Effect of villin on the kinetics of actin polymerization. *Biochemistry* 23: 2613–21

Walsh, T. P., Winzor, D. J., Clarke, F. M., Masters, C. J., Morton, D. J. 1980. Binding of aldolase to actin-containing filaments. *Biochem. J.* 186: 89–98

Wang, K. 1977. Filamin, a new high-molecular-weight protein found in smooth muscle and nonmuscle cells. Purification and properties of chicken gizzard filamin. *Biochemistry* 16: 1857–65

Wang, K., Ash, J. F., Singer, S. J. 1975. Filamin, a new high-molecular weight protein found in smooth muscle and nonmuscle cells. *Proc. Natl. Acad. Sci. USA* 72: 4483–86

Wang, L. L., Bryan, J. 1981. Isolation of calcium-dependent platelet proteins that interact with actin. *Cell* 25: 637–49

Wang, L. L., Spudich, J. A. 1984. A 45,000 mol. wt. protein from unfertilized sea urchin egg severs actin filaments in a calcium-dependent manner and increases the steady-state concentration of nonfilamentous actin. *J. Cell Biol.* 99: 844–51

Wang, Y. L., Bonder, E. M., Mooseker, M. S., Taylor, D. L. 1983. Effects of villin on the polymerization and subunit exchange of actin. *Cell Motil.* 3: 151–65

Wanger, M., Wegner, A. 1985. Equilibrium constant for binding of an actin filament capping protein to the barbed end of actin filaments. *Biochemistry* 24: 1035–40

Weeds, A. 1982. Actin-binding proteins-regulators of cell architecture and motility. *Nature* 296: 811–16

Wegner, A. 1979. Equilibrium of the actin-tropomyosin interaction. *J. Mol. Biol.* 131: 839–53

Wegner, A. 1982a. Kinetic analysis of actin assembly suggests that tropomyosin inhibits spontaneous fragmentation of actin filaments. *J. Mol. Biol.* 161: 217–27

Wegner, A. 1982b. Treadmilling of actin at physiological salt concentrations. *J. Mol. Biol.* 161: 607–15

Wegner, A. 1985. Subtleties of actin assembly. *Nature* 313: 97–98

Wegner, A., Engel, J. 1975. Kinetics of cooperative association of actin to actin filaments. *Biophys. Chem.* 3: 215–25

Wegner, A., Isenberg, G. 1983. 12-fold difference between the critical monomer concentrations of the two ends of actin filaments in physiologic salt concentrations. *Proc. Nat. Acad. Sci. USA* 80: 4922–25

Wegner, A., Savko, P. 1982. Fragmentation of actin filaments. *Biochemistry* 21: 1909–13

Weihing, R. R. 1983. Purification of HeLa cell high molecular weight actin binding protein and its identification in HeLa cell plasma membrane ghosts and intact HeLa cells. *Biochemistry* 22: 1839–47

Woodrum, D. I., Rich, S. A., Pollard, T. D. 1975. Evidence for biased unidirectional polymerization of actin filaments using heavy meromyosin by an improved method. *J. Cell Biol.* 67:231–37

Yamamoto, K., Pardee, J. D., Reidler, J., Stryer, L., Spudich, J. A. 1982. Mechanism of interaction of *Dictyostelium* severin with actin filaments. *J. Cell Biol.* 95:711–19

Yeltman, D. R., Jung, G., Carraway, K. L. 1981. Isolation of a-actinin from sarcoma 180 ascites cell plasma membranes and comparison with smooth muscle a-actinin. *Biochem. Biophys. Acta* 668:201–8

Yin, H. L., Albrecht, J., Fattoum, A. 1981. Identification of gelsolin, a Ca^{2+}-dependent regulatory protein of actin gelsol transformation. Its intracellular distribution in a variety of cells and tissues. *J. Cell Biol.* 91:901–6

Yin, H. L., Kwiatkowski, D. J., Mole, J. E., Cole, F. S. 1984. Structure and biosynthesis of cytoplasmic and secreted variants of gelsolin. *J. Biol. Chem.* 259:5271–76

Yin, H. L., Stossel, T. P. 1979. Control of cytoplasmic actin gel-sol transformation by gelsolin, a calcium-dependent regula-tory protein. *Nature* 281:583–86

Yin, H. L., Stossel, T. P. 1980. Purification and structural properties of gelsolin, a Ca^{2+}-activated regulatory protein of macrophages. *J. Biol. Chem.* 255:9490–93

Yin, H. L., Zaner, K. S., Stossel, T. P. 1980. Ca^+ control of actin gelation. *J. Biol. Chem.* 255:9494–9500

Yoshimura, H., Nishio, T., Mihashi, K., Kinosita, K., Ikegami, A. 1984. Torsional motion of eosin-labelled F-actin as de-tected in the time-resolved anisotropy decay of the probe in the sub-millisecond time range. *J. Mol. Biol.* 179:453–67

Zaner, K. S., Fotland, R., Stossel, T. P. 1981. A low-shear, small volume viscoelasto-meter. *Rev. Sci. Instr.* 52:85–87

Zaner, K. S., Stossel, T. P. 1982. Some perspectives on the viscosity of actin fila-ments. *J. Cell Biol.* 93:987–91

Zaner, K. S., Stossel, T. P. 1983. Physical basis of the rheologic properties of F-actin. *J. Biol. Chem.* 258:11004–9

Zeece, M. G., Robson, R. M., Bechtel, P. J. 1979. Interaction of alpha-actinin, filamin and tropomyosin with F-actin. *Biochem. Biophys. Acta* 581:365–70

Ann. Rev. Cell Biol. 1985. 1:403–45

USING RECOMBINANT DNA TECHNIQUES TO STUDY PROTEIN TARGETING IN THE EUCARYOTIC CELL

Henrik Garoff

European Molecular Biology Laboratory, Postfach 10.2209, D-6900 Heidelberg, Federal Republic of Germany

CONTENTS

INTRODUCTION

The eucaryotic cell has localized many of its metabolic processes into specific membrane-limited compartments or organelles. Since most cellular polypeptides are synthesized on ribosomes in the cytoplasmic compart-

403

0743–4634/85/1115–0403$02.00

ment, it follows that the cell must possess specific transport mechanisms in order to deliver newly synthesized proteins to their site of function.

Recent studies have shown that proteins destined for the extracellular space, the plasma membrane (PM), or lysosomes are all first cotranslationally inserted into or transferred across the membrane of the ER (Blobel & Dobberstein 1975; Walter et al 1984; Hortsch & Meyer 1984), and then transported to their correct destinations via the Golgi complex (Jamieson & Palade 1967a,b; Farquhar & Palade 1981; Novick et al 1981; Green et al 1981a). Proteins specific for the ER and the Golgi complex use only the initial parts of this same transport pathway (Anderson et al 1983; Strous et al 1983a). In contrast, those proteins destined for the nucleus (De Robertis 1983), mitochondria (Schatz & Butow 1983), or chloroplasts (Ellis et al 1980; Schmidt et al 1980) are routed directly in completed (posttranslational) form to the proper organelle.

Evidentally, intracellular protein traffic must be well controlled so as to maintain the functional and structural identity of each compartment. How then does the cell direct a newly synthesized protein to its site of function? It seems reasonable to assume that all proteins carry some specific structural information, that is a signal, that will be recognized by receptor structures on the target organelle (Blobel 1980; Sabatini et al 1982). Using classical biochemical approaches it has been possible to demonstrate the involvement of signal and receptor structures in the process of protein targeting into the ER (Walter et al 1984; Hortsch & Meyer 1984), lysosomes (Neufeld & Ashwell 1980; Creek & Sly 1984), and mitochondria (Zimmermann et al 1981, Gasser et al 1982). However, the actual demonstration of functional peptide signals using the techniques of classical biochemistry has been very difficult. The recent development of cDNA cloning, engineering, and expression technologies has created completely new possibilities for the localization and characterization of those structural features of a protein that act as targeting signals. Most significantly, this approach has made it possible to test directly the signal activity of a proposed peptide by fusing it to another protein and analyzing the behavior of this chimera. In this review I discuss recombinant DNA approaches to the study of protein targeting within the eucaryotic cell. I first present some of the basic principles of this approach and some practical aspects of the manipulation technology itself. Later I discuss the various intracellular transport pathways in relation to recent results obtained using this approach.

PRINCIPLES OF THE RECOMBINANT DNA APPROACH

The approach taken by researchers in this field is to manipulate the cloned gene or cDNA encoding a given protein in such a way that when expressed in vivo or in vitro it will provide information about the signals involved in directing that protein to its correct destination. Inherent in this approach is the assumption that the signal is represented by a continuous stretch of amino acid residues that functions autonomously from the surrounding sequences. This method can be divided into three phases. In the first step, the signal is roughly located on the polypeptide chain by the analysis of various deletion mutants of the protein. More specifically, 5′, 3′, and internal deletions of different sizes are introduced into the cloned gene or cDNA, the altered proteins are expressed in eucaryotic cells or by using in vitro systems, and their targeting is analyzed. In the second step, the putative signal region is fused to a marker protein that normally functions in another compartment. The redirecting of the marker protein into a given compartment is then used as a positive test for the activity of the proposed signal. By following the properties of the marker, the fused targeting segment can be further deleted to give a more precise location of the active signal. In the third step, this peptide is characterized in detail through oligonucleotide-directed mutagenesis to determine which residues of the signal region are crucial for its function.

The success of the scheme outlined above in locating a targeting signal depends on whether or not the particular signal is autonomous. In this respect, it is important to distinguish between cotranslational and post-translational targeting events. For cotranslational events functional signals are represented by contiguous stretches of amino acid residues present on the growing polypeptide chain. These sequences are amenable to manipulation, which explains why protein insertion into the ER has been analyzed so successfully (see below). In posttranslational targeting events signals function as part of the folded protein. From an operational point of view it is useful to discriminate among signals that arise from: (a) a contiguous stretch of amino acid residues that can function irrespective of its location on its own or a foreign polypeptide chain; (b) a contiguous amino acid sequence that can function irrespective of the rest of the sequence only when it is located at either the N- or C-terminus of its own or a foreign polypeptide; (c) a contiguous amino acid sequence that can function only in its proper environment in the protein; and (d) discontinuous stretches of amino acid residues that are juxtaposed by protein folding to create a signal. It is evident that in the first two cases it is possible to transfer a putative signal region to a marker protein and test for its activity. In cases (c) and (d) attempts at locating the signal region are limited to mutagenic analyses.

Sometimes a putative targeting signal can be tentatively located by virtue of its characteristic features (e.g. a cleavable signal peptide or a hydrophobic membrane-spanning peptide), and used in a fusion experiment without the preceding step of deletion mutant analysis. The possible involvement of posttranslational covalent protein modifications, such as polypeptide phosphorylation or sulfation, in the targeting of a given protein can be studied directly by the precise mutagenesis approach (step three). The modified amino acid residue is likely to be located on the surface of the protein molecule, forming part of the recognition signal for the modifying enzyme. It is therefore possible to test the importance of that particular amino acid modification for protein transport using site-directed mutagenesis of the respective codon on the gene or cDNA.

GENETIC MANIPULATION

Genetic manipulation involves the generation of deleted genomes or cDNAs, their fusion to suitable marker protein genes, and the expression of the truncated or chimeric proteins in such cells or in vitro systems where the specific protein targeting can be analyzed. While the engineering is generally straightforward, using standard cloning and recombination technology, attention must be given to (a) the selection of the marker protein, (b) preservation of functional coding units in the manipulated DNA, and (c) the expression system.

Choice of Marker Protein

For a protein to be useful as a marker (also called indicator, reporter, or carrier) protein it should preferably have a monomeric and stable structure, be water soluble, and have an enzymatic activity that is retained even if foreign sequences are added to its C- or N-terminus. Numerous investigators have also found it advantageous to use procaryotic proteins as markers because they are probably devoid of eucaryotic targeting signals. Table 1 lists some of the commonly used proteins.

The 1021 amino acid long β-galactosidase encoded by the *lac Z* gene of *Escherichia coli* has very often been used in procaryotic (reviewed in Silhavy et al 1983) and eucaryotic systems (Guarente & Ptashne 1981; Rose et al 1981; Osley & Hereford 1982; Martinez-Arias & Casadaban 1983; Finkelstein & Strausberg 1983) because it retains its enzymatic activity following fusion to other proteins. However, its usefulness as a marker to study protein traffic in eucaryotic cells has several limitations. First, the polypeptide chain of β-galactosidase cannot be translocated across the ER membrane when fused to an appropriate targeting signal (Nielsen et al 1983; Emr et al 1984). This resembles the situation in *E. coli* in which this enzyme becomes stuck in the inner membrane when fused to a periplasmic protein (Silhavy et al 1979). Secondly, it enters the nuclear compartment on its own (Kalderon et al 1984b). Thirdly, the animal cell has an endogenous β-galactosidase activity in the lysosome (Nielsen et al 1983; Emr et al 1984).

The chimpanzee α-globin (132 amino acids long) appears to be a more useful probe than β-galactosidase for studying protein targeting in the eucaryotic cell. Although it is a cytoplasmic protein of eucaryotes, it is normally present only in cells of the erythroid lineage. In contrast to β-galactosidase, α-globin can be translocated across the ER membrane if equipped with appropriate signals, and it does not enter the nucleus on its own (Lingappa et al 1984; Davey et al 1985).

The 263–amino acid bacterial enzyme neomycin phosphotransferase is another interesting marker protein. It has been used to study the import of proteins into plant chloroplasts (Van den Broeck et al 1985). The enzyme conveys resistance to the antibiotics kanamycin and G418 (Reiss 1982; Southern & Berg 1982). These resistances can be used to select for correct (in frame) fusion proteins in bacteria and eucaryotes, respectively. The phosphotransferase enzyme can also be detected by a sensitive in situ assay (Reiss et al 1984).

The β-lactamase of the *E. coli* plasmid pBR322 and the nuclease of *Staphylococcus aureus* are candidates for use as markers for the exocytic pathways of the eucaryotic cell. Both proteins can be monitored by a sensitive enzymatic assay. The β-lactamase is secreted in frog oocytes, and the nuclease has been fused to the Fc portion of a secreted immunoglobulin (Wiedmann et al 1984; Neuberger et al 1984).

Table 1 Some useful marker proteins for targeting experiments

Protein	Species of origin	Normal cellular location	References
α-globin	chimpanzee	cytoplasm of red blood cell	Lingappa et al 1984 Tabe et al 1985 Davey et al 1985
pyruvate kinase	chicken	cytoplasm of muscle cell	Kalderon et al 1984b
dihydrofolate reductase	mouse	cytoplasm	Hurt et al 1984
β-galactosidase	Escherichia coli	cytoplasm	Kalderon et al 1984b Hase et al 1984 Hall et al 1984 Douglas et al 1984
neomysin phospho-transferase II	bacterial Transposon Tn 5	cytoplasm	Van den Broeck et al 1985
β-lactamase	bacterial plasmid pBR322	periplasm	Lingappa et al 1984 Wiedmann et al 1984
nuclease	Staphylococcus aureus	periplasm	Neuberger et al 1984

Engineering within Coding Sequences

The genetic construction of deleted and fused proteins using DNA-modifying enzymes has two inherent problems in relation to protein translation: (a) the maintenance of efficient translational control signals, and (b) the preservation of the correct reading frame. When performing exonuclease digestions from either the 5' or the 3' end of cloned genes or cDNAs, one cannot avoid also removing the translation initiation or termination signal, respectively. The latter signal can relatively easily be added to the deleted gene in the form of a synthetic oligonucleotide that contains stop codons in all three reading frames (Pettersson et al 1983; Kalderon et al 1984b), or through an in-frame ligation to a commercially available XbaI or HpaI linker molecule that contains a single stop codon (Rose & Bergmann 1983). In some fortunate cases the deleted gene is closely followed at the 3' end by a stop codon in frame with the coding sequence. Restoration of an in-frame ATG triplet with its flanking consensus sequence (Kozak 1981) can be done using the same approaches as described for the termination signal. However, this does not necessarily restore efficient initiation of translation because this involves, in addition to the AUG-region, some as yet undefined features at the 5' end of the RNA molecule.

When chimeric genes are constructed, or internal gene fragments are deleted, the correct reading frame is easily lost in the 3' fragment. An easy way to preserve the frame is to include some randomization at the point of DNA ligation, and screen a number of the recombinants directly for protein expression. In some marker protein fusions this can be done at the level of bacterial cloning, provided that the coding region is controlled by appropriate procaryotic transcription and translation elements. For instance, the in-frame fusion of a DNA fragment encoding the N-terminal portion of a polypeptide to the 5' end of the β-galactosidase coding sequence will, in transformed E. coli cells, result in β-galactosidase production, which can be monitored using a sensitive color assay (Amann 1984). It is even simpler to use the bacterial neomycin phosphotransferase for the same purpose (see above). Expression screening in eucaryotic systems can most conveniently be done using the in vitro or SV40-Cos cell systems described below. When genomic sequences are being analyzed, the reading frame can be maintained by restricting the manipulation to intron sequences, leaving the exons intact. It follows that when a relationship exists between the exons and protein domains of a particular protein then exon deletion, or exon shuffling between different genomes, is a very straightforward way to localize a specific targeting signal to one of the protein domains. This approach has been used to localize those protein domains of mouse class I histocompatability antigens that are recognized by allospecific and virus-specific cytolytic T lymphocytes (Reiss et al 1983, Arnold et al 1984).

Expression Systems

To analyze the intracellular routing of a mutant protein encoded by an engineered gene, the gene must be expressed in an appropriate cell or in vitro system, preferably in such a way that the transport can be followed quantitatively. For this purpose a large number of DNA expression systems have been developed. A detailed treatise of these techniques is, however, beyond the scope of this review, and therefore I will only briefly discuss some general aspects of those systems most often used in the targeting experiments described later. For a thorough discussion of animal, yeast, and plant cell expression systems the reader is referred to Rigby (1982); Hicks et al (1982); Zambryski et al (1984), respectively.

Protein import into ER, mitochondria, and chloroplasts has most successfully been studied using in vitro reconstitution systems. These same systems can now also be used to analyze protein expressed by cloned cDNA molecules. The cDNA molecule is first transcribed using a procaryotic polymerase (the *E. coli*, T7, or SP6 polymerase), and then translated in a eucaryotic in vitro system supplemented with the appropriate target organelle (Paterson & Rosenberg 1979; Contreras et al 1982; Rubenstein & Chappell 1983; Krieg & Melton 1984; Stueber et al 1984; Ahlquist & Janda 1984). Crucial to the success of these systems is the ability to cap the "procaryotic" mRNA either posttranscriptionally using the *Vaccinia* virus capping enzymes (Moss 1977), or cotranscriptionally using 7-methyl GpppA or GpppG for transcription initiation (Contreras et al 1982; Konarska et al 1984).

Since the pioneering experiments by Gurdon (1970) the frog oocyte has been the model system for the study of protein transport to the nucleus. Coleman et al (1983) and Krieg et al (1984) have introduced the use of cloned cDNA molecules in this system. The cDNA molecules are engineered downstream from a eucaryotic promotor/enhancer element, and microinjected into the oocyte nuclei. Alternatively, the cDNA can be transcribed into RNA in vitro, and introduced into the cytoplasm of the oocyte (Krieg & Melton 1984). A few injected oocytes express enough protein to be followed in a pulse-chase experiment.

Protein transport from the ER to the Golgi, and the pathways from the Golgi apparatus to the cell surface and lysosomes have been studied extensively in animal tissues, primary cell cultures, and continuous cell lines. Of these, the established animal cell lines are most convenient for DNA expression studies. Foreign genetic material can be introduced permanently into these cells at a low frequency through DNA transfection. When this is done in combination with a gene that conveys a selectable phenotype to the cell, it is possible to obtain cotransformants that can grow in selective medium (Wigler et al 1977, 1979; Mulligan & Berg 1981; Southern & Berg 1982; Blochlinger & Diggelmann 1984). Cell clones expressing the nonselective gene are then screened by using, for example, a radioimmunoassay or immunofluorescent techniques. The efficiency of protein expression in the transformed cell clone will generally (but not always) be sufficient for quantitative biochemical analysis. The variation in efficiency of protein expression appears to depend on how compatible a particular cellular system is with (a) the promotor/enhancer elements of the foreign gene (for review on the species and tissue specificity of these elements see Gruss 1984), and (b) the continuous production of a foreign soluble or membrane-bound protein. In the second instance the production of a particular protein may be toxic for the cell, and therefore only cell clones that express at a very low level can survive. In principle, this problem can be solved by using an inducible promotor with a low basal level of expression. Such promotors have been described for the β-interferon gene (Mitrani-Rosenbaum et al 1983; McCormick et al 1984), the human and mouse metallothionein genes (Pavlakis & Hamer 1983; Karin et al 1984), and the mouse mammary tumor virus gene (Ringold et al 1977). By maintaining the promoter in the uninduced state, thus keeping the level of the toxic protein low, it should be possible to clone and raise transformants that upon induction express enough protein for quantitative targeting experiments.

Monkey cell lines make possible the use of the well-characterized virus SV40 as an expression system. Infection of monkey cells with SV40 proceeds in two phases. In the first phase large T antigen is produced, which in the second phase supports SV40 DNA replication and late transcription (reviewed in Tooze 1981). Several research groups have engineered their gene of interest into the late region of the SV40 DNA, and used a temperature-sensitive helper virus in early gene functions to generate a higher titer virus stock containing both helper and recombinant genomes (Mulligan et al 1979; Hamer & Leder 1979). This system was most successfully used by Gething & Sambrook (1982), who reported the synthesis and secretion of 300 μg of a soluble form of the influenza hemagglutinin (HA) from a 10 cm dish of monkey CV1 cells that had been infected with recombinant SV40 virus. This is by far the highest level of protein expression in a eucaryotic cell from engineered DNA reported. However, a virus stock with a higher titer of recombinant virus is not always obtained even if the relatively stringent DNA size limits for encapsidation (Rigby 1982) are met. The reasons for this are poorly understood.

The SV40 monkey cell system is not restricted to the use of viral stocks. When naked SV40 recombinant DNA is introduced into monkey cells by means of transfection (Graham & van der Eb 1973; Lopata et al 1984; Sussman & Milman 1984) and allowed to replicate, a level of protein expression can be obtained that is sufficient for pulse chase experiments. The large T antigen can either be provided by the vector DNA (Gruss et al 1982), or by a monkey cell line that has been transformed with the gene for the T antigen (Cos cells) (Gluzman 1981, Rio et al 1985). As virus packaging is not involved in this simple expression system there are fewer constraints on the size of the recombinant DNA molecule.

A major difficulty in using the recombinant DNA approach to study protein traffic in animal cells is that there is no single expression system that works in all cells. Therefore, it is interesting to note that expression systems based on retrovirus infection have recently been developed that should be compatible with cells from practically any species (Mann et al 1983; Watanabe & Temin 1983; Cepko et al 1984; Sorge et al 1984; Cone & Mulligan 1984). Unlike SV40 the retrovirus infection is not lytic to the cell, and usually only one copy of the recombinant gene is introduced into the host chromosome. However, this single copy can yield an expression level severalfold higher than that which results when the same construct is introduced by transfection (Sherry Hwang & Gilboa 1984).

INSERTION OF PROTEINS INTO AND ACROSS THE MEMBRANE OF THE ER

Cotranslational Signals for Protein Topogenesis in the ER

Protein import into the ER involves the targeting of the nascent polypeptide to the outer surface of the ER membrane, and the translocation of the polypeptide chain either completely (secretory proteins) or partially (membrane-bound proteins) across the membrane. The mechanisms responsible for these processes have been better characterized than any other protein targeting event in the eucaryotic cell. This is mostly due to the development of an in vitro system for protein translocation (Blobel & Dobberstein 1975). In this assay in vitro–synthesized secretory and membrane proteins become inserted into and across the membrane of microsomes derived from the rough ER of dog pancreas. Studies using this system, as well as others made in vivo, have shown that polypeptides imported into the ER generally contain an N-terminal "signal peptide."

This signal is thought to direct the ribosome to the ER membrane, where continued polypeptide synthesis is coupled with direct transfer of the nascent chain into the lumen of ER (reviewed in Sabatini et al 1982). In most cases the signal peptide is removed by a proteolytic cleavage before chain completion.

More recent studies have shown that the binding of the polysome to the ER membrane is mediated by a ribonucleoprotein complex (Walter et al 1981; Walter & Blobel 1981a,b), termed the signal recognition particle (SRP). SRP specifically recognizes those polysomes translating polypeptides with signal peptides, and transfers them to the rough ER by binding to an integral membrane protein, the docking protein, that is restricted to this organelle (Meyer et al 1982; Gilmore et al 1982a,b; Hortsch et al 1985). At the rough ER membrane SRP is displaced from the polysome, and the latter remains attached to the membrane by other means (Gilmore & Blobel 1983).

Apparently, the information required for translocation of secreted proteins across the ER membrane resides in the signal peptide. The location of the insertion signal (Figure 1a) at the N-terminus of the polypeptide could serve several purposes. For instance, it might ensure an immediate cotranslational chain transfer across the membrane, and allow the signal to be cleaved by proteolysis after initiation of translocation. The signal peptide cleavage appears to be obligatory for the secretion of at least some secretory proteins (Hortin & Boime 1981a,b; Haguenauer-Tsapis & Hinnen 1984).

However, the topogenesis of various kinds of transmembrane proteins in the ER cannot be explained by an N-terminal insertion signal alone. One has to postulate the existence of additional signals that are decoded on the growing polypeptide chain (Blobel 1980; Sabatini et al 1982). As their mode of function has been inferred from the known topology of various transmembrane proteins, it is helpful to consider this before discussing the biosynthesis of these proteins. Transmembrane proteins can be classified on a topological basis into three groups. Group A proteins span the lipid bilayer once and have their N-termini oriented towards the ER lumen and their C-termini exposed on the cytoplasmic side. This group includes viral spike glycoproteins, such as the G protein of vesicular stomatitis virus (VSV), the EI and E2 proteins of Semliki Forest virus (SFV), and the HA of influenza virus; histocompatibility antigens; membrane-bound immunoglobulins; and the glycophorin molecule (reviewed in Warren 1981). Group B proteins span the membrane in the opposite orientation; their C-termini are in the ER lumen and their N-termini are in the cytoplasm. Examples of this type are the neuraminidase of influenza virus (Blok et al 1982), the invariant chain of class II histocompatibility antigens (Claesson et al 1983;

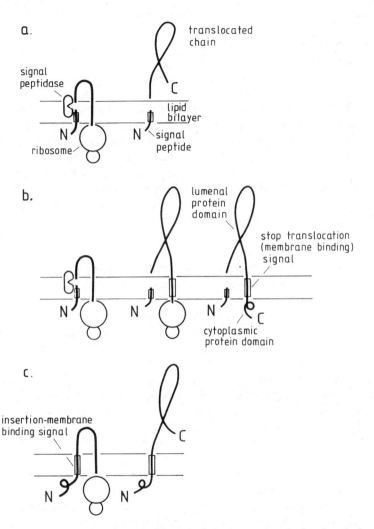

Figure 1   Postulated modes of protein biosynthesis in the rough endoplasmic reticulum. (*a*) Synthesis of a secretory protein. (*b*) Synthesis of a Group A transmembrane protein. (*c*) Synthesis of a Group B transmembrane protein. The boxed segments of the polypeptide chains represent hydrophobic regions of insertion and membrane-binding signals. The complete chain is shown on the right side of each panel.

Strubin et al 1984; Singer et al 1984), the human liver transferrin receptor (Schneider et al 1984), and the rat liver asialoglycoprotein receptor (Holland et al 1984). Group C proteins span the membrane several times (reviewed in Sabatini et al 1982). Group B proteins typically have a permanent insertion signal, whereas most Group A proteins possess a cleavable signal peptide.

Translocation of a membrane polypeptide from the N-terminus is thought to be mediated by an N-terminal insertion signal, as represented by the cleavable signal peptide of the secretory proteins (Figure 1b). For Group A transmembrane proteins the translocation process is postulated to be arrested by a specific "halt" or "stop" transfer sequence (Blobel 1980; Sabatini et al 1982) (Figure 1b). The membrane-spanning segment of these proteins is thought to function as such a sequence. A mechanism for the generation of proteins with the opposite disposition in the membrane (Group B proteins) has been more difficult to envisage. One can postulate that these proteins contain a permanent insertion signal that overlaps with their membrane-spanning segment. This "insertion membrane-binding signal" would intiate the translocation of the polypeptide chain by looping into the lipid bilayer with the N-terminus remaining on the cytoplasmic side (Figure 1c). Such a mechanism was originally suggested by Inouye & Halegoua (1979) for the cleavable signal peptide. By postulating the existence of specific N-terminal insertion signals, stop transfer signals, and insertion membrane-binding signals every known kind of polypeptide disposition across the membrane can be explained. For instance, a protein that spans the membrane three times with its N-terminus within the ER lumen and its C-terminus exposed to the cytoplasm could be generated by the sequential and cotranslational expression of an N-terminal insertion signal, a halt transfer signal, an insertion membrane-binding signal, and a second halt transfer signal, in that order.

Most of the theories regarding the signal peptide and the additional signals required to generate transmembrane proteins with different dispositions in the membrane are amenable to testing by the recombinant DNA approach. Gene deletion and fusion experiments can be designed to demonstrate (a) whether or not a signal peptide carries all the information required to transfer a given polypeptide chain into the ER; (b) the structural specificity of the signal peptide; (c) whether or not a signal peptide can function if it is displaced further away from the N-terminus; (d) the location of a permanent insertion signal on the polypeptide chain; (e) whether or not the spanning peptide of Group B transmembrane proteins functions as an insertion signal; (f) the exact nature of the stop transfer signal; (g) whether or not one can create a Group C protein by inserting the proposed signals into a given polypeptide in the appropriate order.

The Signal Peptide

ER INSERTION ACTIVITY IN THE SIGNAL PEPTIDE   The existence of transient amino-terminal extensions on proteins destined to be inserted into the ER strongly suggests that these peptides act as targeting signals. This was supported by the finding that protein translocation in vitro could be inhibited competitively by the addition of preproteins or synthetic signal peptides to the assay (Prehn et al 1980; Majzoub et al 1980; Koren et al 1983). Genetic evidence for the importance of the signal peptide in translocation was first obtained in procaryotes. Using an elegant selection system based on the conditionally lethal phenotype of some export protein–*lac Z* gene fusions in *E. coli*, it was possible to isolate several translocation-deficient fusion protein mutants that accumulated in the cytoplasm (reviewed in Silhavy et al 1983). Their characterization showed that the mutations were located in the gene segment encoding the signal peptide region.

More recently, recombinant DNA approaches were used to study the signal peptides of both procaryotes and eucaryotes. Inouye and his colleagues have used oligonucleotide-directed mutagenesis of the signal peptide region of the *E. coli* prolipoprotein gene to demonstrate its importance for protein translocation (Inouye et al 1982, 1983, 1984; Vlasuk et al 1984). Gething & Sambrook (1982) deleted the entire signal peptide region in an influenza virus hemagglutinin cDNA molecule. The mutant protein was expressed in vivo in monkey CV1 cells using recombinant SV40 infection. In contrast to the wild type protein, which was glycosylated and transported to the cell surface, the mutant protein was located intracellularly, presumably in the cell cytoplasm, in an unglycosylated form. Sekikawa & Lai (1983) reported similar results with another HA cDNA molecule.

The definitive targeting experiment with a cleavable signal peptide was made by Lingappa et al (1984) (Figure 2a). They inserted the coding sequence for the chimpanzee α-globin (a cytoplasmic protein) into the β-lactamase gene of pBR322 (which codes for a periplasmic protein) such that the globin sequence was preceded by a portion of the lactamase gene coding for the 28 N-terminal amino acid residues of the enzyme, including its 25-residue signal peptide. The hybrid gene was transcribed in vitro into RNA, which upon translation in an in vitro wheat germ system directed the synthesis of the protein chimera. The protein was shown to be translocated into dog pancreas microsomes, and cleaved at the site of normal β-lactamase signal peptide processing. This experiment demonstrated unequivocally that a cleavable signal peptide contains all of the information required to target a protein into the lumen of the ER.

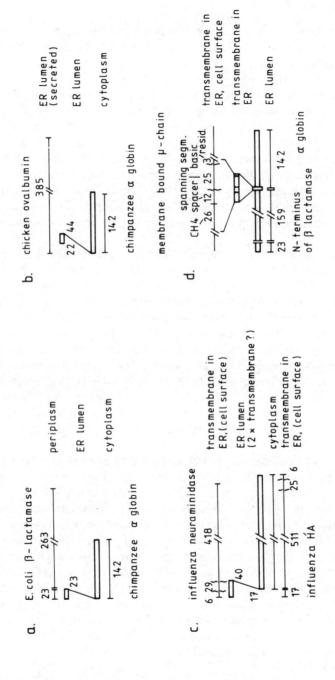

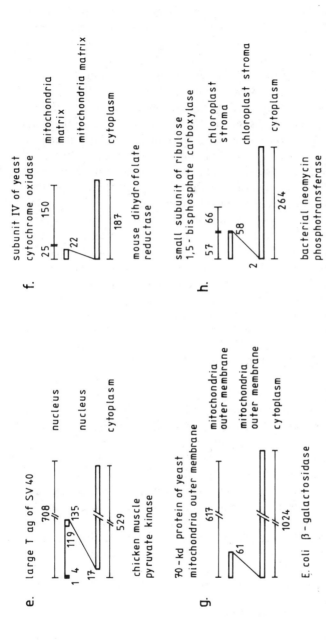

Figure 2 Construction schemes for some chimeric proteins used in retargeting experiments. In each panel the two wild type proteins are represented by thin horizontal lines at the top and bottom of the channel, and the chimeric constructs are represented by elongated boxes joined together at points of fusion. The cellular destination of each protein is given on the right side of each figure. Labels: thin vertical bars limit complete polypeptide chains and topological regions of these; double thin vertical bars indicate signal peptide cleavage sites; wavy vertical bars indicate the boundaries of transmembrane peptide segments; thick bars indicate fusion points in chimeric proteins; and double slanting bars indicate that the part of the polypeptide chain between the bars is not represented in the figure. Numbers above or below lines indicate amino acid residues for polypeptide chains or limited regions of them. Numbers before or after boxes represent amino acids at the fusion point, as numbered from the N-terminus of the corresponding wild type protein.

STRUCTURAL AND TOPOLOGICAL SPECIFICITY OF THE SIGNAL PEPTIDE   At present, more than 200 procaryotic and eucaryotic signal peptides have been sequenced (Watson 1984). Comparison shows that most extensions comprise 20–40 amino acid residues; the longest is that of the Rous sarcoma virus membrane glycoprotein with 62 residues. There is no homology between the sequences, but a characteristic distribution of amino acid side chains is observed. Most prominent is a stretch of 9 or more hydrophobic residues, often flanked by basic residues on the N-terminal side (Watson 1984). The importance of these general features for signal activity is supported by (a) the kind of amino acid alterations found in export-defective E. coli protein mutants and their suppressor mutations (Bedouelle et al 1980; Emr et al 1980; Emr & Silhavy 1983; Bankaitis et al 1984); and (b) by the phenotypes of the various site-directed signal peptide mutations made by Inouye & colleagues (see above).

Until now only a few gene manipulation attempts were made to determine whether or not a signal peptide can function if displaced from its N-terminal location. Eighteen and 22 amino acids were added to the N-terminus of the preproinsulin signal peptide by genetic engineering, with no apparent effect on its normal in vivo processing (Talmadge et al 1981; Kozak 1983). However, when 68 residues were added there was almost no cleavage into the proform (Kozak 1983). The intracellular location of this mutant was not determined, but one possibility is that it remained unprocessed in the cell cytoplasm because of an inactivated signal peptide. In another case, the addition of 76 amino acids to the N-terminal end of the hepatitis surface antigen inhibited its secretion (Liu et al 1984).

A third example of the influence of neighboring sequences on signal peptide function comes from the mode of biosynthesis of the structural proteins of α viruses (Garoff et al 1982). In these viruses, the structural proteins are translated from a single mRNA molecule, and are released by cleavage of the nascent chain. The cytoplasmically located capsid protein is synthesized before the two glycoproteins that will be inserted into the membrane of ER. If the cleavage of the capsid protein is inhibited (e.g. in a conditionally defective Semliki Forest virus mutant) no insertion of the membrane protein translated after the capsid protein is observed (Hashimoto et al 1981). It appears from these examples that amino-terminal signal peptides can function only in their proper N-terminal location. However, this should be confirmed by more detailed experiments using in vitro transcription/translation/translocation system.

Permanent ER Insertion Sequences

Ovalbumin (Palmiter et al 1978) and several membrane proteins (Garoff et al 1978; Meek et al 1982; Braell & Lodish 1982; Anderson et al 1983;

Sakaguchi et al 1984; Rottier et al 1985) are translocated across the ER membrane in a cotranslational fashion without a transient amino-terminal peptide. Different in vitro translation strategies have been used to localize the signal regions of these proteins, but no strategy has been very exact (see references above). One type of experiment investigates how long after synchronous initiation of translation microsomes can be added and still result in proper compartmentalization of the completed chains.

In another type of experiment SRP is used in a similar in vitro assay, where its ability to induce a translation arrest (Walter & Blobel 1981b) is exploited. The time point beyond which the signal is no longer recognized is then used to estimate the length of polypeptide synthesized, and thus to tentatively localize the signal. The problem with such approaches to locating internal signals is that one can never eliminate the possibility that an N-terminal signal remains active even after extensive polypeptide synthesis. More recently, the recombinant DNA approach has been used to localize permanent insertion signals of a secretory protein, the ovalbumin molecule, and a membrane glycoprotein, the neuraminidase molecule of influenza virus.

Tabe & colleagues (1984) first analyzed the insertion of a series of N-terminal deletion mutants of ovalbumin. These mutants were constructed by exonuclease digestions of the 5' end of the ovalbumin cDNA, and then expressed in frog oocytes using the SV40 early gene region, or the herpes simplex thymidine kinase promotor. They found that while removal of amino acids 1–8 had no effect on ovalbumin sequestration and secretion, elimination of amino acids 2–21 inhibited its secretion but not translocation. In contrast, removal of amino acids 1–40 inhibited both processes. In another experiment Tabe & colleagues fused various portions of the 5' coding sequence of the ovalbumin cDNA to the chimpanzee α-globin sequence, and studied subsequent protein expression in oocytes. A protein chimera containing amino acids 9–41 of ovalbumin at its amino terminus was sequestered into the ER and secreted, whereas another chimera containing amino acids 22–41 was sequestered but not secreted (Figure 2b). These authors also deleted a region comprising amino acids 229–279 of ovalbumin, which had been proposed as a possible internal insertion signal (Lingappa et al 1979), and found that the mutant protein was still inserted into the ER. Thus these results localize the ovalbumin ER insertion signal within residues 22–41.

The neuraminidase of influenza virus is a membrane glycoprotein belonging to the Group B proteins previously discussed. It is synthesized and inserted into ER membranes without a transient amino-terminal peptide (Blok et al 1982; Hiti & Nayak 1982), and it becomes anchored in the lipid bilayer by a hydrophobic peptide located close to its N-terminus.

Two groups (Markoff et al 1984; Bos et al 1984) recently tested whether or not the hydrophobic membrane-binding region of neuraminidase actually contains a region with insertion activity. Both groups approached the problem by using cDNA engineering and the infectious SV40 expression system. Markoff & colleagues produced a series of internal deletions in the N-terminal region of this molecule, removing various portions of its 29-residue hydrophobic segment. Removal of 7 residues from the C-terminal side of the membrane-spanning segment resulted in normal cell surface appearance of the molecule, whereas the removal of 20 or more residues resulted in the synthesis of an unglycosylated chain which was probably in the cytosolic compartment. Bos & colleagues fused the amino-terminal region of the neuraminidase, including the total hydrophobic region, to a hemagglutinin molecule from which the normal cleavable signal peptide had been removed, and showed that this fusion protein was glycosylated (Figure 2c), and therefore was probably also inserted across the ER membrane. It was not, however, transported to the cell surface. Thus these studies localize the insertion signal of neuraminidase to its amino-terminal hydrophobic region.

The Stop Transfer Signal

Many cell surface and membrane proteins of animal and procaryotic viruses are bound to the lipid bilayer by a membrane-spanning hydrophobic peptide close to the C-terminus of the polypeptide (reviewed in Warren 1981; Armstrong et al 1981). For the influenza hemagglutinin (Gething & Sambrook 1982; Sveda et al 1982), the VSV G protein (Rose & Bergmann 1982), and the minor coat protein of phage f1 (Boeke & Model 1982) this has been shown most convincingly by the expression of secreted forms of these proteins by 3'-deleted cDNA molecules. The deletions included the gene regions encoding the cytoplasmic and membrane-spanning protein domains of each cDNA.

The hydrophobic membrane-spanning peptide of these proteins is thought to be the essential component of the cotranslational signal that effects the arrest of chain transfer across the rough ER (RER) membrane during synthesis. Some currently undefined sequences on either the luminal or cytoplasmic side of the spanning peptide could also be involved in this process (see Figure 1b). Interestingly, one or more basic residues are always found at the cytoplasmic side of transmembrane polypeptide segments (reviewed in Sabatini et al 1982; Garoff et al 1982). These could well serve some function in the generation of transmembrane polypeptides. Furthermore, support for this view is given by an examination of the exon organization of membrane-bound immunoglobulins and class I histocompatibility antigens. It shows that the membrane exons always include a gene region that encodes the total membrane-spanning segment with at least one flanking basic residue at the 3' end (Alt et al 1980; Rogers et al 1980; Cheng et al 1982; Yamawake-Kataska et al 1982; Tyler et al 1982; Komaromy et al 1983).

Yost et al (1983) directly demonstrated that a transmembrane peptide segment with its flanking sequences can cause translocation arrest during polypeptide synthesis, thereby generating a membrane-spanning polypeptide. Using the same approach described earlier (Lingappa et al 1984), these authors first showed that a β-lactamase chimpanzee α-globin protein chimera was efficiently transferred into the lumen of microsomes in vitro. In a second engineering step a segment of a cloned membrane-bound IgM μ-chain cDNA, which included the region encoding the transmembrane segment of the μ-chain, was inserted between the two genes of the original chimera (Figure 2d). When the new construct was expressed and the protein tested for translocation in vitro, it was arrested and anchored in the membrane at the place of the inserted μ-chain segment.

Guan & Rose (1984) tested the importance of the luminal domain of the VSV G protein for translocation arrest. They constructed a gene encoding a chimeric protein composed of the growth hormone (a secreted protein), and the transmembrane and cytoplasmic domains of the VSV G protein, in such a way that the hormone sequence was joined directly to the first amino acid residue of the 20-residue transmembrane segment of the G protein. When the chimera was expressed in monkey Cos cells, and its topology relative to the membrane was analyzed by protease treatment, the same transmembrane configuration was found as in the wild type VSV G protein. Therefore, at least for the G protein, the luminal sequences flanking its spanning peptide are not required for the generation of its transmembrane configuration.

TARGETING OF PROTEINS TO THE CELL SURFACE

Possible Protein Signals for Transport Along the Constitutive Cell Surface Pathway

Following insertion into the RER membrane during synthesis, plasma membrane proteins are transported via the Golgi complex to the cell surface along a constitutive pathway (Green et al 1981a; Bergmann & Singer 1983; Saraste & Kuismanen 1984; Rothman et al 1984). This pathway is probably also used by all those proteins constitutively secreted by the cell (Strous et al 1983b). It is generally believed that the protein traffic from the ER to the Golgi, between the individual stacks of the Golgi, and from the Golgi to the cell surface is mediated by vesicle carriers. The mechanisms by which proteins become included into and released from these transport vesicles are completely unknown. One possibility is that routing to the cell surface involves transport signals present on the membrane and secretory proteins. The signals could be represented either by features of the polypeptide chain or by covalent modifications of it (e.g. glycosylation, phosphorylation, or sulfation). Another possible explanation of the constitutive transport of proteins to the cell surface is that all proteins translocated into the ER lumen are routed to the cell surface in a nonselective way by a unidirectional traffic of vesicular carriers. Only for proteins specific to the ER and Golgi compartments would it be necessary to postulate the presence of protein signals that mediate their retention in the proper organelle.

Several observations have been used in the past to support the existence of cell surface transport signals in the polypeptide or carbohydrate part of various (glyco)protein molecules. These are (a) the observed differences in the kinetics by which various plasma membrane (PM) and secretory proteins are routed to the cell surface or extracellular space (Coleman & Morser 1979; Cutler et al 1981; Fitting & Kabat 1982; Lodish et al 1983; Novick & Scheckman 1983; Fries et al 1984; Patzer et al 1984; Scheele & Tartakoff 1985); (b) the demonstration of transport-defective protein mutants (Yoshida et al 1976; Katona et al 1978; Krangel et al 1982; Arias et al 1983; Wu et al 1983; Gallione & Rose 1985); and (c) the inhibition or blocking of surface transport of some proteins when their normal glycosylation has been prevented by the presence of drugs, such as tunicamycin or 1-deoxynojirimycin during synthesis (Hickman & Kornfeld 1978; Elbein 1981; Novick & Schekman 1983; Lemansky et al 1984; Lodish & Hong 1984).

However, each of these observations can be explained in a way that obviates the need for transport signals. For instance, the observed nonuniformity in transport rates could as well be due to differences in the general properties of the individual proteins (e.g. solubility inside the ER), and the way in which this influences their inclusion in transport vesicles. The amino acid changes found in the transport-defective mutants might cause aberrant folding of the protein in the ER so that transport might be inhibited for physical reasons rather than because of an inactivated signal. The same is true of the effect obtained by inhibiting a normally occurring cotranslational glycosylation process. In fact, altered solubility properties have been demonstrated for unglycosylated VSV and Sindbis glycoproteins made in vivo (Leavitt et al 1977; Gibson et al 1979). Moreover, yeast invertase can be transported to the cell surface in an unglycosylated form when synthesized at 25°C, but not when made at 37°C, which suggests a stabilizing effect of the sugars on the structure of this protein (Ferro-Novick et al 1984). This is supported by the earlier results of Chu et al (1978), who showed that invertase that had been deglycosylated in vitro became less stable to a variety of denaturing conditions such as heat, freezing and thawing, trypsin digestion, and acidic conditions.

On the other hand, it has been directly demonstrated in several cases that the sugar units of proteins, and various posttranslational covalent protein modifications, are not essential for the process of cell surface transport. For instance: (a) the routing of many secretory and PM proteins are little, if at all, affected by the absence of their sugar units (Hickman & Kornfeld 1978; Green et al 1981b; Novick & Schekman 1983; Lodish & Hong 1984); (b) PM proteins have been shown to be normally transported to the cell surface of Chinese hamster ovary cell mutants that are defective in oligosaccharide processing (Briles et al 1977; Li & Kornfeld 1978; Rothman et al 1984); (c) the VSV G protein (Indiana serotype) reaches the cell surface in monkey cells in an apparently normal fashion even when it is devoid of its usual fatty acid modification (Rose et al 1984); and (d) proalbumin is secreted by hepatocytes in the unprocessed form in a natural human variant (Brennan & Carrell 1978). Another indication of the nonselective nature of the constitutive cell surface transport of proteins is the effect of the drug chloroquine on protein traffic in the cell. This drug, which increases intracellular pH, inhibits routing of proteins to the lysosome (Gonzales-Noriega et al 1980), and along the regulated pathway of protein secretion (Moore et al 1983a) but not along the constitutive pathway.

The analysis of transport signals on proteins routed constitutively to the cell surface by the recombinant DNA approach is met with some difficulties. As there are no useful marker proteins in this pathway with which to test the activity of a postulated signal, the approach is restricted to studies on proteins altered by deletions or point mutations. Such studies can give unequivocal information only about polypeptide regions that are not involved in transport signaling. When the transport to the cell surface is inhibited or blocked, the same problem appears as for transport defective protein mutants generated in vivo, as discussed earlier. Because of these difficulties investigators in the field are trying to introduce only those alterations, such as the deletion of a complete structural domain, that are least likely to cause protein denaturation. Special attention must be paid to oligomeric proteins when studying their cell surface transport. This is so because the unassociated subunits of many oligomeric proteins, especially membrane-bound ones, have been shown to be retarded in the ER, e.g. the unassociated heavy chains of membrane-bound and free immunoglobulins (Burrows et al 1979; Ochi et al 1983); the heavy chain of class I histo-compatibility antigens (Arce-Gomez et al 1978; Severinsson & Peterson 1984); the subunits of class II histocompatibility antigens (Kaufman et al

1984); and the E1 protein of SFV (Hashimoto et al 1981). This possibility must be considered whenever the alteration of an oligomeric protein results in transport inhibition.

Several research groups have specifically altered the cytoplasmic portion of various plasma membrane–spanning glycoproteins by deletions and/or insertions, and analyzed the effects on cell surface transport. This protein domain could in principle contain a transport signal that would interact with cytoplasmic structures. The proteins studied were a class I histocompatibility antigen (Murre et al 1984), the p62 protein of SFV virus (Garoff et al 1983), the VSV G protein (Rose & Bergmann 1983), and the Rous sarcoma virus envelope glycoprotein, pr95[env] (Wills et al 1984). These proteins have cytoplasmic protein domains comprising 31, 31, 29, and 22 amino acid residues, respectively. The proteins analyzed included H2, p62, and pr95[env] mutants with deletions extending up to or close to the basic residues flanking the membrane-spanning segment of each protein, and G mutants with deletions extending approximately halfway through the cytoplasmic domain. In addition, one G protein mutant studied had its basic residue flanking the membrane-spanning region removed, along with the whole cytoplasmic domain. It should be noted that in all mutants the C-terminal region contained one or several aberrant amino acids as a result of the engineering.

Using various expression systems it was shown that all deletions of cytoplasmic protein domains that did not remove the basic residues flanking the transmembrane region were transported to the cell surface. However, when the kinetics of cell surface transport were studied for the G protein and pr95[env] deletion mutants by biochemically monitoring their posttranslational processing, the rate of transport was found to be much slower than that of the wild type proteins. One G protein mutant that had retained almost the complete cytoplasmic domain, and contained an additional short aberrant stretch of amino acids at its C-terminus, remained unprocessed even after the maximum chase time used in the experiment (6 hr). This mutant was, however, detected by immunofluorescent techniques on the cell surface in about 10% of the cells expressing this mutant. In the H2 antigen and the p62 protein the kinetics of cell surface transport were not determined. The G mutant lacking the basic residue and the entire cytoplasmic domain did not appear on the cell surface.

Thus there is probably not an obligatory transport signal in the amino acid sequences between the basic amino acid cluster and the C-terminus of the H2 antigen, the p62 protein, and the pr95[env] protein. This is also true of the C-terminal half of the G protein's cytoplasmic sequence. Furthermore, the kinetic analyses of the pr95[env] and the G protein mutants show clearly that perturbations of the respective cytoplasmic polypeptide sequences in the form of deletions and/or insertions can slow the rate of cell surface transport significantly. One explanation for this phenotype could be that the mutations have altered some general features of the cytoplasmic domain that are required for its inclusion in transport vesicles. We cannot, however, tell whether this effect is caused directly by the deletion or alteration of the cytoplasmic domain, or indirectly by a defect in a possible protein oligomerization process which these proteins might normally undergo and which might be a requirement for efficient transport. It should be noted that the subunit structure of the retrovirus spike is unknown. The G protein has been found to be a monomer (Crimmins et al 1983), but cross-linking studies suggest that it is a trimer (Dubovi & Wagner 1977).

The involvement of the basic amino acids that flank the spanning segment of transmembrane proteins on the cytoplasmic side in cell surface transport has been addressed by our group (Cutler & Garoff, unpublished). The wild-type sequence (Arg-Ser-Lys) of the p62 protein of SFV has been changed into a neutral (Met-Ser-Gly) or an acidic (Met-Ser-Glu) one using oligonucleotide-directed mutagenesis. Expression analyses of this protein showed that it reached the cell surface in the presence or absence of the other SFV glycoprotein, E1. When the protein was expressed with E1 it was possible to demonstrate acid-inducible cell-cell fusion activity, which is characteristic of the viral spike glycoprotein complex of SFV (Kondor-Koch et al 1983). Thus, the basic amino acids of the p62 protein are not obligatory for cell surface transport.

Possible Protein Signals for Polarized Transport in Epithelial Cells

Epithelial cells display a strong structural and functional polarization, a major aspect of which is the division of its PM into apical and basolateral domains. Studies on the generation of PM polarity in these cells have been made considerably easier by the finding that animal enveloped viruses mature in a polarized fashion in infected epithelial cells (Rodriguez-Boulan & Sabatini 1978). For example, during viral infection it has been shown that the influenza virus HA molecule is transported to the apical surface, and the VSV G protein goes to the basolateral surface of a polarized epithelial cell line, MDCK (see review by K. Simons & S. Fuller, this volume). Using infected MDCK cells grown on a permeable support, Matlin & Simons (1984), Misek et al (1984), and Pfeiffer et al (1985) demonstrated that the viral glycoproteins are sorted intracellularly, probably in some late Golgi compartment.

For polarized epithelial cells, which display two kinds of constitutive cell surface pathways for membrane proteins, one must postulate transport or sorting signals, at least for those proteins routed along one of the two pathways. Using classical biochemical techniques it has been shown that the N-glycosidically linked sugar units of the influenza HA molecule and the VSV G protein are not involved in their polar transport in MDCK cells (Roth et al 1979; Green et al 1981b). The recombinant DNA approach has been applied to investigate the possibility that the cytoplasmic domain is required for this same process. In principle, the polar system should enable protein retargeting experiments to be performed by constructing appropriate protein chimeras between the apical HA molecule and the basolateral G protein. Such experiments would provide direct evidence for an active transport signal.

Roth et al (1983) studied the expression of a cloned HA cDNA molecule in a primary monkey kidney cell culture. When the monkey cells were seeded at high density on a solid support and allowed to grow, there arose islands of polarized cells, as judged by polar maturation of control VSV and influenza virus, and by the presence of tight junction-like structures. Infection of such cell cultures with SV40 recombinant virus harboring the HA cDNA resulted in HA protein expression at the apical cell surface, as seen during influenza virus infection. This demonstrates that the HA molecule rather than some other viral component present during an influenza infection carries the information for polar transport.

In a second study, McQueen et al (1984) constructed a protein chimera that contained the N-terminal portion of the HA polypeptide fused to the C-terminal part of the G protein to be used in a retargeting experiment in the monkey cells. In theory, such a chimera should be directed to the basolateral surface if the transport signal resides in the cytoplasmic domain of these proteins. In practice, however, this chimera failed to leave the ER when tested in a nonpolar monkey cell line (CV1). The inability of the chimeric molecule to be transported from the ER might be related to problems with the normal trimerization process of the HA molecule (Wilson et al 1981).

In our laboratory we have approached the problem of signals for polar protein transport in epithelial cells by preparing stable transformants of the MDCK cell line. So far several MDCK lines expressing viral membrane glycoproteins have been established, but the level of protein expression is very low, making quantitative biochemical analyses difficult. This does not appear to result from an incompatible promotor/enhancer sequence in our DNA constructs, as the same SV40 early gene region was very successful in the transformation of MDCK cells when combined with the chicken oviduct lysozyme cDNA (Kondor-Koch et al 1985). We suspect that the viral proteins are toxic to the cells, and we are presently trying to circumvent this problem by using inducible promotor elements with a very low basal level of protein expression.

New HA-G chimeras will be constructed in the future, with emphasis on the preservation of the trimeric structure of the HA (e.g. an exact replacement of the HA cytoplasmic domain with that of G). If these chimeras are expressed on the cell surface then analysis of their transport in either the monkey cell system using recombinant SV40 virus, or preferably in the MDCK cells using inducible promotor elements, should indicate whether or not the cytoplasmic domain of these surface proteins is involved in polarized transport.

Possible Signals for Directing Proteins Along the Regulated Pathway of Protein Secretion

Cells of tissues specialized for either endocrine or exocrine secretion have evolved a way of routing the vast majority of their secretory products from the Golgi into secretory granules. The secretory products are stored in these granules until the cell receives an appropriate stimulation, which effects discharge of the content through fusion of the granule membrane with that of the cell surface. The mechanisms by which the secretory proteins are segregated into granules are unknown. However, from the work of Moore et al (1983b) it appears that this process is conserved among secretory endocrine cells, at least between pancreas β cells and corticotropic cells of the anterior pituitary. For these studies Moore and colleagues prepared a corticotropic tumor cell line (AtT20) stably transformed with the insulin gene linked to the SV40 early promotor. They demonstrated that the proinsulin was cleaved into mature protein, and secreted along the same regulated pathway as taken by endogenous ACTH (Gumbiner & Kelly 1982). In the future it will be interesting to determine whether exocrine secretory proteins (e.g. pancreas zymogens) also share the same transport mechanism. The propeptides (reviewed in Docherty & Steiner 1982; Douglas et al 1984), which are characteristic of most endocrine hormones, represent potential targeting regions that could be studied by appropriate engineering of cloned hormone cDNAs and subsequent expression in AtT20 cells, for example.

THE MANNOSE 6-PHOSPHATE SIGNAL OF LYSOSOMAL HYDROLASES

In animal cells the mannose 6-phosphate residue represents the essential component of a recognition marker involved in the routing of hydrolases to the lysosomes. (Creek & Sly 1984; Goldberg et al 1984). The phosphorylation itself represents a modification of a cotranslational glycosylation event. It has recently been studied in detail by Lang et al (1984) using an in vitro assay. They found that the phosphorylation of lysosomal enzymes could be competitively and specifically inhibited by the addition of deglycosylated lysosomal enzymes. Furthermore, proteolytic fragments or denatured forms of these hydrolases did not inhibit phosphorylation. This suggests that lysosomal enzymes contain a peptide signal or determinant recognized by the phosphorylating enzyme, and that this determinant is active only on the native enzyme. Further localization and characterization of this signal, e.g. by a recombinant DNA approach, may prove difficult for several reasons. First, a tentative localization of the determinant at or around the glycosylation site on the polypeptide chain might be very erroneous because the phosphorylated mannose residue(s) is on the outer branches of large oligosaccharides, which are a considerable distance from the linkage asparagine. Second, if the determinant is present only on the properly folded protein it could be made up of discontinuous stretches of amino acids.

TARGETING OF PROTEINS INTO THE NUCLEUS

The nucleoplasm is separated from the cytoplasm by a double membrane, the nuclear envelope. This membrane is perforated by specialized openings (pore complexes) which facilitate the extensive RNA and protein transport between the nucleus and cytosol. Studies with extranuclear soluble proteins have shown that the pores act as a molecular sieve with an exclusion limit of about 70 kilodaltons (kDa) (Bonner 1975a,b). However, the entry of proteins into the nucleus cannot be entirely explained by diffusion through the pore complexes because some of these proteins are of very large molecular weight (M_r) (Dingwall et al 1982), and many proteins with small M_r enter the nucleus with much faster kinetics than expected from molecular diffusion only (Bonner 1975a,b). Nuclear proteins are therefore believed to carry specific structural features or signals that facilitate or mediate their passage through the pore complex. A second signal needed for the accumulation of the nuclear proteins in the nucleoplasm has also been postulated. Such a signal might mediate binding to some nondiffusible elements of the nucleus, e.g. the chromatin and the nuclear lamina.

Several groups have recently used the recombinant DNA approach to demonstrate the presence of targeting peptides on various nuclear proteins. The proteins studied are the large T antigen of SV40, the nuclear protein (NP) of influenza virus, and the $\alpha2$ protein of yeast.

The T antigen is a well characterized protein involved in cell transformation, tumorigenicity, and SV40 DNA replication (Tooze 1981). Following synthesis in transformed cells, or during SV40 virus infection, the T antigen accumulates efficiently in the cell nucleus. A preliminary localization of the signal involved in its targeting was aided by several observations (Lewis et al 1969; Lewis & Rowe 1971, 1973; Seif et al 1979; Pipas et al 1983; Lanford & Butel 1984), the most significant of which was that of Lanford & Butel. They showed that an SV40 mutant that accumulated T antigen in the cytoplasm had amino acid 128 (Lys) of the T antigen replaced with Asn.

Kalderon et al (1984a,b) used a recombinant DNA approach to show that the peptide Pro (126)-Lys-Lys-Lys-Arg-Lys-Val (132) acts as a nuclear location signal. The T antigen polypeptide region around Lys (128) was first mapped extensively for its targeting function by introducing deletions internally into the T sequence, and expressing the respective mutants in Rat 1 cells using microinjection. The cytoplasmic (and) or nuclear accumulation of the mutant proteins was evaluated from the pattern and intensity of immunofluorescent staining. They then fused the putative signal region to two carrier proteins (chicken muscle type M1 pyruvate kinase and β-galactosidase) to test for targeting activity into the nucleus, and to further minimize the size of the active signal region (Figure 2e). The fact that the nuclear targeting signal was found to be active when transposed from its normal location around Lys (128) to the amino terminus of the T antigen, or to the two carrier proteins demonstrates that it possesses a great degree of autonomy regarding its location within a protein.

The nuclear targeting signal of the T antigen represents the first characterized example of a signal that is not located at either end of the polypeptide chain, and that is active in the folded protein. It will be most important to further test its autonomy in other protein sequences so as to understand the functional basis for it. Interestingly, the signal region is predicted to form a short α-helix comprising amino acids 127–131 (Lanford & Butel 1984). On each side of this region there are Pro residues which may be important to the autonomy of the α-helical region. All of the active constructs tested by Kalderon & colleagues maintained at least one flanking pro residue.

Davey et al (1985) studied the nuclear targeting of the influenza virus nuclear protein (NP) in frog oocytes. By analyzing the transport behavior of C- and N-terminal NP deletion mutants the authors defined a polypeptide region (amino acids 327–345) important for nuclear accumulation of NP. The role of this region in targeting NP was supported by the finding that only those C-terminal deletion mutants that contained the proposed location signal could direct the chimpanzee α-globin to the nucleus when fused to the N-terminus of the latter. The proposed signal region does not have a tract of basic residues as that of T antigen, but it contains a sequence Arg (342)-Val-Leu-Ser-Phe-Ile-Arg (348) that has some similarity (a short stretch of hydrophobic amino acids flanked by basic ones) with the targeting sequence proposed by Hall et al (see below).

A third nuclear targeting peptide has been proposed by Hall et al (1984) for the presumptive nuclear protein α2 in yeast. The authors fused 3-, 13-, 25- and 67-residue N-terminal peptides of this protein, as well as the complete α2 chain, to the β-galactosidase, and expressed the chimeras in yeast cells. Immunofluorescence analyses of the transformed yeast cell lines showed that the β-galactosidase was preferentially localized in the nuclear compartment when fused to more than 13 N-terminal amino acid residues of the α2 protein. However, these results should be interpreted with some caution since the β-galactosidase itself appears to enter the nucleus in animal cells (Kalderon et al 1984b). Hall & colleagues propose that the sequence Lys(3)-Ile-Pro-Ile-Lys(7) constitutes a nuclear targeting sequence because a similar one (two positively charged residues flanking three hydrophobic ones, one of which is Pro) is found in several other nuclear proteins, although not always in an N-terminal position. Certainly many more nuclear targeting sequences will be characterized in the near future, and it will be most interesting to see whether these kinds of generalizations can be substantiated.

TARGETING OF PROTEINS INTO MITOCHONDRIA

In contrast to nearly all other cell organelles (except the chloroplast, see below) mitochondria have the capacity to synthesize their own proteins. However, these account for only a minor fraction of the protein species present in this organelle. Most of the mitochondrial proteins are coded for by nuclear genes, and synthesized on cytosolic ribosomes. The import of these proteins necessitates the recognition of specific structural features in the proteins by receptor molecules on the outer membrane surface of the mitochondria. Recent studies have shown that many nuclear-encoded proteins destined for the matrix, inner-membrane, or intermembrane space of the mitochondria are synthesized as preproteins that have an N-terminal extension (Schatz & Butow 1983; Hay et al 1984). This peptide is cleaved upon import into the mitochondria by a matrix protease. The enzyme has been solubilized from mitochondria, and used to convert preproteins into their mature form (Gasser et al 1982). In vitro reconstitution experiments have shown that proteins lacking the prepiece are unable to enter the mitochondria. This argues strongly for a targeting function of the prepiece.

Hurt et al (1984) have fused the 22 N-terminal amino acids of the 25-residue prepiece of the yeast cytochrome C oxidase subunit IV protein (a matrix protein) to the mouse dihydrofolate reductase (DHFR) (a cytosolic protein), and demonstrated the targeting of DHFR into mitochondrial matrix using in vitro systems (Figure 2f). This shows that the prepiece of this mitochondrial protein has all the information necessary for routing a protein into the mitochondrial matrix. Surprisingly, the prepiece was cleaved even though the natural cleavage site had been deleted. It was postulated that the peptidase specifically recognizes the prepiece structure, but cleaves relatively nonspecifically at a nearby peptide bond.

In another study, the N-terminal fragment (amino acids 1–350) of the yeast F_1-ATPase α-subunit preform has been used to import the *E. coli* β-galactosidase protein into mitochondria (Douglas et al 1984). The pre form of the α-subunit is 479 residues long, and is normally imported into the mitochondrial matrix with the concomitant release of a 2 kDa prepiece. Analysis of the chimeric protein in transformed yeast cells showed that it was protected from proteinase K degradation, but its exact intra-mitochondrial location was not determined. Most important, however, was the finding that the transformed cells became respiration negative. One explanation for this phenotype is that the hybrid protein interferes with the ATPase. The transformant might be useful in the future for the isolation of mutants in the protein-import machinery of mitochondria, in the same way as the conditionally lethal phenotype of some export protein-*lac Z* gene fusions was exploited to characterize the protein translocation event in procaryotes.

A 70 kDa protein of the yeast mitochondrial outer membrane has also been studied with respect to targeting signals, using a gene deletion and fusion approach (Hase et al 1984). This protein lacks a cleavable prepiece. Analyses of a large number of internal deletion mutants suggest that both the targeting and membrane-binding functions reside in the 41 N-terminal amino acids. This region includes a stretch of 28 uncharged residues, previously proposed to function as a membrane-anchoring peptide (Riezman et al 1983a,b; Hase et al 1983). The dual action of the amino terminal peptides was directly demonstrated by the binding of a 70 kDa protein–β-galactosidase chimera to the outer mitochondria membrane (Figure 2*g*). In this chimera the 61 N-terminal amino acids from the 70 kDa protein were fused to the β-galactosidase.

TARGETING OF PROTEINS INTO CHLOROPLASTS

The chloroplasts of plant cells contain three distinct membranes (the outer, inner, and thylakoid membranes) that delimit three separate compartments: the intermembrane space, the stroma, and the thylakoid space. As in the mitochondria, a small fraction of chloroplast proteins are synthesized within the organelle, whereas the majority are imported from the cytosol by a posttranslational mechanism (Ellis et al 1980; Schmidt et al 1980). One example of an imported protein is the small subunit (ss) of the ribulose 1,5-bisphosphate carboxylase. It is synthesized in the cytosol with a 57-amino acid prepiece, which is cleaved in the chloroplast stroma by a protease (Bedbrook et al 1980). Van den Broeck et al (1985) have demonstrated that the localization signal is confined to this prepiece.

In the Van den Broeck et al experiment the gene region encoding the entire prepiece and the first amino acid (Met) of the small carboxylase subunit was fused via a linker molecule to the second codon of the neomycin phosphotransferase gene (Figure 2h). In this way the prepiece cleavage site was preserved. The fused gene was inserted downstream from the normal ss promoter, and then used for an *Agrobacterium*-mediated transformation of tobacco cells. Biochemical analyses demonstrated that nearly all the phosphotransferase expressed from the fused gene was cleaved into "mature" protein. This strongly suggests that the preprotein entered the chloroplast stroma. Direct proof was obtained by protease digestion and suborganelle fractionation experiments performed with isolated chloroplasts.

CONCLUSIONS AND FUTURE DIRECTIONS

The recombinant DNA approach has already demonstrated its usefulness in the localization and characterization of signals that are involved in the process of cotranslational insertion of proteins into the ER, and in the targeting of proteins into the nucleus, mitochondria, and chloroplasts. As more nuclear, chloroplast, and mitochondrial targeting signals are characterized it will be interesting to see whether any common features are revealed among signals for the same organelle. This could be used as an indication of the number of different kinds of signal-receptor interactions involved in each of these transport events. These common features may be found in the predicted secondary structure of the signal sequences, or as a characteristic distribution of particular amino acid residues, like the tract of basic residues in the SV40 T antigen signal. Another way to study the same question would be to use the sequence information in a targeting signal for the chemical synthesis of a homologous peptide, which could then be used in an in vitro system as an inhibitor of targeting, to test for the generality of a particular targeting mechanism. The peptide could also be used to elicit an antibody that could replace the peptide in such an inhibition assay.

In the case of cotranslational signals, it will be interesting to find out whether the different ways by which polypeptides become positioned in the ER membrane can be explained by the existence of a subset of only three types of signals: an N-terminal insertion signal, an insertion membrane-binding signal, and a stop-transfer signal (see Figure 1). This will require the localization and the characterization of the permanent insertion signals for many more membrane proteins belonging to Groups A, B, and C. The definitive experiment will be one in which multiple gene regions encoding insertion and stop-transfer signals are engineered into a given gene, and the expressed polypeptide is shown to become dispositioned back and forth in the ER membrane in the expected way. In the future it may be possible to develop fast and convenient biological assays that selectively screen for functional cotranslational signals. For instance, one could possibly devise a translocation signal "trap" by exploiting the dependence of the bacterio-phage f1 replication on the biosynthesis of its minor coat protein (Marvin & Hohn 1969; Webster & Lopez 1984). The minor coat protein (424 amino acid residues long) has a cleavable signal peptide (18 residues) that mediates the translocation of the major part of the polypeptide chain across the *E. coli* inner membrane during infection. Close to its C-terminus there is a stop translocation signal (a stretch of 23 uncharged residues) that binds the chain to the membrane, leaving 5 residues on the cytoplasmic side. If the authentic signal sequence of the minor coat protein is deleted on the double-stranded replicative form of the f1 DNA, and this is used for *E. coli* transformation, then the expressed chain is not expected to be translocated, and infectious phage should not be formed. In the proposed assay, a cDNA molecule encoding a protein with a permanent insertion signal is frag-mented into small pieces, and then used for ligation into the site of the deletion in the f1 genome. Only when a functional insertion sequence becomes incorporated will replicative phage be produced.

The use of the recombinant DNA approach to the study of transport signals for proteins routed along the constitutive cell surface pathway has been difficult for several reasons. First, we must consider the possibility that receptor-mediated mechanisms are not required to include proteins in cell surface transport vesicles at all. The lack of such mechanisms might be tested for by introducing the genes for procaryotic periplasmic (e.g. *Staphylococcus aureus* nuclease and *E. coli* β-lactamase) and inner mem-brane proteins (e.g. the minor coat protein) into eucaryotic cells, and analyzing the behavior of these proteins in the exocytic pathway. Prokaryotic proteins of this kind are known to carry insertion signals that can mediate their translocation across the ER membrane (see for instance Lingappa et al 1984), but they should be devoid of any transport signals needed for exocytosis. The first experiment along these lines has already been done by Wiedmann et al. (1984). They injected β-lactamase RNA, which had been synthesized in vitro, into frog oocytes, and showed that this periplasmic protein was secreted into the medium in an enzymatically active form. Clearly, if procaryotic proteins are generally routed to the cell surface with similar efficiency and kinetics as normal eucaryotic proteins, then it would cause serious doubt as to the existence of transport signals for constitutive exocytosis. The procaryotic proteins could also be tested in polarized epithelial cells. In this case the question would be whether only one of the two constitutive cell surface pathways present in these cells requires a specific transport or sorting signal, and the other pathway follows the same nonselective process postulated for nonpolar cells.

A second problem in this field has been the apparent ease with which altered proteins are trapped in the ER (see for instance constructs made by McQueen et al 1984; Bos et al 1984; Sveda et al 1984). The same phenotype is also typical of mutants isolated by selection in vivo or found as natural variants (Krangel et al 1982; Arias et al 1983; Wu et al 1983; Gallione & Rose 1985). Furthermore, proteins that differ in their cell surface transport kinetics have been shown to vary in the rate at which they are removed from the ER. This is true for both intact proteins and proteins whose transport to the surface has been inhibited by perturbations of their carbohydrate or polypeptide structure (Hickman & Kornfeld 1978; Elbein 1981; Novick & Schekman 1983; Lemansky et al 1984; Lodish & Hong 1984). Thus it appears that protein export from the ER is very sensitive to some physical features of the protein (e.g. solubility), and that these same features can easily be disturbed by structural alterations of the protein. For future experiments it is therefore important to plan the genetic engineering so as to leave both protein oligomerization and domain structure as intact as possible. The problem of protein oligomerization can be avoided by working with monomeric protein probes. However, among cell surface proteins it is not easy to find any whose monomeric state has been unequivocally demonstrated. The class I histocompatibility antigens could be considered for engineering purposes that exclusively involve the transmembrane and cytoplasmic protein domains. The heavy chain of this protein binds the $\beta2$ microglobulin on its lumenal (exterior) part, whereas it is "monomeric" in its spanning and cytoplasmic portions (Henning et al 1976).

If proteins do not require any transport signal for routing along the constitutive cell surface pathway, then it is important to study those features of the ER- and Golgi-specific proteins that are responsible for their retention in their respective organelles. The alteration or deletion of the "retention signals" by genetic manipulation should result in the release of these proteins into the exocytic pathway. The cDNAs for two ER-specific proteins were recently characterized; they are the E1 glycoprotein of the mouse Corona virus (Armstrong et al 1984), and the 3-hydroxy 3-methyl-glutaryl coenzyme A reductase of a Chinese hamster ovary cell line (Chin et al 1984). The deduced amino acid sequences for these proteins, together with earlier biochemical data, suggest that they both have a small N-terminal lumenal domain, a multispanning membrane-binding domain, and a large cytoplasmic domain, part of which includes the C-terminus of the chain. There are as yet no cloned genes or cDNAs for Golgi-specific proteins. An attractive candidate for cloning is the galactosyl transferase molecule. This enzyme is an integral membrane protein with most of its mass on the lumenal side of Golgi membranes (Fleischer & Smigel 1978; Fleischer 1981; Strous & Berger 1982; Strous et al 1983a). It has been localized using immunocytochemical techniques in trans-Golgi cisternae (Roth & Berger 1982). Apart from the interest in its possible retention signal, the expression of this molecule from cloned cDNA might provide investigators with the means for specifically isolating the distal part of the Golgi complex. The latter compartment is especially interesting because many protein sorting events are throught to occur there (e.g. those of apical and basolateral proteins, proteins for regulated secretion, and lysosomal hydrolases).

Another interesting field of protein transport that can be explored using recombinant DNA technology is receptor-mediated endocytosis. The receptor for the low density lipoprotein (LDL), and possibly the one for the transferring molecule, is clustered into coated pits at the cell surface, internalized, and recycled back to the PM, even without a bound ligand. This suggests that the signals involved are expressed by the receptor molecule alone, rather than by the ligand-receptor complex (Anderson et al 1982; Hopkins & Trowbridge 1983; Hopkins 1985; Watts 1985). The LDL and the transferrin receptors are transmembrane proteins belonging to Group A and B, respectively, and their mRNAs have been cloned as cDNA molecules. Thus it seems a straightforward task to use recombinant DNA techniques to test the involvement of their cytoplasmic and transmembrane protein domains in the process of receptor inclusion into coated pits at the cell surface. The characterization of a defective LDL receptor of a human patient has already demonstrated that the complete lack of these domains result in partially secreted receptor molecules (Lehrman et al 1985). Those receptors that remain membrane-bound were shown to bind ligand but not to cluster into coated pits. Using the recombinant DNA approach it should be possible to replace the transmembrane and cytoplasmic domains of cell surface marker proteins normally not included into coated pits with those of the receptor molecule, and test for endocytosis and recycling activity.

ACKNOWLEDGMENT
I would like to thank Joyce de Bruyn, Annie Steiner, and Heide Seifert for excellent secretarial assistance, and Laurie Roman, Kai Simons, Paul Melancon, Claudia Kondor-Koch, Daniel Cutler, David Meyer, and Graham Warren for critical reading of the manuscript.

Literature Cited

Ahlquist, P., Janda, M. 1984. cDNA cloning and in vitro transcription of the complete brome mosaic virus genome. *Mol. Cell. Biol.* 4:2876–82

Alt, F. W., Bothwell, A. L. M., Knapp, M., Siden, E., Mather, E., et al. 1980. Synthesis of secreted and membrane-bound immunoglobulin Mu heavy chains is directed by mRNAs that differ at their 3′ ends. *Cell* 20:293–301

Amann, E., Bröker, M., Worm, F. 1984. Expression of Herpes Simplex virus type 1 glycoprotein C antigens in *Eschericia coli*. *Gene* 32:203–15

Anderson, D. J., Mostov, K. E., Blobel, G. 1983. Mechanisms of integration of *de novo*-synthesized polypeptides into membranes: Signal-recognition particle is required for integration into microsomal membranes of calcium ATPase and of lens MP26 but not of cytochrome b_5. *Proc. Natl. Acad. Sci. USA* 80:7249–53

Anderson, R. G. W., Brown, M. S., Beisiegel, U., Goldstein, J. L. 1982. Surface distribution and recycling of the low density lipoprotein receptor as visualized with antireceptor antibodies. *J. Cell Biol.* 93:523–31

Arce-Gomez, B., Jones, E. A., Barnstable, C. J., Solomon, E., Bodmer, W. F. 1978. The genetic control of HLA-A and B antigens in somatic cell hybrids: Requirement for β_2 microglobulin. *Tissue Antigens* 11:96–112

Arias, C., Bell, J. R., Lenches, E. M., Strauss, E. G., Strauss, J. H. 1983. Sequence analysis of two mutants of Sindbis virus defective in the intracellular transport of their glycoproteins. *J. Mol. Biol.* 168:87–102

Armstrong, J., Niemann, H., Smeekens, S., Rottier, P., Warren, G. 1984. Sequence and topology of a model intracellular membrane protein, E1 glycoprotein, from a coronavirus. *Nature* 308:751–52

Armstrong, J., Perham, R. N., Walker, J. E. 1981. Domain structure of bacteriophage fed adsorption protein. *FEBS Lett.* 135:167–72

Arnold, B., Burgert, H.-G., Hamann, U., Hämmerling, G., Kees, U. 1984. Cytolytic T cells recognize the two amino-terminal domains of H-2 antigens in tandem in influenza A infected cells. *Cell* 38:79–87

Bankaitis, V. A., Rasmussen, B. A., Bassford, P. J. Jr. 1984. Intragenic suppressor mutations that restore export of maltose binding protein with a truncated signal peptide. *Cell* 37:243–52

Bedbrook, J. R., Smith, S. M., Ellis, R. J. 1980. Molecular cloning and sequencing of cDNA encoding the precursor to the small subunit of chloroplast ribulose-1,5-bisphosphate carboxylase. *Nature* 287:692–97

Bedouelle, H., Bassford, P. J., Jr. Fowler, A. V., Zabin, I., Beckwith, J. 1980. Mutations which alter the function of the signal sequence of the maltose binding protein of *Escherichia coli*. *Nature* 285:78–81

Bergmann, J. E., Singer, S. J. 1983. Immunoelectron microscopic studies of the intracellular transport of the membrane glycoprotein (G) of vesicular stomatitis virus in infected chinese hamster ovary cells. *J. Cell Biol.* 97:1777–87

Blobel, G. 1980. Intracellular protein topogenesis. *Proc. Natl. Acad. Sci. U.S.A.* 77(3):1496–1500

Blobel, G., Dobberstein, B. 1975. Transfer of proteins across membranes. II. Reconstitution of functional rough microsomes from heterologous components. *J. Cell Biol.* 67:852–62

Blochlinger, K., Diggelmann, H. 1984. Hygromycin B phosphotransferase as a selectable marker for DNA transfer experiments with higher eucaryotic cells. *Mol. Cell Biol.* 4:2929–31

Blok, J., Air, G. M., Laver, W. G., Ward, C. W., Lilley, G. G. 1982. Studies on the size, chemical composition and partial sequence of the neuraminidase from type A influenza viruses show that the N-terminal region of the NA is not processed and serves to anchor the NA in the viral membrane. *Virology* 119:109–21

Boeke, J. D., Model, P. 1982. A prokaryotic membrane anchor sequence: Carboxyl terminus of bacteriophage f1 gene III protein retains it in the membrane. *Proc. Natl. Acad. Sci. USA* 79:5200–4

Bonner, W. M. 1975a. Protein migration into nuclei. I. Frog oocyte nuclei accumulate microinjected histones, allow entry to small proteins, and exclude large proteins. *J. Cell Biol.* 64:421–30

Bonner, W. M. 1975b. Protein migration into nuclei. II. Frog oocyte nuclei accumulate a class of microinjected oocyte nuclear proteins and exclude a class of oocyte cytoplasmic proteins. *J. Cell Biol.* 64:431–37

Bos, T. J., Davis, A. R., Nayak, D. P. 1984. NH_2-terminal hydrophobic region of influenza virus neuraminidase provides the signal function in translocation. *Proc. Natl. Acad. Sci. USA* 81:2327–31

Braell, W. A., Lodish, H. F. 1982. Ovalbumin utilizes an NH_2-terminal signal sequence. *J. Biol. Chem.* 257:4578–82

Brennan, S. O., Carrell, R. N. 1978. A circulating variant of human proalbumin. *Nature* 274:908–9

Briles, E. B., Li, E., Kornfeld, S. 1977. Isolation of wheat germ agglutinin-resistant clones of Chinese hamster ovary cells deficient in membrane sialic acid and galactose. *J. Biol. Chem.* 252:1106–16

Burrows, P., LeJeune, M., Kearney, I. F. 1979. Evidence that murine pre-B cells synthesise μ heavy chains but no light chains. *Nature* 280:838–40

Cepko, C. L., Roberts, B. E., Mulligan, R. C. 1984. Construction and applications of a highly transmissible murine retrovirus shuttle vector. *Cell* 37:1053–62

Cheng, H.-L., Blattner, F. R., Fitzmaurice, L., Mushinski, J. F., Tucker, P. W. 1982. Structure of genes for membrane and secreted murine IgD heavy chains. *Nature* 296:410–15

Chin, D. J., Gil, G., Russell, D. W., Liscum, L., Luskey, K. L., et al. 1984. Nucleotide sequence of 3-hydroxy-3-methyl-gluaryl coenzyme A reductase, a glycoprotein of endoplasmic reticulum. *Nature* 308:613–17

Chu, F. K., Trimble, R. B., Maley, F. 1978. The effect of carbohydrate depletion on the properties of yeast external invertase. *J. Biol. Chem.* 253:8691–93

Claesson, L., Larhammar, D., Rask, L., Peterson, P. A. 1983. cDNA clone for the human invariant chain of class II histocompatibility antigens and its implications for the protein structure. *Proc. Natl. Acad. Sci. USA* 80:7395–99

Colman, A., Cutler, D., Krieg, P., Valle, G. 1983. The oocyte as a secretory cell. In *Molecular Biology of Egg Maturation. Ciba Symp.* 98:248–67

Colman, A., Morser, J. 1979. Export of proteins from oocytes of *Xenopus laevis. Cell* 17:517–26

Cone, R. D., Mulligan, R. C. 1984. High-efficiency gene transfer into mammalian cells: Generation of helper-free recombinant retrovirus with broad mammalian host range. *Proc. Natl. Acad. Sci. USA* 81:6349–53

Contreras, R., Cheroutre, H., Degrave, W., Fiers, W. 1982. Simple, efficient in vitro synthesis of capped RNA for direct expression of cloned eukaryotic genes. *Nucleic Acids Res.* 10:6353–62

Creek, K. E., Sly, W. S. 1984. *Lysosomes in Pathology and Biology,* ed. J. T. Dingle, R. T. Dean, W. Sly, pp. 63–82. New York: Elsevier/North-Holland

Crimmins, D. L., Mehard, W. B., Schlesinger, S. 1983. Physical properties of a soluble form of the glycoprotein of vesicular stomatitis virus at neutral and acidic pH. *Biochemistry* 22:5790–96

Cutler, D., Lane, C., Colman, A. 1981. Nonparallel kinetics and the role of tissue-specific factors in the secretion of chicken ovalbumin and lysozyme from *Xenopus* oocytes. *J. Mol. Biol.* 153:917–31

Davey, J., Dimmock, N. J., Colman, A. 1985. Identification of the sequence responsible for the nuclear accumulation of the influenza virus nucleoprotein in *Xenopus* oocytes. In press

De Robertis, E. M. 1983. Nucleocytoplasmic segregation of proteins and RNAs. *Cell* 32:1021–25

Dingwall, C., Sharnick, S. V., Laskey, R. A. 1982. A polypeptide domain that specifies migration of nucleoplasmin into the nucleus. *Cell* 30:449–58

Docherty, K., Steiner, D. F. 1982. Post-translational proteolysis in polypeptide hormone biosynthesis. *Ann. Rev. Physiol.* 44:625–38

Douglas, M. G., Geller, B. L., Emr, S. D. 1984. Intracellular targeting and import of an F_1-ATPase β-subunit-β-galactosidase hybrid protein into yeast mitochondria. *Proc. Natl. Acad. Sci. USA* 81:3983–87

Douglass, J., Civelli, O., Herbert, E. 1984. Polyprotein gene expression: Generation of diversity of neuroendocrine peptides. *Ann. Rev. Biochem.* 84:665–715

Dubovi, E. J., Wagner, R. R. 1977. Spatial relationships of the proteins of vesicular stomatitis virus: Induction of reversible oligomers by cleavable protein cross-linkers and oxidation. *J. Virol.* 22:500–9

Elbein, A. D. 1981. The tunicamycins—useful tools for studies on glycoproteins. *Trends Biochem. Sci.* 6:219–21

Ellis, R. J., Smith, S. M., Barraclough, R. 1980. In *Genome Organization and Expression in Plants,* ed. C. J. Leaver, pp. 321–35. New York: Plenum

Emr, S. D., Hedgpeth, J., Clement, J.-M., Sillhavy, T. J., Hofnung, M. 1980. Sequence analysis of mutations that prevent export of λ receptor, an *Escherichia coli* outer membrane protein. *Nature* 285:82–85

Emr, S. D., Schauer, I., Hansen, W., Esmon, P., Schekman, R. 1984. Invertase β-galactosidase hybrid proteins fail to be transported from the endoplasmic reticulum in *Saccharomyces cerevisiae. Mol. Cell Biol.* 4:2347–55

Emr, S. D., Silhavy, T. J. 1983. Importance of secondary structure in the signal sequence for protein secretion. *Proc. Natl. Acad. Sci. USA* 80:4599–4603

Farquhar, M. G., Palade, G. E. 1981. The Golgi apparatus (complex)—(1954–1981)—from artifact to center stage. *J. Cell Biol.* 91:77s–103s

Ferro-Novick, S., Novick, P., Field C., Schekman, R. 1984. Yeast secretory mutants that block the formation of active cell surface enzymes. *J. Cell Biol.* 98:35–43

Finkelstein, D. B., Strausberg, S. 1983. Heat shock-regulated production of *Escherichia coli β*-galactosidase in *Saccharomyces cerevisiae*. *Mol. Cell Biol.* 3:1625–33

Fitting, T., Kabat, D. 1982. Evidence for a glycoprotein "signal" involved in transport between subcellular organelles. *J. Biol. Chem.* 257:14011–17

Fleischer, B. 1981. Orientation of glycoprotein galactosyltransferase and sialyltransferase enzymes in vesicles derived from rat liver Golgi apparatus. *J. Cell Biol.* 89:246–55

Fleischer, B., Smigel, M. 1978. Solubilization and properties of galactosyltransferase and sulfotransferase activities of Golgi membranes in Triton X-100. *J. Biol. Chem.* 253:1632–38

Fries, E., Gustafsson, L., Peterson, P. A. 1984. Four secretory proteins synthesized by hepatocytes are transported from endoplasmic reticulum to Golgi complex at different rates. *EMBO J.* 3:147–52

Gallione, C. J., Rose, J. K. 1985. A single amino acid substitution in a hydrophobic domain causes temperature-sensitive cell-surface transport of a mutant viral glycoprotein. Submitted to *J. Virol.*

Garoff, H., Kondor-Koch, C., Pettersson, R., Burke, B. 1983. Expression of Semliki Forest virus proteins from cloned complementary DNA. II. The membrane-spanning glycoprotein E2 is transported to the cell surface without its normal cytoplasmic domain. *J. Cell Biol.* 97:652–58

Garoff, H., Kondor-Koch, C., Riedel, H. 1982. Structure and assembly of alpha viruses. In *Current Topics in Microbiology and Immunology*, ed. M. Cooper, W. Henle, P. H. Hofschneider, H. Koprowski, F. Melchers, et al, Vol. 99, pp. 1–50. Berlin: Springer-Verlag

Garoff, H., Simons, K., Dobberstein, B. 1978. Assembly of the Semliki Forest virus membrane glycoproteins in the membrane of the endoplasmic reticulum in vitro. *J. Mol. Biol.* 124:587–600

Gasser, S. M., Daum, G., Schatz, G. 1982. Import of proteins into mitochondria. *J. Biol. Chem.* 257:13034–41

Gething, M.-J., Sambrook, J. 1982. Construction of influenza haemagglutinin genes that code for intracellular and secreted forms of the protein. *Nature* 300:598–603

Gibson, R., Schlesinger, S., Kornfield, S. 1979. The nonglycosylated glycoprotein of vesicular stomatitis virus is temperature-sensitive and undergoes intracellular aggregation at elevated temperatures. *J. Biol. Chem.* 254:3600–7

Gilmore, R., Blobel, G. 1983. Transient involvement of signal recognition particle and its receptor in the microsomal membrane prior to protein translocation. *Cell* 35:677–85

Gilmore, R., Blobel, G., Walter, P. 1982a. Protein translocation across the endoplasmic reticulum. I. Detection in the microsomal membrane of a receptor for the signal recognition particle. *J. Cell Biol.* 95:463–69

Gilmore, R., Walter, P., Blobel, G. 1982b. Protein translocation across the endoplasmic reticulum. II. Isolation and characterization of the signal recognition particle receptor. *J. Cell Biol.* 95:470–77

Gluzman, Y. 1981. SV40-transformed simian cells support the replication of early SV-40 mutants. *Cell* 23:175–82

Goldberg, D., Gabel, C., Kornfeld, S. 1984. *Lysosomes in Pathology and Biology*, ed. J. T. Dingle, R. T. Dean, W. Sly, pp. 45–62. New York: Elsevier/North-Holland

Gonzales-Noriega, A., Grubb, J. H., Talkad, V., Sly, W. S. 1980. Chloroquine inhibits lysosomal enzyme pinocytosis and enhances lysosomal enzyme secretion by impairing receptor recycling. *J. Cell Biol.* 85:839–52

Graham, F. L., van der Eb, A. J. 1973. A new technique for the assay of infectivity of human adenovirus 5 DNA. *Virology* 52:456–67

Green, J., Griffiths, G., Louvard, D., Quinn, P., Warren, G. 1981a. Passage of viral membrane proteins through the Golgi complex. *J. Mol. Biol.* 152:663–98

Green, R. F., Meiss, H. K., Rodriguez-Boulan, E. 1981b. Glycosylation does not determine segregation of viral envelope proteins in the plasma membrane of epithelial cells. *J. Cell Biol.* 89:230–39

Gruss, P. 1984. Magic enhancers? *DNA* 3:1–5

Gruss, P., Rosenthal, N., König, M., Ellis, R. W., Shih, T. Y., et al. 1982. The expression of viral and cellular p21 *ras* genes using SV40 as a vector. In *Eukaryotic Viral Vectors*, ed. Y. Gluzman, pp. 13–18. Cold Spring Harbor, NY: Cold Spring Harbor Lab.

Guan, J.-L., Rose, J. K. 1984. Conversion of a secretory protein into a transmembrane protein results in its transport to the Golgi complex but not to the cell surface. *Cell* 37:779–87

Guarente, L., Ptashne, M. 1981. Fusion of *E. coli* lacZ to the cytochrome c gene of *S. cerevisiae*. *Proc. Natl. Acad. Sci. USA* 78:2199–2203

Gumbiner, B., Kelly, R. B. 1982. Two distinct intracellular pathways transport secretory and membrane glycoproteins to the surface of pituitary tumor cells. *Cell* 28:51–59

Gurdon, J. B. 1970. Nuclear transplantation and the control of gene activity in animal development. *Proc. R. Soc. London Ser. B* 176:303–14

Haguenauer-Tsapis, R., Hinnen, A. 1984. A deletion that includes the signal peptidase cleavage site impairs processing, glycosylation, and secretion of cell surface yeast acid phosphatase. *Mol. Cell Biol.* 4:2668–75

Hall, M. N., Hereford, L., Herskowitz, I. 1984. Targeting of *E. coli* β-galactosidase to the nucleus in yeast. *Cell* 36:1057–65

Hamer, D. H., Leder, P. 1979. Expression of the chromosomal mouse β^{maj}-globin gene cloned in SV40. *Nature* 281:35–40

Hase, T., Müller, U., Riezman, H., Schatz, G. 1984. A 70-kd protein of the yeast mitochondrial outer membrane is targeted and anchored via its extreme amino terminus. *EMBO J.* 3:3157–61

Hase, T., Riezman, H., Suda, K., Schatz, G. 1983. Import of proteins into mitochondria: nucleotide sequence of the gene for a 70-kd protein of the yeast mitochondrial outer membrane. *EMBO J.* 2:2169–72

Hashimoto, K., Erdel, S., Keränen, S., Saraste, J., Kääriäinen, L. 1981. Evidence for a separate signal sequence for the carboxy-terminal envelope glycoprotein E1 of Semliki Forest virus. *J. Virol.* 38:34–40

Hay, R., Böhni, P., Gasser, S. M. 1984. *Biochim. Biophys. Acta* 779:65–87

Henning, R., Milner, R. J., Reske, K., Cunningham, B. A., Edelman, G. M. 1976. Subunit structure, cell surface orientation, and partial amino-acid sequences of murine histocompatibility antigens. *Proc. Natl. Acad. Sci. USA* 73:118–22

Hickman, S., Kornfeld, S. 1978. Effect of tunicamycin on IgM, IgA, and IgG secretion by mouse plasmacytoma cells. *J. Immunol.* 121:990–6

Hicks, J. B., Strathern, J. N., Klar, A. J. S., Dellaporta, S. L. 1982. Cloning by complementation in yeast: The mating type genes. In *Genetic Engineering: Principles and Methods*, Vol. 4, pp. 219–48, ed. J. K. Setlow, A. Hollaender. London: Plenum

Hiti, A. L., Nyak, D. P. 1982. Complete nucleotide sequence of the neuraminidase gene of human influenza virus A/WSN33. *J. Virol.* 41:730–34

Holland, E. C., Leung, J. O., Drickamer, K. 1984. Rat liver asialoglycoprotein receptor lacks a cleavable NH$_2$-terminal signal sequence. *Proc. Natl. Acad. Sci. USA* 81:7338–42

Hopkins, C. R. 1985. The appearance and internalization of transferrin receptors at the margins of spreading human tumor cells. *Cell* 40:199–208

Hopkins, C. R., Trowbridge, I. S. 1983. Internalization and processing of transferrin and the transferrin receptor in human carcinoma A431 cells. *J. Cell Biol.* 97:508–21

Hortin, G., Boime, I. 1981a. Miscleavage at the presequence of rat preprolactin synthesized in pituitary cells incubated with a threonine analog. *Cell* 24:453–61

Hortin, G., Boime, I. 1981b. Transport of an uncleaved preprotein into the endoplasmic reticulum of rat pituitary cells. *J. Biol. Chem.* 256:1491–94

Hortsch, M., Griffiths, G., Meyer, D. 1985. The docking protein is restricted to the rough endoplasmic reticulum. Submitted for publication.

Hortsch, M., Meyer, D. 1984. Pushing the signal hypothesis: What are the limits? *Biol. Cell* 52:1–8

Hurt, E. C., Pesold-Hurt, B., Schatz, G. 1984. The cleavable prepiece of an imported mitochondrial protein is sufficient to direct cytosolic dihydrofolate reductase into the mitochondrial matrix. *FEBS Lett.* 178:306–10

Inouye, M., Halegoua, S. 1979. Secretion and membrane localization of proteins in *Escherichia coli*. *CRC Crit. Rev. Biochem.* 7:339–71

Inouye, S., Hsu, C.-P., Itakura, K., Inouye, M. 1983. Requirement for signal peptide cleavage of *Escherichia coli* prolipoprotein. *Science* 221:59–61

Inouye, S., Soberon, X., Franceschini, T., Nakamura, K., Itakura, K., et al. 1982. Role of positive charge on the aminoterminal region of the signal peptide in protein secretion across the membrane. *Proc. Natl. Acad. Sci. USA* 79:3438–41

Inouye, S., Vlasuk, G. P., Hsiung, H., Inouye, M. 1984. Effects of mutations at glycine residues in the hydrophobic region of the *Escherichia coli* prolipoprotein signal peptide on the secretion across the membrane. *J. Biol. Chem.* 259:3729–33

Jamieson, J., Palade, G. 1967a. Intracellular transport of secretory proteins in the pancreatic exocrine cell. I. Role of peripheral elements of the Golgi complex. *J. Cell Biol.* 34:577–96

Jamieson, J., Palade, G. 1967b. Intracellular transport of secretory proteins in the pancreatic exocrine cell. II. Transport to condensing vacuoles and zymogen granules. *J. Cell Biol.* 34:597–615

Kalderon, D., Richardson, W. D., Markham, A. F., Smith, A. E. 1984a. Sequence requirements for nuclear location of simian virus 40 large-T antigen. *Nature* 311:33–38

Kalderon, D., Roberts, B. L., Richardson, W. D., Smith, A. E. 1984b. A short amino acid sequence able to specify nuclear location. *Cell* 39:499–509

Karin, M., Haslinger, A., Holtgreve, H., Richards, R. I., Krauter, P., et al. 1984. Characterization of DNA sequences through which cadmium and glucocorticoid hormones induce human metallothionein-II$_A$ gene. *Nature* 308:513–19

Katona, A. E., Cuts, E., Wilson, J., Barton, M. 1978. α_1-Antitrypsin: The presence of excess mannose in the Z variant isolated from liver. *Science* 201:1229–32

Kaufman, J. F., Auffray, C., Korman, A. J., Shackelford, D. A., Strominger, J. 1984. The class II molecules of the human and murine major histocompatibility complex. *Cell* 36:1–13

Komaromy, M., Clayton, L., Rogers, J., Robertson, S., Kettman, J. 1983. The structure of the mouse immunoglobulin in γ_3 membrane gene segment. *Nucleic Acids Res.* 11:6775–85

Konarska, M. M., Padgett, R. A., Sharp, P. A. 1984. Recognition of cap structure in splicing in vitro of mRNA presursors. *Cell* 38:731–36

Kondor-Koch, C., Bravo, R., Fuller, S., Cutler, D., Garoff, H. 1985. Protein secretion in the polarized epithelial cell line, MDCK. Submitted for publication

Kondor-Koch, C., Burke, B., Garoff, H. 1983. Expression of Semliki Forest virus proteins from clonal complementary DNA. I. The fusion activity of the spike glycoprotein. *J. Cell Biol.* 97:644–51

Koren, R., Burstein, Y., Soreq, H. 1983. Synthetic leader peptide modulates secretion of proteins from microinjected *Xenopus* oocytes. *Proc. Natl. Acad. Sci. USA* 80:7205–9

Kozak, M. 1981. Mechanism of mRNA recognition by eukaryotic ribosomes during initiation of protein synthesis. *Curr. Top. Microbiol. Immunol.* 93:81–123

Kozak, M. 1983. Translation of insulin-related polypeptides from messenger RNAs with tandemly reiterated copies of the ribosome binding site. *Cell* 34:971–78

Krangel, M. S., Pious, D., Strominger, J. L. 1982. Human histocompatibility antigen mutants immunoselected in vitro. *J. Biol. Chem.* 257:5296–5305

Krieg, P. A., Melton, D. A. 1984. Functional messenger RNAs are produced by SP6 in vitro transcription of cloned cDNAs. *Nucleic Acids Res.* 12:7057–70

Krieg, P., Strachnan, R., Wallis, E., Tabe, L., Colman, A. 1984. Efficient expression of cloned complementary DNAs for secretory proteins after injection into *Xenopus* oocytes. *J. Mol. Biol.* 180:615–43

Lanford, E. R., Butel, J. S. 1984. Construction and characterization of an SV40 mutant defective in nuclear transport of T antigen. *Cell* 37:801–13

Lang, L., Reitman, M. R., Tang, J., Roberts, R. M., Kornfeld, S. 1984. Lysosomal Enzyme Phosphorylation. Recognition of a protein-dependent determinant allows specific phosphorylation of oligosaccharides present on lysosomal enzymes. *J. Biol. Chem.* 259:14663–71

Leavitt, R., Schlesinger, S., Kornfield, S. 1977. Impaired intracellular migration and altered solubility of nonglycosylated glycoproteins of vesicular stomatitis virus and Sindbis virus. *J. Biol. Chem.* 252:9018–23

Lehrman, M. A., Schneider, W. J., Südhof, T. C., Brown, M. S., Goldstein, J. L., et al. 1985. Mutation in LDL receptor: Alu-Alu recombination deletes exons encoding transmembrane and cytoplastic domains. *Science* 227:140–46

Lemansky, P., Gieselmann, V., Hasilik, A., von Figura, K. 1984. Cathepsin D and β-hexosaminidase synthesized in the presence of 1-deoxynojirimycin accumulate in the endoplasmic reticulum. *J. Biol. Chem.* 259:10129–35

Lewis, A. M. Jr., Levin, M. J., Wiese, H. W., Crumpacker, C. S., Henry, P. H. 1969. A nondefective (competent) adenovirus-SV40 hybrid isolated from the Ad2-SV40 hybrid population. *Proc. Natl. Acad. Sci. USA* 63:1128–35

Lewis, A. M. Jr., Rowe, W. P. 1971. Studies on nondefective adenovirus-simian virus 40 hybride viruses. I. A newly characterized simian virus 40 antigen induced by the Ad2+ND$_1$ virus. *J. Virol.* 7:189–97

Lewis, A. M. Jr., Rowe, W. P. 1973. Studies on nondefective adenovirus 2-simian virus 40 hybrid viruses. VIII. Association of simian virus 40 transplantation antigen with a specific region of the early viral genome. *J. Virol.* 12:836–40

Li, E., Kornfeld, S. 1978. Structure of the altered oligosaccharide present in glycoproteins from a clone of Chinese hamster ovary cells deficient in N-acetylglucosaminyltransferase activity. *J. Biol. Chem.* 253:6426–31

Lingappa, V. R., Chaidez, J., Yost, C. S., Hedgpeth, J. 1984. Determinants for protein localization: β-lactamase signal sequence directs globin across microsomal membranes. *Proc. Natl. Acad. Sci. USA* 81:456–60

Lingappa, V. R., Lingappa, J. R., Blobel, G. 1979. Chicken ovalbumin contains an internal signal sequence. *Nature* 281:117–21

Liu, C.-C., Simonsen, C. C., Levinson, A. D. 1984. Initiation of translation at internal AUG codons in mammalian cells. *Nature* 309:82–85

Lodish, H. F., Hong, N. 1984. Glucose removal from N-linked oligosaccharides is required for efficient maturation of certain secretory glycoproteins from the rough endoplasmic reticulum to the Golgi complex. *J. Cell Biol.* 98:1720–29

Lodish, H. F., Hong, N., Snider, M., Strous, G. J. A. M. 1983. Hepatoma secretory proteins migrate from rough endoplasmic reticulum to Golgi at characteristic rates. *Nature* 304:80–83

Lopata, M. A., Cleveland, D. W., Sollner-Webb, B. 1984. High level transient expression of a cholramphenicol acetyl transferase gene by DEAE-dextran mediated DNA transfection coupled with a dimethyl sulfoxide or glycerol shock treatment. *Nucleic Acids Res.* 12:5707–17

Majzoub, J. A., Rosenblatt, M., Fennick, B., Maunus, R., Kronenberg, H. M., et al. 1980. Synthetic pre-proparathyroid hormone leader sequence inhibits cell-free processing of placental, parathyroid, and pituitary prehormones. *J. Biol. Chem.* 255:11478–83

Mann, R., Mulligan, R. D., Baltimore, D. 1983. Construction of a retrovirus packaging mutant and its use to produce helper-free defective retrovirus. *Cell* 33:153–59

Markoff, L., Lin, B.-C., Sveda, M. M., Lai, C.-J. 1984. Glycosylation and surface expression of the influenza virus neuraminidase requires the N-terminal hydrophobic region. *Mol. Cell. Biol.* 4:8–16

Martinez-Arias, A. E., Casadaban, M. J. 1983. Fusion of the saccharomyces cerevisiea LEU2 gene to an *Escherichia coli* β-galactosidase gene. *Mol. Cell. Biol.* 3:580–86

Marvin, D. A., Hohn, B. 1969. Filamentous bacterial viruses. *Bacteriol. Rev.* 33:172–209

Matlin, K. S., Simons, K. 1984. Sorting of an apical plasma membrane glycoprotein occurs before it reaches the cell surface in cultured epithelial cells. *J. Cell Biol.* 99:2131–39

McCormick, F., Trahey, M., Innis, M., Dieckmann, B., Ringold, G. 1984. Inducible expression of amplified human beta interferon genes in CHO cells. *Mol. Cell. Biol.* 4:166–72

McQueen, N. L., Nayak, D. P., Jones, L. V., Compans, R. W. 1984. Chimeric influenza virus hemagglutinin containing either the NH_2 terminus or the COOH terminus of G protein of vesicular stomatitis virus is defective in transport to the cell surface. *Proc. Natl. Acad. Sci. USA* 81:395–99

Meek, R. L., Walsh, K. A., Palmiter, R. D. 1982. The signal sequence of ovalbumin is located near the NH_2 terminus. *J. Biol. Chem.* 257:12245–51

Meyer, D. I., Krause, E., Dobberstein, B. 1982. Secretory protein translocation across membranes—the role of the "docking protein." *Nature* 297:647–50

Misek, D. E., Bard, E., Rodriguez-Boulan, E. 1984. Biogenesis of epithelial cell polarity: Intracellular sorting and vectorial exocytosis of an apical plasma membrane glycoprotein. *Cell* 39:537–46

Mitrani-Rosenbaum, S., Maroteaux, L., Mory, Y., Revel, M., Howley, P. M. 1983. Inducible expression of the human interferon β_1 gene linked to a bovine papilloma virus DNA vector and maintained extrachromosomally in mouse cells. *Mol. Cell. Biol.* 3:223–40

Moore, H.-P. H., Gumbiner, B., Kelly, R. B. 1983a. Chloroquine diverts ACTH from a regulated to a constitutive pathway in AtT-20 cells. *Nature* 302:434–36

Moore, H.-P. H., Walker, M. D., Lee, F., Kelly, R. B. 1983b. Expressing a human proinsulin cDNA in a mouse ACTH-secreting cell. Intracellular storage, proteolytic processing, and secretion on stimulation. *Cell* 35:531–38

Moss, B. 1977. Utilization of the guanylyltransferase amd methyltransferase of vaccinia virus to modify and identify the 5′-terminals of heterologous RNA species. *Biochem. Biophys. Res. Commun.* 74:374–83

Mulligan, R. C., Berg, P. 1981. Selection for animal cells that express the *Escherichia coli* gene coding for xanthine-guanine phosphoribosyltransferase. *Proc. Natl. Acad. Sci. USA* 78:2072–76

Mulligan, R. C., Howard, B. H., Berg, P. 1979. Synthesis of rabbit β-globin in cultured monkey cells following infection with a SV40 β-globin recombinant genome. *Nature* 277:108–14

Murre, C., Reiss, C. S., Bernabeu, C., Chen, L. B., Burakoff, S., et al. 1984. Construction, expression and recognition of an H-2 molecule lacking its carboxyl terminus. *Nature* 307:432–36

Neuberger, M. S., Williams, G. T., Fox, R. O. 1984. Recombinant antibodies possessing novel effector functions. *Nature* 312:604–8

Neufeld, E. F., Ashwell, G. 1980. In *The Biochemistry of Glycoproteins and Proteoglycans*, ed. W. J. Lennarz, pp. 252–57, New York: Plenum

Nielsen, D. A., Chou, J., MacKrell, A. J., Casadaban, M. J., Steiner, D. F. 1983. Expression of a preproinsulin-β-galactosidase gene fusion in mammalian cells. *Proc. Natl. Acad. Sci. USA* 80:5198–5202

Novick, P., Ferro, S., Schekman, R. 1981. Order of events in the yeast secretory pathway. *Cell* 25:461–69

Novick, P., Schekman, R. 1983. Export of major cell surface proteins is blocked in yeast secretory mutants. *J. Cell Biol.* 96:541–47

Ochi, A., Hawley, R. G., Shulman, M. J., Hozumi, N. 1983. Transfer of a cloned immunoglobulin light-chain gene to mutant hybridoma cells restores specific antibody production. *Nature* 302:340–42

Osley, M. A., Hereford, L. 1982. Identification of a sequence responsible for periodic synthesis of yeast histone 2A mRNA. *Proc. Natl. Acad. Sci. USA* 79:7689–93

Palmiter, D. R., Gagnon, J., Walsh, K. A. 1978. Ovalbumin: a secreted protein without a transient hydrophobic leader sequence. *Proc. Natl. Acad. Sci. USA* 75:94–98

Paterson, B. M., Rosenberg, M. 1979. Efficient translation of prokaryotic mRNAs in eukaryotic cell-free system requires addition of a cap structure. *Nature* 919:692–96

Patzer, E. J., Nakamura, G. R., Yaffe, A. 1984. Intracellular transport and secretion of hepatitis B surface antigen in mammalian cells. *J. Virol.* 51:346–53

Pavlakis, G. N., Dean, H., Hamer, D. H. 1983. Regulation of a metallothionein-growth hormone hybrid gene in bovine papilloma virus. *Proc. Natl. Acad. Sci. USA* 80:397–401

Pettersson, R. F., Lundström, K., Chattopadhyaya, J. B., Josephson, S., Philipson, L., et al. 1983. Chemical synthesis and molecular cloning of a STOP oligonucleotide encoding an UGA translation terminator in all three reading frames. *Gene* 24:15–27

Pfeiffer, S., Fuller, S. D., Simons, K. 1985. Intracellular sorting and basolateral appearance of the G protein of vesicular stomatitis virus in MDCK cells. *J. Cell Biol.* In press

Pipas, J. M., Peden, K. W. D., Nathans, D. 1983. Mutational analysis of simian virus 40 T antigen: Isolation and characterization of mutants with deletions in the T-antigen gene. *Mol. Cell. Biol.* 3:203–13

Prehn, S., Tsamaloukas, A., Rapoport, T. A. 1980. Demonstration of specific receptors of the rough endoplasmic membrane for the signal sequence of carp pre-proinsulin. *Eur. J. Biochem.* 107:185–95

Reiss, B. 1982. PhD thesis. Univ. Heidelberg, Federal Republic of Germany

Reiss, C. S., Evans, G. A., Marqulies, D. H., Seidman, J. G., Burakoff, S. J. 1983. Allospecific and virus specific cytolytic T lymphocytes are restricted to the N or C1 domain of H-2 antigens expressed on L cells after DNA-mediated gene transfer. *Proc. Natl. Acad. Sci. USA* 80:2709–12

Reiss, B., Sprengel, R., Will, H., Schaller, H. 1984. A new sensitive method for qualitative and quantitative assay of neomycin phosphotransferase in crude cell extracts.

Gene 30:211–18

Riezman, H., Hay, R., Gasser, S., Daum, G., Schneider, G., et al. 1983a. The outer membrane of yeast mitochondria: Isolation of outside-out sealed vesicles. *EMBO J.* 2:1105–11

Riezman, H., Hay, R., Witte, C., Nelson, N., Schatz, G. 1983b. Yeast mitochondrial outer membrane specifically binds cytoplasmically-synthesized precursors of mitochondrial proteins. *EMBO J.* 2:1113–18

Rigby, P. W. J. 1982. Expression of cloned genes in eukaryotic cells using vector systems derived from viral replicons. In *Genetic Engineering 3*, ed. R. Williamson, pp. 83–141. New York: Academic

Ringold, G. M., Yamamoto, K. R., Bishop, J. M., Varmus, H. E. 1977. Glucocorticoid-stimulated accumulation of mouse mammary tumor virus RNA: Increased rate of synthesis of viral RNA. *Proc. Natl. Acad. Sci. USA* 74:2879–83

Rio, D. C., Clark, S. G., Tjian, R. 1985. A mammalian host-vector system that regulates expression and amplification of transfected genes by temperatures induction. *Science* 227:23–28

Rodriguez-Boulan, E., Sabatini, D. D. 1978. Asymmetric budding of viruses in epithelial monolayers: A model system for study of epithelial polarity. *Proc. Natl. Acad. Sci. USA* 75:5071–75

Rogers, J., Early, P., Carter, C., Calame, K., Bond, M., et al. 1980. Two mRNAs with different 3′ ends encode membrane-bound and secreted forms of immunoglobulin μ chain. *Cell* 20:303–12

Rose, J. K., Adams, G. A., Gallione, C. J. 1984. The presence of cysteine in the cytoplasmic domain of the vesicular stomatitis virus glycoprotein is required for palmitate addition. *Proc. Natl. Acad. Sci. USA* 81:2050–54

Rose, J. K., Bergmann, J. E. 1982. Expression from cloned cDNA of cell-surface and secreted forms of the glycoprotein of vesicular stomatitis virus in eucaryotic cells. *Cell* 30:753–62

Rose, J. K., Bergmann, J. E. 1983. Altered cytoplasmic domains affect intracellular transport of the vesicular stomatitis virus glycoprotein. *Cell* 34:513–24

Rose, M., Casadaban, M., Botstein, D. 1981. Yeast genes fused to β-galactosidase in *E. coli* can be expressed normally in yeast. *Proc. Natl. Acad. Sci. USA* 78:2460–64

Roth, J., Berger, E. G. 1982. Immunocytochemical localization of galactosyltransferase in HeLa cells: Codistribution with thiamine pyrophosphatase in trans-Golgi cisternae. *J. Cell Biol.* 92:223–29

Roth, M. G., Compans, R. W., Giusti, L., Damis, A. R., Nayak, D. P., et al. 1983. Influenza virus hemagglutinin expression is polarized in cells infected with recombinant SV40 viruses carrying cloned hemagglutinin DNA. *Cell* 33:435–43

Roth, M. G., Fitzpatrick, J. P., Compans, R. W. 1979. Polarity of influenza and vesicular stomatitis maturation in MDCK cells: Lack of requirement for glycosylation of viral glycoproteins. *Proc. Natl. Acad. Sci. USA* 76:6430–34

Rothman, J. E., Miller, R. L., Urbani, L. J. 1984. Intercompartmental transport in the Golgi complex is a dissociative process: Facile transfer of membrane protein between two Golgi populations. *J. Cell Biol.* 99:260–71

Rottier, P., Armstrong, J., Meyer, D. I. 1985. Signal recognition particle-dependent insertion of coronavirus, E1 an intracellular membrane glycoprotein. *J. Biol. Chem.* In press

Rubenstein, J. L. R., Chappell, T. G. 1983. Construction of a synthetic messenger RNA encoding a membrane protein. *J. Cell Biol.* 96:1464–69

Sabatini, D. D., Kreibich, G., Morimoto, T., Adesnik, M. 1982. Mechanisms for the incorporation of proteins in membranes and organelles. *J. Cell Biol.* 92:1–22

Sakaguchi, M., Mihara, K., Sato, R. 1984. Signal recognition particle is required for co-translational insertion of cytochrome P-450 into microsomal membranes. *Proc. Natl. Acad. Sci. USA* 81:3361–64

Saraste, J., Kuismanen, E. 1984. Pre- and post-Golgi vacuoles operate in the transport of Semliki Forest virus membrane glycoproteins to the cell surface. *Cell* 38:535–49

Schatz, G., Butow, R. A. 1983. How are proteins imported into mitochondria? *Cell* 32:316–18

Scheele, G., Tartakoff, A. 1985. Exit of non-glycosylated secretory proteins from the rough endoplastic reticulum is asynchronous in the exocrine pancreas. *J. Biol. Chem.* 260:926–31

Schmidt, G. W., Bartlett, S. G., Grossman, A. R., Cashmore, A. R., Chua, N.-H. 1980. In *Genome Organization and Expression in Plants*, ed. C. J. Leaver, pp. 337–51. New York: Plenum

Schneider, C., Owen, M. J., Banville, D., Williams, J. G. 1984. Primary structure of human transferrin receptor deduced from the mRNA sequence. *Nature* 311:675–78

Seif, I., Khoury, G., Dhar, R. 1979. The genome of human papovavirus BKV. *Cell* 18:963–77

Sekikawa, K., Lai, C.-J. 1983. Defects in functional expression of an influenza virus hemagglutinin lacking the signal peptide sequences. *Proc. Natl. Acad. Sci. USA* 80:3563–67

Severinsson, L., Peterson, P. A. 1984. β_2-microglobulin induces intracellular transport of human class I transplantation antigen heavy chains in *Xenopus laevis* oocytes. *J. Cell Biol.* 99:226–32

Sherry Hwang, L.-H., Gilboa, E. 1984. Expression of genes introduced into cells by retroviral infection is more efficient than that of genes introduced into cells by DNA transfection. *J. Virol.* 50:417–24

Silhavy, T. J., Bassford, P. J. Jr., Beckwith, J. R. 1979. A genetic approach to the study of protein localization in *Escherichia coli*. In *Bacterial Outer Membranes: Biogenesis and Functions*, pp. 203–54, ed. M. Inouye. New York: Wiley

Silhavy, T. J., Benson, S. A., Emr, S. D. 1983. Mechanisms of protein localization. *Microbiol. Rev.* 47:313–44

Singer, P. A., Lauer, W., Dembic, Z., Mayer, W. E., Lipp, J., et al. 1984. Structure of the murine Ia-associated invariant (Ii) chain as deduced from a cDNA clone. *EMBO J.* 3:873–77

Sorge, J., Wright, D., Erdman, V. D., Cutting, A. E. 1984. Amphotropic retrovirus vector system for human cell gene transfer. *Mol. Cell. Biol.* 4:1730–37

Southern, P. J., Berg, P. 1982. Transformation of mammalian cells to antibiotic resistance with a bacterial gene under control of the SV40 early region promoter. *J. Mol. Appl. Genet.* 1:327–41

Strous, G. J. A. M., Berger, E. G. 1982. Biosynthesis, intracellular transport, and release of the Golgi enzyme galactosyltransferase (lactose synethetase A protein) in HeLa cells. *J. Biol. Chem.* 257:7623–28

Strous, G. J., van Kerkof, P., Willemsen, R., Geuze, H. J., Berger, E. G. 1983a. Transport and topology of galactosyltransferase in endomembranes of HeLa cells. *J. Cell Biol.* 97:723–27

Strous, G. J. A. M., Willemsen, R., van Kerkhof, P., Slot, J. W., Geuze, H. J. 1983b. Vesicular stomatitis virus glycoprotein, albumin, and transferrin are transported to the cell surface via the same Golgi vesicles. *J. Cell Biol.* 97:1815–22

Strubin, M., Mach, B., Long, E. O. 1984. The complete sequence of the mRNA for the HLA-DR-associated invariant chain reveals a polypeptide with an unusual transmembrane polarity. *EMBO J.* 3:869–72

Stueber, D., Ibrahimi, I., Cutler, D., Dobberstein, B., Bujard, H. 1984. A novel in vitro transcription-translation system: Accurate and efficient synthesis of single proteins from cloned DNA sequences. *EMBO J.* 3:3143–48

Sussman, D. J., Milman, G. 1984. Short-term, high-efficiency expression of transfected DNA. *Mol. Cell. Biol.* 4:1641–43

Sveda, M. M., Markoff, L. J., Lai, C.-J. 1982. Cell surface expression of the influenza virus haemagglutinin requires the hydrophobic carboxy-terminal sequences. *Cell* 30:649–56

Sveda, M. M., Markoff, L. J., Lai, C.-J. 1984. Influenza virus hemagglutinin containing an altered hydrophobic carboxy terminus accumulates intracellularly. *J. Virol.* 49:223–28

Tabe, L., Krieg, P., Strachan, R., Jackson, D., Wallis, E., et al. 1984. Segregation of mutant ovalbumins and ovalbumin-globin fusion proteins in *Xenopus* oocytes: Identification of an ovalbumin signal sequence. *J. Mol. Biol.* 180:645–66

Talmadge, K., Brosius, J., Gilbert, W. 1981. An "internal" signal sequence directs secretion and processing of proinsulin in bacteria. *Nature* 294:176–78

Tooze, J. 1981. DNA tumor viruses. *Molecular Biology of Tumor Viruses*, pp. 799–841. Cold Spring Harbor, New York: Cold Spring Harbor Lab. 2nd. ed.

Tyler, B. M., Cowman, A. F., Gerondakis, S. D., Adams, J. M., Bernard, O. 1982. mRNA for surface immunoglobulin chains encodes a highly conserved transmembrane sequence and a 28-residue intracellular domain. *Proc. Natl. Acad. Sci. USA* 79:2008–12

Van den Broeck, G., Timko, M. P., Kausch, A. P., Cashmore, A. R., Van Montagu, M., et al. 1985. Targeting of a foreign protein to chloroplasts by fusion to the transit peptide from the small subunit of ribulose 1,5-bisphosphate carboxylase. *Nature* 313:358–63

Vlasuk, G. P., Inouye, S., Inouye, M. 1984. Effects of replacing serine and threonine residues within the signal peptide on the secretion of the major outer membrane lipoprotein of *Escherichia coli*. *J. Biol. Chem.* 259:6195–200

Walter, P., Blobel, G. 1981a. Translocation of proteins across the endoplasmic reticulum. II. Signal recognition protein (SRP) mediates the selective binding to microsomal membranes of in vitro assembled polysomes synthesizing secretory protein. *J. Cell Biol.* 91:551–56

Walter, P., Blobel, G. 1981b. Translocation of proteins across the endoplasmic reticulum. III. Signal recognition protein (SRP) causes signal sequence-dependent and site-specific arrest of chain elongation that is released by microsomal membranes. *J. Cell Biol.* 91:557–61

Walter, P., Gilmore, R., Blobel, G. 1984. Protein translocation across the endoplasmic reticulum. *Cell* 38:5–8

Walter, P., Ibrahimi, I., Blobel, G. 1981. Translocation of proteins across the endoplasmic reticulum. I. Signal recognition protein (SRP) binds to in vitro assembled polysomes synthesizing secretory protein. *J. Cell Biol.* 91:545–50

Warren, G. 1981. Membrane proteins: Structure and assembly. In *Membrane Structure*, ed. J. B. Finean, R. H. Michell, 1:215–257. New York: Elsevier

Watanabe, S., Temin, H. M. 1983. Construction of a helper cell line for avian reticuloendotheliosis virus cloning vectors. *Mol. Cell. Biol.* 3:2241–49

Watson, M. E. E. 1984. Compilation of published signal sequences. *Nucleic Acids Res.* 12:5144–64

Watts, C. 1985. Rapid endocytosis of the transferrin receptor in the absence of bound transferrin. *J. Cell Biol.* 100:633–37

Webster, R. E., Lopez, J. 1984. In *Virus Structure and Assembly*, ed. S. Casjen. Boston: Jones & Bartlett. In press

Wiedmann, M., Huth, A., Rapoport, T. A. 1984. *Xenopus* oocytes can secrete bacterial β-lactamase. *Nature* 309:637–39

Wigler, M., Silverstein, S., Lee, L.-S., Pellicer, A., Cheng, Y., et al. 1977. Transfer of purified herpes virus thymidine kinase gene to cultured mouse cells. *Cell* 11:223–32

Wigler, M., Sweet, R., Kee Sim, G., Wold, B., Pellicer, A., et al. 1979. Transformation of mammalian cells with genes from procaryotes and eucaryotes. *Cell* 16:777–85

Wills, J. W., Srinivas, R. V., Hunter, E. 1984. Mutations of the Rous sarcoma virus *env* gene that affect the transport and subcellular location of the glycoprotein products. *J. Cell Biol.* 99:2011–23

Wilson, I. A., Skehel, J. J., Wiley, D. C. 1981. Structure of the haemagglutinin membrane glycoprotein of influenza virus at 3 Å resolution. *Nature* 289:366–73

Wu, G. E., Hozumi, N., Murialdo, H. 1983. Secretion of a λ_2 immunoglobulin chain is prevented by a single amino acid substitution in its variable region. *Cell* 33:77–83

Yamawake-Kataoka, Y., Nakai, S., Miyata, T., Honjo, T. 1982. Nucleotide sequences of gene segments encoding membrane domains of immunoglobulin chains. *Proc. Natl. Acad. Sci. USA* 79:2623–27

Yoshida, A., Lieberman, J., Gaidulis, L., Ewing, C. 1976. Molecular abnormality of human alpha$_1$-antitrypsin variant (Pi-ZZ) associated with plasma activity deficiency. *Proc. Natl. Acad. Sci. USA* 73:1324–28

Yost, C. S., Hedgpeth, J., Lingappa, V. R.

1983. A stop transfer sequence confers predictable transmembrane orientation to a previously secreted protein in cell-free systems. *Cell* 34:759–66

Zambryski, P., Herrera-Estrella, L., De Block, M., Van Montagu, M., Schell, J. 1984. The use of the Ti plasmid of *Agrobacterium* to study the transfer and expression of foreign DNA in plant cells: New vectors and methods. In *Genetic Engineering, Principles and Methods*, ed. J. K. Setlow, A. Hollaender, Vol. 6, pp. 253–78. New York: Plenum

Zimmermann, R., Hennig, B., Neupert, W. 1981. Different transport pathways of individual precursor proteins in mitochondria. *Eur. J. Biochem.* 116:455–60

Ann. Rev. Cell Biol. 1985.1:447–88

PROGRESS IN UNRAVELING PATHWAYS OF GOLGI TRAFFIC

Marilyn Gist Farquhar

Department of Cell Biology, Yale University School of Medicine, 333 Cedar Street, New Haven, Connecticut 06510

CONTENTS

447

0743–4634/85/1115–0447$02.00

INTRODUCTION

It has become apparent that the Golgi complex is the intracellular site where biosynthetic, endocytic, and recycling membrane traffic converges and is sorted and directed to its correct intracellular or cell surface destinations. Indeed, sorting and intracellular traffic control appear to represent two of the major functions of this organelle. The realization of the key role played by the Golgi complex in intracellular traffic has led to an explosion of interest, and has stimulated considerable new research of a varied nature, which has enhanced greatly our knowledge and understanding of this once mysterious organelle.

In this review I outline recent progress in unraveling the pathways of Golgi traffic, paying particular attention to what is known about functional compartmentalization within this organelle. Emphasis is placed on findings obtained since 1980, and on earlier findings that made major contributions to our current concepts. The historical background and earlier work were summarized in previous reviews (Farquhar 1978b; Farquhar & Palade 1981). For different perspectives on the Golgi complex the interested reader can also refer to several other excellent recent reviews (Goldfischer 1982; Tartakoff 1980, 1983b).

BACKGROUND INFORMATION AND OVERVIEW

Biosynthetic Monopolies of the Golgi Complex

The best understood functions—in biochemical terms—of the Golgi complex are those connected with the transport and modification in transit of newly synthesized proteins. This organelle has an exclusive or nearly exclusive monopoly on terminal N-glycosylation, O-glycosylation, synthesis of glycosaminoglycans, sulfation, the phosphorylation of lysosomal enzymes, and concentration of secretory products. In addition, fatty acid acylation and proteolytic processing of biosynthetic products sometimes occurs there. Thus, it follows that (a) the transport systems and enzymes that effect these operations are resident proteins of Golgi membranes, and (b) any product that undergoes these modifications must traverse the appropriate subcompartment of the Golgi complex where the modifying proteins reside. The list of substrates that are channelled through the Golgi complex and successively modified in transit now includes secretory proteins, membrane proteins, proteoglycans, and lysosomal enzymes.

Organization of the Golgi Complex

Our current concept of the topography and organization of the Golgi complex based on morphological data is depicted in Figure 1. The key features of this organelle that have been known for some time are: (1) *Its characteristic stack* of 3–8 flattened cisternae; (2) *its compositional heterogeneity*, i.e. resident enzymes, detected by enzyme cytochemical or immunocytochemical procedures, are differentially distributed across the stack; and (3) *its polarity*—one side, the cis side, faces the endoplasmic reticulum (ER) and the other, the trans side, faces secretion granules or centrioles. Unfortunately, the sidedness (cis versus trans) of the stack is not always evident unless a cytochemical marker is used because some differentiated cells (e.g. fibroblasts, macrophages, plasma cells) and most cell lines in culture lack morphologically recognizable secretion granules, and their centrioles are not readily visible. The cytochemical markers localized so far and their common or "expected" distribution in the Golgi

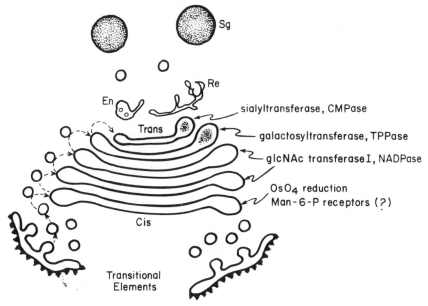

Figure 1   Stationary cisternae model of the Golgi complex. The membrane components that have been localized in situ, together with their most frequent localization in either cis, middle, or trans cisternae, are indicated on the *right*. Note that exceptions and functional modulations in these distributions have also been reported. The flow of biosynthetic products through the Golgi complex is diagrammed on the *left*. The main features of this model are that (1) each cisterna represents a separate subcompartment with a distinctive membrane composition and internal milieu; (2) products move vectorially from rough ER to transitional elements (part rough, part smooth) located on the cis side of the Golgi complex, and then unidirectionally across the stacks (cis to trans), traversing the cisternae one by one; (3) transport along the route is effected by vesicular carriers; and (4) the main flow of traffic is to the dilated rims of the cisternae. (See text for details.) (*Sg* = secretory granules; *En* = endosome; *Re* = reticular element.)

stack are as follows (Figure 1): Thiamine pyrophosphatase (TPPase), acid phosphatase (CMPase) (Novikoff 1976; Novikoff & Novikoff 1977), galactosyltransferase (Roth & Berger 1982), and sialyltransferase (Roth et al 1984) are most often detected in 1–2 trans cisternae; nicotinamide adenine dinucleotide phosphatase (NADPase) (Smith 1980), and GlcNAc transferase I (Dunphy et al 1985) are usually attributed to the middle cisternae; and strong reducing agents, detected by OsO_4 reduction (Friend & Murray 1965; Friend 1969) are most often found in 1–2 cis cisternae. However, exceptions have also been reported (see below).

Flow of Biosynthetic Traffic to and through the Golgi Complex

Several models have been put forth to explain Golgi traffic flow. Those that have received the most attention are the "membrane flow" or "cisternal progression" model, the "distillation tower" hypothesis, and the "stationary cisternae" model.

"MEMBRANE FLOW" OR "CISTERNAL PROGRESSION" MODEL   According to this model both the membrane and contents of each cisterna form on the cis side of the Golgi stack and move in concert across the stack from cis to trans, with the cisternae being "used up" in packaging on the trans side. This idea can be traced to Grassé (1957) who, based on electron microscope findings, proposed that the formation of the peripheral (cis) Golgi saccules balances the conversion of central (trans) Golgi saccules into secretion granules. This concept was further developed and extended by Morré and his co-workers (Morré et al 1971; Morré & Ovtracht 1977; Morré et al 1979) under the name of "membrane flow." It prevailed until relatively recently (Morré et al 1979),[1] and was the framework upon which most investigators interpreted their findings for nearly thirty years. However, the validity of this model was questioned (Farquhar 1978b) because it was difficult to reconcile with (1) compositional differences detected by cytochemical procedures in Golgi cisternae; (2) the demonstration that protein components of secretion granule membranes turn over much more slowly than their contents (Meldolesi 1974; Wallach et al 1975; Castle et al 1975); and (3) evidence that granule membranes are recovered and returned to the trans Golgi cisternae, presumably to be reutilized in packaging (Herzog & Farquhar 1977; Farquhar 1978a, 1981, 1982; Herzog & Miller 1979; Herzog & Reggio 1980; Ottosen et al 1980). Accordingly, this model has few supporters at present and has been replaced by a modified version (see below).

---

[1] See also discussion in *Ciba Found. Symp. Membrane Recycling*, 1982, Vol. 92, pp. 174–82.

DISTILLATION TOWER HYPOTHESIS   This model was proposed by Rothman (1981) to explain and rationalize the organization of the Golgi stack. According to him, the main function of the Golgi stack was to act sequentially to refine the protein export of the ER by removing "escaped" ER proteins. The Golgi complex was viewed as two separate compartments, cis and trans, and refinement was seen as a multistage process akin to fractional distillation. The cis Golgi cisternae were viewed as plates in this distillation tower, and the trans Golgi cisternae were believed to be the receivers that collect refined fractions from the rest of the Golgi complex. This model implies that there is a gradient of ER proteins across the cis Golgi cisternae. However, it has been shown that ER proteins are present in very low concentrations in Golgi membranes (Matsura et al 1978; Aoi et al 1981), and that some are effectively excluded (Lucocq et al 1984; Tashiro et al 1984). In many cases their presence in Golgi fractions can probably be accounted for by contamination with ER membranes. In the single case where an ER component (NADPH-cytochrome P450 reductase) has been clearly detected in bona fide Golgi elements (Ito & Palade 1978), its presence can probably be accounted for by recycling (Farquhar & Palade 1981). This model has recently been withdrawn by Rothman (personal communication) because he and his collaborators were unable to detect an ER marker (cytochrome P-450) in Golgi stacks by immunocytochemistry. Moreover, they did not find any evidence of Golgi-associated oligosaccharide processing of ER proteins (glucosidase II and hexose-6-phosphate dehydrogenase), as might be expected if these ER components reached Golgi cisternae (Brands et al 1985).

STATIONARY CISTERNAE MODEL   This model was originally formulated in 1978 (Farquhar 1978b) and later elaborated on (Farquhar & Palade 1981) to take into account the distinctive composition of individual Golgi cisternae and the existence of membrane recycling. It assumes that (1) only the products move vectorially across the stacks; (2) the individual cisternae remain stationary; (3) transport between the cisternae is effected by specific vesicular carriers; and (4) the main flow of vesicular traffic is to the dilated rims of the Golgi cisternae (Figure 1).

At present this model appears to be widely accepted, and efforts are currently directed toward localizing different functions within the stack, defining the properties of individual Golgi subcompartments, and determining the nature and routing of traffic to these subcompartments.

CURRENT FLOW DIAGRAM   Thus, after some years of controversy and conflicting models we now begin to have a general, widely accepted idea of how biosynthetic traffic flows to and through the Golgi complex, and how the Golgi processing enzymes are distributed along the route (Figure 1).

The key features of the flow diagram of the stationary cisternae model are that (1) intracellular transport is vectorial, proceeding from rough ER to transitional elements on the cis side of the Golgi to cis Golgi cisternae, followed by sequential movement across the stack in the cis to trans direction; (2) the transport route consists of a number of separate compartments of distinctive composition arranged in polarized series; and (3) the transport of newly synthesized products along the route is effected by vesicles that shuttle back and forth between donor and recipient compartments, i.e. from ER to Golgi cisternae, between adjacent cisternae within the Golgi stack, from Golgi cisternae to various destinations (secretion granules, lysosomes, different domains of the plasmalemma, etc).

DIVERSE SOURCES OF EVIDENCE FOR THE CURRENT FLOW DIAGRAM   Due to the inherent limitations of any given technique and the complicated topography of this organelle, the basis of our present flow diagram necessarily comes from convergent information obtained using a variety of approaches, including autoradiography, cytochemistry, cell fractionation and, most recently, in vitro systems and somatic cell fusions. Each of these approaches, used singly or in combination, has provided crucial information on Golgi structure and function. Autoradiography provided the first evidence that addition of terminal sugars (galactose) to glycoproteins takes place in this organelle, and gave the first indication that Golgi cisternae are involved in the transport of secretory proteins (Neutra & Leblond 1966; Whur et al 1969).

In the 1960s, enzyme cytochemistry provided the first demonstrations of compositional heterogeneity between different Golgi cisternae (see Farquhar & Palade 1981; Goldfischer 1982), and later on, within individual Golgi cisternae (Farquhar et al 1974). It also provided the first evidence that secretory products traverse the Golgi cisternae (Bainton & Farquhar 1970; Herzog & Miller 1970), and that the middle Golgi cisternae have a membrane composition distinct from that of cis or trans cisternae (Smith 1980). So far, immunocytochemistry has provided the only direct evidence for the cis to trans movement of products across the Golgi stacks (Bergmann et al 1981; Bergmann & Singer 1983; Saraste & Hedman 1983). Additionally, with the demonstration that galactosyltransferase is confined to 1–2 trans Golgi cisternae (Roth & Berger 1982), immunocytochemistry provided a key piece of evidence that has made possible the integration of cell fractionation and morphological data.

Cell fractionation coupled with autoradiography provided the first evidence that secretory proteins are transported vectorially along the secretory route [ER → Golgi → condensing vacuoles → secretion granules

(Jamieson & Palade 1967a,b)]. Analysis of subcellular fractions provided the initial indication that many glycosyltransferases and the enzymes involved in the processing of lysosomal proteins are more concentrated in Golgi than in ER membranes: first galactosyltransferase (Fleischer et al 1969), and subsequently sialyltransferase and GlcNAc transferase I (Schachter et al 1970), fucosyltransferase (Munro et al 1975), mannosidase I (Tabas & Kornfeld 1979), lysosomal enzyme N-acetylglucosamine-1-phosphotransferase (Waheed et al 1982), and phosphodiester glycosidase (Varki & Kornfeld 1980; Waheed et al 1982) were found to be more concentrated in smooth than in rough microsomal fractions. It should be mentioned parenthetically that for years galactosyltransferase has been the standard Golgi marker against which all other activities have been compared, even though it marks only certain (trans) Golgi elements.

More recently, results obtained by analysis of microsomal fractions on analytical sucrose density gradients, an approach introduced by Rothman and his colleagues (Rothman 1981; Dunphy et al 1981), provided the first evidence that different glycoprotein processing enzymes are present in Golgi subfractions of different density, and that newly synthesized membrane proteins (VSV G protein) pass sequentially from smooth membrane fractions of high density to those of lower density.

Quite recently, use of cell-free systems (Balch et al 1984a,b; Braell et al 1984), and cell hybrids obtained by fusing wild-type cells with mutants lacking specific carbohydrate processing enzymes (Rothman et al 1984a,b), has provided new information on sequential transport between Golgi subcompartments and the conditions required to effect it.

To summarize, convergent information obtained by many investigators using a variety of approaches made possible the construction of our current flow diagram for biosynthetic traffic through the Golgi complex.

Vesicular Transport

The concept of vesicular transport along the biosynthetic route had its origins in the discovery of exocytosis (Palade 1959), and has been extended to include traffic from the ER to Golgi elements (Jamieson & Palade 1967a; Rothman et al 1980), traffic from the Golgi to lysosomes (Friend & Farquhar 1967; Sly & Fischer 1982; Campbell & Rome 1983; Brown & Farquhar 1984a,b,c; Brown et al 1984), transport of membrane proteins to the plasmalemma (Leblond & Bennett 1976; Rothman et al 1980; Wehland et al 1982; Rindler et al 1982), and, most recently, intra-Golgi traffic (Farquhar 1978b; Farquhar & Palade 1981; Rothman et al 1984a,b; Balch et al 1984a,b). Inherent to the concept of transport by shuttling vesicles is that of extensive reutilization or recycling of vesicular membranes, and there is now indirect evidence for recycling along several of these routes, i.e. Golgi to plasmalemma (Farquhar 1981, 1982, 1983) and Golgi to lysosomes (Brown et al 1984; Brown & Farquhar 1984c).

N-Glycosylation Processing Steps Used to Determine Compartmentation Along the Secretory Pathway

Until recently it was very difficult, if not impossible, to relate cell fractionation data and the multiplicity of functions ascribed to the Golgi complex to its complicated structure. As information has become available on the steps involved in N-linked glycosylation, and with the identification and purification of many of the enzymes involved, a cohesive body of cytochemical and immunocytochemical evidence as well as biochemical data derived from analysis of cell fractions, has accumulated and has provided a basis for correlation of Golgi structure and function. Because synthesis of N-linked oligosaccharides is a sequential, multistep process involving a whole battery of co- and posttranslational modifications (Hubbard & Ivatt 1981; Lennarz 1983), the modifications provide markers for the passage of proteins through the cell compartments that contain the modifying enzymes (Dunphy et al 1981; Dunphy & Rothman 1983; Goldberg & Kornfeld 1983; Dautry-Varsat & Lodish 1983).

STEPS IN N-GLYCOSYLATION   The sequence of events and the presumed intracellular location of these events is as follows (Figure 2): Addition of a preformed oligosaccharide (containing three glucose, nine mannose, and two GlcNAc residues) via a lipid-linked donor to the nascent peptide occurs cotranslationally in the rough ER, producing polymannose chains. Subsequently, all three glucose residues are trimmed, ostensibly while the protein is still in the rough ER because the enzymes involved (glucosidases I and II) are more concentrated in rough and smooth microsome than in Golgi fractions (Grinna & Robbins 1979; Goldberg & Kornfeld 1983). Moreover, glucosidase II has recently been found in rough and smooth ER but not in Golgi elements of rat hepatocytes by immunocytochemistry (Lucocq et al 1984). A series of further modifications of the newly synthesized glycoprotein then takes place: one to four mannoses are trimmed [by the combined action of an ER mannosidase and Golgi mannosidase I (Bischoff & Kornfeld 1983)], a GlcNAc is added (by GlcNAc transferase I), two additional mannose residues are trimmed (by mannosidase II), two more GlcNAc residues are added (by GlcNAc transferases II and IV), and galactose, sialic acid, and sometimes fucose residues are added (by appropriate fucosyl-, galactosyl-, and sialyltransferases).

All these steps except trimming of glucose and mannose residues are believed to occur exclusively in the Golgi complex because all the enzymes involved have been found to be more enriched in Golgi and/or smooth microsome fractions than in rough microsome fractions.

GLYCOPROTEIN PROCESSING ENZYMES CAN BE SEPARATED ON DENSITY GRADIENTS   An important finding was the discovery by Rothman and his associates (Rothman 1981; Dunphy et al 1981) that in CHO cells several of the glycoprotein processing enzymes are separable on analytical sucrose density gradients. Initially, kinetically early (mannosidase I) and kinetically late (galactosyltransferase, sialyltransferase) activities were found in heavy and light membrane fractions, respectively, which were suggested to represent two distinct Golgi subcompartments corresponding to cis and trans Golgi cisternae in situ. It was subsequently shown that in mouse lymphoma cells the order of fractionation (heavy to light) on sucrose gradients of a number of enzymes involved in glycoprotein processing is

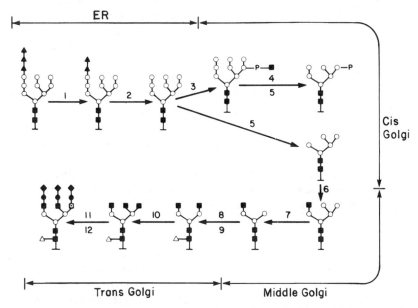

Figure 2   Steps in the processing of asparagine-linked oligosaccharides and their presumptive intracellular site. Steps for addition of Man-6-P to lysosomal enzymes are indicated in the side branch (3–5). 1 = glucosidase I; 2 = glucosidase II; 3 = lysosomal enzyme, N-acetylglucosaminylphosphotransferase; 4 = lysosomal enzyme, phosphodiester glycosidase; 5 = mannosidase I; 6 = GlcNAc transferase I; 7 = mannosidase II; 8 = GlcNAc transferase II; 9 = fucosyltransferase; 10 = GlcNAc transferase IV; 11 = galactosyltransferase; 12 = sialyltransferase. (Symbols: *dark triangles* = glucose; *squares* = GlcNAc; *open circles* = mannose; *dark circles* = galactose; *open triangles* = fucose; *diamonds* = sialic acid; P = phosphate.) Note: ER mannosidase is not indicated in the diagram. (From Goldberg & Kornfeld 1983.)

identical to the sequence in which these enzymes act in vivo (Goldberg & Kornfeld 1983). N-acetylglucosaminyl-1-phosphotransferase, which is involved in the modification of lysosomal enzymes, was present in the densest membranes; the lysosomal enzyme phosphodiester glycosidase was found in the next most dense region; GlcNAc transferases I and IV, mannosidase II, and fucosyltransferase were concentrated in fractions of intermediate density; and galactosyltransferase was present in the lightest membranes. These findings have led to the conclusion (Goldberg & Kornfeld 1983; Dunphy et al 1983; Quinn et al 1983; Dunphy et al 1985) that glycoprotein processing enzymes are spatially distributed sequentially across the stacks, with kinetically earlier processing steps (mannosidase I, fatty acid acylation) located in cis Golgi cisternae and kinetically later steps (terminal transferases) in the trans Golgi cisternae. The basis for the density differences is presumed to be differences in the cholesterol content of the corresponding membranes. Membrane thickness (Morré 1976), electron microscopic findings using filipin as a cholesterol probe (Orci et al 1981), and density shift experiments with digitonin (Borgese & Meldolesi 1980) all suggest that cholesterol is distributed in a gradient across the Golgi stacks, and that it increases from cis to trans.

Although this framework is logical and reasonable, it was initially largely hypothetical because up until very recently the only enzyme that had been localized in situ by immunocytochemistry was galactosyltransferase (Roth & Berger 1982). Even now only three more of the enzymes involved have been directly localized in situ [mannosidase II (Novikoff et al 1983), sialyltransferase (Roth et al 1984), and GlcNAc transferase I (Dunphy et al 1985)]. The assumed intra-Golgi localizations of sialyltransferase and GlcNAc transferase I (trans and medial cisternae, respectively) have been corroborated. In the case of mannosidase II, however, the enzyme was found to be distributed throughout the Golgi stack.

Clearly, it will be useful in the future to localize other transferases and trimming enzymes, as well as the specific transporters that serve to import sugar nucleotides into the Golgi stack (Fleischer 1983; Capasso & Hirschberg 1984). The latter are also considered to be Golgi transmembrane glycoproteins, and Capasso & Hirschberg (1984) have obtained evidence suggesting they serve as antiporters that mediate exchange of sugar nucleotides for their corresponding nucleoside monophosphate product. However, none of the antiporter proteins has yet been identified, characterized, and purified.

ACQUISITION OF ENDO H INSENSITIVITY USED TO MONITOR GOLGI TRANSPORT   Based on the distribution of the glycoprotein processing enzymes in cell fractions, i.e. glucosidases in rough microsomes and mannosidase I in smooth microsomes (Goldberg & Kornfeld 1983; Dunphy & Rothman 1983), and on results obtained with inhibitors of ER to Golgi transport (Godelaine et al 1981), it is believed that newly synthesized glycoproteins exit the ER and are transported to the Golgi complex shortly after removal of their three glucose residues (by glucosidases I and II) and one to three mannose residues (by ER mannosidase), and prior to removal of additional mannose residues by Golgi mannosidase I. Because the oligosaccharides remain sensitive to digestion with endoglycosidase H (Endo H) until acted on by GlcNAc transferase I and mannosidase II, loss of Endo H sensitivity is commonly used operationally in biosynthetic studies to indicate that the glycoprotein has been transported to the Golgi. More precisely, it indicates transport to the middle Golgi cisternae, where these two enzymes are presumed to reside (Dunphy et al 1985).

At present, acquisition of Endo H resistance and addition of galactose are being used to determine the temporal and thereby presumably the spatial relationships between the events that occur during N-glycosylation and other Golgi functions (O-glycosylation, synthesis of glycosamino-glycans, sulfation), about which we know much less (see below). In addition, various inhibitors of specific N-glycosylation steps, such as swainsonine which inhibits mannosidase II (Tulsiani et al 1982), and deoxynojirimycin which inhibits glucosidase II (Peyrieras et al 1983), are also being used to assess the effects of eliminating these processing steps on the fate of specific glycoproteins and on membrane traffic in general.

PATHWAYS TO AND THROUGH THE GOLGI COMPLEX

Traffic from the ER to the Golgi Complex

There is but a single known portal of entry into the Golgi complex, that is from the rough ER. As already mentioned, autoradiographic and cell fractionation data generated on the exocrine pancreas some time ago (Jamieson & Palade 1967a,b, 1968; Palade 1975) indicated (a) that the transport pathway is from the rough ER to transitional elements (part rough and part smooth) to swarms of vesicles found on the cis side of the Golgi complex; and (b) that these vesicles pinch off the transitional elements of the rough ER and carry secretory products to the Golgi complex (see Figure 1). The finding that secretory products can be detected in transitional elements and in peripheral (cis) Golgi vesicles by enzyme cytochemical (Bainton & Farquhar 1970) or immunocytochemical (Geuze et al 1979) procedures confirmed that these elements lie along the ER-to-Golgi transport pathway.

It was further shown that ER-to-Golgi transport is an energy requiring step (Jamieson & Palade 1968): When ATP synthesis was inhibited by respiratory inhibitors (N_2, cyanide, or antimycin A), or by inhibitors of oxidative phosphorylation (dinitrophenol or oligomycin), the secretory proteins remained in rough microsomes and did not enter smooth microsomal fractions, which consist primarily of Golgi-derived vesicles. Autoradiography demonstrated that in the presence of these inhibitors, secretory proteins accumulated in the transitional elements or peripheral (cis) Golgi vesicles, and did not reach Golgi cisternae or condensing vacuoles, which suggests that the energy requiring step is· vesicular transport (Jamieson & Palade 1968). Similar findings were also obtained later in plasma cells and myeloma cells using CCCP (carbonyl cyanide m-chlorophenylhydrazone), another inhibitor of oxidative phosphorylation (Tartakoff & Vassalli 1977). At present, CCCP is sometimes used operationally to distinguish ER from Golgi functions by inhibiting transport out of the ER (Godelaine et al 1981; Geetha-Habib et al 1984).

The assumption that the entry point is the cis, "forming" or "immature," face of the Golgi is an old one inherent in the cisternal progression or membrane flow model, and was based on inferential or indirect evidence until recently when the pathway taken by viral membrane glycoproteins through the Golgi was directly demonstrated by immunocytochemistry. Bergmann and co-workers (Bergmann et al 1981; Bergmann & Singer 1983) showed that when a temperature-sensitive VSV mutant was grown in CHO cells, VSV G protein was arrested in the ER at nonpermissive temperature, but when the cells were shifted to a permissive temperature, the G protein appeared first on the cis side of the Golgi and then moved sequentially across the stack, reaching the trans side quite rapidly (within 5–10 min). Similar results were also obtained by Saraste & Kuismanen (1984) on the membrane proteins of Semliki Forest virus. The latter authors also found that transport across the Golgi is arrested at 15°C; the viral glycoproteins accumulated in vacuoles on the cis side of the Golgi at 15, but not at 20°C.

Recently evidence has been obtained that newly synthesized membrane and secretory proteins exit the ER at different rates (Strous & Lodish 1980; Fitting & Kabat 1982; Lodish et al 1983; Fries et al 1984; Scheele & Tartakoff 1985). In one study, the half time of transport to the Golgi (based on acquisition of Endo H insensitivity) was found to vary from 14 to 137 min (Fries et al 1984).

To explain the finding that different membrane proteins are exported from the ER at different rates, it was postulated by Fitting & Kabat (1982) that the ER to Golgi transport step is selective, that newly synthesized proteins contain an amino acid signal required for their export from the ER, and that transport depends on their binding to a carrier protein. Similar conclusions were reached by others for secretory proteins (Lodish et al 1983; Fries et al 1984; Scheele & Tartakoff 1985). Lodish & Kong (1984) have suggested that glucose removal provides the transport signal in some cases, because inhibition of glucose removal with 1-deoxynojirimycin caused certain hepatoma glycoproteins (α-1 antitrypsin and α-2 chymotrypsin), but not others (α-1 trypsin and transferrin), to accumulate in rough microsome fractions. Clearly, glucose removal cannot constitute a universal ER exit signal because many proteins, primarily secretory proteins, are not glycosylated, and nonglycosylated proteins also exit the rough ER asynchronously (Scheele & Tartakoff 1985). Glucose removal cannot even be the common signal for transport of N-linked glycoproteins out of the ER because not all of the latter were affected by deoxynojirimycin treatment.

More work is needed to determine whether or not transport from the ER to the Golgi is mediated by a specific receptor or carrier and, if so, to determine what signals and receptors are involved and to rule out other possibilities, e.g. that binding of proteins to ER components or its content proteins, rather than binding to receptors in transport vesicles, controls retention in the ER.

Intra-Golgi Traffic

THE STACKED CISTERNAE PROVIDE SEPARATE SUBCOMPARTMENTS   Upon leaving the rough ER and being transported to the cis Golgi cisternae, newly synthesized proteins are assumed to be transported unidirectionally (cis to trans), also by vesicular transport, across the stacked cisternae in which the enzyme and transport systems responsible for posttranslational modifications of proteins are distributed in polarized series. Because of their different composition (based on cytochemical markers) and the fact that connections between cisternae are rarely if ever seen, each cisterna (or small group of cisternae) can be assumed to represent a separate subcompartment (Farquhar & Palade 1981). Thus, the pile of stacked but separate cisternae, the characteristic organizational feature of this organelle, can now be rationalized as the solution the cell has devised to provide a series of subcompartments of distinctive composition and function whose succession in space corresponds to the succession of modifying operations in time. The advantage of this arrangement is that the internal environments of each cisterna (pH, ionic composition, substrate concentration) can vary as required by different resident enzyme and transport functions, and be maintained separate. That the internal milieu of individual Golgi cisternae differs was already suggested some time ago by the finding that reduced OsO_4 is deposited exclusively in some (cis) Golgi cisternae but not others (middle and trans) (Friend & Murray 1965; Friend 1969), and more recently by the demonstration that the ionophore monesin affects some but not all Golgi functions (Tartakoff 1983a; Griffiths et al 1983; Quinn et al 1983), and that some Golgi elements, namely trans cisternae and secretory vacuoles, have a lower pH than others (Anderson & Pathak 1985).

It was proposed some time ago (Farquhar 1978b; Farquhar & Palade 1981) that transport between individual Golgi cisternae comprising distinctive subcompartments is effected by vesicular carriers that shuttle from one subcompartment to another. Biochemical evidence for vesicular transport has been provided by the novel experiments of Rothman & co-workers (1984a,b) in which glycosylation of VSV G protein was studied in somatic cell hybrids produced by fusion of glycosylation-defective mutant CHO cells lines. When mutant cells capable of incorporating GlcNAc but unable to add galactose were infected with VSV and fused with another mutant deficient in GlcNAc transferase I (unable to incorporate GlcNAc but able to add galactose), galactosylation of G protein took place normally. This indicates that there had been inter-Golgi transport, i.e. the G protein that has just received GlcNAc in the Golgi complex of one cell was effectively transported to that of the other cell. Remarkably, similar results were obtained in a cell-free system: when postnuclear supernatants of VSV-infected mutants lacking GlcNAc transferase I were mixed with postnuclear supernatants from wild-type cells in the presence of ATP, an ATP-generating system, UDP-[³H]GlcNAc, and cytosolic proteins, incorporation of [³H]GlcNAc into G protein took place (Balch et al 1984a). When the donor (mutant) fraction that had been primed with ATP and cytosol was examined by electron microscopy, many "budding" vesicular profiles were seen in continuity with Golgi cisternae, suggesting that vesicles may be responsible for transport from the donor to the recipient cisternae (Balch et al 1984b). These results are intriguing and promising because the availability of an in vitro cell-free system for study of Golgi transport should facilitate the elucidation of the nature of and requirements for intra-Golgi transport.

DEFINITION OF THREE BIOCHEMICALLY DISTINCT GOLGI SUBCOMPARTMENTS   Based on the separation of early acting (mannosidase I, fatty acid acylation) and late acting (galactosyl- and sialyltransferases) Golgi processing enzymes on analytical gradients, Dunphy et al (1981) proposed that these activities define two biochemically distinct Golgi subcompartments, cis and trans, assumed to correspond to cis and trans cisternae. Later a third or "medial" Golgi subcompartment was defined by Griffiths and co-workers based on the effects of monensin treatment on the transport and processing of Semliki Forest virus (Griffiths et al 1983; Quinn et al 1983). The enzymes that have been ascribed to the cis Golgi subcompartment are mannosidase I, fatty acylase (Dunphy et al 1981), and the two enzymes that add the mannose-6-phosphate recognition marker to lysosomal enzymes (Goldberg & Kornfeld 1983); those ascribed to the "medial"[2] subcompartment are mannosidase II and GlcNAc transferase I (Dunphy et al 1985); and the terminal (fucosyl, galactosyl, and sialyl) transferases are ascribed to the trans subcompartment (Dunphy & Rothman 1983; Goldberg & Kornfeld 1983; Dunphy et al 1985). Given the morphological and cytochemical heterogeneity of the Golgi complex, it seems likely that additional biochemically distinct subcompartments will be defined in the future (see below).

MONENSIN PERTURBS PRIMARILY TRANS GOLGI FUNCTIONS   Monensin is a monovalent cation ionophore that exchanges Na^+, K^+, and protons. It was introduced by Tartakoff & Vassalli (1977) for the study of Golgi traffic and has proved very useful because it (1) slows or arrests intra-Golgi transport (Tartakoff 1983a); (2) inhibits or prevents late Golgi functions, such as terminal N-glycosylation, sulfation of proteoglycans (Tartakoff 1983a; Ledger & Tanzer 1984), and proprotein processing (Orci et al 1984); and (3) leads to the accumulation of some membrane and secretory protein precursors in middle (Griffiths et al 1983) or trans (Strous et al 1983) Golgi cisternae, or in condensing vacuoles (Orci et al 1984). Morphologically the ionophore causes a dramatic dilation and vacuolization of some or all Golgi cisternae. The extent of these effects varies with the dosage used, the cell type, and the protein under study. The simplest explanation for all these effects is that monensin disrupts a proton gradient maintained by an ATP-dependent proton pump believed to be present in Golgi membranes (Zhang & Schneider 1983; Glickman et al 1983; Barr et al 1984), causing a rise in the pH of the Golgi subcompartments affected. The finding that trans Golgi elements and condensing vacuoles have a low pH (Yamashiro et al 1984; Anderson & Pathak 1985) along with the fact that trans Golgi functions, such as proprotein processing (Docherty et al 1984; Steiner et al 1984; Chang & Loh 1983; Hook & Loh 1984) and CMPase activity (Novikoff & Novikoff 1977), require a low pH and are the Golgi functions inhibited by monensin, all point to the trans 1–2 cisternae and condensing vacuoles as the main site of monensin's effects.

It should be noted, however, that other effects of monensin treatment, e.g. inhibition of galactosyltransferase activity, have also been reported (Wilcox et al 1982) or suspected (Strous et al 1983). With the information available a direct effect of lowering Na^+ or K^+ levels on specific Golgi functions cannot be ruled out. In addition to effects on Golgi functions, monensin

[2] We prefer the term "middle" cisternae to "medial" or central since anatomically it is more accurate.

also causes an increase in intralysosomal and intraendosomal pH (Tarta-koff 1983a), and thereby perturbs endocytic traffic and plasmalemmal receptor recycling (Brown et al 1983; Wilcox et al 1982; Wileman et al 1984). Thus, although there is no doubt that monensin has been a very useful tool for the investigation of Golgi functions, these additional effects must be considered when interpreting the results obtained after treating cells with the ionophore.

CYTOCHEMICAL CORRELATES   As indicated earlier, only a few of the glycoprotein processing enzymes have been directly localized in the Golgi complex in situ by immunocytochemistry. However, in most of those cases in which information is available, i.e. galactosyltransferase (Roth & Berger 1982), and more recently sialyltransferase (Roth et al 1984) and GlcNAc transferase I (Dunphy et al 1985), the predictions based on biochemical data have been borne out. Galactosyltransferase and sialyltransferase were localized in trans cisternae and GlcNAc transferase I in medial Golgi cisternae. One of the processing enzymes, mannosidase II, expected to be concentrated in the middle cisternae, has been detected in all cisternae across the stack in hepatocytes (Novikoff et al 1983). However, there may be some variation in the distribution of mannosidase II from one cell type to another; in exocrine pancreatic cells and anterior pituitary cells this enzyme is concentrated in 2–3 middle Golgi cisternae (Farquhar et al, unpublished data). Besides immunocytochemical observations, the fact that lectins specific for terminal sugars, such as galactose (ricin) and sialic acid (wheat germ agglutinin), bind specifically to trans Golgi cisternae (Griffiths et al 1982; Tartakoff & Vassalli 1983) corroborates the conclusion that terminal glycosylation occurs in this location.

Variations from the "expected" distribution of a given enzyme in different cell types, or in the same cell type under different physiologic conditions, have been reported for many Golgi resident proteins. For example, TPPase was found in up to 4–5 stacked cisternae in epididymal cells (Friend & Farquhar 1967), and in all cisternae of intestinal epithelial cells (Pavelka & Ellinger 1983); NADPase activity was found in several trans cisternae in hepatocytes and Kupffer cells (Angermuller & Fahimi 1984), and galacto-syltransferase was recently detected by immunocytochemistry at the surface of intestinal epithelial cells (Roth et al 1985). Functional modulations in the amount and distribution of TPPase and acid phosphatase in Golgi and other structures have also been reported (Smith & Farquhar 1966, 1970; Paavola 1978; Hand & Oliver 1984). [See Goldfischer (1982) and Hand & Oliver (1984) for additional discussion of this problem.] These examples serve to illustrate that the distribution of Golgi enzymes varies from one cell type to another, and in a given cell type depending on its functional state. It follows that before using Golgi enzymes as biochemical markers for cell fractionation, their distribution in situ should be checked wherever possible by appropriate cytochemical procedures.

Until recently the functions of many of the cytochemical Golgi markers was unclear. However, we have begun to understand the biochemical basis for at least some of these enzyme activities that have served as valuable Golgi cytochemical markers. In particular, the presence of TPPase in Golgi cisternae and its co-localization with galactosyltransferase (Roth & Berger 1982) are now understood: The TPPase activity detected is due to a nucleoside diphosphatase (UDPase) that functions to degrade UDP released as a result of the transfer of galactose from UDP-gal to the nascent glycoprotein (Kuhn & White 1977; Brandan & Fleischer 1982).

In summary, the current framework and flow diagram for intra-Golgi traffic is logical, and is supported by a number of pieces of converging evidence. However, data generated on the direct localization of specific Golgi components in situ is still sparse. Not only have relatively few of the glycoprotein processing enzymes been localized, but also, as mentioned earlier, none of the sugar nucleotide transporter proteins have yet been isolated, purified, and localized. Because so few of the resident Golgi proteins have been directly localized in situ, we may expect some surprises when the details are elucidated. We may also expect some variations from one cell type to another, especially in highly specialized cell types.

How Many Golgi Subcompartments Are There?

There are already clear indications that the proposed existence of three biochemically distinct Golgi subcompartments represents an oversimplification of the true situation. In all likelihood there are many more than three. First, there is already evidence that the condensing vacuole compartment along with the analogous dilated rims of the trans cisternae where concentration of secretory products takes place (Farquhar & Palade 1981) represent a distinct compartment, because the membrane composition of the dilated cisternal rims differs from that of their centers (Farquhar et al 1974; Cheng & Farquhar 1976). Moreover, the condensing vacuoles appear to have a lower pH than other Golgi subcompartments (Anderson & Pathak 1985).

Secondly, recent immunocytochemical evidence obtained by Roth et al (1984), along with enzyme cytochemical data obtained previously by Novikoff (1976), suggests that there may be compositional heterogeneity among trans Golgi cisternae. Galactosyl- and sialyltransferases were found to reside in morphologically distinct trans Golgi compartments (Roth et al 1984). The former were detected in one or two cisternae also marked by TPPase, and the latter were detected in the transmost cisternae also marked by acid phosphatase (CMPase). This observation identifies the trans-most Golgi cisterna or GERL as the most distal station in the glycoprotein processing pathway. It also raises the intriguing possibility that, in analogy to the relationship between TPPase and galactosyltransferase, the acid phosphatase (CMPase) detected in trans Golgi or GERL cisternae may not be a lysosomal enzyme in transit to lysosomes (Novikoff 1977), but instead may function in situ to degrade the CMP released after transfer of sialic acid from CMP-sialic acid to the nascent oligosaccharide chain.

The transitional elements with their vesicular protrusions and the transporting vesicles at the periphery of the Golgi complex may also be considered another potentially distinctive subcompartment for which there are as yet no known markers. Conceivably, some of the functions attributed to cis Golgi cisternae may reside in this compartment.

Finally, if each of the transport operations between Golgi subcompartments is effected by vesicular carriers, it can be assumed that there are as many specific, compositionally distinct populations of vesicular carriers as there are Golgi subcompartments. Indirect evidence already suggests that the vesicles involved in the transport of phosphorylated lysosomal enzymes (from the cis Golgi to lysosomes) represent a distinct subpopulation of coated vesicles (Brown & Farquhar 1984b; Brown et al 1984).

Thus, although the precise number of Golgi subcompartments of distinctive composition is unknown at present, there may be as many as six or even more. Information is needed on this point, and specific markers are required for the characterization of each subcompartment.

Intra-Golgi Location of Other Posttranslational Modifications

O-GLYCOSYLATION   Much less information is available on the localization of events involved in the glycosylation of proteins O-glycosidically linked (to serine or threonine) through GalNAc; however, the bulk of the evidence available indicates that it occurs in the Golgi complex by direct transfer of GalNAc from nucleotide sugars without requiring lipid intermediates (Hubbard & Ivatt 1981; Lennarz 1983). The conclusion that this operation occurs in the Golgi rests on several findings: (1) kinetically O-glycosylation coincides with addition of terminal sugars to N-linked oligosaccharides (Hanover et al 1982; Johnson & Spear 1983); (2) two of the enzymes involved (GalNAc transferase and GalNAc galactosyltransferase) are more concentrated in smooth than in rough microsomal fractions (Elhammer & Kornfeld 1984); and (3) GalNAc-containing glycoproteins can be detected in Golgi, but not in ER cisternae, after *Helix pomatia* lectin-gold labeling (Roth 1984).

Attempts to refine the localization of enzymes involved in O-glycosylation (Elhammer & Kornfeld 1984) have revealed that in BW 5147 lymphocytes GalNAc transferase activity is highest in membrane fractions of intermediate density between ER and trans Golgi markers, and GalNAc-galactosyltransferase is highest in light (trans) Golgi fractions, coincidental with the galactosyltransferase that acts on N-linked oligosaccharides. Detailed analysis of the N- and O-glycosylation patterns of the LDL receptor (Cummings et al 1983) suggests that transfer of GalNAc occurs very soon after the receptor leaves the ER, and that assembly of O-linked oligosaccharides, like that of N-linked oligosaccharides, occurs in more than one region of the Golgi complex. No information is available, however, on whether the enzymes involved in N- and O-glycosylation are localized in the same or different Golgi membranes.

ADDITION OF GLYCOSAMINOGLYCANS TO PROTEOGLYCANS   Proteoglycans hold the record among glycoproteins for posttranslational modifications because they contain, in addition to N- and O-linked oligosaccharides, glycosaminoglycan (GAG) chains O-glycosidically linked (to serine or threonine) through xylose (Roden 1980; Dorfman 1981), and proteoglycan synthesis involves the addition to the protein core of up to ten times its weight in GAG polysaccharides. The assembly of these GAG chains is a complex process that varies in detail from one type of proteoglycan to another, but in general can be resolved into several basic operations: transfer of xylose (by xylosyltransferase) from UDP-xylose, stepwise transfer of monosaccharide units (usually hexosamine or uronic acid) from the corresponding nucleotide sugars by specific glycosyltransferases to form polymers of increasing size, followed by extensive modification of the GAG (e.g. deacetylation, epimerization, and N- and O-sulfation). The magnitude of the processing that can take place is illustrated by the fact that the chondroitin sulfate proteoglycan of cartilage is estimated to undergo more than 200 posttranslational modifications (Kimura et al 1984).

Even less is known about intracellular sites of proteoglycan synthesis than about O-glycosylation. Most of the studies that have provided information on this problem have been done on cultured cartilage or chondrosarcoma cells in which the combined immunocytochemical, cell fractionation, and kinetic analyses suggest that all these steps, with the possible exception of xylose transfer, occur in the Golgi complex. GalNAc transferase, glucuronosyltransferase, and sulfotransferase have been shown to be more concentrated in smooth than in rough microsomes (Geetha-

Habib et al 1984), chondroitin sulfate GAG chains have been localized to the Golgi region by immunofluorescence (Vertel & Barkman 1984), and GAG addition has been found to occur late in biosynthesis at about the same time as N-glycosylation (Kimura et al 1984). The site of xylose coupling, whether ER or Golgi, is controversial at present. Early cell fractionation data (Horwitz & Dorfman 1968) obtained on embryonic chick cartilage indicated that xylosyltransferase activity is higher in rough than in smooth microsomes, and by immunocytochemistry xylosyltransferase has been localized to ER, not Golgi, cisternae (Hoffman et al 1984). Recently, however, Kimura et al (1984) found that incorporation of xylose into cartilage proteoglycan in chondrosarcoma cells occurs within five minutes of galactose addition to N-linked oligosaccharides, which is a late (trans) Golgi event. In addition, in rat liver the xylosyltransferase–specific activity has been found to be five to ten times higher in Golgi than in rough or smooth microsome fractions (Hirschberg, personal communication).

In summary, there is agreement that all the steps in GAG synthesis after xylose transfer take place largely or exclusively in the Golgi complex, but none of the enzymes involved has yet been localized in situ. The evidence concerning the site of xylose transfer is conflicting at present, and further studies are required, not only on cartilage cells but also on other cell types that make different proteoglycans, to resolve this issue.

SULFATION   Sulfation of proteoglycans (and of glycoproteins and glycolipids as well) has long been known to be a Golgi-associated event. Using autoradiography it was determined some time ago that the initial uptake of [^{35}S]sulfate is largely or exclusively into this organelle (Young 1973). Moreover, sulfotransferases have been known for some time to be concentrated in Golgi fractions (Fleischer & Zambrano 1974; Fleischer & Smigel 1978). A detailed kinetic analysis of cartilage proteoglycan revealed that sulfation of GAG chains is virtually the last modification that occurs before their discharge (Kimura et al 1984). Therefore biochemically sulfation is a late—presumably a trans—Golgi event. To date none of the sulfotransferases has been localized directly in situ, so their distribution in trans Golgi elements relative to other glycoprotein processing enzymes has not yet been determined.

Sorting and Transport of Lysosomal Enzymes

Considerable progress has been made during the last five years in understanding the mechanisms by which lysosomal enzymes are sorted intracellularly and targeted to lysosomes. In fact, lysosomal enzymes are the only biosynthetic product for which an intra-Golgi sorting signal has been identified. It is now known that in many cell types intracellular targeting is effected by a receptor-mediated process involving mannose-6-phosphate (Man-6-P) residues on lysosomal enzymes and intracellular Man-6-P receptors (Sly & Fischer 1982; Creek & Sly 1984).

SYNTHESIS OF LYSOSOMAL ENZYMES   The steps involved in lysosomal enzyme biosynthesis and their presumed intracellular location have been worked out in some detail. Lysosomal enzymes are known to be synthesized as typical high mannose–type glycoproteins and partially trimmed (all glucose residues and at least one mannose residue are removed), presumably in the rough ER. They are then transported to smooth membranes of high density where, in contrast to other N-linked glycoproteins, Man-6-P residues are added to the oligosaccharide chains (Goldberg & Kornfeld 1983; Goldberg et al 1984). Addition of the recognition signal is assumed to take place in cis Golgi cisternae (Figure 2) because the two enzymes involved, N-acetylglucosamine-1-phospho-transferase and phosphodiester glycosidase ("uncovering enzyme"), are concentrated in membrane fractions of lower density than ER markers and of higher density than those in which galactosyltransferase is found (Pohlmann et al 1982; Goldberg & Kornfeld 1983).

SITE OF SORTING OF LYSOSOMAL ENZYMES   The sorting and diversion of lysosomal enzymes from the secretory pathway has been assumed to take place after addition of the recognition marker in the Golgi complex, and a 215 kDa Man-6-P receptor has been identified, purified (Sahagian et al 1981; Steiner & Rome 1982; Sahagian 1984), and localized by immunocytochemistry to Golgi cisternae (Willingham et al 1983; Brown & Farquhar 1984a; Geuze et al 1984).

Regarding the intra-Golgi site of sorting, it has been commonly assumed that lysosomal enzymes follow the same cis to trans pathway across the stack as secretory proteins and membrane glycoproteins, and that all these products are sorted and packaged for delivery to their respective desti-nations in trans Golgi or GERL cisternae. This concept is based on (1) the finding that acid phosphatase (CMPase) activity, which is assumed to represent a lysosomal marker enzyme, is concentrated in GERL cisternae (Novikoff 1976); and (2) the finding that lysosomal enzymes, especially those secreted by cultured cells (Hasilik & von Figura 1981; Rosenfeld et al 1982; Goldberg et al 1984), contain complex carbohydrates, which implies that they must pass through the trans-most Golgi cisternae where the terminal transferases reside.

If sorting of lysosomal enzymes is a trans Golgi event, one would expect the receptors to be concentrated in trans Golgi cisternae. So far conflicting evidence has been obtained on the localization of the 215-kDa Man-6-P receptor. Willingham et al (1983) localized the 215-kDa receptor in ER and

Golgi cisternae, in "receptosomes" (endosomes), and in coated pits at the cell surface in CHO cells. It was found in highest concentration in peripheral Golgi cisternae interpreted as GERL, and was occasionally detected in Golgi stacks. Geuze et al (1984) found the receptors in CURL (a "system of anastomosing tubules that extend inwards up to the Golgi areas"), in coated vesicles, and distributed all across the Golgi stack in rat hepatocytes. They also found the receptor along the entire plasma membrane and in coated pits at the cell surface.

Somewhat different findings were obtained by Brown & Farquhar (1984a), who found the receptors concentrated in cis Golgi cisternae, in coated vesicles located near the cis-most Golgi cisternae, and in endosomes or lysosomes in a number of cell types, namely hepatocytes, exocrine pancreatic cells, and epididymal epithelia, as well as several cell lines (Clone 9 hepatocytes, BHK cells) in culture (Brown et al 1984). Some cell types (hepatocytes, cultured cells) also had receptors concentrated at the cell surface in coated pits. The cis Golgi cisternae, coated vesicles, and endosomes/lysosomes were assumed to represent the sorting site, carrier vesicles, and delivery sites, respectively, for lysosomal enzymes. Subsequent cell fractionation data on Clone 9 hepatocytes demonstrated that the 215 kDa Man-6-P receptors are concentrated in sucrose density gradients at a higher density than middle (mannosidase II) and trans (galactosyl- and sialyltransferase) Golgi markers (Brown & Farquhar 1985). Based on the localization of the 215-kDa Man-6-P receptor in cis Golgi elements, it was suggested (Brown & Farquhar 1984a) that: (1) In some cell types, lysosomal enzymes are sorted and removed from the secretory pathway for delivery to lysosomes in cis Golgi cisternae; and (2) if sorting does not take place in the cis Golgi, due to lack of appropriate receptors or ligands, the lysosomal enzymes would be expected to continue along the secretory pathway, be terminally glycosylated and either be secreted or sorted intracellularly by some alternative means. The fact that in Clone 9 cells intracellular lysosomal enzyme precursors were found to have high-mannose type (Endo H sensitive) oligosaccharides, and secreted enzymes complex-type (Endo H insensitive) oligosaccharides (Rosenfeld et al 1982) lends credence to this suggestion.

The localization of the receptor to the cis Golgi was not found to apply universally to all cell types studied because the receptors were distributed all across the Golgi stack in the osteoclast (Baron et al 1984), which is a specialized cell type that secretes lysosomal enzymes into extracellular pockets that are the functional equivalent of lysosomes where bone resorption takes place. This finding suggests that variations in lysosomal enzyme traffic occur among different cell types.

The explanation for the different results obtained by different investigators on the localization of the 215-kDa Man-6-P receptor is not immediately apparent, and is complicated by the different antibodies and immunocytochemical techniques used and, in some cases, the different cell types studied. It is also complicated by the recent discovery (Hoflack & Kornfeld 1985) of a second Man-6-P receptor, a 46-kDa protein with somewhat different properties. The new receptor also has a specificity for Man-6-P, but it differs from the 215-kDa receptor in two ways: First, it requires divalent cations, particularly $MnCl_2$, for ligand binding whereas the 215-kDa receptor is cation independent. Second, it has a subtle difference in carbohydrate binding specificity: The new receptor binds only phosphomonoesters, whereas the 215-kDa receptor binds methylphosphomannosyl diesters in addition to phosphomonoesters. Most cell types studied so far appear to have both types of receptors, but a few murine tissue culture cell lines (e.g. mouse $P338D_1$ macrophages) are deficient in the 215-kDa receptor (Gabel et al 1983). As yet neither the intracellular distribution of the 46 kDa receptor nor its role in lysosomal enzyme traffic have been determined.

In summary, the steps involved in biosynthesis and processing of lysosomal enzymes have been elucidated, and the role of Man-6-P as a sorting signal is firmly established. However, a good deal of work is still required to fully understand the mechanisms and sites of lysosomal enzyme sorting, to determine the comparative roles of the two Man-6-P receptors (cation-dependent and cation-independent), and to chart pathways of lysosomal enzyme traffic in different cell types.

TRAFFIC FROM THE GOLGI TO THE PLASMA MEMBRANE

Sorting and Packaging of Secretory Proteins: The regulated versus constitutive pathways

Exocrine and endocrine glandular cells typically concentrate their secretory products [up to 200 times that of the ER (Salpeter & Farquhar 1981)], and package them into secretory granules that are stored in the cytoplasm until released by exocytosis (Palade 1975). In such cases concentration and packaging occur in the dilated rims of the trans-most Golgi cisternae or in condensing vacuoles (Farquhar & Palade 1981), and secretion granule discharge is "regulated" by an appropriate signal, such as Ca^{++} or cAMP (Figure 3). Nonglandular cell types, such as fibroblasts and plasma cells, as well as most cultured cell lines have a "nonregulated" or "constitutive" type of secretion (Figure 3): Concentration and storage in granules does not occur, and instead the cells package their products in

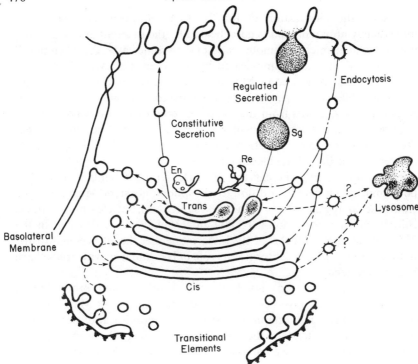

Figure 3 Current flow diagram depicting the major routes for biosynthetic (--- →), exocytic (——→), endocytic (— · →), and lysosomal (-- →) traffic. Only traffic that is directed to or through the Golgi complex is illustrated. The *biosynthetic pathway* from ER to Golgi and across the Golgi stack is utilized for the transport of secretory proteins and membrane glycoproteins. *Lysosomal enzymes* follow the same pathway as secretory proteins from the ER to the cis Golgi cisterna, where the Man-6-P recognition marker is believed to be added. The pathway taken from there on is not yet resolved. Two possible routes for transport of lysosomal enzymes to lysosomes via coated vesicles are indicated because opinions vary as to whether sorting via Man-6-P receptors occurs in cis or trans Golgi cisternae or in both of these locations. Several different *exocytosis* routes are indicated. *Secretory proteins* can be discharged by either the *regulated* or *constitutive* (nonregulated) pathways. Products that follow the regulated pathway are concentrated and stored in granules until discharge is stimulated by an appropriate secretagogue; those that follow the constitutive pathway are packaged in dilute form into vesicles that are continuously discharged. *Membrane glycoproteins* are also delivered to the plasmalemma in vesicles discharged by exocytosis. In polarized cells, two separate pathways exist for delivery of membrane proteins to the apical and basolateral plasmalemmal domains, and sorting is believed to occur at the time of exit from the Golgi (as diagrammed here) or shortly thereafter. Whether membrane and secretory proteins destined for the apical plasmalemma are ferried in the same or in different vesicles is not yet known. Endocytic vesicles are depicted as capable of fusing with all Golgi elements. In the majority of those cases in which endocytic traffic to Golgi elements was investigated, the incoming vesicles were found to fuse with reticular (*Re*) or tubular elements on the trans side of the Golgi stack. In some cases, primarily in secretory cells, the incoming vesicles were traced to the stacked cisternae. (*En* = endosome; *Re* = reticular element.)

dilute form into vesicles that pinch off of trans Golgi cisternae and are rapidly and continually released by exocytosis (Tartakoff & Vassalli 1977, 1978).

It is now evident from the studies of Kelly and co-workers (Gumbiner & Kelly 1982; Kelly et al 1985) that two separate exocytosis pathways, regulated and nonregulated, can coexist in a single cell type. The cell type they studied is the AtT-20 cell, a mouse pituitary tumor cell line that produces pro-opiomelanocortin (POMC), the precursor of ACTH and endorphins. The main secretory product, ACTH, and several acidic polypeptides and sulfated macromolecules were found in a distinct secretory granule fraction of dense–cored granules whose discharge was stimulated by cAMP. However, unprocessed precursor (POMC), a completely different set of polypeptides and sulfated molecules, gp70 (a viral membrane protein used as a marker for endogenous membrane glyco-proteins), and laminin were found in a separate membrane fraction continuously discharged by a constitutive pathway (Gumbiner & Kelly 1982; Moore et al 1983a; Burgess et al 1985). Since concentration and packaging of secretory products are known to occur in the dilated rims of the trans-most Golgi cisternae and in immature granules in anterior pituitary cells (Farquhar 1971), these observations imply that sorting of the products secreted by the two different routes must occur after addition of terminal sugars in the trans Golgi and prior to, or concomitant with, packaging and concentration, i.e. either in the rims of the trans-most cisternae or in immature granules.

These observations raise questions as to the mechanisms involved in directing products along the regulated and constitutive pathways. Kelly et al (1985) have addressed this issue by transfecting AtT-20 cells with cloned DNA encoding for several heterologous proteins, and determining the pathway taken by the expressed proteins. When DNA coding for proinsulin (Moore et al 1983c), trypsinogen (Burgess et al 1985), or growth hormone (Moore & Kelly 1985) was added to the cells, all of these proteins were secreted by the regulated pathway and, interestingly, proinsulin was proteolytically processed to insulin along with POMC. In contrast, when AtT-20 cells were transfected with DNA sequences encoding a secreted form of VSV "G" protein, a membrane protein from which the trans-membrane and cytoplasmic domains had been deleted, the truncated G protein was preferentially secreted by a constitutive pathway (Moore & Kelly 1985). Thus, the endogenous peptide hormone (ACTH), two hetero-logous peptide hormones (human insulin, growth hormone), and an exocrine protein (rat trypsinogen)—but not the secreted G protein—follow the regulated route. These observations led Kelly & co-workers (Burgess et al 1985; Moore & Kelly 1985) to the conclusion that exocrine and endocrine secretory products must share a common recognition signal (a "sorting domain") that lies in the protein structure of these molecules and is responsible for their being sorted and directed along the regulated pathway. The nature of the signal is unknown, since these proteins have no obvious common structural features or primary amino acid sequence homology (Moore & Kelly 1985).

Does traffic directed along the regulated and constitutive pathways require two different sorting signals, or is segregation of one class of protein mediated by a specific receptor and the other carried to the cell surface in the content of the transporting vesicles by bulk flow? Several findings led Moore & Kelly (1985) to suggest that only entry into the regulated pathway is receptor-mediated. These were (1) the fact that glycosaminoglycans produced under conditions in which they are not coupled to protein (treatment with B-D-xyloside) follow the constitutive pathway (Burgess & Kelly 1984); (2) the fact that chloroquine diverts ACTH from the regulated to the constitutive pathway (Moore et al 1983b); and (3) the fact that the truncated G protein, which would not be expected to contain a natural sorting signal, follows the constitutive pathway. The presumptive signal(s) involved remain unknown at present, but recombinant DNA technology involving gene transfer and in vitro mutagenesis is being utilized to explore these questions, and holds great promise for eventually providing clues as to their nature.

Are Membrane and Secretory Proteins Sorted into Separate Carrier Vesicles by the Golgi Complex?

The question of whether membrane and secretory proteins are sorted into the same or separate carrier vesicles cannot be answered with certainty at present, because the studies bearing on this problem are too limited, and the evidence obtained so far is conflicting. The biochemical results obtained by Kelly and his co-workers discussed above on AtT-20 cells indicate that secretory proteins can follow either the regulated or constitutive route, whereas the membrane proteins, at least those studied up to now (gp70, VSV G), follow the constitutive pathway. They leave open the questions of whether or not any membrane proteins follow the regulated pathway, and whether or not the secretory proteins and membrane proteins released constitutively are carried in the same vesicle. To answer these questions requires colocalization of a secretory protein and a membrane protein in the same vesicle by immunoelectron microscopy, and only a single study of this type has been reported so far. In that case albumin and transferrin, which are both secretory proteins, and VSV G protein, a membrane protein, were detected in the same vesicle (Strous et al 1983). This finding suggests that both membrane and secreted proteins can be ferried to the cell surface together by a constitutive pathway.

Little information is available on the question of whether or not newly synthesized membrane proteins can be delivered to the cell surface via the regulated pathway. In PMN leukocytes there is indirect evidence that N-formulated peptide receptors are inserted into the plasmalemma upon exocytosis of the cell's specific granules (Fletcher & Gallin 1983). In this case, the event has considerable functional significance because it serves to upregulate surface receptors. However, the receptor has not yet been directly localized in situ within the granules. Clearly, more information is needed on this issue.

Membrane Proteins Destined for Delivery to Different Plasmalemmal Domains Are Sorted Intracellularly

Epithelial cells, such as those lining the intestine, kidney tubule, and many other organs, are well known to have apical and basolateral plasmalemmal domains of distinctive protein composition, e.g. Na^+K^+-ATPase and peptide hormone receptors are associated with the basolateral domain, and alkaline phosphatase, aminopeptidases, and other digestive enzymes are associated with the apical membrane. It follows that these cells must have a means of sorting and specifically targeting membrane proteins to their domain of residence. This targeting problem has been investigated by a number of workers in different systems, and different conclusions have been reached concerning the site of sorting. It has been proposed that sorting occurs in the Golgi, that it occurs in post-Golgi vesicles, or that it occurs after delivery to the cell surface (see Rindler et al 1982; Rodriguez-Boulan 1983; Simons & Fuller, this volume). The most convincing data to date were obtained by studying the delivery of viral membrane glycoproteins in cultured MDCK cells, a stable cell line that grows as a polarized epithelium in culture. When these cells are infected with enveloped viruses they distribute newly synthesized viral membrane proteins preferentially into one or the other surface domain (Rodriguez-Boulan & Sabatini 1978; Rodriguez-Boulan & Pendergast 1980), e.g. VSV G protein to the basolateral and influenza virus HA protein to the apical cell surface.

There is now both morphologic (Rindler et al 1984) and biochemical (Fuller et al 1985) data indicating that viral membrane proteins destined for different cell surfaces pass through the same Golgi cisternae, are sorted intracellularly, and delivered directly to the appropriate plasmalemmal domain by polarized exocytosis (Matlin & Simons 1984; Rodriguez-Boulan et al 1984; Misek et al 1984; Rindler et al 1985; Pfeiffer et al 1985). This means that sorting must occur somewhere between the Golgi and the cell surface. Investigations of this problem have taken advantage of the fact that transport to the plasmalemma is blocked or retarded at 20°C (Matlin et al 1983; Saraste & Kuismanen 1984). Griffiths et al (1985) recently demonstrated by immunocytochemistry that under these conditions VSV G protein accumulates in a system of tubules and vesicles located on the trans side of the Golgi, which stains for acid phosphatase activity and therefore corresponds to the trans-most Golgi or GERL cisternae. When the temperature was raised so that transport to the cell surface could take place, G protein was seen within small vesicles, but was not detected in endosomes (identified by prelabeling with horseradish peroxidase). These observations indicate that sorting of membrane proteins does not take place in endosomes, and suggest that it must occur at the time of exit from the trans Golgi cisternae or in some other post-Golgi compartment. As in many other cases, the mechanism of sorting and the nature of the signals involved is unknown. The individual viral glycoproteins must contain all the necessary information to direct sorting because when genes encoding for influenza virus HA protein (Roth et al 1983) or VSV G protein (Gottlieb et al 1985) are introduced into cultured cells, the viral glycoproteins are expressed and are directed to the correct plasmalemmal domain. Once again, recombinant DNA technology involving construction and transfection of chimeric genes, in which the role of specific polypeptide segments on sorting and polarized delivery can be assessed (Rizzolo et al 1985), offers great promise of providing insights into the mechanisms and signals involved in this type of targeting. For a more detailed review of this topic the reader is referred to several recent reviews (Rodriguez-Boulan 1983; Simons & Fuller, this volume).

Nature of the Carriers

The only vesicular carriers that have been isolated in pure enough form to warrant biochemical analysis are secretory granules, which have been prepared from several regulated secretory cell types: exocrine pancreatic (Meldolesi & Cova 1972; MacDonald & Ronzio 1972), parotid (Castle et al 1975; Cameron 1984; Cameron & Castle 1984), and adrenal chromaffin (Winkler et al 1970) cells. Comparison of the gel electrophoretic patterns obtained for the secretory granule membranes of these and other (Cameron 1984) cell types indicates that membranes specialized for storage of exportable products have a distinct membrane composition that is simpler than that of other compartments along the secretory pathway (e.g. Golgi, plasmalemma).

None of the vesicular carriers involved in other transport operations have been isolated and characterized. Based on cell fractionation data, clathrin-coated vesicles have been said to be involved in several types of traffic, namely the transport of membrane proteins from ER to the Golgi, and from the Golgi to the plasmalemma (Rothman et al 1980); however, the latter point is still controversial (Wehland et al 1982; Saraste & Hedman

1983; Rodriguez-Boulan et al 1984). To solve this problem, and to characterize the carriers further, will require their isolation in reasonably pure form. Since separation based on physical properties (size, density) is unlikely to yield highly purified subpopulations of vesicular carriers, the preparation of such fractions is likely to require more refined approaches, such as immunoaffinity absorption based on differences in their constituent membrane proteins (Merisko et al 1982). This approach was recently successfully applied to the fractionation of brain coated vesicles (Pfeffer & Kelly 1985).

PATHWAY FROM THE PLASMA MEMBRANE TO THE GOLGI COMPLEX

As recently as 1978 the existence of endocytic traffic to the Golgi complex was not recognized because all such traffic was believed to converge on lysosomes (see Farquhar 1981, 1982). With the demonstration that electron-dense tracers taken up by endocytosis can reach the stacked Golgi cisternae (Herzog & Farquhar 1977; Farquhar 1978a; Herzog & Miller 1979; Ottosen et al 1980; Herzog & Reggio 1980) or GERL cisternae (Gonatas et al 1980; Gonatas 1982), the existence of plasmalemma-to-Golgi traffic has been increasingly acknowledged. However, endocytic uptake into Golgi elements is often claimed but seldom unambiguously demonstrated. Uptake of various ligands into Golgi elements is often described by immunofluorescence based on their localization in a juxta-nuclear fluorescent spot. The latter undoubtedly corresponds to the Golgi region, but this region houses not only Golgi cisternae, but also many other cell structures and compartments, including endosomes and lysosomes. Electron microscopy is required to distinguish between localization in bona fide Golgi elements and other structures present in the same general location.

Transport to Reticular Elements on the Trans Side of the Golgi Stacks

There have been repeated demonstrations of uptake of various electron-dense tracers, exogenous ligands, or surface membrane constituents into reticular elements located on the trans side of the Golgi stack, variously described as "vesicles or tubules located on the trans side of the Golgi stack" (Yanashiro et al 1984), GERL cisternae (Gonatas 1982), CURL or "anastomosing tubules that extend inwards up to the Golgi areas" (Geuze et al 1983, 1984), "juxtanuclear endosomes with tubular extensions" (Hopkins & Trowbridge 1983; Hopkins 1983), or "reticular portion of the Golgi" (Willingham & Pastan 1982, 1984; Willingham et al 1984). Hence, there is general agreement that there is endocytic uptake of surface ligands into a vesicular or tubular system located on the trans side of the Golgi stacks, but there is no agreement and great confusion concerning the nature of the compartment(s), i.e. whether they correspond to endosomes, GERL cisternae, or Golgi cisternae. The confusion is compounded by the following: (1) there are wide variations in the morphology of all these structures in different cell types at steady state; (2) the extent and topography of any one of these structures can vary dramatically in individual cell types undergoing functional modulations; (3) Golgi cisternae can appear fenestrated or tubular in grazing section (Rambourg et al 1979); and (4) endosomes often have long tubular extensions (Hopkins & Trowbridge 1983; Geuze et al 1983). To unravel this complicated endocytic traffic and to identify the trans Golgi compartments more precisely will require the simultaneous localization of an exogenous tracer and enzyme markers. This has not usually been done in the past, but should be done regularly in the future.

For other perspectives on the organization of the trans Golgi region and traffic to it, the reader can refer to recent reviews by Goldfischer (1982) and Pastan & Willingham (1985).

Transport to the Golgi Stacks

In only a few cases has access of cell surface markers to the piled cisternae in the Golgi stacks been described (Herzog & Farquhar 1977; Farquhar 1978a; Herzog & Miller 1979; Ottosen et al 1980; Herzog & Reggio 1980; Woods & Farquhar 1984, 1985). All of these studies were done on secretory cells producing large amounts of proteins for export in which this type of traffic appears to be amplified. Indirect evidence suggests that this traffic is connected with the recycling of secretory granule containers (Farquhar 1981, 1982, 1983; Meldolesi & Ceccarelli 1981). Earlier work demonstrated the uptake of nonspecific tracers (dextrans, cationized ferritin), and their delivery to Golgi cisternae. More recently evidence has been obtained that the transferrin receptor, a specific plasmalemmal membrane protein, can reach the stacked Golgi cisternae upon internalization into cultured myeloma cells (Woods & Farquhar 1984, 1985). When transferrin receptors were tagged at the cell surface with an anti-transferrin receptor antibody, the antibody—presumably still bound to the receptor—was subsequently detected in all the stacked Golgi cisternae as well as in reticular elements on the trans side of the Golgi. These observations clearly indicate that this surface membrane constituent traverses cis and middle as well as trans Golgi cisternae in its cycling in and out of these secretory cells.

The discovery that exogenous markers and, more recently, plasmalemma receptors gain access to bona fide Golgi cisternae, where Golgi processing enzymes reside, has far reaching implications, among them is the possibility suggested some time ago (Farquhar 1981, 1982) that biosynthetic repair of ligands and membrane constituents may take place. Initial evidence for the existence of such a repair mechanism was obtained by Regoeczi and co-workers (1982), whose findings suggest that when asialotransferrin is injected into rats it can be resialylated, presumably by liver cells. Quite recently, Snider & Rogers (1985) found that the transferrin receptor can also be resialylated by cultured hepatoma cells after deglycosylation of surface components by neuraminidase treatment. In view of the finding that in some cell types terminal transferases (specifically galactosyltransferase) can occur at the cell surface (Roth et al 1984), the distribution of sialotransferase in these hepatoma cells needs to be checked. However, regardless of the site, whether cell surface or Golgi, these findings establish that biosynthetic modification and repair of ligands and plasmalemmal constituents are possible. In the future it will be of considerable interest to obtain additional information on this important problem in order to find out how limited or extensive this repair mechanism is, and to determine other possible metabolic consequences of plasmalemmal-to-Golgi traffic.

PAUCITY OF INFORMATION ON GOLGI SORTING SIGNALS

It is clear that the Golgi complex must sort both biosynthetic and recycling membrane traffic and direct each molecule to its appropriate destination. Receptor-mediated sorting has been proposed to explain ER-to-Golgi transport, intra-Golgi transport, direction along the regulated and/or constitutive secretory pathways, and the packaging of membrane proteins for delivery to different plasmalemmal domains. With the exception of the Man-6-P sorting signal for lysosomal enzymes, not a single clue has been obtained so far as to what any of the sorting signals might be, in spite of the many efforts devoted to the problem. It seems unlikely that any of the other Golgi sorting events that occur during biosynthesis of glycoproteins is mediated by a carbohydrate recognition marker, due to lack of uniform effects of agents (e.g. swainsonine, tunicamycin) that modify glycosylation (Elbein 1983). However, since carbohydrate recognition signals (e.g. mannose, galactose) are well known to mediate binding and uptake at the cell surface, they could be involved in directing recycling membrane traffic from the cell surface to and through the Golgi complex.

It also seems unlikely that the recognition signals involve homologies in the amino acid sequences of the proteins to be sorted, because of the large number and diversified nature of the products that follow any one of these pathways and the fact that no such homologies have yet been detected in spite of numerous attempts to do so. More likely, in analogy to the signal peptides of secretory, lysosomal, and some membrane proteins that recognize a common signal recognition particle (Walter et al 1984), there must be certain common features contained in the secondary or tertiary structure of these peptides that allow recognition by a membrane receptor or carrier protein of a whole class of transported products. Genetic manipulations involving construction of chimeric genes with specific modifications and deletions, as well as reconstituted in vitro systems, offer great hope for finally deciphering the sorting codes.

What Has Been Learned About Golgi Traffic from Studies on Man-6-P–Mediated Sorting of Lysosomal Enzymes?

Because it constitutes our only known Golgi sorting mechanism, it is worth pausing to consider what we have learned up to now from the study of lysosomal enzyme traffic. The first lesson, learned some time ago, is that coated vesicles can be involved in intra-Golgi sorting, as well as in sorting at the cell surface, because there is good evidence based on enzyme cytochemistry (Friend & Farquhar 1967), cell fractionation (Campbell & Rome 1983), and immunocytochemistry (Brown & Farquhar 1984a,b; Brown et al 1984) that coated vesicles are involved in the sorting and transport of lysosomal enzymes to lysosomes.

A second lesson is that perturbants that modify the milieu of a given compartment can disturb and reroute traffic by causing redistribution of receptors from one site and their accumulation at another. For example, in cells treated with agents such as NH_4Cl and chloroquine that cause an increase in intralysosomal and intraendosomal pH, the Man-6-P receptors become trapped in endosomes or lysosomes because lysosomal enzymes are unable to dissociate from their receptors when the pH rises above 6.0 (Gonzalez-Noriega et al 1980; Brown et al 1984). This results in depletion of receptors from the loading site in the Golgi (Brown et al 1984), which prevents sorting of lysosomal enzymes and results in their secretion (Gonzalez-Noriega et al 1980; Rosenfeld et al 1982).

We have also learned in the case of the 215-kDa Man-6-P receptor that ligand loading (Brown et al 1984) and unloading (Brown & Farquhar 1984c) trigger receptor relocation, via coated vesicles, to the delivery and sorting sites, respectively. The effect of occupation of the receptors by ligand was particularly striking in cells placed in normal culture medium after treating them with pH perturbants. As already mentioned, pH perturbants cause depletion of receptors from the sorting site in the Golgi, and their accumulation at the delivery site in lysosomes. Upon placing the cells in normal (NH_4Cl-free) medium, the pH drops, allowing dissociation of lysosomal enzymes, and the receptors immediately return via coated vesicles to the Golgi (Brown & Farquhar 1984c). These findings indicate that ligand unloading triggers movement of vesicles, and imply that the ligand must induce conformational changes in the receptor or proteins closely associated with it. The rapid return of the receptors to the Golgi complex tells us that recycling of intracellular receptors, like that of cell surface receptors (Ciechanover et al 1983), is very rapid : Man-6-P receptors return to the Golgi within 2–5 minutes after placing NH_4Cl-treated cells in normal medium (Brown & Farquhar 1984c).

Finally, from the recent studies by Kornfeld and his co-workers (Lang et al 1984; Kornfeld et al 1985) we have learned something about how Golgi processing enzymes recognize their specific substrates. The determinant that allows the selective recognition of acid hydrolases by N-acetylglucosaminylphosphotransferase was found to be on the protein rather than the oligosaccharide portion of lysosomal enzymes, because deglycosylated enzymes (i.e. Endo H–treated) proved to be specific inhibitors of phosphorylation of intact enzymes. Moreover, the intact protein is required because even large proteolytic fragments did not retain the ability to be recognized.

Thus, the study of lysosomal enzyme traffic has already provided a number of insights into some of the important events in this type of traffic. Whether or not these same principles apply to other Golgi recognition and sorting events awaits the discovery and study of other sorting signals.

CONCLUSIONS

In the last five years we have witnessed an enormous surge of interest in the Golgi complex, and in that time rapid progress has been made in understanding the functions and organization of the organelle, as well as the routing of traffic to and through it. Figure 3 summarizes in diagrammatic form our current concepts of the flow of biosynthetic, exocytic, lysosomal, and endocytic traffic to and from the Golgi complex. A key development in this period was the working out of the kinetics and sequential steps in N-glycosylation, and the purification of many of the enzymes involved, which made possible the integration of biochemical and morphological data. Improvements in techniques for immunofluorescence and especially immunocytochemical localization at the electron microscope level have been equally important. This progress made possible the localization of some of the resident Golgi enzymes in situ, and the tracking of the traffic of individual proteins. As a result of these new developments and a large volume of earlier work we now have developed a logical framework for the flow diagram of biosynthetic and other traffic through the Golgi, which will be useful in interpretation of future data.

The euphoria associated with having achieved the current, enhanced level of understanding should not blind us to the reality that the facts are still sparse, the framework is still sketchy, and enormous questions remain to be answered. First, as far as Golgi organization alone is concerned, few of the total resident Golgi proteins have been identified, purified, and localized. Second, direct knowledge of the composition of any individual subcompartment is totally lacking because techniques for isolation of Golgi fractions and subfractions based on physical parameters always yield fractions of mixed composition; as a result, none of the subfractions has been isolated and characterized. Third, we have little idea at present of how all this complicated Golgi traffic is controlled. Signals and sorting mechanisms are assumed to be essential at every step for proper traffic regulation but, with the exception of the Man-6-P sorting signal for lysosomal enzymes, not a single sorting signal has been identified as yet. Even in the case of lysosomal enzymes, although we know the signal involved, we have no idea how the vesicular carrier traffic is guided to its appropriate destinations. All this remains to be worked out in the future.

Based on past experience, given the inherent complexity of the Golgi and of its impinging traffic, it can be safely predicted that important and lasting insights into the problems remaining will come only from an approach combining biochemical, cell biological, and molecular biological evidence. An interesting period of Golgi research lies ahead.

ACKNOWLEDGMENT

The author's own work cited in this chapter has been supported by NIH research grant AM17780. I would like to thank all my collaborators over the last twenty years of Golgi research, plus the many colleagues who provided me with preprints of unpublished work. I would also like to thank M. Lynne Wootton and Leah D'Eugenio for their patience and assistance with the editing and word processing of the manuscript, and Virginia Simon for preparation of the diagrams.

Literature Cited

Anderson, R. G. W., Pathak, R. K. 1985. Vesicles and cisternae in the trans Golgi apparatus of human fibroblasts are acidic compartments. *Cell* 40:635–43

Angermüller, S., Fahimi, D. H. 1984. Cytochemical localization of β-NADPase in rat hepatocytes and Kupffer cells. Comparison with thiamine pyrophosphatase (TPPase). *J. Histochem. Cytochem.* 32: 541–46

Aoi, K., Fujii-Kuriyama, Y., Tashiro, Y. 1981. Intracellular distribution of NADPH-cytochrome C reductase in rat hepatocytes studied by direct ferritin-immunoelectron microscopy. *J. Cell. Sci.* 50:181–98

Bainton, D. F., Farquhar, M. G. 1970. Segregation and packaging of granule enzymes in eosinophilic leukocytes. *J. Cell Biol.* 45:54–73

Balch, W. E., Dunphy, W. G., Braell, W. A., Rothman, J. E. 1984a. Reconstitution of the transport of protein between successive compartments of the Golgi measured by the coupled incorporation of N-acetyl-glucosamine. *Cell* 39:405–16

Balch, W. E., Glick, B. S., Rothman, J. E. 1984b. Sequential intermediates in the pathway of intercompartmental transport in a cell-free system. *Cell* 39:525–36

Baron, R., Neff, L., Brown, E., Louvard, D., Courtoy, P., Farquhar, M. G. 1984. Redistribution of a 100 Kd membrane protein, the mannose-6-phosphate receptor and lysosomal enzymes during the inactivation of osteoclasts by calcitonin. *J. Cell Biol.* 99:362a

Barr, R., Safranski, K., Sun, I. L., Crane, F. L., Morré, D. J. 1984. An electrogenic proton pump associated with the Golgi apparatus of mouse liver driven by NADH and ATP. *J. Biol. Chem.* 259:14064–67

Bergmann, J. E., Singer, S. J. 1983. Immuno-electron microscopic studies of the intra-cellular transport of the membrane glyco-protein (G) of vesicular stomatitis virus in infected chinese hamster ovary cells. *J. Cell Biol.* 97:1777–87

Bergmann, J. E., Tokuyasu, K. T., Singer, S. J. 1981. Passage of an integral membrane protein, the vesicular stomatitis virus glycoprotein, through the Golgi apparatus en route to the plasma membrane. *Proc. Natl. Acad. Sci. USA* 78:1746–50

Bischoff, J., Kornfeld, R. 1983. Evidence for an α-mannosidase in endoplasmic reti-culum of rat liver. *J. Biol. Chem.* 258:7907–10

Borgese, N., Meldolesi, J. 1980. Localization and biosynthesis of NADH-cytochrome b_5 reductase, an integral membrane pro-tein, in rat liver cells. I. Distribution of the enzyme activity in microsomes, mito-chondria, and Golgi complex. *J. Cell Biol.* 85:501–15

Braell, W. A., Balch, W. E., Dobbertin, D. C., Rothman, J. E. 1984. The glycoprotein that is transported between successive com-partments of the Golgi in a cell-free system resides in stacks of cisternae. *Cell* 39:511–24

Brandan, E., Fleisher, B. 1982. Orientation and role of nucleosidediphosphatase and 5′-nucleotidase in Golgi vesicles from rat liver. *Biochemistry* 21:4640–45

Brands, R., Snider, M. D., Hino, Y., Park, S. S., Gelboin, H. V., Rothman, J. E. 1985. The retention of membrane proteins by the endoplasmic reticulum. *J. Cell Biol.* In press

Brown, M. S., Anderson, R. G. W., Goldstein, J. L. 1983. Recycling receptors: The round-trip itinerary of migrant membrane pro-teins. *Cell* 32:663–67.

Brown, W. J., Constantinescu, E., Farquhar, M. G. 1984. Redistribution of mannose-6-phosphate receptors induced by tunica-mycin and chloroquine treatments. *J. Cell Biol.* 99:320–26

Brown, W. J., Farquhar, M. G. 1984a. The mannose-6-phosphate receptor for lyso-somal enzymes is concentrated in cis Golgi cisternae. *Cell* 36:295–307

Brown, W. J., Farquhar, M. G. 1984b. Accu-mulation of coated vesicles bearing man-nose-6-phosphate receptors for lysosomal enzymes in the Golgi region of I cell fibroblasts. *Proc. Natl. Acad. Sci. USA* 81:5135–39

Brown, W. J., Farquhar, M. G. 1984c. Mannose-6-phosphate receptor recycling during recovery from chloroquine or NH$_4$Cl treatment. *J. Cell Biol.* 99:78a

Brown, W. J., Farquhar, M. G. 1985. Man-nose-6-phosphate receptors (215 Kd) are concentrated in heavy (cis) Golgi fractions in clone 9 hepatocytes. *J. Cell Biol.* In press

Burgess, T. L., Craik, C. S., Kelly, R. B. 1985. The exocrine protein trypsinogen is tar-geted into the secretory granules of an endocrine cell line: studies by gene trans-fer. *J. Cell Biol.* 101:639–45

Burgess, T. L., Kelly, R. B. 1984. Sorting and secretion of adrenocorticotropin in a pitui-tary tumor cell line after perturbation of the level of a secretory granule-specific proteoglycan. *J. Cell Biol.* 99:2223–30

Cameron, R. S. 1984. Parotid secretion granule membranes as a compositional probe: Identification of membrane poly-peptides potentially related to secretory function. PhD thesis. Yale University, New Haven, Conn. 172 pp.

Cameron, R. S., Castle, J. D. 1984. Isolation and compositional analysis of secretion granules and their membrane subfraction from the rat parotid gland. *J. Memb. Biol.* 79:127–44

Campbell, C. H., Rome, L. H. 1983. Coated vesicles from rat liver and calf brain con-tain lysosomal enzymes bound to mannose 6-phosphate receptors. *J. Biol. Chem.* 258:13347–52

Capasso, J. M., Hirschberg, C. B. 1984. Mechanisms of glycosylation and sulfation in the Golgi apparatus: Evidence for nucleotide sugar/nucleoside monophos-phate and nucleotide sulfate/nucleoside monophosphate antiports in the Golgi apparatus membrane. *Proc. Natl. Acad. Sci. USA* 81:7051–55

Castle, J. D., Jamieson, J. D., Palade, G. E. 1975. Secretion granules of the rabbit parotid gland. Isolation, subfractionation, and characterization of membrane and content subfractions. *J. Cell Biol.* 64:182–210

Chang, T.-L., Loh, Y. P. 1983. Characterization of proopiocortin converting activity in rat anterior pituitary secretory granules. *Endocrinology* 112:1832–38

Cheng, H., Farquhar, M. G. 1976. Presence of adenylate cyclase activity in Golgi and other fractions from rat liver. II. Cytochemical localization within Golgi and ER membranes. *J. Cell Biol.* 70:671–84

Ciechanover, A., Schwartz, A. L., Dautry-Varsat, A., Lodish, H. 1983. Kinetics of internalization and recycling of transferrin and the transferrin receptor in a human hepatoma cell line. *J. Biol. Chem.* 258:9681–89

Creek, K. E., Sly, W. 1984. The role of the phosphomannosyl receptor in the transport of acid hydrolases to lysosomes. In *Lysosomes in Biology and Pathology*, ed. J. T. Dingle, R. T. Dean, W. Sly, pp. 63–82. The Netherlands: Elsevier

Cummings, R. D., Kornfeld, S., Schneider, W. J., Hobgood, K. K., Tolleshauge, M., et al. 1983. Biosynthesis of N- and O-linked oligosaccharides of the low density lipoprotein receptor. *J. Biol. Chem.* 258:15261–73.

Dautry-Varsat, A., Lodish, H. F. 1983. The Golgi complex and the sorting of membrane and secreted proteins. *Trends Neurosci.* 6:484–90

Docherty, K., Hutton, J. C., Steiner, D. F. 1984. Cathepsin B-related proteases in the insulin secretory granule. *J. Biol. Chem.* 259:6041–44

Dorfman, A. 1981. Proteoglycan biosynthesis. In *Cell Biology of Extracellular Matrix*, ed. E. D. Hay, pp. 115–38. New York/London: Plenum

Dunphy, W. G., Brands, R., Rothman, J. E. 1985. Attachment of terminal N-acetylglucosamine to asparagine-linked oligosaccharides occurs in central cisternae of the Golgi stack. *Cell* 40:463–72

Dunphy, W. G., Fries, E., Urbani, L. J., Rothman, J. E. 1981. Early and late functions associated with the Golgi apparatus reside in distinct compartments. *Proc. Natl. Acad. Sci. USA* 78:7453–57

Dunphy, W. G., Rothman, J. E. 1983. Compartmentation of asparagine-linked oligosaccharide processing in the Golgi apparatus. *J. Cell Biol.* 97:270–75

Elbein, A. D. 1983. Inhibitors of glycoprotein synthesis. *Methods Enzymol.* 98:135–54

Elhammer, A., Kornfeld, S. 1984. Two enzymes involved in the synthesis of O-linked oligosaccharides are localized on membranes of different densities in mouse lymphoma BW5147 cells. *J. Cell Biol.* 98:327–31

Farquhar, M. G. 1971. Processing of secretory products by cells of the anterior

pituitary gland. In *Subcellular Structure and Function in Endocrine Organs*, ed. H. Heller, K. Lederis, pp. 79–122. Cambridge, England: Cambridge Univ. Press

Farquhar, M. G. 1978a. Recovery of surface membrane in anterior pituitary cells. Variations in traffic detected with anionic and cationic ferritin. *J. Cell Biol.* 77:R35–R42

Farquhar, M. G. 1978b. Traffic of products and membranes through the Golgi complex. In *Transport of Macromolecules in Cellular Systems*, ed. S. Silverstein, pp. 341–62. Berlin: Dahlem Konferenzen

Farquhar, M. G. 1981. Membrane recycling in secretory cells: Implications for traffic of products and specialized membranes within the Golgi complex. *Methods Cell Biol.* 23:399–427

Farquhar, M. G. 1982. Membrane recycling in secretory cells: Pathway to the Golgi complex. *Ciba Found. Symp.* 92:157–74

Farquhar, M. G. 1983. Multiple pathways of exocytosis, endocytosis and membrane recycling: Validation of a Golgi route. *Fed. Proc.* 42:2407–13

Farquhar, M. G., Bergeron, J. J. M., Palade, G. E. 1974. Cytochemistry of Golgi fractions prepared from rat liver. *J. Cell Biol.* 60:8–25

Farquhar, M. G., Palade, G. E. 1981. The Golgi apparatus (complex)—(1954–1981) —from artifact to center stage. *J. Cell Biol.* 77s–103s.

Fitting, T., Kabat, D. 1982. Evidence for a glycoprotein "signal" involved in transport between subcellular organelles. Two membrane glycoproteins encoded by murine leukemia virus reach the cell surface at different rates. *J. Biol. Chem.* 257:14011–17

Fleischer, B. 1983. Mechanism of glycosylation in the Golgi apparatus. *J. Histochem. Cytochem.* 31:1033–40

Fleischer, B., Fleischer, S., Ozawa, H. 1969. Isolation and characterization of Golgi membranes from bovine liver. *J. Cell Biol.* 43:59–79

Fleischer, B., Smigel, M. 1978. Solubilization and properties of galactosyltransferase and sulfotransferase activities of Golgi membranes in Triton X-100. *J. Biol. Chem.* 253:1632–38.

Fleischer, B., Zambrano, F. 1974. Golgi apparatus of rat kidney. Preparation and role in sulfatide formation. *J. Biol. Chem.* 249:5995–6003

Fletcher, M. P., Gallin, J. I. 1983. Human neutrophils contain an intracellular pool of putative receptors for the chemoattractant N-formyl-methionyl-leucyl-phenylalanine. *Blood* 62:792–99

Friend, D. S. 1969. Cytochemical staining of multivesicular body and Golgi vesicles. *J. Cell Biol.* 41:269–79

Friend, D. S., Farquhar, M. G. 1967. Functions of coated vesicles during protein absorption in the rat vas deferens. *J. Cell Biol.* 35:357–76

Friend, D. S., Murray, M. J. 1965. Osmium impregnation of the Golgi apparatus. *Am. J. Anat.* 117:135–50

Fries, E., Gustafsson, L., Peterson, P. A. 1984. Four secretory proteins synthesized by hepatocytes are transported from endoplasmic reticulum to Golgi complex at different rates. *EMBO J.* 3:147–52

Fuller, S. D., Bravo, R., Simons, K. 1985. An enzymatic assay reveals that proteins destined for the apical or basolateral domains of an epithelial cell line share the same late Golgi compartment. *EMBO J.* 4:297–308

Gabel, C. A., Goldberg, D. E., Kornfeld, S. 1983. Identification and characterization of cells deficient in the mannose 6-phosphate receptor: Evidence for an alternate pathway for lysosomal enzyme targeting. *Proc. Natl. Acad. Sci. USA* 80:775–79

Geetha-Habib, M., Campbell, S. C., Schwartz, N. B. 1984. Subcellular localization of the synthesis and glycosylation of chondroitin sulfate proteoglycan core protein. *J. Biol. Chem.* 259:7300–10

Geuze, H. J., Slot, J. W., Strous, G. J. A. M., Hasilik, A., von Figura, K. 1984. Ultrastructural localization of the mannose 6-phosphate receptor in rat liver. *J. Cell Biol.* 98:2047–54

Geuze, H. J., Slot, J. W., Strous, G. J. A. M., Lodish, H. F., Schwartz, A. L. 1983. Intracellular site of asialoglycoprotein receptor-ligand uncoupling: Double-label immunoelectron microscopy during receptor-mediated endocytosis. *Cell* 32:277–87

Geuze, H. J., Slot, J. W., Tokuyasu, K. T. 1979. Immunocytochemical localization of amylase and chymotrysinogen in the exocrine pancreatic cell with special attention to the Golgi complex. *J. Cell Biol.* 82:697–707

Glickman, J., Croen, K., Kelly, S., Al-Awqati, Q. 1983. Golgi membranes contain an electrogenic H^+ pump in parallel to a chloride conductance. *J. Cell Biol.* 97:1303–8

Godelaine, D., Spiro, M. J., Spiro, R. G. 1981. Processing of the carbohydrate units of thyroglobulin. *J. Biol. Chem.* 256:10161–68

Goldberg, D., Gabel, C., Kornfeld, S. 1984. Processing of lysosomal enzyme oligosaccharide units. In *Lysosomes in Biology and Pathology*, ed. J. T. Dingle, R. T. Dean, W. Sly, pp. 45–62. The Netherlands: Elsevier

Goldberg, D. E., Kornfeld, S. 1983. Evidence for extensive subcellular organization of asparagine-linked oligosaccharide processing and lysosomal enzyme phosphorylation. *J. Biol. Chem.* 258:3159–65

Goldfischer, S. 1982. The internal reticular apparatus of Camillo Golgi: A complex, heterogeneous organelle, enriched in acid, neutral, and alkaline phosphatases, and involved in glycosylation, secretion, membrane flow, lysosome formation, and intracellular digestion. *J. Histochem. Cytochem.* 30:717–33

Gonatas, J., Stieber, A., Olsnes, S., Gonatas, N. K. 1980. Pathways involved in fluid phase and adsorptive endocytosis in neuroblastoma. *J. Cell Biol.* 87:579–88

Gonatas, N. K. 1982. The role of the neuronal Golgi apparatus in a centripetal membrane vesicular traffic. *J. Neuropath. Exp. Neurol.* 41:6–17

Gonzalez-Noriega, A., Grubb, J. H., Talkad, V., Sly, W. S. 1980. Chloroquine inhibits lysosomal enzyme pinocytosis and enhances lysosomal enzyme secretion by impairing receptor recycling. *J. Cell Biol.* 85:839–52

Gottlieb, T. A., Gonzalez, A., Rizzolo, L., Rindler, M. J., Adesnik, M., Sabatini, D. D. 1985. Sorting and endocytosis of viral glycoproteins in transfected polarized epithelial cells. *J. Cell Biol.* Submitted for publication

Grassé, P. P. 1957. Ultrastructure polarité et réproduction de l'appareil de Golgi. *C. R. Acad. Sci.* 245:1228–81

Griffiths, G., Brands, R., Burke, B., Louvard, D., Warren, G. 1982. Viral membrane proteins acquire galactose in trans Golgi cisternae during intracellular transport. *J. Cell Biol.* 95:781–92

Griffiths, G., Pfeiffer, S., Simons, K., Matlin, K. 1985. Exit of newly synthesized membrane proteins from the trans cisterna of the Golgi complex to the plasma membrane. *J. Cell Biol.* In press

Griffiths, G., Quinn, P., Warren, G. 1983. Dissection of the Golgi complex I. Monensin inhibits the transport of viral membrane proteins from medial to trans Golgi cisternae in baby hamster kidney cells infected with Semliki forest virus. *J. Cell Biol.* 96:835–50

Grinna, L. S., Robbins, P. W. 1979. Glycoprotein biosynthesis. Rat liver microsomal glucosidases which process oligosaccharides. *J. Biol. Chem.* 254:8814–18

Gumbiner, B., Kelly, R. B. 1982. Two distinct intracellular pathways transport secretory and membrane glycoproteins to the surface of pituitary tumor cells. *Cell* 28: 51–59

Hand, A. R., Oliver, C. 1984. Effects of secretory stimulation on the Golgi apparatus and GERL of rat parotid acinar cells. *J. Histochem. Cytochem.* 32: 403–12

Hanover, J. A., Elting, J., Mintz, G. R., Lennarz, W. J. 1982. Temporal aspects of the N- and O-glycosylation of human chorionic gonadotropin. *J. Biol. Chem.* 257: 10172–77

Hasilik, A., von Figura, K. 1981. Oligosaccharides in lysosomal enzymes. Distribution of high-mannose and complex oligosaccharides in cathepsin D and β-hexosaminidase. *Eur. J. Biochem.* 121: 125–29

Herzog, V., Farquhar, M. G. 1977. Luminal membrane retrieved after exocytosis reaches most Golgi cisternae in secretory cells. *Proc. Natl. Acad. Sci. USA* 74: 5073–77

Herzog, V., Miller, F. 1970. Die lokalisation endogener peroxydase in der glandular parotis de ratte. *Z. Zell. Mikrosk. Ant.* 107: 403–20

Herzog, V., Miller, F. 1979. Membrane retrieval in epithelial cells of isolated thyroid follicles. *Eur. J. Cell Biol.* 19: 203–15

Herzog, V., Reggio, H. 1980. Pathways of endocytosis from luminal plasma membrane in rat exocrine pancreas. *Eur. J. Cell Biol.* 21: 141–50

Hoffmann, H.-P., Schwartz, N. B., Roden, L., Prockop, D. J. 1984. Location of xylosyltransferase in the cisternae of the rough endoplasmic reticulum of embryonic cartilage cells. *Connect. Tissue Res.* 12: 151–64

Hoflack, B., Kornfeld, S. 1985. Lysosomal enzyme binding to mouse P388D$_1$ macrophage membranes lacking the 215Kd mannose 6-phosphate receptor: Evidence for the existence of a second mannose 6-phosphate receptor. *Proc. Natl. Acad. Sci. USA* 80: 4428–32

Hook, V. Y. H., Loh, Y. P. 1984. Carboxypeptidase B-like converting enzyme activity in secretory granules of rat pituitary. *Proc. Natl. Acad. Sci. USA* 81: 2776–80

Hopkins, C. R. 1983. Intracellular routing of transferrin and transferrin receptors in epidermoid carcinoma A431 cells. *Cell* 35: 321–30

Hopkins, C. R., Trowbridge, I. S. 1983. Internalization and processing of transferrin and the transferrin receptor in human carcinoma A431 cells. *J. Cell Biol.* 97: 508–21

Horwitz, A., Dorfman, A. 1968. Subcellular sites for synthesis of chondromucoprotein of cartilage. *J. Cell Biol.* 38: 358–69

Hubbard, S. C., Ivatt, R. J. 1981. Synthesis and processing of asparagine-linked oligosaccharides. *Ann. Rev. Biochem.* 50: 555–83

Ito, A., Palade, G. E. 1978. Presence of NADPH-cytochrome P-450 reductase in rat liver Golgi membranes. *J. Cell Biol.* 79: 590–97

Jamieson, J. D., Palade, G. E. 1967a. Intracellular transport of secrctory proteins in the pancreatic exocrine cell. I. Role of the peripheral elements of the Golgi complex. *J. Cell Biol.* 34: 577–96

Jamieson, J. D., Palade, G. E. 1967b. Intracellular transport of secretory proteins in the pancreatic exocrine cell. II. Transport to condensing vacuoles and zymogen granules. *J. Cell Biol.* 34: 597–616

Jamieson, J. D., Palade, G. E. 1968. Intracellular transport of secretory proteins in the pancreatic exocrine cell. IV. Metabolic requirements. *J. Cell Biol.* 39: 589–603

Johnson, D. C., Spear, P. G. 1983. O-linked oligosaccharides are acquired by Herpes simplex virus glycoproteins in the Golgi apparatus. *Cell* 32: 987–97

Kelly, R. B., Burgess, T. L., Moore, H.-P. H. 1985. Sorting of secretory proteins in an endocrine cell line studied by DNA transfection. In *Protein Transport and Secretion*, ed. M. J. Gething, pp. 156–161. Cold Spring Harbor, NY: Cold Spring Harbor

Kimura, J. H., Lohmander, L. S., Hascall, V. C. 1984. Studies on the biosynthesis of cartilage proteoglycan in a model system of cultured chondrocytes from the swarm rat chondrosarcoma. *J. Cell Biochem.* 26: 261–78

Kornfeld, S., Lang, L., Hoflack, B. 1985. Lysosomal enzyme targeting. *Fed. Proc.* 44(3): ix

Kuhn, N. J., White, A. 1977. The role of nucleoside diphosphatase in an uridine nucleotide cycle associated with lactose synthesis in rat mammary gland Golgi apparatus. *Biochem. J.* 168: 423–33

Lang, L., Reitman, M., Tang, J., Roberts, R. M., Kornfeld, S. 1984. Lysosomal enzyme phosphorylation. Recognition of a protein-dependent determinant allows specific phosphorylation of oligosaccharides present on lysosomal enzymes. *J. Biol. Chem.* 259: 14663–71

Leblond, C. P., Bennett, G. 1976. Role of the Golgi apparatus in terminal glycosylation. In *International Cell Biology*, ed. B. R. Brinkley, K. R. Porter, pp. 326–36. New York: Rockefeller Univ. Press

Ledger, P. W., Tanzer, M. L. 1984. Monensin—a perturbant of cellular physiology. *TIBS* 9: 313–19

Lennarz, W. 1983. Overview: Role of intracellular membrane systems in glycosylation of proteins. *Methods Enzymol.* 98:91–97

Lodish, H. F., Kong, N. 1984. Glucose removal from N-linked oligosaccharides is required for efficient maturation of certain secretory glycoproteins from the rough endoplasmic reticulum to the Golgi complex. *J. Cell Biol.* 98:1720–29

Lodish, H. F., Kong, N., Snider, M., Strous, G. J. A. M. 1983. Hepatoma secretory proteins migrate from rough endoplasmic reticulum to Golgi at characteristic rates. *Nature* 304:80–83

Lucocq, J. M., Brada, D., Roth, J. 1984. Immunolocalization of glucosidase II in pig liver. *J. Cell Biol.* 99 (No. 4, Pt. 2):354a

MacDonald, R. J., Ronzio, R. A. 1972. Comparative analysis of zymogen granule membrane polypeptides. *Biochem. Biophys. Res. Commun.* 49:377–82

Matlin, K., Simons, K. 1984. Sorting of a plasma membrane glycoprotein occurs before it reaches the cell surface in cultured epithelial cells. *J. Cell Biol.* 99:2131–39

Matlin, K. S., Simons, K. 1983. Reduced temperature prevents transfer of a membrane glycoprotein to the cell surface but does not prevent terminal glycosylation. *Cell* 34:233–43

Matsura, S., Fujii-Kuriyama, Y., Tashiro, Y. 1978. Immunoelectron microscope localization of cytochrome P-450 on microsomes and other membrane structures of rat hepatocytes. *J. Cell Biol.* 78:503–19

Meldolesi, J. 1974. Dynamics of cytoplasmic membranes in guinea pig pancreatic acinar cells. I. Synthesis and turnover of membrane proteins. *J. Cell Biol.* 61:1–13

Meldolesi, J., Ceccarelli, B. 1981. Exocytosis and membrane recycling. *Philos. Trans. R. Soc. London B* 296:55–65

Meldolesi, J., Cova, D. 1972. Composition of cellular membranes in the pancreas of the guinea pig. IV. Polyacrylamide gel electrophoresis and amino acid composition of membrane proteins. *J. Cell Biol.* 55:1–18

Merisko, E. M., Farquhar, M. G., Palade, G. E. 1982. Coated vesicle isolation by immunoadsorption on *Staphylococcus aureus* cells. *J. Cell Biol.* 92:846–58

Misek, D. E., Bard, E., Rodriguez-Boulan, E. 1984. Biogenesis of epithelial cell polarity: Intracellular sorting and vectorial exocytosis of an apical plasma membrane glycoprotein. *Cell* 39:537–46

Moore, H.-P., Gumbiner, B., Kelly, R. B. 1983a. A subclass of proteins and sulfated macromolecules secreted by AtT-20 (mouse pituitary tumor) cells is sorted with adrenocorticotropin into dense secretory granules. *J. Cell Biol.* 97:810–17

Moore, H.-P., Gumbiner, B., Kelly, R. B. 1983b. Chloroquine diverts ACTH from a

regulated to a constitutive secretory pathway in AtT-20 cells. *Nature* 302:434–36

Moore, H.-P., Kelly, R. B. 1985. Secretory protein targeting in a pituitary cell line. Differential transport of foreign secretory proteins to distinct secretory pathways. *J. Cell Biol.* In press

Moore, H.-P., Walker, M. D., Lee, F., Kelly, R. B. 1983c. Expressing a human proinsulin cDNA in a mouse ACTH-secreting cell. Intracellular storage, proteolytic processing, and secretion on stimulation. *Cell* 35:531–38

Morré, D. J. 1976. Membrane differentiation and the control of secretion: A comparison of plant and animal Golgi apparatus. In *International Cell Biology*, ed. B. R. Brinkley, K. R. Porter, pp. 293–303. New York: Rockefeller Univ. Press

Morré, D. J., Franke, W. W., Deumling, B., Nyquist, S. E., Ovtracht, L. 1971. Golgi apparatus function in membrane flow and differentiation: Origin of plasma membranes from endoplasmic reticulum. *Biomembranes* 2:95–104

Morré, D. J., Kartenbeck, J., Franke, W. W. 1979. Membrane flow and interconversions among endomembranes. *Biochim. Biophys. Acta* 559:71–152

Morré, D. J., Ovtracht, L. 1977. Dynamics of the Golgi apparatus: Membrane differentiation and membrane flow. *Int. Rev. Cytol.* 5:61–188

Munro, J. R., Narasimhan, S., Wetmore, S., Riordan, J. R., Schachter, H. 1975. Intracellular localization of GDP-L-fucose: Glycoprotein and CMP-sialic acid: apolipoprotein glycosyltransferases in rat and pork livers. *Arch. Biochem. Biophys.* 169:269–77

Neutra, M., Leblond, C. P. 1966. Radioautographic comparison of the uptake of galactose-^{3}H and glucose-^{3}H in the Golgi region of various cells secreting glycoproteins or mucopolysaccharides. *J. Cell Biol.* 30:137–50

Novikoff, A. B. 1976. The endoplasmic reticulum: A cytochemist's view (a review). *Proc. Natl. Acad. Sci. USA* 73:2781–87.

Novikoff, A. B., Novikoff, P. M. 1977. Cytochemical contributions to differentiating GERL from the Golgi apparatus. *Histochem. J.* 9:525–51

Novikoff, P. M., Tulsiani, D. R. P., Touster, O., Yam, A., Novikoff, A. B. 1983. Immunocytochemical localization of α-D-mannosidase II in the Golgi apparatus of rat liver. *Proc. Natl. Acad. Sci. USA* 80:4364–68

Orci, L., Halban, P., Amherdt, M., Ravazzola, M., Vassalli, J.-D., Perrelet, A. 1984. A clathrin-coated, Golgi-related compartment of the insulin secreting cell accumulates proinsulin in the presence of monensin. *Cell* 39:39–47

Orci, L., Montesano, R., Meda, P., Malaisse-Lagae, F., Brown, D., et al. 1981. Heterogeneous distribution of filipin-cholesterol complexes across the cisternae of the Golgi apparatus. *Proc. Natl. Acad. Sci. USA* 78:293–97

Ottosen, P. D., Courtoy, P. J., Farquhar, M. G. 1980. Pathways followed by membrane recovered from the surface of plasma cells and myeloma cells. *J. Exp. Med.* 152:1–19

Paavola, L. G. 1978. The corpus luteum of the guinea pig. III. Cytochemical studies on the Golgi complex, GERL and lysosomes in luteal cells during maximal progesterone secretion. *J. Cell Biol.* 79:45–58

Palade, G. 1959. Functional changes in the structure of cell components. In *Subcellular Particles*, ed. T. Hayashi, pp. 64–80. New York: Ronald

Palade, G. 1975. Intracellular aspects of the process of protein secretion. *Science* 189:347–58

Pastan, I., Willingham, M. C. 1985. The pathway of endocytosis. In *Endocytosis*, ed. I. Pastan, M. C. Willingham, pp. 1–44. New York: Plenum

Pavelka, M., Ellinger, A. 1983. The trans Golgi face in rat small intestinal absorptive cells. *Eur. J. Cell Biol.* 29:253–61

Peyrieras, N., Bause, E., Legler, G., Vasilov, R., Claesson, L., et al. 1983. Effects of the glucosidase inhibitors nojirimycin and deoxynojirimycin on the biosynthesis of membrane and secretory glycoproteins. *EMBO J.* 2:823–32

Pfeffer, S. R., Kelly, R. B. 1985. The subpopulation of brain coated vesicles that carries synaptic vesicle proteins contains two unique polypeptides. *Cell* 40:949–57

Pfeiffer, S., Fuller, S. D., Simons, K. 1985. Intracellular sorting and basolateral appearance of the G protein of vesicular stomatitis virus in MDCK cells. *J. Cell Biol.* In press

Pohlmann, R., Waheed, A., Hasilik, A., von Figura, K. 1982. Synthesis of phosphorylated recognition marker in lysosomal enzymes is located in the cis part of Golgi apparatus. *J. Biol. Chem.* 257:5323–25

Quinn, P., Griffiths, G., Warren, G. 1983. Dissection of the Golgi complex. II. Density separation of specific Golgi functions in virally infected cells treated with monensin. *J. Cell Biol.* 96:851–56

Rambourg, A., Clermont, Y., Hermo, L. 1979. Three-dimensional architecture of the Golgi apparatus in Sertoli cells of the rat. *Am. J. Anat.* 154:455–70

Regoeczi, E., Chindemi, P. A., DeBanne, M. T., Charlwood, P. A. 1982. Partial resialylation of human asialotransferrin

type 3 in the rat. *Proc. Natl. Acad. Sci. USA* 79:2226–30

Rindler, M. J., Ivanov, I. E., Plesken, H., Rodriguez-Boulan, E. J., Sabatini, D. D. 1984. Viral glycoproteins destined for apical or basolateral plasma membrane domains traverse the same Golgi apparatus during their intracellular transport in doubly infected Madin-Darby canine kidney cells. *J. Cell Biol.* 98:1304–19

Rindler, M. J., Ivanov, I. E., Plesken, H., Sabatini, D. D. 1985. Polarized delivery of viral glycoproteins to the apical and basolateral plasma membranes of Madin-Darby canine kidney cells infected with temperature-sensitive viruses. *J. Cell Biol.* 100:136–51

Rindler, M. J., Ivanov, I. E., Rodriguez-Boulan, E. J., Sabatini, D. D. 1982. Biogenesis of epithelial cell plasma membranes. *Ciba Found. Symp.* 92:184–208

Rizzolo, L. J., Gonzalez, A., Gottlieb, T. A., Finidori, J., Ivanov, I. E., et al. 1985. Intracellular sorting and distinct recycling patterns of viral glycoproteins in polarized epithelial cells. In *Current Communications in Molecular Biology. Protein Transport and Secretion*, ed. M. J. Gething, pp. 147–51. Cold Spring Harbor, NY: Cold Spring Harbor Lab.

Roden, L. 1980. Structure and metabolism of connective tissue proteoglycans. In *The Biochemistry of Glycoproteins and Proteoglycans*, ed. W. J. Lennarz, pp. 267–371. New York: Plenum

Rodriguez-Boulan, E. 1983. Membrane biogenesis, enveloped RNA viruses, and epithelial polarity. In *Modern Cell Biology*, ed. B. Satir, pp. 119–70. New York: Liss

Rodriguez-Boulan, E., Paskiet, K. T., Salas, P. J. I., Bard, E. 1984. Intracellular transport of influenza virus hemagglutinin to the apical surface of Madin-Darby canine kidney cells. *J. Cell Biol.* 98:308–19

Rodriguez-Boulan, E., Pendergast, M. 1980. Polarized distribution of viral envelope proteins in the plasma membrane of infected epithelial cells. *Cell* 20:45–54

Rodriguez-Boulan, E., Sabatini, D. D. 1978. Asymmetric budding of viruses in epithelial monolayers: a model system for study of epithelial polarity. *Proc. Natl. Acad. Sci. USA* 75:5071–75

Rosenfeld, M. G., Kreibich, G., Popov, D., Kato, K., Sabatini, D. D. 1982. Biosynthesis of lysosomal hydrolases: Their synthesis in bound polysomes and the role of co- and post-translational processing in determining their subcellular distribution. *J. Cell Biol.* 93:135–43

Roth, J. 1984. Cytochemical localization of terminal N-acetyl-D-galactosamine residues in cellular compartments of intestinal goblet cells: Implications for the topology of O-glycosylation. *J. Cell Biol.* 98:399–406

Roth, J., Berger, E. G. 1982. Immunocytochemical localization of galactosyltransferase in HeLa cells: Codistribution with thiamine pyrophosphatase in trans-Golgi cisternae. *J. Cell Biol.* 93:223–29

Roth, J., Lentze, M. J., Berger, E. G. 1985. Immunocytochemical demonstration of ecto-galactosyltransferase in absorptive intestinal cells. *J. Cell Biol.* 100:118–25

Roth, J., Lucocq, J. M., Berger, E. G., Paulson, J. C., Watkins, W. M. 1984. Terminal glycosylation is compartmentalized in the Golgi apparatus. *J. Cell Biol.* 99(No. 4, Pt. 2):229a

Roth, M. D., Compans, R. W., Giusti, L., Davi, A. R., Mayak, D. P., et al. 1983. Influenza virus hemagglutinin expression is polarized in cells infected with recombinant SV40 viruses carrying cloned hemagglutinin DNA. *Cell* 33:435–43

Rothman, J. E. 1981. The Golgi apparatus: Two organelles in tandem. *Science* 213:1212–19

Rothman, J. E., Miller, R. L., Urbani, L. J. 1984a. Intercompartmental transport in the Golgi complex is a dissociative process: Facile transfer of membrane protein between two Golgi populations. *J. Cell Biol.* 99:260–71

Rothman, J. E., Pettegrew, H. G., Fine, R. E. 1980. Transport of the membrane glycoprotein of the vesicular stomatitis virus to the cell surface in two stages by clathrin-coated vesicles. *J. Cell Biol.* 86:162–71

Rothman, J. E., Urbani, L. J., Brands, R. 1984b. Transport of protein between cytoplasmic membranes of fused cells: Correspondence to processes reconstituted in a cell-free system. *J. Cell Biol.* 99:248–59

Sahagian, G. G. 1984. The mannose 6-phosphate receptor: Function, biosynthesis and translocation. *Biol. Cell* 51:207–14

Sahagian, G. G., Distler, J., Jourdian, G. W. 1981. Characterization of a membrane-associated receptor from bovine liver that binds phosphomannosyl residues of bovine testicular β-galactosidase. *Proc. Natl. Acad. Sci. USA* 78:4289–93

Salpeter, M., Farquhar, M. G. 1981. High resolution autoradiographic analysis of the secretory pathway in mammotrophs of the rat anterior pituitary. *J. Cell Biol.* 91:240–46

Saraste, J., Hedman, K. 1983. Intracellular vesicles involved in the transport of Semliki Forest virus membrane proteins to the cell surface. *EMBO J.* 2:2001–6

Saraste, J., Kuismanen, E. 1984. Pre- and post-Golgi vacuoles operate in the transport of Semliki Forest virus membrane glycoproteins to the cell surface. *Cell* 38:535–49

Schachter, H., Jabbal, I., Hudgin, R. L., Pinteric, L., McGuire, E. J., Roseman, S. 1970. Intracellular localization of liver sugar nucleotide glycoprotein glycosyltransferases in a Golgi-rich fraction. *J. Biol. Chem.* 245:1090–1110

Scheele, G., Tartakoff, A. 1985. Exit of nonglycosylated secretory proteins from the rough endoplasmic reticulum is asynchronous in the exocrine pancreas. *J. Biol. Chem.* 260:926–31

Sly, W. S., Fischer, H. D. 1982. The phosphomannosyl recognition system for intracellular and intercellular transport of lysosomal enzymes. *J. Cell Biochem.* 18:67–85

Smith, C. E. 1980. Ultrastructural localization of nicotinamide adenine dinucleotide phosphatase (NADPase) activity to the intermediate saccules of the Golgi apparatus in rat incisor ameloblasts. *J. Histochem. Cytochem.* 28:16–26

Smith, R. E., Farquhar, M. G. 1966. Lysosome function in the regulation of the secretory process in cells of the anterior pituitary gland. *J. Cell Biol.* 31:319–47

Smith, R. E., Farquhar, M. G. 1970. Modulation in nucleoside diphosphatase activity of mammotrophic cells of the rat adenohypophysis during secretion. *J. Histochem. Cytochem.* 18:237

Snider, M. D., Rogers, O. C. 1985. Intracellular movement of cell surface receptors after endocytosis: Resialyation of asialotransferrin receptor in human erythroleukemia cells. *J. Cell Biol.* 100:826–34

Steiner, A. W., Rome, L. H. 1982. Assay and purification of a solubilized membrane receptor that binds the lysosomal enzyme α-L-iduronidase. *Arch. Biochem. Biophys.* 214:681–87

Steiner, D. F., Kicherty, K., Carroll, R. 1984. Golgi/granule processing of peptide hormone and neuropeptide precursors: A minireview. *J. Cell Biochem.* 24:121–30

Strous, G. J. A. M., Lodish, H. F. 1980. Intracellular transport of secretory and membrane proteins in hepatoma cells infected by vesicular stomatitis virus. *Cell* 22:709–17

Strous, G. J. A. M., Willemsen, R., van Kerkhof, P., Slot, J. W., Geuze, H. J., Lodish, H. F. 1983. Vesicular stomatitis virus glycoprotein, albumin, and transferrin are transported to the cell surface via the same Golgi vesicles. *J. Cell Biol.* 97:1815–22

Tabas, I., Kornfeld, S. 1979. Purification and characterization of a rat liver Golgi α-mannosidase capable of processing asparagine-linked oligosaccharides. *J. Biol. Chem.* 254:11655–63

Tartakoff, A. M. 1980. The Golgi complex: crossroads for vesicular traffic. *Int. Rev. Exp. Pathol.* 22:227–51

Tartakoff, A. M. 1983a. Perturbation of vesicular traffic with the carboxylic ionophore monensin. *Cell* 32:1026–28

Tartakoff, A. M. 1983b. The confined function model of the Golgi complex: Center for ordered processing of biosynthetic products of the rough endoplasmic reticulum. *Int. Rev. Cytol.* 85:221–52

Tartakoff, A. M., Vassalli, P. 1977. Plasma cell immunoglobulin secretion. Arrest is accompanied by alterations of the Golgi complex. *J. Exp. Med.* 146:1332–45

Tartakoff, A., Vassalli, P. 1978. Comparative studies of intracellular transport of secretory proteins. *J. Cell Biol.* 79:694–707

Tartakoff, A. M., Vassalli, P. 1983. Lectin-binding sites as markers of Golgi subcompartments: Proximal-to-distal maturation of oligosaccharides. *J. Cell Biol.* 97:1243–48

Tashiro, Y., Nakada, H., Masaki, R., Yamamoto, Y. 1984. Transport of membrane proteins in rat liver Golgi apparatus. In *International Cell Biology*, ed. S. Sem, Y. Okado, p. 98. Tokyo, Japan: Jpn. Soc. Cell Biol.

Tulsiani, D. R. P., Harris, T. M., Touster, O. 1982. Swainsonine inhibits the biosynthesis of complex glycoproteins by inhibition of Golgi mannosidase II. *J. Biol. Chem.* 257:7936–39

Varki, A., Kornfeld, S. 1980. Identification of a rat liver α-N-acetylglucosaminyl phosphodiesterase capable of removing "blocking" α-N-acetylglucosamine residues from phosphorylated high mannose oligosaccharides of lysosomal enzymes. *J. Biol. Chem.* 255:8398–8401

Vertel, B. M., Barkman, L. L. 1984. Immunofluorescence studies of chondroitin sulfate proteoglycan biosynthesis. The use of monoclonal antibodies. *Collagen Rel. Res.* 4:1–20

Waheed, A., Hasilik, A., von Figura, K. 1982. UDP-N-acetylglucosamine: lysosomal enzyme precursor N-acetylglucosamine-1-phosphotransferase. *J. Biol. Chem.* 257:12322–31

Wallach, D., Kirshner, N., Schramm, M. 1975. Non-parallel transport of membrane proteins and content proteins during assembly of the secretory granule in rat parotid gland. *Biochim. Biophys. Acta* 375:87–105

Walter, P., Gilmore, R., Blobel, G. 1984. Protein translocation across the endoplasmic reticulum. *Cell* 38:5–8

Wehland, J., Willingham, M. C., Gallo, M. G., Pastan, I. 1982. The morphologic pathway of exocytosis of the vesicular stomatitis virus G protein in cultured fibroblasts.

Cell 28:831–41

Whur, P., Herscovics, A., Leblond, C. P. 1969. Radioautographic visualization of the incorporation of galactose-^{3}H and mannose-^{3}H by rat thyroids in vitro in relation to the stages of thyroglobulin synthesis. *J. Cell Biol.* 43:289–311

Wilcox, D. K., Kitson, R. P., Widnell, C. C. 1982. Inhibition of pinocytosis in rat embryo fibroblasts treated with monensin. *J. Cell Biol.* 92:859–64

Wileman, T., Boshans, R. L., Schlesinger, P., Stahl, P. 1984. Monensin inhibits recycling of macrophage mannose-glycoprotein receptors and ligand delivery to lysosomes. *Biochem. J.* 220:665–75

Willingham, M. C., Hanover, J. A., Dickson, R. B., Pastan, I. 1984. Morphologic characterization of the pathway of transferrin endocytosis and recycling in human KB cells. *Proc. Natl. Acad. Sci. USA* 81:175–79

Willingham, M. C., Pastan, I. H. 1982. Transit of epidermal growth factor through coated pits of the Golgi system. *J. Cell Biol.* 94:207–12

Willingham, M. C., Pastan, I. H. 1984. Ultrastructural immunocytochemical localization of the transferrin receptor using a monoclonal antibody in human KB cells. *J. Histochem. Cytochem.* 33:59–64

Willingham, M. C., Pastan, I. H., Sahagian, G. G. 1983. Ultrastructural immunocytochemical localization of the phosphomannosyl receptor in chinese hamster ovary (CHO) cells. *J. Histochem. Cytochem.* 31:1–11

Winkler, H., Hortnagl, H., Smith, A. D. 1970. Membrane of the adrenal medulla. Behavior of insoluble proteins of chromaffin granules on gel electrophoresis. *Biochem. J.* 118:303–10

Woods, J. W., Farquhar, M. G. 1984. The transferrin receptor cycles through the Golgi complex in myeloma cells. *J. Cell Biol.* 99(4, pt. 2):375a

Woods, J. W., Farquhar, M. G. 1985. Ligand binding redirects transferrin receptor recycling among Golgi subcompartments. *J. Cell Biol.* In press

Yamashiro, D. J., Tycko, B., Fluss, S. R., Maxfield, F. R. 1984. Segregation of transferrin to a mildly acidic (pH 6.5) para-Golgi compartment in the recycling pathway. *Cell* 37:789–800

Young, R. W. 1973. The role of the Golgi complex in sulfate metabolism. *J. Cell Biol.* 57:175–89

Zhang, F., Schneider, D. L. 1983. The bioenergetics of Golgi apparatus function: Evidence for an ATP-dependent proton pump. *Biochem. Biophys. Res. Comm.* 114:620–25

Ann. Rev. Cell Biol. 1985. 1:489–530
Copyright © 1985 by Annual Reviews Inc. All rights reserved

BIOGENESIS OF PEROXISOMES

P. B. Lazarow and Y. Fujiki[1]

The Rockefeller University, New York, New York 10021

CONTENTS

PERSPECTIVES AND SUMMARY

The peroxisome is nearly ubiquitous in eukaryotic cells, and it plays diverse metabolic roles, depending on cell type. Many peroxisomes in animal, plant, and fungal cells share at least two biochemical capabilities:

---

[1] Dr. Fujiki's current address is the Meiji Institute of Health Science, 540 Naruda, Odawara, 250 Japan.

0743–4634/85/1115–0489$02.00

respiration (based on H_2O_2 metabolism) and fatty acid oxidation. These and other common features, discussed later, justify treating peroxisomes (including the specialized peroxisomes in germinating seeds called glyoxysomes) as one kind of organelle.

Examination of the past five years' literature on the biogenesis of individual, well-characterized peroxisomal proteins, together with recent morphological observations and other experiments, reveals an unexpectedly simple pattern. All such proteins are synthesized on free polyribosomes; these include matrix and crystalloid core proteins, as well as one major integral membrane protein. Most of them are made at their final size. Posttranslational import in vivo has been observed for many: They move from the cytosolic to peroxisomal fraction with half-times ranging from 1 to 15 min or more. Posttranslational import has been reconstituted in vitro for several proteins, which in the future will help elucidate the details of the import mechanism. The proteolytic processing of those exceptional proteins that are synthesized as large precursors appears not to be tightly coupled to import; rather, processing occurs slowly after import (in at least one case, over hours). Endoplasmic reticulum is presumed to synthesize the membrane phospholipids of peroxisomes (as it is presumed to do for mitochondria), but endoplasmic reticulum plays no known role in the formation of peroxisomal (glyoxysomal) proteins, including membrane proteins.

In yeasts, peroxisomes are observed to form by division. In several types of animal and plant cells serial sectioning has revealed extensive interconnections among peroxisomes, which sometimes form elaborate compartments consisting of branched, tortuous tubules and larger cup-shaped structures. Parts of these compartments resemble endoplasmic reticulum. In many other cells peroxisomes apparently exist as individual entities, but the presence of clustered, dumbbell, and tail forms suggests that transient interconnections among peroxisomes are formed via fission and fusion. Biochemical evidence for a uniform peroxisomal pool supports the concept of a "peroxisomal reticulum." The former practice of interpreting peroxisomal tails as connections to endoplasmic reticulum presently appears unjustified.

These experimental observations afford a simple interpretation: Peroxisomes grow by the posttranslational incorporation of new content and membrane proteins into preexisting peroxisomes, which then divide to form daughter peroxisomes. In this respect peroxisomes are like mitochondria and chloroplasts.

This model of peroxisome biogenesis raises interesting questions concerning the evolutionary origin of the organelle, and makes experimentally testable predictions. Moreover, this model, together with regulation of peroxisomal gene expression at the level of transcription (which has been demonstrated in several cases), explains simply many aspects of peroxisome regulation, including induction by drugs and the "glyoxysome-peroxisome transition."

Further experimental testing of this model is essential, especially of the biogenesis of well-characterized integral membrane proteins, for which only limited information is presently available. DNA cloning and sequencing projects currently under way in several laboratories promise to help elucidate the details of the biogenetic mechanism(s).

INTRODUCTION

Functions and Inducibility

Peroxisomes carry out a diverse set of metabolic functions, which vary with cell type. These include respiration (based on H_2O_2-forming oxidases and catalase); fatty acid β-oxidation; the initial reactions in ether glycerolipid (plasmalogen) biosynthesis; the glyoxalate cycle; photorespiration; alcohol oxidation; transaminations; and purine and polyamine catabolism (reviewed by Tolbert 1981). Their morphological appearance also varies. They are often spherical, ranging from 0.2 to 1.0 μm in diameter, and sometimes contain a crystalline core. However, peroxisomes in the form of small tubules and elaborate pleomorphic cell compartments are also observed (see below). Peroxisomes are found in nearly all eukaryotic cells, but their abundance varies from one or two to hundreds per cell.

One striking feature of peroxisome biology is that the organelle is inducible. Proliferation of rat liver peroxisomes is induced five- to ninefold by plasticizers and other compounds (reviewed by Reddy et al 1982). Peroxisomes may be induced in some *Candida* species and other yeasts by growth on methanol, alkanes, or fatty acids (reviewed by Fukui & Tanaka 1979; Veenhuis et al 1983). Glyoxysomes are induced in fatty seeds by germination. These means of inducing peroxisome proliferation have been exploited in biogenesis studies.

Nomenclature

Four names have been used to describe this organelle: peroxisome, glyoxysome, microbody, and microperoxisome. The initial biochemical characterization of peroxisomes came from the pioneering work of de Duve and his colleagues in subfractionating the mitochondrial fraction of rat liver, which led first to the discovery of the lysosome (de Duve 1975). "Peroxisome" was defined by de Duve as an organelle containing at least one oxidase to form H_2O_2 and catalase to decompose it (de Duve & Baudhuin 1966). The name refers to hydrogen peroxide metabolism, not to peroxidases. Although catalase is capable of physiologically significant peroxidation (Chance & Oshino 1971) in addition to its catalatic reaction (Nicholls & Schonbaum 1963), simple peroxidases (e.g. glutathione, horse-radish) are nonperoxisomal. The early peroxisome research was elegantly reviewed by de Duve & Baudhuin (1966).

The term glyoxysome was introduced by Breidenbach & Beevers (1967) to describe a plant organelle found by them in castor bean endosperm that contained the five enzymes of the glyoxalate cycle. Subsequently, this organelle was found to contain the oxidases and catalase characteristic of peroxisomes (Beevers 1969).

Both plant glyoxysomes, e.g. in castor bean (Cooper & Beevers 1969; Hutton & Stumpf 1969), and peroxisomes, e.g. in liver (Lazarow & de Duve 1976), *Candida tropicalis* (Kawamoto et al 1978a), and *Tetrahymena pyriformis* (Blum 1973), contain a set of enzymes that catalyze the β-oxidation of fatty acids. The properties of the β-oxidation enzymes (discussed in detail later) are similar in these several distant species. The first enzyme is an H_2O_2-forming acyl-CoA oxidase, as opposed to the dehydrogenase found in mitochondria. The second and third enzymes (enoyl-CoA hydratase and β-hydroxyacyl-CoA dehydrogenase) are found together on a bi- or trifunctional protein, as opposed to the individual enzymes found in mitochondria.

Isocitrate lyase and malate synthetase, key enzymes of the glyoxalate cycle, have been found in peroxisomes of *T. pyriformis* (Müller et al 1968), *C. tropicalis* (Fukui & Tanaka 1979), and even in toad bladder (Jones et al 1982), although they are believed to be absent from most animal tissues. Peroxisomes and glyoxysomes both have high equilibrium densities in sucrose gradients (~ 1.23 g/cm^3). Taken together, these facts suggest an evolutionary relationship between animal, plant, fungal, and protozoan peroxisomes.

The biochemical properties of peroxisomes change strikingly during the development of fat-storing dicotyledons. At the beginning of germination, when fat is converted to carbohydrate, abundant "glyoxysomes" containing an active β-oxidation system and glyoxalate cycle are closely associated with lipid droplets. Later, after greening, the cotyledons contain abundant "peroxisomes" that carry out photosynthetic glycolate catabolism and are often closely associated with chloroplasts. During an intermediate stage both sets of enzymatic activities are present in the cotyledons. Beevers (1979) considered that there are two separate particle populations during the transition: one containing the glyoxalate cycle enzymes and the other the glycolate pathway enzymes. Trelease et al (1971) suggested that the transition occurred within one population of existing particles by a change in the complement of enzymes. The latter idea has been supported by cytochemical, biochemical, and morphometric analyses (Burke & Trelease 1975; Schopfer et al 1976; Köller & Kindl 1978). Recent experiments by Titus & Becker (1985) have unequivocally demonstrated, by immuno-electron microscopy with two sizes of protein A–gold, that both types of enzymes are present within the same organelle during the transition.

"Microbody" is a morphological term for a subcellular structure bounded by one membrane with a homogeneous matrix and a diameter generally between 0.2 and 1.0 μm. It was first used by Rhodin (1954) for rat kidney particles, and by Rouiller & Bernhard (1956) for similar structures in rat liver. In this tissue they contain a dense paracrystalline core structure (or "crystalloid"), which makes them easy to recognize. The term microbody has been used to describe similar structures in other cell types, many of which lack cores (Hruban & Rechcigl 1969). Cell fractionation and cyto-chemistry have established that microbodies are usually, but not always, peroxisomes.

One exception is the hydrogenosome of trichomonads (reviewed by Müller 1980). Although this organelle resembles most other microbodies, it is biochemically entirely different (it forms H_2 and ATP), and recent evidence indicates that it is bounded by two membranes (Honigberg et al 1984).

Another apparent exception is the glycosome of the parasitic Try-panosomatids, which catalyzes the reactions of the glycolytic pathway up to 3-phosphoglycerate formation (Opperdoes et al 1984). The glycosome has the characteristic appearance of a microbody, including in some cases a crystalline core, and an equilibrium density of 1.23 g/cm^3. Although biochemically the glycosome differs greatly from the peroxisome, both contain the first enzymes in the pathway of ether lipid biosynthesis (Hajra & Bishop 1982; Opperdoes 1984).

"Microperoxisome" has been redefined twice; it is a morphological term meaning a particle that has a positive cytochemical reaction for catalase and that is somewhat small (0.2 instead of 0.5 μm diameter), or coreless, or near the endoplasmic reticulum, or some combination of the above (Novikoff et al 1973). Those that have been investigated biochemically have turned out to be bona fide peroxisomes, often with β-oxidation activity, e.g. in intestine (Small et al 1980), adrenal cortex (Russo & Black 1982), and heart (Connock & Perry 1983). In mouse liver, serial sectioning revealed that small coreless peroxisomes are connected to larger core-containing peroxisomes (Gorgas 1985).

Historical Studies on Biogenesis

Historically, peroxisomes were thought to form by budding from the endoplasmic reticulum. This was based upon electron micrographs interpreted by Novikoff & Shin (1964) as showing connections between peroxisomes and endoplasmic reticulum, and on an early report of Higashi & Peters (1963) that newly synthesized catalase is found in rat liver microsomes. De Duve & Baudhuin (1966) included this hypothesis in their review. Many subsequent electron micrographs have been interpreted in a fashion consistent with the budding theory.

Plant glyoxysomes also were thought at one time to form by budding from the endoplasmic reticulum. Gonzalez & Beevers (1976) found considerable malate synthetase activity in castor bean microsomes at the time of maximal glyoxysome formation. Bowden & Lord (1976a,b) reported that the membranes of glyoxysomes and endoplasmic reticulum had similar polypeptide compositions and that ^{35}S-labeled proteins were synthesized in endoplasmic reticulum and transferred to glyoxysomes. These and subsequent data supporting the budding hypothesis (summarized by Beevers 1982) are controversial, and have been disputed and discussed in several recent reviews (Kindl 1982a; Lord & Roberts 1983; Huang et al 1983; Trelease 1984). It is not our intention to re-review the older literature. We simply point out that much of this work was done in the days before SDS-PAGE and fluorography became routine, and that without these tools immunoprecipitates may be contaminated with small amounts of highly radioactive impurities. Kruse & Kindl (1983a) have shown that malate synthetase forms large aggregates that happen to cosediment with microsomes, and serious questions have been raised about possible cross-contamination of fractions. In the next section we review the more recent literature, focusing on those papers in which modern molecular tools have been used to investigate the synthesis of well-characterized proteins, and adequate controls have been carried out. Older papers will be included that meet these criteria and that contributed significantly to our current understanding.

The first experiment that critically tested a prediction of the budding hypothesis was carried out by Brian Poole. Rats were labeled with [^{3}H]leucine and the incorporation of radioactive catalase into small and large peroxisomes was measured as a function of time. The experiment was designed to look for the gradual enlargement of peroxisomes forming as dilations of the endoplasmic reticulum membrane (Poole et al 1970). The results were uniformly negative. This did not disprove the budding hypothesis, but in retrospect, it was the first of many unsuccessful attempts to find compelling evidence for it.

BIOGENESIS OF INDIVIDUAL PEROXISOMAL PROTEINS

In this section we review the experimental data concerning the manner of synthesis, intracellular transport, and assembly of individual, well characterized proteins. Some of their molecular properties are summarized in Table 1. For those readers not requiring all the details, the main results are summarized in Table 1 and under the heading "Conclusions."

Integral Membrane Proteins

Prominent among the polypeptides of the rat liver peroxisomal membrane is one with an M_r of 22,000 (Fujiki et al 1982a). It is an integral membrane protein, as judged by the criterion of resistance to extraction with 0.1 M Na_2CO_3, which has been shown to separate peripheral from integral membrane proteins (Fujiki et al 1982b), and it is rich in hydrophobic amino acids (Fujiki et al 1984). It is exposed on the cytosolic face of the membrane (proteolytically nicked to M_r 21,000); whether it is also exposed on the luminal face is not presently known. The mRNA encoding this polypeptide is found in free polyribosomes. Its cell-free translation product (in the reticulocyte lysate protein-synthesizing system) is the same size as the mature protein (Fujiki et al 1984).

Catalase

Catalase is a tetrameric hemoprotein (Nicholls & Schonbaum 1963) with known sequence (Schroeder et al 1982; Korneluk et al 1984) and three-dimensional structure (Murthy et al 1981). Two catalases contain NADPH (Kirkman & Gaetani 1984; Fita & Rossman 1985).

LIVER CATALASE   Catalase is the most abundant protein in the peroxisomes of normal rat liver (Lazarow et al 1982a). It is located in the matrix space of the organelle, from which it readily leaks out if the peroxisomes are osmotically or mechanically damaged (Alexson et al 1985).

Rat liver catalase is synthesized in vivo as a heme-less precursor, which is far more labile than the mature enzyme to denaturation with the organic solvents used for catalase purification (Lazarow & de Duve 1971, 1973a). The precursor cosediments in sucrose gradients with hemoglobin (M_r 68,000 native size) (Lazarow & de Duve 1973b), and in SDS-PAGE it has an M_r of 65–66,000 (Robbi & Lazarow 1978), thereby identifying it as an apomonomer. There are 13 μg of this apomonomer per gram of liver, and it has a half-life of 34 min. Heme addition produces a second biosynthetic intermediate, presumed to be the monomer, which amounts to 4 μg/g liver, and turns over with a half-life of 12 min (Lazarow & de Duve 1973a). Upon

Table 1 Biosynthesis of peroxisomal proteins

Enzyme / Cell Type	Subunit size[a] ($M_r \times 10^{-3}$)	In vitro size difference[b]	Ribosomes (F/B)[c]	Initial site in vivo[d]	Import in vitro[e]	Oligomer	Prosthetic group	pI
Integral membrane protein								
Rat liver	22	0	F	—	—	—	—	—
Catalase								
Rat liver	65–66	0	F	cytosol	yes	4	heme	6.5–6.7
C. tropicalis	54–57	0	—	—	—	4	heme	—
H. polymorpha	56	0	—	—	—	4	heme	—
Pumpkin cotyledon, active[f]	55	4	—	—	—	4	heme	—
Acyl-CoA oxidase								
Rat liver, subunit A	72–77[g]	0	F	cytosol	yes	2[g]	FAD	9.2
C. tropicalis	72–76	0	—	—	—	8	FAD	5.2–5.5
Hydratase-dehydrogenase								
Rat liver	70–81	0	F	cytosol	—	1	—	9.5–9.9
Cucumber cotyledon	75	0	—	cytosol	—	1	—	9.8
C. tropicalis (& epimerase)	102	0	—	—	—	2	—	—
Thiolase								
Rat liver	40–41	3–6	F	cytosol	—	2	—	—
Isocitrate lyase								
Cucumber cotyledon	62–64	0–1.5?[h]	—	cytosol	—	4	—	5.9
Castor bean endosperm	62	0	F	cytosol	—	4	—	—
N. crassa[i]	67	0	F	—	yes	4	—	—

Enzyme / source	$M_r \times 10^{-3}$ [a]	Translation product difference [b]	mRNA (F/B) [c]	Location [d]	Import [e]	No. of subunits	Cofactor	mRNA (kb)
Malate synthetase								
Cucumber cotyledon	57–63	0	—	cytosol	yes	8	—	9
Castor bean endosperm	64	0	—	cytosol	—	8	—	—
N. crassa[i]	59	0	F	—	—	—	—	—
Malate dehydrogenase								
Watermelon cotyledon	33	8	—	—	yes	2	—	8.9
Cucumber cotyledon	33–34	5[j]	—	—	—	2	—	—
Glycolate oxidase								
Cucumber cotyledon	43	0	—	cytosol	—	4[k]	FMN	8.7
Lentil leaf	43	0	—	—	—	—	—	—
Carnitine acetyltransferase								
C. tropicalis, large subunit	64	7	—	—	—	?	—	} 5.1
C. tropicalis, small subunit	57	0	—	—	—	?	—	
Carnitine octanoyltransferase								
Rat liver	64–66	0	F	—	—	1	—	—
Urate oxidase								
Rat liver	31–37	0	F	—	—	3, 6[l]	—	—
Alcohol oxidase								
C. boidinii	73–74	0	—	cytosol	—	8	FAD	—
H. polymorpha	75–83	0	—	—	—	8	FAD	—

[a] By SDS-PAGE. Range of values reported in different laboratories indicated. [b] (M_r of cell-free translation product) − (M_r of enzyme subunit). [c] mRNA found in free (F) or membrane-bound (B) polyribosomes. [d] Location of newly synthesized protein after pulse-labeling (cytosol or endoplasmic reticulum). [e] Post-translational import of translation product into organelle in a cell-free system. [f] Pumpkin cotyledons also contain an inactive catalase tetramer with a subunit M_r of 59,000, which probably lacks heme. [g] Subunits B (M_r 45–52,000) and C (19–22,000) are derived from subunit A by slow proteolytic cleavage within peroxisomes; oligomers are thought to be a mixture of A_2, ABC, and B_2C_2. [h] Conflicting results, see text. [i] The microbody containing isocitrate lyase and malate synthetase in N. crassa has unusual properties (see text). [j] In the wheat germ system. The difference for the watermelon cell-free product was also 5000 in the wheat germ system, but 8000 in the reticulocyte system. [k] At an early stage in the purification the enzyme behaves as a hexadecamer. [l] The purified enzyme behaves as a trimer or hexamer; in situ the protein forms the large crystalloid core.

cell fractionation 8 min after pulse-labeling of rats, the apomonomer is found predominantly in the high speed supernatant fraction (Redman et al 1972; Lazarow & de Duve 1973b; Lazarow et al 1980). It then enters peroxisomes (showing clear precursor-product kinetics) with a half-time of 14 min; the bulk of heme addition occurs inside the peroxisome (Lazarow & de Duve 1973b). The 8-min old extraperoxisomal apomonomeric precursor made in vivo and the subunit of mature peroxisomal catalase are indistinguishable by SDS-PAGE (Robbi & Lazarow 1978) or by one- and two-dimensional peptide mapping (Robbi & Lazarow 1982).

Messenger RNA encoding liver catalase is found to the extent of 83–89% in a free polyribosome fraction both in normal rats (Goldman & Blobel 1978) and in rats treated with clofibrate to cause peroxisome proliferation (Rachubinski et al 1984). Its cell-free translation product comigrates in SDS-PAGE with the subunit of mature catalase (~ 66 kDa) (Robbi & Lazarow 1978; Goldman & Blobel 1978). The cell-free translation product and the mature enzyme subunit are indistinguishable by one- and two-dimensional peptide mapping (Robbi & Lazarow 1982). When in vitro translation is performed with N-formyl-[^{35}S]Met-tRNA$_f^{Met}$ the labeled catalase is the same size (~ 66 kDa). Normally the (nonformylated) initiator methionine is removed, presumably cotranslationally (Housman et al 1970; Palmiter 1983), because the amino-terminal residue of the bovine liver enzyme is a blocked alanine (Schroeder et al 1982).

These experiments on catalase biogenesis in vivo and in vitro establish that this rat liver enzyme is imported into peroxisomes by a posttranslational mechanism without detectable modification of the primary structure. Most binding of the prosthetic heme group and oligomerization are observed to occur within the peroxisomes. Posttranslational import of the catalase apomonomer into purified peroxisomes has been reproduced in vitro; a cytosolic factor(s) stimulates the import (Fujiki & Lazarow 1985).

In rats treated with allylisopropylacetamide, which disrupts heme metabolism and prevents the synthesis of active catalase, monomeric apocatalase accumulates within peroxisomes (Sugita et al 1982). This confirms that apomonomers enter peroxisomes, and that heme addition precedes tetramerization. However, intraperoxisomal catalase assembly is not obligatory, because in Zellweger syndrome, a rare human disease in which hepatic peroxisomes are missing (Goldfischer et al 1973), active catalase is present in normal amounts in the liver but it is located in the cytosol (Wanders et al 1984; Lazarow et al 1985).

Higashi and his coworkers advocate a role of the endoplasmic reticulum in catalase biogenesis. They argue that mRNA encoding catalase destined for peroxisomes is translated on membrane-bound polyribosomes (Tobe &

Higashi 1980), and that newly synthesized catalase precursor of high specific radioactivity is found in the microsomal fraction (Sugita et al 1982). These data are unconvincing because the radioactivity was measured on immunoprecipitates without benefit of SDS-PAGE and fluorography. They further suggest that catalase is synthesized on membrane-bound polyribosomes, then released into the cytosol, later taken up posttranslationally into the lumen of endoplasmic reticulum, and finally translocated to peroxisomes. This view cannot be reconciled with our current understanding that protein translocation into the endoplasmic reticulum lumen is a cotranslational event mediated by a signal recognition particle and docking protein (recently reviewed by Walter et al 1984).

YEAST CATALASE   Catalase activity is induced, together with a proliferation of peroxisomes, in *Candida tropicalis* by growth on n-alkanes or oleic acid, and in *Hansenula polymorpha* by growth on methanol (Fukui & Tanaka 1979). Catalase mRNA in both species encodes a cell-free translation product that comigrates in SDS gels with the subunit of the mature enzyme (Yamada et al 1982b; Roa & Blobel 1983; Fujiki et al 1985b). One-dimensional peptide maps of the *C. tropicalis* translation product and purified enzyme give similar patterns, arguing against appreciable posttranslational modification of the primary structure (Yamada et al 1982b).

The presence of catalase in *Saccharomyces cerevisiae* peroxisomes was described by Szabo & Avers (1969) and Hoffmann et al (1970). However, Susani et al (1976) reported that catalase is located in the vacuole and cytosol. In any event, peroxisomes are rare in *S. cerevisiae* and, unlike those in other yeast species, have not responded to attempts to induce them. The synthesis of catalase in *S. cerevisiae* follows the same biochemical steps as in rat liver, i.e. it is synthesized at its final subunit size (Ammerer & Ruis 1979; Ammerer et al 1981), and is assembled via heme-less and heme-containing intermediates (Zimniak et al 1975, 1976).

PUMPKIN CATALASE   Catalase biogenesis in pumpkins has unusual features (Yamaguchi & Nishimura 1984; Yamaguchi et al 1984). It is synthesized in vivo and in vitro with an M_r of 59,000, and is imported into peroxisomes (glyoxysomes) posttranslationally in vivo without change in its size or one-dimensional peptide map. Inside, two things can happen. It may be processed proteolytically to M_r 55,000, bind heme, and aggregate to an active tetrameric enzyme with a native M_r of 230,000 (the sequence of these steps is unknown). Or it can escape proteolysis and assemble to a tetramer with little or no enzymatic activity. This tetramer is apparently slightly smaller (more compact?) by gel filtration (M_r 215,000) and it may lack heme. The preponderance of these two maturation pathways depends on the developmental stage. In etiolated (dark-grown) cotyledons, which have typical glyoxysome-type peroxisomes, most of the catalase is processed to the active form. After greening of the cotyledons, 60% of the catalase remains unprocessed within "peroxisomes." The two types of tetramers are antigenically indistinguishable and copurify (except for a slight separation in gel filtration). The result is that peroxisomal catalase appears to have only 40% the specific activity of glyoxysomal catalase, whereas it actually consists of active and inactive forms.

The mechanism of catalase synthesis in other plant species has not been fully clarified. Size differences between cell-free product and mature enzyme subunit have been observed in cucumber (Kindl 1982b; Riezman et al 1980). Lentil leaf catalase mRNA encodes a cell-free product that has the same M_r as the largest of several catalase species observed in the leaves (Kindl 1982b).

β-Oxidation Enzymes

Peroxisomes contain an enzyme system for the β-oxidation of fatty acids (Cooper & Beevers 1969; Hutton & Stumpf 1969; Lazarow & de Duve 1976; Lazarow 1978; Kawamoto et al 1978a). The reactions are similar to those of the mitochondrial β-oxidation system, but the enzymes catalyzing them are different (Cooper & Beevers 1969; Hashimoto 1982; Moreno de la Garza et al 1985), and their regulation and specificity are also different (Lazarow 1982; Osmundsen 1982; Mannaerts & Debeer 1982; Leighton et al 1982). The peroxisomal β-oxidation system can be induced in liver by hypolipidemic drugs such as clofibrate (Lazarow 1977), and plasticizers like di(2-ethylhexyl)phthalate (Hashimoto 1982); these and other compounds cause massive peroxisome proliferation (Reddy et al 1982). The system is induced in C. tropicalis by growth on n-alkanes or oleic acid (Kawamoto et al 1978a).

Although two of the liver β-oxidation enzymes have been attributed to the peroxisomal core (Hayashi et al 1981), other biochemical (Huang & Beevers 1973; Hüttinger et al 1980; Frevert & Kindl 1980b; Alexson et al 1985) and immunoelectron microscopic (Yokota & Fahimi 1982; Bendayan & Reddy 1982) evidence places all these enzymes in the matrix in both liver and seedlings.

ACYL-CoA OXIDASE   The first enzyme of the β-oxidation spiral is an H_2O_2-forming acyl-CoA oxidase, which in liver is rate limiting for, and defines the specificity of, the overall spiral pathway (Osumi & Hashimoto 1979b; Inestrosa et al 1979). The rat liver enzyme purified by Osumi et al (1980) consists of three polypeptides, A, B, and C, with sizes of 72–75,000, 50–52,000, and 19–21,000, respectively (Table 1). The amino acid composition of subunits B and C together is similar to that of subunit A. Subunits A and

B comigrate with major clofibrate-induced peroxisomal polypeptides (Lazarow et al 1982a). The oxidase purified by Inestrosa et al (1980) consists mainly of 45,000 and 22,000 m_r subunits with some 77,000 polypeptides.

Acyl-CoA oxidase is synthesized in liver on free polyribosomes, both in clofibrate-treated rats (92% free, 8% bound) (Rachubinski et al 1984) and in di(2-ethylhexyl)phthalate-treated rats (82% free, 18% bound) (Miura et al 1984). A single cell-free translation product was found in both cases. It was first reported to be slightly larger than the A subunit (Furuta et al 1982a), but later was shown to be identical in size (Rachubinski et al 1984; Miura et al 1984). The cell-free translation product also has the same isoelectric point and one-dimensional peptide map as the native A subunit (Miura et al 1984). In isotopic labeling experiments in vivo or in isolated hepatocytes, radioactivity appears at first only in the A subunit; after an hour label also can be detected in the B (Miura et al 1984) or B plus C subunits (Furuta et al 1982b). These data suggest that subunit A is slowly cleaved in vivo to the two smaller subunits.

The acyl-CoA oxidase cell-free translation product has been imported into liver peroxisomes in vitro. During this process it did not undergo proteolytic processing (Fujiki & Lazarow 1985).

C. tropicalis acyl-CoA oxidase differs from the rat liver enzyme by being an octamer of a single subunit with a lower pI (Table 1). It is similar to the mammalian enzyme in that it is an H_2O_2-forming oxidase, and the size and amino acid composition of its subunit resemble those of the liver subunit A (Shimizu et al 1979; Coudron & Frerman 1982; Coudron et al 1983; Jiang & Thorpe 1983). It is synthesized in vitro at its final subunit size (M_r 76,000) (Fujiki et al 1985b).

BIFUNCTIONAL ENOYL-CoA HYDRATASE/β-HYDROXYACYL-CoA DEHYDRO-GENASE   The second and third reactions in the β-oxidation spiral are cata-lyzed by a single, bifunctional protein in rat liver (Osumi & Hashimoto 1979a) and cucumber cotyledons (Frevert & Kindl 1980). In C. tropicalis, a larger trifunctional protein catalyzes these reactions as well as β-hydroxyacyl-CoA epimerization (Moreno de la Garza et al 1985).

The bifunctional hydratase-dehydrogenase is synthesized at its final size on free polyribosomes both in clofibrate-treated rats (91% free, 9% bound) (Rachubinski et al 1984), and in di(2-ethylhexyl)phthalate-treated rats (75% free, 25% bound) (Miura et al 1984). The cell-free translation product of hydratase-dehydrogenase mRNA also had the same pI and one-dimensional peptide map as the mature protein (Miura et al 1984). In an experiment in which isolated hepatocytes were labeled for 1 hr with [^{35}S]methionine, 60% of the labeled bifunctional protein was found in a cytosolic fraction at 10 min, whereas at 60 min 80% of a much larger amount of labeled protein was in a particulate fraction (presumably peroxisomes) (Miura et al 1984). Because the experimental design involved continuous rather than pulse-chase labeling, the observed kinetics support, but do not rigorously prove, the posttranslational import of hydratase-dehydrogenase into peroxisomes in the isolated cells.

The cucumber bifunctional protein (Kindl 1982b) and the *C. tropicalis* trifunctional protein (Fujiki et al 1985b) are likewise synthesized in vitro at their final sizes. The cucumber enzyme has been shown to pass through a cytosolic pool en route to the peroxisomes (Frevert et al 1980).

THIOLASE   The biogenesis of the fourth β-oxidation spiral enzyme, thiolase, has been investigated only in liver. Thiolase mRNA is found in free polyribosomes in di(2-ethylhexyl)phthalate-treated rats (Muira et al 1984), and in normal and clofibrate-treated rats (Fujiki et al 1985a). The cell-free translation product had an M_r that was 3000 (Furuta et al 1982a) or 7000 (Fujiki et al 1985a) greater than the purified enzyme. A similar discrepancy had been found for rat liver catalase translation product and purified enzyme subunit, but this was due to the artifactual proteolysis of catalase during its purification (Robbi & Lazarow 1978). Immunoblotting of thiolase antigen in liver (homogenized in the presence of protease inhibitors) and of the purified enzyme demonstrated that both have an M_r of 41,000 (Fujiki et al 1985a). This excludes artifactual proteolysis and confirms that thiolase, unlike other rat liver peroxisomal proteins studied thus far, is synthesized as a larger precursor.

The proteolytic processing of thiolase was investigated in the hepatocyte-labeling experiment described above (Miura et al 1984). Labeling of both the larger precursor and the processed subunit increased throughout the 1 hr incubation. The same ratio of processed to unprocessed thiolase was found in the cytosolic fraction as in the peroxisomal fraction at each time point studied, leading the investigators to conclude that the "result is not consistent with proteolytic processing being linked with transport into the peroxisome."

Glyoxalate Cycle Enzymes

Isocitrate lyase, malate synthetase, malate dehydrogenase, citrate synthetase, and aconitase catalyze the conversion of two acetyl-CoA to succinate. All five enzymes are found in plant glyoxysomes, whereas in other cells, e.g. *T. pyriformis* (Müller et al 1968) and *C. tropicalis* (Kawamoto et al 1977), isocitrate lyase and malate synthetase are in peroxisomes and the other three enzymes are mitochondrial.

PLANTS   Malate synthetase of cucumber and castor bean is synthesized in vivo and in vitro at its final subunit size (Riezman et al 1980; Kruse et al

1981; Lord & Roberts 1982). In vivo studies in cucumber showed that malate synthetase passes through a cytosolic pool en route to the glyoxysomes (Köller & Kindl 1980; Frevert et al 1980). The 5s monomer is amphipathic, and has a strong tendency to bind lipid or to oligomerize to a 19s octamer or aggregate to a 100s particle (Kruse & Kindl 1983b). This 100s form cosediments with endoplasmic reticulum (Kruse & Kindl 1983b), which explains previous observations of "microsomal malate synthetase" (Gonzalez 1982). Posttranslational import of malate synthetase into glyoxysomes has been demonstrated in vitro (Kruse et al 1981; Kruse & Kindl 1983a).

The isocitrate lyase cell-free translation product of castor bean endo-sperm mRNA comigrated in SDS-PAGE with the glyoxysomal enzyme purified in the presence of phenylmethylsulfonyl fluoride (M_r 62,000), as well as with the in vivo-labeled protein immunoprecipitated from total cell homogenate (Roberts & Lord 1981). Isocitrate lyase was not cotransla-tionally segregated into dog pancreatic microsomes under conditions that resulted in segregation of control plant proteins. In vivo labeling experiments suggested a slow translocation of isocitrate lyase from cytosol to glyoxysomes.

Pulse-chase labeling experiments in ripening cucumber seeds (3 hr pulse, 15 hr chase) followed by cell fractionation demonstrated that newly made isocitrate lyase was present first in a cytosolic pool (at its final size) and later in glyoxysomes (Frevert et al 1980). In contrast, Riezman et al (1980) observed two cucumber isocitrate lyase subunits, M_r 63,000 and 61,500, in the purified enzyme and in the in vivo-labeled protein. These forms were immunologically cross-reactive and yielded similar one-dimensional peptide maps after limited proteolysis. They are said to be regulated differ-ently in development. Two cell-free translation products were observed with M_r 61,500 and 60,000. The relationship between the smaller in vitro products and the two mature forms is still uncertain, but glycosylation was excluded (Riezman et al 1980).

Malate dehydrogenase of watermelon, which has a subunit M_r of 33,000, is exceptional in that it is synthesized in the wheat germ cell-free protein synthesizing system as a larger precursor with M_r 38,000 (Walk & Hock 1978; Riezman et al 1980), and in the reticulocyte system as a still larger 41,000 form (Gietl & Hock 1982). It is suggested that the size difference between the two translation systems is the result of proteolysis in the plant system. A precursor-product relationship between a 41,000 and a 33,000 form in vivo was demonstrated by pulse-chase labeling (1 hr pulse, 1–3 hr chase) (Gietl & Hock 1982). The cell-free translation product is imported posttranslationally into glyoxysomes in vitro in a process that is reported to depend upon the presence of some exogenous protease (Gietl & Hock

1984). The authors speculate that processing may immediately precede import in vivo, which would be highly unusual.

NEUROSPORA CRASSA   *Neurospora crassa* (grown on acetate) is exceptional in that it has two types of microbodies (Wanner & Theimer 1982). One contains catalase and urate oxidase, and has an equilibrium density in sucrose gradients of 1.23–1.24. The other contains isocitrate lyase, malate synthetase, and a small amount of catalase, and has a density of 1.22. When grown on oleate, enzymes catalyzing β-oxidation appear in the lighter particle, together with the glyoxalate cycle enzymes, but catalase is absent (Kionka & Kunau 1985). Surprisingly, the first of these β-oxidation enzymes is an acyl-CoA dehydrogenase, previously found only in mitochondria. The second and third enzymes reside together on a single protein, as in peroxisomes. The relationship of this microbody to other glyoxysomes and peroxisomes requires further study.

Both isocitrate lyase and malate synthetase are synthesized in *N. crassa* on free polyribosomes at their final sizes (Zimmerman & Neupert 1980; Desel et al 1982). One-dimensional peptide maps of the in vivo– and in vitro–synthesized isocitrate lyase are identical. The isocitrate lyase translation product was imported posttranslationally in vitro into a particulate fraction of acetate-grown *N. crassa* (by the criterion of protease resistance) (Zimmerman & Neupert 1980). No such import occurred with a similar fraction from glucose-grown cells, which contain few of these microbodies. Thus the mechanism of the biogenesis of isocitrate lyase and malate synthetase in *N. crassa* is similar to that in higher plants, despite the fact that the microbody that houses them has unusual properties.

Other Matrix Proteins

GLYCOLATE OXIDASE   Cucumber glycolate oxidase has a subunit M_r of 43,000 (Behrends et al 1982). The cell-free translation product also has an M_r of 43,000, and exists as the monomer (Gerdes et al 1982). When cotyledons were labeled in vivo with [^{35}S]methionine, radioactivity was found first in the cytosolic fraction, and only much later in peroxisomes. Oligomerization occurred to some extent in the cytosol, and to a larger extent in the peroxisomes (Gerdes et al 1982). Lentil glycolate oxidase is likewise synthesized in vitro at its final size of 43,000 (Kindl 1982b).

CARNITINE ACYLTRANSFERASES   In rat liver, acyltransferases are found in both peroxisomes and mitochondria. Peroxisomes contain a carnitine octanoyltransferase (Markwell et al 1976; Miyazawa et al 1983b) that is synthesized on free polyribosomes at its final size of 66,000 (Ozasa et al 1983). In addition they contain a carnitine acetyltransferase, but the biogenesis of this enzyme has been studied only in mitochondria (Miya-zawa et al 1983a).

In *C. tropicalis*, carnitine acetyltransferases are also found in both peroxisomes and mitochondria (Kawamoto et al 1978b). They are believed to function in shuttling acetyl groups generated by peroxisomal β-oxidation to mitochondria for condensation with oxalacetate, forming citrate (see glyoxalate cycle enzymes, above). The peroxisomal and mitochondrial carnitine acetyltransferases have similar properties and are immunochemically indistinguishable (Ueda et al 1982, 1984a). However, the peroxisomal enzyme contains subunits of 64,000 and 57,000 M_r, whereas the mitochondrial enzyme contains subunits of 64,000 and 52,000 M_r. The enzymes also differ in pI and in chromatographic behavior on DEAE-Sephacel.

Two mRNAs were distinguished in *C. tropicalis* that encode cell-free translation products (M_r 57,000 and doublet at 71,000) recognized by antiserum against carnitine acetyltransferase (Ueda et al 1984b). The relationship between translation products and enzyme subunits cannot be unambiguously decided because the antiserum does not distinguish between the four subunits (two peroxisomal and two mitochondrial). Ueda & coworkers (1984a,b) suggest that the peroxisomal and mitochondrial subunits are formed by differential posttranslational modifications of common gene products (ignoring the doublet nature of the larger translation product). Gene duplication and independent transcription could also account for the experimental data, including the doublet.

Crystalloid Core Proteins

RAT LIVER URATE OXIDASE   Urate oxidase in rat liver forms the large, electron-dense paracrystalline cores seen in electron micrographs of peroxisomes (Figure 1, *arrow*) (Hruban & Swift 1964; Tsukada et al 1966; Leighton et al 1968). Rat liver mRNA encoding urate oxidase is located in free polyribosomes. The protein is synthesized in vitro at its final size, and it is not cotranslationally imported into dog pancreas microsomes under conditions in which pre-proalbumin is so transported (Goldman & Blobel 1978). In vivo, urate oxidase must enter peroxisomes posttranslationally with a half-time less than 1–2 min, because in pulse-chase labeling studies transport of urate oxidase into peroxisomes was complete at the earliest time point investigated (4 min) (Lazarow et al 1982b).

YEAST ALCOHOL (METHANOL) OXIDASE   Several yeasts can be grown in methanol; this causes the formation of abundant peroxisomes, which contain an induced alcohol oxidase (Fukui & Tanaka 1979; van Dijken et

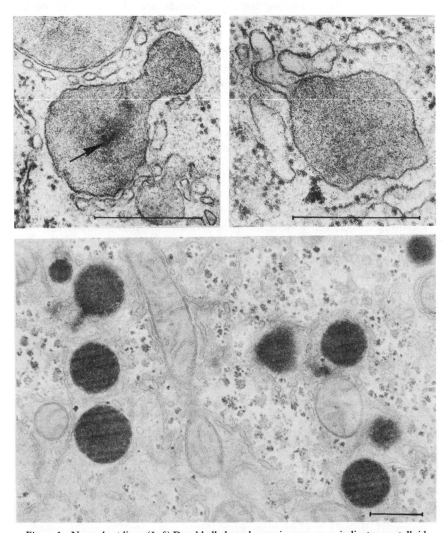

Figure 1   Normal rat liver. (*Left*) Dumbbell-shaped peroxisome; *arrow* indicates crystalloid core. Close proximity to endoplasmic reticulum. (*Right*) Rare peroxisome with tail; (*bottom*) diaminobenzidine cytochemical reaction for catalase; cluster of 4 peroxisomes at left. Bar = 0.5 μm. Electron microscopy by Ms. Helen Shio.

al 1982; Veenhuis et al 1983). These yeasts include *Candida boidinii* (Roggenkamp et al 1975), *Kloeckera* sp. no. 2201 (Fukui et al 1975) [subsequently this yeast was identified as a strain of *C. boidinii* (Fukui & Tanaka 1979)], and *Hansenula polymorpha* (Veenhuis et al 1976). In spheroplasts fixed in glutaraldehyde-OsO_4, alcohol oxidase is visible as a large crystal within the peroxisome, whereas such crystals are not seen in $KMnO_4$-fixed cells (Veenhuis et al 1976, 1978a). Osumi et al (1979) reported that in *Kloeckera* sp. 2201 alcohol oxidase and catalase are cocrystallized. Veenhuis et al (1981) found that the *H. polymorpha* crystals consisted only of alcohol oxidase, but reported that spaces in the crystal lattice were large enough to allow soluble catalase to enter.

Alcohol oxidases of *H. polymorpha* and *C. boidinii* are synthesized in vitro at their final sizes (73–75 kDa) (Roa & Blobel 1983; Roggenkamp et al 1984; Goodman et al 1984). The *H. polymorpha* cell-free product differs from the mature protein in that it sediments as a monomer, and is very sensitive to trypsinolysis (Roa & Blobel 1983). Alcohol oxidase of *C. boidinii* is synthesized in vivo (5 min pulse labeling) as a monomer (it cosediments in sucrose gradients with bovine serum albumin, which has an M_r of 66,000), and is then converted to an octamer with a half-time of 20 min (Goodman et al 1984). In spheroplasts, the newly synthesized monomer is found first in the cytosolic fraction. It is imported into peroxisomes posttranslationally with kinetics that are at least as fast as, if not faster than, those of oligomerization. Thus octamers are formed inside the peroxisomes (Goodman et al 1984).

MORPHOLOGICAL STUDIES

Connections to Endoplasmic Reticulum?

It was for many years an accepted dogma that peroxisomes are directly connected to the endoplasmic reticulum, and this in turn suggested that peroxisomes form from it by budding. Many published images have been interpreted as connections, beginning with the earliest papers of Rhodin (1963) and Novikoff & Shin (1964). This view was strongly questioned by Lazarow et al (1980), who pointed out the conflict between this morphological view and the growing body of biochemical and molecular evidence concerning peroxisome biogenesis. They suggested that this conflict could be reconciled by a reinterpretation of the experimental evidence. This evidence was of two sorts: proximity and tails.

TAILS   Peroxisomes occasionally have tails, i.e. narrow tubular extensions (Figure 1, *top right*), which have been interpreted as endoplasmic reticulum. There is no way to tell by looking at a micrograph whether the membrane of the tail contains those proteins characteristic of the peroxisomal membrane (Fujiki et al 1982a) or the quite different proteins of the endoplasmic reticulum membrane (Fujiki et al 1982b). Hence it is unjustified to label such tails "endoplasmic reticulum." It is equally likely on morphological grounds, and more probable on biochemical grounds, that they are connections to other peroxisomes.

PROXIMITY   Peroxisomes are often found in close physical proximity to endoplasmic reticulum. Indeed, peroxisomes are sometimes seen wrapped up in a strand of endoplasmic reticulum, as are mitochondria. Arrowheads (denoting connections) often appear on micrographs where the ER and peroxisome membranes touch, or are broken, or are fuzzy (perhaps because of tangential sectioning or structures overlapping within the plane of the section). However, the issue is whether or not a direct luminal connection exists between the two structures. Minimal criteria for demonstrating such a continuity include a sufficiently high magnification to visualize the tri-laminar unit membrane, a section cut (or tilted) such that the membrane is parallel to the electron beam so that its profile is narrow and sharp, and an image of the membrane running continuously and uninterrupted from peroxisome to endoplasmic reticulum. Such images are rare, if they exist at all.

Several morphological investigations reported that direct peroxisomal-endoplasmic reticulum connections could not be found in liver (Legg & Wood 1970; Rigatuso et al 1970), in pneumocytes (Schneeberger 1972), or in bean leaves (Gruber et al 1973). Fahimi et al (1976) carried out serial sectioning of the coreless peroxisomes (0.1–0.3 μm diameter) of rat liver sinusoidal cells, and found "images suggestive of direct continuity between the two organelles only when the limiting membrane of the particles and/or the endoplasmic reticulum were tangentially sectioned." Gulyas & Yuan (1975) reported a lack of connections in corpus luteum of the rhesus monkey. Novikoff et al (1973), in a paper concluding that there are pervasive connections between peroxisomes and endoplasmic reticulum, published electron micrographs showing that when the section was tilted with a goniometer stage such that the membrane images were sharp, there were no connections visible (their figures 6 and 7). Shio & Lazarow (1981) looked for diffusion of reaction products of catalase and glucose-6-phosphatase cytochemistry through the putative connections, but their results were uniformly negative.

Interconnections Among Peroxisomes

In rat liver examined by random sectioning, most peroxisomes apparently exist as individual entities. Nevertheless, evidence suggests that a dynamic fission/fusion process may occur among these peroxisomes. Morphologically, dumbbell-shaped interconnected peroxisomes are seen (although rarely) in normal liver (Figure 1, *top left*), and such images are more common when peroxisome proliferation is induced by partial hepatectomy (Rigatuso et al 1970) or by hypolipidemic drugs such as clofibrate (Legg & Wood 1970; Reddy & Svoboda 1971, 1973; Leighton et al 1975). Dumbbell-shaped peroxisomal cross-sections may also be found in the published electron micrographs of other tissues, including adrenal cortex (Beard 1972), corpus luteum (Gulyas & Yuan 1975), and myocardium (Hicks & Fahimi 1977). Peroxisomes tend to be found in clusters (Figure 1, *bottom*), as pointed out for liver by Novikoff & Shin (1964), Essner (1970), Legg & Wood (1970), and Reddy & Svoboda (1971), among others. Finally, peroxisomes sometimes have tails, as discussed above.

Based on these results and on considerable biochemical evidence, Lazarow et al (1980) suggested that peroxisomes may be interconnected transiently by fission and fusion, if not permanently, into a "peroxisomal reticulum." This idea is strongly supported by recent observations of Gorgas (1985), who carried out serial sectioning on mouse liver that had been subjected to the diaminobenzidine cytochemical procedure in order to visualize catalase. The several types of diaminobenzidine-positive profiles seen in random sections (large core-containing peroxisomes, smaller coreless peroxisomes, and tubular forms) appear in serial sections to be interconnected.

In mouse preputial gland, a sebaceous tissue that synthesizes ether glycerolipids and waxes (Sansone & Hamilton 1969), even more elaborate interconnections have been found by serial sectioning (Gorgas 1982, 1984). Peroxisomes do not exist as individual spherical entities, but form a large contorted space, more than 4 μm in length, with tortuously branched tubular elements and an enlarged terminal cup-shaped structure (Figure 2). Peroxisomal profiles in individual sections vary; their shapes include small and round, rod-shaped, dumbbells, and ring-like cisternae. Some resemble endoplasmic reticulum cisternae. Elements of the contorted peroxisomal compartment, mitochondria, and endoplasmic reticulum were sometimes very close to each other, but no interorganellar continuities were observed.

Pais & Carrapico (1979, 1982) described a "microbodial compartment" consisting of branched diaminobenzidine-positive tubules in chlorophyllian spores of the moss *Byrum capillare*. In *Neurospora crassa* a large pleomorphic peroxisomal compartment occurs in the early stationary phase of growth (Wanner & Theimer 1982).

The size and shape of the elements of the peroxisomal compartment are affected by hormones and drugs. For example, thyroid hormone treatment of rats causes a 60% increase in the total volume of liver peroxisomes. This

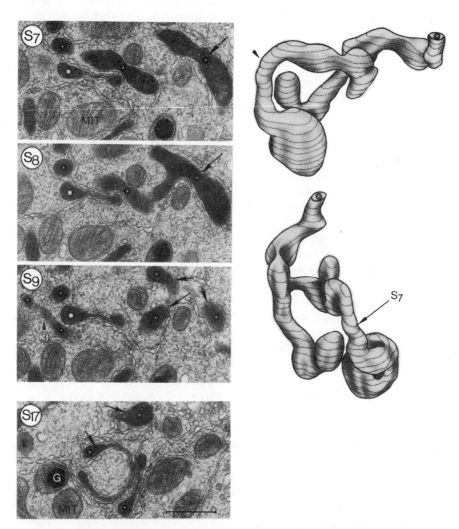

Figure 2 Serial section analysis of mouse preputial gland with cytochemical reaction for catalase. (*Left*) Sections 7, 8, 9, and 17 from a series of 25. Bar = 0.5 μm. Arrows indicate some parts of one peroxisome. MIT = mitochondria; G = mitochondrial inclusion body. (*Right*) Reconstruction of peroxisomal compartment or "peroxisomal reticulum." Arrow and arrowhead indicate the plane of section 7. Reprinted with permission from Gorgas 1984.

occurs by a threefold decrease in the mean volume of individual peroxisomes, accompanied by a quintupling of the number of peroxisomes and an increase in apparent clustering (Fringes & Reith 1982). The size distribution of peroxisomes is reversibly altered by nafenopin, a hypolipidemic drug (Staubli et al 1977). Animals treated with hypolipidemic and other drugs often have elongated and/or bizarrely shaped peroxisomes (Hruban et al 1966, 1974).

Peroxisome Budding in Yeast

In *C. tropicalis*, the size of the peroxisomal compartment increased 12 times when the cells were shifted from growth on malt extract to growth on alkanes (Osumi et al 1975). One hr after the shift one or two peroxisomes with a diameter of ~ 0.1 μm were observed per sectioned cell. By 6–8 hr the peroxisome diameter had increased to ~ 0.5 μm, and there were ~ 10 peroxisomes per sectioned cell. Larger oblate peroxisomes ($\sim 0.5 \times 1.0$–1.5 μm) were also seen, some of which contained septa dividing the peroxisome unequally. At 46 hr the cytoplasm was full of peroxisomes, most with a somewhat reduced diameter of 0.3–0.6 μm. Endoplasmic reticulum was sometimes wrapped around peroxisomes, but "no direct continuity was observed" (Osumi et al 1975). The authors conclude that *C. tropicalis* peroxisomes form by enlargement of preexisting peroxisomes followed by division.

Peroxisome fission also was observed when *H. polymorpha* grown to the stationary phase on methanol were diluted into fresh methanol medium (Veenhuis et al 1978). After a lag of 3 hr, the cells began to form small buds. During this process the peroxisomes also divided into unequally sized progeny, often in the narrow neck between mother and daughter yeast cells (Figure 3, *left*). Sometimes this peroxisomal budding occurred in the mother cell; in these cases the small new peroxisomes apparently migrated into the yeast bud, because the observed end products were small new *H. polymorpha* cells containing one or two small peroxisomes, and larger mother cells filled with large peroxisomes. The mother cells could be identified by the thickness of their cell walls and the presence of bud scars. In a second growth phase, the peroxisomes in the daughter cells gradually enlarged, and then divided into a large and a small organelle, which remained attached to each other (Figure 3, *bottom right*). Enlargement of peroxisomes (from 0.14 to 0.63 μm diameter) also occurred when *H. polymorpha* grown on glucose were transferred to methanol (Veenhuis et al 1979). A close association with ER, but no direct continuities, was observed during this growth.

In summary, there is in our opinion no convincing evidence that peroxisomes are directly connected to the endoplasmic reticulum with

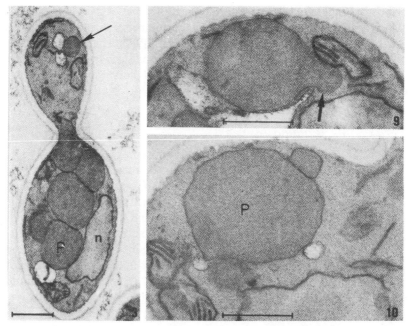

Figure 3 Hansenula polymorpha growing on methanol. (*Left*) Dividing cell with one small peroxisome in the bud (*arrow*) and a larger one in the neck. (*Top*) Enlarged peroxisome budding in daughter cell. (*Bottom*) Large and small peroxisome still apposed in daughter cell. P = peroxisome; n = nucleus. Bar = 0.5 μm. Reprinted with permission from Veenhuis et al 1978.

continuity of the lumens. Of course, absence of proof does not disprove the possibility. Nevertheless, recent evidence strongly suggests that peroxisomes are interconnected only with each other. Physical proximity with endoplasmic reticulum probably facilitates biochemical cooperation between the organelles.

OTHER INVESTIGATIONS

Membrane Properties

Some of the properties of the membrane of peroxisomes (glyoxysomes) are briefly summarized here; they have recently been reviewed in detail (Lazarow 1984). The membrane is 6.5–7 nm thick with a typical trilaminar appearance. The phospholipid/protein ratio is 140–200 nmol/mg (Donaldson et al 1981; Fujiki et al 1982a). Membranes have been isolated from purified peroxisomes by osmotic shock (Huang & Beevers 1973), sometimes at pH 9 (Donaldson et al 1972). This yields "ghosts" with apparently sealed membranes and some residual content material, including cores in the case of rat liver (Alexson et al 1985). Unsealed membrane sheets, stripped of their peripheral membrane proteins, but retaining phospholipids and integral membrane proteins, may be isolated by treatment with Na_2CO_3 at pH 11.5 (Fujiki et al 1982b).

Rat liver peroxisomal membranes prepared by the latter procedure and analyzed by SDS-PAGE reveal a pattern of proteins that differs greatly from that of the endoplasmic reticulum (Fujiki et al 1982a). Prominent among the peroxisomal polypeptides are three with M_r of 22,000, 68,000, and 70,000. Glyoxysomal membrane proteins are reported to resemble those of the endoplasmic reticulum in castor bean (Goldberg & Gonzalez 1982), but to be quite different in cucumber (Kruse & Kindl 1982).

The only uniquely peroxisomal integral membrane protein whose biogenesis has been studied thus far is the M_r 22,000 protein of rat liver, which is made on free polyribosomes at its final size (Fujiki et al 1984). In cucumber glyoxysomes, a portion of the amphipathic protein malate synthetase appears to be integrally embedded in the membrane. This conclusion is based on one-dimensional peptide mapping (Kruse & Kindl 1982), which showed identity between malate synthetase and a M_r 63,000 integral membrane protein (Ludwig & Kindl 1976). Malate synthetase is made at its final size and passes through the cytosol on its way to peroxisomes (see above).

A few proteins appear to be present in small amounts in peroxisomal membranes, and in much larger amounts in the endoplasmic reticulum in rat liver and castor bean endosperm. In rat liver these include cytochrome b_5, NADH cytochrome b_5 reductase, NADH-cytochrome c reductase (Fowler et al 1976; Remacle 1978; Hüttinger et al 1981), and acyl-CoA synthetase (Mannaerts et al 1982; Alexson et al 1985). The biogenesis of the first two of these has been studied; they are made on free polyribosomes at their final sizes (Rachubinski et al 1980; Borgese et al 1960; Borgese & Gaetani 1980). Therefore, this overlap in polypeptide composition is not evidence that the peroxisome forms by budding from the endoplasmic reticulum. On the contrary, the available data supports the posttranslational assembly mechanism for peroxisomal membrane proteins.

Phosphatidyl choline and phosphatidyl ethanolamine are the principal phospholipids in the peroxisomal membrane of rat liver (Fujiki et al 1982a) and castor bean (Donaldson & Beevers 1977). Rat liver and brain peroxisomes contain enzymes that catalyze the initial reactions in the formation of plasmalogens (Hajra & Bishop 1982). However, peroxisomes lack the other enzymes of phospholipid biosynthesis (Hajra & Bishop 1982; Ballas et al 1984). These enzymes are found mainly in the endoplasmic reticulum in both animals (Bell & Coleman 1980) and plants (Beevers 1979;

Moore 1982). Thus endoplasmic reticulum appears to be the site of formation of phospholipids for peroxisomes (as it does for mitochondria and chloroplasts). It is not known how these phospholipids are transferred to their final sites; speculations focus on phospholipid carrier proteins (Wirtz 1982) and putative shuttle vesicles.

Peroxisomal Diseases: Zellweger Syndrome

Our understanding of peroxisome function and biogenesis has increased through recent studies of a growing group of peroxisomal disorders (Goldfischer & Reddy 1984). Prominent among these is Zellweger Syndrome, a rare autosomal recessive disease characterized by profound neuropathology, hepatic and renal dysfunction, and death within the first weeks or months after birth. Peroxisomes are morphologically absent in liver and kidney (Goldfischer et al 1973). The biochemical symptoms include accumulation of very long chain fatty acids (Moser et al 1984), deficient plasmalogens (Heymans et al 1983), and abnormal bile acids, all of which can be plausibly attributed to the absence of peroxisomes (reviewed by Kelley 1983; Goldfischer & Reddy 1984). In Zellweger patients some peroxisomal enzymes are deficient, but others, such as catalase, are present in normal amounts and are found in the cell cytosol (Wanders et al 1984; Lazarow et al 1985). This is exactly what one would expect if there is no preexisting peroxisome into which to package the newly made enzymes. We suspect that Zellweger Syndrome may be due to a failure of peroxisome membrane assembly or protein import.

Enzymes in Multiple Cell Compartments

The presence of some enzymatic activities in more than one cell compartment raises questions about the mechanism(s) by which this is accomplished in the cell. In some instances, the same activity is catalyzed by entirely different proteins in the two compartments; this poses no conceptual problems for intracellular protein sorting—presumably the two proteins carry different targeting information. When the enzymes in the two spaces appear very similar, or indistinguishable, one must ask how they are targeted to different intracellular places.

It is clear from the work of Hashimoto and his colleagues that the proteins responsible for the β-oxidation reactions, which occur in both peroxisomes and in mitochondria, are entirely different in the two organelles (Hashimoto 1982). Watermelon cotyledons contain five isozymes of malate dehydrogenase: one in the glyoxysomes, one in the mitochondria and three in the cytosol. The glyoxysomal isozyme is biochemically and immunochemically different from the others (Walk & Hock 1977; Walk et al 1977; Sautter & Hock 1982). The glyoxysomal and mitochondrial isozymes also differ in their subunit molecular weights (33,000 and 38,000, respectively). Both organelle isozymes are synthesized in vitro as larger precursors, M_r 41,000 in each case (Gietl & Hock 1982). Hock (1984) reported that the mature forms lose their antigenic difference after denaturation, and raised the possibility that both were derived from a common M_r 41,000 precursor. Ueda et al (1984b) made a similar suggestion for the peroxisomal and mitochondrial carnitine acetyltransferases of *C. tropicalis*. These enzymes are immunochemically indistinguishable, but have different subunit sizes, pIs, and chromatographic behaviors. One can imagine that proteolytic processing machinery in peroxisomes and mitochondria might cleave the same precursor at different sites, but how the identical precursor is targeted to two different places has not yet been satisfactorily explained. *Immunochemical* identity is weak evidence for *molecular* identity. Therefore, it appears more likely at present that these similar isozymes arose by gene duplication followed by divergent evolution of the targeting information. Hybridization of genomic clones to more than one peroxisomal mRNA in *C. tropicalis* led Kamiryo & Okazaki (1984) to suggest gene duplication to explain their results.

Peroxisomes are fragile structures, so leakage may explain the presence of some peroxisomal proteins found in soluble form in cell homogenates. In female rat liver, about 30% of catalase and several oxidases are present in the supernatant fraction, and free cores are detected when the light mitochondrial fraction is sedimented into a sucrose gradient. These are thought to have been released from peroxisomes that were partly damaged during cell homogenization (Leighton et al 1968; Lazarow & de Duve 1973b). This conclusion is supported by recent observations that dilution and recentrifugation of isolated peroxisomes caused further loss of catalase and certain other enzymes (Alexson et al 1985).

In sheep liver, on the other hand, a genuine cytosolic catalase has been demonstrated cytochemically (Roels 1976). In sheep, there is about the same absolute amount of sedimentable liver catalase as in rat, but the total liver activity is ten times higher, and the excess is soluble. The molecular properties of the soluble and peroxisomal catalase have not been compared, and nothing is known concerning their biogenesis.

There are a number of reports of charge heterogeneity in rat and mouse liver catalase, which have been interpreted to imply that soluble and peroxisomal catalases are different proteins (Higashi & Shibata 1965; Holmes & Masters 1970, 1972; Sugita et al 1982). Sando et al (1984) attached significance to minor differences in nearly identical amino acid compositions of purified soluble and sedimentable rat liver catalases. In the case of mouse, where there is genetic evidence for a single catalase gene (Feinstein 1970; Holmes 1972), the charge heterogeneity was attributed to the presence of different amounts of sialic acid. This conclusion was reached solely on the basis of one experiment in which liver homogenates were incubated for one week with a commercial neuraminidase preparation, or with a microsomal extract plus sialic acid and CTP, which produced changes in catalase charge (Jones & Masters 1975). Sialic acid was not demonstrated chemically, proteolysis was not excluded, and the presence of sialic acid has never been confirmed. The physiological significance of the charge heterogeneity has been called into question by experiments demonstrating that lysosomal proteases cause artifactual changes in pI, accompanied by decreases in M_r, that mimic the observed charge differences (Mainferme & Wattiaux 1976, 1982; Crane et al 1982).

The existence of multiple catalase isozymes in maize has been firmly established on genetic grounds (Scandalios 1975; Scandalios et al 1980a), but their biogenesis has not been studied.

Glycoproteins?

There is little new information on whether or not any peroxisomal proteins are glycosylated. Reports conflict on whether cucumber isocitrate lyase is glycosylated (Frevert & Kindl 1978; Riezman et al 1980); the castor bean enzyme is not (Bergner & Tanner 1981). Malate synthetase, which was initially reported to be glycosylated in both cucumber and castor beans (Mellor et al 1978; Riezman et al 1980) has subsequently been found to lack sugar (Bergner & Tanner 1981; Lord & Roberts 1982; Kruse & Kindl 1983a).

Catalase appears to lack sugar in rat liver (Sugita et al 1982), *S. cerevisiae* (Ammerer et al 1981), and cucumber (Becker et al 1982). (For discussion of the situation in mouse liver see previous section.) Carnitine acetyl-transferase of *C. tropicalis* (Ueda et al 1984a) and cucumber malate dehydrogenase (Becker et al 1982) are reportedly not glycosylated. Volkl & Lazarow (1982) failed to detect any glycoproteins among the major rat liver peroxisomal proteins, tested by binding to lectin columns.

A monoclonal antibody against rat liver 3-hydroxymethylglutaryl-CoA reductase (a glycosylated integral membrane protein of endoplasmic reticulum) cross-reacts with some matrix protein of peroxisomes (Keller et al 1985). Nothing is known of the molecular properties of this peroxisomal protein, including whether or not it is glycosylated.

There are two recent claims for glyoxysomal glycoproteins based on labeling of unidentified bands with radioactive sugars (Hanson et al 1983; Lord & Roberts 1983). This approach seems dangerous to us, because the great sensitivity of this method (like the sensitivity of iodinated lectin binding assays) easily permits the detection of trace contaminant glycoproteins in the cell fractions.

In conclusion, there is little evidence of glycosylated proteins in peroxisomes. If future experiments provide such evidence, it will imply either that these proteins are exceptional in that they pass through the endoplasmic reticulum, or that there is another hitherto unsuspected glycosylation site.

DNA in Peroxisomes?

The presence of DNA in *Candida tropicalis* peroxisomes was suggested on the basis of electron microscopic (Osumi 1976) and physicochemical data (Osumi et al 1978). Kamiryo et al (1982) reinvestigated this question with highly purified peroxisomes from oleic acid–grown cells. A very small amount of DNA was found, but restriction-mapping demonstrated that this was due to traces of contaminating mitochondrial DNA. Peroxisomal DNA was not detected in CsCl density gradients by ethidium bromide–staining or by staining with the fluorescent dye 4′,6-diamidino-2-phenylindole (DAPI). Neither was peroxisomal DNA found in the peroxisomes of cells grown in the presence of [^{3}H]adenine. In all these experiments mitochondrial DNA was readily detected. These results make it very unlikely that DNA is present in peroxisomes of *C. tropicalis*. Douglass et al (1973) similarly could not find unique glyoxisomal DNA in castor bean endosperm. Leighton et al (1968) labeled rat DNA with [^{14}C]thymidine and then carried out liver cell fractionations at various intervals after labeling. The distribution of radioactivity in the isopycnic sucrose gradient subfractionation of a mitochondrial plus light mitochondrial fraction coincided with that of cytochrome oxidase, a marker enzyme for mitochondria. No labeled DNA was found in peroxisomes.

Cloning and Genetics

There is genetic evidence for a single catalase gene in the mouse (Feinstein 1970; Holmes 1972), which has been mapped to chromosome 2 (Holmes & Duley 1975). It is closely linked to a gene encoding a peroxisomal L-α-hydroxyacid oxidase isozyme, but is not linked to the gene for peroxisomal D-amino acid oxidase (Holmes 1976).

In maize, three catalase genes with multiple alleles have been identified that encode glyoxysomal catalases expressed differently during development (Scandalios 1975; Scandalios et al 1980b). A regulatory gene, located on chromosome 1S together with, but not adjacent to, one of the structural genes, controls the developmental expression (Scandalios et al 1980a).

In order to investigate the biogenesis and regulation of peroxisomes at the molecular level, several groups have created cDNA libraries for cells in which peroxisome proliferation was induced in a variety of ways. cDNA libraries were made for di(2-ethylhexyl)phthalate-induced rat liver (Osumi et al 1984a), alkane-grown *Candida tropicalis* (Rachubinski et al 1985a,b), and methanol-grown *Pichia pastoris* (Ellis et al 1985). A genomic library was constructed for oleate-grown *C. tropicalis* (Kamiryo & Okazaki 1984). The yeast libraries were screened by selecting alkane- or oleate- or methanol-induced sequences and then carrying out hybrid-selection translation.

Korneluk et al (1984) identified a human catalase cDNA clone by screening a fibroblast cDNA library with synthetic oligonucleotides. Catalase clones were identified in the rat liver (Osumi et al 1984b) and *C. tropicalis* (Rachubinski et al 1985b) cDNA libraries. The two mammalian catalase mRNAs have 700–800 3' noncoding bases, in contrast to the *C. tropicalis* catalase mRNA which has only about 200 total untranslated bases (Rachubinski et al 1985b). An even greater difference occurs for the acyl-CoA oxidase mRNAs: rat liver has ~ 1800 untranslated bases (Osumi et al 1984a) whereas *C. tropicalis* has ~ 100 (Rachubinski et al 1985a).

Ellis et al (1985) made use of cDNA clones for alcohol oxidase and two other methanol-induced sequences to identify and clone the corresponding genes. DNA subfragments of two of these have been identified as containing the control regions involved in methanol regulation.

CONCLUSIONS

All peroxisomal (including glyoxysomal) proteins whose site of synthesis has been investigated are encoded by mRNAs found in free polyribosomes and/or are initially located in the cytosolic fraction, as shown by in vivo pulse-labeling experiments. Of the nearly 30 polypeptides studied, most are made at their final size (Table 1). Malate dehydrogenase (watermelon and cucumber), rat liver thiolase, pumpkin catalase, and *C. tropicalis* carnitine acetyltransferase are synthesized with molecular weights that are 3000–8000 larger than the mature subunits. In one exceptional case the cucumber isocitrate lyase cell-free product appeared to be slightly smaller than the mature protein, but in other experiments it was the same size.

The posttranslational import of many of these proteins has been observed in vivo, and has been reproduced in vitro for cucumber malate synthetase, *N. crassa* isocitrate lyase, watermelon malate dehydrogenase, and rat liver catalase and acyl-CoA oxidase. The rate of import in vivo varies greatly, from 2–3 minutes (e.g. several rat liver proteins) to 15–20 min (e.g. rat liver catalase and *C. boidinii* alcohol oxidase) or more. These transit times may reflect the varying sizes of cytosolic pools of newly-synthesized proteins. Remarkably, the proteolytic processing of the exceptional larger precursors thus far does *not* appear to be tightly coupled to their import into the organelle. Significant amounts of unprocessed thiolase accumulate within peroxisomes, while some thiolase is apparently cleaved before entry. Unprocessed pumpkin catalase aggregates within peroxisomes to an inactive tetramer, and at certain developmental stages this inactive form is more abundant than the active enzyme. Rat liver acyl-CoA oxidase, which is initially synthesized at the final size of the A subunit, undergoes proteolysis within peroxisomes over a period of hours or days to form the smaller B and C subunits.

In several cases it has been demonstrated that oligomerization occurs after import (rat liver catalase and *C. boidinii* alcohol oxidase). It is widely expected, but not yet established, that this is a general rule. The site of prosthetic group addition (inside peroxisomes) has been determined thus far only for rat liver catalase, but this may be a kinetic phenomenon because active catalase assembles in the cytosol in Zellweger patients who lack peroxisomes. Heme addition precedes tetramerization for catalase in rat liver and *S. cerevisiae*, but in pumpkin, inactive tetramers form that are thought to be lacking in heme. Further work is needed to establish the detailed relationships between import, prosthetic group addition, and oligomerization for the various proteins.

Molecular analysis of peroxisomal membrane formation has begun. In rat liver, the peroxisomal and endoplasmic reticulum membrane have dissimilar polypeptide compositions, and a major peroxisomal integral membrane protein (M_r 22,000) is made on free polysomes at its final size. In cucumber, it appears that a glyoxysomal integral membrane protein of M_r 63,000 is malate synthetase, which is also made on free polysomes at its final size. However, most malate synthetase is not integral to the membrane. Of those proteins that may be present in membranes of both peroxisomes and endoplasmic reticulum, the two whose biosynthesis has been studied, cytochrome b_5 and cytochrome b_5 reductase, are made on free polysomes. Thus this overlap in composition does *not* imply that the peroxisomal membrane forms from endoplasmic reticulum. The available data are limited, but those we have indicate that newly made membrane proteins insert posttranslationally into preexisting peroxisomal membranes.

Despite some hints, there is as yet no proof that any one well-characterized peroxisomal protein is glycosylated. No peroxisomal protein is known to pass through the endoplasmic reticulum, where those glyco-sylation reactions that have been characterized occur.

CURRENT UNDERSTANDING OF PEROXISOME BIOGENESIS

The literature reviewed has a very simple interpretation which we take as a working model (Figure 4). Peroxisomal content and crystalloid core proteins are made on free polyribosomes, and imported posttranslationally into preexisting peroxisomes. Peroxisomal membrane proteins are made on free polyribosomes and inserted posttranslationally into preexisting peroxisomal membranes. Thus the peroxisomes grow and undergo fission to form new peroxisomes. In the case of interconnected peroxisomes, the peroxisomal reticulum expands, and the formation of new peroxisomes may appear as budding from the contorted peroxisomal compartment, rather than fission from a sphere. The biogenesis of peroxisomes appears to fall into the same general category as mitochondria and chloroplasts; thus there appear to be three rather than two cell organelles that form by fission.

This model has a number of implications: one is that peroxisomes never form de novo. Therefore most cells, including germ cells, must contain at least one. In the case of yeast, this prediction has been confirmed. Despite some initial reports to the contrary, yeast grown on glucose, which represses peroxisome formation, still contain at least one or two small peroxisomes. This model also implies a simple result if the expression of genes encoding peroxisomal proteins is altered. The synthesis of peroxisomal proteins will be directed by the new population of mRNAs, and the altered mixture of new proteins will be incorporated into the preexisting peroxisomes. Since peroxisomes are destroyed by autophagy, probably mostly at random, the peroxisomal enzyme complement will come to reflect the pattern of proteins synthesized after the differential change in gene expression, once a few peroxisomal half-lives have elapsed. This may

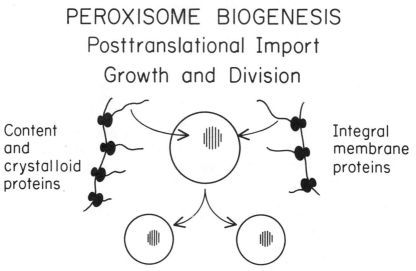

Figure 4   Current scheme of peroxisome biogenesis.

explain the glyoxysome-peroxisome transition, and many of the effects of peroxisomal inducers.

There is no proof that the endoplasmic reticulum does not play a role in peroxisomal protein formation, but the accumulated evidence supports the posttranslational assembly mechanism. Further work is required to test the generality of the results, especially insofar as the integral membrane proteins are concerned. We note that at present there are just two established mechanisms for organelle formation: (a) cotranslational segregation and derivation from the secretory system (lysosomes, Golgi apparatus, secretion granules, etc); and (b) posttranslational accretion of new proteins and division (mitochondria and chloroplasts). Theories that would require parts of both of these mechanisms are inherently more complicated.

If peroxisomes form by fission, where did the first one come from? Could peroxisomes be the descendants of ancient endosymbionts, as mitochondria and chloroplasts are supposed to be? Peroxisomes differ from mitochondria and chloroplasts in two fundamental respects: they have only one membrane, and they lack DNA. These issues have been considered by de Duve (1983), who pointed out that mitochondria have lost at least 90% of their DNA to the nucleus, so why not 100%? He also observed that some endocytic parasites escape into the cell cytosol and thus are bounded by only one membrane. Therefore the speculation is not entirely implausible.

Work in progress on peroxisomal genes should yield exciting data concerning peroxisome biogenesis, and may also shed some light on these evolutionary questions.

ACKNOWLEDGMENT

We thank Ms. Karrie Polowetzky, Judiann McGhee, Laurie McGhee, and Martha Wemple for extraordinary help with word processing and information gathering; Dr. Gillian Small for critical reading of the manuscript; and Ms. Helen Shio for preparation of the figures. The work of the authors is supported by NIH grant AM-19394, and NSF grants PCM82 08315 and PCM83 09222. Paul B. Lazarow is the recipient of an Established Fellowship of the New York Heart Association.

Literature Cited

Alexson, S. E. H., Fujiki, Y., Shio, H., Lazarow, P. B. 1985. Partial disassembly of peroxisomes. *J. Cell Biol.* 101:294–305

Ammerer, G., Richter, K., Hartter, E., Ruis, H. 1981. Synthesis of *Saccharomyces cere-* *visiae* catalase A in vitro. *Eur. J. Biochem.* 113:327–31

Ammerer, G., Ruis, H. 1979. Cell-free synthesis of *Saccharomyces cerevisiae* catalase T. *FEBS Lett.* 99:242–46

522   LAZAROW & FUJIKI

Antonenkov, V. D., Panchenko, L. F. 1978. Organization of urate oxidase in peroxisomal nucleoids. *FEBS Lett.* 88:151–54

Ballas, L. M., Lazarow, P. B., Bell, R. M. 1984. Glycerolipid synthetic capacity of rat liver peroxisomes. *Biochem. Biophys. Acta* 795:297–300

Beard, M. E. 1972. Identification of peroxisomes in the rat adrenal cortex. *J. Histochem. Cytochem.* 20:173–79

Becker, W. M., Riezman, H., Weir, E. M., Titus, D. E., Leaver, C. J. 1982. In vitro synthesis and compartmentalization of glyoxysomal enzymes from cucumber. *Ann. NY Acad. Sci.* 386:329–49

Beevers, H. 1969. Glyoxysomes of castor bean endosperm and their relation to gluconeogenesis. *Ann. NY Acad. Sci.* 168:313–24

Beevers, H. 1979. Microbodies in higher plants. *Ann. Rev. Plant Physiol.* 30:159–93

Beevers, H. 1982. Glyoxysomes in higher plants. *Ann. NY Acad. Sci.* 386:243–53

Behrends, W., Rausch, U., Loffler, H.-G., Kindl, H. 1982. Purification of glycollate oxidase from greening cucumber cotyledons. *Planta* 156:566–71

Bell, R. M., Coleman, R. A. 1980. Enzymes of glycerolipid synthesis in eukaryotes. *Ann. Biochem.* 49:459–87

Bendayan, M., Reddy, J. K. 1982. Immunocytochemical localization of catalase and heat-labile enoyl-CoA hydratase in the livers of normal and peroxisome proliferator-treated rats. *Lab. Invest.* 47:364–69

Bergner, U., Tanner, W. 1981. Occurrence of several glycoproteins in glyoxysomal membranes of castor beans. *FEBS Lett.* 131:68–72

Blum, J. J. 1973. Localization of some enzymes of β-oxidation of fatty acids in the peroxisomes of Tetrahymena. *J. Protozool.* 20:688–92

Borgese, N., Pietrini, G., Meldolesi, J. 1980. Localization and biosynthesis of NADH-cytochrome b_5 reductase, an integral membrane protein, in rat liver cells. III. Evidence for the independent insertion and turnover of the enzyme in various subcellular compartments. *J. Cell Biol.* 86:38–45

Borgese, N., Gaetani, S. 1980. Site of synthesis of NADH-cytochrome b_5 reductase, an integral membrane protein, in rat liver cells. *FEBS Lett.* 112:216–20

Bowden, L., Lord, J. M. 1976a. Similarities in the polypeptide composition of glyoxysomal and endoplasmic-reticulum membranes from castor-bean endosperm. *Biochem. J.* 154:491–99

Bowden, L., Lord, J. M. 1976b. The cellular origin of glyoxysomal proteins in germinating castor-bean endosperm. *Biochem. J.* 154:501–6

Bowden, L., Lord, J. M. 1978. Purification and comparative properties of microsomal and glyoxysomal malate synthase from castor bean endosperm. *Plant Physiol.* 61:259–65

Breidenbach, R. W., Beevers, H. 1967. Association of the glyoxylate cycle enzymes in a novel subcellular particle from castor bean endosperm. *Biochem. Biophys. Res. Commun.* 27:462–69

Burke, J. J., Trelease, R. N. 1975. Cytochemical demonstration of malate synthase and glycolate oxidase in microbodies of cucumber cotyledons. *Plant Physiol.* 56:710–17

Chance, B., Oshino, N. 1971. Kinetics and mechanisms of catalase in peroxisomes of the mitochondrial fraction. *Biochem. J.* 122:225–33

Connock, M. J., Perry, S. R. 1983. Detection of acyl-CoA β-oxidation enzymes in peroxisomes (microperoxisomes) of mouse heart. *Biochem. Int.* 6:545–51

Cooper, T. G., Beevers, H. 1969. β-Oxidation in glyoxysomes from castor bean endosperm. *J. Biol. Chem.* 244:3514–20

Coudron, P., Frerman, F. 1982. Chemical and catalytic properties of an acyl-CoA oxidase from *Candida tropicalis*. *Ann. NY Acad. Sci.* 386:397–400

Coudron, P. E., Frerman, F. E., Schowalter, D. B. 1983. Chemical and catalytic properties of the peroxisomal acyl-coenzyme A oxidase from *Candida tropicalis*. *Arch. Biochem. Biophys.* 226:324–36

Crane, D., Holmes, R., Masters, C. 1982. Proteolytic modification of mouse liver catalase. *Biochem. Biophys. Res. Commun.* 104:1567–72

de Duve, C., Baudhuin, P. 1966. Peroxisomes (microbodies and related particles). *Physiol. Rev.* 46:323–57

de Duve, C. 1975. Exploring cells with a centrifuge. *Science* 189:186–94

de Duve, C. 1983. Microbodies in the living cell. *Sci. Amer.* 248:74–84

Desel, H., Zimmermann, R., Janes, M., Miller, F., Neupert, W. 1982. Biosynthesis of glyoxysomal enzymes in *Neurospora crassa*. *Ann. NY Acad. Sci.* 386:377–89

Donaldson, R. P., Beevers, H. 1977. Lipid composition of organelles from germinating castor bean endosperm. *Plant Physiol.* 59:259–63

Donaldson, R. P., Tolbert, N. E., Schnarrenberger, C. 1972. A comparison of microbody membranes with microsomes and mitochondria from plant and animal tissue. *Arch. Biochem. Biophys.* 152:199–215

Donaldson, R. P., Tully, R. E., Young, O. A., Beevers, H. 1981. Organelle membranes from germinating castor bean endosperm. II. Enzymes, cytochromes, and permeability of the glyoxysome membrane. *Plant Physiol.* 67:21–25

Douglass, S. A., Criddle, R. S., Breidenbach, R. W. 1973. Characterization of deoxyribonucleic acid species from castor bean endosperm. *Plant Physiol.* 51:902–6

Ellis, S. B., Brust, P. F., Koutz, P. J., Waters, A. F., Harpold, M. M., Gingeras, T. R. 1985. The isolation of alcohol oxidase and two other methanol regulatable genes from the yeast *Pichia pastoris*. *Mol. Cell. Biol.* 5:1111–21

Essner, E. 1970. Observations on hepatic and renal peroxisomes (microbodies) in the developing chick. *J. Histochem. Cytochem.* 18:80–92

Fahimi, H. D., Gray, B. A., Herzog, V. K. 1976. Cytochemical localization of catalase and peroxidase in sinusoidal cells of rat liver. *Lab. Invest.* 34:192–201

Feinstein, R. N. 1970. Acatalasemia in the mouse and other species. *Biochem. Genet.* 4:135–55

Fita, I., Rossmann, M. G. 1985. The NADPH binding site on beef liver catalase. *Proc. Natl. Acad. Sci. USA* 82:1604–8

Fowler, S., Remacle, J., Trouet, A., Beaufay, H., Berthet, J., et al. 1976. Analytical study of microsomes and isolated subcellular membranes from rat liver. V. Immunological localization of cytochrome b_5 by electron microscopy: methodology and application to various subcellular fractions. *J. Cell Biol.* 71:535–50

Frevert, J., Kindl, H. 1978. Plant microbody proteins. Purification and glycoprotein nature of glyoxysomal isocitrate lyase from cucumber cotyledons. *Eur. J. Biochem.* 92:35–43

Frevert, J., Kindl, H. 1980. A bifunctional enzyme from glyoxysomes. Purification of a protein possessing enoyl-CoA hydratase and 3-hydroxyacyl-CoA dehydrogenase activities. *Eur. J. Biochem.* 107:79–86

Frevert, J., Koller, W., Kindl, H. 1980. Occurrence and biosynthesis of glyoxysomal enzymes in ripening cucumber seeds. *Hoppe-Seyler's Z. Physiol. Chem.* 361:1557–65

Fringes, B., Reith, A. 1982. Time course of peroxisome biogenesis during adaptation to mild hyperthyroidism in rat liver. *Lab. Invest.* 47:19–26

Fujiki, Y., Fowler, S., Shio, H., Hubbard, A. L., Lazarow, P. B. 1982a. Polypeptide and phospholipid composition of the membrane of rat liver peroxisomes. Comparison with endoplasmic reticulum and mitochondrial membranes. *J. Cell Biol.* 93:103–10

Fujiki, Y., Hubbard, A. L., Fowler, S., Lazarow, P. B. 1982b. Isolation of intracellular membranes by means of sodium carbonate treatment. Application to endoplasmic reticulum. *J. Cell Biol.* 93:97–102

Fujiki, Y., Lazarow, P. B. 1985. Post-translational import of fatty acyl-CoA oxidase and catalase into peroxisomes of rat liver in vitro. *J. Biol. Chem.* 260:5603–9

Fujiki, Y., Rachubinski, R. A., Lazarow, P. B. 1984. Synthesis of a major integral membrane polypeptide of rat liver peroxisomes on free polysomes. *Proc. Natl. Acad. Sci. USA* 81:7127–31

Fujiki, Y., Rachubinski, R. A., Mortensen, R. M., Lazarow, P. B. 1985a. Synthesis of 3-ketoazyl-CoA thiolase of rat liver peroxisomes on free polyribosomes as a larger precursor. Induction of thiolase mRNA activity by clofibrate. *Biochem. J.* 226:697–704

Fujiki, Y., Rachubinski, R. A., Zentella-Dehesa, A., Lazarow, P. B. 1985b. Induction, identification, and cell-free translation of mRNAs coding for peroxisomal proteins in *Candida tropicalis*. *J. Biol. Chem.* In press

Fukui, S., Kawamoto, S., Yasuhara, S., Tanaka, A. 1975. Microbody of methanolgrown yeasts: localization of catalase and flavin-dependent alcohol oxidase in the isolated microbody. *Eur. J. Biochem.* 59:561–66

Fukui, S., Tanaka, A. 1979. Peroxisomes of alkane- and methanol-grown yeasts. *J. Appl. Biochem.* 1:171–201

Furuta, S., Hashimoto, T., Miura, S., Mori, M., Tatibana, M. 1982a. Cell-free synthesis of the enzymes of peroxisomal β-oxidation. *Biochem. Biophys. Res. Commun.* 105:639–46

Furuta, S., Miyazawa, S., Hashimoti, T. 1982b. Biosynthesis of enzymes of peroxisomal β-oxidation. *J. Biochem.* 92:319–26

Furuta, S., Miyazawa, S., Osumi, T., Hashimoto, T., Ui, N. 1980. Properties of mitochondrial and peroxisomal enoyl-CoA hydratases from rat liver. *J. Biochem.* 88:1059–70

Gerdes, H.-H., Behrends, W., Kindl, H. 1982. Biosynthesis of a microbody matrix enzyme in greening cotyledons. *Planta* 156:572–78

Gietl, C., Hock, B. 1982. Organelle-bound malate dehydrogenase isoenzymes are synthesized as higher molecular weight precursors. *Plant Physiol.* 70:483–87

Gietl, C., Hock, B. 1984. Import of in-vitro–synthesized glyoxysomal malate dehydrogenase into isolated watermelon *Citrullus vulgaris* glyoxysomes. *Planta* 162:261–67

Goldberg, D. B., Al-Marayati, S., Gonzalez,

E. 1982. A comparison of castor bean endoplasmic reticulum and glyoxysome intrinsic membrane proteins. *Ann. NY Acad. Sci.* 386:502–3

Goldfischer, S., Reddy, J. K. 1984. Peroxisomes (microbodies) in cell pathology. *Int. Rev. Exp. Pathol.* 26:45–84

Goldfischer, S., Moore, C. L., Johnson, A. B., Spiro, A. J., Valsamis, M. P., et al. 1973. Peroxisomal and mitochondrial defects in the cerebro-hepato-renal syndrome. *Science* 182:62–64

Goldman, B. M., Blobel, G. 1978. Biogenesis of peroxisomes: Intracellular site of synthesis of catalase and uricase. *Proc. Natl. Acad. Sci. USA* 75:5066–70

Gonzalez, E. 1982. Aggregated forms of malate and citrate synthase are localized in endoplasmic reticulum of endosperm of germinating castor bean. *Plant Physiol.* 69:83–87

Gonzalez, E., Beevers, H. 1976. Role of the endoplasmic reticulum in glyoxysome formation in castor bean endosperm. *Plant Physiol.* 57:406–9

Goodman, J. M., Scott, C. W., Donahue, P. N., Atherton, J. P. 1984. Alcohol oxidase assembles post-translationally into peroxisomes of *Candida boidinii. J. Biol. Chem.* 259:8485–93

Gorgas, K. 1982. Serial section analysis of peroxisomal shape and membrane relationships in the mouse preputial gland. *Ann. NY Acad. Sci.* 386:519–22

Gorgas, K. 1984. Peroxisomes in sebaceous glands. V. Complex peroxisomes in the mouse preputial gland: Serial sectioning and three-dimensional reconstruction studies. *Anat. Embryol.* 169:261–70

Gorgas, K. 1985. Serial section analysis of mouse hepatic peroxisomes. *Anat. Embryol.* 172:21–32

Gruber, P. J., Becker, W. M., Newcomb, E. H. 1973. The development of microbodies and peroxisomal enzymes in greening bean leaves. *J. Cell Biol.* 56:500–18

Gulyas, B. J., Yuan, L. C. 1975. Microperoxisomes in the late pregnancy corpus luteum of Rhesus monkeys (*Macaca mulatta*). *J. Histochem. Cytochem.* 23:359–68

Hajra, A. K., Bishop, J. E. 1982. Glycerolipid biosynthesis in peroxisomes via the acyl dihydroxyacetone phosphate pathway. *Ann. NY Acad. Sci.* 386:170–82

Harson, M. M., Conder, M. J., Lord, J. M. 1983. Endoplasmic reticulum and glyoxysomal membranes from castor-bean endosperm: interaction between membrane glycoproteins and organelle matrix proteins. *Planta* 157:143–49

Hashimoto, T. 1982. Individual peroxisomal β-oxidation enzymes. *Ann. NY Acad. Sci.* 386:5–12

Hayashi, H., Hino, S., Yamasaki, F. 1981. Intraparticulate localization of some peroxisomal enzymes related to fatty acid β-oxidation. *Eur. J. Biochem.* 120:47–51

Hayashi, H., Taya, K., Suga, T., Niinobe, S. 1976. Studies on peroxisomes. VI. Relationship between the peroxisomal core and urate oxidase. *J. Biochem.* 79:1029–34

Heymans, H. S. A., Schutgens, R. B. H., Tan, R., van den Bosch, H., Borst, P. 1983. Severe plasmalogen deficiency in tissues of infants without peroxisomes (Zellweger syndrome). *Nature* 306:69–70

Hicks, L., Fahimi, H. D. 1977. Peroxisomes (microbodies) in the myocardium of rodents and primates. *Cell Tiss. Res.* 175:467–81

Higashi, T., Peters, T. Jr. 1963. Studies on rat liver catalase. II. Incorporation of C14-leucine into catalase of liver cell fractions in vivo. *J. Biol. Chem.* 238:3952–54

Higashi, T., Shibata, Y. 1965. Studies on rat liver catalase. IV. Heterogeneity of mitochondrial and supernatant catalase. *J. Biochem.* 58:530–37

Hock, B. 1984. Processing and organelle import of malate dehydrogenase isoenzymes. Is there a common precursor for the glyoxysomal and mitochondrial forms? *Physiol. Veg.* 22:333–40

Hoffmann, H.-P., Szabo, A., Avers, C. J. 1970. Cytochemical localization of catalase activity in yeast peroxisomes. *J. Bacteriol.* 104:581–84

Holmes, R. S. 1972. Catalase multiplicity in normal and acatalasemic mice. *FEBS Lett.* 24:161–64

Holmes, R. S. 1976. Genetics of peroxisomal enzymes in the mouse: Nonlinkage of D-amino acid oxidase locus (Dao) to catalase (Cs) and L-α-hydroxyacid oxidase (Hao-1) loci on chromosome 2. *Biochem. Genet.* 14:981–87

Holmes, R. S., Duley, J. A. 1975. Biochemical and genetic studies of peroxisomal multiple enzyme systems: α-hydroxyacid oxidase and catalase. *Isozymes I. Molecular Structure*, ed. C. L. Markert, pp. 191–211. New York: Academic

Holmes, R. S., Masters, C. J. 1970. Epigenetic interconversions of the multiple forms of mouse liver catalase. *FEBS Lett.* 11:45–48

Holmes, R. S., Masters, C. J. 1972. Species specific features of the distribution and multiplicity of mammalian liver catalase. *Arch. Biochem. Biophys.* 148:217–23

Honigberg, B. M., Volkmann, D., Entzeroth, R., Scholtyseck, E. 1984. A freeze-fracture electron microscope study of *Trichomonas vaginalis* donné and *Tritrichomonas foetus* (Reidmuller). *J. Protozool.* 31:116–31

Housman, D., Jacobs-Lorena, M., Rajbhandary, U. L., Lodish, H. F. 1970. Initiation of haemoglobin synthesis by methionyl-tRNA. *Nature* 227:913–18

Hruban, Z., Swift, H., Slesers, A. 1966. Ultrastructural alterations of hepatic microbodies. *Lab. Invest.* 15:1884–1901

Hruban, Z., Gotoh, M., Slesers, A., Chou, S.-F. 1974. Structure of hepatic microbodies in rats treated with acetylsalicylic acid, clofibrate, and dimethrin. *Lab. Invest.* 30:64–75

Hruban, Z., Rechcigl, M. Jr. 1969. *Microbodies and Related Particles.* New York: Academic. 296 pp.

Hruban, Z., Swift, H. 1964. Uricase: localization in hepatic microbodies. *Science* 146:1316–18

Huang, A. H. C., Beevers, H. 1973. Localization of enzymes within microbodies. *J. Cell Biol.* 58:379–89

Huang, A. H. C., Trelease, R. N., Moore, T. S. Jr. 1983. *Plant Peroxisomes*, pp. 157–200. New York: Academic

Hüttinger, M., Goldenberg, H., Kramar, R. 1980. Intraparticulate localization of the peroxisomal fatty acid β-oxidation system in rat liver. *Hoppe-Seyler's Z. Physiol. Chem.* 361:1125–28

Hüttinger, M., Pavelka, M., Goldenberg, H., Kramar, R. 1981. Membranes of rat liver peroxisomes. *Histochemistry* 71:259–67

Hutton, D., Stumpf, P. K. 1969. Fat metabolism in higher plants. XXXVII. Characterization of the β-oxidation systems from maturing and germinating castor bean seeds. *Plant Physiol.* 44:508–16

Inestrosa, N. C., Bronfman, M., Leighton, F. 1979. Detection of peroxisomal fatty acylcoenzyme A oxidase activity. *Biochem. J.* 182:779–88

Inestrosa, N. C., Bronfman, M., Leighton, F. 1980. Purification of the peroxisomal fatty acyl-CoA oxidase from rat liver. *Biochem. Biophys. Res. Commun.* 95:7–12

Jiang, Z., Thorpe, C. 1983. Acyl-CoA oxidase from *Candida tropicalis*. *Biochemistry* 22:3752–58

Johanson, R. A., Hill, J. M., McFadden, B. A. 1974. Isocitrate lyase from *Neurospora crassa*. II. Composition, quaternary structure, C-terminus, and active-site modification. *Biochim. Biophys. Acta* 364:341–52

Jones, G. L., Masters, C. J. 1975. On the nature and characteristics of the multiple forms of catalase in mouse liver. *Arch. Biochem. Biophys.* 169:7–21

Jones, R. G., Davis, W. L., Goodman, D. B. P. 1982. The role of peroxisomes in the response of the toad bladder to aldosterone. *Ann. NY Acad. Sci.* 386:165–69

Kamiryo, T., Abe, M., Okazaki, K., Kato, S., Shimamoto, N. 1982. Absence of DNA in peroxisomes of *Candida tropicalis*. *J. Bacteriol.* 152:269–74

Kamiryo, T., Okazaki, K. 1984. High-level expression and molecular cloning of genes encoding *Candida tropicalis* peroxisomal proteins. *Mol. Cell. Biol.* 4:2136–41

Kato, N., Omori, Y., Tani, Y., Ogata, K. 1976. Alcohol oxidases of *Kloeckera* sp. and *Hansenula polymorpha*. Catalytic properties and subunit structures. *Eur. J. Biochem.* 64:341–50

Kawamoto, S., Nozaki, C., Tanaka, A., Fukui, S. 1978a. Fatty acid β-oxidation system in microbodies of *n*-alkane-grown *Candida tropicalis*. *Eur. J. Biochem.* 83:609–13

Kawamoto, S., Tanaka, A., Yamamura, M., Teranishi, Y., Fukui, S., Osumi, M. 1977. Microbody of *n*-alkane-grown yeast. Enzyme localization in the isolated microbody. *Arch. Microbiol.* 112:1–8

Kawamoto, S., Ueda, M., Nozaki, C., Yamamura, M., Tanaka, A., Fukui, S. 1978b. Localization of carnitine acetyltransferase in peroxisomes and in mitochondria of *n*-alkane-grown *Candida tropicalis*. *FEBS Lett.* 96:37–40

Keller, G.-A., Barton, M. C., Shapiro, D. J., Singer, S. J. 1985. 3-Hydroxy-3-methylglutaryl-coenzyme A reductase is present in peroxisomes in normal rat liver cells. *Proc. Natl. Acad. Sci. USA* 82:770–74

Kelley, R. I. 1983. The cerebrohepatorenal syndrome of Zellweger, morphologic and metabolic aspects. *Am. J. Med. Genet.* 16:503–17

Kindl, H. 1982a. The biosynthesis of microbodies (peroxisomes, glyoxysomes). *Int. Rev. Cytol.* 80:193–229

Kindl, H. 1982b. Glyoxysome biogenesis via cytosolic pools in cucumber. *Ann. NY Acad. Sci.* 386:314–28

Kindl, H., Kruse, C. 1983. Biosynthesis of glyoxysomal proteins. *Methods Enzymol.* 96:700–15

Kindl, H., Lazarow, P. B., eds. 1982. Peroxisomes and glyoxysomes. *Ann. NY Acad. Sci.* 386:1–550

Kionka, C., Kunau, W.-H. 1985. Inducible β-oxidation pathway in *Neurospora crassa*. *J. Bacteriol.* 161:153–57

Kirkman, H. N., Gaetani, G. F. 1984. Catalase: a tetrameric enzyme with four tightly bound molecules of NADPH. *Proc. Natl. Acad. Sci. USA* 81:4343–47

Köller, W., Kindl, H. 1977. Glyoxylate cycle enzymes of the glyoxysomal membrane from cucumber cotyledons. *Arch. Biochem. Biophys.* 181:236–48

Köller, W., Kindl, H. 1978. Studies supporting the concept of glyoxyperoxisomes as intermediary organelles in transformation of glyoxysomes into peroxisomes. *Z. Naturforsch.* 33c:962–68

Köller, W., Kindl, H. 1980. 19S Cytosolic malate synthase. A small pool characterized by rapid turnover. *Hoppe-Seylers Z. Physiol. Chem.* 361:1437–44

Korneluk, R. G., Quan, F., Lewis, W. H., Guise, K. S., Willard, H. F., et al. 1984. Isolation of human fibroblast catalase cDNA clones. Sequence of clones derived from spliced and unspliced mRNA. *J. Biol. Chem.* 259:13819–23

Kruse, C., Frevert, J., Kindl, H. 1981. Selective uptake by glyoxysomes of in vitro translated malate synthase. *FEBS Lett.* 129:36–38

Kruse, C., Kindl, H. 1982. Integral proteins of the glyoxysomal membranes. *Ann. NY Acad. Sci.* 386:499–501

Kruse, C., Kindl, H. 1983a. Oligomerization of malate synthase during glyoxysome biosynthesis. *Arch. Biochem. Biophys.* 223:629–38

Kruse, C., Kindl, H. 1983b. Malate synthase: aggregation, deaggregation, and binding of phospholipids. *Arch. Biochem. Biophys.* 223:618–28

Lamb, J. E., Riezman, H., Becker, W. M., Leaver, C. L. 1978. Regulation of glyoxysomal enzymes during germination of cucumber. 2. Isolation and immunological detection of isocitrate lyase and catalase. *Plant Physiol.* 62:754–60

Lazarow, P. B. 1977. Three hypolipidemic drugs increase hepatic palmitoyl-coenzyme A oxidation in the rat. *Science* 197:580–81

Lazarow, P. B. 1978. Rat liver peroxisomes catalyze the β-oxidation of fatty acids. *J. Biol. Chem.* 253:1522–28

Lazarow, P. B. 1982. Compartmentation of β-oxidation of fatty acids in peroxisomes. In *Metabolic Compartmentation*, ed. H. Sies, pp. 317–29. New York: Academic

Lazarow, P. B. 1984. The peroxisomal membrane. In *Membrane Structure and Function*, Vol. 5, ed. E. E. Bittar, pp. 1–31. New York: Wiley

Lazarow, P. B., Black, V., Shio, H., Fujiki, Y., Hajra, A. K., et al. 1985. Zellweger syndrome: biochemical and morphological studies on two patients treated with clofibrate. *Pediatr. Res.* In press

Lazarow, P. B., de Duve, C. 1971. Intermediates in the biosynthesis of peroxisomal catalase in rat liver. *Biochem. Biophys. Res. Commun.* 45:1198–1204

Lazarow, P. B., de Duve, C. 1973a. The synthesis and turnover of rat liver peroxisomes. IV. Biochemical pathway of catalase synthesis. *J. Cell Biol.* 59:491–506

Lazarow, P. B., de Duve, C. 1973b. The synthesis and turnover of rat liver peroxisomes. V. Intracellular pathway of catalase synthesis. *J. Cell Biol.* 59:507–24

Lazarow, P. B., de Duve, C. 1976. A fatty acyl-CoA oxidizing system in rat liver peroxisomes; enhancement by clofibrate, a hypolipidemic drug. *Proc. Natl. Acad. Sci. USA* 73:2043–46

Lazarow, P. B., Fujiki, Y., Mortensen, R., Hashimoto, T. 1982a. Identification of β-oxidation enzymes among peroxisomal polypeptides: increase in Coomassie Blue-stainable protein after clofibrate treatment. *FEBS Lett.* 150:307–10

Lazarow, P. B., Robbi, M., Fujiki, Y., Wong, L. 1982b. Biogenesis of peroxisomal proteins in vivo and in vitro. *Ann. NY Acad. Sci.* 386:285–300

Lazarow, P. B., Shio, H., Robbi, M. 1980. Biogenesis of peroxisomes and the peroxisome reticulum hypothesis. In *31st Mosbach Colloq. Biol. Chem. Organelle Formation*, ed. T. Bucher, W. Sebald, H. Weiss, pp. 187–206. New York: Springer-Verlag

Legg, P. G., Wood, R. L. 1970. New observations on microbodies: A cytochemical study on CPIB-treated rat liver. *J. Cell Biol.* 45:118–29

Leighton, F., Brandan, E., Lazo, O., Bronfman, M. 1982. Subcellular fractionation studies on the organization of fatty acid oxidation by liver peroxisomes. *Ann. NY Acad. Sci.* 386:62–78

Leighton, F., Coloma, L., Koenig, C. 1975. Structure, composition, physical properties, and turnover of proliferated peroxisomes. A study of the trophic effects of Su-13437 on rat liver. *J. Cell Biol.* 67:281–309

Leighton, F., Poole, B., Beaufay, H., Baudhuin, P., Coffey, J. W., et al. 1968. The large scale separation of peroxisomes, mitochondria, and lysosomes from the livers of rats injected with Triton WR-1339. *J. Cell Biol.* 37:482–513

Lord, J. M., Roberts, L. M. 1982. Glyoxysome biogenesis via the endoplasmic reticulum in castor bean endosperm? *Ann. NY Acad. Sci.* 386:362–76

Lord, J. M., Roberts, L. M. 1983. Formation of glyoxysomes. *Int. Rev. Cytol.* (Suppl.) 15:115–56

Ludwig, B., Kindl, H. 1976. Plant microbody proteins. II. Purification and characterization of the major protein component (SP-63) of peroxisome membranes. *Hoppe-Seyler's Z. Physiol. Chem.* 357:177–86

Mainferme, F., Wattiaux, R. 1976. Behavior of rat liver catalase during electrophoresis in a pH gradient. *Cancer Biochem. Biophys.* 1:313–16

Mainferme, F., Wattiaux, R. 1982. Effect of lysosomes on rat-liver catalase. *Eur. J. Biochem.* 127:343–46

Mannaerts, G. P., Debeer, L. J. 1982. Mitochondrial and peroxisomal β-oxidation of fatty acids in rat liver. *Ann. NY Acad. Sci.* 386:30–38

Mannaerts, G. P., Van Veldhoven, P., Van Broekhoven, A., Vandebroek, G., Debeer, L. J. 1982. Evidence that peroxisomal acyl-CoA synthetase is located at the cytoplasmic side of the peroxisomal membrane. *Biochem. J.* 204:17–23

Markwell, M. A. K., Tolbert, N. E., Beiber, L. L. 1976. Comparison of the carnitine acyltransferase activities from rat liver peroxisomes and microsomes. *Arch. Biochem. Biophys.* 176:479–88

Mellor, R. B., Bowden, L., Lord, J. M. 1978. Glycoproteins of the glyoxysomal matrix. *FEBS Lett.* 90:275–78

Miura, S., Mori, M., Takiguchi, M., Tatibana, M., Furuta, S., et al. 1984. Biosynthesis and intracellular transport of enzymes of peroxisomal β-oxidation. *J. Biol. Chem.* 259:6397–6402

Miyazawa, S., Furuta, S., Osumi, T., Hashimoto, T., Ui, N. 1981. Properties of peroxisomal 3-ketoacyl-CoA thiolase from rat liver. *J. Biochem.* 90:511–19

Miyazawa, S., Ozasa, H., Furuta, S., Osumi, T., Hashimoto, T., et al. 1983a. Biosynthesis and turnover of carnitine acetyltransferase in rat liver. *J. Biochem.* 93:453–59

Miyazawa, S., Ozasa, H., Osumi, T., Hashimoto, T. 1983b. Purification and properties of carnitine octanoyltransferase and carnitine palmitoyltransferase from rat liver. *J. Biochem.* 94:529–42

Moore, T. S. Jr. 1982. Phospholipid biosynthesis. *Ann. Rev. Plant Physiol.* 33:235–59

Moreno de la Garza, M., Schultz-Borchard, U., Crabb, J. W., Kunau, W.-H. 1985. Peroxisomal β-oxidation system of *Candida tropicalis*. Purification of a multifunctional protein possessing enoyl-CoA hydratase, 3-hydroxyacyl-CoA dehydrogenase, and 3-hydroxyacyl-CoA epimerase activities. *Eur. J. Biochem.* 148:285–91

Moser, A. E., Singh, I., Brown, F. R. III, Solish, G. I., Kelley, R. I., et al. 1984. The cerebrohepatorenal (Zellweger) syndrome. Increased levels and impaired degradation of very-long-chain fatty acids and their use in prenatal diagnosis. *New Eng. J. Med.* 310:1141–46

Müller, M. 1975. Biochemistry of protozoan microbodies: peroxisomes, α-glycerophosphate oxidase bodies, hydrogenosomes. *Ann. Rev. Microbiol.* 29:467–83

Müller, M. 1980. The hydrogenosome. In *The Eukaryotic Microbial Cell*, ed. G. W. Gooday, D. Lloyd, A. P. J. Trinci, pp. 127–42. Cambridge: Cambridge Univ. Press

Müller, M., Hogg, J. F., de Duve, C. 1968.

Distribution of tricarboxylic acid cycle enzymes and glyoxylate cycle enzymes between mitochondria and peroxisomes in *Tetrahymena pyriformis*. *J. Biol. Chem.* 243:5385–95

Murthy, M. R. N., Reid, T. J. III, Sicignano, A., Tanaka, N., Rossmann, M. G. 1981. Structure of beef liver catalase. *J. Mol. Biol.* 152:465–99

Nicholls, P., Schonbaum, G. R. 1963. Catalases. *Enzymes* 8:147–225

Novikoff, A. B., Shin, W.-Y. 1964. The endoplasmic reticulum in the Golgi zone and its relations to microbodies, Golgi apparatus and autophagic vacuoles in rat liver cells. *J. Micros.* Oxford 3:187–206

Novikoff, P. M., Novikoff, A. B., Quintana, N., Davis, C. 1973. Studies on microperoxisomes. III. Observations on human and rat hepatocytes. *J. Histochem. Cytochem.* 21:540–58

Opperdoes, F. R. 1984. Localization of the initial steps in alkoxyphospholipid biosynthesis in glycosomes (microbodies) of *Trypanosoma brucei*. *FEBS Lett.* 169:35–39

Opperdoes, F. R., Baudhuin, P., Coppens, I., de Roe, C., Edwards, S. W., et al. 1984. Purification, morphometric analysis, and characterization of the glycosomes (microbodies) of the protozoan hemoflagellate *Trypanosoma brucei*. *J. Cell Biol.* 98:1178–84

Osmundsen, H. 1982. Peroxisomal β-oxidation of long fatty acids: effects of high fat diets. *Ann. NY Acad. Sci.* 386:13–27

Osumi, M. 1976. Possible existence of DNA in yeast microbody. *J. Electron Microsc.* 25:43–47

Osumi, M., Fukuzumi, F., Teranishi, Y., Tanaka, A., Fukui, S. 1975. Development of microbodies in *Candida tropicalis* during incubation in a n-alkane medium. *Arch. Microbiol.* 103:1–11

Osumi, M., Kazama, H., Sato, S. 1978. Microbody-associated DNA in *Candida tropicalis* pK 233 cells. *FEBS Lett.* 90:309–12

Osumi, M., Sato, M., Sakai, T., Suzuki, M. 1979. Fine structure of crystalloid in the yeast *Kloeckera* microbodies. *J. Electron Microsc.* 28:295–300

Osumi, T., Hashimoto, T. 1979a. Peroxisomal β-oxidation system of rat liver. Copurification of enoyl-CoA hydratase and 3-hydroxyacyl-CoA dehydrogenase. *Biochem. Biophys. Res. Commun.* 89:580–84

Osumi, T., Hashimoto, T. 1979b. Subcellular distribution of the enzymes of the fatty acyl-CoA β-oxidation system and their induction by di(2-ethylhexyl)phthalate in rat liver. *J. Biochem.* 85:131–39

Osumi, T., Hashimoto, T. 1980. Purification and properties of mitochondrial and peroxisomal 3-hydroxyacyl-CoA dehydrogenase from rat liver. *Arch. Biochem. Biophys.* 203:372–83

Osumi, T., Hashimoto, T., Ui, N. 1980. Purification and properties of acyl-CoA oxidase from rat liver. *J. Biochem.* 87:1735–46

Osumi, T., Ozasa, H., Hashimoto, T. 1984a. Molecular cloning of cDNA for rat acyl-CoA oxidase. *J. Biol. Chem.* 259:2031–34

Osumi, T., Ozasa, H., Miyazawa, S., Hashimoto, T. 1984b. Molecular cloning of cDNA for rat liver catalase. *Biochem. Biophys. Res. Commun.* 122:831–37

Ozasa, H., Miyazawa, S., Osumi, T. 1983. Biosynthesis of carnitine octanoyltransferase and carnitine palmitoyltransferase. *J. Biochem.* 94:543–49

Pais, M. S. S., Carrapico, F. 1982. Microbodies—a membrane compartment. *Ann. NY Acad. Sci.* 386:510–13

Palmiter, R. D. 1983. Identifying primary translation products: use of N-formylmethionyl-tRNA and prevention of NH$_2$-terminal acetylation. *Methods Enzymol.* 96:150–57

Poole, B., Higashi, T., de Duve, C. 1970. The synthesis and turnover of rat liver peroxisomes. III. The size distribution of peroxisomes and the incorporation of new catalase. *J. Cell Biol.* 45:408–15

Rachubinski, R. A., Fujiki, Y., Lazarow, P. B. 1985a. Cloning of cDNA coding for peroxisomal acyl-CoA oxidase from the yeast *Candida tropicalis* pK233. *Proc. Natl. Acad. Sci. USA* 82:3973–77

Rachubinski, R. A., Fujiki, Y., Lazarow, P. B. 1985b. Cloning of cDNAs coding for peroxisomal proteins of the yeast *Candida tropicalis* pK233: identification of a clone coding for catalase. Submitted for publication

Rachubinski, R. A., Fujiki, Y., Mortensen, R. M., Lazarow, P. B. 1984. Acyl-CoA oxidase and hydratase-dehydrogenase, two enzymes of the peroxisomal β-oxidation system, are synthesized on free polysomes of clofibrate-treated rat liver. *J. Cell Biol.* 99:2241–46

Rachubinski, R. A., Verma, D. P. S., Bergeron, J. J. M. 1980. Synthesis of rat liver microsomal cytochrome b_5 by free ribosomes. *J. Cell Biol.* 84:705–16

Reddy, J., Svoboda, D. 1971. Microbodies in experimentally altered cells. VIII. Continuities between microbodies and their possible biologic significance. *Lab. Invest.* 24:74–81

Reddy, J., Svoboda, D. 1973. Further evidence to suggest that microbodies do not exist as individual entities. *Am. J. Pathol.* 70:421–32

Reddy, J. K., Warren, J. R., Reddy, M. K., Lalwani, N. D. 1982. Hepatic and renal effects of peroxisome proliferators: biological implications. *Ann. NY Acad. Sci.* 386:81–110

Reddy, M. K., Qureshi, S. A., Hollenberg, P. F., Reddy, J. K. 1981. Immunochemical identity of peroxisomal enoyl-CoA hydratase with the peroxisome-proliferation-associated 80,000 mol wt polypeptide in rat liver. *J. Cell Biol.* 89:406–17

Redman, C. M., Grab, D. J., Irukulla, R. 1972. The intracellular pathway of newly formed rat liver catalase. *Arch. Biochem. Biophys.* 152:496–501

Remacle, J. 1978. Binding of cytochrome b_5 to membranes of isolated subcellular organelles from rat liver. *J. Cell Biol.* 79:291–313

Rhodin, J. 1954. Correlation of ultrastructural organization and function in normal and experimentally changed proximal convoluted tubule cells of the mouse kidney. PhD thesis. Aktiebolaget Godvil, Stockholm. 76 pp.

Rhodin, J. A. G. 1963. In *An Atlas of Ultrastructure*. Philadelphia: Saunders

Riezman, H., Weir, E. M., Leaver, C. J., Titus, D. E., Becker, W. M. 1980. Regulation of glyoxysomal enzymes during germination of cucumber. 3. In vitro translation and characterization of four glyoxysomal enzymes. *Plant Physiol.* 65:40–46

Rigatuso, J. L., Legg, P. G., Wood, R. L. 1970. Microbody formation in regenerating rat liver. *J. Histochem. Cytochem.* 18:893–900

Roa, M., Blobel, G. 1983. Biosynthesis of peroxisomal enzymes in the methylotrophic yeast *Hansenula polymorpha*. *Proc. Natl. Acad. Sci. USA* 80:6872–76

Robbi, M., Lazarow, P. B. 1978. Synthesis of catalase in two cell-free protein-synthesizing systems and in rat liver. *Proc. Natl. Acad. Sci. USA* 75:4344–48

Robbi, M., Lazarow, P. B. 1982. Peptide mapping of peroxisomal catalase and its precursor: comparison to the primary wheat germ translation product. *J. Biol. Chem.* 257:964–70

Roberts, L. M., Lord, J. M. 1981. Synthesis and posttranslational segregation of glyoxysomal isocitrate lyase from castor bean endosperm. *Eur. J. Biochem.* 119:43–49

Roels, F. 1976. Cytochemical demonstration of extraperoxisomal catalase. I. Sheep liver. *J. Histochem. Cytochem.* 24:713–24

Roggenkamp, R., Janowicz, Z., Stanikowski, B., Hollenberg, C. P. 1984. Biosynthesis and regulation of the peroxisomal methanol oxidase from the methylotrophic yeast *Hansenula polymorpha*. *Mol. Gen. Genet.* 194:489–93

Roggenkamp, R., Sahm, H., Hinkelmann, W., Wagner, F. 1975. Alcohol oxidase and catalase in peroxisomes of methanol-grown *Candida boidinii. Eur. J. Biochem.* 59:231–36

Rouiller, C., Bernhard, W. 1956. Microbodies and the problem of mitochondrial regeneration in liver cells. *J. Biophys. Biochem. Cytol.* 2 Suppl:355–59

Russo, J. J., Black, V. H. 1982. Hormone-dependent changes in peroxisomal enzyme activity in guinea pig adrenal. *J. Biol. Chem.* 257:3883–89

Sahm, H., Wagner, F. 1973. Microbial assimilation of methanol. The ethanol- and methanol-oxidizing enzymes of the yeast *Candida boidinii. Eur. J. Biochem.* 36:250–56

Sando, T., Konno, K., Takei, N., Sakamoto, T., Higashi, T. 1984. Purification and characterization of rat liver cytosol catalase. *Cell Struct. Funct.* 9:125–33

Sansone, G., Hamilton, J. G. 1969. Glyceryl ether, wax ester and triglyceride composition of the mouse preputial gland. *Lipids* 4:435–40

Sautter, C., Hock, B. 1982. Fluorescence immunohistochemical localization of malate dehydrogenase isoenzymes in watermelon cotyledons. *Plant Physiol.* 70:1162–68

Scandalios, J. G. 1975. Differential gene expression and biochemical properties of catalase isozymes in maize. In *Isozymes. III. Developmental biology*, ed. C. L. Markert, pp. 213–38. New York: Academic

Scandalios, J. G., Chang, D. Y., McMillin, D. E., Tsaftaris, A., Moll, R. H. 1980a. Genetic regulation of the catalase developmental program in maize scutellum: Identification of a temporal regulatory gene. *Proc. Natl. Acad. Sci. USA* 77:5360–64

Scandalios, J. G., Tong, W. F., Roupakias, D. G. 1980b. Cat 3, a third gene locus coding for a tissue-specific catalase in maize: Genetics, intracellular location, and some biochemical properties. *Mol. Gen. Genet.* 179:33–41

Schneeberger, E. E. 1972. Development of peroxisomes in granular pneumocytes during pre- and postnatal growth. *Lab. Invest.* 27:581–89

Schopfer, D., Bajracharya, D., Bergfeld, R., Falk, H. 1976. Phytochrome-mediated transformation of glyoxysomes into peroxisomes in the cotyledons of mustard (*Sinapis alba L.*) seedlings. *Planta* 133:73–80

Schroeder, W. A., Shelton, J. R., Shelton, J. B., Robberson, B., Apell, G., et al. 1982. The complete amino acid sequence of bovine liver catalase and the partial sequence of bovine erythrocyte catalase. *Arch. Biochem. Biophys.* 214:397–421

Shimizu, S., Yasui, K., Tani, Y., Yamada, H. 1979. Acyl-CoA oxidase from *Candida tropicalis. Biochem. Biophys. Res. Commun.* 91:108–13

Shio, H., Lazarow, P. B. 1981. Relationship between peroxisomes and endoplasmic reticulum investigated by combined catalase and glucose-6-phosphatase cytochemistry. *J. Histochem. Cytochem.* 29:1263–72

Small, G. M., Brolly, D., Connock, M. J. 1980. Palmityl-CoA oxidase: Detection in several guinea pig tissues and peroxisomal localisation in mucosa of small intestine. *Life Sci.* 27:1743–51

Staübli, W., Schweizer, W., Suter, J., Weibel, E. R. 1977. The proliferative response of hepatic peroxisomes of neonatal rats to treatment with SU-13 437 (Nafenopin). *J. Cell Biol.* 74:665–89

Sugita, Y., Tobe, T., Sakamoto, T., Higashi, T. 1982. Immature precursor catalase in subcellular fractions of rat liver. *J. Biochem.* 92:509–15

Susani, M., Zimniak, P., Fessl, F., Ruis, H. 1976. Localization of catalase A in vacuoles of *Saccharomyces cerevisiae*: Evidence for the vacuolar nature of isolated "yeast peroxisomes". *Hoppe-Seyler's Z. Physiol. Chem.* 357:961–70

Szabo, Á. S., Avers, C. J. 1969. Some aspects of regulation of peroxisomes and mitochondria in yeast. *Ann. NY Acad. Sci.* 168:302–12

Titus, D. E., Becker, W. M. 1985. Investigation of the glyoxysome-peroxisome transition in germinating cucumber cotyledons using double-label immunoelectron microscopy. *J. Cell Biol.* 101:1288–99

Tobe, T., Higashi, T. 1980. Studies on rat liver catalase. XI. Site of synthesis and segregation by stripped ER membranes. *J. Biochem.* 88:1341–47

Tolbert, N. E. 1981. Metabolic pathways in peroxisomes and glyoxysomes. *Ann. Rev. Biochem.* 50:133–57

Trelease, R. N. 1984. Biogenesis of glyoxysomes. *Ann. Rev. Plant Physiol.* 35:321–47

Trelease, R. N., Becker, W. M., Gruber, P. J., Newcomb, E. H. 1971. Microbodies (glyoxysomes and peroxisomes) in cucumber cotyledons. Correlative biochemical and ultrastructural study in light- and dark-grown seedlings. *Plant Physiol.* 48:461–75

Tsubouchi, J., Tonomura, K., Tanaka, K. 1976. Ultrastructure of microbodies of methanol-assimilating yeasts. *J. Gen. Appl. Microbiol.* 22:131–42

Tsukada, H., Mochizuki, Y., Fujiwara, S. 1966. The nucleoids of rat liver cell microbodies. Fine structure and enzymes. *J. Cell Biol.* 28:449–60

Ueda, M., Tanaka, A., Fukui, S. 1982. Peroxisomal and mitochondrial carnitine acetyltransferases in alkane-grown yeast *Candida tropicalis*. *Eur. J. Biochem.* 124 : 205–10

Ueda, M., Tanaka, A., Fukui, S. 1984a. Characterization of peroxisomal and mitochondrial carnitine acetyltransferases purified from alkane-grown *Candida tropicalis*. *Eur. J. Biochem.* 138 : 445–49

Ueda, M., Tanaka, A., Horikawa, S., Numa, S., Fukui, S. 1984b. Synthesis *in vitro* of precursor-type carnitine acetyltransferase with messenger RNA from *Candida tropicalis*. *Eur. J. Biochem.* 138 : 451–57

Van Dijken, J. P., Veenhuis, M., Harder, W. 1982. Peroxisomes of methanol-grown yeast. *Ann. NY Acad. Sci.* 386 : 200–16

Veenhuis, M., Van Dijken, J. P., Harder, W. 1976. Cytochemical studies on the localization of methanol oxidase and other oxidases in peroxisomes of methanol-grown *Hansenula polymorpha*. *Arch. Microbiol.* 111 : 123–35

Veenhuis, M., Van Dijken, J. P., Harder, W. 1983. The significance of peroxisomes in the metabolism of one-carbon compounds in yeasts. In *Advances in Microbial Physiology*, ed. A. H. Rose, J. G. Morris, D. W. Tempest, Vol. 24, pp. 1–82. New York : Academic

Veenhuis, M., Van Dijken, J. P., Pilon, S. A. F., Harder, W. 1978. Development of crystalline peroxisomes in methanol-grown cells of the yeast *Hansenula polymorpha* and its relation to environmental conditions. *Arch. Microbiol.* 117 : 153–63

Veenhuis, M., Harder, W., Van Dijken, J. P., Mayer, F. 1981. Substructure of crystalline peroxisomes in methanol-grown *Hansenula polymorpha*: evidence for an in vivo crystal of alcohol oxidase. *Mol. Cell. Biol.* 1 : 949–57

Veenhuis, M., Keizer, I., Harder, W. 1979. Characterization of peroxisomes in glucose-grown *Hansenula polymorpha* and their development after the transfer of cells into methanol-containing media. *Arch. Microbiol.* 120 : 167–75

Völkl, A., Lazarow, P. B. 1982. Affinity chromatography of peroxisomal proteins on lectin-sepharose columns. *Ann. NY Acad. Sci.* 386 : 504–6

Walk, R.-A., Hock, B. 1977. Glyoxysomal malate dehydrogenase of watermelon cotyledons: de novo synthesis on cytoplasmic ribosomes. *Planta* 124 : 277–85

Walk, R.-A., Hock, B. 1978. Cell-free synthesis of glyoxysomal malate dehydrogenase. *Biochem. Biophys. Res. Commun.* 81 : 636–43

Walk, R.-A., Michaeli, S., Hock, B. 1977.

Glyoxysomal and mitochondrial malate dehydrogenase of watermelon (*Citrullus vulgaris*) cotyledons. I. Molecular properties of the purified isoenzymes. *Planta* 136 : 211–20

Walter, P., Gilmore, R., Blobel, G. 1984. Protein translocation across the endoplasmic reticulum. *Cell* 38 : 5–8

Wanders, R. J. A., Kos, M., Roest, B., Meijer, A. J., Schrakamp, G., et al. 1984. Activity of peroxisomal enzymes and intracellular distribution of catalase in Zellweger syndrome. *Biochem. Biophys. Res. Commun.* 123 : 1054–61

Wanner, G., Theimer, R. R. 1982. Two types of microbodies in *Neurospora crassa*. *Ann. NY Acad. Sci.* 386 : 269–84

Wirtz, K. W. A. 1982. Phospholipid transfer proteins. In *Lipid-Protein Interactions*, ed. P. Jost, O. H. Griffith, pp. 151–231. New York : Wiley

Yamada, T., Tanaka, A., Fukui, S. 1982a. Properties of catalase purified from whole cells and peroxisomes of n-alkane-grown *Candida tropicalis*. *Eur. J. Biochem.* 125 : 517–21

Yamada, T., Tanaka, A., Horikawa, S., Numa, S., Fukui, S. 1982b. Cell-free translation and regulation of *Candida tropicalis* catalase messenger RNA. *Eur. J. Biochem.* 129 : 251–55

Yamaguchi, J., Nishimura, M. 1984. Purification of glyoxysomal catalase and immunochemical comparison of glyoxysomal and leaf peroxisomal catalase in germinating pumpkin cotyledons. *Plant Physiol.* 74 : 261–67

Yamaguchi, J., Nishimura, M., Akazawa, T. 1984. Maturation of catalase precursor proceeds to a different extent in glyoxysomes and leaf peroxisomes of pumpkin cotyledons. *Proc. Natl. Acad. Sci. USA* 81 : 4809–13

Yokota, S., Fahimi, H. D. 1982. Immunoelectron microscopic localization of acyl-CoA oxidase in rat liver and kidney. *Ann. NY Acad. Sci.* 386 : 491–94

Zimmermann, R., Neupert, W. 1980. Biogenesis of glyoxysomes. Synthesis and intracellular transfer of isocitrate lyase. *Eur. J. Biochem.* 112 : 225–33

Zimniak, P., Hartter, E., Woloszczuk, W., Ruis, H. 1976. Catalase biosynthesis in yeast: Formation of catalase A and catalase T during oxygen adaptation of *Saccharomyces cerevisiae*. *Eur. J. Biochem.* 71 : 393–98

Zimniak, P., Hartter, E., Ruis, H. 1975. Biosynthesis of catalase T during oxygen adaptation of *Saccharomyces cerevisiae*. *FEBS Lett.* 59 : 300–4

Ann. Rev. Cell Biol. 1985. 1:531–61

STABILIZING INFRASTRUCTURE OF CELL MEMBRANES

V. T. Marchesi

Department of Pathology, Yale University School of Medicine, 310 Cedar Street, Post Office Box 3333, New Haven, Connecticut 06510

CONTENTS

OVERVIEW

Cell membranes are composed of phospholipid bilayers punctuated by transmembrane proteins, parts of which span the hydrophobic regions of the bilayer as α-helically coiled polypeptides. The proteins that span different types of surface membranes can be grouped into two distinct classes on the basis of their membrane-spanning segments. One type has a single membrane-spanning segment, which appears to be in an α-helical conformation and contains 22–23 uncharged amino acids, just long enough to cross the nonpolar region of the membrane. This class of transmembrane proteins includes glycophorin, the principal sialic acid–containing glyco-

531

0743–4634/85/1115–0531$02.00

protein of the human red blood cell membrane (1); histocompatibility antigens and membrane-bound immunoglobulins (2); and a variety of receptor molecules, including those for transferrin (3), asialoglycoproteins (4), low density lipoproteins (5), polyimmunoglobulins (6), epidermal growth factor (7), and insulin (8). The second class of transmembrane proteins is arranged in a distinctly different manner: multiple polypeptide chains cross the lipid bilayer rather than a single membrane-spanning loop. The intramembranous segments of these molecules also appear to be largely α-helical, although in most cases this interpretation is more inferential than the result of direct experimental evidence. The membrane-spanning segments of this class of molecules also differ from the first type in that charged amino acids exist within the hydrophobic regions of the bilayer, in contrast to the completely nonpolar complement of amino acids in those proteins with a single spanning loop.

Since the proteins with multiple membrane-spanning segments are also known to be involved in different transport functions, it is generally believed that the intramembranous segments create "channels" across the bilayer. The mechanisms by which adjacent polypeptide chains create aqueous channels across a lipid bilayer is still largely obscure, but associations between specific amino acid side chains and pairing of charged amino acids probably play important roles. A simplistic model depicting the possible arrangement of these two classes of transmembrane glyco-proteins within the bilayer context is illustrated in Figure 1.

It has been realized for some time that models that only deal with integral membrane proteins do not explain how cell surface functions are regulated nor how membranes can undergo dynamic changes in their shape, deformability, and stability. It is now clear that a complex set of proteins is attached to the undersurfaces of cell membranes, and is responsible for

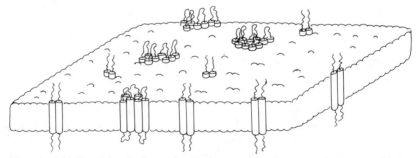

Figure 1   A highly schematic picture showing how transmembrane proteins interact with the lipid bilayer. The intramembranous polypeptide segments may all be in an α-helical conformation, depicted here as cylinders.

stabilizing membranes, and regulating the topography and mobility of the different transmembrane proteins. This submembranous network, often referred to as a membrane skeleton, is composed of filaments of actin, actin-binding proteins such as spectrin and its analogues, and a set of connecting proteins that link the stabilizing infrastructure to the overlying membrane.

The stabilizing infrastructure of mammalian red blood cells has received the most attention over the past decade, and as a result we know most about the proteins involved in its construction. The methods used to analyze the red cell system are now being applied to the study of other cell types with promising results. This review attempts to highlight some of the principles that are thought to govern this submembranous infrastructure. Most of the generalizations are derived from studies of human red blood cells, since this has proved to be the most accessible system for study so far.

PROPERTIES OF ERYTHROCYTE SPECTRIN

The properties of erythrocyte spectrin have been reviewed extensively in recent years (9–14), and I will not attempt to provide yet another historical account of the study of this interesting molecule. Instead, I will present generalities about spectrin's structure and function that may be relevant to understanding how spectrin subserves its principal supportive role in the submembranous skeleton.

Spectrin, when isolated from erythrocyte ghost membranes by low-salt extraction techniques and purified using conventional chromatography methods, is soluble in aqueous buffers, and exists as flexible, rod-like forms of varying length and complexity. The anatomy of the spectrin molecule has been determined by low-angle rotary shadowing and electron microscopy (15), and our conception of its molecular substructure is based largely on images revealed by this technique. The smallest forms of purified erythrocyte spectrin seen are flexible rods, approximately 100 nm composed of two nonidentical subunits, each of which is a long peptide that extends the length of the dimeric unit. Erythrocyte spectrin subunits, referred to as α and β forms, are both sizable polypeptide chains; the α subunit has an approximate molecular weight of 240, and the β subunit is close to 225 kDa. These two subunits associate together noncovalently, probably at multiple contact sites, since peptide complexes containing both α and β peptides can be produced by proteolytic cleavage of spectrin molecules under non-denaturing conditions (16). Erythrocyte spectrin dimers can be dissociated into monomers in neutral salt solutions containing 3 M urea, and the isolated subunits reassociate back into dimers under nondenaturing conditions (17). The reassociation of α and β subunits occurs extremely rapidly and displays properties that suggest that the α-β interactions are characterized by cooperativity and extremely high affinity. Isolated α subunits associate into homopolymers of variable size and shape in the absence of β forms. The addition of purified β subunits to α homopolymers converts the loosely associated α subunits into α, β heterodimers with a 1 : 1 stoichiometry (17). This event is probably comparable to the assembly of heterodimers that occurs following spectrin biosynthesis.

Spectrin is Composed of Multiple Triple-Helical Segments

Recent amino acid sequencing studies of human erythrocyte spectrin have provided some remarkable results (18, 19). A segment of 850 amino acid residues extending from the amino terminus of the α subunit consists of 8 subsegments, each 106 amino acids long, that contain a number of identical residues at corresponding sites. These subsegments appear to be homologous to one another to the degree usually displayed by other polypeptides of common genetic origin. This 106 amino acid repeating subsegment also seems to extend throughout the length of both subunits (19), and is believed to represent the predominant structural feature of the entire spectrin molecule.

The 106 amino acid sequences of each of the repeating subsegments have another noteworthy feature. Certain predictions about their conformations can be made based on the arrangement of similar sequences in proteins of known structures (20). Such an analysis suggests that each 106 amino acid subsegment of erythrocyte spectrin could form 3 α-helical units linked by short segments of nonhelical peptide to produce multiple, closely packed triple-helical domains, as illustrated diagramatically in Figure 2. If the entire spectrin molecule is made up largely of these 106 amino acid subsegments, the molecule could contain approximately 38 triple-helical domains. This repeating-helices concept agrees very well with the overall size and shape of spectrin, as revealed by electron microscopy and other physical techniques. However, this model does not provide any explanation for the many different functional sites that spectrin displays. The associations between the α and β subunits are likely to involve interactions between adjacent helices, probably involving the invariant tryptophans that seem to occur in every repeating segment (19). A special peptide segment seems to be responsible for associations between the amino terminus of the α subunit and the carboxyl terminus of the β subunit (16). In addition, there are specific binding sites on human erythrocyte spectrin for ankyrin, protein 4.1 and actin, and calmodulin. It is likely that each of these sites contains a unique amino acid sequence distinct from that characteristic of the triple-helical conformation.

Another remarkable and still incompletely explored feature of the

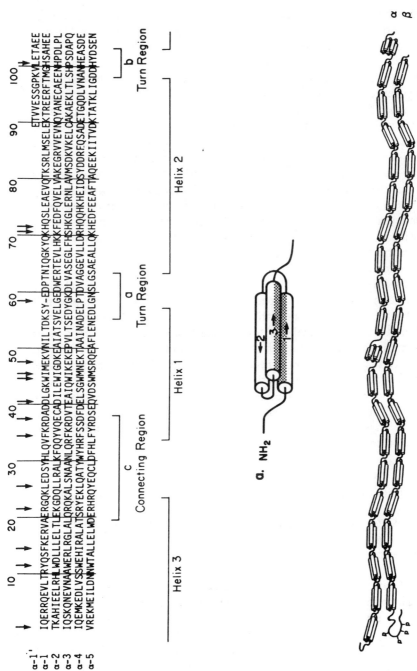

Figure 2 A model of the human erythrocyte spectrin dimer deduced from amino acid sequences (18, 19). The basic folding unit appears to be a 106 amino acid segment that has the capacity to fold into multiple triple helical segments (a). Both spectrin subunits are composed largely of the 106 amino acid segments, five of which are presented at the top of the figure.

erythrocyte spectrin molecule is its capacity to self-assemble into large oligomeric networks (21–23). These spectrin-spectrin interactions depend upon the concentration of protein in solution, require energy (24), and involve specific binding sites at one particular end of the spectrin dimer. Self-assembly of spectrin is initiated when two dimeric units join together to form a tetramer; this complex can then accommodate a third dimer to form a hexamer. Eventually, large oligomeric forms can be generated, apparently by the same dimer addition mechanism. The evidence now available suggests that higher oligomeric forms are also created by head-to-head association of additional spectrin units, but the precise way that additional units join together is still unclear. A highly schematic diagram depicting this self-assembly process is shown in Figure 3. Although this process has only been studied in vitro on isolated spectrin molecules, there is no reason to believe that the relevant observations do not reflect the capacity of spectrin to self-associate into large oligomeric structures in situ. The high local concentration of spectrin in the submembranous region of the red cell membrane created by the association of spectrin units with high affinity binding sites in the ankyrin molecule, as described below, will certainly promote this self-assembly process to a degree far greater than that which can be achieved under in vitro conditions. Freshly extracted spectrin prepared under the usual low ionic strength conditions also appears in oligomeric forms, even though the conditions of extraction are likely to influence the distribution of spectrin forms released from the membrane. Until a better understanding of the mechanism of spectrin extraction under these conditions is developed, it is not possible for us to predict how accurately the state of freshly isolated spectrin molecules in solution reflects the true physical state of spectrin when attached to the inner surface of the red cell membrane.

ANALOGUES OF THE RED CELL MEMBRANE SKELETON IN NON-ERYTHROID CELLS

There is now abundant evidence that spectrin and spectrin-like molecules are not confined solely to red blood cells, but instead are distributed widely over the cells of the animal kingdom. Cells as diverse as neurons and amoeba have spectrin-like molecules.

Early attempts to detect spectrin-like molecules in non-erythroid cells using immunochemical probes produced negative results, which we can now ascribe to the complex nature of spectrin's immunologic reactivity. Rabbits and other species produce antibodies to spectrin that may display remarkably narrow reactivities, sometimes reacting with only one of the two subunits, or showing restricted reactivity or low-affinity binding to

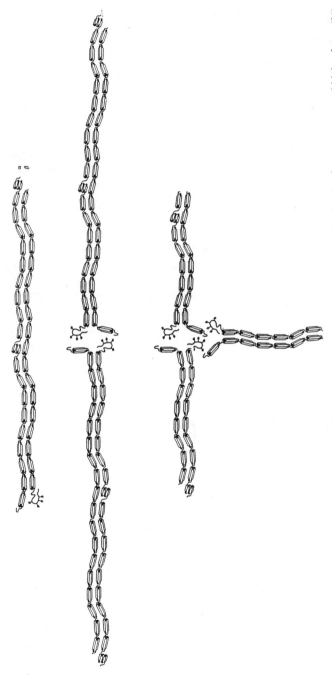

Figure 3 A model for self-assembly of human erythrocyte spectrin. Dimers of spectrin join head-to-head to form tetramers and higher forms (21). Actin and protein 4.1 are believed to associate with the free ends of the spectrin tetramers and oligomers.

spectrins derived from other species. The first positive evidence for the presence of spectrin in non-erythroid cells came from studies describing a cross-reaction between proteins of non-erythroid cells and antibodies raised against erythrocyte spectrin (25). There was some understandable skepticism in response to these early reports, both because of the previous negative findings using the same approach, and the feeling expressed trenchantly by Baines (26) that to qualify as spectrin a protein must do more than simply react with anti-spectrin antibodies.

Most investigators in this field would probably agree to the following criteria for a spectrin-like molecule: (a) spectrin-like forms should be composed of two nonidentical subunits (α and β in the human erythrocyte form), ranging in mass from 225 to 250 kDa. (b) Such molecules should have the capacity to self-associate at least to tetrameric forms, if not to higher oligomers. (c) Spectrin-like molecules should be able to interact with F-actin, preferably in the presence of protein 4.1. This interaction should take place in isotonic salt solutions and be of reasonably high affinity. (d) Spectrin-like molecules should have the capacity to bind ankyrin. (e) Finally, spectrin-like molecules should be elongated rod-like forms with a highly α-helical conformation. It would be even more convincing if all spectrin-like molecules were composed of repeating subsegments of the type described above, but there is no reason to expect each member of the spectrin family to have identical 106 amino acid homologous sub-segments.

The first convincing example of a spectrin-like molecule in non-erythroid cells was the protein called fodrin, which was first identified in neuronal tissue and was believed to be transported along axonal segments (27). This protein appeared to lie along the inner surfaces of the axonal extensions, and it was hypothesized that it might be attached to the overlying plasma membrane, perhaps in company with actin filaments, an arrangement reminiscent of the spectrin-actin network previously described in human red blood cells. The existence of fodrin was subsequently confirmed by work done in other laboratories (28–29). Immunochemically similar material, believed to be spectrin-like, was also identified in a wide variety of cell types more or less at the same time that fodrin was identified in nerve tissue (30). More detailed studies of fodrin produced further interesting results. Brain fodrin is composed of two subunits with molecular weights close to those of erythrocyte spectrin, although both proteins can be readily distinguished by SDS-PAGE. The two fodrin subunits differ from the erythrocyte spectrin subunits when compared by peptide mapping tech-niques (28, 31), but fodrin molecules also form tetrameric forms, which are remarkably similar to erythrocyte spectrin tetramers when examined by rotary shadowing electron microscopy (31, 32). Brain fodrin also binds both ankyrin and actin (28, 32). On the basis of these properties it seems clear that brain fodrin and erythrocyte spectrin are closely related members of a common spectrin family.

Similar observations have been made on proteins extracted from intestinal brush borders, which also show remarkable similarities to erythrocyte spectrin. A protein extracted from chicken intestine epithelium is composed of two subunits with molecular weights of 260 and 240 kDa, and has been designated TW 260/240 (33). TW 260/240 binds to F-actin and has a rod-like structure with a double-stranded morphology similar to the appearance of erythrocyte spectrin tetramers. Comparative studies show that TW 260/240 and brain fodrin are similar but distinct entities (34). The rotary shadowed images of these molecules are shown in Figure 4. An immunofluorescent analysis using antibodies directed against TW 260/240 showed that these molecules are concentrated at the base of the microvilli (33), and high-resolution immunoelectron microscopic studies confirm this localization (35, 36).

Immunochemical studies using well characterized antibodies, often affinity purified against specific spectrin subunits, confirm the widespread distribution of spectrin-like forms. Using both polyclonal and monoclonal antibodies, spectrin-like proteins have been found in muscle cells [both normal (37) and dystrophic (38)], skeletal muscles (39), and developing myoblasts (40). Similar immunologically reactive material is found in avian lens (41), fibroblasts (42), sperm cells (43), pre-implantation mouse embryos (44), and even *Acanthamoeba* (45). Studies with monoclonal antibodies directed against specific domains of the α subunit of human erythrocyte spectrin show that spectrin-like proteins exist in rat Sertoli cells in culture (46). In these cells, a fibrillar network is seen that is distinct from the intermediate filament pattern revealed using appropriate antibodies.

Studies of the structure and function of spectrin-like molecules in non-erythroid cells have been extended far beyond simple immunochemical identification. For example, brain fodrin seems to have almost all the characteristics of erythrocyte spectrin (47). Brain fodrin can be phosphorylated by protein kinases, and can form specific complexes with both F-actin and protein 4.1, as does erythrocyte spectrin.

Further evidence that non-erythroid spectrins such as fodrin have a membrane skeletal role comes from studies of lymphocyte capping, both natural and induced. Under conditions of artificial capping, induced by either lectin or anti-immunoglobin treatments (48, 49), fodrin concentrates under the capped regions of the lymphocyte membranes. This behavior could be explained by fodrin attachment to the cytoplasmic segments of the transmembrane proteins that are involved in the formation of the cap. It also appears that a sizable fraction of lymphocytes are naturally capped in

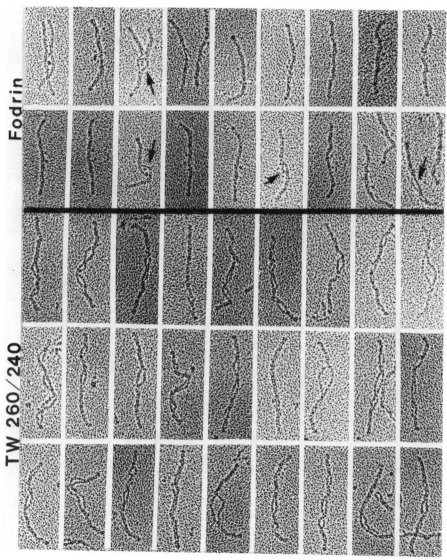

Figure 4 Electron micrographs of rotary shadowed non-erythrocyte spectrin forms, TW 260/240 and fodrin. Both exist as tetramers. (Kindly provided by John Glenney, Jr.)

situ (50); these cells also have fodrin concentrated beneath the patched surfaces. The presence of a sizable number of naturally capped lymphocytes presents a biological puzzle, and its significance remains to be explored.

In spite of this interesting observation that shows fodrin's apparent connection with the overlying surface membrane, and clear evidence that fodrin and other spectrin-like molecules share many of the functional properties of erythrocyte spectrin, there is still some uncertainty as to the precise role that spectrin-like molecules play in non-erythrocytes (51). In an attempt to gain some direct evidence of a functional role for these molecules, antispectrin molecules have been microinjected into isolated cells in hopes of detecting some telling effects. To the surprise and perhaps disappointment of many, anti-spectrin antibodies injected into a number of different isolated epithelial cells did not produce dramatic changes in cell shape or membrane stability (52). A clear cut and almost total precipitation of spectrin was observed inside antibody-treated cells, yet there seemed to be no obvious effect on the shape of the cells, or on the distribution of coated vesicles, microtubules, or the arrangement of the actin cytoskeleton. Only one clear change was noted : The distribution of intermediate filaments was distinctly modified, which raises the possibility that non-erythroid spectrins may be involved in linking intermediate filaments to the plasma membrane.

The two non-erythrocytic forms of spectrin about which most is known, brain fodrin and TW 260/240 show one interesting feature which deserves comment. Both fodrin and TW 260/240 exist in solution as tetrameric forms, similar to the tetrameric forms of erythrocyte spectrin, but in contrast to erythrocyte spectrin neither fodrin nor TW 260/240 seems to have the capacity to self-assemble into higher oligomers. Although we can not rule out the possibility that isolation of these molecules has in some way impaired their capacity to oligomerize, it is perhaps likelier that the capacity of erythrocyte spectrin for more extensive polymerization may be a special adaptation needed to provide greater stability for the red cell membrane than nucleated cells require. Red cells experience a turbulent journey within the intravascular circulation, and it is generally felt that their membrane skeletons are uniquely designed to insure their four-month lifespan.

Red cell membranes of patients with several hereditary hemolytic anemias are clearly less stable than those of normal individuals. In at least two conditions, hereditary pyropoikilocytosis (HPP) and hereditary elliptocytosis (HE), spectrin molecules isolated from affected red cells have a significantly decreased capacity to self-assemble into higher oligomeric forms (53–55). This is consistent with the idea that the capacity to oligomerize may confer greater membrane stability than can be achieved by tetrameric forms of the same molecules. Perhaps spectrin-like tetrameric forms have other roles inside the cell cytoplasm rather than simply serving as membrane stabilizers.

ACTIN IS PART OF THE MEMBRANE SKELETON

The critical role that actin filaments play in muscle structure and function has been evident to biologists for decades, but only recently has it been appreciated that actin also plays an important role in nonmuscle cells. Actin is present within the cytoplasm of most cells in the form of thin filaments of considerable length, but recent studies indicate that an oligomeric form of actin, much shorter than the actin filaments in skeletal muscle, is also associated with the membrane skeletons of many cells, including human red blood cells. It has been difficult for technical reasons to study the molecular structure of actin as it exists in the red blood cell membrane skeleton. The conditions used to extract spectrin from red blood cell ghost membranes also coextract actin, and the two appear in solution in multiple molecular forms.

As described earlier, freshly isolated spectrin exists as dimers, tetramers, and higher oligomers of varying proportions, which depend upon whether the membranes are extracted at 37°C or at a lower temperature. A greater proportion of oligomers and tetramers exists if the extraction procedure is carried out at a low temperature; higher temperatures favor the conversion of both forms to the dimer. Actin isolated by the same low ionic strength buffers containing EDTA is modified unless measures are taken to stabilize it before extraction. Unfortunately, treatments that effectively stabilize actin filaments also render them unextractable by the mild methods commonly used to extract spectrin and actin from membranes, thereby making it difficult to analyze the lengths of the actin filaments using simple physical and microscopic methods.

Actin filaments that are stabilized by phalloidin can be isolated from red blood cell ghost membranes if the membranes are first subjected to mild tryptic digestion (56). The actin filaments released from the membrane by this treatment seem to be relatively short oligomers, composed of 15–20 monomers on the average. The interpretation of these results is complicated since we do not know what influence phalloidin has on the natural actin forms on the membrane, nor do we know whether longer filaments of actin also exist on membranes but are either not released by tryptic digestion or are partially degraded during the isolation steps.

The state of actin on red blood cell membranes was recently re-investigated by Pinder & Gratzer (57). They determined that ADP not ATP is the predominant nucleotide bound to red blood cell membrane skeletal complexes. Using a quantitative DNase-1 assay these investigators also found that less than 10% of the actin in lysed red blood cells is in the monomeric form. Both findings strongly support the earlier suggestion that the bulk of the actin attached to the red blood cell membranes is in a filamentous form.

The actin filaments that make up the red blood cell membrane skeleton differ from simple actin filaments prepared from muscle actin in their stability and nucleating activity, in a manner consistent with their being coupled with other proteins that may have capping properties. Interactions between spectrin and protein 4.1, known to occur in vitro (58), may also be responsible for this modified activity. The total amount of actin present in red blood cells has also been re-evaluated by quantitative analysis of SDS gels and DNase-1 assays (57). The average erythrocyte is reported to have approximately 500,000 actin molecules which are almost all polymerized into short (15–20 monomer) filaments; thus there should be approximately 25,000 such filaments per cell. Since each red blood cell contains approximately 100,000 spectrin tetramers (57), each actin oligomer could accommodate an average of 4 spectrin tetramers.

Short actin filaments on the red cell membrane may also be stabilized by a membrane-bound form of tropomyosin (59), and possibly by an actin-bundling protein that was recently described (60). There appears to be enough tropomyosin per red cell to associate with the actin oligomers by binding to the grooves between helically coiled filaments, as postulated for muscle sarcomere organization. Spectrin molecules also bind to actin filaments in vitro, both alone and together with protein 4.1 molecules (61, 62). Spectrin-actin interactions seem to be of relatively low affinity, and involve the association of only one end of the spectrin dimer (or tetramer) with unspecified sites along the actin filament. These interactions have been seen most clearly by studying spectrin-actin preparations using low-angle shadowing and electron microscopy (61, 62). Since both dimers and tetramers of spectrin bind to actin filaments with equal facility, the association sites must involve the ends of spectrin molecules that are opposite the tetramer-oligomer binding site. The latter is also the end of the spectrin molecule that binds protein 4.1. The addition of protein 4.1 to mixtures of spectrin and actin greatly augments complex formation (58, 63). Spectrin-actin-protein 4.1 complexes are far more stable than spectrin-actin alone, and it is the general feeling that this ternary complex plays a special role in stabilizing the spectrin-actin lattice, in addition to the role protein 4.1 plays in linking the spectrin-actin network to the overlying membrane. A model incorporating these recent observations has been provided by Fowler & Bennett, and it is presented in Figure 5.

Although spectrin and actin clearly interact to form a two-dimensional lattice that is attached to the inner surface of the red cell membrane, the precise arrangement of this network is still unclear. Two alternative models have been proposed. Since the smallest functional unit of spectrin seems to be the tetramer, and actin binds to spectrin tetramers at their free ends, a simple model consistent with both findings would consist of extended chains of spectrin tetramers linked together at their free ends by short actin filaments. A spectrin-actin lattice created by this arrangement could extend symmetrically and without interruption over the entire inner surface of the red cell membrane. This arrangement would be consistent with fluorescent antibody findings and immunoelectron microscopic data which show that both spectrin and actin exist in an uninterrupted fashion throughout the entire inner surface of the membrane (64, 65).

Another model of spectrin-actin assembly in situ emphasizes the importance of spectrin oligomers generated by the head-to-head self-assembly mechanism described earlier. Neighboring spectrin oligomers could be joined together at some of their free ends by short actin filaments. The connecting segments of actin might be distributed more irregularly under the membrane surface, and different regions of the membrane might be stabilized to a varied extent by the local underlying lattice. This model allows for relaxation and redistribution of the underlying supporting

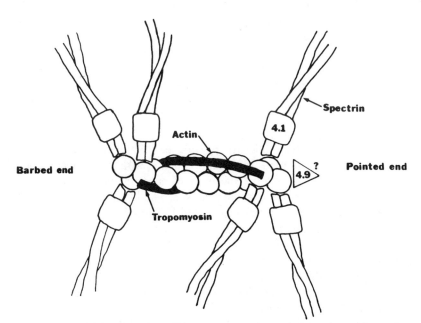

Figure 5   A model showing one possible arrangement of spectrin-4.1-actin complexes. These may be stabilized by tropomyosin and possibly protein 4.9. (Kindly provided by Velia Fowler and Vann Bennett.)

segments through regulation of the factors that control spectrin-actin interactions. Potential regulators include calcium and other molecules capable of modulating tropomyosin-actin interactions.

Both models, or a combination of the two, are consistent with the behavior of spectrin-actin under in vitro conditions. Unfortunately, the picture of the spectrin-actin lattice in intact red cells that is revealed by electron microscopy is consistent with either model. High resolution electron micrographs of carefully prepared red cell ghost membranes demonstrate the existence of a fibrillar network immediately under the lipid bilayer (66, 67). Although these electron micrographs show an anastomosing fibrillar network with exquisite detail, it is not possible to decide whether the microfibrils seen in the electron micrographs are arranged as spectrin tetramers that are connected by actin filaments, or whether they are spectrin oligomers joined at their edges by an actin-tropomyosin complex. More definitive morphological studies are clearly needed, particularly if we hope to detect subtle structure variations in genetically defective red cell membranes using this approach.

Actomyosin May Be Part of the Membrane Skeleton

The co-isolation of spectrin and actin from red cell membranes, first described more than a decade ago, prompted the suggestion that spectrin might be a myosin-like protein attached to the membrane as an actomyosin-like complex capable of contractile activity. A role in membrane shape changes was postulated for this complex that seemed appropriate to explain the effects of ATP on membrane deformability. Subsequent studies have shown that spectrin is not at all like myosin. However, two laboratories have recently described the existence of authentic myosin in red blood cells that may be part of the membrane skeleton (68, 69). A relatively small number of copies of myosin filaments are present in red cells, if all the myosin is arranged as it exists in muscle, and it is not clear how they interact with the spectrin-actin-protein 4.1-tropomyosin network described above. If these myosin molecules are part of a sliding filament–type contractile system they may serve as a supplemental mechanism for modulating membrane shape changes, perhaps by mobilizing segments of the membrane skeleton to promote more global changes in cell shape, such as may be involved in echinocyte formation.

LINKING PROTEINS CONNECT SPECTRIN-ACTIN TO THE MEMBRANE

For the submembranous skeleton to play a significant role in supporting the overlying cell membrane it must be attached in some way to the lipid bilayer, either directly to specific phospholipid molecules, or indirectly via linking or attachment proteins. Early studies, largely using in vitro techniques, suggested that perhaps specific associations existed between spectrin and phospholipids of artificial liposomes. Based on these preliminary findings it was proposed that the submembranous skeleton might coordinate the activities of the overlying membrane through direct contacts with specific polar groups of the phospholipid molecules. Several problems were apparent with this conception. The associations between spectrin and different liposomal preparations appeared to be of low affinity, and no clear indication for phospholipid specificity was evident. These studies also did not identify any particular domains of the spectrin molecule that carry the lipid-binding specificity, nor were attempts made to study lipid binding by more complex forms of the membrane skeleton such as spectrin-actin or spectrin-actin-protein 4.1 complexes. The studies described below present an alternative mechanism for the attachment of the membrane skeleton to the membrane through the action of linking proteins that bind both to specific segments of the spectrin-actin complex and link them, not directly to the phospholipid bilayer, but indirectly to it through associations with the cytoplasmic segments of two different transmembrane glycoproteins.

Human erythrocyte membranes contain two principal linking proteins that connect the submembranous skeleton to the overlying membrane. These are now referred to as ankyrin and protein 4.1. Both proteins remain bound to the membrane under the low ionic strength conditions used to extract the spectrin-actin network, and they retain their capacity to rebind spectrin and actin back on the membrane when isotonicity is restored. Both proteins can be quantitatively detached from the residual ghost membranes by hypertonic neutral salt solutions (1.0 mM KCl is usually used), and they can be purified by DEAE ion exchange chromotography (70). Since high concentrations of neutral salt solutions can be used to extract these proteins from the membrane, rather than detergents or strong chaotropic agents, the interactions between these linking proteins and the membranes are not thought to involve direct insertion of any part of either protein directly into the lipid bilayer, implying further that these linking molecules are not integral membrane proteins.

Ankyrin

Ankyrin was the first protein to be identified as a high-affinity binding site for spectrin on the inner surface of the human red blood cell membrane (71, 72). Ankyrin binds to one specific segment of the β subunit, and links it to a site on the cytoplasmic segment of band 3 (73), the major transmembrane glycoprotein. The ankyrin molecule is itself quite large (approximately 210 kDa), it is multiply phosphorylated, and appears to be slightly asymmetrical. One end of ankyrin binds to spectrin and the other links to band 3 (74–76). The association between ankyrin and spectrin is disrupted at low ionic strength conditions, hence the capacity of low ionic strength buffers to extract spectrin and actin from the membrane. In contrast, the ankyrin-band 3 connection is dissociated only at high ionic strength or at strongly alkaline pH.

Although ankyrin is phosphorylated at several sites along the poly-peptide chain, we know little about the factors that regulate its capacity to bind to either spectrin or band 3. It was originally thought that ankyrin bound only to a subset of band 3 molecules, but recent studies have not supported this idea.

Ankyrin has a number of puzzling features which deserve to be explored further. Although it binds to isolated β chains of spectrin as well as to native α, β dimers, it seems to bind to spectrin tetramers and oligomers with higher affinity (76). This surprising result implies that the ankyrin binding sites on spectrin are modified in some way when multiple spectrin dimers are assembled into higher forms. It is also unclear why the ankyrin molecule is so large. If this molecule's sole purpose is to connect parts of the β subunit of spectrin to a peptide sequence on band 3, one wonders why approximately 2000 amino acids are needed to establish this connection. Ankyrin does not seem to be an extended rod of the spectrin type, but its substructure has not yet been analyzed in a definitive way.

Not surprisingly, analogues of ankyrin are found in many different cell types (77–80). In some cases the capacity of non-erythroid ankyrin to bind spectrin has been demonstrated (78), but in most the functional properties of non-erythroid ankyrin have hardly been explored. It may perhaps be misleading to assume that non-erythroid ankyrin analogues behave in an identical fashion to the erythroid counterpart.

Protein 4.1

Protein 4.1 derives its name from the position it assumes on SDS-PAGE when normal red blood cell membranes are electrophoresed under standard conditions. Protein 4.1 migrates to a position between band 3, the anion channel protein, and band 5, an erythrocyte form of actin. Protein 4.1 remains bound to red cell membrane fragments after spectrin and actin are released, but it is solubilized almost quantitatively by high salt solutions along with most of the membrane-bound ankyrin (70). Protein 4.1 is easily separated from ankyrin by DEAE-ion exchange chromatography. Protein 4.1 is soluble in neutral salt solutions without the need for detergents, but it appears to aggregate easily, particularly after it has been radio-iodinated in the presence of strong oxidizing agents (T. Leto, unpublished observations). Sometimes these treatments destroy its natural binding properties, and at the same time render it prone to nonspecific binding to unnatural substrates.

Protein 4.1 is composed of two almost identical polypeptide chains of 80 and 78 kDa. These two have essentially identical peptide maps (82), the larger form differs by having an extension on its carboxy terminus (83). A preliminary structural model of protein 4.1 has been proposed (83), based on restricted proteolytic digestion and chemical cleavage experiments, and is presented in Figure 6.

The substructure of both protein 4.1 polypeptides is noteworthy in that the amino terminal end has a higher concentration of basic amino acids than other parts of the molecule. A 30 kDa peptide that is resistant to further degradation is easily generated by proteolytic digestion of protein 4.1; this peptide has a relatively basic isoelectric point, and contains a cluster of cysteines. The carboxy-terminal region of protein 4.1 has more acidic amino acids than the 30 kDa peptide, and is easily degraded by even mild proteolytic treatment. Peptides that connect the two terminal segments are phosphorylated at multiple sites, and one 67 amino acid segment represents the site on protein 4.1 that interacts with spectrin and actin (84). Figure 6 shows its location. It is interesting that this site is immediately adjacent to several different phosphopeptides, particularly since several observers have reported changes in the capacity of protein 4.1 to link spectrin and actin together as a function of the state of phosphorylation of the entire ensemble (85). As mentioned earlier, the capacity of protein 4.1 to enhance spectrin and actin associations is impressive. Spectrin and actin associate together only weakly, but an extraordinarily stable complex is created when protein 4.1 is added to the system. This stabilizing effect is also mimicked by the 67 amino acid peptide described above.

Protein 4.1 differs from ankyrin in that it does not bind to a specific segment of one of the spectrin subunits, but instead seems to require the presence of both subunits to generate a stable complex (86). Electron microscopic studies and other considerations indicate that protein 4.1 binds to the tail region of spectrin, the same end that binds to filamentous actin. The precise way that protein 4.1 actually augments the association between spectrin and actin is still a mystery. We do not know whether protein 4.1 binds to both spectrin and actin together, or whether its association with spectrin enhances the latter's capacity to bind to actin.

In addition to enhancing the associations between spectrin and actin (an association that also involves tropomyosin and band 4.9, the actin bundling protein), protein 4.1 also links the entire ensemble to the overlying membrane. In this capacity protein 4.1 serves as the second high affinity link between the skeleton and the membrane.

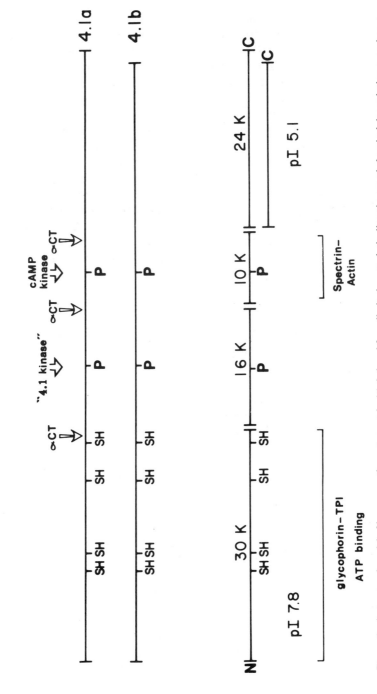

Figure 6 A structural model of human erythrocyte protein 4.1 deduced from limited proteolytic digestions and chemical degradation experiments (83). Functional sites have been determined by analyzing the properties of individual peptides. αCT denotes sites of cleavage by α-chymotrypsin.

Like ankyrin, protein 4.1 also binds to the cytoplasmic segment of a transmembrane glycoprotein, but one distinctly different from band 3. Protein 4.1 binds to the glycophorins (87, 88), which are the principal sialoglycoproteins of the human red blood cell membrane (89). Glycophorin A, the predominate species, is a 131 amino acid polypeptide that extends across the lipid bilayer. It has a heavily glycosylated segment outside the membrane, and a relatively short polypeptide segment inside the cell, with a single helically coiled membrane-spanning region of 23 amino acids joining the two (1). Protein 4.1 binds to some part of the 35 amino acid segment of glycophorin that extends into the cytoplasm of the erythrocyte, but it does so only if the glycophorin molecules are associated with a specific phospholipid, phosphatidylinositol 4,5 bisphosphate (90). This phospholipid, commonly referred to as triphosphoinositide or TPI, comprises a relatively small fraction of the total phospholipid of the red cell membrane, but it appears to be concentrated entirely within the inner or cytoplasmic leaflet of the lipid bilayer. TPI co-isolates with glycophorin when the latter is extracted from red blood cells by the LIS-phenol method (91), and its capacity to promote the association of protein 4.1 with glycophorin cannot be substituted by any of the other phospholipids commonly found in cell membranes (92).

The capacity of TPI to promote one of the links between the skeleton and the overlying membrane is particularly interesting since polyphosphoinositide turnover seems to play an important role in transmembrane signalling events in a wide variety of cell types (93). Activation of many cells by one or another physiologic ligand can cause the immediate breakdown of TPI, which is accompanied by the generation of diacyl-glycerol, a protein kinase activator, and inositol triphosphate, an agent capable of mobilizing intracellular calcium (94).

If TPI is an essential cofactor needed to promote one of the links between the skeleton and the overlying membrane its turnover could have important consequences for membrane organization, in addition to the metabolic effects produced by its breakdown products. The hydrolysis of TPI bound to glycophorin might be expected to cause the detachment of protein 4.1 and its partners from the membrane, allowing both the liberated glycophorin and a segment of the membrane skeleton to relocate elsewhere in the cell. Glycophorin would be free to move within the plane of the bilayer while the protein 4.1 complex could reattach to some other internal site.

Protein 4.1 has a second binding site on the inner surface of the red cell membrane that deserves comment. Protein 4.1 can also associate with the cytoplasmic segment of band 3, at a site close to but distinct from the ankyrin binding site (95). The binding of protein 4.1 to band 3 is of lower affinity than its association with glycophorin-TPI, but if glycophorin loses its lipid cofactor, protein 4.1 can bind preferentially to band 3. The physiologic significance of this secondary binding site is unclear, since there is at present no direct evidence that protein 4.1 is ever linked to band 3 when ATP levels are adequate to insure maximum phosphorylation of the phosphoinositides. A reduction in TPI levels, due possibly to ATP depletion or other metabolic rearrangements, significantly reduces the capacity of glycophorin to bind protein 4.1, and under these conditions band 3 binding could play an important role in stabilizing the membrane skeleton.

Analogues of protein 4.1 are present in a wide variety of cell types (96–100). These findings, based primarily on immunologic cross-reactivity with antibodies directed against erythrocyte protein 4.1, have been confirmed by analyzing immunoreactive forms with more specific biochemical assays, such as NTCB cleavage or restricted proteolytic digestion (100). First identified in other blood cells, a wider search revealed an almost ubiquitous distribution of protein 4.1 analogues. One of the surprising findings was the recent demonstration that synapsin-1, a protein originally thought to be confined to neurons and associated with synaptic vesicle membranes (101–103), is probably a homologue of protein 4.1 (104).

Another interesting aspect of protein 4.1 was uncovered by studies of its analogues in avian erythrocytes. Mature avian red cells contain multiple molecular forms of protein 4.1 not found in red cell precursors of avian embryos at different stages of development (105). Protein 4.1 molecules of variable molecular mass ranging from 87 to 175 kDa are characteristic of terminally differentiated cells. It is hard to conceive of a simple explanation for this finding if the sole function of these different protein 4.1 molecules is to attach the skeleton to the overlying membrane. It is perhaps more likely that different protein 4.1 isoforms have different functional capacities, related in some way to the role they play in cell development.

At the present time we know little about the functional roles that protein 4.1 analogues play in non-erythroid cells. Some may be involved in linking the membrane skeleton to transmembrane glycoproteins, but other possible functions should not be overlooked.

ASSEMBLY OF THE MEMBRANE SKELETON

In the mature mammalian erythrocyte all of the components of the membrane skeleton are bound to the inner surface of the membrane. In other cells the precise disposition of all the different components is less certain, but it is likely that the bulk of the proteins that are synthesized to serve a skeletal function eventually wind up on membrane surfaces. It is generally agreed that the most important links between the membrane skeleton and the overlying membrane are forged by high-affinity associations between parts of the membrane skeleton and specific transmembrane glycoproteins. Thus it is reasonable to suspect that assembly of the developing skeleton first requires the presence of appropriate transmembrane glycoprotein anchoring sites. In developing erythrocytes, band 3, the attachment site for ankyrin-spectrin, is synthesized and inserted into membranes cotranslationally (106). Membrane units having such band 3 molecules presumably serve as initiating sites for skeletal assembly, but it is not known whether critical concentrations of band 3 molecules are needed to initiate assembly, or whether other posttranslational modifications must occur before band 3 molecules acquire the competence to direct the assembly process.

The pathway of assembly of the remaining components of the skeleton is still uncertain. A multicomponent system such as this can assemble in at least two ways. Each of the components could add on stepwise, one at a time, from pools of precursors synthesized elsewhere in the cell. Alternatively, assembly could proceed along modular lines, with preformed units of several different molecules joining the membrane. Spectrin-actin complexes are perhaps the most likely candidates for preformed units of the human erythrocyte skeleton, but the association between the two proteins is weak in the absence of protein 4.1, and a complex of the two alone may not be stable. Complexes of spectrin, 4.1, and actin would be extremely stable, and they could attach to membranes containing either band 3, occupied by ankyrin, or those containing glycophorin-TPI. Complexes of the entire ensemble, spectrin-4.1-actin-ankyrin, would be the most complete functional units, but also the most complicated to assemble and distribute to the different membrane surfaces. Would these units be too large and cumbersome to be transported from their sites of assembly to appropriate anchoring sites on the surface membrane? Would they be likely to attach to inappropriate membrane surfaces?

Lazarides and co-workers have studied the assembly of the membrane skeleton of avian erythrocytes in great detail (107–110), and their findings provide clear support for the stepwise addition model. They propose that skeletal assembly in these cells is mediated by the availability of high affinity receptors for each of the components individually. A membrane containing competent band 3 attracts ankyrin molecules, and once these are bound, new high-affinity sites become available on the membrane for β-spectrin to bind. Since β-spectrin binds to α-spectrin with very high affinity, the α, β complex joins ankyrin on the membrane surface. Although no experimental data are yet available for the assembly of 4.1 and actin into the membrane skeleton it is presumed that they join the assembled complexes by the same route.

Ankyrin and the two spectrin subunits are synthesized in excess of the amounts that eventually become part of the membrane skeleton. To deal with this potential oversupply avian red cells have mechanisms for the selective degradation of unbound forms, which appear to be different for each of the spectrin subunits. β-spectrin turns over rapidly and is degraded by a non-ATP-dependent cytosolic system, while the α subunit seems to be processed within the lysosomal system (110).

Not all of the components of the skeleton behave in the same way, since these investigators also noted that almost all the protein 4.1 that is synthesized becomes incorporated into the membrane skeleton. There is also no detectable turnover of membrane-bound band 3. It will be important to see whether this pathway of assembly for the membrane skeleton of avian erythrocytes is shared by mammalian red cells.

One might also expect the ankyrin content of cells to influence the rate of attachment of spectrin to the membrane, and this seems to be the case. Another interesting finding has been reported for avian cells: Ankyrin seems to influence both the amount of spectrin bound to membranes, and the type of spectrin localized at different sites within the cell (111). Avian nervous tissue contains at least two isoforms of spectrin, one composed of α, β subunits of the erythroid type, and the other composed of the same α subunit but a different β form (often referred to as γ), the type associated with fodrin. Careful immunofluorescent analysis of the localization of these two spectrins in the developing embryo shows that the α, β form is associated exclusively with the membranes of nerve cell bodies and dendrites; in contrast to the α, γ form which concentrates along axonal membranes and nerve endings (112–114). Some cells also seem to have at least two different forms of ankyrin (α and β), and it has been suggested that each of these ankyrin isoforms links a specific type of spectrin to a specific subregion of the membrane.

The site-specific localization of ankyrin isoforms to different populations of band 3 could be a simple way to segregate different classes of spectrin throughout the cell, but this would only be effective if different classes of transmembrane glycoproteins were themselves segregated into distinct membrane domains. How can this be achieved? In the steady-state, "mature" surface membrane, the membrane skeleton is thought to play a critical role in regulating the topography of the transmembrane glycoproteins through specific attachments with their cytoplasmic domains. Without these attachments the transmembrane glycoproteins are presumably free to diffuse within the plane of the lipid bilayer. How then can they be localized to specific domains without the constraining influence of an underlying membrane skeleton? What could bring about the segregation of ankyrin binding sites on the membranes of the nerve cell body, as opposed to the accumulation of the β ankyrin sites along the axonal membrane?

Two reasonable possibilities may be proposed. Different populations of transmembrane glycoproteins may be sequestered within particular regions of the plasma membrane in response to signals from the external environment. For example, in developing axons a special class of transmembrane molecules may be concentrated at the growing regions in response to trophic stimuli from the surrounding medium (in vitro) or matrix molecules (in vivo). Comparable influences may be generated by contacts with neighboring cells, or in response to high concentrations of physiologic ligands, neurotransmitters, or other substances.

Site-specific localizations of specific populations of transmembrane proteins may also occur in response to signals generated from within the cell itself. Other cytoskeletal elements such as intermediate filaments may play some regulatory role. It is interesting in this context to point out that vimentin filaments can attach to the inner surface of erythrocyte membranes through a high-affinity association with ankyrin (115).

What is the Driving Force for Assembly of the Membrane Skeleton?

In at least one cell type, synthesis of membrane skeletal parts seems to take place in the neighborhood of the preexisting skeletal complex, which suggests that assembly of nascent chains destined to be parts of the skeleton is facilitated by their synthesis immediately proximal to the assembled skeleton in a cotranslational assembly arrangement. But this seems not to be the case with developing erythroblasts and other cells. Instead, newly synthesized components become attached to the membrane through an affinity-modulated mechanism (21) driven by both the concentration of unbound components and the concentration of available receptor sites (114). No evidence now exists for the action of "carrier molecules" that transport nascent skeletal proteins to appropriate membrane sites, but it would be a mistake to dismiss this possibility without further investigation. Still unresolved is the mechanism by which segregation of the transmembrane proteins at specific sites is maintained without a fully assembled membrane skeleton in place.

Rearrangements of the membrane skeleton can certainly be induced by forces imposed upon the cells from the external environment, as countless demonstrations of patching and capping attest. Experimentally induced caps of lymphocytes cause selective accumulation of the fodrin form of spectrin in lymphocytes (48, 49), and this even occurs in certain populations of lymphocytes that are found to be naturally capped in situ (50). These findings suggest some physical role for the segregated spectrin.

PROSPECTS FOR THE FUTURE

Our understanding of the molecular features of the proteins that provide a supporting infrastructure for mammalian erythrocytes is increasing at a reasonably brisk pace. The main components have been identified, and we have provisional ideas about their general structures. The application of modern gene technology to this field has been slower than for others, but initial successes are now being reported, and it is likely that detailed information about the structures of the principal proteins will soon be available.

More insight into the functional arrangements of different membrane skeletons may not be so easy to acquire. At present, most of the widely advertised provisional models are based almost entirely on the behavior of isolated skeletal components in vitro. Our ultimate goal, understanding the structural and functional arrangements that characterize the membrane skeleton in situ, is still beyond reach. Conventional microscopic analysis has been less revealing than expected. Techniques that rely on chemical fixation, heavy metal treatments, and other processing steps seem to create artifacts too complex to relate back to the native structure. Freeze-fracture and other attempts to avoid tissue processing have not yet progressed much beyond their initial applications, but more imaginative applications still offer promise.

The correlation of structure with function may soon be possible by using modern genetic approaches that involve site-directed mutagenesis, transfection into appropriate hosts, and analysis of the functional consequences. The success of this approach does not depend solely on the ability to modify any one protein, but also requires new ways to measure quantitatively the function of complex structural units attached to membranes. The analysis of naturally occurring mutant forms of membrane skeletons have not been as informative as one would like. We still have no acceptable explanation for the fragility of the membranes of erythrocytes of people with hereditary spherocytosis, one of the most common hereditary hemolytic anemias that is thought to be due to an alteration of the membrane skeleton.

Most cell biologists would probably regard the analysis of the membrane skeletons of nucleated cells an even more formidable challenge. How does one identify the components of the membrane skeleton of a more complex cell? Do all cells have one? Do the membrane skeletons of other cells function in ways analogous to their red cell counterparts?

One approach to the study of potential skeletal complexes in nucleated cells involves the use of nonionic detergents such as Triton X-100. This method, first used by Steck & co-workers to dissect the human red blood cell membrane, has been widely applied with very promising results. The success of this approach is based on the apparent insolubility of skeletal complexes in nonionic detergents when such agents are added to cells at concentrations high enough to solubilize most of the membrane phospholipids and most (but not all) of their integral proteins. Monomer forms of the different skeletal components are also soluble under these conditions. The insoluble residues that result when intact cells are treated in this way are often referred to as Triton shells. This method offers the possibility of isolating protein complexes from a wide variety of cell types (116–122), and can be used to study the biosynthesis and assembly of the skeletal components (107–110). Rearrangements of membrane skeletal complexes brought about by the activation of surface receptors have also been identified by this method (123–126). The skeletal complexes exposed by this treatment are accessible to study by many different ultrastructural techniques, and it represents a promising way to study how the different components fit together (127).

Literature Cited

1. Tomita, M., Furthmayr, H., Marchesi, V. T. 1978. Primary structure of human erythrocyte glycophorin A. Isolation and characterization of peptides and complete amino acid sequence. *Biochemistry* 17:4756–70
2. Kaufman, J. F., Auffray, C., Korman, A. J., Shackelford, D. A., Strominger, J. 1984. The class II molecules of the human and murine major histocompatibility complex. *Cell* 36:1–13
3. McClelland, A., Kuhn, L. C., Ruddle, F. H. 1984. The human transferrin receptor gene: genomic organization, and the complete primary structure of the receptor deduced from a cDNA sequence. *Cell* 39:267–74
4. Drickamer, K., Mamon, J. F., Binns, G., Leung, J. O. 1984. Primary structure of the rat liver asialoglycoprotein receptor. *J. Biol. Chem.* 259:770–78
5. Yamamoto, T., Davis, C. G., Brown, M. S., Schneider, W. J., Casey, M. L., et al. 1984. The human LDL receptor: a cysteine-rich protein with multiple Alu sequences in its mRNA. *Cell* 39:27–38
6. Mostov, K. E., Friedlander, M., Blobel, G. 1984. The receptor for transepithelial transport of IgA and IgM

contains multiple immunoglobulin-like domains. *Nature* 308:37–43
7. Ullrich, A., Coussens, L., Hayflick, J. S., Dull, T. J., Gray, A., et al. 1984. Human epidermal growth factor receptor cDNA sequence and aberrant expression of the amplified gene in A431 epidermoid carcinoma cells. *Nature* 309:418–25
8. Ullrich, A., Bell, J. R., Chen, E. Y., Herrera, R., Petruzzelli, L. M., et al 1985. Human insulin receptor and its relationship to the tyrosine kinase family of oncogenes. *Nature* 313:756–61
9. Branton, D., Cohen, C. M., Tyler, J. 1981. Interaction of cytoskeletal proteins on the human erythrocyte membrane. *Cell* 24:24–32
10. Marchesi, V. T. 1983. The red cell membrane skeleton: recent progress. *Blood* 61:1–11
11. Lux, S. 1979. Dissecting the red cell membrane skeleton. *Nature* 281:426
12. Cohen, C. M. 1983. The molecular organization of the red cell membrane skeleton. *Semin. Hematol.* 20:141–58
13. Goodman, S. R., Shiffer, K. 1983. The spectrin membrane skeleton of normal

and abnormal human erythrocytes: a review. *Am. Physiol. Soc.* C121–C133

14. Morrow, J. S. 1984. Spectrins: Mediators of cytoskeletal function. *Am. J. Dermatol.* 6:573–81

15. Shotton, D., Burke, B., Branton, D. 1978. The shape of spectrin molecules from human erythrocyte membranes. *Biochim. Biophys. Acta* 536:313

16. Morrow, J. S., Speicher, D. W., Knowles, W. J., Hsu, C. J., Marchesi, V. T. 1980. Identification of functional domains of human erythrocyte spectrin. *Proc. Natl. Acad. Sci. USA* 77: 6592–96

17. Yoshino, H., Marchesi, V. T. 1984. Isolation of spectrin subunits and reassociation in vitro: Analysis by fluorescence polarization. *J. Biol. Chem.* 259:4496–4500

18. Speicher, D. W., Davis, G., Marchesi, V. T. 1983. The structure of human erythrocyte spectrin: II. The sequence of alpha-I domain. *J. Biol. Chem.* 258: 14938–47

19. Speicher, D. W., Marchesi, V. T. 1984. Erythrocyte spectrin is comprised of many homologous triple helical segments. *Nature* 311:177–80

20. Chou, P. Y., Fasman, G. D. 1978. Empirical predictions of protein conformation. *Ann. Rev. Biochem.* 47:251–76

21. Morrow, J. S., Marchesi, V. T. 1981. Self-assembly of spectrin oligomers in vitro: a basis for a dynamic cytoskeleton. *J. Cell. Biol.* 88:463–68

22. Morris, M., Ralston, G. B. 1984. A reappraisal of the self-association of human spectrin. *Biochim. Biophys. Acta* 788:132–37

23. Liu, S.-C., Windisch, P., Kim, S., Palek, J. 1984. Oligomeric states of spectrin in normal erythrocyte membranes: biochemical and electron microscopic studies. *Cell* 37:587–94

24. Ungewickell, E., Gratzer, W. 1978. Self-association of human spectrin. A thermodynamic and kinetic study. *Eur. J. Biochem.* 88:379

25. Goodman, S. R., Zagon, I. S., Kulikowski, R. R. 1981. Identification of a spectrin-like protein in nonerythroid cells. *Proc. Natl. Acad. Sci. USA* 78: 7570–74

26. Baines, A. J. 1983. The spread of spectrin. *Nature* 301:377

27. Levine, J., Willard, M. 1981. Fodrin: axonally transported polypeptides associated with the internal periphery of many cells. *J. Cell. Biol.* 90:631–43

28. Bennett, V., Davis, J., Fowler, W. E. 1982. Brain spectrin, a membrane-associated protein related in structure and function to erythrocyte spectrin. *Nature* 299:126–31

29. Glenney, J. R. Jr., Glenney, P. 1983. Fodrin is the general spectrin-like protein found in most cells whereas spectrin and the TW protein have a restricted distribution. *Cell* 34:503–12

30. Repasky, E. A., Granger, B. L., Lazarides, E. 1982. Widespread occurrence of avian spectrin in non-erythroid cells. *J. Cell. Biol.* 29:821–33

31. Davis, J., Bennett, V. 1983. Brain spectrin. Isolation of subunits and formation of hybrids with erythrocyte spectrin subunits. *J. Biol. Chem. USA* 258:7757–66

32. Glenney, J. R. Jr., Glenney, P., Weber, K. 1982. F-actin-binding and cross-linking properties of porcine brain fodrin, a spectrin-related molecule. *J. Biol. Chem.* 257:9781–87

33. Glenney, J. R. Jr., Glenney, P., Osborn, M., Weber, K. 1982. An F-actin and calmodulin-binding protein from isolated intestinal brush borders has a morphology related to spectrin. *Cell* 28:843–54

34. Glenney, J. R. Jr., Glenney, P., Weber, K. 1982. Erythroid spectrin, brain fodrin, and intestinal brush border proteins (TW-260/240) are related molecules containing a common calmodulin-binding subunit bound to a variant cell type-specific subunit. *Proc. Natl. Acad. Sci. USA* 79:4002–5

35. Glenney, J. R. Jr., Glenney, P., Weber, K. 1983. The spectrin-related molecule, TW-260/240, cross-links the actin bundles of the microvillus rootlets in the brush borders of intestinal epithelial cells. *J. Cell. Biol.* 96:1491–96

36. Hirokawa, N., Cheney, R. E., Willard, M. 1983. Location of a protein of the fodrin-spectrin-TW 260/240 family in the mouse intestinal brush border. *J. Cell. Biol.* 32:953–65

37. Kasturi, K., Fleming, J., Harrison, P. 1983. A monoclonal antibody against erythrocyte spectrin reacts with both α- and β-subunits and detects spectrin-like molecules in non-erythroid cells. *Exp. Cell. Res.* 144:241–47

38. Appleyard, S. T., Dunn, M. J., Dubowitz, V., Scott, M. L., Pittman, S. J., Shotton, D. M. 1984. Monoclonal antibodies detect a spectrin-like protein in normal and dystrophic human skeletal muscle. *Proc. Natl. Acad. Sci. USA* 81:776–80

39. Nelson, W. J., Lazarides, E. 1983. Ex-

pression of the β subunit of spectrin in non-erythroid cells. *Proc. Natl. Acad. Sci. USA* 80:363–67

40. Nelson, W. J., Lazarides, E. 1983. Switching of subunit composition of muscle spectrin during myogenesis in vitro. *Nature* 304:364–68

41. Nelson, W. J., Granger, B. L., Lazarides, E. 1983. Avian lens spectrin: Subunit composition compared with erythrocyte and brain spectrin. *J. Cell. Biol.* 97:1271–76

42. Lehto, V.-P., Virtanen, I. 1983. Immunolocalization of a novel, cytoskeletal-associated polypeptide of M_r 230,000 daltons. *J. Cell. Biol.* 96:703–16

43. Virtanen, I., Badley, R. A., Paasivuo, R., Lehto, V.-P. 1984. Distinct cytoskeletal domains revealed in sperm cells. *J. Cell. Biol.* 99:1083–91

44. Sobel, S. J., Alliegro, W. 1985. Changes in the distribution of a spectrin-like protein during development of the pre-implantation mouse embryo. *J. Cell. Biol.* 100:333–36

45. Pollard, T. D. 1984. Purification of a high molecular weight actin filament gelatin protein from acanthamoeba that shares antigenic determinants with vertebrate spectrins. *J. Cell. Biol.* 99:1970–80

46. Ziparo, E., Zani, B. M., Marchesi, V. T., Stefanini, M. 1984. Analogues of the red cell membrane skeleton are present in rat sertoli cells. *J. Cell. Biol.* 99:296a

47. Burns, N. R., Ohanian, V., Gratzer, W. B. 1983. Properties of brain spectrin (fodrin). *Cell* 153:165–68

48. Levine, J., Willard, M. 1983. Redistribution of fodrin (a component of the cortical cytoplasm) accompaying capping of cell surface molecules. *Proc. Natl. Acad. Sci. USA* 80:191–95

49. Nelson, W. J., Calaco, C. A. L. S., Lazarides, E. 1983. Involvement of spectrin in cell-surface receptor capping in lymphocytes. *Proc. Natl. Acad. Sci. USA* 80:1626–30

50. Repasky, E. A., Symer, D. E., Bankert, R. B. 1984. Spectrin immunofluorescence distinguishes a population of naturally capped lymphocytes in situ. *J. Cell. Biol.* 99:350–55

51. Burridge, K., Kelly, T., Mangeat, P. 1982. Non-erythrocyte spectrins: actin-membrane attachment proteins occurring in many cell types. *J. Cell Biol.* 95:478–86

52. Mangeat, P. H., Burridge, K. 1984. Immunoprecipitation of non-erythrocyte spectrin within live cells following microinjection of specific antibodies:

relation to cytoskeletal structures. *J. Cell Biol.* 98:1363–77

53. Knowles, W. J., Morrow, J. S., Speicher, D. W., Zarkowsky, H. S., Mohandas, N., et al. 1983. Molecular and functional changes in spectrin from patients with hereditary pyropoikilocytosis. *J. Clin. Invest.* 71:1867–77

54. Palek, J., Liu, S.-C., Liu, P.-Y., Prchal, J., Castleberry, R. 1981. Altered assembly of spectrin in red cell membranes in hereditary pyropoikilocytosis. *Blood* 57:130

55. Marchesi, S. L., Knowles, W. J., Morrow, J. S., Bologna, M., Marchesi, V. T. 1985. Abnormal spectrin in hereditary elliptocytosis. *Blood.* In press

56. Atkinson, M. A. L., Morrow, J. S., Marchesi, V. T. 1982. The polymeric state of actin in the human erythrocyte cytoskeleton. *J. Cell. Biochem.* 18:493–505

57. Pinder, J. C., Gratzer, W. B. 1983. Structural and dynamic states of actin in the erythrocyte. *J. Cell Biol.* 96:768–75

58. Ungewickell, E., Bennett, P. M., Calvert, R., Ohanian, V., Gratzer, W. B. 1979. In vitro formation of a complex between cytoskeletal proteins of the human erythrocyte. *Nature* 280:811–14

59. Fowler, V. M., Bennett, V. 1984. Erythrocyte membrane tropomyosin. *J. Biol. Chem.* 259:5978–89

60. Siegel, D. L., Branton, D. 1985. Partial purification and characterization of an actin-bundling protein, band 4.9, from human erythrocytes. *J. Cell Biol.* 100:775–85

61. Fowler, V. M., Luna, E. J., Hargreaves, W. R., Taylor, D. L., Branton, D. 1981. Spectrin promotes the association of F-actin with the cytoplasmic surface of the human erythrocyte membrane. *J. Cell Biol.* 88:388

62. Cohen, C. M., Tyler, J. M., Branton, D. 1980. Spectrin-actin associations studied by electron microscopy of shadowed preparations. *Cell* 21:875

63. Ohanian, V., Wolfe, L. C., John, K. M., Pinder, J. C., Lux, S. E., Gratzer, W. B. 1984. Analysis of the ternary interaction of the red cell membrane skeletal proteins spectrin, actin and 4.1. *Biochemistry* 23:4416–20

64. Geiduschek, J. B., Singer, S. J. 1979. Molecular changes in the membranes of mouse erythroid cells accompanying differentiation. *Cell* 16:149

65. Ziparo, E., Lemay, A., Marchesi, V. T. 1978. The distribution of spectrin along the membranes of normal and echino-

cytic human erythrocytes. *J. Cell. Sci.* 34:91–101

66. Tsukita, S., Tsukita, S., Ishikawa, H. 1980. Cytoskeletal network underlying the human erythrocyte membrane. Thin section electron microscopy. *J. Cell Biol.* 85:507–76

67. Tsukita, S., Tsukita, S., Ishikawa, H., Sato, S., Nakao, M. 1981. Electron microscopic study of reassociation of spectrin and actin with the human erythrocyte membrane. *J. Cell Biol.* 90:70–77

68. Fowler, V. M., Davis, J. Q., Bennett, V. 1985. Human erythrocyte myosin: identification and purification. *J. Cell Biol.* 100:47–55

69. Wong, A. J., Kiehart, D. P., Pollard, T. D. 1985. Myosin from human erythrocytes. *J. Cell. Biol. Chem.* 260:46–49

70. Tyler, J. M., Hargreaves, W. R., Branton, D. 1979. Purification of two spectrin-binding proteins. Biochemical and electron microscopic evidence for site-specific reassociation between spectrin and bands 2.1 and 4.1. *Proc. Natl. Acad. Sci. USA* 76:5192

71. Bennett, V., Stenbuck, P. J. 1979. Identification and partial purification of ankyrin, the high affinity membrane attachment site for human erythrocyte spectrin. *J. Biol. Chem.* 254:2533–41

72. Bennett, V., Stenbuck, P. J. 1980. Human erythrocyte ankyrin. *J. Biol. Chem.* 255:2540–48

73. Hargreaves, W. R., Giedd, K. N., Verkleij, A., Branton, D. 1980. Reassociation of ankyrin with band 3 in erythrocyte membranes and in lipid vesicles. *J. Biol. Chem.* 255:11965–72

74. Bennett, V., Stenbuck, P. J. 1979. The membrane attachment protein for spectrin is associated with band 3 in human erythrocyte membranes. *Nature* 280:468–73

75. Weaver, D. C., Marchesi, V. T. 1984. The structural basis of ankyrin function. I. Identification of two structural domains. *J. Biol. Chem.* 259:6165–69

76. Weaver, D. C., Pasternack, G. R., Marchesi, V. T. 1984. The structural basis of ankyrin function. II. Identification of two functional domains. *J. Biol. Chem.* 259:6170–75

77. Davis, J. Q., Bennett, V. 1984. Brain ankyrin. A membrane-associated protein with binding sites for spectrin, tubulin, and the cytoplasmic domain of the erythrocyte anion channel. *J. Biol. Chem.* 259:13550–59

78. Davis, J. Q., Bennett, V. 1984. Brain ankyrin. Purification of a 72,000 M_r spectrin-binding domain. *J. Biol. Chem.*

259:1874–81

79. Alper, S. L., Bean, K. G., Greengard, P. 1980. Hormonal control of Na^+-K^+ cotransport in turkey erythrocytes. Multiple sites of phosphorylation of goblin, a high molecular weight protein of the plasma membrane. *J. Biol. Chem.* 225:4864–71

80. Moon, R. T., Ngai, J., Wold, B. J., Lazarides, E. 1985. Tissue-specific expression of distinct spectrin and ankyrin transcripts of erythroid and nonerythroid cells. *J. Cell Biol.* 100:152–60

81. Deleted in proof

82. Goodman, S. R., Yu, J., Whitefield, C. F., Culp, E. N., Posna, E. J. 1982. Erythrocyte membrane skeletal protein bands 4.1 a and b are sequence-related phosphoproteins. *J. Biol. Chem.* 257:4564–69

83. Leto, T. L., Marchesi, V. T. 1984. A structural model of human erythrocyte protein 4.1. *J. Biol. Chem.* 259:4603–8

84. Correas, I., Leto, T. L., Speicher, D. W., Marchesi, V. T. 1984. Identification of a proteolytic fragment of erythrocyte protein 4.1 that promotes spectrin-actin associations in vitro. *J. Cell Biol.* 99:300a

85. Pinder, J. C., Ungewickell, E., Bray, D., Gratzer, W. B. 1978. The spectrin-actin complex and erythrocyte shape. *J. Supramol. Struct.* 8:439

86. Cohen, C. M., Langley, R. C. Jr. 1984. Functional characterization of human erythrocyte spectrin α and β chains: association with actin and erythrocyte protein 4.1. *Biochemistry* 23:4488–95

87. Anderson, R. A., Lovrien, R. E. 1984. Glycophorin is linked by band 4.1 protein to the human erythrocyte membrane skeleton. *Nature* 307:655–58

88. Mueller, T. J., Morrison, M. 1981. In *Erythrocyte Membranes 2: Recent Clinical and Experimental Advances,* ed. W. C. Kruckenberg, J. W. Eaton, G. J. Brewer, pp. 95–112. New York: Liss

89. Furthmayr, H., Marchesi, V. T. 1984. Glycophorins: isolation, orientation, and localization of specific domains. *Methods Enzymol.* 96:268–80

90. Anderson, R. A., Marchesi, V. T. 1983. Polyphosphoinositide modulation of glycophorin-protein 4.1 interactions: A possible mechanism for regulation of the red cell membrane skeleton. *J. Cell Biol.* 97:297a

91. Armitage, I., Shapiro, D. L., Furthmayr, H., Marchesi, V. T. 1977. p^{31} nuclear magnetic resonance evidence for polyphosphoinositide associated with the hydrophobic segment of glycophorin A. *Biochemistry* 16:1317–20

92. Anderson, R. A., Marchesi, V. T. 1985. Associations between glycophorin and protein 4.1 are modulated by polyphosphoinositides: A mechanism for membrane skeletal regulation. *Nature.* In press

93. Nishizuka, Y. 1984. Turnover of inositol phospholipids and signal transduction. *Science* 225:1365–70

94. Berridge, M. J., Irvine, R. F. 1984. Inositol triphosphate, a novel second messenger in cellular signal transduction. *Nature* 312:315–21

95. Pasternack, G. R., Anderson, R. A., Leto, T. L., Marchesi, V. T. 1985. Interactions between protein 4.1 and band 3. An alternative binding site for an element of the membrane skeleton. *J. Biol. Chem.* 260:3676–83

96. Cohen, C. M., Foley, S. F., Korsgren, C. 1982. A protein immunologically related to erythrocyte band 4.1 is found on stress fibres of nonerythroid cells. *Nature* 299:648–50

97. Aster, J. C., Welsh, M. J., Brewer, G. J., Maisel, H. 1984. Identification of spectrin and protein 4.1-like proteins in mammalian lens. *Biochem. Biophys. Res. Comm.* 119:726–34

98. Goodman, S. R., Casoria, L. A., Coleman, D. B., Zagon, I. S. 1984. Identification and localization of brain protein 4.1. *Science* 224:1433–36

99. Spiegel, J. E., Beardsley, D. S., Southwick, F. S., Lux, S. E. 1984. An analogue of the erythroid membrane skeletal protein 4.1 in non-erythroid cells. *J. Cell Biol.* 99:886–93

100. Leto, T. L., Pratt, B. M., Madri, J. A. 1985. Mechanisms of cytoskeletal regulation: modulation of aortic endothelial cell protein band 4.1 by the extracellular matrix. *Fed. Proc.* 44:1600

101. De Camilli, P., Cameron, R., Greengard, P. 1983. Synapsin I (protein 1), a nerve terminal-specific phosphoprotein. I. Its general distribution in synapses of the central and peripheral nervous system demonstrated by immunofluorescence in frozen and plastic sections. *J. Cell Biol.* 96:1337–54

102. De Camilli, P., Harris, S. M. Jr., Huttner, W. B., Greengard, P. 1983. Synapsin I (protein I), a nerve terminal-specific phosphoprotein. II. Its specific association with synaptic vesicles demonstrated by immunocytochemistry in agarose-embedded synaptosomes. *J. Cell Biol.* 96:1355–73

103. Huttner, W. B., Schiebler, W., Greengard, P., De Camilli, P. 1983. Synapsin I (protein 1), a nerve terminal-specific phosphoprotein. III. Its association with synaptic vesicles studied in a highly purified synaptic vesicle preparation. *J. Cell Biol.* 96:1374–88

104. Baines, A. J., Bennett, V. 1985. Synapsin I is a spectrin-binding protein immunologically related to erythrocyte protein 4.1. *Nature* 315:410–13

105. Granger, B. L., Lazarides, E. 1985. Appearance of new variants of membrane skeletal protein 4.1 during terminal differentiation of avian erythroid and lenticular cells. *Nature* 313:238–41

106. Braell, W. A., Lodish, H. F. 1982. The erythrocyte anion transport protein is co-translationally inserted into microsomes. *Cell* 28:23–31

107. Blikstad, I., Nelson, W. J., Moon, R. T., Lazarides, E. 1983. Synthesis and assembly of spectrin during avian erythropoiesis: Stoichiometric assembly but unequal synthesis of α and β spectrin. *Cell* 32:1081–91

108. Moon, R. T., Lazarides, E. 1984. Biogenesis of the avian erythroid membrane skeleton: receptor mediated assembly and stabilization of ankyrin (goblin) and spectrin. *J. Cell Biol.* 98:1899–1904

109. Lazarides, E., Moon, R. T. 1984. Assembly and topogenesis of the spectrin-based membrane skeleton in erythroid development. *Cell* 37:354–56

110. Woods, C. M., Lazarides, E. 1985. Degradation of unassembled α- and β-spectrin by distinct intracellular pathways: regulation of spectrin topogenesis by β-spectrin degradation. *Cell* 40:959–69

111. Nelson, W. J., Lazarides, E. 1984. The patterns of expression of two ankyrin isoforms demonstrate distinct steps in the assembly of the membrane skeleton in neuronal morphogenesis. *Cell* 39:309–20

112. Lazarides, E., Nelson, W. J. 1983. Erythrocyte and brain forms of spectrin in cerebellum: distinct membrane-cytoskeletal domains in neurons. *Science* 220:1295–96

113. Lazarides, E., Nelson, W. J. 1983. Erythrocyte form of spectrin in cerebellum: appearance at a specific stage in the terminal differentiation of neurons. *Science* 222:931–33

114. Lazarides, E., Nelson, W. J., Kasamatsu, T. 1984. Segregation of two spectrin forms in the chicken optic system: a mechanism for establishing restricted membrane-cytoskeletal domains in neurons. *Cell* 36:269–78

115. Georgatos, S. D., Marchesi, V. T. 1985. The binding of vimentin to human

erythrocyte membranes: a model system for the study of intermediate filament-membrane interactions. *J. Cell Biol.* 100: 1955–61

116. Tarone, G., Ferracini, R., Galetto, G., Comoglio, P. 1984. A cell surface integral membrane glycoprotein of 85,000 mol wt (gp85) associated with triton x-100-insoluble cell skeleton. *J. Cell Biol.* 99: 512–19

117. Ranscht, B., Moss, D. J., Thomas, C. 1984. A neuronal surface glycoprotein associated with the cytoskeleton. *J. Cell Biol.* 99: 1803–13

118. Braun, J., Unanue, E. R. 1983. The lymphocyte cytoskeleton and its control of surface receptor functions. *Semin. Hemetol.* 20: 322–33

119. Gonen, A., Weisman-Shomer, P., Fry, M. 1979. Cell adhesion and acquisition of detergent resistance by the cytoskeleton of cultured chick fibroblasts. *Biochim. Biophys. Acta* 552: 307–21

120. Schliwa, M., Van Blerkom, J. 1981. Structural interaction of cytoskeletal components. *J. Cell Biol.* 90: 222–35

121. Herrmann, H., Wiche, G. 1983. Specific in situ phosphorylation of plectin in detergent-resistant cytoskeletons from cultured chinese hamster ovary cells. *J.*

Biol. Chem. 258: 14610–18

122. Carroll, R. C., Butler, R. G., Morris, P. A. 1982. Separable assembly of platelet pseudopodal and contractile cytoskeletons. *Cell* 30: 385–93

123. Phillips, D. R., Jennings, L. K., Edwards, H. H. 1980. Identification of membrane proteins mediating the interaction of human platelets. *J. Cell Biol.* 86: 77–86

124. Painter, R. G., Ginsberg, M. 1982. Concanavalin A induces interactions between surface glycoproteins and the platelet cytoskeleton. *J. Cell Biol.* 92: 565–73

125. Loftus, J. C., Choate, J., Albrecht, R. M. 1984. Platelet activation and cytoskeletal reorganization: high voltage electron microscopic examination of intact and Triton-extracted whole mounts. *J. Cell Biol.* 98: 2019–25

126. Jennings, L. K., Fox, J. E. B., Edwards, H. H., Phillips, D. R. 1981. Changes in the cytoskeletal structure of human platelets following thrombin activation. *J. Biol. Chem.* 256: 6927–32

127. Shen, B. W., Josephs, R., Steck, T. L. 1984. Ultrastructure of unit fragments of the skeleton of the human erythrocyte membrane. *J. Cell Biol.* 99: 810–21

SUBJECT INDEX

A

Absorption
  and brush border cytoskeleton,
    214
Acanthamoeba
  amino acid sequence in, 362
  capping protein nucleating
    actin, 372
Accessory proteins
  and actin, 354
Acetylcholine binding sites
  structure of high affinity, 334-
    37
Acetylcholine receptor
  Ach release, 318
  in *Torpedo*, 325
  mechanisms of, 319
  sequence homology, 339
Acetylcholine receptor sequences
  amphipathic fourier
    figure, 333
Acetyltransferase
  of *C. tropicalis*, 516
Acid hydrolases
  and the Golgi complex, 479
Acidic endosome
  and receptor-mediated endocy-
    tosis, 6
Acidic keratins
  and intermediate filaments, 45
  as Type 1 IF subunits, 58
Acidic polypeptide
  villin as, 217
Acid phosphatase (CMPase)
  and the Golgi cisternae, 464
  and organization of the Golgi
    complex, 450
  and protein localization in
    yeast, 118
Acid phosphate
  and protein localization in
    yeast, 129
Actin
  assembly of, 355-59
  cross-linking proteins
    figure, 376-78
  interaction of fimbrin with,
    216
  as part of the membrane
    skeleton, 542-45
  and spectrin biosynthesis, 534-
    36
  see also Nonmuscle actin-
    binding proteins

Actin assembly
  effects of villin on, 218
  regulation of, 373-75
  and the BB cytoskeleton,
    236
Actin-binding ability
  and acetylcholine receptor,
    322
Actin-binding protein
  and filamin, 380-82
  and brush border cytoskeleton,
    215
Actin bundling protein
  and spectrin, 548
Actin cytoskeleton
  and non-erythroid spectrin,
    541
Actin filament
  cross-linking proteins, 375-88
  end blocking proteins
    figure, 366
Actin filament bundling proteins
  and cross-linking, 385
Actin filaments
  arrowhead morphology of,
    217
  and brush border cytoskeleton,
    213
  proteins binding sides of, 388-
    89
  proteins blocking, 364-65
  the spectrin-like, 539
Actin gelling proteins
  and isotropic networks, 387-
    88
Actin-gelsolin complex
  and elution by EGTA, 368
Actin monomers
  inhibiting self-assembly into
    filaments, 355
  proteins that sequester, 359-73
Actin-myosin interaction
  and function of tropomyosin,
    389
Actin polymerization
  and monomer sequestering
    proteins, 374-75
  and profilins, 363
Actin subunit
  binding sites, 364
Actomyosin
  as part of the membrane
    skeleton, 545
Acumentin
  binding to actin filaments, 372

Acyl-CoA oxidase
  and biogenesis of per-
    oxisomes, 500-1
Acyltransferases
  and biogenesis of per-
    oxisomes, 504-5
Adenine dinucleotide phospha-
    tase
  and organization of the Golgi
    complex, 450
Adenine nucleotides
  binding by actin monomers,
    356
Adenocarcinoma
  and brush border cytoskeleton,
    234
Adenosine conversion
  and receptor-mediated endocy-
    tosis, 31
Adenylate cyclase
  and *Caulobacter* cell differen-
    tiation, 194
ADP
  binding by actin monomers,
    356
ADP monomer
  and brush border cytoskeleton,
    218
Adrenal cortex
  and biogenesis of per-
    oxisomes, 509
Affinity binding
  in fibronectin, 78
Affinity purification
  and acetylcholine receptor,
    318
Aggregation
  and embryonic cell migration,
    101
Alanine
  and *Caulobacter* cell differen-
    tiation, 188
Alcohol oxidases
  and biogenesis of per-
    oxisomes, 507
Aldolase
  adsorption to actin filaments,
    390-91
Alkaline phosphatase
  and protein localization in
    yeast, 136
Allylisopropylacetamide
  and biogenesis of per-
    oxisomes, 498
Alpha-actinin

563

Annual Reviews Inc.   ORDER FORM
A NONPROFIT SCIENTIFIC PUBLISHER
4139 El Camino Way, Palo Alto, CA 94306-9981, USA • (415) 493-4400

Annual Reviews Inc. publications are available directly from our office by mail or telephone (paid by credit card or purchase order), through booksellers and subscription agents, worldwide, and through participating professional societies. Prices subject to change without notice.

- **Individuals:** Prepayment required on new accounts by check or money order (in U.S. dollars, check drawn on U.S. bank) or charge to credit card — American Express, VISA, MasterCard.
- **Institutional buyers:** Please include purchase order number.
- **Students:** $10.00 discount from retail price, per volume. Prepayment required. Proof of student status must be provided (photocopy of student I.D. or signature of department secretary is acceptable). Students must send orders direct to Annual Reviews. Orders received through bookstores and institutions requesting student rates will be returned.
- **Professional Society Members:** Members of professional societies that have a contractual arrangement with Annual Reviews may order books through their society at a reduced rate. Check with your society for information.

Regular orders: Please list the volumes you wish to order by volume number.
Standing orders: New volume in the series will be sent to you automatically each year upon publication. Cancellation may be made at any time. Please indicate volume number to begin standing order.
Prepublication orders: Volumes not yet published will be shipped in month and year indicated.
California orders: Add applicable sales tax.
Postage paid (4th class bookrate/surface mail) **by Annual Reviews Inc.** Airmail postage extra.

ANNUAL REVIEWS SERIES		Prices Postpaid per volume USA/elsewhere	Regular Order Please send:	Standing Order Begin with:
			Vol. number	Vol. number
Annual Review of ANTHROPOLOGY (Prices of Volumes in brackets effective until 12/31/85)				
[Vols. 1-10	(1972-1981)	**$20.00/$21.00**		
[Vol. 11	(1982)	**$22.00/$25.00**]		
[Vols. 12-14	(1983-1985)	**$27.00/$30.00**]		
Vols. 1-14	(1972-1985)	**$27.00/$30.00**		
Vol. 15	(avail. Oct. 1986)	**$31.00/$34.00**	Vol(s). _____	Vol. _____
Annual Review of ASTRONOMY AND ASTROPHYSICS (Prices of Volumes in brackets effective until 12/31/85)				
[Vols. 1-2, 4-19	(1963-1964; 1966-1981)	**$20.00/$21.00**		
[Vol. 20	(1982)	**$22.00/$25.00**]		
[Vols. 21-23	(1983-1985)	**$44.00/$47.00**]		
Vols. 1-2, 4-20	(1963-1964; 1966-1982)	**$27.00/$30.00**		
Vols. 21-23	(1983-1985)	**$44.00/$47.00**		
Vol. 24	(avail. Sept. 1986)	**$44.00/$47.00**	Vol(s). _____	Vol. _____
Annual Review of BIOCHEMISTRY (Prices of Volumes in brackets effective until 12/31/85)				
[Vols. 30-34, 36-50	(1961-1965; 1967-1981)	**$21.00/$22.00**		
[Vol. 51	(1982)	**$23.00/$26.00**]		
[Vols. 52-54	(1983-1985)	**$29.00/$32.00**]		
Vols. 30-34, 36-54	(1961-1965; 1967-1985)	**$29.00/$32.00**		
Vol. 55	(avail. July 1986)	**$33.00/$36.00**	Vol(s). _____	Vol. _____
Annual Review of BIOPHYSICS AND BIOPHYSICAL CHEMISTRY (Prices of Vols. in brackets effective until 12/31/85)				
(Formerly Annual Review of Biophysics and Bioengineering)				
[Vols. 1-10	(1972-1981)	**$20.00/$21.00**		
[Vol. 11	(1982)	**$22.00/$25.00**]		
[Vols. 12-14	(1983-1985)	**$47.00/$50.00**]		
Vols. 1-11	(1972-1982)	**$27.00/$30.00**		
Vols. 12-14	(1983-1985)	**$47.00/$50.00**		
Vol. 15	(avail. June 1986)	**$47.00/$50.00**	Vol(s). _____	Vol. _____
Annual Review of CELL BIOLOGY				
Vol. 1	(1985)	**$27.00/$30.00**		
Vol. 2	(avail. Nov. 1986)	**$31.00/$34.00**	Vol(s). _____	Vol. _____
Annual Review of COMPUTER SCIENCE				
Vol. 1	(avail. late 1986)	**Price not yet established**	Vol. _____	Vol. _____
Annual Review of EARTH AND PLANETARY SCIENCES (Prices of Volumes in brackets effective until 12/31/85)				
[Vols. 1-9	(1973-1981)	**$20.00/$21.00**		
[Vol. 10	(1982)	**$22.00/$25.00**]		
[Vols. 11-13	(1983-1985)	**$44.00/$47.00**]		
Vols. 1-10	(1973-1982)	**$27.00/$30.00**		
Vols. 11-13	(1983-1985)	**$44.00/$47.00**		
Vol. 14	(avail. May 1986)	**$44.00/$47.00**	Vol(s). _____	Vol. _____